W9-DBY-528

Introduction to Electromagnetic Compatibility

Introduction to Electromagnetic Compatibility

CLAYTON R. PAUL
Department of Electrical Engineering
University of Kentucky, Lexington

A WILEY-INTERSCIENCE PUBLICATION

JOHN WILEY & SONS, INC.

NEW YORK / CHICHESTER / BRISBANE / TORONTO / SINGAPORE

In recognition of the importance of preserving what has been
written, it is a policy of John Wiley & Sons, Inc., to have books
of enduring value published in the United States printed on
acid-free paper, and we exert our best efforts to that end.

Library of Congress Cataloging in Publication Data:

Paul, Clayton R.
  Introduction to electromagnetic compatibility/Clayton R. Paul.
    p. cm.—(Wiley series in microwave and optical engineering)
  "A Wiley Interscience publication."
  Includes index.
    1. Electronic circuits—Noise. 2. Electromagnetic compatibility.
3. Digital electronics. 4. Shielding (Electricity) I. Title.
II. Series.
TK7867.5.P38 1992
621.382'24—dc20   91–16016
ISBN 0-471-54927-4   CIP

Printed in the United States of America

10 9 8 7 6 5 4 3

To the humane and compassionate treatment of animals

"For every difficult problem there is always
a simple answer and most of them are wrong"

Contents

Preface

This text is intended for a university course in electromagnetic compatibility (EMC). It may also be used as a reference for industrial professionals interested in EMC design. The prerequisites are the completion of the basic undergraduate electrical engineering courses in circuits, signals and systems, electronics and electromagnetic fields. The text builds on these basic concepts and principles and applies them to the design of electronic systems so that they will operate compatibly with other electronic systems and also comply with various governmental regulations on radiated and conducted emissions. In essence, EMC deals with *interference* in electronic systems. Earlier designations of the subjects referred to it as EMI (an acronym for electromagnetic interference) or RFI (an acronym for radio frequency interference). It is now referred to in more positive terms by replacing "interference" with "compatibility".

The subject of EMC has grown in importance in the United States in the last ten years primarily due to the Federal Communications Commission's (FCC's) imposition of restrictions on radiated and conducted emissions of *digital devices. Digital devices having clock frequencies that exceed 9 kHz cannot be legally sold in the US unless they have been tested and found not to exceed limits on radiated and conducted emissions set by the FCC.* Prior to the imposition of regulations by the FCC, digital devices were subject to similarly stringent legal restrictions imposed by other countries in which the device was to be marketed. It can now be said that the subject of EMC is of worldwide concern to manufacturers of digital products. Until the imposition of the FCC regulations on digital products, interference in electronic systems designed for the US market was only controlled in military systems by military standards, in commercial products by voluntary compliance to requirements of the product manufacturer, and by the FCC in certain systems that interfere with radio and wire communications.

Industries throughout the world are now keenly interested in EMC because of the cost-competitive nature of today's markets. If a product fails the regulatory limits, it cannot be sold, no matter how innovative the design. In order to modify the design to comply with the requirements, additional suppression components may need to be added and/or the physical configuration of the product might require modification. These costs are borne by the manufacturer and may cause the product's price to be noncompetitive in the marketplace. Also, the additional schedule delays required to fix the problem may cause the product announcement to miss the window of optimum marketability, resulting in reduced sales. There are additional areas of the EMC profile that affect the quality of that product. A good example is the susceptibility to electrostatic discharge (ESD). If a digital product is particularly sensitive to ESD such that it will fail to operate reliably in dry climates, customers will soon cease to purchase it. In summary, failure to consider EMC in product design adversely affects a product's profitability as well as the manufacturer's reputation in the industry.

Present electrical engineering (EE) curricula do not explicitly address EMC. This text fills that gap. Chapter 1 provides examples of interference, a history of the evolution of EMC, and an introduction to the decibel. Chapter 2 supplies additional "motivation" for the study of EMC by discussing governmental regulations as well as the various issues of quality product design.

The remainder of the text is divided into two parts. Part I, on basic electromagnetic principles, constitutes a review of the important concepts of electromagnetic field theory (Chapter 3), transmission lines (Chapter 4) and antennas (Chapter 5).

Part II, on applications to EMC design, explains how to use the basic principles to design electronic systems and, in particular, digital systems, so they will satisfy governmental regulations and perform reliably in the presence of interference sources. An important aspect in the ability of an engineer to effectively incorporate EMC into a design is an understanding of the *nonideal behavior of electronic components* as discussed in Chapter 6. Typical suppression elements, such as ferrite cores, are also considered. Chapter 7 analyzes the correlation between the time-domain and frequency-domain representation of signals. Most EMC regulations are written in the frequency domain, so it is important that a designer be able to view a signal with equal flexibility in either domain. The important relationships between rise/fall time of digital signals and the signal's spectral content are emphasized. Measurement devices, such as spectrum analyzers, are also discussed. Chapter 8 provides simple emission and susceptibility models that give first-order predictions. These models also highlight, in a qualitative sense, the important factors that influence the radiated emissions and susceptibility of a product. Of equal importance are the product's noise emissions passed out through the unit's power cord. This important aspect of EMC is discussed in Chapter 9. Chapter 10 considers a frequently overlooked aspect of EMC: the potential of a product to interfere with itself. This is analyzed in relation to crosstalk between wires and a printed circuit board's lands. Simple crosstalk models are developed and the often misunderstood concepts of using shielded wires or twisted pairs to reduce crosstalk are clarified. Chapter 11 covers the subject of shielding enclosures. This aspect of EMC is often relied upon to save a poor design when in actuality the ideal performance of a shield is often not realized. Chapter 12 discusses electrostatic discharge. A basic understanding of the event, as well as mitigation design techniques, are set forth. Finally, Chapter 13 brings together all the previous topics in the design of systems. The often misunderstood subject of grounding is discussed along with the physical configuration of the system. Printed circuit board layout and design is a critical aspect in the ability of digital systems to comply with EMC regulations and is also discussed in this chapter. This final chapter is the capstone of the text. Upon completion of the text, the reader will understand the basic fundamentals of EMC design. Although the emphasis has been placed on digital systems, the concepts apply to other electronic systems.

Problems have been provided at the end of each chapter. Answers are given in brackets, [ ], at the end of each problem. Some emphasize theory while others stress applications. A detailed solutions manual is also available. Each chapter contains a list of references for additional details, noted where appropriate.

Conducting experiments is an important aspect in comprehension of the material. It also gives the student the opportunity to understand the limitations of current measurement equipment and a realistic appreciation of the "numbers" one should expect. For this reason, several "mini-experiments" that illustrate the basic principles should be incorporated in the course. Present-day EMC test equipment is quite sophisticated (and expensive) and not all universities have access to such equipment. Local industries that practice EMC may provide students an opportunity, in a "field trip" format, to meet their industrial counterparts, use the industry's test equipment and gain a feeling of "relevance" of the material.

Many people in the field of EMC have contributed to the author's ability to produce this text. Insights gained from Mr Henry W. Ott significantly impacted the author's EMC perspective. Numerous helpful discussions with Messrs Robert F. German and Keith B. Hardin are also gratefully acknowledged. Many other individuals have had a similar impact on the author's EMC perspective, and to them, the author extends his appreciation. The editing of the manuscript by Carol Gage as well as her encouragement are also gratefully acknowledged.

Clayton R. Paul

Lexington, Kentucky, July 1991

CHAPTER ONE

Introduction to Electromagnetic Compatibility (EMC)

Since the early days of radio and telegraph communications, it has been known that a spark gap generates electromagnetic waves rich in spectral content (frequency components) and that these waves can cause interference or noise in various electronic and electrical devices such as radio receivers and telephone communications. Numerous other sources of electromagnetic emissions such as lightning, relays, dc electric motors and fluorescent lights also generate electromagnetic waves that are rich in spectral content and which can cause interference in those devices. There are also sources of electromagnetic emissions that contain only a narrow band of frequencies. High-voltage power transmission lines generate electromagnetic emissions at the power frequency (60 Hz). Radio transmitters transmit desired emissions by encoding information (voice, music, etc.) on a carrier frequency. Radio receivers intercept these electromagnetic waves, amplify them and extract the information that is encoded in the wave. Radar transmitters also transmit pulses of a single-frequency carrier. As this carrier frequency is pulsed on and off, these pulses radiate outward from the antenna, strike a target and return to the radar antenna. The total transit time of the wave is directly related to the distance of the target from the radar antenna. The spectral content of this radar pulse is distributed over a larger band of frequencies around the carrier than are radio transmissions. Another important and increasingly significant source of electromagnetic emissions is associated with digital computers in particular and digital electronic devices in general. These digital devices utilize pulses to signify a binary number, 0 (off) or 1 (on). Numbers and other symbols are represented as sequences of these binary digits. The transition time of the pulse from off to on and vice versa is perhaps the most important factor in determining the spectral content of the pulse. Fast (short) transition times generate a wider range of frequencies than do slower (longer) transition times. The spectral content of digital devices

1

generally occupies a wide range of frequencies and can also cause interference in electrical and electronic devices.

This text is concerned with the ability of these types of electromagnetic emissions to cause *interference* in electrical and electronic devices. The reader has no doubt experienced noise produced in an AM radio by nearby lightning discharges. The lightning discharge is rich in frequency components, some of which pass through the input filter of the radio, causing noise to be superimposed on the desired signal. Also, even though a radio may not be tuned to a particular transmitter frequency, the transmission may be received, causing the reception of an unintended signal. These are examples of interference produced in *intentional receivers*. Of equal importance is the interference produced in *unintentional receivers*. For example, a strong transmission from an FM radio station or TV station may be picked up by a digital computer, causing the computer to interpret it as data or a control signal resulting in incorrect function of the computer. Conversely, a digital computer may create emissions that couple into a TV, causing interference.

This text is also concerned with the design of electronic systems such that interference from or to that system will be minimized. The emphasis will be on *digital* electronic systems. An electronic system that is able to function compatibly with other electronic systems and not produce or be susceptible to interference is said to be *electromagnetically compatible* with its environment. The objective of this text is to learn how to design electronic systems for *electromagnetic compatibility (EMC)*. A system is electromagnetically compatible if it satisfies three criteria:

1. *It does not cause interference with other systems.*
2. *It is not susceptible to emissions from other systems.*
3. *It does not cause interference with itself.*

Designing for EMC is not only important for the desired functional performance, but the device must also meet *LEGAL* requirements in virtually all countries of the world before it can be sold. Designing an electronic product to perform a new and exciting function is a waste of effort if it cannot be placed on the market!

EMC design techniques and methodology have become as integral a part of design as, for example, digital design. Consequently the material in this text has become a fundamental part of an electrical engineer's background. This will no doubt increase in importance as the trend toward increased clock speeds and data rates of digital systems continues.

This text is intended for a university course in electromagnetic compatibility in an undergraduate/graduate curriculum in electrical engineering. There are textbooks available that concern EMC, but these are primarily designed for the industrial professional. Consequently, we will draw on a number of sources for reference material. These will be given at the end of each chapter and their reference will be denoted in the text by brackets, e.g., [xx]. Numerous trade journals, EMC conference proceedings and the *Institute of Electrical and Electronics Engineers* (*IEEE*) *Transactions on Electromagnetic Compatibility* contain useful tutorial articles on various aspects of EMC that we will discuss, and these will similarly be referenced where appropriate. The most important aspect in successfully dealing with EMC design is to have a sound understanding of the basic principles of electrical engineering—circuits, electronics, signals, electromagnetics, linear system theory, digital system design, etc. We will therefore review these basics so that the fundamentals will be understood and can be used effectively and correctly by the reader in solving the EMC problem. A representative set of such basic texts is [1–3]. A representative but not exhaustive list of texts that cover the general aspects of EMC is represented by [4–10]. The text by Ott [4] will form our primary EMC text reference. Other texts and journal articles that cover aspects of EMC will be referenced in the appropriate chapters. For a discussion of the evolution of this EMC course see [11, 12].

1.1 ASPECTS OF EMC

As illustrated above, EMC is concerned with the *generation, transmission* and *reception* of electromagnetic energy. These three aspects of the EMC problem form the basic framework of any EMC design. This is illustrated in Fig. 1.1. A *source* (also referred to as an *emitter*) produces the emission, and a *transfer* or *coupling path* transfers the emission energy to a *receptor* (*receiver*), where it is processed, resulting in either desired or undesired behavior. *Interference occurs if the received energy causes the receptor to behave in an undesired manner.* Transfer of electromagnetic energy occurs frequently via unintended coupling modes. However, the unintentional transfer of energy causes interference only if the received energy is of sufficient magnitude and/or spectral content at the receptor input to cause the receptor to behave in an undesired fashion. *Unintentional transmission or reception of electromagnetic energy is not necessarily detrimental; undesired behavior of the receptor constitutes interference.* So the *processing of the received energy* by the receptor is an important part of the question of whether interference will occur. Quite often it is difficult to determine, a priori, whether a signal that is incident on a receptor will cause interference in that receptor. For example, clutter on a radar scope may cause a novice radar operator to incorrectly interpret the desired data, whereas the clutter may not create problems for an operator who has considerable experience. In one

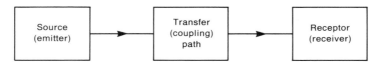

FIGURE 1.1 The basic decomposition of the EMC coupling problem.

case we have interference and in the other we do not, although one could argue that the receptor is the radar operator and not the radar receiver. This points out that it is often difficult to uniquely identify the three aspects of the problem shown in Fig. 1.1!

It is also important to understand that a source or receptor may be classified as intended or unintended. In fact, a source or receptor may behave in both modes. Whether the source or the receptor is intended or unintended *depends on the coupling path as well as the type of source or receptor.* As an example, an AM radio station transmitter whose transmission is picked up by a radio receiver that is tuned to that carrier frequency constitutes an intended emitter. On the other hand, if the same AM radio transmission is processed by another radio receiver that is not tuned to the carrier frequency of the transmitter, then the emission is unintended. (Actually the emission is still intended but the coupling path is not.) There are some emitters whose emissions can serve no intended purpose. An example is the (nonvisible) electromagnetic emission from a fluorescent light.

This suggests that there are three ways to prevent interference:

1. *Suppress the emission at its source.*
2. *Make the coupling path as inefficient as possible.*
3. *Make the receptor less susceptible to the emission.*

As we proceed through the examination of the EMC problem, these three alternatives should be kept in mind. The "first line of defense" is to suppress the emission as much as possible at the source. For example, we will find that fast (short) rise/fall times of digital pulses are the primary contributors to the high-frequency spectral content of these signals. In general, the higher the frequency of the signal to be passed through the coupling path, the more efficient the coupling path. So we should slow (increase) the rise/fall times of digital signals. However, the rise/fall times of digital signals can be increased only to a point at which the digital circuitry malfunctions. This is not sufficient reason to use digital signals having 1 ns rise/fall times when the system will properly function with 10 ns rise/fall times. Remember that reducing the high-frequency spectral content of an emission tends to inherently reduce the efficiency of the coupling path and hence reduces the signal level at the receptor. There are " brute force" methods of reducing the efficiency of the coupling path that we will discuss. For example, placing the receptor in a metal enclosure (a shield) will serve to reduce the efficiency of the coupling path. But shielded enclosures are more expensive than reducing the rise/fall time of the emitter, and, more often than not, their actual performance in an installation is far less than ideal. Reducing the susceptibility of the receptor is quite often difficult to implement and still preserve the desired function of the product. An example of implementing reduced susceptibility of a receptor to noise would be the use of error-correcting codes in a digital receptor. Although undesired electromagnetic energy is incident on the receptor, the error-correcting codes may allow the receptor to function properly in the presence of a potentially troublesome signal. If the reader will think in terms of reducing the coupling by working from left to right in Fig. 1.1, success will usually be easier to achieve and with less additional cost to the system design. Minimizing the cost added to a system to make it electromagnetically compatible will continue to be an important consideration in EMC design. One can put all electronic products in metallic enclosures and power them with internal batteries, but the product appearance, utility and cost would be unacceptable to the customer.

We may further break the transfer of electromagnetic energy (with regard to the prevention of interference) into four subgroups: *radiated emissions, radiated susceptibility, conducted emissions,* and *conducted susceptibility,* as illustrated in Fig. 1.2. A typical electronic system usually consists of one or more subsystems that communicate with each other via cables (bundles of wires). A means for providing power to these subsystems is usually the commercial ac power system of the installation site. A power supply in a

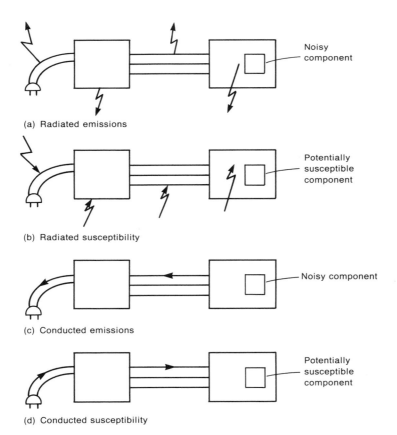

(a) Radiated emissions

(b) Radiated susceptibility

(c) Conducted emissions

(d) Conducted susceptibility

FIGURE 1.2 The four basic EMC subproblems: (a) radiated emissions; (b) radiated susceptibility; (c) conducted emissions; and (d) conducted susceptibility.

particular electronic system converts this ac 120 V, 60 Hz voltage (240 V, 50 Hz in Europe) to the various dc voltage levels required to power the internal electronic components of the system. For example, 5 V dc is required to power the digital logic, +12 V, and −12 V dc voltages are required to power analog electronics. Other dc voltages are required to power devices such as motors. Sometimes the 60 Hz (50 Hz) ac power is required to power other components such as small cooling fans. The 60 Hz, 120 V ac system power is obtained from the commercial power net via a line cord. Other cables are required to interconnect subsystems so that functional signals can be passed between them. All of these cables have the potential for emitting and/or picking up electromagnetic energy, and are usually quite efficient in doing so. Generally speaking, the longer the cable, the more efficient it is in emitting or picking up electromagnetic energy. Interference signals can also be passed directly between the subsystems via direct conduction on these cables. If the subsystems are enclosed in metallic enclosures, currents may be induced on these enclosures by internal signals or external signals. These induced currents can then radiate to the external environment or to the interior of the enclosure. It is becoming more common, particularly in low-cost systems, to use nonmetallic enclosures, usually plastic. The electronic circuits contained in these nonmetallic enclosures are, for the most part, completely exposed to electromagnetic emissions, and as such can directly radiate or be susceptible to them. The four aspects of the EMC problem, *radiated emissions*, *radiated susceptibility*, *conducted emissions*, and *conducted susceptibility*, illustrated in Fig. 1.2, reflect these considerations.

Electromagnetic emissions can occur from the ac power cord, a metallic enclosure containing a subsystem, a cable connecting subsystems or from an electronic component within a nonmetallic enclosure as Fig. 1.2(a) illustrates. Throughout the text we will be trying to replace certain misconceptions that prevent an understanding of the problem. An example is the notion that the ac power cord carries only 60 Hz signals. Although the primary intent of this cable is to transfer 60 Hz commercial power to the system, it is important to realize that *other much higher-frequency signals may and usually do exist on the ac power cord!* These are coupled to the ac power cord from the internal subsystems via a number of coupling paths that we will discuss. Once these high-frequency currents appear on this long (1 m or more) cable, they will radiate quite efficiently. Also, this long cable may function as an efficient "antenna" and pick up radiated emissions from other nearby electronic systems as shown in Fig. 1.2(b). Once these external signals are induced on this cable as well as any cables connecting the subsystems, they may be transferred to the internal components of the subsystems, where they may cause interference in those circuits. To summarize, undesired signals may be radiated or picked up by the ac power cord, interconnection cables, metallic cabinets or internal circuitry of the subsystems, even though these structures or wires are not intended to carry the signals.

Emissions of and susceptibility to electromagnetic energy occur not only by electromagnetic waves propagating through air but also by direct conduction on metallic conductors as illustrated in Figs. 1.2(c) and (d). Usually this coupling path is inherently more efficient than the air coupling path. Electronic system designers realize this, and intentionally place barriers, such as filters, in this path to block the undesired transmission of this energy. It is particularly important to realize that the interference problem often extends beyond the boundaries shown in Fig. 1.2. For example, currents conducted out the ac power cord are placed on the power distribution net of the installation. This power distribution net is an extensive array of wires that are directly connected and as such may radiate these signals quite efficiently. In this case, a *conducted emission* produces a *radiated emission*. Consequently, restrictions on the emissions conducted out the product's ac power cord are intended to reduce the radiated emissions from this power distribution system.

Our primary concern will be the design of electronic systems so that they will comply with the *legal* requirements imposed by governmental agencies. However, there are also a number of other important EMC concerns that we will discuss. Some of these are depicted in Fig. 1.3. Fig. 1.3(a) illustrates an increasingly common susceptibility problem for today's small-scale integrated circuits, *electrostatic discharge* (*ESD*). Walking across a nylon carpet with rubber-soled shoes can cause a buildup of static charge on the body. If an electronic device such as a keyboard is touched, this static charge may be transferred to the device, and an arc is created between the finger tips and the device. The direct transfer of charge can cause permanent destruction of electronic components such as integrated circuit chips. The arc also bathes the device in an electromagnetic wave that is picked up by the internal circuitry. This can result in system malfunction. ESD is such a pervasive problem today that we will devote Chapter 12 to its discussion.

After the first nuclear detonation in the mid-1940s, it was discovered that the semiconductor devices (a new type of amplifying element) in the electronic systems that were used to monitor the effects of the blast were destroyed. This was not due to the direct physical effects of the blast but was caused by an intense electromagnetic wave created by the charge separation and movement within the detonation as illustrated in Fig. 1.3(b). Consequently, there is significant interest within the military communities in regard to "hardening" communication and data processing facilities against the effect of this *electromagnetic pulse* (*EMP*). The concern is not with the physical effects of the blast but with the inability to direct retaliatory action if the communication and data processing facilities are rendered nonfunctional by the EMP. This represents a radiated susceptibility problem. We will find that the same principles used to reduce the effect of radiated emissions from neighboring electronic systems also apply to this problem, but with larger numbers.

Lightning occurs frequently and direct strikes illustrated in Fig. 1.3(c) are obviously important. However, the indirect effects on electronic systems can be equally devastating. The "lightning channel" carries upwards of 50,000 A of current. The electromagnetic fields from this intense current can couple to electronic systems either by direct radiation or by coupling to the commercial

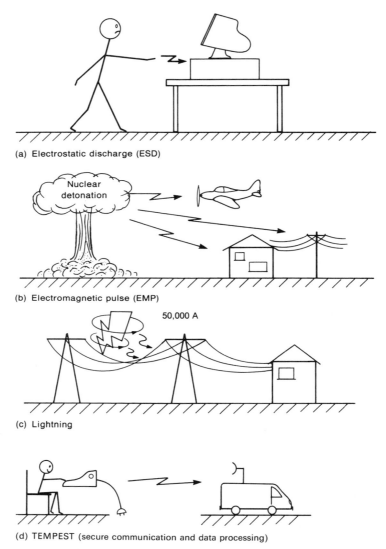

FIGURE 1.3 Other aspects of EMC: (a) electrostatic discharge (ESD), (b) electro-magnetic pulse (EMP), (c) lightning, (d) TEMPEST (secure communication and data processing).

power system and subsequently being conducted into the device via the ac power cord. Consequently, it is important to design and test the product for its immunity to transient voltages on the ac power cord. Most manufacturers inject "surges" onto the ac power cord and design their products to withstand these and other undesired transient voltages.

It has also become of interest to prevent the interception of electromagnetic emissions by unauthorized persons. It is possible, for example, to determine what is being typed on an electronic typewriter by monitoring its electromagnetic emissions as illustrated in Fig. 1.3(d). There are also other instances of direct interception of radiated emissions from which the content of the communications or data can be determined. Obviously, it is imperative for the military to contain this problem, which it refers to as TEMPEST. The commercial community is also interested in this problem from the standpoint of preserving trade secrets, the knowledge of which could affect the competitiveness of the company in the marketplace.

There are several other related problems that fit within the purview of the EMC discipline. However, it is important to realize that these can be viewed in terms of the four basic subproblems of radiated emissions, radiated susceptibility, conducted emissions, and conducted susceptibility shown in Fig. 1.2. Only the context of the problem changes.

The primary vehicle used to understand the effects of interference is a *mathematical model*. A mathematical model quantifies our understanding of the phenomenon and also may bring out important properties that are not so readily apparent. An additional, important advantage of a mathematical model is its ability to aid in the design process. *The criterion that determines whether the model adequately represents the phenomenon is whether it can be used to predict experimentally observed results.* If the predictions of the model do not correlate with experimentally observed behavior of the phenomenon, it is useless. However, our ability to solve the equations resulting from the model and extract insight from them quite often dictates the approximations used to construct the model. For example, we often model nonlinear phenomena with linear, approximate models.

Calculations will be performed quite frequently, and correct *unit conversion* is essential. Although the trend in the international scientific community is toward the metric or SI system of units, there is still the need to use other systems. One must be able to convert a unit in one system to the equivalent in another system, as in an equation where certain constants are given in another unit system. A simple and flawless method is to multiply by unit ratios between the two systems *and* cancel the unit names to insure that the quantity should be multiplied rather than divided and vice versa. For example, the units of distance in the English system are inches, feet, miles, yards, etc. Some representative conversions are 1 inch = 2.54 cm, 1 foot = 12 inches, 1 m = 100 cm, 1 mile = 5280 feet, 1 yard = 3 feet, etc. For example, suppose we wish to convert a distance of 5 miles to kilometers. Multiply

$$5 \text{ miles} \times \frac{5280 \text{ feet}}{1 \text{ mile}} \times \frac{12 \text{ inches}}{1 \text{ foot}} \times \frac{2.54 \text{ cm}}{1 \text{ inch}} \times \frac{1 \text{ m}}{100 \text{ cm}} \times \frac{1 \text{ km}}{1000 \text{ m}} = 8.047 \text{ km}.$$

Cancellation of the unit names in this conversion avoids the improper multiplication (division) of a unit ratio when division (multiplication) should be used. The inability to properly convert units is a leading reason for numerical errors.

1.2   HISTORY OF EMC

It may be said that interference and its correction arose with the first spark-gap experiment of Marconi in the late 1800s. In 1901 he provided the first transatlantic transmission using an array of copper wires. The only receptors of significance at that time were radio receivers. These were few and widely separated, so that the correction of an interference problem was relatively simple. However, technical papers on radio interference began to appear in various technical journals around 1920. The radio receivers and antennas were rather crude and were prone to interference either from external sources or from within as with self-induced oscillations. Improvements in design technology cured many of these problems. Radio interference from electrical apparatus such as electric motors, electric railroads, and electric signs soon began to appear as a major problem around 1930.

During World War II, the use of electronic devices, primarily radios, navigation devices, and radar, accelerated. Instances of interference between radios and navigational devices on aircraft began to increase. These were usually easily corrected by reassignment of transmitting frequencies in an uncrowded spectrum or physically moving cables away from noise emission sources to prevent the cables from picking up those emissions. Because the density of the electronics (primarily vacuum tube electronics) was considerably less than it is today, these interference remedies could be easily implemented on a case-by-case basis in order to correct any electromagnetic interference (EMI) problem. However, the most significant increases in the interference problem occurred with the inventions of high-density electronic components such as the bipolar transistor in the 1950s, the integrated circuit (IC) in the 1960s, and the microprocessor chip in the 1970s. The frequency spectrum also became more crowded due to the increased demand for voice and data transmission. This required considerable planning with regard to *spectrum utilization* and continues today.

Perhaps the primary event that brought the present emphasis on EMC to the forefront was the introduction of digital signal processing and computation. In the early 1960s digital computers used vacuum tubes as switching elements. These were rather slow (by today's standards) and required large power consumption and considerable "real estate." In the 1970s the integrated circuit allowed the construction of computers that consumed far less power and required much less physical space. Towards the end of the 1970s the trend toward replacing *analog signal processing* with *digital signal processing* began to accelerate. Almost all electronic functions were being implemented digitally because of the increased switching speed and miniaturization of the ICs. The implementation of various tasks ranging from computation to word processing to digital control became widespread, and continues today. This meant that the density of noise sources rich in spectral content (switching waveforms) was becoming quite large. Consequently, the occurrence of EMI problems began to rise.

Because of the increasing occurrence of digital system interference with wire and radio communication, the Federal Communications Commission (FCC) in the US published a regulation in 1979 that required the electromagnetic emissions of all "digital devices" to be below certain limits. The intent of this rule was to try to limit the "electromagnetic pollution" of the environment in order to prevent, or at least reduce, the number of instances of EMI. Because no "digital device" could be sold in the US unless its electromagnetic emissions met these limits imposed by the FCC, the subject of EMC generated intense interest among the manufacturers of commercial electronics ranging from digital computers to electronic typewriters.

This is not intended to imply that the US was at the forefront of "cleaning up the electromagnetic environment" in mandating limits on electromagnetic emissions. Countries in Europe imposed similar requirements on digital devices well before the FCC issued its rule. In 1933 a meeting of the International Electrotechnical Commission (IEC) in Paris recommended the formation of the International Special Committee on Radio Interference (CISPR) to deal with the emerging problem of EMI. The committee produced a document detailing measurement equipment for determining potential EMI emissions. The CISPR reconvened after World War II in London in 1946. Subsequent meetings yielded various technical publications, which dealt with measurement techniques as well as recommended emission limits. Some European countries adopted versions of CISPR's recommended limits. The FCC rule was the first regulation for digital systems in the US, and the limits follow the CISPR recommendations with variations peculiar to the US environment. Most manufacturers of electronic products within the US already had internal limits and standards imposed on their products in order to prevent "field problems" associated with EMI. However, the FCC rule made what had been voluntary a matter of legal compliance.

The military community in the US also imposed limits on the electromagnetic emissions of electronic systems to prevent EMI through MIL–STD–461 prior to the FCC issuing its rule. These had been in effect from the early 1960s and were imposed to insure "mission success." All electronic and electrical equipment ranging from hand drills to sophisticated computers were required to meet the emission limits of these standards. Another aspect of the military's regulations is the imposition of a *susceptibility requirement*. Interfering signals are purposely injected into the equipment, which must then operate properly in the presence of these signals. At the time of writing this book, the CISPR and FCC requirements only regulate the *emissions* of the equipment. It is expected that the various regulatory bodies will also impose susceptibility requirements. Even though an electronic product complies with the emission requirements, it could cause interference with or be susceptible to the emissions of another electronic device in close proximity. The emission requirements simply attempt to limit electromagnetic pollution. Susceptibility requirements, used by the military, go one step further in attempting to insure electromagnetically compatible operation of all equipment.

These regulations have made EMC a critical aspect in the marketability of an electronic product. If the product does not comply with these regulations for a particular country, it cannot be sold in that country. The fact that the product performs some very desirable task and customers are willing to purchase it is unimportant if it does not comply with the regulatory requirements. Throughout this text the reader should keep in mind that the evolution of technology has caused the subject of EMC design to be as critical a part of electronic design as any of the traditional aspects.

1.3   EXAMPLES

There are numerous examples of EMI, ranging from the commonplace to the catastrophic. In this section we will mention a few of these.

Probably one of the more common examples is the occurrence of "lines" across the face of a television screen when a blender, vacuum cleaner, or other household device containing a dc motor is turned on. This problem results from the arcing at the brushes of the dc motor. As the commutator makes and breaks contact through the brushes, the current in the motor windings (an inductance) is being interrupted, causing a large voltage ($L\, di/dt$) across the contacts. This voltage is similar to the Marconi spark-gap generator and is rich in spectral content. The problem is caused by the direct radiation of this signal to the TV antenna and the passage of this noise signal out through the ac power cord of the device. This places the interference signal on the common power net of the household. As mentioned earlier, this common power distribution system is a large array of wires. Once the signal is present on this efficient "antenna," it radiates to the TV antenna, creating the interference.

A manufacturer of office equipment placed its first prototype of a new copying machine in its headquarters. An executive noticed that when someone made a copy, the hall clocks would sometimes reset or do strange things. The problem turned out to be due to the silicon-controlled rectifiers (SCR) in the power conditioning circuitry of the copier. These devices turn on and off to "chop" the ac current to create a regulated dc current. These signals are also rich in spectral content due to the abrupt change in current, and were coupled out through the copier's ac power cord onto the common power net in the building. Clocks in hallways are often set and synchronized by use of a modulated signal imposed on the 60 Hz ac power signal. The "glitch" caused by the firing of the SCRs in the copier coupled into the clocks via the common ac power net and caused them to interpret it as a signal to reset.

A new version of an automobile had a microprocessor-controlled emission and fuel monitoring system installed. A dealer received a complaint that when the customer drove down a certain street in the town, the car would stall. Measurement of the ambient fields on that street revealed the presence of an illegal FM radio transmitter. The signals from that transmitter coupled onto the wires leading to the processor and caused it to shut down.

Certain trailer trucks had electronic breaking systems installed. Keying a citizens band (CB) transmitter in a passing automobile would sometimes cause the brakes on the truck to "lock up." The problem turned out to be the coupling of the CB signal into the electronic circuitry of the braking system. Shielding the circuitry cured the problem.

A large computer system was installed in an office complex near a commercial airport. At random times the system would lose or store incorrect data. The problem turned out to be synchronized with the sweep of the airport surveillance radar as it illuminated the office complex. Extensive shielding of the computer room prevented any further interference.

In 1982 the United Kingdom lost a destroyer, the HMS *Sheffield*, to an Exocet missile during an engagement with Argentinian forces in the battle of the Falkland Islands. The destroyer's radio system for communicating with the UK would not operate properly while the ship's anti-missile detection system was being operated due to interference between the two systems. To temporarily prevent interference during a period of communication with the UK, the anti-missile system was turned off. Unfortunately, this coincided with the enemy launch of the Exocet missile.

The US Army purchased an attack helicopter designated as the UH-60 Black Hawk. On Sunday, November 8, 1988, various news agencies reported that the helicopter was susceptible to electromagnetic emissions. Evidence was revealed that indicated most of the crashes of the Black Hawk since 1982, which killed 22 servicemen, were caused by flying too close to radar transmitters, radio transmitters, and possibly even a citizens band (CB) transmitter. The susceptibility of the helicopter's electronically controlled flight control system to these electromagnetic emissions was thought to have caused these crashes.

On July 29, 1967, the US aircraft carrier *Forrestal* was deployed off the coast of North Vietnam. The carrier deck contained numerous attack aircraft that were fueled and loaded with 1000-pound bombs, as well as air-to-air and air-to-ground missiles. One of the aircraft missiles was inadvertently deployed, striking another aircraft and causing an explosion of its fuel tanks and the subsequent death of 134 servicemen. The problem was thought to be caused by the generation of radio-frequency (RF) voltages across the contacts of a shielded connector by the ship's high-power search radar.

These are a few of the many instances of EMI in our dense electronic world. The life-threatening results clearly demand remedies. The occurrences that merely result in annoyance or loss of data in a computer are not as dramatic, but still create considerable disruption and also require resolution. We will discuss design principles that solve many of these problems.

1.4 ELECTRICAL DIMENSIONS

Perhaps the most important concept that the reader should grasp is that of the *electrical dimensions* of an electronic circuit or electromagnetic radiating structure. Although there are other possibilities, we will consider an electromagnetic radiating structure to be any structure that is composed of current-carrying conductors. *Physical dimensions are not important, per se, in determining the ability of a source to couple to a receptor. Electrical dimensions are more significant in determining this efficiency.*

Electrical dimensions are measured in *wavelengths.* (This will be discussed in more detail in subsequent sections of the text.) A wavelength represents the distance an electromagnetic wave must travel in order to change phase by 360°. Strictly speaking, this applies to one type of wave: the uniform plane wave. However, other types of waves have similar characteristics, so that this concept is generally applicable. This phase-shift property is also related to the time of propagation of the wave between two points, as we shall see in later chapters. For the present it is sufficient to simply determine the electrical dimensions of a circuit or electromagnetic structure in terms of wavelengths. Wavelengths are denoted by the symbol λ. We will first consider electromagnetic waves traveling in *nonconductive (lossless) media.* In later sections we will consider this property for waves that travel in conductive media. For *lossless media* a wavelength can be calculated from [1]

$$\lambda = \frac{v}{f} \tag{1.1}$$

where v is the velocity of propagation of the wave and f is the frequency of the wave. A structure whose physical dimension (in meters) is $\mathscr{L}$ has electrical dimensions in wavelengths of

$$k = \frac{\mathscr{L}}{\lambda} \tag{1.2}$$

$$= \frac{\mathscr{L} f}{v}$$

The electrical dimensions of a structure depend on the physical dimensions, the frequency of excitation, and the velocity of propagation of the wave in the medium in which the structure is immersed. *An electronic circuit or electromagnetic radiating structure is said to be electrically small if its largest dimension $\mathscr{L}$ is much smaller than a wavelength, $k \ll 1$, or, $\mathscr{L} \ll \lambda$.* Although only an approximate criterion, we will assume that a circuit or electromagnetic structure is electrically small if $\mathscr{L} < \frac{1}{10}\lambda$.

The velocity of wave propagation in free space is denoted by v_o and is given approximately by

$$v_o = 3 \times 10^8 \text{ m/s}$$

Broadly speaking, the velocity of propagation of a wave in a nonconductive medium other than free space is determined by the *permittivity* ϵ and *permeability* μ of the medium. For free space these are denoted as ϵ_o and μ_o and are given by

$$\epsilon_o = \frac{1}{36\pi} \times 10^{-9} \text{ F/m} \quad \text{(approximate)}$$

$$\mu_o = 4\pi \times 10^{-7} \text{ H/m} \quad \text{(exact)}$$

The units of ϵ are farads per meter or a capacitance per distance. The units of μ are henrys per meter or an inductance per distance. We will see these combinations of units several times in later portions of this text and in a different context. The velocity of propagation is given in terms of these as

$$v_o = \frac{1}{\sqrt{\epsilon_o \mu_o}} \tag{1.3}$$

$$= 3 \times 10^8 \text{ m/s} \quad \text{(approximate)}$$

The exact value for the speed of light (or the velocity of propagation of any other electromagnetic wave) in free space is $2.997925\cdots \times 10^8$ m/s, but the approximate value of 3×10^8 m/s is sufficiently accurate for most calculations. Other media through which the wave may propagate are characterized in terms of their permittivity and permeability *relative to that of free space*, ϵ_r and μ_r, so that $\epsilon = \epsilon_r \epsilon_o$ and $\mu = \mu_r \mu_o$. For example, Teflon has $\epsilon_r = 2.1$ and $\mu_r = 1.0$. Note that the permeability μ is the same as in free space. This is an important property of *nonferrous or nonmagnetic* materials. On the other hand, the permeability of sheet steel (a ferrous or magnetic material) is 2000 times that of free space, $\mu_r = 2000$, whereas it has a relative permittivity of $\epsilon_r = 1.0$. For nonconductive media, other than free space, the velocity of wave propagation is

$$v = \frac{1}{\sqrt{\epsilon \mu}} \tag{1.4}$$

$$= \frac{v_o}{\sqrt{\epsilon_r \mu_r}}$$

For example, a wave propagating in Teflon ($\epsilon_r = 2.1$, $\mu_r = 1$) has a velocity of propagation of

$$v = \frac{v_o}{\sqrt{\epsilon_r \mu_r}}$$

$$= \frac{3 \times 10^8 \text{ m/s}}{\sqrt{2.1 \times 1}}$$

$$= 207{,}019{,}667.8 \text{ m/s}$$

$$= 0.69 \, v_o$$

Dielectric materials ($\mu_r = 1$) have relative permittivities (ϵ_r) typically between 2 and 5, so that velocities of propagation range from $0.70v_o$ to $0.45v_o$ in dielectrics. Table 1.1 gives ϵ_r for various dielectric materials.

It is very important for the reader to be able to *correctly* calculate the electrical dimensions of a structure at a particular frequency. The key to doing this is to realize that *a dimension of one meter in free space (air) is one wavelength at a frequency of 300 MHz.* Wavelengths in free space can be easily calculated at another frequency by appropriately scaling the dimension, remembering that one wavelength at 300 MHz is 1 m. To do this, it is important to realize that *as frequency increases, a wavelength decreases and vice versa.* For example, a wavelength at 50 MHz is $1 \text{ m} \times 300 \text{ MHz}/50 \text{ MHz} = 6 \text{ m}$. A wavelength at

TABLE 1.1

Material	ϵ_r
Bakelite	4.74
Mica	5.4
Neoprene	5–7
Nylon	3.5
Paper	3.0
Plexiglas	3.45
Polyethylene	2.27–2.50
Polypropylene	2.50–2.65
Polystyrene	2.56
Polyurethane	5.6–7.6
Polyvinyl chloride	3.0–8.0
Quartz	3.8
Rubber	2.5–3
Silicon	11.8
Silicone	3.0–3.5
Styrofoam	1.03
Teflon	2.1
Wood (dry)	1.5–4

TABLE 1.2   Wavelengths at Different Frequencies (in Air).

f	λ
60 Hz	3107 miles (5000 km)
3 kHz	100,000 m
30 kHz	10,000 m
300 kHz	1000 m
3 MHz	100 m
30 MHz	10 m
300 MHz	*1 m*
3 GHz	10 cm
30 GHz	1 cm
300 GHz	0.1 cm

2 GHz (1 GHz = 1000 MHz) is $300/2000 = 0.15$ m $= 15$ cm. Table 1.2 gives some representative values.

The electrical dimensions of a circuit or other electromagnetic structure need to be calculated to determine whether it is electrically small ($\mathscr{L} < \frac{1}{10}\lambda$) or not. If it is electrically small, we can apply simpler concepts and calculations than would be necessary if it were electrically large ($\mathscr{L} > \frac{1}{10}\lambda$). For example, Kirchhoff's voltage and current laws along with the lumped-circuit modeling of elements *are only applicable if the largest dimension of the circuit is electrically small*! If the circuit is electrically large, we have no other recourse but to use Maxwell's equations (or some appropriate simplification of them) in order to describe the problem. Clearly then, it is important to determine the electrical dimensions of a circuit. One can determine this by first calculating the wavelength at the *highest frequency of interest* and then computing k in relation to $\mathscr{L} = k\lambda$ by writing

$$k = \frac{\mathscr{L}}{\lambda} \tag{1.5}$$

For example, a circuit or radiating structure whose maximum dimension is 3.6 m and is operated at a frequency of 86 MHz is $3.6/3.49 = 1.03$ wavelengths because a wavelength in free space at 86 MHz is $300/86 = 3.49$ m. If this structure were immersed in a polyvinyl chloride (PVC) dielectric ($\epsilon_r = 4$, $\mu_r = 1$), its maximum dimension of 3.6 m would be 2.06 wavelengths since the wavelength of 86 MHz in PVC is

$$\lambda = \frac{v}{f}$$

$$= \frac{v_o}{f\sqrt{\epsilon_r\mu_r}} = \frac{\lambda_o}{\sqrt{\epsilon_r\mu_r}}$$

$$= \frac{3.49}{\sqrt{4 \times 1}} = 1.744 \text{ m}$$

1.5   DECIBELS AND COMMON EMC UNITS

The primary quantities of interest in the EMC problem are conducted emissions (voltage in volts (V), and current in amperes (A)) and radiated emissions (electric field in volts per meter (V/m) and magnetic field in amperes per meter (A/m)). Associated with these primary quantities are the quantities of power in watts (W) or power density in watts per square meter (W/m²). The numerical range of these quantities can be quite large. For example, electric fields can have values ranging from 1 μV/m to 200 V/m. This represents a *dynamic range* of over eight orders of magnitude (10^8). Because these wide ranges in units are common in the EMC community, EMC units are expressed in *decibels (dB)*. Decibels have the property of compressing data, e.g., a range of voltages of 10^8 is 160 dB. There are also other reasons for expressing these quantities in dB, as we will see. In order to be effective in EMC, we must be able to express and manipulate units that are expressed in dB. It is also important to conceptualize the values of various EMC units when they are expressed in dB. This is somewhat similar to the conversion from the English system of units (inches, feet, gallons,...) to the metric system of units (meters, centimeters, liters,...). Those accustomed to the English system have a feel for the length of, for example, 100 yards (the length of a football field in the US) but may have difficulty visualizing a length of 100 meters although both dimensions are approximately the same. In order to be effective in EMC, it is imperitive to convert, understand and use units expressed in dB. This section is devoted to that objective.

The decibel was originally developed in the telephone industry to describe the effect of noise in telephone circuits [4]. The ear tends to hear logarithmically so describing the effect of noise in dB is natural. To begin the discussion, consider the amplifier circuit shown in Fig. 1.4. A source consisting of an open-circuit voltage V_S and source resistance R_S delivers a signal to an amplifier whose load is represented by R_L. The input resistance to the amplifier is denoted by R_{in} and the power delivered to the amplifier is

$$P_{in} = \frac{v_{in}^2}{R_{in}} \tag{1.6}$$

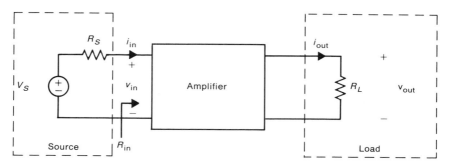

FIGURE 1.4   An illustration of the definition and use of the decibel (dB).

The power delivered to the load is

$$P_{out} = \frac{v_{out}^2}{R_L} \qquad (1.7)$$

The *power gain* of the amplifier is

$$\text{Power Gain} \equiv \frac{P_{out}}{P_{in}}$$

$$= \frac{v_{out}^2}{v_{in}^2} \frac{R_{in}}{R_L} \qquad (1.8)$$

The power gain expressed in decibels is *defined* as

$$\text{Power Gain}_{dB} \equiv 10 \log_{10} \left(\frac{P_{out}}{P_{in}} \right) \qquad (1.9)$$

where $\log_{10}$ is the logarithm to the base 10.

The related quantities of interest are the *voltage gain* and *current gain* of the amplifier. These are defined by

$$\text{Voltage Gain} \equiv \frac{v_{out}}{v_{in}} \qquad (1.10)$$

$$\text{Current Gain} \equiv \frac{i_{out}}{i_{in}} \qquad (1.11)$$

In dB these are *defined* as

$$\text{Voltage Gain}_{dB} \equiv 20 \log_{10} \left(\frac{v_{out}}{v_{in}} \right) \qquad (1.12)$$

$$\text{Current Gain}_{dB} \equiv 20 \log_{10} \left(\frac{i_{out}}{i_{in}} \right) \qquad (1.13)$$

Note that power gain in dB is defined as $10 \log_{10}$ of the ratio of the two quantities, whereas voltage gain and current gain in dB are defined as $20 \log_{10}$ of the ratio of the two quantities! Although this could be taken as simply definition, there is a rationale for it. To see this, observe that *if* the input resistance to the amplifier equals the load resistance, $R_L = R_{in}$, then (1.8) becomes

$$\text{Power Gain} = \left(\frac{v_{out}}{v_{in}} \right)^2_{R_L = R_{in}} \qquad (1.14)$$

In dB the power gain becomes, for $R_{in} = R_L$,

$$\text{Power Gain}_{dB} = 20 \log_{10} \left(\frac{v_{out}}{v_{in}} \right)_{R_L = R_{in}} \tag{1.15}$$

from which we obtain the defining relation for expressing the voltage gain in dB as in (1.12). In summary, *the ratio of two quantities in dB is given by*

$$dB \equiv 10 \log_{10} \left(\frac{P_2}{P_1} \right) \quad \text{(power)} \tag{1.16}$$

$$dB \equiv 20 \log_{10} \left(\frac{v_2}{v_1} \right) \quad \text{(voltage)} \tag{1.17}$$

$$dB \equiv 20 \log_{10} \left(\frac{i_2}{i_1} \right) \quad \text{(current)} \tag{1.18}$$

Note that *decibels are the ratio of two quantities*. Absolute power, voltage or current levels are expressed in dB by giving their value *above or referenced to some base quantity*. For example, voltages are commonly expressed relative to 1 μV as dBμV:

$$dB\mu V \equiv 20 \log_{10} \left(\frac{\text{volts}}{1 \, \mu V} \right) \tag{1.19}$$

For example, a voltage of 1 V is 120 dBμV since

$$20 \log_{10} \left(\frac{1 \, V}{1 \, \mu V \equiv 10^{-6} \, V} \right) = 20 \log_{10} 10^6$$

$$= 120 \, dB\mu V$$

This is commonly stated as "1 V is 120 dB above a microvolt." As another example, 1 mV is 60 dBμV. Similarly, 350 mV is

$$20 \log_{10} \left(\frac{350 \times 10^{-3} V}{10^{-6} V} \right) = 20 \log_{10} (350 \times 10^3)$$

$$= 110.88 \, dB\mu V$$

Conversely, 0.1 μV is -20 dBμV, or, 0.1 μV is 20 dB *below* a microvolt.

Some other standard units are dBmV (dB above a millivolt), dBμA (dB above a microamp) and dBmA (dB above a milliamp), where

$$dBmV \equiv 20 \log_{10} \left(\frac{volts}{1 \text{ mV}} \right) \tag{1.20}$$

$$dB\mu A \equiv 20 \log_{10} \left(\frac{amps}{1 \text{ } \mu A} \right) \tag{1.21}$$

$$dBmA \equiv 20 \log_{10} \left(\frac{amps}{1 \text{ mA}} \right) \tag{1.22}$$

Powers are also expressed relative to a microwatt, dBμW, and dB above a milliwatt, dBmW, or more commonly dBm as

$$dB\mu W \equiv 10 \log_{10} \left(\frac{watts}{1 \text{ } \mu W} \right) \tag{1.23}$$

$$dBm \equiv dBmW \tag{1.24}$$

$$\equiv 10 \log_{10} \left(\frac{watts}{1 \text{ mW}} \right)$$

Note that the pattern for the names follow that for voltage and current except that the very common *dB above a milliwatt is usually denoted as dBm*. Some examples are

$$15 \text{ mV} = 15{,}000 \text{ } \mu V$$
$$= 83.52 \text{ dB}\mu V$$
$$= 23.52 \text{ dBmV}$$

$$630 \text{ mA} = 630{,}000 \text{ } \mu A$$
$$= 115.99 \text{ dB}\mu A$$
$$= 55.99 \text{ dBmA}$$

$$250 \text{ mW} = 250{,}000 \text{ } \mu W$$
$$= 53.98 \text{ dB}\mu W$$
$$= 23.98 \text{ dBm}$$

Note the use of the designation dBm in the last example to designate dBmW.

Radiated electromagnetic fields are given in terms of electric field intensity in units of volts per meter, V/m, or in terms of magnetic field intensity in units of amperes per meter, A/m. The common EMC units reference these to 1 μV/m, 1 mV/m, 1 μA/m, or 1 mA/m as dBμV/m, dBmV/m, dBμA/m or dBmA/m, respectively. For example, one of the legal limits on radiated electric field is 100 μV/m. This translates to 40 dBμV/m. So these units translate in the same fashion as voltage and current:

$$dB\mu V/m \equiv 20 \log_{10} \left(\frac{V/m}{1\ \mu V/m} \right) \tag{1.25}$$

$$dB\mu A/m \equiv 20 \log_{10} \left(\frac{A/m}{1\ \mu A/m} \right) \tag{1.26}$$

It is also important to be able to convert a unit given in dB to its absolute value. To do this we use the definition of the logarithm of a number to the base m:

$$\log_m A = n \tag{1.27}$$

This denotes the power to which the base m must be raised to give A:

$$m^n = A \tag{1.28}$$

Therefore we may convert a number given in dB to its absolute value by performing the operation given in (1.28). For example, 108 dBμV is

$$108\ dB\mu V = 20 \log_{10} \left(\frac{V}{10^{-6}} \right)$$

Thus the absolute value of V in this expression is

$$V = 10^{108\ dB\mu V/20} \times 10^{-6}$$

$$= 0.2512\ V$$

The common conversions are

$$\text{volts} = 10^{dB\mu V/20} \times 10^{-6} \tag{1.29}$$

$$\text{volts} = 10^{dBmV/20} \times 10^{-3} \tag{1.30}$$

$$\text{watts} = 10^{dB\mu W/10} \times 10^{-6} \tag{1.31}$$

$$\text{watts} = 10^{dBm/10} \times 10^{-3} \tag{1.32}$$

The exact value is given in parentheses. Some other examples are

$$20 \log_{10} 360 = 20 \log_{10}(3 \times 2 \times 3 \times 2 \times 10)$$
$$= 10 + 6 + 10 + 6 + 20$$
$$= 52 \text{ dB} \quad (51.126)$$

$$10 \log_{10} \tfrac{1}{180} = 10 \log_{10} 1 - 10 \log_{10} 180$$
$$= 0 - 10 \log_{10}(2 \times 3 \times 3 \times 10)$$
$$= -3 - 5 - 5 - 10$$
$$= -23 \text{ dB} \quad (-22.52)$$

With these observations, the reader should be able to estimate a number in dB and convert a number expressed in dB to its absolute value. For example, an electric field intensity of $86 \text{ dB}\mu\text{V/m}$ is $86 = 20 + 20 + 20 + 20 + 6$. Thus $86 \text{ dB}\mu\text{V/m}$ represents (approximately) $10 \times 10 \times 10 \times 10 \times 2 = 2 \times 10^4 = 20{,}000 \ \mu\text{V/m}$ or 20 mV/m or 0.02 V/m. The exact value is $19{,}952.62 \ \mu\text{V/m}$.

The ability of the dB (and the logarithm) to compress large numbers into smaller ones means that we can make some relatively crude approximations and still arrive at a reasonable estimate of the number in dB. Another example of the utility of expressing EMC units in dB is given in Fig. 1.5. The power gain of the amplifier is the ratio of the output and input power

$$\text{Gain} = \frac{P_{\text{out}}}{P_{\text{in}}} \tag{1.36}$$

Thus the output power, given the input power, is

$$P_{\text{out}} = \text{Gain} \times P_{\text{in}} \tag{1.37}$$

Taking $10 \log_{10}$ of both sides of (1.37) and using (1.33) gives

$$P_{\text{out dB}} = \text{Gain}_{\text{dB}} + P_{\text{in dB}} \tag{1.38}$$

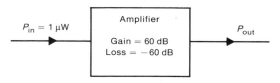

FIGURE 1.5   An illustration of the use of the decibel in computing amplifier performance.

The reference quantity for P_{out} and P_{in} used to convert them to dB can be any convenient base such as 1 mW. So (1.38) can be written in a number of ways as

$$P_{out\ dBm} = \text{Gain}_{dB} + P_{in\ dBm} \qquad (1.39a)$$

$$P_{out\ dB\mu W} = \text{Gain}_{dB} + P_{in\ dB\mu W} \qquad (1.39b)$$

Note that Gain in dB is the same in both cases. It is the ratio of two powers and as long as the two powers are expressed in the same units (dBμW, dBm, etc.), Gain is unchanged. This makes the computation of powers in a system simple, since outputs of signal sources are usually rated in terms of power (more about this later) and are typically given in dBm. For the example given in Fig. 1.5, $P_{out\ dBm} = 60$ dB $+ -30$ dBm $= 30$ dBm. Also, $P_{out\ dB\mu W} = 60$ dB $+ 0$ dBμW $= 60$ dBμW. So *the products of transfer functions become sums when the transfer functions are expressed in dB*. The same holds true when $R_{in} = R_L$ in Fig. 1.5 and the transfer function is a ratio of two voltages or two currents, or a ratio of a voltage and a current. Because of the way we defined the dB (10 log for power and 20 log for voltage and current), output and input quantities are similarly related with the same Gain in dB that was used for power in (1.38) and (1.39) (assuming $R_{in} = R_L$):

$$v_{out\ dB\mu V} = \text{Gain}_{dB} + v_{in\ dB\mu V} \qquad (1.40a)$$

$$v_{out\ dBmV} = \text{Gain}_{dB} + v_{in\ dBmV} \qquad (1.40b)$$

$$i_{out\ dB\mu A} = \text{Gain}_{dB} + i_{in\ dB\mu A} \qquad (1.40c)$$

$$i_{out\ dBmA} = \text{Gain}_{dB} + i_{in\ dBmA} \qquad (1.40d)$$

1.5.1   Power Loss in Cables

Computing the power loss in long connection cables is another example that illustrates the utility of expressing quantities in dB. To begin that discussion, we need to briefly review the topic of transmission lines. (This topic will be covered in more detail in Chapter 4.) Consider the transmission line of length $\mathscr{L}$ shown in Fig. 1.6. The line is usually expressed in terms of its *characteristic impedance* $\hat{Z}_C$ and *velocity of propagation of waves on the line v*. Although we may be interested in the behavior of the line when arbitrary time-domain pulses are applied to it, we are usually concerned with its *sinusoidal steady-state behavior*, i.e., for single frequency, sinusoidal excitation after all transients have died out. The equations for the voltage and current on the line at position z for sinusoidal steady-state excitation are [1]

$$\hat{V}(z) = \hat{V}^+ e^{-\alpha z} e^{-j\beta z} + \hat{V}^- e^{\alpha z} e^{j\beta z} \qquad (1.41a)$$

$$\hat{I}(z) = \frac{\hat{V}^+}{\hat{Z}_C} e^{-\alpha z} e^{-j\beta z} - \frac{\hat{V}^-}{\hat{Z}_C} e^{\alpha z} e^{j\beta z} \qquad (1.41b)$$

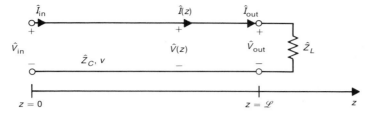

FIGURE 1.6 The basic transmission line problem with symbols defined.

The quantities $\hat{V}(z)$ and $\hat{I}(z)$ are the *phasor* line voltage and current, respectively, and are functions of position z on the line. The quantities $\hat{V}^+$ and $\hat{V}^-$ are undetermined constants that will be determined by the source and load attached to the line. We will denote all complex-valued quantities such as phasor voltages and currents with a caret ($\hat{\ }$). Each complex quantity will have a magnitude and a phase angle, e.g., $\hat{V} = V\ \underline{/\theta_V}$, $\hat{I} = I\ \underline{/\theta_I}$ and $\hat{Z} = Z\ \underline{/\theta_Z}$. The quantity α is the *attenuation constant* caused by losses in the line (in the line conductors and in the surrounding medium). If the line is lossless, $\alpha = 0$. The quantity β is the *phase constant* and gives the phase shift incurred in the wave as it moves down the line. The units of β are radians/m; whereas the units of α are nepers/m.

The expressions in (1.41) can be written as

$$\hat{V}(z) = \hat{V}_f(z) + \hat{V}_b(z) \tag{1.42a}$$

$$\hat{I}(z) = \frac{\hat{V}_f(z)}{\hat{Z}_C} - \frac{\hat{V}_b(z)}{\hat{Z}_C} \tag{1.42b}$$

where

$$\hat{V}_f(z) = \hat{V}^+\ e^{-\alpha z}\ e^{-j\beta z} \tag{1.43a}$$

$$\hat{V}_b(z) = \hat{V}^-\ e^{\alpha z}\ e^{j\beta z} \tag{1.43b}$$

The quantities $\hat{V}_f(z)$ and $\hat{V}_b(z)$ are referred to as the *forward- and backward-traveling waves*, respectively. This is because the phasor forms in (1.41) converted to the time domain become [1]

$$v(z, t) = \mathscr{R}e[\hat{V}(z)e^{j\omega t}]$$
$$= V^+\ e^{-\alpha z}\cos(\omega t - \beta z + \theta^+) + V^-\ e^{\alpha z}\cos(\omega t + \beta z + \theta^-) \tag{1.44a}$$

$$i(z, t) = \mathscr{R}e[\hat{I}(z)e^{j\omega t}] \tag{1.44b}$$

$$= \frac{V^+}{Z_C}\ e^{-\alpha z}\cos(\omega t - \beta z + \theta^+ - \theta_{Z_C})$$

$$- \frac{V^-}{Z_C}\ e^{\alpha z}\cos(\omega t + \beta z + \theta^- - \theta_{Z_C})$$

where t is the time variable and the complex quantities are written as $\hat{V}^+ = V^+ \; \underline{/\theta^+}$, $\hat{V}^- = V^- \; \underline{/\theta^-}$ and $\hat{Z}_C = Z_C \; \underline{/\theta_{Z_C}}$. The symbol $\mathscr{R}e[\quad]$ denotes the real part of the enclosed complex quantity. The forward-traveling waves contain $\cos(\omega t - \beta z + \theta)$. *As t increases, we must increase z in order to track the movement of a point on the waveform, i.e., to keep the argument of* $\cos(\omega t - \beta z + \theta)$ *constant.* Therefore this wave is moving in the positive z direction, a forward-traveling wave. Similarly, the term containing $\cos(\omega t + \beta z + \theta)$ represents a backward-traveling wave, since, as t increases, z must decrease in order to track a point on the waveform.

It is common to define the *voltage reflection coefficient* $\hat{\Gamma}(z)$ as the ratio of the backward- and forward-traveling phasor voltages as [1]

$$\hat{\Gamma}(z) = \frac{\hat{V}_b(z)}{\hat{V}_f(z)} \tag{1.45}$$

$$= \frac{\hat{V}^-}{\hat{V}^+} e^{2\alpha z} e^{j2\beta z}$$

The reflection coefficient at the load is [1]

$$\hat{\Gamma}_L = \frac{\hat{Z}_L - \hat{Z}_C}{\hat{Z}_L + \hat{Z}_C} \tag{1.46}$$

If $\hat{Z}_L = \hat{Z}_C$, the line is said to be *matched* and the reflection coefficient at the load is zero, $\hat{\Gamma}_L = 0$. The reflection coefficient at any point on the line can be related to the load reflection coefficient as [1]

$$\hat{\Gamma}(z) = \hat{\Gamma}_L \, e^{2\alpha(z - \mathscr{L})} \, e^{j2\beta(z - \mathscr{L})} \tag{1.47}$$

The general phasor expressions in (1.41) can be written in terms of the reflection coefficient as [1]

$$\hat{V}(z) = \hat{V}^+ \, e^{-\alpha z} \, e^{-j\beta z}[1 + \hat{\Gamma}(z)] = \hat{V}_f(z)[1 + \hat{\Gamma}(z)] \tag{1.48a}$$

$$\hat{I}(z) = \frac{\hat{V}^+}{\hat{Z}_C} e^{-\alpha z} \, e^{-j\beta z}[1 - \hat{\Gamma}(z)] = \frac{\hat{V}_f(z)}{\hat{Z}_C}[1 - \hat{\Gamma}(z)] \tag{1.48b}$$

The input impedance at any point on the line can be obtained as the ratio of (1.48a) and (1.48b) as

$$\hat{Z}_{in}(z) = \frac{\hat{V}(z)}{\hat{I}(z)} \tag{1.49}$$

$$= \hat{Z}_C \frac{1 + \hat{\Gamma}(z)}{1 - \hat{\Gamma}(z)}$$

It is important to realize that *if the line is matched, i.e.,* $\hat{Z}_L = \hat{Z}_C$, *the reflection coefficient at the load and anywhere on the line is zero and there are no backward-traveling waves on the line.* Therefore the phasor expressions for a *matched line* simplify to

$$\hat{V}(z) = \hat{V}^+ \, e^{-\alpha z} \, e^{-j\beta z} = \hat{V}_f(z) \tag{1.50a}$$

$$\hat{I}(z) = \frac{\hat{V}^+}{\hat{Z}_C} e^{-\alpha z} \, e^{-j\beta z} = \frac{\hat{V}_f(z)}{\hat{Z}_C} \tag{1.50b}$$

with

$$\hat{Z}_L = \hat{Z}_C$$

Thus the input impedance to a matched line at any point on the line is

$$\hat{Z}_{in}(z) = \frac{\hat{V}(z)}{\hat{I}(z)} \tag{1.51}$$

$$= \hat{Z}_C$$

with

$$\hat{Z}_C = \hat{Z}_L$$

The average power delivered to the right at any position z on the line is given by [1]

$$P_{av}(z) = \tfrac{1}{2} \mathscr{R}e[\hat{V}(z)\hat{I}^*(z)] \tag{1.52}$$

where the asterisk (*) denotes the complex conjugate of the quantity. This leads to the concept and characterization of power loss in cables that are used to interconnect measurement equipment. Typical interconnection cables are of the *coaxial* type, consisting of a cylindrical shield with an inner wire located on its interior axis. The waves travel in the space interior to the overall shield and that space is usually filled with a dielectric characterized by ϵ_r and $\mu_r = 1$ such as RG–58U, whose interior is filled with Teflon ($\epsilon_r = 2.1$). The voltage and current waves travel within this with a velocity of

$$v = \frac{v_0}{\sqrt{\epsilon_r \mu_r}} \tag{1.53}$$

Cable manufacturers usually specify the coaxial cable by giving (1) the magnitude of the characteristic impedance Z_C, assuming small losses ($Z_C = 50 \, \Omega$ for RG–58U), (2) the velocity of propagation as a percentage of free-space velocity

($v = 0.69\, v_o$ for RG–58U), and (3) the loss per 100 feet at a selected set of frequencies. It is this latter parameter, loss, that we need to understand.

Loss occurs in a transmission line via loss in the conductors as well as in the surrounding dielectric [1]. In the normal frequency range of use the primary loss mechanism is due to the loss in the conductors. The resistance of the conductors increases at a rate proportional to $\sqrt{f}$ due to the skin effect [1]. Nevertheless, the cable loss must be specified at each frequency of interest. Normally the cable manufacturers specify this at a few selected frequencies. For example, the loss of RG–58U coaxial cable is specified at 100 MHz as 4.5 dB/100 feet. Specification of the loss *assumes that the cable is matched,* $\hat{Z}_L = \hat{Z}_C$! In this case only forward-traveling waves exist on the line and are given by (1.50). For example, consider the expression for power given in (1.52). If the cable is matched, the reflection coefficient is zero ($\hat{\Gamma}(z) = 0$), and the average power delivered to the right at any point on the line is obtained by substituting (1.50) into (1.52) [1]:

$$P_{av}(z) = \frac{1}{2}\frac{V^{+2}}{Z_C}e^{-2\alpha z}\cos\theta_{Z_C} \tag{1.54}$$

with

$$\hat{Z}_L = \hat{Z}_C$$

The input power to the cable is

$$P_{av}(z = 0) = \frac{1}{2}\frac{V^{+2}}{Z_C}\cos\theta_{Z_C} \tag{1.55}$$

and the power delivered to the load is

$$P_{av}(z = \mathscr{L}) = \frac{1}{2}\frac{V^{+2}}{Z_C}e^{-2\alpha\mathscr{L}}\cos\theta_{Z_C} \tag{1.56}$$

The power loss in the cable is defined by

$$\text{Power Loss} = P_{av}(z = 0) - P_{av}(z = \mathscr{L}) \quad \text{(in W)} \tag{1.57}$$

Rather than stating this loss as in (1.57), cable manufacturers specify loss as the ratio of the input and output powers:

$$\text{Power Loss} = \frac{P_{av}(z = 0)}{P_{av}(z = \mathscr{L})} \tag{1.58}$$

$$= \frac{P_{in}}{P_{out}}$$

$$= e^{2\alpha\mathscr{L}}$$

as substitution of (1.55) and (1.56) will show. Cable manufacturers give the loss in dB/length. By this they mean

$$\text{Cable Loss}_{dB} = 10 \log_{10} e^{2\alpha\mathscr{L}} \tag{1.59}$$

$$= 20\alpha\mathscr{L} \log_{10} e$$

$$= 8.686\alpha\mathscr{L}$$

where $\mathscr{L}$ is chosen to be some length, e.g., 100 feet. This is obtained by *measuring the power delivered to that length of cable and the output power for a matched load* so that for those quantities expressed in dB we have, by converting (1.58) to dB,

$$\text{Cable Loss}_{dB} = P_{\text{in dBx}} - P_{\text{out dBx}} \tag{1.60}$$

where dBx denotes the power referenced to some level. Typically, dBm is used.

Given the manufacturer's specification of the cable loss, we can obtain the attenuation constant at that frequency from (1.59) as

$$\alpha = \frac{\text{Power Loss in dB/length}}{8.686\mathscr{L}} \tag{1.61}$$

where $\mathscr{L}$ in (1.61) is the length used to specify the loss by the manufacturer. For example, RG–58U coaxial cable is specified as having 4.5 dB/100 feet loss at 100 MHz. So the attenuation constant *at 100 MHz* is

$$\alpha = \frac{4.5}{8.686 \times 100}$$

$$= 5.18 \times 10^{-3} \text{ nepers/foot}$$

It is very important to realize that *the specification of cable loss as defined above assumes that the cable is matched, $\hat{Z}_L = \hat{Z}_C$. If the cable is not matched, the specification has nothing to do with the cable loss!*

1.5.2 Signal Source Specification

Signal sources (pulse and sinusoidal) can be characterized in terms of a Thévenin equivalent as shown in Fig. 1.7. The quantity V_{OC} is the open-circuit voltage and R_S is the source resistance. *Virtually all signal sources today have $R_S = 50 \, \Omega$!* Also, *the vast majority of instruments used to measure signals have an input resistance of 50 Ω and can be characterized as shown in Fig. 1.8, where $C_{in} = 0$ and $R_{in} = 50 \, \Omega$.* There are exceptions to this latter statement, notably voltmeters and some oscilloscopes. However, it can be said that if the input resistance is not designed to be 50 Ω, it will be designed to be very large, and its input

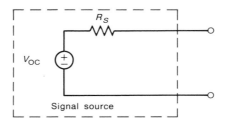

FIGURE 1.7 Specification of a signal source as a Thévenin equivalent circuit.

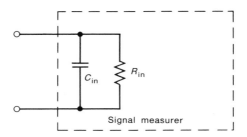

FIGURE 1.8 An equivalent circuit for the input to a signal measurer.

circuitry can generally be represented as a capacitance in parallel with a large resistance. It is very easy to determine the input representation of a particular signal measurer since the manufacturer will clearly state these parameters near the input connector. For example, typical spectrum analyzers used to display the frequency spectrum of a signal have $C_{in} = 0$ and $R_{in} = 50\ \Omega$. The high-impedance plug-in for an oscilloscope typically has $C_{in} = 47$ pF and $R_{in} = 1$ MΩ. However, there are also other plug-ins available that have $C_{in} = 0$ and $R_{in} = 50\ \Omega$.

With these concepts in mind, let us examine the connection of a signal source to a signal measurer with a length of coaxial cable as shown in Fig. 1.9. Suppose

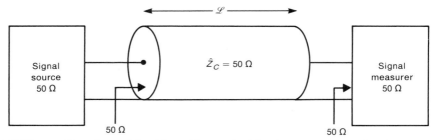

FIGURE 1.9 Use of coaxial cables with matched loads to connect a signal source to a signal measurer.

the signal measurer has an input impedance that is 50 Ω (purely resistive) and the coaxial connection cable also has a characteristic resistance of $\hat{Z}_C = 50\,\Omega$ (approximately real due to small losses of the cable). Because the load on this cable equals $\hat{Z}_C$, the cable is *matched* and the input impedance, *at any frequency* and *for any length of the cable*, is $\hat{Z}_{\text{in}} = 50\,\Omega = \hat{Z}_C$. This shows why signal measurers typically have input resistances of 50 Ω and coaxial cables have $\hat{Z}_C = 50\,\Omega$. Any other choice of characteristic impedance other than 50 Ω would be suitable, but 50 Ω has become the *industry standard. If the terminal resistance of the cable, which is the input resistance to the signal measurer, had not equalled $\hat{Z}_C$ of the cable, the input impedance to the cable as seen by the signal source would not be 50 Ω for all frequencies and all lengths $\mathscr{L}$ of the cable but would vary with frequency and cable length.* Consequently, it would be very difficult to determine the impedance presented to the source, and furthermore this impedance would vary with frequency and line length so that the source output would vary. (Although the open-circuit voltage of the source in Fig. 1.7 may be stable, the output voltage will depend on the source resistance R_S and the load resistance placed across its terminals). It is frequently important to be able to perform *swept-frequency measurements* in which the frequency of the source is swept over a band. If we could not rely on the output being constant with frequency, this swept measurement would be useless since we would not know the output at a particular frequency! This illustrates why *pieces of modern EMC test equipment have input and source impedances of pure 50 Ω and are connected by 50 Ω coaxial cables*!

Another important outcome of this practicality is that it is a trivial matter to compute the output of the source and the signal level delivered to the signal measurer. The output of a signal source is usually displayed on a meter of that instrument in terms of output power *to a matched load* in dBm. For example, consider Fig. 1.10, where a signal source is terminated in a load R_L either directly as the input to a device or via the input to a connection cable. If $R_L = R_S$ then the output voltage at the terminals of the source, V_{out}, is simply one-half of V_{OC}:

$$V_{\text{out}} = \frac{R_L}{R_S + R_L}\, V_{OC} \tag{1.62}$$

$$= \tfrac{1}{2} V_{OC}$$

with

$$R_L = R_S$$

Outputs of signal sources typically assume $R_S = R_L = 50\,\Omega$ and are given in terms of the power delivered to the $R_L = 50\,\Omega$ load in dBm:

$$P_{\text{out}} = \frac{V_{\text{out}}^2}{R_L = 50\,\Omega} \tag{1.63}$$

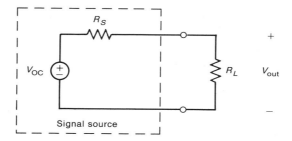

FIGURE 1.10   Calculation of a signal source output for a mismatched load.

Note the absence of the factor of $\frac{1}{2}$ in this power expression. This is because the voltage is assumed to be given in its *RMS* value, where $V_{\text{out peak}} = \sqrt{2} V_{\text{out RMS}}$ [2]. This is a typical assumption throughout industry; *voltages and currents are specified in their RMS value and no factor of $\frac{1}{2}$ is then required in power expressions.* The output power *into an assumed 50 Ω load* is given via a meter reading in dBm as

$$P_{\text{out dBm}} = 10 \log_{10}\left(\frac{P_{\text{out}}}{1\,\text{mW}}\right) \tag{1.64}$$

For example, if the output voltage across a 50 Ω load is $V_{\text{out}} = 120\,\mu\text{V} = 41.6\,\text{dB}\mu\text{V}$, the power delivered to this 50 Ω load is

$$P_{\text{out}} = \frac{(120 \times 10^{-6}\,\text{V})^2}{50\,\Omega} \times \frac{1000\,\text{mW}}{1\,\text{W}}$$

$$= 2.88 \times 10^{-7}\,\text{mW}$$

$$= -65.4\,\text{dBm}$$

Similarly, suppose the meter reading gives the output power (into an assumed 50 Ω load) of -37 dBm. This can be converted to give V_{out} (in RMS) by the following:

$$-37\,\text{dBm} = 2 \times 10^{-4}\,\text{mW}$$

$$= 2 \times 10^{-7}\,\text{W}$$

Thus

$$2 \times 10^{-7}\,\text{W} = V_{\text{out}}^2/50$$

giving $V_{\text{out}} = 3.162\,\text{mV}$ (RMS) $= 70\,\text{dB}\mu\text{V}$.

It is very important to realize that *if the load attached to the source is not 50 Ω then the meter reading does not give the output power across this load.* We can, however, determine the actual output voltage from the meter reading, but this requires some calculation. For example, suppose a $(50\,\Omega)$ source is set to put out -26 dBm, *but* is connected across a $150\,\Omega$ load. The simplest way to determine the actual output voltage is to (1) determine the V_{OC} of the source and then (2) compute V_{out}. First determine V_{OC} *assuming a load of 50 Ω* (which is what the meter reading is calibrated to assume). The output of -26 dBm *into a 50 Ω load* gives a power in that load of 0.002512 mW or 2.512×10^{-6} W. The voltage across a $50\,\Omega$ load that would dissipate this power is

$$V_{out} = \sqrt{50 \times P_{out}} \quad (R_L = 50\,\Omega)$$
$$= 11.2\text{ mV}$$
$$= 80.98\text{ dB}\mu\text{V}$$

So the open-circuit voltage (since all this assumes $R_S = R_L$) is

$$V_{OC} = 2 \times V_{out(R_L = 50\,\Omega)}$$
$$= 22.4\text{ mV}$$
$$= 87\text{ dB}\mu\text{V}$$

Now the actual output voltage can be computed from Fig. 1.10 using $R_S = 50\,\Omega$ and $R_L = 150\,\Omega$ as

$$V_{out} = \frac{150}{50 + 150} \times 22.4\text{ mV}$$
$$= 16.8\text{ mV}$$
$$= 84.5\text{ dB}\mu\text{V}$$

This could be directly obtained in dB. The open-circuit voltage can be immediately obtained as twice the output voltage when the load is $50\,\Omega$:

$$V_{OC\text{ dB}\mu V} = 6\text{ dB} + V_{out\text{ dB}\mu V}|_{R_L = 50\,\Omega}$$
$$= 6\text{ dB} + 80.98\text{ dB}\mu\text{V}$$
$$= 87\text{ dB}\mu\text{V}$$

Therefore the output voltage is

$$V_{out\text{ dB}\mu V} = \underbrace{20\log_{10}\left(\frac{150}{50 + 150}\right)}_{-2.5\text{ dB}} + V_{OC\text{ dB}\mu V}$$
$$= -2.5\text{ dB} + 87\text{ dB}\mu\text{V}$$
$$= 84.5\text{ dB}\mu\text{V}$$

Many signal measurers such as spectrum analyzers also have their responses specified *assuming a 50 Ω input impedance to the instrument*. For example, a level of -25 dBm means that -25 dBm or 0.003162 mW of power is dissipated in the 50 Ω input resistance of the instrument. The voltage across the input terminals can then be computed from the above basic power relations as

$$V_{in} = \sqrt{50 \times 0.003162 \text{ mW}}$$

$$= 12.574 \text{ mV}$$

$$= 82 \text{ dB}\mu\text{V}$$

The maximum signal that may be applied to the input of an instrument is often given in dBm. This is the maximum power that can be safely dissipated in the input resistance (assumed 50 Ω) of the instrument—above which you send the instrument back to the manufacturer for repair! This maximum input level is usually clearly marked near the input connector of the instrument, so there should be no excuse for "blowing out the front-end" of the instrument. For example, a spectrum analyzer has a maximum input rating of -30 dBm or 1 μW. Therefore the maximum input voltage to the instrument (in RMS) is

$$V_{in \text{ max}} = \sqrt{50 \times 1 \text{ μW}}$$

$$= 7.07 \text{ mV (RMS)}$$

Finally, we can now use these principles to compute the signal measured by the signal measurer of Fig. 1.9 given the cable loss (at the appropriate frequency) and the output reading of the meter on the signal source. Throughout this we assume a 50 Ω source, cable and signal measurer. If this is not the case, none of the following makes sense, and, furthermore, the actual level measured by the signal measurer would be extremely difficult to determine *if at all* (without some other measurements being made). Assume that the signal source meter shows the source is putting out a 100 MHz signal with a level of -30 dBm. Suppose the cable (RG–58U) is 150 feet in length. The cable loss at 100 MHz is 4.5 dB/100 feet. The received power is computed from

$$P_{rec} = \frac{P_{out} \text{ of cable}}{P_{in} \text{ to cable}} \times P_{source} \tag{1.65}$$

Taking $10 \log_{10}$ of both sides of this expression gives

$$P_{rec \text{ dBm}} = \text{Cable Gain}_{dB} + P_{source \text{ dBm}} \tag{1.66}$$

$$= -\frac{4.5 \text{ dB}}{100 \text{ feet}} \times 150 \text{ feet} + (-30 \text{ dBm})$$

$$= -36.75 \text{ dBm}$$

This converts to a power of 0.2113 μW or voltage across the 50 Ω load of 3.25 mV or 70.24 dBμV.

Because of the definition of dB as 10 $\log_{10}$ for power ratios and 20 $\log_{10}$ for voltage and current ratios, we can also use the cable loss (a ratio of powers) in dB to convert voltages. However, because this definition for voltages or currents implicitly assumed that $R_{in} = R_L$ in Fig. 1.4, its use in converting voltages requires a matched load as in Fig. 1.9. For example, the output power of the source in the previous problem of -30 dBm translates to a voltage across its output terminals of 7.07 mV or 77 dBμV. Since the cable is matched, the voltage across the input terminals of the signal measurer can also be computed as

$$V_{\text{rec dB}\mu\text{V}} = \text{Cable Gain}_{\text{dB}} + V_{\text{source dB}\mu\text{V}} \qquad (1.67)$$

For this problem we compute

$$V_{\text{rec dB}\mu\text{V}} = -\frac{4.5 \text{ dB}}{100 \text{ feet}} \times 150 \text{ feet} + 77 \text{ dB}\mu\text{V}$$

$$= -6.75 \text{ dB} + 77 \text{ dB}\mu\text{V}$$

$$= 70.24 \text{ dB}\mu\text{V}$$

as we computed using power.

REFERENCES

[1]   C.R. Paul and S.A. Nasar, *Introduction to Electromagnetic Fields*, second edition, McGraw-Hill, NY (1987).

[2]   C.R. Paul, *Analysis of Linear Circuits*, McGraw-Hill, NY (1989).

[3]   C.R. Paul, S.A. Nasar, and L.E. Unnewehr, *Introduction to Electrical Engineering*, McGraw-Hill, NY (1986).

[4]   H.W. Ott, *Noise Reduction Techniques in Electronic Systems*, second edition, John Wiley Interscience, NY (1988).

[5]   R.K. Keenan, *Digital Design for Interference Specifications*, The Keenan Corporation, Pinellas Park, FL (1983).

[6]   B. Keiser, *Principles of Electromagnetic Compatibility*, third edition, Artech House, Dedham, MA (1987).

[7]   N. Violette, D.R.J. White, and M. Violette, *Electromagnetic Compatibility Handbook*, Van Nostrand Reinhold, NY (1987).

[8]   J.R. Barnes, *Electronic System Design: Interference and Noise Control Techniques*, Prentice-Hall, Englewood Cliffs, NJ (1987).

[9]   E.R. Freeman and M. Sachs, Electromagnetic Compatibility Design Guide, Artech House, Dedham, MA (1982).

[10] T.A. Jerse, *Designing for Electromagnetic Compatibility*, Hewlett Packard Company, P/N 11949–60001, Palo Alto, CA (1989).

[11] C.R. Paul, An undergraduate course in electromagnetic compatibility, *1986 IEEE International Symposium on Electromagnetic Compatibility, San Diego, CA, September 1986.*
[12] C.R. Paul, Establishment of a university course in electromagnetic compatibility (EMC), *IEEE Trans. on Education,* **33**, 111–118 (1990).

PROBLEMS

1.1 Reduce each of the following signal transmission situations to a source, coupling path, and receptor:

1. AM radio transmission to the human ear.
2. TV transmission to the human eye.
3. Radar target identification.
4. Transfer of 60 Hz power to an air conditioner.
5. Transfer of digital computer data to a printer.
6. Unintended interference between a radar transmitter and a digital computer.
7. Interference between an automobile ignition system and the car radio.
8. Interference in a radio due to lightning.
9. Several other interference situations of your choosing.

1.2 Scan the topic index of the following publications to get an appreciation for the various topics of EMC:

1. *IEEE Transactions on Electromagnetic Compatibility.*
2. *Proceedings of the IEEE International Symposia on Electromagnetic Compatibility.*
3. Various trade magazines such as *RF Design, EMC Technology, ITEM, Compliance Engineering,* and *EMC Test and Design.*

1.3 Convert the following dimensions to those indicated:

1. 30 miles to km [48.2 km].
2. 1 ft to mils [12,000 mils].
3. 100 yds (length of a football field) to meters [91.44 m].
4. 1 mm to mils [39.37 mils].
5. 1 μm (micron) to mils [0.03937 mils].
6. 880 yd (race distance) to m [804.67 m].

1.4 Determine the wavelength at the following frequencies in metric and English units:

1. 850 MHz, free space [35.3 cm, 13.9 in].
2. 430 MHz, Teflon [48.1 cm, 18.95 in].
3. 250 kHz, air [1200 m, 3937 ft].
4. 20 kHz (RS–232 data transmission), air [15,000 m, 9.32 miles].
5. 450 kHz–1.5 MHz AM radio transmission, air [667 m–200 m, 2187 ft–656 ft].
6. 108 MHz FM transmission, air [2.78 m, 9.11 ft].
7. 20 MHz computer clock fundamental frequency, glass epoxy ($\epsilon_r = 4.7$) [6.92 m, 22.7 ft].
8. 1 GHz microwave signal, air [30 cm, 11.81 in].

1.5 Determine the following physical dimensions in wavelengths:

1. 120 MHz, 18 cm, air [0.072λ].
2. 100 MHz, 6 ft, air [0.61λ].
3. 500 MHz, 10 in, glass–epoxy ($\epsilon_r = 4.7$) [0.918λ].
4. A 6 ft printer cable at 80 MHz in air [0.49λ].
5. The 3 m measurement distance of the FCC regulations at the lower frequency (30 MHz) and upper frequency (1 GHz) of the limit [0.091λ, 3.048λ].
6. A 10 mile length of 60 Hz power transmission line in air [$3.22 \times 10^{-3}\lambda$].
7. An automobile (12 ft) at the lower frequency of the AM band (450 kHz) [$5.5 \times 10^{-3}\lambda$].

1.6 Determine the following voltages in dBµV and dBm:

1. 23 mV [87.2 dBµV, -19.8 dBm].
2. 670 µV [56.5 dBµV, -50.5 dBm].
3. 3.2 V [130 dBµV, 23.1 dBm].
4. 0.1 µV [-20 dBµV, -127 dBm].
5. 1 mV [60 dBµV, -47 dBm].
6. 300 mV [110 dBµV, 2.55 dBm].
7. 21 mV [86.4 dBµV, -20.6 dBm].
8. 30 V [149.5 dBµV, 42.5 dBm].
9. 48 mV [93.6 dBµV, -13.37 dBm].
10. 0.3 V [109.5 dBµV, 2.55 dBm].
11. 0.5 µV [-6.02 dBµV, -113 dBm].
12. 200 mV [106 dBµV, -0.97 dBm].

1.7   Convert the following quantities to V:

1. -26 dBμV [0.05 μV].
2. -35 dBm [4 mV].
3. -16 dBm [35 mV].
4. 36 dBμV [63.1 μV].
5. -28 dBmV [39.8 μV].
6. 20 dBm [2.24 V].

1.8   Determine a simple expression to convert voltage in dBμV to dBm [dBμV = 107 + dBm].

1.9   Determine the following electric field intensity levels in dBμV/m:

1. 100 μV/m [40 dBμV/m].
2. 1 mV/m [60 dBμV/m].
3. 200 V/m [166 dBμV/m].

1.10   Estimate the following ratios of currents or voltages in dB. Give exact values.

1. 200 [46, 46].
2. 640 [56, 56.1].
3. 32×10^{-3} [-30, -29.9].
4. 5.7×10^{-6} [-104, -105].

1.11   A 50 Ω source is connected to a 50 Ω receiver using 30 ft of RG–58U coaxial cable. If the source output is 100 MHz and -30 dBm, determine the voltage at the receiver in mV and dBμV. [6.05 mV, 75.6 dBμV].

1.12   A 50 Ω receiver is attached to an antenna via 200 m of RG–58U coaxial cable. The receiver indicates a level of -20 dBm at 200 MHz. Determine the voltage at the base of the antenna in dBμV if the cable loss at 200 MHz is 8 dB/100 ft. [139.48 dBμV]

1.13   A 50 Ω source is tuned to 100 MHz and attached to a 50 Ω spectrum analyzer with 200 ft of 50 Ω coaxial cable that has a loss of 4.5 dB/100 ft at 100 MHz. The spectrum analyzer reads a level of signal at 100 MHz of 56.5 dBμV. If the cable is removed and the signal source is attached directly to a 100 Ω load, determine the voltage across this load in dBμV. [67.98 dBμV] Determine the reading on the meter of the source in dBm. [-41.5 dBm]

1.14   A 50 Ω source is attached to a 50 Ω receiver with 200 ft of RG–58U coaxial cable (4.5 dB/100 ft loss at 100 MHz). The source is tuned to 100 MHz and the meter indicates that the output is -30 dBm. Determine the voltage at the receiver input in dBμV. [68 dBμV] If the voltage at the output terminals of the source is measured and found to be 30 mV, determine the voltage at the input to the receiver in dBm. [-26.45 dBm] If the received voltage at the terminals of the receiver is measured and found to be -50 dBm, determine the voltage at the output terminals of the source in dBμV. [66 dBμV]

1.15 A 50 Ω source is attached to a 300 Ω receiver with a 100 ft length of Twin Lead ($Z_C = 300\ \Omega$). The Twin Lead has a loss of 10 dB/100 ft at the frequency of the source. Determine the voltage at the input to the receiver in dBμV if the source indicates an output of -30 dBm. [71.7 dBμV]

1.16 A 50 Ω oscillator is putting out a signal level of -20 dBm according to the meter on the instrument. If a 150 Ω load is placed across its terminals, determine the output voltage in dBμV. [90.5 dBμV]

1.17 A 50 Ω oscillator is attached to the high-impedance input of an oscilloscope ($C_{in} = 47$ pF, $R_{in} = 1$ MΩ). The source is tuned to 100 MHz and the level set to -30 dBm. Determine the voltage level (peak) of the sinusoid seen on the oscilloscope face. [11.22 mV]

$100\ ft = 30.48\ m$

$1\ ft = 0.3048\ m$

CHAPTER TWO

EMC Requirements for Electronic Systems

In this chapter we will discuss the *motivation* for studying the subject of EMC. This motivation results from the imposition of additional design objectives for electronic systems over and above those required for the functional performance of the system. These additional design objectives stem from the overall requirement that the system be *electromagnetically compatible* with its environment. There are basically two classes of EMC requirements that are imposed on electronic systems:

1. *Those mandated by governmental agencies.*
2. *Those imposed by the product manufacturer.*

The requirements imposed by governmental agencies are *legal* requirements and generally cannot be waived. These requirements are imposed in order to control the interference produced by the product. However, compliance with these EMC requirements does not guarantee that the product will not cause interference. It only allows the country imposing the requirement to control the amount of "electromagnetic pollution" that the product generates. In order for the product to be marketed (advertised and sold) in a country, the product must comply with these requirements. If a product cannot be sold due to its inability to comply with the governmental EMC requirements, the fact that it may perform a function that gives it significant sales potential is unimportant!

On the other hand, EMC requirements that manufacturers impose on their products are intended to result in customer satisfaction. They are imposed for the purpose of insuring a reliable, quality product. For example, if a new digital computer turns out to be highly susceptible to electrostatic discharge (ESD), the company will obtain a poor reputation from the standpoint of quality control of its product, resulting in loss of future sales on this and other products.

Maintaining a good reputation in the marketplace is clearly of critical importance. We will only be able to discuss these company-mandated requirements in general terms since most are proprietary and not available to the public. Nevertheless, most manufacturers impose a fairly standard set of requirements which we will discuss.

Compliance with both of the above EMC requirements is critical to the success of the product in the marketplace. We will begin this discussion with governmental requirements and then discuss those requirements that are imposed by the manufacturing company.

2.1 GOVERNMENTAL REQUIREMENTS

With the availability of rapid, global transportation and communications, the marketplace today encompasses the entire world. Consequently, the EMC requirements of all countries are of importance to manufacturers of electronic equipment. We will divide these into two sectors: those imposed on products marketed in the United States (US) and those imposed on products marketed outside the US. Furthermore, the regulatory requirements of each country are further subdivided into those for commercial use and those for military use.

In this section we will give a brief overview of the typical regulatory requirements. There are a number of technical nuances in each regulation that will not be discussed so that we can focus on the general requirements without getting involved in minutiae of technical detail. It should be emphasized that the governmental requirements are in a constant state of change. The regulations that we will discuss are those in effect as of this writing. The reader should consult the most current issue of those regulations.

2.1.1 Requirements for Commercial Products Marketed in the United States

In the United States the Federal Communications Commission (FCC) is charged with the regulation of radio and wire communication. A significant part of that responsibility is to control interference from and to wire and radio communication. The FCC Rules and Regulations contained in Title 47 of the Code of Federal Regulations have several parts that apply to nonlicensed electronic equipment. Part 15 applies to radio-frequency devices and will be of primary concern here [1]. The range of frequencies defined by the FCC to be "radio frequencies" extends from 9 kHz to 3000 GHz. A radio-frequency device is any device that is capable of emitting radio-frequency energy by radiation, conduction or other means whether intentionally or not. The purpose of Part 15 is to control the interference from these emitters. Transmitters operating under a radio station license are covered in another part. Some examples of "radio-frequency devices" are dc motors where arcing at the brushes generates a wide spectrum of energy that includes this band of frequencies, digital computers whose "clock" signals generate radiated emissions in this band, electronic typewriters that also employ digital circuitry, etc. Our discussion of the FCC regulations will be brief but will cover the essential points and will concern those parts that affect digital devices. More detailed discussions of the regulations are found in [2–4].

In 1979 the FCC published, under Part 15 of its Rules and Regulations, a requirement that has had and will continue to have considerable impact on the electrical engineering community and the electronics industry. With the increasing proliferation of computers and other digital devices, the FCC realized that some limits on the electromagnetic emissions from these devices was necessary in order to minimize their potential for interfering with radio and wire communications. Numerous instances of such interference had begun to surface with increasing regularity. This resulted in the publication of the above rule, which has the force of law. It basically sets limits on the radiated and conducted emissions of a *digital device*. The FCC defines a digital device as

> *Any unintentional radiator (device or system) that generates and uses timing pulses at a rate in excess of 9000 pulses (cycles) per second and uses digital techniques... .*

Any electronic device that has digital circuitry and uses a clock signal in excess of 9 kHz is covered under the rule, although there are a limited number of exemptions. This rule includes, for example, electronic typewriters, calculators, point-of-sale terminals, printers, modems, etc., as well as personal computers.

It is *illegal* to market a "digital device" in the US unless its radiated and conducted emissions have been measured and do not exceed the limits of the regulation. The FCC considers *marketing* as shipping, selling, offering for sale, etc. Monetary fines and/or jail terms can be imposed for the willful violation of this rule. Companies that manufacture these products are not as much concerned about these consequences as they are the devastating publicity resulting from a "minor infraction." They are also concerned about the financial impact of a potential recall of a product if some units are randomly tested by the FCC and found to exceed the limits. So it is not sufficient to construct one sample that complies with the regulation. All units sold must comply.

The FCC further breaks the digital device class of products into Class A and Class B. Class A digital devices are those that are marketed for use in a commercial, industrial or business environment. Class B digital devices are those that are marketed for use in a residential environment, notwithstanding their use in a commercial, industrial or business environment. The Class B limits are more stringent than the Class A limits under the reasonable assumption that interference from the device in an industrial environment can be more readily corrected than in a residential environment, where the interference source and the susceptible device are likely to be in closer proximity. Further, the owner of the interfering device in a residential environment is not as likely to have the expertise or financial resources to correct the problem as would an industrial user. Hence the potential for interference is more closely controlled for the residential market. Personal computers and their peripherals are a subcategory of Class B digital devices. They must be tested for compliance by the manufacturer and the test data submitted to the FCC for *certification*. The FCC may request a sample to test. For all other "digital devices," the manufacturer must test the device for compliance, and no test data are required to be submitted to the FCC. The FCC employs random sampling to verify compliance. In this subsection we will discuss the limits. The measurement procedure to verify compliance will be discussed in Section 2.1.4.

The FCC limits of this rule concern the conducted and radiated emissions of the digital product. Conducted emissions are those currents that are passed out through the unit's ac power cord and placed on the common power net, where they may radiate more efficiently due to the much larger expanse of this "antenna" and thus cause interference with other devices. The frequency range for conducted emissions extends from 450 kHz to 30 MHz. Compliance is verified by inserting a line impedance stabilization network (LISN) into the unit's ac power cord. Although the emission to be controlled is current passing out the ac line cord, the limits are given in volts. This is because the test device (LISN) described in Section 2.1.4 measures a voltage that is directly related to the interference current. Radiated emissions concern the electric and magnetic fields radiated by the device that may be received by other electronic devices, causing interference in those devices. The FCC, as well as other regulatory agencies, requires measurement of only the radiated electric field, and the regulatory limits are given in terms of that field in dBμV/m. The frequency range for radiated emissions begins at 30 MHz and extends to 40 GHz. Compliance is verified by measuring the radiated electric fields of the product in either a semianechoic chamber or at an open-field test site. The radiated emissions must be measured with the measurement antenna in both the vertical and horizontal polarizations with respect to the ground plane of the test site, and the product must comply for both polarizations.

FCC Part 15 requirements have recently been revised and now include three subparts [5, 6]:

Subpart A: General requirements.

Subpart B: Unintentional radiators.

Subpart C: Intentional radiators.

Subpart A contains general information and Subpart B contains the specific requirements for unintentional radiators, which are primarily radio receivers and digital devices. These new requirements took effect June 23, 1989. One of the many changes is in the upper frequency of applicability for radiated emissions. These are given in Table 2.1. The new emission limits are given in Table 2.2 for Class B devices. The radiated emissions are to be verified at 3 m as for the old requirements and are to be measured with the measurement antenna in horizontal and in vertical polarizations. The new limits for Class A digital devices are given in Table 2.3. The measurement distance for radiated emissions has been changed to 10 m for Class A measurements. The radiated emissions are to be measured at the required distance from the product (3 m for Class B

TABLE 2.1

Highest Frequency Generated or Used in the Device or on Which the Device Operates or Tunes (MHz)	Upper Frequency of Measurement Range (MHz)
< 1.705	30
1.705–108	1000
108–500	2000
500–1000	5000
> 1000	5th harmonic of highest frequency or 40 GHz, whichever is lower

TABLE 2.2   FCC Emission Limits for Class B Digital Devices.

Radiated Emissions (3 m)		
Frequency (MHz)	μV/m	dBμV/m
30–88	100	40
88–216	150	43.5
216–960	200	46
> 960	500	54
Conducted Emissions		
Frequency (MHz)	μV	dBμV
0.45–30	250	48

TABLE 2.3   FCC Emission Limits for Class A Digital Devices.

Radiated Emissions (10 m)		
Frequency (MHz)	μV/m	dBμV/m
30–88	90	39
88–216	150	43.5
216–960	210	46
> 960	300	49.5
Conducted Emissions		
Frequency (MHz)	μV	dBμV
0.45–1.705	1000	60
1.705–30	3000	69.5

and 10 m for Class A), with the measurement antenna in horizontal and in vertical polarizations. The product must also be rotated to obtain the maximum radiation. The Class B conducted emissions are illustrated in Fig. 2.1, whereas the Class A conducted emissions are illustrated in Fig. 2.2. The Class B and Class A radiated emissions are illustrated in Figs. 2.3 and 2.4, respectively.

It is informative to compare the radiated emission limits for Class A and Class B products in order to determine how much less stringent the Class A limits are. However, the radiated emissions for Class B devices are to be measured at a distance of 3 m from the product, whereas the radiated emissions for Class A devices are to be measured at a distance of 10 m. In order to compare these limits, we must determine a method for scaling the measurements from one measurement distance to another. A common way of doing this is to use

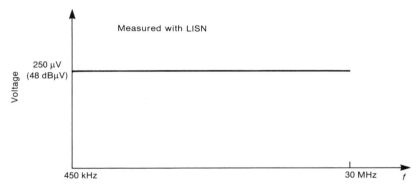

FIGURE 2.1   The FCC conducted emission limits for Class B digital devices.

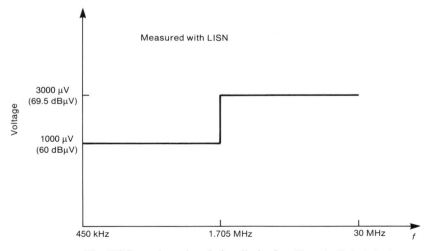

FIGURE 2.2   The FCC conducted emission limits for Class A digital devices.

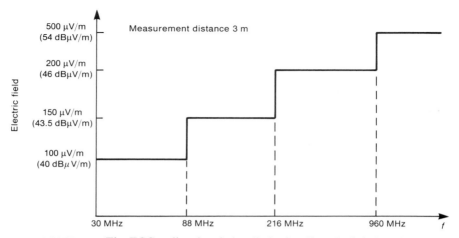

FIGURE 2.3   The FCC radiated emission limits for Class B digital devices.

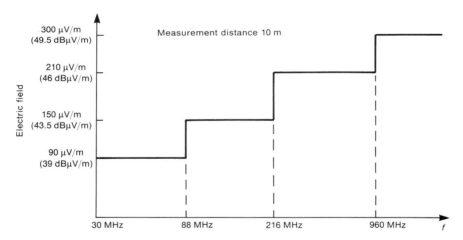

FIGURE 2.4   The FCC radiated emission limits for Class A digital devices.

the *inverse distance method* in which the emissions are assumed to fall off linearly with increasing distance to the measurement antenna. Thus the emissions at 3 m are assumed to be reduced by $\frac{3}{10}$ if the measurement antenna is moved to a further distance of 10 m and vice versa. To translate the Class A limits from a distance of 10 m to 3 m, we *add* $20 \log_{10} \frac{10}{3} = 10.46$ dB $\cong 10$ dB to the Class A limits, since moving the measurement point closer to the source is expected to increase the electric field levels that are measured. These are compared in Fig. 2.5 and show, based on this extrapolation, that the Class A limits are some 10 dB less stringent than the Class B limits. It should be pointed out that, as we shall see in Chapter 5, the emissions from antennas fall off inversely with

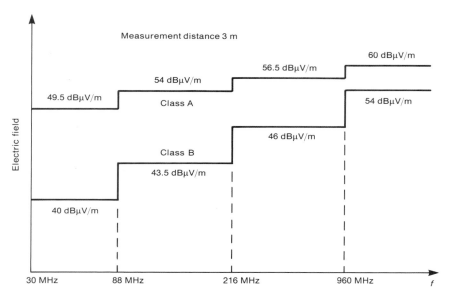

FIGURE 2.5   A comparison of the FCC Class A and FCC Class B radiated emission limits for a measurement distance of 3 m.

distance only if the measurement points are in the *far field* of the emitter. An approximate criterion for the far-field boundary is three wavelengths (3λ). Therefore the near-field–far-field boundary at the lowest measurement frequency of 30 MHz is 30 m, but is 90 cm at 1 GHz. Thus the measurement distance of 3 m is likely in the near field of the product at 30 MHz, and this extrapolation of the Class A limit at 30 MHz to 3 m (or the extrapolation of the Class B limit out to 10 m) is probably not valid. This comparison, although somewhat approximate, nevertheless illustrates the fact that emissions of Class A devices are controlled to a lesser degree than are those of Class B devices.

2.1.2   Requirements for Commercial Products Marketed Outside the United States

We should begin this discussion of the legal requirements for commercial products that are marketed in countries other than the US by stating that, at present, these requirements vary from country to country. Consequently we cannot give a single discussion that covers all countries. However, there are certain common requirements that cover the bulk of these.

In the past each country had its somewhat unique EMC requirements. For example, products intended to be marketed in Germany are required to comply with the VDE (Verband Deutscher Elektrotechniker) regulations [3]. The general requirements in this regulation are quite similar to the FCC requirements (limits on radiated and conducted emissions are specified), but the frequency ranges and limit levels are different. There is a trend toward the adoption of a single standard for international EMC requirements. The primary candidate for this international EMC standard is one developed in 1985 by CISPR (the French translation meaning International Special Committee on Radio Interference). CISPR is a committee of the International Electrotechnical Commission (IEC), which is an international body that promulgates standards in order to facilitate trade between countries. CISPR is not a regulatory body but simply develops standards which when adopted by a government become the standard for that government. CISPR published a set of emission standards in 1985 referred to as Publication 22 that concern information technology equipment (ITE), which includes digital devices [7]. Many countries in Europe and throughout the world have adopted the CISPR 22 standards or some variant of them as their national standard, and it is anticipated that CISPR 22 will become the world standard. Although many countries still hold their own variation of CISPR 22, we will concentrate on this standard as our international standard. Other standards (including the FCC standards) are very similar to the CISPR 22 standards. For a general discussion of the evolution of international EMC requirements see [8].

The radiated emission limits for Class B devices are to be measured at 10 m and are given in Table 2.4. These are shown in Fig. 2.6(a) and compared with the FCC Class B limits extrapolated to a measurement distance of 10 m. (Subtract 10.46 dB or approximately 10 dB from the FCC Class B limits at 3 m to translate them to 10 m.) From this comparison we see that the CISPR 22 Class B limits are somewhat more restrictive than the FCC Class B limits in the frequency range of 88–230 MHz. From 88 to 216 MHz the CISPR 22 limits are 3 dB more restrictive, and from 216 to 230 MHz they are 5.5 dB more restrictive. From 230 to 960 MHz the FCC limits are more restrictive by about 1.5 dB.

The radiated emission limits for Class A devices are shown in Fig. 2.6(b) and given in Table 2.5. These are to be measured at 30 m. The FCC Class A limits are translated to a 30 m measurement distance from their specified measurement distance of 10 m by subtracting 9.54 dB ($20 \log_{10} \frac{10}{30} = -9.54$ dB) or approximately 10 dB and are shown for comparison. Again we see that the CISPR 22 limits for Class A digital devices are more restrictive than the FCC limits in the frequency range of 88–216 MHz by some 4 dB and 6.5 dB in the range of 216–230 MHz. From 230 to 960 MHz the CISPR 22 limits are less restrictive than the FCC limits by some 0.5 dB.

TABLE 2.4   CISPR 22 Radiated Emission Limits Class B Digital Devices (10 m).

Frequency (MHz)	μV/m	dBμV/m
30–230	31.6	30
230–1000	70.8	37

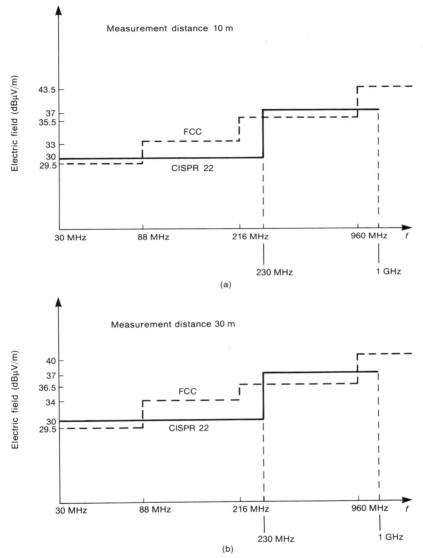

FIGURE 2.6 The CISPR 22 radiated emission limits compared to the FCC radiated emission limits: (a) Class B; (b) Class A.

TABLE 2.5   CISPR 22 Radiated Emission Limits for Class A Digital Devices (30 m).

Frequency (MHz)	μV/m	dBμV/m
30–230	31.6	30
230–1000	70.8	37

The CISPR 22 and FCC limits on conducted emissions are compared in
Fig. 2.7. A significant difference between the CISPR 22 and FCC conducted
emission limits is in the frequency range of applicability. The CISPR 22
conducted emission limits extend down to 150 kHz instead of 450 kHz as for
the FCC limits. Both extend to an upper limit of 30 MHz. Note that the
CISPR 22 limit for Class B devices rises below 500 kHz. This extension was put

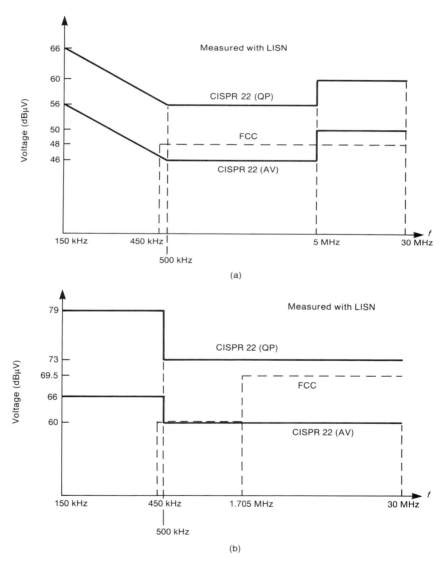

FIGURE 2.7   The CISPR 22 conducted emission limits compared to the FCC conducted
emission limits: (a) Class B; (b) Class A.

TABLE 2.6   CISPR 22 Conducted Emission Limits for Class B Digital Devices.

Frequency (MHz)	μV QP(AV)	dBμV QP(AV)
0.15	1995(631)	66(56)
0.5	631(199.5)	56(46)
0.5–5	631(199.5)	56(46)
5–30	1000(316)	60(50)

TABLE 2.7   CISPR 22 Conducted Emission Limits for Class A Digital Devices.

Frequency (MHz)	μV QP(AV)	dBμV QP(AV)
0.15–0.5	8912.5(1995)	79(66)
0.5–30	4467(1000)	73(60)

in place to cover the emissions of *switching power supplies*, which are growing in importance over *linear power supplies* due to their efficiency and light weight. Switching power supplies generate a dc voltage by "chopping" the ac voltage waveform. (More will be said about this in Chapter 9.) Typical chopping frequencies range from 20 kHz to over 100 kHz. Thus these types of power supplies generate conducted noise at the basic switch rate and its harmonics. The break frequencies of the CISPR 22 Class B limit are at 500 kHz and 5 MHz. The break frequency for the CISPR 22 Class A limit is at 500 kHz. The FCC Class B limits are 2 dB less restrictive than the CISPR 22 limits from 500 kHz to 5 MHz but 2 dB more restrictive from 5 to 30 MHz. The CISPR 22 Class A limits are 9.5 dB more restrictive than the FCC Class A limits from 1.705 to 30 MHz. It should be noted that the VDE limits on conducted emissions for digital products marketed in Germany extend down to 10 kHz! The CISPR 22 limits are given in Table 2.6 for Class B devices and in Table 2.7 for Class A devices. These tables show the limits when the receiver uses a *quasi-peak* detector (QP) and when the receiver uses an *average* detector (AV). These detectors are discussed in Section 2.1.4. Both the FCC and CISPR 22 radiated emission limits and the FCC conducted emission limits apply to the use of a quasi-peak detector.

2.1.3   Requirements for Military Products Marketed in the United States

A large portion of the products of commercial firms in the US and outside the US are produced for military applications. Specification of limits on emissions for the control of interference is obviously more critical for military products than commercial products, since interference can affect mission performance of the system containing the product. In addition, it is important to control the *susceptibility* of the electronic device to emissions from other electronic devices. Both the FCC and CISPR 22 requirements regulate only emissions. However, the CISPR 21 regulations will regulate susceptibility (including ESD) for the European Community after December 31, 1992. The FCC was given the statutory authority to regulate susceptibility of home electronics equipment and systems, but has yet to promulgate such a standard.

The EMC requirements for products that are marketed for use by the US military are contained in a document MIL-STD-461 [9]. The limits and applicability in this document are much more complicated than the FCC or CISPR 22 requirements, so we will only give a brief discussion of these requirements to give the reader the essence of this requirement. A general discussion of the international EMC requirements for military products is given in [10], and the military and commercial EMC requirements are compared in [11]. One key difference between the commercial EMC requirements and MIL-STD-461 are that compliance with the military requirements can be *waived* if the military agency responsible for procurement of the product judge this to be not critical to the mission success of the system in which the product is to be installed. The FCC and CISPR 22 requirements *cannot be waived*. Another distinction is that the military standards cover the EMC of a considerably larger class of products than the FCC or CISPR 22 standards. MIL-STD-461 covers systems ranging from electric hand drills to sophisticated computer systems.

The frequency ranges and types of limits are as follows:

CONDUCTED EMISSIONS

CE01–	dc–15 kHz
CE03–	15 kHz–50 MHz
CE06–	10 kHz–26 GHz
CE07–	Spikes, time domain

CONDUCTED SUSCEPTIBILITY

CS01–	30 Hz–50 kHz
CS02–	50 kHz–400 MHz
CS03–	Intermodulation, 15 kHz–10 GHz
CS04–	Rejection of undesired signals, 30 Hz–20 GHz
CS05–	Crossmodulation, 30 Hz–20 GHz
CS06–	Spikes
CS07–	Squelch circuits
CS09–	Common-mode currents, 60 Hz–100 kHz
CS10–	Damped sinusoidal transients, pins and terminals, 10 kHz–100 MHz
CS11–	Damped sinusoidal transients, cables, 10 kHz–100 MHz

RADIATED EMISSIONS

RE01–	Magnetic field, 30 Hz–50 kHz
RE02–	Electric field, 14 kHz–10 GHz
RE03–	Spurious emissions and harmonics

RS01–   Magnetic field, 30 Hz–50 kHz
RS02–   Magnetic induction, spikes, power frequency
RS03–   Electric field, 14 kHz–40 GHz
RS05–   EMP transients

UM03–   Radiated emissions, tactical and special purpose vehicles and engine-driven equipment
UM04–   Conducted emissions and radiated emissions and susceptibility, engine generators, and associated components, UPS and MEP equipments
UM05–   Conducted and radiated emissions, commercial electrical and electromechanical equipments

There are also a number of other distinctions between the MIL-STD-461 EMC requirements and the FCC or CISPR 22 requirements. Not only do they regulate susceptibility, but the frequency range is considerably larger (30 Hz–40 GHz) for some tests, and some specific tests in the time domain are required (spikes, intermodulation). So it is clear that the military requirements are much more extensive and detailed than are the commercial requirements.

The classes to which some of these requirements apply are as follows:

A—must operate when installed in critical areas
    A1—aircraft
    A2—spacecraft and launch vehicles
    A3—ground facilities (fixed and mobile)
    A4—surface ships
    A5—submarines
B—support Class A systems but will not be physically located in critical areas
C—miscellaneous, general-purpose equipment
    C1—tactical and special-purpose vehicles and engine-driven equipment
    C2—engine generators
    C3—commercial electrical or electromechanical equipment

As an example, the limits for Class A equipment and conducted emission limit CE03 are shown in Fig. 2.8. Note that the units are in dBμA, a current. The conducted emissions for FCC and CISPR 22 limits are voltages even though the intent is to limit noise emission currents exiting the power cord of the product. The military standards directly measure this emission current using a current probe as shown in Fig. 2.9. Also, conducted emissions are to be measured on other cables in addition to the ac power cord.

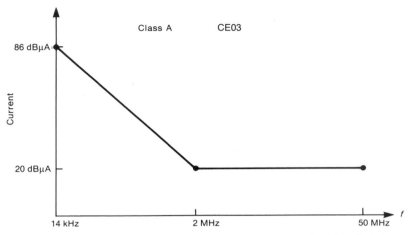

FIGURE 2.8   The MIL-STD-461 CE03 conducted emission limit for Class A equipment.

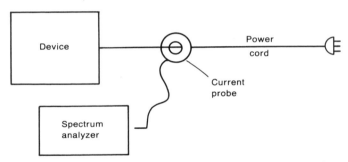

FIGURE 2.9   Use of a current probe to verify the MIL-STD-461 conducted emission limits.

An example of radiated emission limits is given in Fig. 2.10 for Class A equipment RE02. All of the radiated emissions in the military standards are to be measured at a distance of 1 m from the product in a shielded room! This measurement procedure has caused considerable problems in the verification of compliance. A measurement distance of 1 m is certainly in the near field of the product for frequencies below 300 MHz. Also, the requirement to make the measurements in a shielded room can give different results for different test sites. This is examined in [12]. The reader can obtain an approximate comparison of the severity of the narrowband radiated emission limits in MIL-STD-461 to those of the FCC limits by scaling the FCC Class B limits from 3 m to 1 m by adding $20 \log_{10}(3/1) = 9.54$ dB or approximately 10 dB to the FCC limits. These are shown in Fig. 2.10. In doing so one finds that the military radiated emission limits are considerably more stringent than the FCC or CISPR 22 limits [11].

Another distinction between the commercial and military limits is illustrated in Fig. 2.11, in which the UM05 limits for *broadband* conducted emissions for Class C3 equipment are shown. Observe the units of dBμA/MHz. The radiated emission limits also contain *broadband* emission limits. The broadband radiated emission UM05 limits for Class C3 equipment are shown in Fig. 2.12. Observe

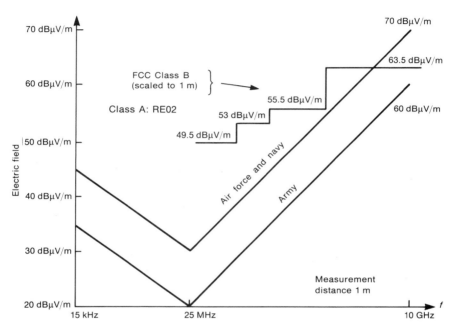

FIGURE 2.10   Comparison of the MIL-STD-461 RE02 radiated emission limit for Class A equipment to the FCC Class B radiated emission limit scaled to a measurement distance of 1 m.

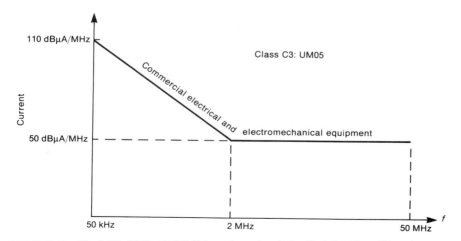

FIGURE 2.11   The MIL-STD-461 UM05 conducted emission limit for Class C3 equipment.

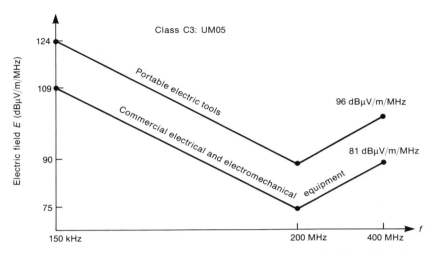

FIGURE 2.12   The MIL-STD-461 UM05 radiated emission limit for Class C3 equipment.

the units of dBμV/m/MHz. The FCC and CISPR 22 limits are narrowband limits, and those standards contain no broadband limit. Essentially, a broadband emission is one that is wider in bandwidth than the bandwidth of the measurement equipment. An example is the arcing from brushes of a dc motor. These emissions are relatively constant over a wide frequency range. The UM05 limit in Fig. 2.12 covers, among other things, electric hand drills that contain dc motors. Therefore it is reasonable to expect to find a broadband limit here.

2.1.4   Measurement of Commercial Product Emissions for Verification of Compliance

It is as important to clearly specify how one is to measure the product emissions when attempting to verify compliance with the limits as it is to clearly specify the limits. Measurement of radiated and conducted emissions is a complex subject. It is fair to say that if the measurement procedures are not clearly spelled out but are left to the interpretation of the measurement personnel, one can obtain different sets of measured data at different measurement sites *for the same product*! Every standard that sets out limits on radiated and conducted emissions (FCC, CISPR 22, and MIL-STD-461) has a related standard that clearly defines how the data are to be measured. This includes test procedure, test equipment, bandwidth, test antennas, etc. Once again, the specification of the method for gathering the data is critically important so that the governing agency can be sure that data gathered on a product at one company's test site can be validly compared to the limits and to data gathered at another test site. Otherwise the governing agency as well as the product manufacturer cannot be assured that the product's emissions comply with the limits.

   In this section we will discuss the measurement procedures that apply to the FCC and CISPR 22 requirements. These are documented in [13–15]. The measurement procedures for the MIL-STD-461 requirements are contained in MIL-STD-462 [16].

2.1.4.1 Radiated Emissions   At present, the test procedure for verifying compliance with the FCC radiated emission limits is specified by the FCC in FCC/OST MP-4 *FCC Methods of Measurement of Radio Noise Emissions from Computing Devices* [13]. The FCC specifies that the measurements of radiated and conducted emissions must be performed on the complete system. All interconnect cables to peripheral equipment must be connected and the system must be in a typical configuration. The cables and the system must also be configured in a representative way such that the *emissions are maximized*. Therefore the measurement personnel are required to determine the configuration that maximizes the emissions (within the bounds of a reasonable configuration) and use that configuration when gathering compliance data.

   The FCC radiated emissions are to be measured at a distance of 10 m for Class A products and 3 m for Class B products. These measurements are to be made over a ground plane using a tuned dipole antenna at an *open-field test site*. Furthermore, the measurements are to be made with the measurement antenna in the vertical position (perpendicular to the ground plane) and in the horizontal position (parallel to the ground plane). This test method preferred by the FCC is difficult to automate. For example, when using the tuned dipole, its length must be adjusted to a half wavelength *at each measurement frequency*. There are *broadband* antennas that do not need to be readjusted in length at each frequency. Consequently *swept-frequency* measurements can be made using these broadband antennas, thereby speeding up the data gathering. Examples of these broadband antennas—the biconical and log-periodic antennas—are discussed in Chapter 5. The other requirement that the test be conducted at an open-field test site also presents measurement difficulties. There are numerous (and quite strong) ambient signals present in addition to the emissions from the product. Consequently, in order to obtain only those frequencies that need to be measured at the open-field site, preliminary screening tests are usually conducted in a *semianechoic chamber* as illustrated in Fig. 2.13. The semianechoic chamber consists of two aspects: (1) a shielded room and (2) radio-frequency absorbing cones lining the walls and the ceiling. An example is shown in Fig. 2.14. The shielded room is intended to prevent external signals from contaminating the test. The absorbing cones are intended to prevent reflections of the emissions from the walls and ceiling. Reflections can and do occur at the ground plane (floor) of the chamber. The absorbing cones are intended to simulate an open-field site. Swept-frequency measurements can be made very rapidly with this arrangement. The FCC specifies that the measurement antenna must also be scanned from a height of 1 m to 4 m above the ground plane and the maximum signal obtained in that scan be recorded for that frequency. Some test sites automate the entire test procedure. Computer-controlled equipment

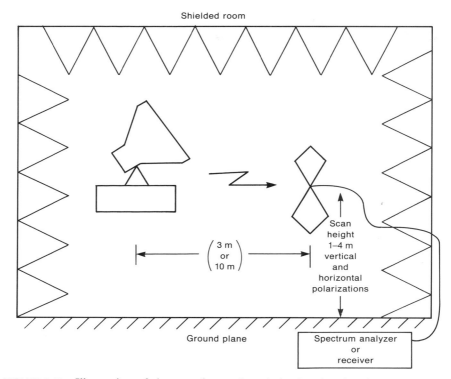

FIGURE 2.13 Illustration of the use of a semianechoic chamber for the measurement of radiated emissions.

moves the test antenna up and down, changes the polarization of the test antenna and records the data. Plotting of the measurement data is also automated. This greatly speeds up the gathering of the data. Any frequencies at which the emissions exceed the limit are recorded, and the product is tested at those frequencies at a valid open-field test site. The CISPR 22 measurement procedure defined in [15] is similar to the FCC test.

One final test requirement needs to be addressed. The FCC and CISPR test procedures specify that the bandwidth (6 dB) of the spectrum analyzer or receiver that is used to measure the radiated emissions must be at least 100 kHz. If the emission at a particular frequency is very narrow, such as a clock harmonic, the bandwidth of the receiver is not of much concern. However, if the emission is due to a fairly broadband source such as arcing at the brushes of a dc motor, the bandwidth of the receiver directly determines the signal level that is measured: the wider the bandwidth, the larger the measured level. Limiting the bandwidth of the receiver to no less than 100 kHz means that certain broadband signals will be measured. (Of course no one would use a bandwidth larger than 100 kHz, since that would increase the received signal level from broadband

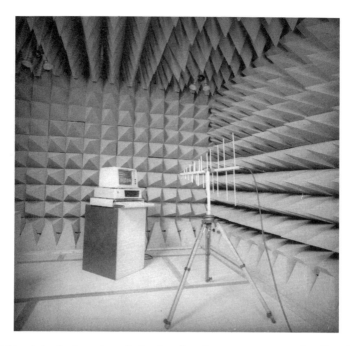

FIGURE 2.14 A typical semianechoic chamber for measurement of radiated emissions (courtesy the Amador Corporation).

sources, making compliance more difficult.) A related issue is the *detector* used in the output stage of the receiver. Typical spectrum analyzers use *peak detectors* that record the maximum signal level at the desired frequency. The FCC and CISPR test procedures require that the receiver use a *quasi-peak* detector. The quasi-peak detector is similar to an $R-C$ maximum-hold circuit having a certain time constant. Fast changing, momentary signals such as randomly occurring "spikes" will not "charge up" the quasi-peak detector to as high a level as periodic signals. In order to appreciate the need for the quasi-peak detector, one must remember the reason these limits are imposed. The FCC has authority to regulate interference to wire and radio communications. Consequently, a randomly occurring, one-time signal that does not reoccur very often will not pose as great an interference to a listener as would a continually occurring tone. Also, the human ear tends to hear logarithmically. The time constant of the quasi-peak detector tends to simulate this phenomenon.

The radiated emission test procedures for verifying compliance with the CISPR 22 requirements are very similar to those dictated for the FCC requirements discussed above. One important difference is that the radiated emissions must be measured at a distance of 10 m for Class B devices and 30 m for Class A devices.

2.1.4.2   Conducted Emissions   The intent of the conducted emission limits is to restrict the noise *current* passing out through the product's ac power cord. The reason for this is that these noise currents will be placed on the common power net of the installation. The common power net of an installation is an array of interconnected wires in the installation walls, and as such represents a large antenna. Noise currents that are placed on this power net will therefore radiate quite efficiently, which can produce interference. An example of this is the lines that appear on a TV set when a blender or other device powered by a dc motor is turned on. The noise generated by the arcing at the brushes of the dc motor pass out through the ac power cord of the blender, are placed on the household ac power system, and are then radiated and picked up by the TV, where they show up as interference.

Therefore the conducted emission that should be measured is the noise current conducted out through the ac power cord of the product. Yet the FCC and CISPR conducted emission limits are given in units of volts. This is because the tests are to be conducted by inserting a line impedance stabilization network (LISN) in series with the ac power cord of the product. In order to understand the performance of this device, we need to discuss the standard ac power distribution system. In the US, ac voltage utilized in residential and business environments has a frequency of 60 Hz and an RMS voltage of 120 V. This power is transmitted to these sites at various other, higher voltages. For example, the distribution wiring entering a typical residence is composed of two wires and a ground wire connected to earth. The voltage between the two wires is 240 V. At the service entrance panel in the home, the 120 V is obtained between one wire and the ground and between the other wire and ground. A third or safety wire (referred to as the *green wire*) is carried throughout the residence along with these two wires that carry the desired 60 Hz power. The two wires that carry the desired 60 Hz power are referred to as the *phase* and *neutral* wires. The currents to be measured are those exiting the product via the phase and the neutral wires. Thus, like the radiated emission measurements, two measurements are needed for conducted emissions, phase and neutral.

The LISN and its use is illustrated in Fig. 2.15. The LISN used for CISPR 22 conducted measurements is similar. There are two purposes of the LISN. The first, like the shielded room of the radiated emission measurements, is to *prevent noise external to the test (on the common ac power net) from contaminating the measurement.* The inductor L_1 and capacitor C_2 are for this purpose: L_1 blocks noise whereas C_2 diverts noise. The value of L_1 is 50 μH, and its impedance ranges from 141 to 9425 Ω over the conducted emission frequency range (450 kHz–30 MHz). The value of C_2 is 1 μF, and its impedance ranges from 0.35 to 0.005 Ω over this frequency range. The second purpose of the LISN is to ensure that measurements made at one test site will be correlatible with measurements at another test site. The possibility of this inconsistency between test sites is in the variability of the ac impedance seen looking into the ac power net from site to site. Measurements of the ac impedance seen looking into the ac power net at different locations show variability from site to site in addition

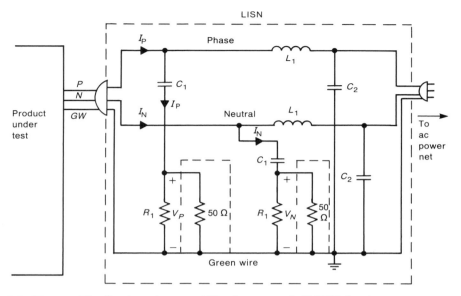

FIGURE 2.15   The line impedance stabilization network (LISN) for the measurement of conducted emissions.

to the variability with frequency [17]. (Remember that our interest in this measurement is not the 60 Hz power frequency but noise signals superimposed on the ac power conductors at frequencies from 450 kHz to 30 MHz.) In order to insure that conducted emissions measured at one site correlate with those measured at another, we must be sure that the impedance seen by the product looking into its power cord is the same from site to site at corresponding frequencies. This is the second purpose of the LISN: *to present a constant impedance in frequency and from site to site to the product between phase and ground between neutral and ground.* The capacitor C_1 and the 50 Ω resistor accomplish this task. The capacitor C_1 is included to prevent any dc from overloading the test receiver, and the $R_1 = 1$ kΩ resistor is used to provide a discharge path for C_1 in the event that the 50 Ω resistor is disconnected. The value of C_1 is 0.1 μF, so that the impedance of C_1 over the conducted emission frequency range (450 kHz–30 MHz) ranges from 3.5 to 0.05 Ω. The inductor L_1 and capacitor C_2 prevent noise on the commercial power distribution system from being measured, but also pass the required 60 Hz power necessary to operate the product. The impedances of L_1 and C_2 at 60 Hz are 0.019 and 2653 Ω, respectively.

Over the frequency range of the regulatory limit (450 kHz–30 MHz), L_1 and C_2 essentially give an open circuit looking into the commercial power distribution system. Thus the impedance seen by the product between phase and green wire (ground) and between neutral and green wire is essentially 50 Ω. Furthermore, this is fairly constant over the frequency range of the conducted

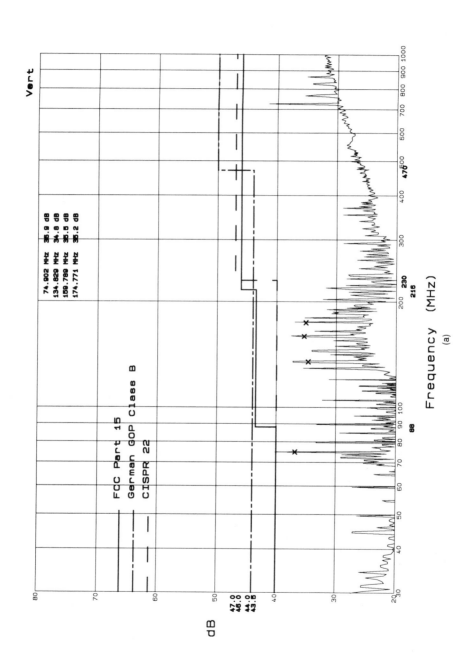

(a)

Frequency (MHz)

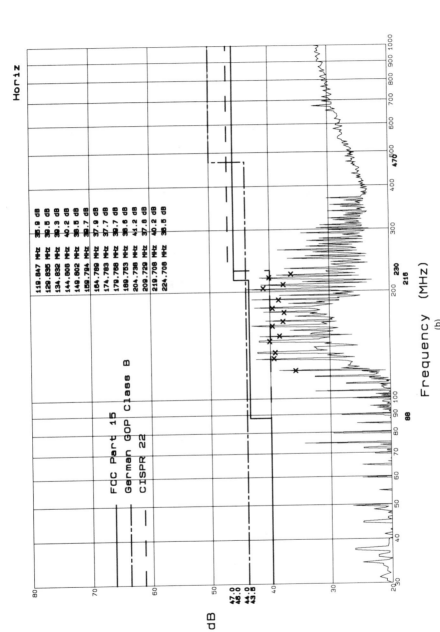

FIGURE 2.16 Radiated emissions of a typical digit product: (a) vertical emissions; (b) horizontal emissions.

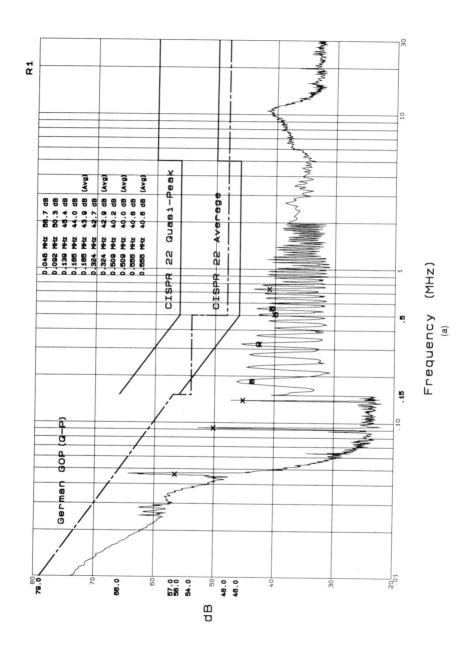

Frequency (MHz)

(a)

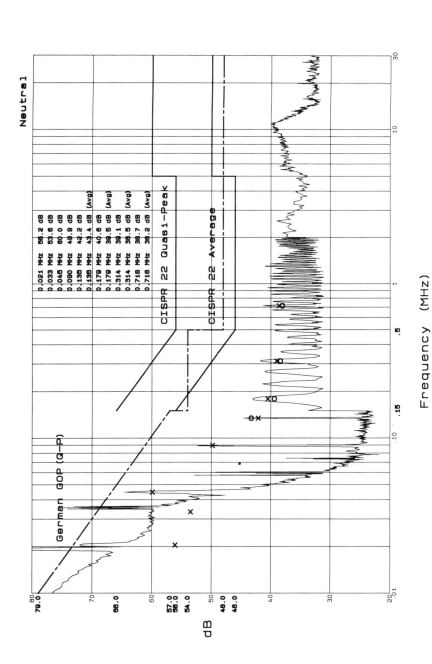

FIGURE 2.17 Conducted emissions of a typical digit product containing a switching power supply: (a) phase (R1); (b) neutral.

emission measurement. The LISN must be modified if the frequency range of measurement is changed. Thus the LISN for CISPR 22 conducted emission measurements has larger values of inductors and smaller values of capacitors since the frequency range extends down to 150 kHz. The 50 Ω resistors represent the standard 50 Ω input impedance to the spectrum analyzer that is used to measure the phase, V_P, and neutral, V_N, voltages. Now it is clear that these measured voltages are directly related to the noise currents passed out the phase and neutral conductors, I_P and I_N:

$$I_P = \tfrac{1}{50} V_P$$
$$I_N = \tfrac{1}{50} V_N$$

Each voltage must be measured over the frequency range of 450 kHz–30 MHz and the emissions must not exceed the limits shown in Fig. 2.1 or 2.2 for either voltage. One of the 50 Ω resistors represents the input impedance of the measurement spectrum analyzer or receiver, and the other 50 Ω termination is a "dummy load." The FCC Class B conducted emission limit of 48 dBμV therefore represents a current of 14 dBμA or 5 μA.

2.1.5   Typical Product Emissions

In this section we will show some examples of radiated and conducted emissions from a typical digital product. These data are included to give the reader an appreciation for the typical range of expected numbers.

The radiated emissions of a typical digital product are shown in Fig. 2.16(a) for vertical emissions and in Fig. 2.16(b) for horizontal emissions. Values of quasi-peak measurements at selected frequencies are also shown on the plots. Although we see individual harmonics of the clocks in the system, we do not see them of the same level but instead see certain regions of the spectrum accentuated such as one sees in a resonance situation. Some of this may be due to "tuning" by system cables or other resonance conditions. The FCC Class B and CISPR 22 Class B limits are drawn on the plots. These emissions were measured in a semianechoic chamber that is used for product compliance testing.

The conducted emissions of this product are shown in Fig. 2.17, with the phase measurement in (a) and the neutral measurement in (b). The VDE conducted emission specifications are drawn on the plot down to 10 kHz. This product employs a 45 kHz switching power supply. Note the peaks at the fundamental (45 kHz) and second (90 kHz) and third (135 kHz) harmonics of the switcher. The switcher frequency was chosen to be 45 kHz rather than a more obvious choise of 50 kHz in order that the third harmonic (135 kHz) occur prior to the break frequency of the CISPR 22 limit (150 kHz). This moves the third harmonic below the lower frequency limit of CISPR 22 than would a choice of a 50 kHz switch frequency, and only marginally impairs the efficiency of the switcher. Therefore, from the standpoint of satisfying the CISPR 22 requirement, the levels of the first three harmonics of the switcher *are not regulated!* This is a good example of the point that simply being aware of the potential problem can lead to *cost-free* solutions in the early design. Recognizing this later in the design after a 50 kHz switch frequency had been chosen would require changes to be made that may be difficult to implement. Although not apparent on these plots, the harmonics of the system clocks that appear on the ac power cord contribute to the conducted emission levels at the higher frequencies. These may contribute to *radiated emissions* above 30 MHz from this "long antenna."

2.1.6   A Simple Example to Illustrate the Difficulty in Meeting the Regulatory Limits

The measured emissions of a typical product shown in Figs. 2.16 and 2.17 seem to indicate that designing a complex electronic system to satisfy the regulatory requirements is a deceptively simple problem. In this section we will show data that indicate that this is not true. If one proceeds through a design with no thought given to EMC, it is highly likely, almost a certainty, that the product will not satisfy the regulatory requirements.

In order to illustrate this, a simple experiment was performed. A PCB consisting of two 7 inch, parallel, 15 mil lands separated by 180 mils shown in Fig. 2.18 was constructed. A 10 MHz DIP oscillator drove a 74LS04 inverter that was connected to one end of the pair of lands, and a 74LS04 inverter served as an active load attached to the other end of the pair of lands. Two regulated and compact 5 V power supplies were constructed to power the devices

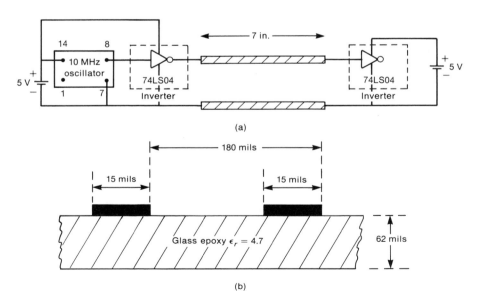

(a)

(b)

FIGURE 2.18   A simple experiment to demonstrate the difficulty in meeting the radiated emission limits: (a) schematic and dimensions of device tested; (b) cross-sectional dimensions of the printed circuit board (PCB).

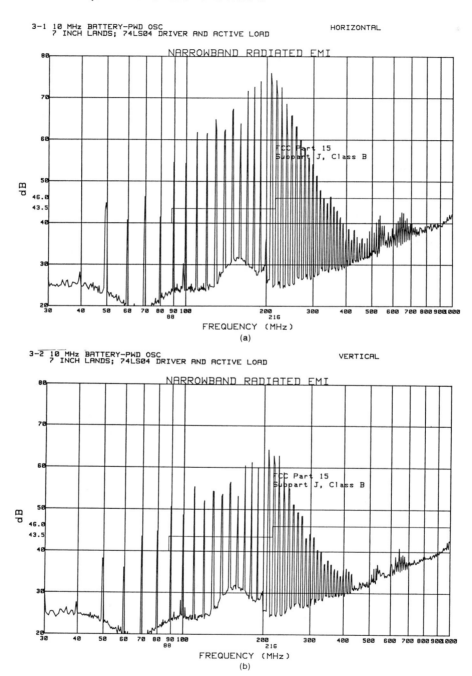

FIGURE 2.19 Radiated emissions at 3 m for the device of Fig. 2.18: (a) horizontal; (b) vertical.

at either end of the line. These consisted of a 9 V battery driving a 7805 regulator chip. Both supplies were very compact (no larger than 2 in × 2 in) and had no connection to the commercial power system. The horizontal radiated emissions are shown in Fig. 2.19(a) and the vertical radiated emissions in Fig. 2.19(b). Observe that the horizontal emissions exceed the FCC Class B limit by as much as 30 dB! Even though the board is placed parallel to the ground plane of the chamber, the vertical emissions also exceed the FCC Class B limit, but only by some 15 dB.

This very simple experiment is a reasonable approximation to circuitry found on PCB's of products, yet it fails the FCC Class B limit by a large amount. Therefore, even if the device performed some useful function that a consumer would desire, it could not be sold in the US (or in many other countries in the world for that matter)! This simple example should serve to illustrate that designing an electronic product to comply with the regulatory limits is a deceptively difficult problem and one that should not be taken lightly.

2.2 ADDITIONAL PRODUCT REQUIREMENTS

As pointed out earlier, it is of no benefit to design a product that performs some new and marketable function if it fails to comply with the EMC regulatory requirements. Similarly, it is also of no benefit to design a product that passes the EMC regulatory requirements imposed by governmental agencies but will not operate satisfactorily when placed near an FM radio transmitter or an airport surveillance radar. Consumers do not appreciate the problems these emitters pose on the proper operation of electronic devices. They expect that a product that was purchased in good faith will operate satisfactorily in any residential installation, and the consumer will not appreciate a warning that states "Caution, this computer will not work if your home is within half a mile of an FM transmitting tower." An equally embarrassing event for the product manufacturer is the inability of the product to function properly if the operator walks across a nylon rug in an office situated in a dry climate and touches the product, causing an electrostatic discharge to reset the machine. These examples illustrate the importance of the manufacturer imposing certain tests over and above those required by governmental agencies in order to insure that the product will perform properly in a wide variety of field installations. We will discuss those requirements in this section. However, these self-imposed requirements are usually proprietary, and the limits and test procedures are not generally made available to the public. Consequently we will only be able to discuss these in qualitative terms.

2.2.1 Radiated Susceptibility

The purpose of these tests is to insure that the product will operate properly when it is installed in the vicinity of high-power transmitters. The common types of such transmitters are AM and FM transmitters and airport surveillance radars. Manufacturers test their products to these types of emitters by illuminating the product with a typical waveform and signal level representing the worse-case exposure of the product and determining whether the product will perform satisfactorily. If the product cannot perform satisfactorily in such installations, this deficiency should be determined prior to its marketing so that "fixes" can be applied to prevent a large number of customer complaints and service calls.

2.2.2   Conducted Susceptibility

Products can be susceptible to a wide variety of interference signals that enter it via the ac power cord. An obvious example is lightning-induced transients. Thunderstorms frequently strike power transmission lines and substations. Circuit breakers are intended to momentarily clear any faults and reclose after a few cycles of the ac waveform. The product must be insensitive to these types of momentary power interuptions as well as the transient spikes that are generated on the power line. Of course, there is little that the manufacturer can do about a complete power "blackout," but consumers consider it reasonable to expect the product to operate so long as only momentary surges occur. Most manufacturers subject their products to these scenarios by intentionally injecting spikes into the product's ac power cord to simulate lightning-induced transients. The ac voltage is also momentarily reduced and/or interrupted to insure that the product will operate through any such event. These types of tests represent conducted susceptibility tests.

2.2.3   Electrostatic Discharge (ESD)

This phenomenon has been mentioned previously and is becoming an increasingly important concern with today's integrated circuit technology. The basic phenomenon is the buildup of static charge on a person's body or furniture with the subsequent discharge to the product when the person or the furniture touches the product. The static voltage can approach 25 kV in magnitude. When the discharge occurs, large currents momentarily course through the product. These currents can cause IC memories to clear, machines to reset, etc. Consumers do not view these events as being normal operation of a well-designed product. Consequently, manufacturers test their products for susceptibility to the ESD phenomenon by subjecting their products to a controlled ESD event that represents a typical field scenario and determining whether the product operates successfully. Typical ESD tests used by a manufacturer are described in [18]. Other aspects of ESD are described in [2]. The phenomenon of ESD is investigated in more detail in Chapter 12.

2.3   DESIGN CONSTRAINTS FOR PRODUCTS

Of course, virtually any product can be made to comply with the governmental regulatory limits and the manufacturer's self-imposed susceptibility requirements. For example, a digital device that is susceptible to disturbances on the ac power cord can be made insensitive to this by simply powering the product with internal batteries. The cost to the manufacturer, not only in terms of added cost to the product but also in customer acceptance, is prohibitive. As another example, suppose that the product fails the FCC Class B limits on radiated emissions at a particular frequency. The product can be made to comply with these limits if it is enclosed entirely in a metallic box that has no openings. However, penetrations in the box must be made in order to provide power as well as data signals from keyboards and other peripheral devices. We will find in later chapters that cables provide efficient antennas for the radiation of high-frequency noise signals. So these penetrations drastically defeat the shielding effect of the enclosure.

A more realistic way of reducing the radiated and conducted emissions of a product is to add suppression components that reduce the levels of emissions. However, these components add extra cost to the product over that required for its functional performance. The manufacturer must add the cost of these components to the manufacturing cost of the product. In today's highly competitive market, any additional cost affects the selling price of the product and the success of the product in the marketplace. Thus an important consideration is *product cost*. Generally, manufacturers are driven by product cost, with all other considerations being secondary.

Another important design constraint is *product marketability*. For example, market surveys determine what the consumer will be likely to purchase. Product appearance and ease of use are important factors here. It may be possible to enclose a typewriter in a contiguous metal box such that radiated emissions will be reduced. However the consumer must be able to easily insert paper, type on the keyboard, etc. The acceptability of a product by the consumer is frequently a top priority, otherwise cost savings are inconsequential.

Another design constraint is imposed by the *manufacturability* of the product. Increasing numbers of manufacturers are using automated assembly methods to manufacture the final product. Electronic components as well as other parts of the product are inserted by robots. Even though today's robots are capable of inserting minute electronic components into printed circuit boards, they are not capable of assembling certain components such as wires that humans can readily place by hand. Thus any EMC suppression components that are to be added must be such that they can be handled easily in the manufacturing process, or else the cost advantage of automated assembly is not fully realized.

And finally, one of the most important considerations is the *product development schedule*. Manufacturers determine the need for a new product by conducting market surveys. These also serve to indicate the trend in products that will be produced by other manufacturers in the future. In order to take advantage of the interest of the consumer, the product must be placed on the market during a certain time frame, otherwise competing products will gain an advantage. Product development schedules are developed to not only gauge the progress of the product development but to also insure that the product will be made available to the consumer when his/her interest in that product is at its peak. Delays in this product development schedule obviously affect the marketability of the product, and any delays can be as serious to the manufacturer as cost overruns. EMC problems can and frequently do cause delays in the product's development schedule. The typical scenario is to wait until the end of the development schedule to test the product for its compliance with the governmental EMC requirements as well as those imposed by the manufacturer. If it is discovered at this late date that the product fails to comply, fixes must be developed. Generally the difficult part of fixing the problem is the *diagnosis* of the *source of the problem*. For example, even though it is obvious that the product's radiated emissions exceed the FCC Class B limit at a particular frequency, the essential question is "how do we reduce the level of that emission in a cost-effective and manufacturable manner?" Although the original source of the emission is obvious (e.g., one of the system processor clocks), the primary radiation point or mechanism is not known, e.g., which land of a printed circuit board or wire of one of the system's cables is the dominant radiator. Determining the primary or dominant source of the emission is important so that (1) a fix can be efficiently made and (2) unnecessary cost is not added to the product in the way of fixes that do not substantially contribute to the reduction of the emission. We will study this important aspect of *diagnostic tools* in later chapters. *Diagnosing the problem correctly and quickly is critical to maintaining the product's development schedule.*

2.4 ADVANTAGES OF EMC DESIGN

The primary advantages of adequate EMC design are (1) minimizing the additional cost required by suppression elements or redesign in order to satisfy the regulatory requirements (minimizing product cost), (2) maintaining the development and product announcement schedule (minimizing development schedule delays), and (3) insuring that the product will operate satisfactorily in the presence of the inevitable external noise sources at its installation location (minimizing customer complaints). The reader may not be sensitive to the importance of these issues until he/she has entered the workplace. However, the author can assure the reader that these issues are as important as the technical (functional) aspects of the design!

Perhaps the most important factor in insuring that the product will satisfy the regulatory requirements at the end of the design and will maintain the development schedule is the *early and continuous application of the EMC design principles throughout the entire development cycle of the product.* The longer we proceed into the development of a product, the more aspects are "set in concrete" and cannot be changed without much additional cost and schedule delays. For example, at the early conceptual stages of the design, we can move cables, change locations of cable connectors on PCBs, reorient PCBs in the product, etc., since the design exists only "on paper." Once the design has proceeded to the development of prototypes, it becomes an increasingly difficult and costly matter to make physical changes, since many other aspects of the design must also be changed. For example, suppose someone notices that the clock oscillator on a PCB is located adjacent to a cable connector of that PCB. It should be abundantly clear to the reader, even without reading further in this text, that there exists a strong potential for the oscillator signal to radiate to the cable wires, proceed out the attached cable and radiate very efficiently from that cable, resulting in serious radiated emission problems. Simply moving the connector away from the oscillator or the oscillator away from the connector may considerably reduce this problem. If this observation is made early in the design, it is *cost-free*! If the change is made later after the PCB has been laid out, it will be extremely costly and will probably result in significant delays in the development of the product, which also represent cost in terms of lost sales. *Early and consistent attention to EMC will minimize cost and schedule delays and will provide the best chance for complying with the regulatory requirements*!

An equally important reason for early and consistent attention to EMC is to make the implementation of any suppression measures that are necessary to comply with the regulatory requirements a much simpler matter. For example, suppose that during the design of a PCB, it is felt that suppression of some of the clock harmonics may be needed when the product is eventually tested for compliance. If the PCB designer will place holes or pads at the output of the clock on the PCB, a capacitor can be easily inserted, if needed, across the clock terminals, thus reducing the emissions of the clock. Also holes may be placed in series with one land of the clock output to provide for later insertion of a series resistor to additionally reduce the clock rise/fall times and further reduce the high-frequency emissions of the clock signal. In the initial design the capacitor holes can be left vacant and the series resistor holes can be "wired across" with a $0\,\Omega$ resistor. If problems with the clock emissions occur during testing, a capacitor can be inserted and/or a series resistor can be inserted and *only the PCB artwork and the product parts list need be changed*! If this had not been done, the entire PCB would need to be relaid out, which would be extremely costly and result in significant schedule delays. This represents one of the most effective EMC design principles: *assume that some EMC suppression will be needed for compliance and provide the ability to implement it if it is needed*! It is doubtful that adherence to the EMC design principles of this text will result in a product that exhibits no EMC problems when it is tested. However, adherence to these design principles and maintaining the necessary EMC insight throughout the design will tend to make any necessary suppression easy to apply and at minimal expense. Waiting until the last minute to consider EMC will generally mean that complying with the necessary EMC requirements will be a difficult, time-consuming, and costly experience.

And finally, every electronic design team should include, as an integral partner, an experienced EMC engineer. This person should be consulted on the potential EMC impact of every aspect of the design *early and continuously* in the product design cycle. He/she should be consulted on every detail of the design, no matter how seemingly innocuous it is. For example, the design of the "package," i.e., the product enclosure(s), is very critical to EMC. Once the product package shape is determined, it cannot usually be changed, yet the shape will determine where PCBs can be placed, where cables can exit the enclosure, where disk drive cables must be routed to the PCB, etc. All these aspects remove some of the EMC designer's "cost-free" fix options. Also, once the first prototype is available, no matter how crude, it should be tested for its EMC problems. Granted, the final product may exhibit less problems than the prototype, but this early test will show where there are significant trouble spots that must be attacked early.

Early and continuous attention to the affect on EMC will give the product the best possible change for minimum cost and schedule delay resulting from EMC. A failure to do this will almost certainly translate to added cost and schedule delay. Managers of products may not be attuned to the fine details of EMC design, but they do understand cost and schedule delay!

REFERENCES

[1] *Code of Federal Regulations, Title 47 (47CFR). Part 15, Subpart B: "Unintentional Radiators."*

[2] H.W. Ott, *Noise Reduction Techniques in Electronic Systems*, second edition, John Wiley Interscience, New York (1988).

[3] R.K. Keenan, *Digital Design for Interference Specifications*, The Keenan Corporation, Pinellas Park, FL (1983).

[4] FCC, Understanding FCC regulations concerning computing devices, *OST Bulletin*, **62** (May 1984).

[5] *FCC General Docket 87–389, First Report and Order* (Released April 1987).

[6] I. Straus, A new part 15, *Compliance Engineering*, **6**, Issue 4 (Summer 1989).

[7] *Limits and Methods of Measurement of Radio Interference Characteristics of Information Technology Equipment*, CISPR Publication 22 (1985).

[8] G.A. Jackson, International EMC cooperation past, present and future, *IEEE Aerospace and Electronic Systems Society Magazine*, pp. 2–5 (April 1987).

[9] Department of Defense (US), *Electromagnetic Emission and Susceptibility Requirements for the Control of Electro-Magnetic Interference*, MIL-STD-461B (April 1, 1980).

[10] N.J. Carter, International EMC cooperation in the military, *IEEE Aerospace and Electronic Systems Society Magazine*, pp. 6–9 (April 1987).

[11] R.B. Cowdell, The relationship between MIL-SPEC and commercial EMI requirements, *IEEE Aerospace and Electronic Systems Society Magazine*, pp. 9–13 (April 1987).

[12] A.C. Marvin, The use of screened (shielded) rooms for the identification of radiated mechanisms and the measurement of free-space emissions from electrically small sources, *IEEE Trans. on Electromagnetic Compatibility*, **EMC-26**, 149 (1984).

[13] FCC, *FCC Methods of Measurement of Radio Noise Emissions from Computing Devices*, FCC/OST MP-4 (July 1987).

[14] R. Fabina and J. Husnay, FCC experiences in measuring computing devices, *1985 IEEE International Symposium on Electromagnetic Compatibility, Wakefield, MA*.

[15] *Specification for Radio Interference Measuring Apparatus and Measurement Methods*, Amendment #1 CISPR Publication 16 (1980).

[16] *Electromagnetic Interference Characteristics, Measurement of*, MIL-STD-462 (July 1967).

[17] J.R. Nicholson and J.A. Malack, RF impedance of power lines and line impedance stabilization networks in conducted interference measurements, *IEEE Trans. Electromagnetic Compatibility*, **EMC-15**, 84–86 (1973).

[18] R.J. Calcavecchio and D.J. Pratt, A standard test to determine the susceptibility of a machine to electrostatic discharge, *1986 IEEE International Symposium on Electromagnetic Compatibility, San Diego, CA, September 1986*.

PROBLEMS

2.1 A product is tested for FCC Class B radiated emission compliance as shown in Fig. P2.1. The distance between the measurement antenna and the product is 20 feet. The spectrum analyzer is connected to the measurement antenna with 30 feet of RG-58U coaxial cable that has a loss of 4.5 dB/100 feet at 100 MHz. The receiving antenna provides an output voltage at 100 MHz of 6.31 V for each V/m of incident electric field. If the spectrum analyzer indicates a level of 53 dBμV at 100 MHz, determine the level of received electric field at the antenna. [38.35 dBμV/m] Determine whether the product will pass or fail the FCC Class B test, and by how much. [No, fails by 1.01 dB]

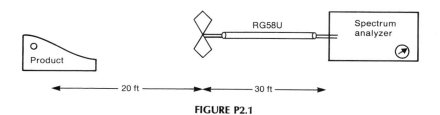

FIGURE P2.1

2.2 The radiated emissions from a product are measured at 50 MHz at 15 m away and are found to be 21 μV/m. Does the product comply with the

FCC Class B limit? [No] By how much does the product pass or fail? [0.424 dB].

2.3 The voltage induced at the terminals of an antenna V_{ant}, is 5 V for every V/m of incident field E_{ant}. What level in dBμV at the base of the antenna would correspond to the FCC Class B limit at 100 MHz? [57.48 dBμV] Determine the reading of the spectrum analyzer if it is connected to the antenna with 200 feet of RG-58U coaxial cable that has 4.5 dB/100 feet of loss at 100 MHz. [48.48 dBμV]

2.4 A product emits a radiated electric field level of 36 dBμV/m at 100 MHz when measured at a distance of 30 feet from the product. Determine the emission level when measured at the FCC Class B distance of 3 m. [45.68 dBμV/m]

2.5 An antenna measures the radiated emissions at 220 MHz from a product as shown in Fig. P2.5. If the receiver measures a level of −93.5 dBm at 220 MHz, determine the voltage at the base of the antenna in dBμV. The cable loss at 220 MHz is 8 dB/100 feet. [29.5 dBμV] If the product providing these emissions is located a distance of 20 m and the antenna provides 1.5 V for every V/m of incident electric field at 220 MHz, determine whether the emissions comply with the CISPR 22 Class B and FCC Class B limits and by how much. [Fails CISPR 22 by 3.5 dB but passes FCC Class B by 2 dB]

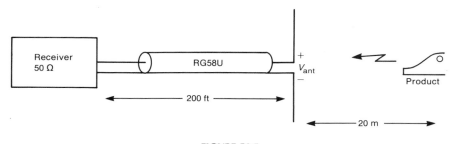

FIGURE P2.5

2.6 The CISPR 22 conducted emission limit at 150 kHz is 56 dBμV when measured with a 50 Ω LISN. Determine the current level corresponding to this limit. [22 dBμA or 12.6 μA]

2.7 The MIL-STD-461 CE03 conducted emission limit is shown in Fig. 2.8. Determine the corresponding FCC Class A limits. [26 dBμA, 450 kHz– 1.705 MHz; 35.5 dBμA, 1.705 MHz–30 MHz]

2.8 Assume the impedance looking into the ac power net between all combinations of wires is infinite (which it isn't). Use SPICE to determine the input impedance between Phase and the Green wire and between

Neutral and the Green wire looking into the FCC LISN shown in Fig. 2.15 over the FCC conducted emission frequency range of 450 kHz– 30 MHz. Determine the frequency where the impedance of L_1 is 1 Ω. [3.183 kHz] Determine the frequency where the impedance of C_1 is 10 kΩ. [159 Hz] Determine the frequency where the impedance of C_2 is 10 kΩ. [16 Hz]

2.9   Determine the distance the PCB of Fig. 2.18 must be moved to in order for its horizontal emissions in Fig. 2.19(a) to be lower than the FCC Class B limit at 100 MHz. [9.5 m]

2.10   Give reasons for using a semianechoic chamber to measure radiated emissions.

2.11   Give reasons for using a LISN to measure conducted emissions.

PART I

BASIC ELECTROMAGNETIC PRINCIPLES

CHAPTER THREE

Electromagnetic Field Theory

In this chapter we will review the essential concepts of electromagnetic field theory needed to understand the important notions of transmission lines (Chapter 4) and antennas (Chapter 5). All of the basic concepts contained in this chapter and the following two chapters are crucial to our ability to understand the basic EMC design concepts discussed in Part II.

Recall that the frequency range of the regulations is rather large: from 150 kHz to 30 MHz for conducted emissions and from 30 MHz to above 1 GHz for radiated emissions. Thus the electrical dimensions of an electronic product and its associated connection cables (as well as the ac power cord) may not be electrically small (much less than a wavelength), in which case the usual lumped-circuit notions and analysis principles such as Kirchhoff's laws do not apply. Attempting to analyze electrically large structures using these lumped-circuit analysis principles will lead to erroneous conclusions and faulty designs. The laws governing the behavior of electrically large structures (Maxwell's equations) are not as simple to use as are the lumped-circuit analysis principles. However, for electrically large structures, we have no other recourse. Some problems are sufficiently small, electrically, so that the simpler lumped-circuit analysis techniques will be applicable (in a reasonably approximate sense). An example is the modeling of small, electronic components. Where it is possible, we will utilize the simpler analysis method.

All macroscopic electromagnetic phenomena are governed by *Maxwell's equations*. We will give a brief discussion of these here, and the reader is referred to [1] or other similar texts for more details. Maxwell's equations are difficult from a mathematical standpoint, although they are quite easy to describe in conceptual terms. These equations describe the *distributed-parameter* nature of electromagnetic fields; that is, the electromagnetic field quantities are distributed throughout space. Thus the differential equations that are referred to as Maxwell's equations are a set of *partial differential equations* since the field quantities are functions of spatial parameters x, y, z in three-dimensional space as well as time t. Where appropriate (electrically small structures) we will use lumped-circuit approximations, and the governing equations become *ordinary differential equations* where the variables are functions of only one parameter, time *t*.

The material in this chapter constitutes a brief review of basic electromagnetic theory. The reader is referred to [1] or similar texts for additional results.

3.1 VECTOR ANALYSIS

Maxwell's equations are described concisely in terms of certain mathematical operations on vector quantities in three-dimensional space. The field quantities are described as *vector* quantities, which will be denoted with an arrow above the symbol, e.g., $\vec{A}$, and a vector is represented as a directed line segment. Vector quantities convey two pieces of information: a magnitude and a direction of effect. The magnitude or length of a vector will be denoted as $|\vec{A}| = A$. In order to quantitatively describe a vector and perform mathematical operations on it as required by Maxwell's equations, we describe the vector in a coordinate system. A *rectangular* or *Cartesian* coordinate system consists of three orthogonal planes, $x = X =$ constant, $y = Y =$ constant, $z = Z =$ constant. The location of a point is described by the intersection of these three planes as $P = [X, Y, Z]$. This results in the coordinate system being represented as three orthogonal axes: x, y, z. The vector $\vec{A}$ has *projections* on each axis: A_x, A_y, A_z. A vector represented in the rectangular coordinate system is written in terms of *unit vectors* $\vec{a}_x, \vec{a}_y, \vec{a}_z$ in terms of its components as

$$\vec{A} = A_x\vec{a}_x + A_y\vec{a}_y + A_y\vec{a}_z \tag{3.1}$$

There are other coordinate systems that will be convenient to use. These are the *cylindrical* and *spherical* coordinate systems, and will be discussed as needed in later chapters. Vectors may be added by adding corresponding components as

$$\vec{A} + \vec{B} = (A_x + B_x)\vec{a}_x + (A_y + B_y)\vec{a}_y + (A_z + B_z)\vec{a}_z \tag{3.2}$$

Vectors may also have their lengths changed by a factor k by multiplying by that constant which multiplies each component by that factor. If the factor k is a negative number, this has the effect of reversing the direction of the vector. This shows how we may subtract two vectors; simply add $\vec{A} + k\vec{B}$ and choose $k = -1$. We will often need differential path lengths $\vec{dl}$, differential surface areas $\vec{ds}$, and differential volumes dv. These become [1]

$$\vec{dl} = dx\,\vec{a}_y + dy\,\vec{a}_y + dz\,\vec{a}_z \tag{3.3a}$$

$$\vec{ds} = dy\,dz\,\vec{a}_x + dx\,dz\,\vec{a}_y + dx\,dy\,\vec{a}_z \tag{3.3b}$$

$$dv = dx\,dy\,dz \tag{3.3c}$$

There are two definitions for the product of two vectors: the *dot product* and the *cross product*. The *dot product* of two vectors is defined as

$$\vec{A} \cdot \vec{B} = |\vec{A}| |\vec{B}| \cos \theta_{AB} \tag{3.4}$$
$$= A_x B_x + A_y B_y + A_z B_z$$

This represents the product of the length of one vector and the *projection* of the other vector on that vector, as illustrated in Fig. 3.1(a). The *cross product* of two vectors is defined as

$$\vec{A} \times \vec{B} = |\vec{A}| |\vec{B}| \sin \theta_{AB} \, \vec{a}_n \tag{3.5}$$
$$= (A_y B_z - A_z B_y)\vec{a}_x + (A_z B_x - A_x B_z)\vec{a}_y + (A_x B_y - A_y B_x)\vec{a}_z$$

where $\vec{a}_n$ is a unit vector perpendicular to the plane formed by the two vectors, as shown in Fig. 3.1(b). The direction of this unit vector is determined by the *right-hand rule*; that is, if the fingers of the right hand are curled from $\vec{A}$ to $\vec{B}$, the thumb will point in the direction of $\vec{a}_n$.

There are a number of *vector calculus* operations that are important to understand. The first is the concept of *line integral* of a vector field $\vec{F}$ along a contour C:

$$\int_C \vec{F} \cdot d\vec{l} = \int_C |\vec{F}| \cos \theta \, dl \tag{3.6}$$
$$= \int_{C_x} F_x \, dx + \int_{C_y} F_y \, dy + \int_{C_z} F_z \, dz$$

as illustrated in Fig. 3.2. The integral in (3.6) symbolizes that we add (integrate) the product of the components of $\vec{F}$ that are tangent to the path, $|\vec{F}| \cos \theta$, and the differential path length dl along the contour C. Perhaps the more obvious use of this result is in the computation of the work required to move an object against a force field from point a to point b. For example, suppose that the force field is given by $\vec{F} = (2y\vec{a}_x + xy\vec{a}_y + z\vec{a}_z)N$, and it is desired to determine the work required to move an object from $[x = 1 \text{ m}, y = 1 \text{ m}, z = 0]$ to $[x = 0,$

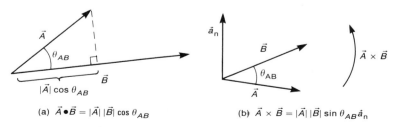

(a) $\vec{A} \bullet \vec{B} = |\vec{A}| |\vec{B}| \cos \theta_{AB}$ (b) $\vec{A} \times \vec{B} = |\vec{A}| |\vec{B}| \sin \theta_{AB} \vec{a}_n$

FIGURE 3.1   Illustration of (a) the dot product, and (b) the cross product.

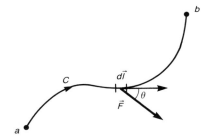

FIGURE 3.2   Illustration of the line integral.

$y = 2$ m, $z = 3$ m]. The work required is

$$W = -\int_a^b \vec{F} \cdot d\vec{l}$$

$$= -\left(\int_{x=1}^0 2y \, dx + \int_{y=1}^2 xy \, dy + \int_{z=0}^3 z \, dz \right)$$

$$= -\left[\int_{x=1}^0 2(-x+2) \, dx + \int_{y=1}^2 (-y+2)y \, dy + \int_{z=0}^3 z \, dz \right]$$

$$= -\tfrac{13}{6} \text{J}$$

and we have substituted the equation of the path, $y = -x + 2$, into the appropriate integrand variables to reduce the integrand to a function of only that variable. The integral symbol with a circle, $\oint_C$, denotes the integral around the *closed path* C. The next integral that we will find useful is the *surface integral* of a vector field over a surface S, defined by

$$\int_S \vec{F} \cdot d\vec{s} = \int_S |\vec{F}| \cos \theta \, ds \tag{3.7}$$

$$= \iint F_x \, dy \, dz + \iint F_y \, dx \, dz + \iint F_z \, dx \, dy$$

where θ is the angle between $\vec{F}$ and a normal to the surface. The surface integral in (3.7) symbolizes that we are to add (integrate) the product of the components of $\vec{F}$ that are perpendicular to the differential surface and the differential surface elements as illustrated in Fig. 3.3. This gives the *flux of the vector $\vec{F}$ through the surface S*. The surface integral with a circle, $\oint_S$, denotes the surface integral over the *closed surface* S. As an example, consider the vector field $\vec{F} = 2x\vec{a}_x + \vec{a}_y - \vec{a}_z$ to be integrated over the planar surface S defined by the three corners [$x = 2$, $y = 1$, $z = 0$], [$x = 2$, $y = 3$, $z = 0$] and [$x = 2$, $y = 3$, $z = 4$]. The surface

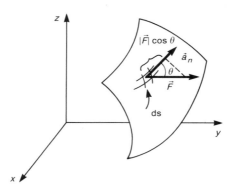

FIGURE 3.3   Illustration of the surface integral in the determination of the flux through a surface.

integral is

$$\int_S \vec{F} \cdot d\vec{s} = \int_{y=1}^3 \int_{z=0}^{z=2y-2} 2x \, dy \, dz + \underbrace{\iint dx \, dz}_{0} - \underbrace{\iint dx \, dy}_{0}$$

$$= 16$$

Since $dx = 0$ over the surface ($x = 2$ at all points on that surface), the second and third integrals are zero. Another way of looking at this is that, because of the orientation of the surface, in the yz plane for $x = 2$, the required flux is that in the positive x direction. The y and z components of the vector are not orthogonal to the surface, and therefore contribute nothing to the integral. Observations such as these are important in simplifying the result.

There are two additional vector calculus operations that will be useful: the *divergence* and the *curl* of a vector field. The *divergence of a vector field* is defined as

$$\text{div } \vec{F} = \nabla \cdot \vec{F} \tag{3.8}$$

$$= \lim_{\Delta v \to 0} \frac{\oint_S \vec{F} \cdot d\vec{s}}{\Delta v}$$

$$= \frac{\partial F_x}{\partial x} + \frac{\partial F_y}{\partial y} + \frac{\partial F_z}{\partial z}$$

where surface S encloses volume Δv. Therefore *the divergence of a vector field is the net outward flux of the vector field per unit volume as the volume shrinks to zero*. Essentially, the divergence of a vector field represents the flux of the field from a point which gives an indication of the *net source* of the field at that point, and is a scalar quantity. An important identity that will prove to be useful is the *divergence theorem* [1]:

$$\oint_S \vec{F} \cdot d\vec{s} = \int_v \nabla \cdot \vec{F} \, dv \qquad (3.9)$$

The divergence theorem shows that *the net flux of a vector field out of a closed surface can be obtained as the integral of the divergence of the vector field throughout the volume bounded by the surface.* This result allows us to interchange surface and volume integrals. The next vector calculus operation that will be required is the *curl of a vector field,* defined

$$\text{curl } \vec{F} \cdot \vec{a}_n = \nabla \times \vec{F} \qquad (3.10)$$

$$= \lim_{\Delta S \to 0} \frac{\oint_c \vec{F} \cdot d\vec{l}}{\Delta S}$$

$$= \left(\frac{\partial F_z}{\partial y} - \frac{\partial F_y}{\partial z} \right) \vec{a}_x + \left(\frac{\partial F_x}{\partial z} - \frac{\partial F_z}{\partial x} \right) \vec{a}_y + \left(\frac{\partial F_y}{\partial x} - \frac{\partial F_x}{\partial y} \right) \vec{a}_z$$

Equation (3.10) symbolizes that *the curl of a vector field is the line integral of a vector field around a contour C that bounds an open surface S as that surface area shrinks to zero.* Essentially, the curl is the net *circulation of a vector field about a point.* The curl is a *vector quantity* symbolizing the circulation of the vector in three orthogonal planes. A related result is *Stokes' theorem,* which is given by [1]

$$\int_S (\nabla \times \vec{F}) \cdot d\vec{s} = \oint_C \vec{F} \cdot d\vec{l} \qquad (3.11)$$

Stokes' theorem symbolizes that *the net flux of the curl of a vector field through an open surface S can be computed by obtaining the line integral of the vector field around the contour C bounding the open surface.* This result allows us to interchange surface and contour integrals.

3.2 MAXWELL'S EQUATIONS

Maxwell's equations are stated concisely in terms of the vector calculus operations of the previous section. Maxwell's equations form the cornerstones of electromagnetic phenomena, and are therefore essential to our understanding of how to design electronic systems such that they will comply with regulatory requirements and will also not cause interference or be interfered with. Solution of Maxwell's equations is not a simple process, yet this practicality does not diminish their fundamental importance. Quite often we will utilize certain approximations of them, such as lumped-circuit models, in order to simplify the solution of a specific problem. This is permissible so long as the dimensions of the problem are *electrically small*. Nevertheless, we should always be cognizant of the fact that Maxwell's equations govern all electromagnetic phenomena, and their complexity does not change this fact.

3.2.1 Faraday's Law

Faraday's law can be stated concisely in *integral form* as

$$\oint_C \vec{E} \cdot d\vec{l} = -\frac{d}{dt} \int_S \vec{B} \cdot d\vec{s} \tag{3.12}$$

The parameter $\vec{E}$ is the *electric field intensity vector*, with units of volts/meter (V/m). The parameter $\vec{B}$ is the *magnetic flux density vector*, with units of webers/square meter (Wb/m²). Faraday's law provides that *the electromotive force (emf) generated around a closed contour C is related to the time rate of change of the total magnetic flux through the open surface S bounded by that contour*. The emf is given by

$$\text{emf} = \oint_C \vec{E} \cdot d\vec{l} \tag{3.13}$$

whose units are volts (V). The total magnetic flux through the open surface S that is bounded by the contour C is

$$\psi_m = \int_S \vec{B} \cdot d\vec{s} \tag{3.14}$$

whose units are webers (Wb). Therefore Faraday's law can be written as

$$\text{emf} = -\frac{d\psi_m}{dt} \tag{3.15}$$

If no magnetic flux penetrates the surface bounded by the contour, $\psi_m = 0$, then emf = 0, which is similar to Kirchhoff's voltage law of lumped-circuit theory. Closed loops that have flux from external sources penetrating the surface bounded by the loop are accounted for in lumped-circuit theory by the use of *mutual inductance*, wherein a voltage source representing this effect is inserted in the loop according to (3.15) so that Kirchhoff's voltage law will handle this case [2].

The contour C and surface S are intimately related by the right-hand rule, as shown in Fig. 3.4. If the fingers of the right hand are directed in the direction of the contour C, the thumb will give the direction of the unit normal to the

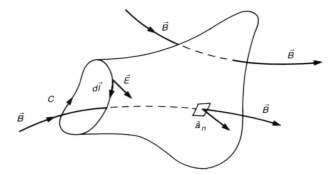

FIGURE 3.4   Illustration of Faraday's law.

enclosed surface. The vector surface is $d\vec{s} = ds\,\vec{a}_n$, where $\vec{a}_n$ is a unit normal to the open surface. The negative sign in (3.12) and (3.15) is referred to as *Lenz's law*, and symbolizes that *the emf induced in the contour is of a polarity that tends to generate an induced current whose magnetic flux tends to oppose any change in (is opposite to) the original magnetic flux*. This is illustrated in Fig. 3.5 for a flat surface S. Consider Fig. 3.5(a). The magnetic flux density vector is directed upward and is decreasing in magnitude. An emf is induced in the contour or loop of a polarity such that it tends to induce (produce) a current i_{induced}, whose magnetic flux (by the right-hand-rule) tends to oppose this decrease in $\vec{B}$. *It is very important to think of this induced emf as being equivalent to a voltage source that is inserted in the loop as shown*. However, the emf is a *distributed-parameter* quantity and cannot be truly localized (lumped). But if the dimensions of the loop are *electrically small*, we can, as a reasonable approximation, think of it as a lumped voltage source. Conversely, if the magnetic flux penetrating the surface and directed upward is increasing in magnitude as shown in Fig. 3.5(b), the induced emf and its associated induced current is oppositely directed such that the flux due to this induced current is (by the right-hand-rule) in a direction that tends to oppose this increase in $\vec{B}$. All of this makes sense because if the reverse were true, the flux due to the induced current would aid the original

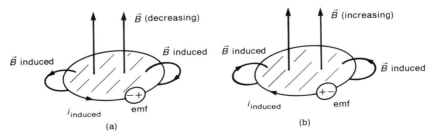

FIGURE 3.5   Further illustration of Faraday's law, giving the relation between the induced emf and the rate-of-change of the magnetic flux density through the surface.

magnetic field, causing it to build up causing a further increase in the induced current, and so forth. Thus energy would not be conserved. Note that the surface S is considered to be flat to simplify the result. *The actual shape of the surface S is not important, since Faraday's law yields the same result for the induced emf for all surfaces that are bounded by the contour C.* This important consideration is illustrated in Fig. 3.4. Any magnetic flux density vector that passes through the mouth of the "balloon-shaped" surface will necessarily exit that surface and contribute to ψ_m for that surface. On the other hand, if that flux enters the surface but does not pass through the mouth, it must necessarily exit the surface also and will not contribute to ψ_m. This is intimately tied to Gauss' law, which states that magnetic flux lines must close on themselves since there are no known isolated sources of the magnetic field.

It is not necessary to have a closed loop in order that a time-changing magnetic field will generate (induce) an emf in that loop. For example, suppose that the loop of Fig. 3.5 is opened at some point as shown in Fig. 3.6. An emf in the form of an induced voltage source is nevertheless induced in the perimeter of the loop. That emf will be present at the terminals of the loop. Although no current will flow because the loop is broken, this voltage will nevertheless appear at the terminals. The polarity of the voltage at the terminals is essentially the *open-circuit voltage* of the loop, which is the induced emf source.

Faraday's law essentially shows that a *time-changing magnetic field will generate (induce) an electric field* in the same fashion as a static charge distribution. However, the induced electric field lines due to a time-changing magnetic field must close on themselves, whereas the electric field lines due to charges originate on positive charge and terminate on negative charge.

The negative sign due to Lenz's law is often a confusing issue. *The simplest way to correctly account for the minus sign is to ignore it and choose the polarity (direction) of the induced emf voltage source such that it will produce the correct direction of the induced current and associated induced magnetic field that will oppose a change in $\vec{B}$.* This takes care of the minus sign in Faraday's law (Lenz's law). To obtain the value of the source, compute the rate-of-change of the total magnetic flux through the loop, which may be positive or negative depending on whether the flux is increasing or decreasing, and place this value (including

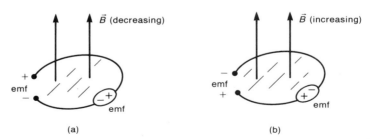

FIGURE 3.6   Illustration of the emf as an open-circuit voltage induced at the terminals of an open-circuited loop.

its sign) on the emf voltage source. This will always result in the correct polarity of the induced emf. For example, in Fig. 3.5(a), because of the *direction of $\vec{B}$* through the surface, the polarity of the source would be placed opposite to that shown. But since $\vec{B}$ is decreasing, its rate-of-change is negative, so that the value of the source is *negative*, which has the effect of reversing this source direction to give that shown in Fig. 3.5(a).

Faraday's law was introduced in its *integral form* in (3.12). The integral form is useful from the standpoint of visualization of the meaning of the result. From a computational standpoint, the *point* or *differential form* is more useful. Applying Stokes' theorem to the left-hand side of (3.12) gives the *point form* of Faraday's law as

$$\nabla \times \vec{E} = -\frac{\partial \vec{B}}{\partial t} \tag{3.16a}$$

Substituting the result for the curl gives

$$\left(\frac{\partial E_z}{\partial y} - \frac{\partial E_y}{\partial z}\right)\vec{a}_x + \left(\frac{\partial E_x}{\partial z} - \frac{\partial E_z}{\partial x}\right)\vec{a}_y + \left(\frac{\partial E_y}{\partial x} - \frac{\partial E_x}{\partial y}\right)\vec{a}_z \tag{3.16b}$$

$$= -\frac{\partial B_x}{\partial t}\vec{a}_x - \frac{\partial B_y}{\partial t}\vec{a}_y - \frac{\partial B_z}{\partial t}\vec{a}_z$$

Matching components gives three equations:

$$\frac{\partial E_z}{\partial y} - \frac{\partial E_y}{\partial z} = -\frac{\partial B_x}{\partial t}$$

$$\frac{\partial E_x}{\partial z} - \frac{\partial E_x}{\partial x} = -\frac{\partial B_y}{\partial t} \tag{3.16c}$$

$$\frac{\partial E_y}{\partial x} - \frac{\partial E_x}{\partial y} = -\frac{\partial B_z}{\partial t}$$

This point form symbolizes that a time-changing magnetic field results in a *circulation* (curl) of the electric field.

3.2.2   Ampère's Law

Faraday's law showed that a time-changing magnetic field can produce (induce) an electric field. Ampère's law shows that the converse is true; that is, *a time-changing electric field can produce (induce) a magnetic field*. Ampère's law is given in integral form as

$$\oint_c \vec{H} \cdot d\vec{l} = \int_s \vec{J} \cdot d\vec{s} + \frac{d}{dt}\int_s \vec{D} \cdot d\vec{s} \tag{3.17}$$

The quantity $\vec{H}$ is the *magnetic field intensity vector*, having units of amperes per meter (A/m). The quantity $\vec{J}$ is the *current density vector*, with units of A/m^2. The quantity $\vec{D}$ is the *electric flux density vector*, with units of coulombs per square meter (C/m^2). Note that the units of the result in (3.17) after integration are amperes (A). The contour C bounds the open surface S, as illustrated in Fig. 3.7. Their directions are related by the right-hand rule. As was the case with Faraday's law, any surface shape is suitable so long as contour C bounds it. Only the $\vec{J}$ and $\vec{D}$ that pass through the opening contribute. The line integral of $\vec{H}$ around the closed contour C is referred to as the *magnetomotive force* or *mmf* around that contour:

$$\text{mmf} = \oint_C \vec{H} \cdot d\vec{l} \quad \text{(in A)} \tag{3.18}$$

This is essentially the dual of the emf of Faraday's law. The first term on the right-hand side of Ampère's law is the *total conduction current that penetrates the surface S bounded by the contour C*:

$$I_c = \int_S \vec{J} \cdot d\vec{s} \quad \text{(in A)} \tag{3.19}$$

This is the total current bounded by the contour that is due to free charges. The second term on the right-hand side of Ampère's law is the *total displacement*

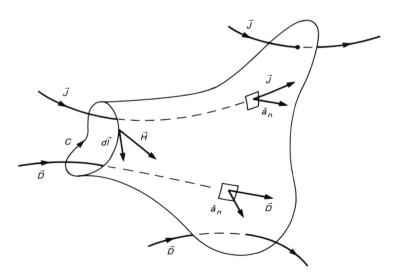

FIGURE 3.7 Illustration of Ampère's law.

current that penetrates the surface S bounded by the contour C:

$$I_d = \frac{d}{dt} \int_S \vec{D} \cdot d\vec{s} \quad \text{(in A)} \tag{3.20}$$

Therefore Ampère's law can be written as

$$\text{mmf} = I_c + I_d \quad \text{(in A)} \tag{3.21}$$

Ampère's law essentially shows that a *time-changing electric flux $\vec{D}$ has the same effect as free current $\vec{J}$ in producing a magnetic field $\vec{H}$.* We already know that a static (dc) current will produce a magnetic field. Ampère's law shows that a time-changing electric field will give the same result.

The *point* or *differential form* of Ampère's law can be obtained by applying Stokes' theorem to (3.17) to give

$$\nabla \times \vec{H} = \vec{J} + \frac{\partial \vec{D}}{\partial t} \tag{3.22a}$$

Substituting the result for the curl and matching components gives three equations;

$$\frac{\partial H_z}{\partial y} - \frac{\partial H_y}{\partial z} = J_x + \frac{\partial D_x}{\partial t}$$

$$\frac{\partial H_x}{\partial z} - \frac{\partial H_z}{\partial x} = J_y + \frac{\partial D_y}{\partial t} \tag{3.22b}$$

$$\frac{\partial H_y}{\partial x} - \frac{\partial H_x}{\partial y} = J_z + \frac{\partial D_z}{\partial t}$$

3.2.3 Gauss' Laws

Gauss' law for the electric field is stated in integral form as

$$\oint_S \vec{D} \cdot d\vec{s} = \int_v \rho_v \, dv \tag{3.23}$$

The quantity ρ_v is the *volume free charge density*, whose units are coulombs per cubic meter (C/m^3). Gauss' law for the electric field provides that the *net flux of the electric flux density vector out of the closed surface S is equivalent to the net positive charge enclosed by the surface.* Fig. 3.8 illustrates this point. Electric field lines that begin on positive charge must terminate on an equal amount of negative charge. Integrating $\vec{D}$ over a closed surface will only reveal the *net*

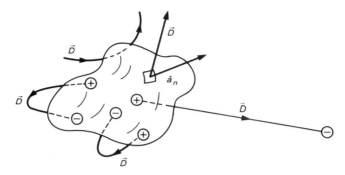

FIGURE 3.8   Illustration of Gauss' law for the electric field. The net flux of $\vec{D}$ through the closed surface is the net positive charge enclosed by the surface.

positive charge enclosed by the surface. Also note that electric field lines can be generated by a time-changing magnetic field, as Faraday's law shows. These form closed paths, and as such enter and leave the closed surface and contribute nothing to Gauss' law.

Equation (3.23) is the integral form of Gauss' law. To obtain the point form, we apply the divergence theorem to (3.23) to yield

$$\nabla \cdot \vec{D} = \rho_v \tag{3.24a}$$

Substituting the result for the divergence gives

$$\frac{\partial D_x}{\partial x} + \frac{\partial D_y}{\partial y} + \frac{\partial D_z}{\partial z} = \rho_v \tag{3.24b}$$

A related result is *Gauss' law for the magnetic field*, stated in integral form as

$$\oint_S \vec{B} \cdot d\vec{s} = 0 \tag{3.25}$$

This result implies that *all magnetic field lines form closed paths; that is, there are no (known) isolated sources of the magnetic field.* This is illustrated in Fig. 3.9. If we try to divide a permanent magnet, we find that new N–S poles are formed at the opposite ends of these pieces. The *point form of Gauss' law for the magnetic field* is obtained by applying the divergence theorem to (3.25) to give

$$\nabla \cdot \vec{B} = 0 \tag{3.26a}$$

Substituting the result for the divergence gives

$$\frac{\partial B_x}{\partial x} + \frac{\partial B_y}{\partial y} + \frac{\partial B_z}{\partial z} = 0 \tag{3.26b}$$

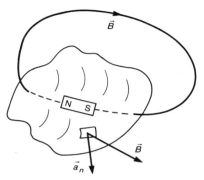

FIGURE 3.9   Illustration of Gauss' law for the magnetic field. The net flux of $\vec{B}$ through the closed surface is zero.

3.2.4   Conservation of Charge

Charge can be neither created nor destroyed. The mathematical statement of this important result is given by

$$\oint_S \vec{J} \cdot d\vec{s} = -\frac{d}{dt} \int_v \rho_v \, dv \tag{3.27}$$

This result is a rather common-sense statement of the fact that *any current leaving a closed surface S implies a decrease of charge within that closed surface.* The point form of this is obtained by applying the divergence theorem to (3.27) to give

$$\nabla \cdot \vec{J} = -\frac{\partial \rho_v}{\partial t} \tag{3.28a}$$

Expanding this gives

$$\frac{\partial J_x}{\partial x} + \frac{\partial J_y}{\partial y} + \frac{\partial J_z}{\partial z} = -\frac{\partial \rho_v}{\partial t} \tag{3.28b}$$

It should be pointed out that Faraday's, Ampère's, and Gauss' laws, along with conservation of charge, are collectively referred to as Maxwell's equations. These five equations are not all independent. It can be shown that Faraday's law, Ampère's law, and conservation of charge are all that are required to completely characterize the electromagnetic field [1].

3.2.5   Constitutive Parameters of the Medium

Maxwell's equations involve five unknown vector field quantities: $\vec{E}$, $\vec{B}$, $\vec{D}$, $\vec{H}$, and $\vec{J}$. The *constitutive relations* relate these quantities. The electromagnetic fields exist in material media and there are various characterizations of those media. The simplest and most common type of medium is the *simple medium*, in which the field vectors are simply related as

$$\vec{D} = \epsilon \vec{E} \tag{3.29a}$$

$$\vec{B} = \mu \vec{H} \tag{3.29b}$$

$$\vec{J_c} = \sigma \vec{E} \tag{3.29c}$$

The current density $\vec{J}$ will also consist of impressed currents $\vec{J_s}$, which can be viewed as the source of the fields. Thus $\vec{J} = \vec{J_c} + \vec{J_s}$ in Ampère's law. The parameters ϵ, μ, and σ in these relations are the *permittivity, permeability,* and *conductivity,* respectively, of the medium. The units of ϵ are farads per meter (F/m) or a capacitance per unit length. The units of μ are henrys per meter (H/m) or an inductance per unit length. The units of σ are siemens per meter (S/m) or a conductance per unit length. Substituting (3.29) into Faraday's and Ampère's laws gives Maxwell's equations for simple media:

$$\nabla \times \vec{E} = -\mu \frac{\partial \vec{H}}{\partial t} \tag{3.30a}$$

$$\nabla \times \vec{H} = \sigma \vec{E} + \epsilon \frac{\partial \vec{E}}{\partial t} + \vec{J_s} \tag{3.30b}$$

Thus (3.30) give six equations in the six components of $\vec{E}$ and $\vec{H}$. Once these are found, we can obtain $\vec{D}$, $\vec{B}$, and $\vec{J_c}$ from (3.29). Our emphasis in this text will be on electromagnetic fields in simple media, and so the equations in (3.30) will be of paramount importance.

Simple media for which the field vectors are related simply by (3.29) are said to be *linear, homogeneous,* and *isotropic*. A *nonlinear medium* is one in which $\vec{D}$ is a function of the *magnitude* of $\vec{E}$, $\vec{B}$ is a function of the *magnitude* of $\vec{H}$, and/or $\vec{J_c}$ is a function of the *magnitude* of $\vec{E}$. An example of a nonlinear medium is a ferromagnetic one, in which $|\vec{B}|$ is related to $|\vec{H}|$ by the familiar nonlinear hysteresis curve. In other words, we may write the parameters as $\epsilon(E)$, $\mu(H)$, and/or $\sigma(E)$. An *inhomogeneous medium* is one in which the medium parameters are functions of position, e.g., $\epsilon(x, y, z)$, $\mu(x, y, z)$, and/or $\sigma(x, y, z)$. Examples of inhomogeneous media are dielectric-insulated wires or printed circuit boards where the electric field exists partly in air ($\epsilon_r = 1$) and partly in the insulating material for which $\epsilon_r \neq 1$. And finally, an *anisotropic medium* is one in which $\vec{E}$ is not parallel to $\vec{D}$, or $\vec{B}$ is not parallel to $\vec{H}$, or $\vec{J_c}$ is not parallel to $\vec{E}$. Ferrites are examples of anisotropic media. These types of materials are used to construct microwave devices such as circulators. Because of the difficulty in solving Maxwell's equations for media that are not *linear, homogeneous,* and *isotropic*, we will try where possible to approximate media as being simple media.

3.3 BOUNDARY CONDITIONS

Maxwell's equations are differential equations, and as such are no different than other differential equations in that they have an infinite number of solutions. Ordinary differential equations such as occur in lumped circuits have an infinite number of solutions and require that certain initial conditions be specified in order to pin down the specific solution. Partial differential equations such as those of Maxwell require the specification of *boundary conditions* in order to pin down which of the infinite number of possible solutions in the particular medium apply. The boundary conditions will be stated, and the reader is referred to [1] for a derivation.

First we consider the boundary between two physical media shown in Fig. 3.10. Medium #1 is characterized by ϵ_1, μ_1, and σ_1, while medium #2 is characterized by ϵ_2, μ_2, and σ_2. It is not necessary that the two media be simple media for the following boundary conditions to hold. The *boundary conditions* provide constraints on the components of the field vectors as they transition across the boundary between the two media. The *tangential components of the electric field intensity vector $\vec{E}$ and the magnetic field intensity vector $\vec{H}$ must be continuous across the boundary between two physical media*:

$$E_{t_1} = E_{t_2} \tag{3.31a}$$

$$H_{t_1} = H_{t_2} \tag{3.31b}$$

Also, *the normal components of the electric flux density vector $\vec{D}$ and the magnetic flux density vector $\vec{B}$ must be continuous across the boundary between two physical media*:

$$D_{n_1} = D_{n_2} \tag{3.32a}$$

$$B_{n_1} = B_{n_2} \tag{3.32b}$$

These boundary conditions are illustrated in Fig. 3.10. The continuity of the normal components of the electric flux density vector $\vec{D}$ given in (3.32a) *assumes that no charge has been intentionally placed and resides on the boundary surface.* This may occur, for example, by rubbing cat's fur against nylon, in which case charges are separated, leaving net charge on the two boundaries. This will be important in the examination of electrostatic discharge (ESD) problems in Chapter 12.

So far we have been discussing *real, physical media*; that is, media that can exist physically. We will have occasion to utilize the concept of certain idealized mathematical media that, although they do not exist physically, serve to simplify the mathematical calculations and, moreover, are reasonable approximations to certain physical media. The primary such idealized medium is the *perfect conductor*, which may be characterized by an infinite conductivity, $\sigma = \infty$. The impact of an infinite conductivity is to render all the fields in that perfect

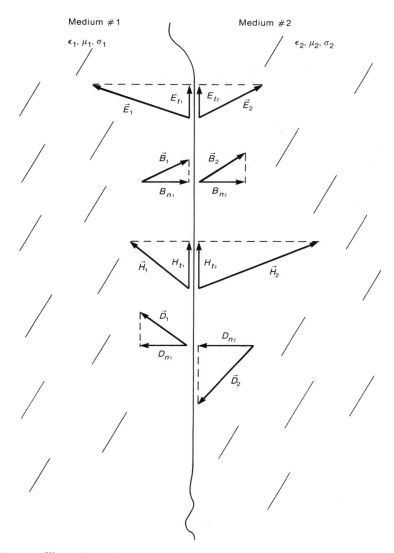

FIGURE 3.10 Illustration of the boundary conditions at an interface between two media.

conductor to be *zero*: $\vec{E}_2 = 0$, $\vec{D}_2 = 0$, $\vec{H}_2 = 0$, and $\vec{B}_2 = 0$. Since all the fields in medium #2 (the perfect conductor) are zero, so are the normal and tangential components at the boundary, as illustrated in Fig. 3.11. This requires that the tangential component of $\vec{E}_1$ must be zero at the boundary:

$$E_{t_1} = 0, \quad \sigma_2 = \infty \tag{3.33a}$$

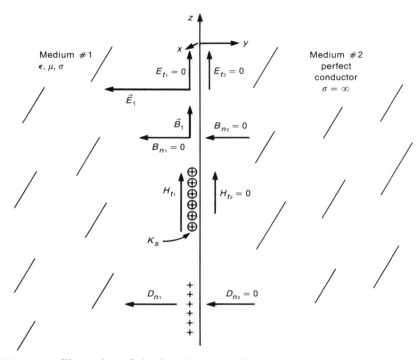

FIGURE 3.11 Illustration of the boundary conditions where one medium is a perfect conductor.

Also the normal component of $\vec{B}_1$ must be zero at the boundary:

$$B_{n_1} = 0, \quad \sigma_2 = \infty \qquad (3.33b)$$

It turns out that we cannot similarly require that H_{t_1} and D_{n_1} be zero, or else the resulting fields in medium #1 would be overspecified [1]. The resulting requirements are

$$H_{t_1} = K_s \quad (\text{in A/m}), \quad \sigma_2 = \infty \qquad (3.34c)$$

and

$$D_{n_1} = \rho_s \quad (\text{in C/m}^2), \quad \sigma_2 = \infty \qquad (3.34d)$$

The quantity K_s is the *surface current density existing on the interface*. The units of K_s are amperes per meter (A/m), and this represents a distribution of current along the boundary surface per unit length along the surface [1]. The surface current density K_s is orthogonal to the tangential component of $\vec{H}$, H_{t_1}. Similarly, the quantity ρ_s given in (3.34d) is the *surface charge density existing on the interface*. The units of ρ_s are coulombs per square meter of the interface (C/m^2), and this represents a distribution of charge (free charge) over the boundary surface [1].

3.4 SINUSOIDAL STEADY STATE

The time variation of the field vectors is not restricted to any particular functional form. However, our main interest in this text will be in the *sinusoidal time variation* of these field vectors. We assume that the fields have been in existence for a sufficient length of time that all transients have decayed to zero, leaving the *sinusoidal steady state*. This has the effect of simplifying the mathematics, and the techniques are virtually no different than for sinusoidal steady-state, lumped-circuit analysis [2]. For example, the phasor form of the x component of the electric field intensity vector is

$$\hat{E}_x(x, y, z) = E_{xm}(x, y, z)e^{j\theta_x(x,y,z)} \qquad (3.35)$$
$$= E_{xm} \underline{/\theta_x}$$

Phasor quantities are complex-valued quantities, and will be denoted by a caret over the symbol. *The time-domain form of phasor quantities may be obtained by multiplying the phasor form by $e^{j\omega t}$ and taking the real part of the result.* This is an important technique, and will be used on numerous occasions. For example,

$$E_x(x, y, z, t) = \mathscr{R}e\{\hat{E}_x(x, y, z)e^{j\omega t}\} \qquad (3.36)$$
$$= \mathscr{R}e\{E_{xm} \underline{/\theta_x} e^{j\omega t}\}$$
$$= \mathscr{R}e\{E_{xm}e^{j(\omega t + \theta_x)}\}$$
$$= \mathscr{R}e\{E_{xm}\cos(\omega t + \theta_x) + jE_{xm}\sin(\omega t + \theta_x)\}$$
$$= E_{xm}\cos(\omega t + \theta_x)$$

where $\mathscr{R}e\{\ \ \}$ denotes the *real part* of the enclosed complex quantity.

In order to solve Maxwell's equations for sinusoidal excitation, we replace the field vectors with their phasor forms multiplied by $e^{j\omega t}$. Differentiation of these forms of the field vectors with respect to time t gives

$$\frac{\partial}{\partial t}\vec{\hat{E}}(x, y, z)e^{j\omega t} = j\omega\vec{\hat{E}}(x, y, z)e^{j\omega t} \qquad (3.37)$$

This very important property allows a considerable simplification in the solution of Maxwell's equations. Substituting the forms of the field vectors and cancelling the $e^{j\omega t}$ that is common to both sides gives the *phasor forms of Maxwell's equations.* If the medium is linear, homogeneous, and isotropic (a simple medium), the phasor Maxwell's equations become

$$\oint_c \vec{E} \cdot d\vec{l} = -j\omega\mu \int_s \vec{H} \cdot d\vec{s}, \qquad \nabla \times \vec{E} = -j\omega\mu\vec{H} \qquad (3.38a)$$

$$\oint_c \vec{H} \cdot d\vec{l} = (\sigma + j\omega\varepsilon) \int_s \vec{E} \cdot d\vec{s} + \int_s \vec{J_s}, \quad \nabla \times \vec{H} = (\sigma + j\omega\varepsilon)\vec{E} + \vec{J_s}$$
$$(3.38b)$$

$$\oint_s \vec{H} \cdot d\vec{s} = 0, \qquad \nabla \cdot \vec{H} = 0 \qquad (3.38c)$$

$$\oint_s \vec{E} \cdot d\vec{s} = \frac{1}{\varepsilon} \int_v \hat{\rho}_v \, dv, \qquad \nabla \cdot \vec{E} = \frac{\hat{\rho}_v}{\varepsilon} \qquad (3.38d)$$

$$\oint_s \vec{J} \cdot d\vec{s} = -j\omega \int_v \hat{\rho}_v \, dv, \qquad \nabla \cdot \vec{J} = -j\omega\hat{\rho}_v \qquad (3.38e)$$

Here the permittivity, permeability, and conductivity may be functions of frequency, i.e., $\varepsilon(f)$, $\mu(f)$, and $\sigma(f)$, as they usually are for material media.

3.5 POWER FLOW

The units of the electric field intensity vector $\vec{E}$ are V/m while those of the magnetic field intensity vector $\vec{H}$ are A/m. Thus the product of these two vectors has the units of power density or watts per square meter (W/m^2). But there are two possibilities for the product of two vectors: dot product and cross product. It turns out that the cross product is more meaningful in that the *power density vector* or *Poynting vector* $\vec{S}$ relates to power flow [1]. The Poynting vector is defined as

$$\vec{S} = \vec{E} \times \vec{H} \qquad (3.39)$$

Using certain vector identities, it is possible to show that [1]

$$-\oint_s \vec{S} \cdot d\vec{s} = \int_v \vec{E} \cdot \vec{J} \, dv + \int_v \left(\vec{E} \cdot \frac{\partial \vec{D}}{\partial t} + \vec{H} \cdot \frac{\partial \vec{B}}{\partial t} \right) dv \quad \text{(in W)} \qquad (3.40)$$

The term on the left-hand side of (3.40) represents the net inward flux of $\vec{S}$ into the volume. The first term on the right-hand side represents *power dissipation* within the volume, while the second represents *the time rate of change of the energy stored within the volume* [1]. The Poynting vector $\vec{S}$ defined above represents *instantaneous power*. For sinusoidal steady-state excitation we are interested in *average power*. To determine this average power flow, we define the *phasor Poynting vector* as

$$\vec{S} = \vec{E} \times \vec{H}^* \tag{3.41}$$

where the *complex conjugate* of a phasor $\hat{A}$ is denoted by $\hat{A}^*$. The *density of average power* is obtained as the *average power density Poynting vector*:

$$\vec{S}_{av} = \tfrac{1}{2} \mathscr{R}e\{\vec{S}\} \quad (\text{in W/m}^2) \tag{3.42}$$
$$= \tfrac{1}{2} \mathscr{R}e\{\vec{E} \times \vec{H}^*\}$$

3.6   UNIFORM PLANE WAVES

We now embark upon a discussion of the simplest type of wave propagation: *uniform plane waves*. We initially consider these not only because they are simple but also because wave propagation on transmission lines and waveguides and waves propagated by antennas bear a striking similarity to uniform plane waves. Therefore the structure and properties of many other forms of wave propagation will be more easily understood.

There are two important terms in the name: *uniform* and *plane*. The term "plane" means that at any point in space the electric and magnetic field intensity vectors lie in a plane and the planes at any two different points are parallel. The term "uniform" means that $\vec{E}$ and $\vec{H}$ are independent of position in each plane. Without any loss of generality, we may assume the electric and magnetic field intensity vectors to lie in the xy plane and may choose the electric field intensity vector to be directed in the x direction, as shown in Fig. 3.12(a):

$$\vec{E} = E_x(z, t)\vec{a}_x \tag{3.43}$$

The criterion that the field vectors be uniform means that they must be independent of x and y:

$$\frac{\partial E_x}{\partial x} = \frac{\partial E_x}{\partial y} = 0 \tag{3.44}$$

and thus the vectors can be a function of only z and, of course time t, as indicated in (3.43). Substituting (3.43) into Faraday's law and using (3.44) shows that the magnetic field has only a y component:

$$\vec{H} = H_y(z, t)\vec{a}_y \tag{3.45}$$

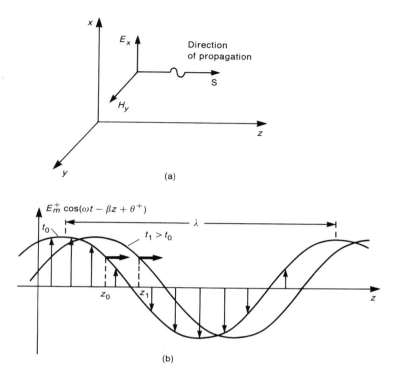

FIGURE 3.12 The uniform plane wave: (a) relation of the fields to each other and the direction of propagation of the wave; (b) general spatial properties of the traveling uniform plane wave.

Thus $\vec{E}$ and $\vec{H}$ are orthogonal and lie in the xy plane. Similarly, because the wave is uniform over the xy plane,

$$\frac{\partial H_y}{\partial x} = \frac{\partial H_y}{\partial y} = 0 \tag{3.46}$$

Substituting these results into Faraday's and Ampère's laws gives the differential equations governing the field vectors for simple media having no impressed sources:

$$\frac{\partial E_x(z, t)}{\partial z} = -\mu \frac{\partial H_y(z, t)}{\partial t} \tag{3.47a}$$

$$\frac{\partial H_y(z, t)}{\partial z} = -\sigma E_x(z, t) - \epsilon \frac{\partial E_x(z, t)}{\partial t} \tag{3.47b}$$

For sinusoidal, steady-state variation of the field vectors, the results in (3.47) simplify to

$$\frac{d\hat{E}_x(z)}{dz} = -j\omega\mu\hat{H}_y(z) \tag{3.48a}$$

$$\frac{d\hat{H}_y(z)}{dz} = -(\sigma + j\omega\epsilon)\hat{E}_x(z) \tag{3.48b}$$

Note that we have replaced partial derivatives with ordinary derivatives since the phasor field vectors are functions of only one variable, z. These are a set of coupled first-order ordinary differential equations. They may be reduced to sets of uncoupled second-order ordinary differential equations by differentiating one with respect to z and substituting the other, and vice versa, resulting in

$$\frac{d^2\hat{E}_x(z)}{dz^2} = \hat{\gamma}^2\hat{E}_x(z) \tag{3.49a}$$

$$\frac{d^2\hat{H}_y(z)}{dz^2} = \hat{\gamma}^2\hat{H}_y(z) \tag{3.49b}$$

The solutions to these equations are of a simple form [1]:

$$\hat{E}_x = \hat{E}_m^+ e^{-\hat{\gamma}z} + \hat{E}_m^- e^{\hat{\gamma}z} \tag{3.50a}$$

$$\hat{H}_y = \frac{\hat{E}_m^+}{\hat{\eta}} e^{-\hat{\gamma}z} - \frac{\hat{E}_m^-}{\hat{\eta}} e^{\hat{\gamma}z} \tag{3.50b}$$

where the *propagation constant* is

$$\hat{\gamma} = \sqrt{j\omega\mu(\sigma + j\omega\epsilon)} \tag{3.51}$$
$$= \alpha + j\beta$$

and the *intrinsic impedance* is

$$\hat{\eta} = \sqrt{\frac{j\omega\mu}{\sigma + j\omega\epsilon}} \tag{3.52}$$

$$= \frac{j\omega\mu}{\hat{\gamma}}$$

$$= \eta \underline{/\theta_\eta}$$

The quantity α in (3.51) is referred to as the *attenuation constant*, with units of nepers per meter (Np/m), and the quantity β is referred to as the *phase constant*, with units of radians per meter (rad/m). The quantity $\hat{\eta}$ in (3.52) is referred to as the *intrinsic impedance of the medium*, with units of ohms (Ω). In terms of these quantities, the solutions in (3.50) can be written as

$$\hat{E}_x = \hat{E}_m^+ e^{-\alpha z} e^{-j\beta z} + \hat{E}_m^- e^{\alpha z} e^{j\beta z} \tag{3.53a}$$

$$\hat{H}_y = \frac{\hat{E}_m^+}{\eta} e^{-\alpha z} e^{-j\beta z} e^{-j\theta_\eta} - \frac{\hat{E}_m^-}{\eta} e^{\alpha z} e^{j\beta z} e^{-j\theta_\eta} \tag{3.53b}$$

Writing the complex undetermined constants $\hat{E}_m^+$ and $\hat{E}_m^-$ as a magnitude and angle as $\hat{E}_m^+ = E_m^+ e^{j\theta^+} = E_m^+ \underline{/\theta^+}$ and $\hat{E}_m^- = E_m^- e^{j\theta^-} = E_m^- \underline{/\theta^-}$ gives

$$\hat{E}_x = E_m^+ e^{-\alpha z} e^{-j\beta z} e^{j\theta^+} + E_m^- e^{\alpha z} e^{j\beta z} e^{j\theta^-} \tag{3.54a}$$

$$\hat{H}_y = \frac{E_m^+}{\eta} e^{-\alpha z} e^{-j\beta z} e^{-j\theta_\eta} e^{j\theta^+} - \frac{E_m^-}{\eta} e^{\alpha z} e^{j\beta z} e^{-j\theta_\eta} e^{j\theta^-} \tag{3.54b}$$

The *time-domain forms* are

$$E_x = \mathcal{R}e\{\hat{E}_x e^{j\omega t}\} \tag{3.55a}$$
$$= E_m^+ e^{-\alpha z} \cos(\omega t - \beta z + \theta^+) + E_m^- e^{\alpha z} \cos(\omega t + \beta z + \theta^-)$$

$$H_y = \mathcal{R}e\{\hat{H}_y e^{j\omega t}\} \tag{3.55b}$$
$$= \frac{E_m^+}{\eta} e^{-\alpha z} \cos(\omega t - \beta z + \theta^+ - \theta_\eta) - \frac{E_m^-}{\eta} e^{\alpha z} \cos(\omega t + \beta z + \theta^- - \theta_\eta)$$

3.6.1 Lossless Media

It is important that we investigate the implications and properties of these equations. To simplify our analysis we will first consider uniform plane waves in *lossless media*, $\sigma = 0$. For this case the propagation constant becomes

$$\alpha = 0 \tag{3.56a}$$

$$\beta = \omega\sqrt{\mu\epsilon} \tag{3.56b}$$

Since $\alpha = 0$, the wave suffers no *attenuation* as it propagates through the medium. The intrinsic impedance becomes

$$\eta = \sqrt{\frac{\mu}{\epsilon}} \qquad (3.57a)$$

$$\theta_\eta = 0 \qquad (3.57b)$$

Thus the field vectors become for lossless **media**

$$E_x = E_m^+ \cos(\omega t - \beta z + \theta^+) + E_m^- \cos(\omega t + \beta z + \theta^-) \qquad (3.58a)$$

$$H_y = \frac{E_m^+}{\eta} \cos(\omega t - \beta z + \theta^+) - \frac{E_m^-}{\eta} \cos(\omega t + \beta z + \theta^-) \qquad (3.58b)$$

Consider the first term of (3.58a), $E_m^+ \cos(\omega t - \beta z + \theta^+)$. This portion of E_x represents a wave traveling in the positive z direction. This can be seen from Fig. 3.12(b), in which $E_m^+ \cos(\omega t - \beta z + \theta^+)$ has been plotted as a function of z for two different instants of time, t_0 and $t_1 > t_0$. Note that *corresponding points on the waveforms occur at positions and times such that the argument of the cosine has the same value*; that is,

$$\omega t_0 - \beta z_0 + \theta^+ = \omega t_1 - \beta z_1 + \theta^+$$

Thus we observe that *a point on the waveform must move in the positive z direction for increasing time, so that*

$$\omega t - \beta z + \theta^+ = \text{constant}$$

Taking the derivative of this expression with respect to time yields the velocity at which the points of constant phase travel, and is referred to as the *phase velocity* of the wave:

$$v = \frac{dz}{dt} \quad \text{(in m/s)} \qquad (3.59)$$

$$= \frac{\omega}{\beta}$$

$$= \frac{1}{\sqrt{\mu\epsilon}}$$

Similarly, we observe that the term $E_m^- \cos(\omega t + \beta z + \theta^-)$ represents a wave traveling in the negative z direction, a *backward-traveling wave*, since, in order to track the movement of a point on the waveform, the argument of the cosine must remain constant. Similar observations hold for the magnetic field intensity vector. However, observe that the magnetic field intensity vector of the backward-traveling wave is in the negative y direction. This is important and sensible because the direction of power flow in the individual waves must be in the same direction as the direction of wave propagation.

The quantity $\beta = \omega \sqrt{\mu \epsilon}$ is referred to as the *phase constant*. The units are radians per meter, so that β is a change in phase of the wave with distance of propagation. The distance between corresponding adjacent points on the wave is known as the *wavelength* and is denoted as λ. From Fig. 3.12(b) we observe that $\beta \lambda = 2\pi$. Since $\beta = \omega \sqrt{\mu \epsilon}$ and $v = 1/\sqrt{\mu \epsilon}$ for this lossless medium, the wavelength becomes

$$\lambda = \frac{2\pi}{\beta} \qquad (3.60)$$

$$= \frac{v}{f}$$

Increased frequencies result in shorter wavelengths. Note that the wavelength is also a function of the properties of the medium since $v = 1/\sqrt{\mu \epsilon}$. For typical materials, $\mu \geq \mu_0$ and $\epsilon \geq \epsilon_0$. Thus *the phase velocity of propagation is slower than in free space and the wavelength is shorter.*

In free space $\mu_0 = 4\pi \times 10^{-7}$ H/m and $\epsilon_0 \cong 1/36\pi \times 10^{-9}$ F/m, so that

$$v_0 \cong 3 \times 10^8 \text{ m/s}$$

At a frequency of 300 MHz

$$\lambda_0 = \frac{v_0}{f}$$

$$= 1 \text{ m} \quad (f = 300 \text{ MHz})$$

Similarly,

$$\lambda_0 = 1 \text{ cm} \quad (f = 30 \text{ GHz})$$

$$\lambda_0 = 3107 \text{ miles} \quad (f = 60 \text{ Hz})$$

The intrinsic impedance becomes

$$\eta_0 = \sqrt{\frac{\mu_0}{\epsilon_0}}$$

$$= 120\pi$$

$$= 377 \ \Omega$$

For any other lossless material medium with $\epsilon = \epsilon_r \epsilon_0$ and $\mu = \mu_r \mu_0$

$$v = \frac{v_0}{\sqrt{\epsilon_r \mu_r}} \quad \text{(in m/s)} \tag{3.61a}$$

$$\eta = \eta_0 \sqrt{\frac{\mu_r}{\epsilon_r}} \quad \text{(in } \Omega\text{)} \tag{3.61b}$$

$$\beta = \beta_0 \sqrt{\mu_r \epsilon_r} \quad \text{(in rad/m)} \tag{3.61c}$$

$$\lambda = \frac{\lambda_0}{\sqrt{\mu_r \epsilon_r}} \quad \text{(in m)} \tag{3.61d}$$

In calculating these quantities in *lossless* material media, one should translate the corresponding results in free space to those in the medium by using these relationships. *One should never again calculate v_0 and η_0!* For example, in order to calculate the wavelength of a 750 MHz signal, we would scale the result of $\lambda = 1$ m at 300 MHz and realize that the wavelength at 750 MHz will be shorter than at 300 MHz:

$$\lambda_{750 \text{ MHz}} = \frac{300}{750} \times \lambda_{300 \text{ MHz}}$$

$$= 0.4 \text{ m}$$

$$= 40 \text{ cm}$$

3.6.2 Lossy Media

There are two important differences between uniform plane waves in lossless media and in *lossy media*. The first difference is that the propagation constant $\hat{\gamma}$ has a nonzero real part, α. This results in the waves for the lossless case being multiplied by the exponentials $e^{-\alpha z}$ and $e^{\alpha z}$, as shown in (3.53). These are obviously still forward- and backward-traveling waves, but the *amplitudes* of the forward-traveling waves are $E_m^+ e^{-\alpha z}$ and $(E_m^+/\eta)e^{-\alpha z}$, which are reduced for increasing z (in the direction of propagation). This is shown for a fixed time as a function of z in Fig. 3.13. Similarly, the amplitudes of the backward-traveling waves, $E_m^- e^{\alpha z}$ and $(E_m^-/\eta)e^{\alpha z}$, are also reduced, since they are propagating in the $-z$ direction (decreasing z). The real part of $\hat{\gamma}$, α, is referred to as the *attenuation constant* for these reasons.

The second difference between lossless and lossy media is that the intrinsic impedance of a lossy medium has a nonzero phase angle, $\theta_\eta \neq 0$. For the lossless case $\theta_\eta = 0$, and we observe that the electric and magnetic fields of the forward-traveling wave are in time phase, as are those of the backward-traveling

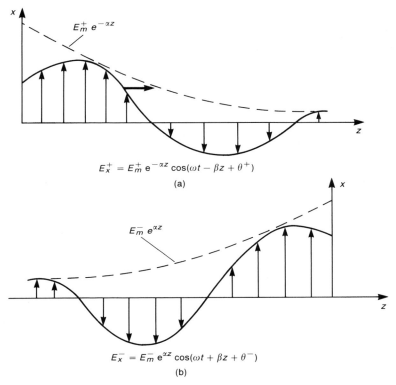

$$E_x^+ = E_m^+ \, e^{-\alpha z} \cos(\omega t - \beta z + \theta^+)$$

(a)

$$E_x^- = E_m^- \, e^{\alpha z} \cos(\omega t + \beta z + \theta^-)$$

(b)

FIGURE 3.13   Spatial properties of a uniform plane wave in a lossy medium: (a) the forward-traveling wave; (b) the backward-traveling wave.

wave. However, for the lossy case the phase angle of the intrinsic impedance, θ_η, results in the electric and magnetic fields of each traveling wave being out of time phase by the phase angle θ_η. Note from (3.52) that $0 \le \theta_\eta \le 45°$. Also the defining relationships for the phase velocity and wavelength are

$$v = \frac{\omega}{\beta} \qquad (3.62)$$

$$\lambda = \frac{2\pi}{\beta} \qquad (3.63)$$

$$= \frac{v}{f}$$

However, β is the imaginary part of the propagation $\hat{\gamma}$ and is no longer equal to $\omega\sqrt{\mu\epsilon}$ for the case of a lossy medium.

3.6.3   Power Flow

The average power density Poynting vector for the uniform plane wave is

$$\vec{S}_{av} = \tfrac{1}{2}\mathscr{R}e\{\vec{E} \times \vec{H}*\} \tag{3.64}$$
$$= \tfrac{1}{2}\mathscr{R}e\{\hat{E}_x\vec{a}_x \times \hat{H}_y^*\vec{a}_y\}$$
$$= \tfrac{1}{2}\mathscr{R}e\{\hat{E}_x\hat{H}_y^*\}\,\vec{a}_z$$

Substituting the phasor forms given in (3.54) gives [1]

$$\vec{S}_{av} = \left[\frac{1}{2}\frac{(E_m^+)^2}{\eta}e^{-2\alpha z}\cos\theta_\eta \right. \tag{3.65}$$
$$-\frac{1}{2}\frac{(E_m^-)^2}{\eta}e^{2\alpha z}\cos\theta_\eta$$
$$\left. -\frac{E_m^+E_m^-}{\eta}\sin\theta_\eta\sin(2\beta z + \theta^- - \theta^+)\right]\vec{a}_z$$

The first term is the average power density in the forward-traveling wave (in the absence of the backward-traveling wave). The second term is the average power density in the backward-traveling wave. The third and final term is a cross-coupling term that disappears for lossless media where $\theta_\eta = 0$. For the case of a lossless medium we may determine the net power transferred in the positive z direction by subtracting the power density of the backward-traveling wave from that of the forward-traveling wave, which is logical to expect.

3.6.4   Conductors versus Dielectrics

Note that the only change in Maxwell's equations introduced by losses in the medium occurs in Ampère's law. For $\sigma \neq 0$ Ampère's law may be written as

$$\nabla \times \vec{H} = (\sigma + j\omega\epsilon)\vec{E} \tag{3.66}$$

On comparing this expression with the lossless case ($\sigma = 0$):

$$\nabla \times \vec{H} = j\omega\epsilon\vec{E} \tag{3.67}$$

we see that we could derive the results for the lossless case and "fix up" those results to include losses by replacing ϵ in the lossless results with a *complex permittivity*:

$$\hat{\epsilon} = \epsilon\left(1 - j\frac{\sigma}{\omega\epsilon}\right) \tag{3.68}$$

The term $\sigma/\omega\epsilon$ is referred to as the *loss tangent* of the material, and is a function of frequency. Values of the loss tangent are experimentally measured for various materials, and are tabulated at several frequencies in handbooks. Note that there are two components of current involved in Ampère's law: a conduction current density $\vec{J}_c = \sigma\vec{E}$ and a displacement current density $\vec{J}_d = j\omega\epsilon\vec{E}$. The conduction current represents an energy loss mechanism, while the displacement current represents energy storage, as we have observed previously. The ratio of these two currents is a measure of the lossy nature of the material. The conduction and displacement current phasors are 90° out of time phase.

The notion of the loss tangent as the ratio of conduction current to displacement current provides a meaningful way of distinguishing between *conductors* and *dielectrics*. Materials are classified as conductors or dielectrics according to whether the conduction current is larger or smaller than the displacement current:

$$\frac{\sigma}{\omega\epsilon} \ll 1 \quad \text{good dielectric}$$

$$\frac{\sigma}{\omega\epsilon} \gg 1 \quad \text{good conductor}$$

Calculations of the propagation constant $\hat{\gamma}$ and the intrinsic impedance $\hat{\eta}$ can be simplified, depending on whether the medium can be classified as a good conductor or a good dielectric according to the above criterion.

First we consider *good dielectrics* where $\sigma/\omega\epsilon \ll 1$. The propagation constant can be written as

$$\hat{\gamma} = \alpha + j\beta \tag{3.69}$$

$$= \sqrt{j\omega\mu(\sigma + j\omega\varepsilon)}$$

$$= j\omega\sqrt{\mu\varepsilon}\sqrt{1 - j\frac{\sigma}{\omega\varepsilon}}$$

For $\sigma/\omega\epsilon \ll 1$ this essentially simplifies to that of free space: $\hat{\gamma} = j\beta = j\omega\sqrt{\mu\epsilon}$. The velocity of propagation is also essentially unchanged from that of a lossless medium. The intrinsic impedance can be similarly simplified:

$$\hat{\eta} = \sqrt{\frac{j\omega\mu}{\sigma + j\omega\epsilon}} \tag{3.70}$$

$$= \sqrt{\frac{\mu}{\epsilon}\frac{1}{1 - j\sigma/\omega\epsilon}}$$

$$\cong \sqrt{\frac{\mu}{\epsilon}}$$

and the intrinsic impedance is virtually unchanged from the lossless medium case.

Next we consider the case of *good conductors*, where $\sigma/\omega\epsilon \gg 1$. The propagation constant can be written as

$$\hat{\gamma} = \sqrt{j\omega\mu(\sigma + j\omega\epsilon)} \tag{3.71}$$

$$= \sqrt{j\omega\mu\sigma\left(1 + j\frac{\omega\epsilon}{\sigma}\right)}$$

$$\cong \sqrt{j\omega\mu\sigma}$$

$$= \sqrt{\omega\mu\sigma} \; \underline{/45^\circ}$$

so that

$$\alpha = \beta \tag{3.72}$$

$$\cong \sqrt{\tfrac{1}{2}\omega\mu\sigma}$$

The velocity of propagation becomes

$$v = \frac{\omega}{\beta} \tag{3.73}$$

$$\cong \sqrt{\frac{2\omega}{\mu\sigma}}$$

The intrinsic impedance becomes

$$\hat{\eta} = \sqrt{\frac{j\omega\mu}{\sigma + j\omega\epsilon}} \tag{3.74}$$

$$= \sqrt{\frac{j\omega\mu/\sigma}{1 + j\omega\epsilon/\sigma}}$$

$$\cong \sqrt{\frac{j\omega\mu}{\sigma}}$$

$$= \sqrt{\frac{\omega\mu}{\sigma}} \; \underline{/45^\circ}$$

$$= \sqrt{\frac{\omega\mu}{2\sigma}}(1 + j1)$$

In addition to conductive loss mechanisms inherent in σ, dielectrics also exhibit another loss mechanism. They are characterized by microscopic dipoles of *bound charge* [1]. As the frequency of the fields is increased, the dipoles of bound charge cannot completely align with the field, and tend to lag behind the changes in field direction. This loss phenomenon is also accounted for by ascribing a complex permittivity to the material, $\hat{\epsilon} = \epsilon' - j\epsilon''$.

TABLE 3.1   Skin Depth of Copper.

f	δ
60 Hz	8.5 mm
1 kHz	2.09 mm
10 kHz	0.66 mm
100 kHz	0.21 mm
1 MHz	2.6 mils
10 MHz	0.82 mils
100 MHz	0.26 mils
1 GHz	0.0823 mils

3.6.5   Skin Depth

The notion of *skin depth* frequently occurs throughout the analysis of electromagnetic field propagation in conductive material media. Consider a forward-traveling wave in a lossy material. As the wave travels through the lossy medium, the amplitude of the wave, $E_m^+ e^{-\alpha z}$, decreases. Over a distance of $\delta = 1/\alpha$, the amplitude will have been reduced by $1/e$ or 37%. The quantity δ is termed the *skin depth* or *depth of penetration* of the material at that frequency. Substituting the relation for the attenuation constant for a *good conductor* given in (3.72) gives

$$\delta = \sqrt{\frac{2}{\omega\mu\sigma}} \tag{3.75}$$

$$= \frac{1}{\sqrt{\pi f \mu\sigma}}$$

Values of the skin depth for copper are given in Table 3.1, where 1 mil = 0.001 inch = 2.54×10^{-5} m. For increasing frequencies the skin depth decreases, and becomes extremely small for frequencies in the radiated emission limit range. For example, the skin depth for copper at 100 MHz is 0.0066 mm = 6.6 μm. For current-carrying conductors we may consider the current to be essentially concentrated in a depth of a few skin depths on the surface of the conductor that is adjacent to the field that caused the current. The remainder of the conductor has essentially no effect.

REFERENCES

[1]   C.R. Paul and S.A. Nasar, *Introduction to Electromagnetic Fields*, second edition, McGraw-Hill, NY (1987).

[2]   C.R. Paul, *Analysis of Linear Circuits*, McGraw-Hill, NY (1989).

PROBLEMS

3.1 Three vectors $\vec{A}$, $\vec{B}$, and $\vec{C}$ are given in a rectangular coordinate system as

$$\vec{A} = 2\vec{a}_x + 3\vec{a}_y - \vec{a}_z$$
$$\vec{B} = \vec{a}_x + \vec{a}_y - 2\vec{a}_z$$
$$\vec{C} = 3\vec{a}_x - \vec{a}_y + \vec{a}_z$$

Compute $\vec{A} + \vec{B}$, $\vec{B} - \vec{C}$, $\vec{A} + 3\vec{B} - 2\vec{C}$, $|\vec{A}|$, $\vec{a}_B$, $\vec{A} \cdot \vec{B}$, $\vec{B} \cdot \vec{A}$, $\vec{B} \times \vec{C}$, $\vec{C} \times \vec{B}$, $\vec{A} \cdot \vec{B} \times \vec{C}$. $[3\vec{a}_x + 4\vec{a}_y - 3\vec{a}_z, -2\vec{a}_x + 2\vec{a}_y - 3\vec{a}_z, -1\vec{a}_x + 8\vec{a}_y - 9\vec{a}_z, \sqrt{14},$ $\sqrt{\frac{1}{6}}\vec{a}_x + \sqrt{\frac{1}{6}}\vec{a}_y - 2\sqrt{\frac{1}{6}}\vec{a}_z$, 7, 7, $-1\vec{a}_x - 7\vec{a}_y - 4\vec{a}_z$, $1\vec{a}_x + 7\vec{a}_y + 4\vec{a}_z$, $-1\vec{a}_x - 7\vec{a}_y - 4\vec{a}_z$, -19]

3.2 If $\vec{A} = \vec{a}_x + 2\vec{a}_y - 3\vec{a}_z$ and $\vec{B} = 2\vec{a}_x - \vec{a}_y + \vec{a}_z$, determine (a) the magnitude of the projection or component of $\vec{B}$ on $\vec{A}$, (b) the angle (smallest) between $\vec{A}$ and $\vec{B}$, (c) the vector projection of $\vec{A}$ onto $\vec{B}$, and (d) a unit vector perpendicular to the plane containing $\vec{A}$ and $\vec{B}$. $[3\sqrt{\frac{1}{14}}, 109.1°,$ $-1\vec{a}_x + 0.5\vec{a}_y - 0.5\vec{a}_z, -\sqrt{\frac{1}{75}}\vec{a}_x - 7\sqrt{\frac{1}{75}}\vec{a}_y - 5\sqrt{\frac{1}{75}}\vec{a}_z]$

3.3 If two vectors are given by $\vec{A} = \vec{a}_x + 2\vec{a}_y - \vec{a}_z$ and $\vec{B} = \alpha\vec{a}_x + \vec{a}_y + 3\vec{a}_z$, determine α such that the two vectors are perpendicular. $[1]$

3.4 If two vectors are given by $\vec{A} = \vec{a}_x + \pi\vec{a}_y + 3\vec{a}_z$ and $\vec{B} = \alpha\vec{a}_x + \beta\vec{a}_y - 6\vec{a}_z$, determine α and β such that the two vectors are parallel. $[\alpha = -2, \beta = -2\pi]$

3.5 Evaluate the line integral of $\vec{F} = x\vec{a}_x + 2xy\vec{a}_y - y\vec{a}_z$ from $[1, -1, 0]$ to $[0, 0, 0]$ along paths consisting of (a) a straight line between the two points, and (b) a two-segment path with segment 1 from $[1, -1, 0]$ to $[1, 0, 0]$ and segment 2 from $[1, 0, 0]$ to $[0, 0, 0]$. $[-\frac{7}{6}, -\frac{3}{2}]$

3.6 If the force exerted on an object is given by $\vec{F}(x, y, z) = (2x\vec{a}_x + 3z\vec{a}_y - 4\vec{a}_z)$ N, determine the work required to move the object in a straight line (a) from $[0, 0, 1$ m$]$ to $[0, 0, -3$ m$]$, (b) from $[1$ m, 1 m, 0$]$ to $[0, 1$ m, 0$]$ and (c) from $[1$ m, 1 m, 1 m$]$ to $[0, 0, 1$ m$]$. $[16$ J, -1 J, -4 J$]$

3.7 Determine the net flux of the vector field $\vec{F}(x, y, z) = 2x^2y\vec{a}_x + z\vec{a}_y + y\vec{a}_z$ leaving the closed surfaces of a cube having unit length sides whose faces are parallel to the three coordinate axes of a rectangular coordinate system when the cube is centered at (a) $[0.5, 0.5, 0.5]$ and (b) $[1.5, 0.5, 0.5]$. $[1, 3]$ Check your result with the divergence theorem.

3.8 Compute the divergence of the vector field $\vec{F} = yz\vec{a}_x + zy\vec{a}_y + xz\vec{a}_z$. $[\nabla \cdot \vec{F} = z + x]$

3.9 Compute the curl of the vector field $\vec{F} = xy\vec{a}_x + 2yz\vec{a}_y - \vec{a}_z$. $[\nabla \times \vec{F} = -2y\vec{a}_x + 0\vec{a}_y - x\vec{a}_z]$

3.10 Verify Stokes' theorem for the flat rectangular surface in the xy plane bounded by $[0, 0, 0]$, $[1, 0, 0]$, $[1, 1, 0]$, and $[0, 1, 0]$ when the vector field is (a)$\vec{F}(x, y, z) = 2\vec{a}_x + \vec{a}_y$ and (b) $\vec{F}(x, y, z) = 2xy\vec{a}_x - y\vec{a}_z$. $[0, -1]$

3.11 A square loop with sides 20 cm by 20 cm is located in free space adjacent to a straight conductor that carries a sinusoidal current of 0.5 A at 5 kHz. Two of the sides of the loop are parallel to the conductor and located at 5 cm and 25 cm from the conductor. If a small gap is introduced into the loop, determine the induced voltage across the gap and its proper polarity. $[1.01\ \text{mV}]$

3.12 A uniform and constant magnetic field of 0.01 Wb/m² is directed along the z axis of a rectangular coordinate system. A circular contour in the xy plane centered at the origin has a radius that is decreasing at a rate of 100 m/s. If the initial radius is 10 cm, determine the induced emf in the path as a function of time. $[0.628\ \text{V}]$

3.13 In the circuit of Fig. P3.13 a rectangular wire loop with resistance 0.02 Ω rotates (as shown) in a constant magnetic field that is given by $\vec{B} = 0.01\vec{a}_y$ Wb/m². One side of the loop lies along the z axis, while the other side rotates at an angular speed of $\omega = 2$ rad/s. Determine the induced current with the direction shown. The loop lies in the xz plane at $t = 0$. $[2 \times 10^{-4} \sin \omega t\ \text{V}]$

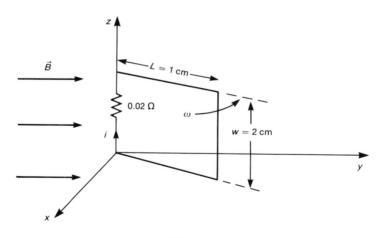

FIGURE P3.13

3.14 A loop of wire is inserted into a time-varying magnetic field as shown in Fig. P3.14. Determine the current circulating around the loop. From this determine the voltages V_2 and V_1 across the resistors. $[V_1 = 0.1/3\ \text{V}, V_2 = -0.2/3\ \text{V}]$ Are they equal? If not, why not?

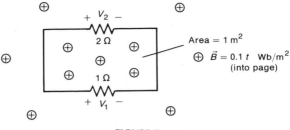

FIGURE P3.14

3.15 A 1 m^2 loop of wire completely encloses a time-varying magnetic field that is uniformly distributed over the loop as shown in Fig. P3.15. A voltmeter that draws neglible current is placed in the three positions shown. Determine (and give reasons for) the voltmeter readings in each of the three positions. $[-0.1/3$ V, $0.2/3$ V, $0.2/3$ V$]$

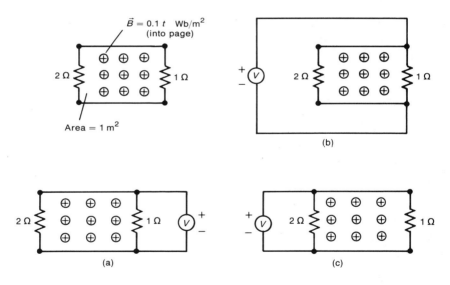

FIGURE P3.15

3.16 A pair of thin, perfectly conducting wires forms a set of rails along which another thin, perfectly conducting wire of length L is moving with velocity u as shown in Fig. P3.16. A magnetic field $\vec{B}$ is perpendicular to this loop and directed into the page. Determine the emf generated across a small gap in the loop with polarity shown when the magnetic field is given by (a) $B = B_o$ and (b) $B = B_o \cos \omega t$. $[-B_o Lu, -B_o Lu(\cos \omega t - \omega t \sin \omega t)]$

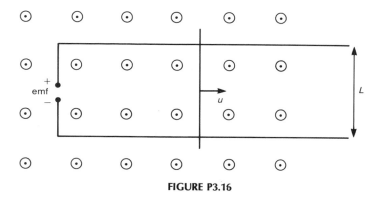

FIGURE P3.16

3.17 A material having a conductivity σ and permittivity ϵ is placed in a sinusoidal, time-varying electric field having a frequency f. At what frequency will the conduction current equal the displacement current? $[f = \sigma/2\pi\epsilon]$ If $\sigma = 10^{-2}$ S/m and $\epsilon = 3\epsilon_o$, determine the frequency. [60 MHz]

3.18 An air-filled parallel-plate capacitor has plates of 10 cm² that are separated by a distance of 2 mm. If the capacitance is connected to a 50 V, 1 MHz sinusoidal voltage source, calculate the magnitude of the displacement current, neglecting fringing. [1.4 mA]

3.19 Which of Maxwell's equations do the following field vectors satisfy? $\vec{E} = E_m \sin x \sin t \, \vec{a}_y$, $\vec{H} = (E_m/\mu_o) \cos x \cos t \, \vec{a}_z$ [Faraday's and the two laws of Gauss] Are they valid electromagnetic fields? [No]

3.20 An electric field in free space is given by $\vec{E} = E_m \sin \alpha x \cos(\omega t - \beta z)\vec{a}_y$. Find the corresponding magnetic field from Faraday's law. $[\vec{H} = -(\beta/\mu_o\omega)E_m \sin \alpha x \cos(\omega t - \beta z)\vec{a}_x - (\alpha/\mu_o\omega)E_m \cos \alpha x \sin(\omega t - \beta z)\vec{a}_z]$ Under what conditions (α, β, ω) do these fields satisfy Maxwell's equations? $[\alpha^2 + \beta^2 = \mu_o\epsilon_o\omega^2]$

3.21 Two regions are separated by a plane defined by $x + y = 1$ as shown in Fig. P3.21. If the electric and magnetic fields at the interface in region 1 are given by $\vec{E}_1 = (2\vec{a}_y + 3\vec{a}_z)$ V/m and $\vec{H}_1 = (0.1\vec{a}_x + 0.2\vec{a}_z)$ A/m, determine $\vec{D}_2$, $\vec{B}_2$, $\vec{E}_2$, $\vec{H}_2$ at the interface in region 2. $[\vec{D}_2 = \epsilon_o(-4\vec{a}_x + 12\vec{a}_y + 24\vec{a}_z), \vec{B}_2 = \mu_o(\vec{a}_x - 0.6\vec{a}_y + 0.8\vec{a}_z), \vec{E}_2 = -0.5\vec{a}_x + 1.5\vec{a}_y + 3\vec{a}_z, \vec{H}_2 = 0.25\vec{a}_x - 0.15\vec{a}_y + 0.2\vec{a}_z)]$

3.22 In Fig. P3.21 suppose that region 2 is a perfect conductor and that region 1 is unchanged. If the fields in region 1 at the interface are given by $\vec{E}_1 = (2\vec{a}_x + \vec{a}_y)$ V/m and $\vec{B}_1 = (-2\vec{a}_x + 2\vec{a}_y + \vec{a}_z)$ Wb/m², determine the surface charge density and vector linear surface current density on

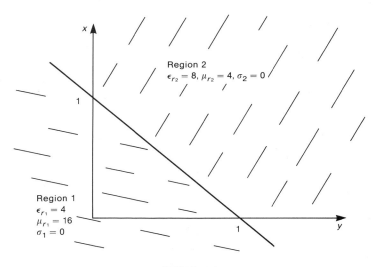

FIGURE P3.21

the surface of the perfect conductor. $[\rho_s = -8\sqrt{2}\epsilon_o\, C/m^2,\ \vec{K}_s = -(1/16\mu_o\sqrt{2})(-\vec{a}_x + \vec{a}_y - 4\vec{a}_z)]$

3.23 The electric and magnetic fields in a region are given in a rectangular coordinate system by $\vec{E} = 10e^{-200x}\cos(\omega t - 200x)\,\vec{a}_z$ V/m and $\vec{H} = -0.5e^{-200x}\cos(\omega t - 200x - \pi\frac{1}{4})\,\vec{a}_y$ A/m. Write the phasor expressions for these field vectors. $[\hat{E} = 10e^{-200x - j200x}\vec{a}_z, \hat{H} = -0.5e^{-200x - j200x - j45°}\vec{a}_y]$ Determine the phasor Poynting vector. $[\hat{S} = 5e^{-400x}e^{j45°}\vec{a}_x]$ Compute the average power dissipated in a cube with sides 1 cm in length. The cube has faces parallel to the coordinate axes and is centered at [0.5 cm, 0.5 cm, 0.5 cm]. [174 μW]

3.24 Suppose that a uniform plane wave is traveling in the x direction in a lossless medium with the 100 V/m electric field in the z direction. If the wavelength is 25 cm and the velocity of propagation is 2×10^8 m/s, determine the frequency of the wave and the relative permittivity of the medium if it has free-space permeability. [800 MHz, 2.25] Write complete time-domain expressions for the electric and magnetic field vectors. $[\vec{E} = 100\cos(1.6\pi \times 19^9 t - 25.13x)\,\vec{a}_z, \vec{H} = -(100/251.33)\cos(1.6\pi \times 10^9 t - 25.13x)\,\vec{a}_y]$

3.25 Write a time-domain expression for the electric field of a uniform plane wave if the magnetic field is given by $\vec{H} = 0.1e^{-200y}\cos(2\pi \times 10^{10}t - 300y)\,\vec{a}_x$ A/m and the medium has free space permeability. $[\vec{E} = 21e^{-200y}\cos(2\pi \times 10^{10}t - 300y + 33.69°)\,\vec{a}_z]$

3.26 If a material has $\sigma = 2$ S/m, $\epsilon_r = 9$, $\mu_r = 16$ at a frequency of 1 GHz, calculate the attenuation constant, phase constant, and intrinsic impedance for the material at this frequency. $[\alpha = 314.06, \; \beta = 402.25, \; \hat{\eta} = 247.55 \; \underline{/38^\circ} \doteq (195.12 + j152.34)\,\Omega.]$

3.27 A 5 GHz uniform plane wave is propagating in a material characterized by $\epsilon_r = 2.53$, $\mu_r = 1$, and $\sigma = 0$. If the electric field is given by $\vec{E} = 10 \cos(10\pi \times 10^9 t - \beta z)\,\vec{a}_x$, determine (a) the phase velocity of propagation, (b) the wavelength, (c) the phase constant, and (d) the magnitude of the magnetic field. $[188.6 \times 10^6 \text{ m/s}, 3.77 \text{ cm}, 166.6 \text{ rad/m}, 0.04 \text{ A/m}]$ Write the time-domain expression for the magnetic field. $[\vec{H} = 0.04 \cos(10\pi \times 10^9 t - 166.6)\,\vec{a}_y]$

3.28 Suppose that a 2 GHz uniform plane wave is propagating in a lossy medium that has a loss tangent of 10^{-2} at 2 GHz. At 2 GHz the relative permittivity of the medium is 2.25 and the relative permeability is 1. Determine the conductivity, the attenuation constant, the phase constant, and the intrinsic impedance at 2 GHz. $[0.25 \times 10^{-2} \text{ S/m}, 0.31, 62.83 \text{ rad/m}, 251.33 \; \underline{/45^\circ}]$ Compute these using approximations and then using exact equations, and then compare the results.

3.29 Repeat Problem 3.28 if the loss tangent is 10^2. $[25 \text{ S/m}, 442.07, 446.52 \text{ rad/m}, 25.13 \; \underline{/44.7^\circ}]$

3.30 Determine the frequency range for which the conduction current exceeds the displacement current by a factor of at least 100 in sea water ($\sigma = 4$ S/m, $\mu_r \cong 1$, $\epsilon_r \cong 81$). $[f \leq 8.89 \text{ MHz}]$

3.31 Determine the phase velocity, attenuation constant, phase constant, skin depth, and intrinsic impedance of a uniform plane wave traveling in wet, marshy soil, ($\sigma \cong 10^{-2}$ S/m, $\epsilon_r \cong 15$, $\mu_r \cong 1$) at 60 Hz (power frequency), 1 MHz (AM radio broadcast frequency), 100 MHz (FM radio broadcast frequency), and 10 GHz (microwave radio relay frequency). [60 Hz: 2.45×10^5 m/s, $\alpha = \beta = 1.539 \times 10^{-3}$, $\delta = 650$ m, $\hat{\eta} = 0.2177 \; \underline{/45^\circ}$; 1 MHz: 3.03×10^7 m/s, $\alpha = 0.1906$, $\beta = 0.2071$, $\delta = 5.25$ m, $\hat{\eta} = 28.1 \; \underline{/42.62^\circ}$; 100 MHz: 7.73×10^7 m/s, $\alpha = 0.4858$, $\beta = 8.127$, $\delta = 2.058$ m, $\hat{\eta} = 96.99 \; \underline{/3.42^\circ}$;   10 GHz: 7.75×10^7 m/s, $\alpha = 0.4867$, $\beta = 811.2$, $\delta = 2.055$ m, $\hat{\eta} = 97.34 \; \underline{/3.44 \times 10^{-2\circ}}]$ Also compute the distance for the wave to travel at each frequency such that its amplitude is reduced by a factor of 20 dB. [1496 m, 12 m, 4.74 m, 4.732 m]

3.32 A 100 MHz uniform plane wave traveling in a lossy dielectric ($\mu_r \cong 1$) has the following phasor expression for the magnetic field intensity vector: $\vec{H} = (1\vec{a}_y + j2\vec{a}_z)e^{-0.2x}e^{-j2x}$ A/m. Write complete time-domain expressions for the electric and magnetic field vectors. $[\vec{E} = -392.8e^{-0.2x}\cos(\omega t - 2x + 5.7^\circ)\,\vec{a}_y + 785.6e^{-0.2x}\sin(\omega t - 2x + 5.7^\circ)\,\vec{a}_z, \; \vec{H} = e^{-0.2x}\cos(\omega t - 2x)\,\vec{a}_y - 2e^{-0.2x}\sin(\omega t - 2x)\,\vec{a}_z]$

CHAPTER FOUR

Transmission Lines

A *transmission line* is a system of two or more closely spaced, parallel conductors. In this chapter we will review the concepts of transmission lines that consist of only two conductors. Transmission lines that consist of more than two parallel conductors admit the possibility of *crosstalk* or the unintentional coupling between the circuits that the conductors interconnect. This will be considered in Chapter 10. In order to understand crosstalk, it is important that we first have a sound understanding of the two-conductor case.

Some common examples of two-conductor transmission lines are illustrated in Fig. 4.1. A two-wire transmission line consisting of two parallel wires (circular, cylindrical conductors) is shown in Fig. 4.1(a). A source, shown as a Thévenin equivalent, consisting of an open-circuit voltage $V_S(t)$ and source resistance R_S is connected via these wires to a load that is represented by a resistance R_L. The second common example of a two-conductor line is the case of a wire above and parallel to an infinite ground plane, shown in Fig. 4.1(b). The terminations at the ends of the line are connected to the ground plane, which serves as the return for the signal on the wire. The third common example is the case of a coaxial transmission line, shown in Fig. 4.1(c). An overall, circular–cylindrical shield encloses an interior wire that is located on the axis of the shield. Thus the shield serves as the return for the signal on the interior wire.

Figure 4.2 shows some other common forms of two-conductor transmission lines, which are often referred to as printed circuit boards (PCB's). Figure 4.2(a) shows a rectangular–cross-section conductor (land) on the surface of a dielectric substrate such as glass–epoxy with a ground plane beneath the substrate. This is commonly used in microwave circuit design and is referred to as a *microstrip* transmission line. Figure 4.2(b) shows lands etched on the same side of the PCB, whereas Fig. 4.2(c) shows lands etched on opposite sides of the board.

The basic problem involved in the analysis of all these two-conductor lines is to determine the currents on the conductors and the voltage between the two conductors at all points along the line. Usually we will only be interested in these currents and voltages at the two ends of the line where the source and

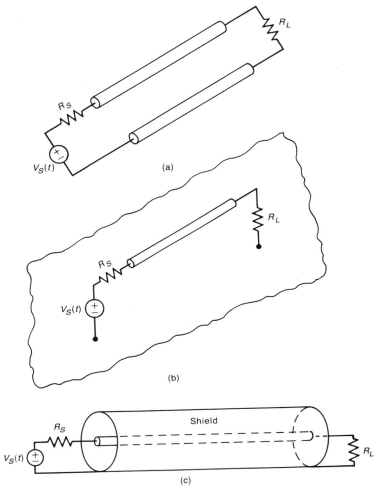

FIGURE 4.1 Illustration of wire-type transmission lines in homogeneous media: (a) a two-wire line; (b) a wire above a ground plane; (c) a coaxial cable.

load are attached. There is an important difference between the configurations of Fig. 4.1 and those of Fig. 4.2 with regard to the analysis of the structures. The surrounding medium for the configurations of Fig. 4.2 is *inhomogeneous* (in permittivity), since the fields developed between the two conductors will lie partially in air, $\epsilon_r = 1$, and partially in the board dielectric, $\epsilon_r \neq 1$. The configurations in Fig. 4.1 represent two-conductor lines in *homogeneous media*. If the wires in Figs. 4.1(a) and (b) had cylindrical dielectric insulations surrounding them then they would also represent lines in inhomogeneous media, since the electric fields would lie partly in the insulation and partly in the surrounding air. For the purposes of simplifying our results, we will ignore any

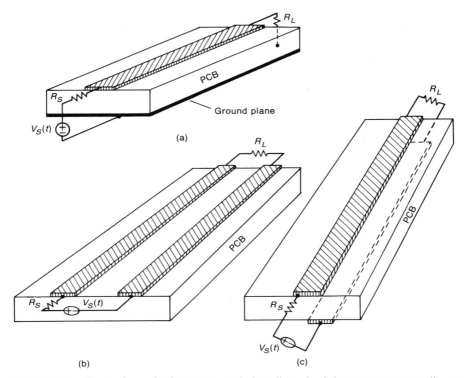

FIGURE 4.2 Illustration of planar transmission lines in inhomogeneous media as occur with printed circuit boards (PCBs): (a) the microstrip line; (b) strips on the same side of the PCB; (c) strips on opposite sides of the PCB.

insulations around the wires in Figs. 4.1(a) and (b) and will assume free space: $\epsilon_r = \mu_r = 1$.

4.1   THE TRANSMISSION-LINE EQUATIONS

The differential equations that govern the voltage and current of a transmission line are referred to as the *transmission-line equations*. These equations constitute the *model* of the transmission line. The fundamental structure of the electromagnetic fields surrounding the conductors of a transmission line is the *transverse electromagnetic (TEM)* field structure in which the electric and magnetic field intensity vectors at each point in space have no component parallel to the line conductors; that is, the field vectors are transverse or perpendicular to the line axis [1]. It turns out that waves will propagate along the line, so that this represents the TEM mode of propagation where the field vectors are transverse to the direction of propagation. The uniform plane wave considered in Chapter 3 is an example of TEM propagation in that the field vectors in those waves are also orthogonal to the direction of propagation of the wave. However, TEM waves on transmission lines differ from uniform plane waves in that the field vectors of the transmission line fields are not independent of position in the planes perpendicular to the line axis. In other words, waves on transmission lines are plane waves, but are not *uniform*, plane waves. We will also confine our discussions to *uniform transmission lines*. A uniform transmission line is one in which the cross-sectional dimensions of the line, i.e., conductor cross sections as well as dielectric cross sections, are the same at any two points along the line.

In order to derive the transmission-line equations and quantify the results, we will orient the two conductors along the z axis of a rectangular coordinate system. For illustration we will consider the two-wire line shown in Fig. 4.1(a), although the results we obtain will also apply to all the lines of Figs. 4.1 and 4.2. Consider applying a dc voltage and current to the parallel-wire line. The voltage between the two wires will cause charge to appear on the wire surfaces, which creates an electric field $\vec{E}_T(z, t)$ directed from one wire to the other. The current flowing down one wire and returning on the other creates a magnetic field $\vec{H}_T(z, t)$ that encircles each wire. If the line is infinitely long, so that there can be no fringing of the fields at the endpoints, the fields will lie in the xy plane and are *transverse* to the line axis, the z axis. This is the meaning of the subscript T on the field vectors. As we increase the frequency of the voltage and current, it is evident that the field vectors will maintain the transverse structure up to some higher-frequency limit where higher-order, waveguide-type modes begin to propagate [2]. For typical line dimensions, these higher order modes will become significant above a very high frequency, typically in the GHz range for typical line dimensions, so that for our interest we can ignore the higher-order modes and assume that the TEM mode of propagation is the only significant mode of propagation on the line.

Assumption of the TEM mode allows us to uniquely define voltage and current for the line. In general, voltage and current can be uniquely defined only for dc excitation, but are approximately defined so long as the structure is *electrically small*. This is the assumption inherent in the analysis of lumped-circuit models for frequencies other than dc. Voltage and current can be uniquely defined for transmission lines so long as the cross-sectional dimensions (conductor separations) are electrically small and for any line length (electrically short or electrically long). In order to define the voltage and current of the line, consider Faraday's and Ampère's laws in integral form,

$$\oint_c \vec{E} \cdot d\vec{l} = -\mu \frac{d}{dt} \int_s \vec{H} \cdot d\vec{s} \tag{4.1a}$$

$$\oint_c \vec{H} \cdot d\vec{l} = \int_s \vec{J} \cdot d\vec{s} + \epsilon \frac{d}{dt} \int_s \vec{E} \cdot d\vec{s} \tag{4.1b}$$

where we have assumed a simple medium characterised by $\vec{B} = \mu \vec{H}$ and $\vec{D} = \epsilon \vec{E}$. If we take the contour C and surface S to lie completely in the xy plane transverse to the line axis, we obtain

$$\oint_{C_{xy}} (E_x \, dx + E_y \, dy) = -\mu \frac{d}{dt} \int_{S_{xy}} H_z \, dx \, dy \tag{4.2a}$$

$$\oint_{C_{xy}} (H_x \, dx + H_y \, dy) = \int_{S_{xy}} J_z \, dx \, dy + \epsilon \frac{d}{dt} \int_{S_{xy}} E_z \, dx \, dy \tag{4.2b}$$

But for the TEM mode $E_z = H_z = 0$! Thus we obtain

$$\oint_{C_{xy}} (E_x \, dx + E_y \, dy) = 0 \tag{4.3a}$$

$$\oint_{C_{xy}} (H_x \, dx + H_y \, dy) = \int_{S_{xy}} J_z \, dx \, dy \tag{4.3b}$$

which are precisely the equations we obtain for the static-field case (dc). Equation (4.3a) allows us to uniquely define voltage and current as is done for static fields as [1]

$$V(z, t) = -\int_{C_V} \vec{E}_T \cdot \vec{dl} \tag{4.4}$$

where the contour C_V is *any path* from the lower wire to the upper wire so long as that path lies in the transverse xy plane. So the line integral in (4.4) is independent of path in this transverse plane, which makes the voltage unique. Similarly, equation (4.3b) allows us to uniquely define current on the conductors as the line integral of H_T around *any closed contour C_I in the transverse xy plane that encircles the wire*; that is [1],

$$I(z, t) = \oint_{C_I} \vec{H}_T \cdot \vec{dl} \tag{4.5}$$

The currents on the top and bottom wires at any cross section are equal in magnitude but oppositely directed. This is the usual notion of *differential-mode current*. We will find that another component of current, *common-mode current*, may exist on the wires, but the transmission-line equations will only predict the differential-mode component! Note that the voltage and current are shown to be functions of position z along the line, as well as time t.

Equations (4.3) being identical to the field equations for static (dc) fields shows another important point: *the transverse electromagnetic field structure for the TEM mode is identical to that for dc excitation.* Therefore we can derive the required per-unit-length parameters solely using static field methods, which greatly simplifies their determination [1]. It is therefore interesting to observe that *the field structure for the TEM mode is identical to the static (dc) field structure for higher frequencies so long as the TEM mode of propagation is the only mode of propagation on the line.*

Now that we can uniquely define voltage and current we can model an electrically small section of the line of length Δz with a lumped-circuit model as shown in Fig. 4.3(a). The electric field surrounding the line conductors gives rise to a capacitance between the conductors. Because the line is assumed to be uniform, we can obtain a capacitance per unit length, c, that applies at every point along the line. If the line were a nonuniform line due to a varying cross section, this per-unit-length parameter would be a function of the place along the line where it is defined, i.e., $c(z)$. The total capacitance of a section of length Δz is the product of the per-unit-length capacitance and the length of the section, $c\Delta z$. If, in addition, the surrounding dielectric has a nonzero conductivity so that it is lossy, there is an additional per-unit-length conductance parameter g, giving an element $g\Delta z$ between the two conductors. It is important to note that the per-unit-length capacitance represents displacement current flowing between

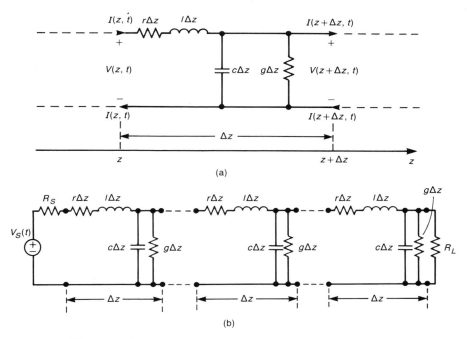

(a)

(b)

FIGURE 4.3 The per-unit-length equivalent circuit of a two-conductor line for the TEM mode of propagation: (a) the equivalent circuit for a Δz section; (b) modeling the entire line as a cascade of Δz sections from which the transmission line equations are derived in the limit as $\Delta z \to 0$.

the conductors and is present whether the medium is lossy or lossless. The per-unit-length conductance represents conduction current flowing between the two conductors. It is not present if the surrounding medium is lossless. In addition, the current flowing along the conductors creates a magnetic field that penetrates the space between the two conductors. Viewing the loop formed by the two conductors as an inductance, this suggests that a per-unit-length inductance l be inserted in series with the line conductors. The total inductance formed by two conductors of length Δz is $l\Delta z$. Observe that it is unimportant whether we place all of this total inductance of the loop in one conductor or divide it between the two conductors, since it represents the inductance of the loop formed between the two conductors. In addition, if the conductors are not perfect conductors but have finite, nonzero conductivity then we include a per-unit-length resistance r to represent the loss associated with the conductors. The total resistance inserted for a Δz section is $r\,\Delta z$ and represents the loss of both conductors.

We now wish to derive the equations relating the voltages and currents at the two ends of this Δz section. These can be straightforwardly derived, from the per-unit-length circuit in Fig. 4.3(a), by using Kirchhoff's voltage and current laws as [1]

$$V(z + \Delta z, t) - V(z, t) = -r\,\Delta z\,I(z, t) - l\,\Delta z\,\frac{\partial}{\partial t}\,I(z, t) \tag{4.6a}$$

$$I(z + \Delta z, t) - I(z, t) = -g\,\Delta z\,V(z + \Delta z, t) - c\,\Delta z\,\frac{\partial}{\partial t}\,V(z + \Delta z, t) \tag{4.6b}$$

In order for this lumped-circuit model to be valid, its length Δz must be electrically short at the highest frequency of use of the model. In order to make the model valid for all frequencies, we model the line with additional subsections as shown in Fig. 4.3(b) and require that the section length go to zero: $\Delta z \to 0$. Dividing both sides of (4.6) by Δz gives

$$\frac{V(z + \Delta z, t) - V(z, t)}{\Delta z} = -rI(z, t) - l\frac{\partial}{\partial t}I(z, t) \tag{4.7a}$$

$$\frac{I(z + \Delta z, t) - I(z, t)}{\Delta z} = -gV(z + \Delta z, t) - c\frac{\partial}{\partial t}V(z + \Delta z, t) \tag{4.7b}$$

Taking the limit as $\Delta z \to 0$ gives the *transmission-line equations*:

$$\frac{\partial V(z, t)}{\partial z} = -rI(z, t) - l\frac{\partial}{\partial t}I(z, t) \tag{4.8a}$$

$$\frac{\partial I(z, t)}{\partial z} = -gV(z, t) - c\frac{\partial}{\partial t}V(z, t) \tag{4.8b}$$

These equations are a set of *coupled, first-order, partial differential equations.*

The above derivation of the transmission-line equations is for the general case. Most of our analyses will be for *lossless transmission lines*, where we assume $r = g = 0$. *A lossless transmission line is one in which (1) the conductors are perfect conductors with infinite conductivity, and (2) the medium surrounding the conductors has zero conductivity and therefore no losses associated with it.* This will yield reasonable approximations to the true results and will simplify the mathematics considerably, which is the primary reason for making the assumption of a lossless line.

4.2   THE PER-UNIT-LENGTH PARAMETERS

Key ingredients in the transmission-line equations are the per-unit-length parameters r, l, g, and c. The form of transmission-line equations derived above is identical for all two-conductor lines. The only aspects of them that distinguish between two different cross sections, for example, between the case of two wires and the case of a coaxial line, are the per-unit-length parameters. *All cross-sectional dimensions unique to the particular line are contained in these per-unit-length parameters and only in these parameters.* Consequently it is as important to be able to determine these parameters as it is to be able to solve the transmission-line equations. The per-unit-length parameters for the common two-conductor transmission lines will be obtained in this section.

The external per-unit-length parameters are due to the electromagnetic fields that are external to the line conductors. In the previous section we also showed that *the field structure for the TEM mode of propagation is identical to a static (dc) field structure.* Therefore *the external per-unit-length parameters can be determined using static field analysis techniques*, which considerably simplifies their determination [1].

Consider the two-wire line of Fig. 4.1(a). The transverse magnetic field $\vec{H}_T$ contributes to the per-unit-length inductance of the line. Figure 4.4 shows the magnetic field intensity of a current-carrying wire internal and external to the wire. The portion of the magnetic field internal to the wire contributes to a per-unit-length *internal inductance* l_i. The portion of the magnetic flux external to the wires contributes to a portion of the total per-unit-length inductance of the line that is referred to as the external inductance and is denoted by l_e. The total per-unit-length inductance is the sum of the internal and external inductances, $l = l_e + l_i$. The external inductance is much larger than the internal inductance for typical line dimensions, $l_e > l_i$, so that the total per-unit-length inductance is approximated by the external inductance, $l \cong l_e$, as we will generally assume. The transverse electric field $\vec{E}_T$ between the two wire surfaces of a two-wire line contributes to a per-unit-length capacitance c and a per-unit-length conductance g between the two wires. The capacitance represents displacement current flowing between the two wires, $\vec{J}_d = j\omega\epsilon\vec{E}_T$, whereas the conductance represents conduction current, $\vec{J}_c = \sigma\vec{E}_T$, flowing between the two wires through the surrounding (lossy) medium. It is possible to show that *for*

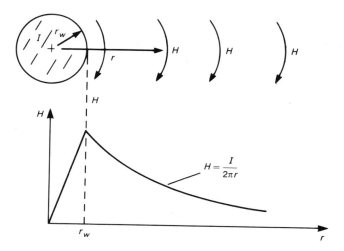

FIGURE 4.4 The magnetic field about a current-carrying wire.

the TEM mode of propagation and a homogeneous medium characterized by σ, μ and ϵ the per-unit-length external inductance l_e, capacitance c, and conductance g are related by [1]

$$l_e g = \mu \sigma \tag{4.9a}$$

$$l_e c = \mu \epsilon \tag{4.9b}$$

We will not repeat this derivation, but the reader is instead referred to [1] or other texts for a proof. Equations (4.9) are very important, and show that *we only need to determine one of the three external parameters for a particular line.* For example, suppose the per-unit-length external inductance l_e is known either from measurement or from a direct derivation. The other external parameters can be found from this as $c = \mu\epsilon/l_e$ and $g = \mu\sigma/l_e$.

The external per-unit-length parameters of inductance and capacitance are derivable in closed form for the three cases of lines in a homogeneous medium shown in Fig. 4.1. Derivations of the inductance and capacitance associated with wires (circular cylindrical conductors) rely on the following two *basic subproblems.* Consider an isolated, current-carrying wire shown in Fig. 4.4. *An important assumption implicit in the following result is that the current is uniformly distributed around the wire periphery.* If we bring another current-carrying wire in close proximity to this wire, the magnetic fields of the two wires will interact, causing the current distributions around them to be nonuniform; the current density will be greatest on the sides of the wires facing each other. This is referred to as *proximity effect,* and the following result disregards this. Under the assumption of a uniform distribution of current around the wire periphery, symmetry shows that the magnetic field intensity vector $\vec{H}_T$ is transverse to the wire axis, is constant for a fixed radius, and is circumferentially directed according to the right-hand rule. An equation for this magnetic field can be easily derived from Ampère's law that was discussed in Chapter 3. Recall that the problem at hand is a static one (dc), so we can omit the displacement current term of Ampère's law. In doing so we arrive at Ampère's law for the static case given in (4.5):

$$\oint_C \vec{H}_T \cdot d\vec{l} = I_{\text{enclosed}} \tag{4.10}$$

Choosing a circular contour C at a radius r from the wire, we observe that due to symmetry the magnetic field is tangent to the contour and the dot product can therefore be replaced by the ordinary product and the vector directions removed. Furthermore, it is evident, from symmetry, that the magnetic field will be the same at all points on this constant–radius contour and so can be removed from the integral. Thus Ampère's law in (4.10) simplifies to

$$H_T = \frac{I}{\oint_C dl} \tag{4.11}$$

$$= \frac{I}{2\pi r} \quad (\text{in A/m})$$

where the direction of H_T is circumferential about the wire. The transverse magnetic flux density vector is related to this by $\vec{B}_T = \mu_o \vec{H}_T$, where we assume the surrounding medium is free space or is not ferromagnetic. The first basic subproblem is to determine the total magnetic flux ψ_e external to the wire and penetrating a surface S of unit length along the wire direction that lies between a radius R_1 and a radius R_2 from the wire, as shown in Fig. 4.5(a). This is obtained by integrating the magnetic flux density vector over the surface, resulting in

$$\psi_e = \int_S \vec{B}_T \cdot d\vec{s} \tag{4.12}$$

$$= \int_{S_1} \vec{B}_T \cdot d\vec{s} + \underbrace{\int_{S_2} \vec{B}_T \cdot d\vec{s}}_{0} + \underbrace{\int_{S_{\text{end}}} \vec{B}_T \cdot d\vec{s}}_{0}$$

$$= \int_{r=R_1}^{R_2} \frac{\mu_o I}{2\pi r} dr$$

$$= \frac{\mu_o I}{2\pi} \ln\left(\frac{R_2}{R_1}\right) \quad (\text{in Wb})$$

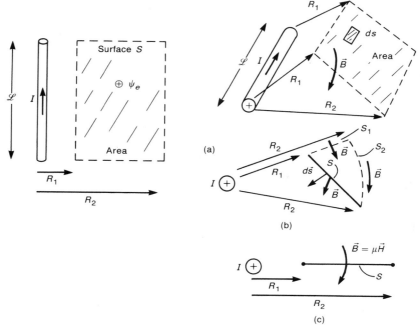

FIGURE 4.5   Illustration of a basic subproblem of determining the flux of a current through a surface: (a) dimensions of the problem; (b) use of Gauss' law; (c) an equivalent but simpler problem.

with

$$R_2 \geq R_1$$

This simple result is easy to see. Consider the surface S as being, for generality, "tilted" as shown in Fig. 4.5(a). Now, consider the *closed* surface consisting of the "wedge-shaped" surface that consists of the original tilted surface S and two other surfaces along with end caps illustrated in Fig. 4.5(b). Surface S_1 is the flat surface that is formed by moving radially outward from radius R_1 to R_2. Surface S_2 is formed by moving from the end of this previously constructed surface along a constant radius R_2 down to the edge of the original surface that is at a radius R_2 from the wire. This somewhat wedge-shaped surface along with the end caps forms a closed surface to which we apply Gauss' law, which states that *the total magnetic flux leaving a closed surface is zero; that is, whatever magnetic flux enters a closed surface must leave that closed surface.* The magnetic flux vector is parallel to the end caps, and so no net flux leaves the closed surface through the end caps. Also the magnetic flux vector is tangent to S_2 (the surface at a constant radius R_2 from the wire), so that no flux enters or leaves this surface. This leaves us with the fact that *magnetic flux that penetrates S_1 must also penetrate the original surface S*. Therefore *the total magnetic flux* *penetrating the original (tilted) open surface S shown in Fig. 4.5(a) is the same as the total magnetic flux penetrating the (untilted) surface S shown in Fig. 4.5(c).* This important observation considerably simplifies the determination of the flux penetrating the original (tilted) surface S. Evaluating (4.12) for the untilted surface gives the desired result. It is important to note the direction of the resulting flux through the surface. The direction of the magnetic field is determined by the right-hand rule. For $R_2 > R_1$ the flux is directed through the surface as shown in Fig. 4.5(a), so that ψ_e given in (4.12) is a positive quantity. The result in (4.12) should be committed to memory and the resulting direction of the flux through the surface should be firmly understood by the reader, since they will be used on numerous occasions. This is a basic result that will be used to derive the per-unit-length parameters for multiconductor lines.

The second basic subproblem concerns the voltage between two points due to a wire that is carrying a per-unit-length charge distribution of q C/m that is uniformly distributed around the wire periphery as shown in Fig. 4.6(a). As was the case with the previous derivations, an important assumption that is implicit in the following result is that the charge is uniformly distributed around the wire periphery. If we bring another charge-carrying wire in close proximity to this wire, the interaction between the fields of the two charge distributions will cause the charge distributions to be largest on the facing sides. This is again referred to as *proximity effect*, and the following result ignores this. We will assume that the medium surrounding the wire is free space with $\epsilon = \epsilon_o \cong 1/36\pi \times 10^{-9}$ F/m. Because of symmetry, the electric field due to this charge distribution, $\vec{E}_T$, is transverse to the wire, is directed away from the wire, and is constant at a constant distance away from the wire. The electric field is obtained by using Gauss' law for the electric field

$$\oint_S \epsilon_o \vec{E}_T \cdot d\vec{s} = Q_{enclosed} \tag{4.13}$$

Choosing the closed surface S to be a cylinder of radius r and unit length, with the wire centered on its axis as shown in Fig. 4.6(b), we observe that the electric field is parallel to the end caps and so contributes nothing to Gauss' law over the ends. Therefore we may simply evaluate (4.13) over the sides of the cylinder. But over the sides the electric field is perpendicular to the sides, and so the dot product can be removed, as can the vector directions. Similarly, because the electric field is, by symmetry, the same at all points on this surface, the electric field can be removed from the integral, leaving

$$E_T = \frac{q \times 1 \text{ m}}{\epsilon_o \oint_S ds} \tag{4.14}$$

$$= \frac{q}{2\pi \epsilon_o r} \quad \text{(in V/m)}$$

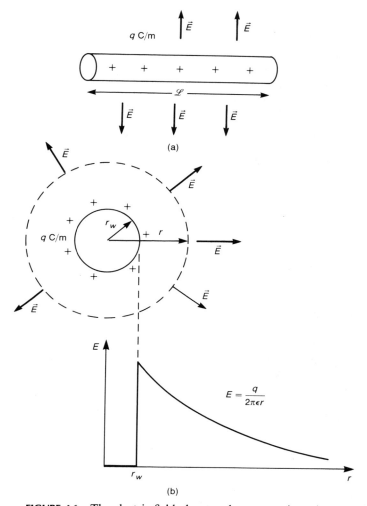

FIGURE 4.6   The electric field about a charge-carrying wire.

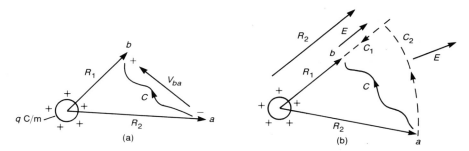

FIGURE 4.7   Illustration of a basic subproblem of determining the voltage between two points: (a) dimensions of the problem; (b) an equivalent but simpler problem.

where the field is directed in the radial direction. We now wish to obtain the voltage between two points that are at radial distances R_1 and R_2 from the wire as shown in Fig. 4.7(a). The voltage is related to the line integral between point a, which is at the larger radius R_2, and point b, which is at the smaller radius R_1. The result is

$$V = -\int_C \vec{E}_T \cdot \vec{dl} \qquad (4.15)$$

$$= -\int_{C_1} \vec{E}_T \cdot \vec{dl} - \underbrace{\int_{C_2} \vec{E}_T \cdot \vec{dl}}_{0}$$

$$= -\int_{r=R_2}^{R_1} \frac{q}{2\pi\epsilon_o r} \, dr$$

$$= \frac{q}{2\pi\epsilon_o} \ln\left(\frac{R_2}{R_1}\right) \quad \text{(in V)}$$

with

$$R_2 \geq R_1$$

Again this simple result is due to our choice of the contour over which to perform the integral in (4.15). We choose to integrate along the two contours C_1 and C_2 shown in Fig. 4.7(b) instead of some general contour between the two points. Because of symmetry, the electric field is perpendicular to the contour C_2, which is at a constant radius R_2 from the wire, and so nothing is contributed to the integral along this *equipotential* contour. Thus the voltage can be obtained by simply integrating from a distance R_2 to a distance R_1 along a radial line between these two points. Observe that the point closer to the positively charged wire is at the higher voltage. This is sensible, since voltage is the work required to move a unit positive charge between the two points. In this case, with the wire positively charged and $R_2 > R_1$, work will be required to move the unit positive charge from R_2 to R_1, so that the closer point will be at the higher potential. Equation (4.15) confirms this observation. It is important for the reader to commit the fundamental result in (4.15) to memory and to be able to perform these types of simple checks on the result, since this result will be used on numerous occasions. Also observe the duality between this result in (4.15) and the result in (4.12).

The above two fundamental subproblems and results allow the straightforward derivation of the per-unit-length parameters for the structures of Fig. 4.1. First consider the two-wire line of Fig. 4.1(a). Suppose one wire carries a current I directed into the page and the other wire carries a current of the same magnitude but directed out of the page. Denote the radius of one wire as r_{w1} and the radius of the other as r_{w2}. If the wires are separated by a distance s, the total magnetic flux between the wires is obtained from Fig. 4.8(a) using (4.12) as

$$\psi_e = \frac{\mu_o I}{2\pi} \ln \left(\frac{s - r_{w2}}{r_{w1}} \right) + \frac{\mu_o I}{2\pi} \ln \left(\frac{s - r_{w1}}{r_{w2}} \right) \tag{4.16}$$

$$= \frac{\mu_o I}{2\pi} \ln \left[\frac{(s - r_{w2})(s - r_{w1})}{r_{w2} r_{w1}} \right]$$

In order to use the basic result in (4.12), the currents must be uniformly distributed around the periphery of each wire, which essentially requires that the wires be widely separated [1]. Thus implicit in (4.16) is the requirement that $s \gg r_{w2}, r_{w1}$. Therefore the result simplifies to

$$\psi_e = \frac{\mu_o I}{2\pi} \ln \left(\frac{s^2}{r_{w2} r_{w1}} \right) \tag{4.17}$$

with

$$s \gg r_{w1}, r_{w2}$$

Ordinarily, if the ratio of the wire separation to wire radius is greater than approximately 5, i.e., $s/r_{w1} > 5$ and $s/r_{w2} > 5$, equation (4.17) gives accuracy within approximately 10% [1]. The per-unit-length external inductance is

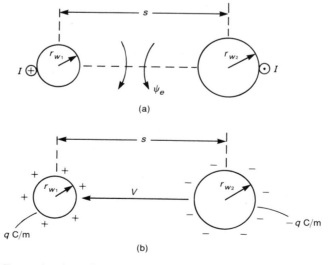

(a)

(b)

FIGURE 4.8   Determination of the per-unit-length parameters of a two-wire line: (a) inductance; (b) capacitance.

defined as the ratio of the flux penetrating a unit length surface between the wires. Using (4.17), we obtain

$$l_e = \frac{\psi_e}{I} \tag{4.18}$$

$$= \frac{\mu_o}{2\pi} \ln\left(\frac{s^2}{r_{w2} r_{w1}}\right) \quad \text{(in H/m)}$$

with

$$s \gg r_{w1}, r_{w2}$$

Generally, both wires will have the same radii. For this case the result in (4.18) becomes

$$l_e = \frac{\mu_o}{\pi} \ln\left(\frac{s}{r_w}\right), \quad \text{with } r_{w1} = r_{w2} = r_w \tag{4.19}$$

$$= 0.4 \ln\left(\frac{s}{r_w}\right) \quad \text{(in } \mu\text{H/m)}$$

$$= 10.16 \ln\left(\frac{s}{r_w}\right) \quad \text{(in nH/inch)}$$

Although this result gives reasonable approximations for widely separated wires, the exact result is derived in [1] and is

$$l_e = \frac{\mu_o}{\pi} \cosh^{-1}\left(\frac{s}{2r_w}\right) \quad \text{(in H/m)} \tag{4.20}$$

where $\cosh^{-1} x = \ln(x + \sqrt{x^2 - 1})$. For $x \gg 1$, $\cosh^{-1} x \cong \ln 2x$. As an example, consider two #28 gauge wires 7×36 ($r_w = 7.5$ mils) of a typical ribbon cable with center-to-center spacing of 50 mils. The exact expression in (4.20) gives the per-unit-length inductance of this cable as $l_e = 7.5 \times 10^{-7}$ H/m $= 19.04$ nH/inch. The approximate expression in (4.19) gives $l_e = 7.6 \times 10^{-7}$ H/m $= 19.3$ nH/inch, which is reasonably close to the exact value. The error is of order 1%. The ratio of separation to wire radius is 6.7.

The next per-unit-length parameter required for the two-wire line is the capacitance. This can be obtained from the per-unit-length inductance using (4.9b) and the exact value of the per-unit-length inductance from (4.20) as

$$c = \frac{\mu_o \epsilon_o}{l_e} \tag{4.21}$$

$$= \frac{1}{v_o^2 l_e}$$

$$= \frac{\pi \epsilon_o}{\cosh^{-1}(s/2r_w)} \quad \text{(in F/m)}$$

If the ratio of wire separation to wire radius is sufficiently large such that the charge distribution around the peripheries of the wires is essentially constant then (4.21) simplifies to

$$c \cong \frac{\pi \epsilon_o}{\ln (s/r_w)} \quad \text{(in F/m)} \tag{4.22}$$

with

$$s \geq r_w$$

This approximate result can be derived using the result for the second basic subproblem for the voltage between two points due to a charged wire given in (4.15). In order to use that result, the charge distributions must be uniform around the wire peripheries so that we must implicitly rule out proximity effect and consider widely-spaced wires. Consider the two wires of radii r_{w1} and r_{w2} and separated by distance s shown in Fig. 4.8(b). The voltage between the wires is given, according to (4.15), by

$$V = \frac{q}{2\pi \epsilon_o} \ln \left(\frac{s - r_{w2}}{r_{w1}} \right) + \frac{q}{2\pi \epsilon_o} \ln \left(\frac{s - r_{w1}}{r_{w2}} \right) \tag{4.23}$$

$$= \frac{q}{2\pi \epsilon_o} \ln \left[\frac{(s - r_{w2})(s - r_{w1})}{r_{w2} r_{w1}} \right]$$

$$\cong \frac{q}{2\pi \epsilon_o} \ln \left(\frac{s^2}{r_{w2} r_{w1}} \right)$$

For equal radii wires (4.23) reduces to

$$V = \frac{q}{\pi \epsilon_o} \ln \left(\frac{s}{r_w} \right), \quad \text{with } r_{w1} = r_{w2} = r_w \tag{4.24}$$

The per-unit-length capacitance is the ratio of the per-unit-length charge to the voltage between the two wires:

$$c = \frac{q}{V} \tag{4.25}$$

$$= \frac{\pi \epsilon_o}{\ln (s/r_w)}$$

$$= \frac{27.78}{\ln (s/r_w)} \quad \text{(in pF/m)}$$

$$= \frac{0.706}{\ln (s/r_w)} \quad \text{(in pF/inch)}$$

For the case of two #28 gauge 7×36 wires ($r_w = 7.5$ mils) of a ribbon cable separated by 50 mils, the exact expression in (4.21) gives $c = 14.82$ pF/m = 0.3765 pF/inch. The approximate expression in (4.22) and (4.25) gives $c = 14.64$ pF/m = 0.372 pF/inch. Again, the approximate expression valid for widely separated wires gives an error of only 1%, which is a reasonable approximation error.

The per-unit-length conductance g represents conduction current flowing through the surrounding medium from one wire to another. If the conductivity of the surrounding medium is zero then so is the per-unit-length conductance. This is generally a reasonable approximation for other typical dielectrics for frequencies below the GHz range, so that *in all our future analyses we will ignore the per-unit-length conductance parameter*; i.e., $g \cong 0$.

The per-unit-length external parameters for the case of one wire above a ground plane shown in Fig. 4.1(b) can be easily derived from the results for the case of two parallel wires using the method of images [1]. First consider the per-unit-length capacitance. The wire is at a height h above the ground plane, and the latter can be replaced with the image of the wire, as shown in Fig. 4.9. From this result we can see that the desired capacitance between the wire and the ground plane is twice the capacitance of two wires separated a distance of $2h$, since capacitors in series add like resistors in parallel. Thus, using the previous results, we obtain

$$c = \frac{2\pi\epsilon_o}{\cosh^{-1}(h/r_w)} \quad \text{(in F/m)} \tag{4.26}$$

For wires that are sufficiently far from the ground plane this simplifies to

$$c \cong \frac{2\pi\epsilon_o}{\ln(2h/r_w)} \quad \text{(in F/m)} \tag{4.27}$$

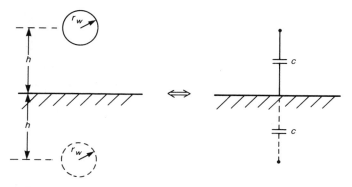

FIGURE 4.9 Determination of the per-unit-length capacitance of a wire above a ground plane with the method of images.

with

$$h \gg r_w$$

As an example, consider a #20 gauge solid wire ($r_w = 16$ mils) at a height of 1 cm above an infinite ground plane. The exact expression in (4.26) gives $c = 14.26$ pF/m $= 0.36$ pF/inch. The approximate expression in (4.27) gives $c = 14.26$ pF/m $= 0.36$ pF/inch also. The per-unit-length inductance can be obtained from the per-unit-length capacitance in (4.26) using (4.9b) as

$$l_e = \frac{\mu_o \epsilon_o}{c} \qquad (4.28)$$

$$= \frac{\mu_o}{2\pi} \cosh^{-1}\left(\frac{h}{r_w}\right) \quad \text{(in H/m)}$$

If the wires are widely separated, this simplifies to

$$l_e \cong \frac{\mu_o}{2\pi} \ln\left(\frac{2h}{r_w}\right) \quad \text{(in H/m)} \qquad (4.29)$$

with

$$h \gg r_w$$

For the case of a #20 gauge solid wire 1 cm above a ground plane the exact result in (4.28) gives $l_e = 0.779$ μH/m $= 19.79$ nH/inch. The approximate result in (4.29) gives essentially the same value.

The remaining structure in Fig. 4.1, the coaxial cable shown in (c), is derived in a similar fashion [1]. Because of symmetry, the electric field is radially directed and the magnetic field is circumferentially directed, as shown in Fig. 4.10(a). Symmetry also shows that if we place a per-unit-length positive charge distribution q C/m on the inner wire and a negative charge distribution $-q$ C/m on the inner surface of the shield, the resulting charge distributions will be uniform around the peripheries of these conductors regardless of the wire radius r_w or the shield interior radius r_s. Similarly, if current I flows along the inner wire surface and returns along the interior surface of the shield, these currents will also be uniformly distributed around the peripheries of the conductors. In other words, the proximity effect is not a factor for this structure. Consequently the *exact* equations for the per-unit-length parameters can be easily derived using the two basic subproblem results in (4.12) and (4.15). For example, the transverse magnetic flux density is circumferentially directed, and

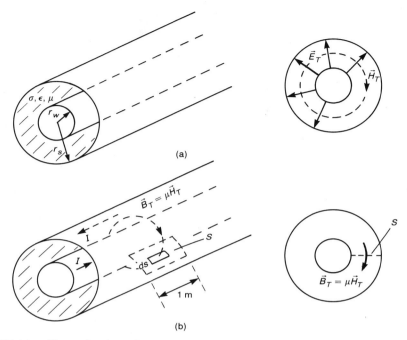

FIGURE 4.10 Determination of the per-unit-length parameters of a coaxial cable: (a) symmetry of the cross-sectional fields; (b) inductance.

is obtained from (4.11) as

$$B_T = \mu_o H_T \tag{4.30}$$

$$= \frac{\mu_o I}{2\pi r}$$

where the dielectric interior to the shield is assumed to be nonferromagnetic, so that $\mu = \mu_o$. The per-unit-length inductance can be found by determining the total magnetic flux penetrating a unit-length surface between the wire and the shield interior surface shown in Fig. 4.10(b) as

$$\psi_e = \int_s \vec{B}_T \cdot d\vec{s} \tag{4.31}$$

$$= \int_{r = r_w}^{r_s} \frac{\mu_o I}{2\pi r} \, dr$$

$$= \frac{\mu_o I}{2\pi} \ln\left(\frac{r_s}{r_w}\right)$$

Observe that this could have been directly obtained using the fundamental subproblem result given in (4.12). Therefore the *exact* value of the per-unit-length external inductance is

$$l_e = \frac{\psi_e}{I} \tag{4.32}$$

$$= \frac{\mu_o}{2\pi} \ln\left(\frac{r_s}{r_w}\right)$$

$$= 0.2 \ln\left(\frac{r_s}{r_w}\right) \quad (\text{in } \mu H/m)$$

$$= 5.08 \ln\left(\frac{r_s}{r_w}\right) \quad (\text{in } nH/inch)$$

The per-unit-length capacitance can be obtained using the fundamental result in (4.14) to give

$$E_T = \frac{q}{2\pi \epsilon r} \tag{4.33}$$

where the dielectric interior to the shield has permittivity ϵ. The voltage between the inner wire and the interior surface of the shield is (assuming the wire is at the higher potential) obtained as

$$V = -\int_{r_s}^{r_w} \vec{E}_T \cdot d\vec{l} \tag{4.34}$$

$$= \frac{q}{2\pi \epsilon} \ln\left(\frac{r_s}{r_w}\right)$$

The per-unit-length capacitance is obtained as

$$c = \frac{q}{V} \tag{4.35}$$

$$= \frac{2\pi \epsilon}{\ln(r_s/r_w)}$$

$$= \frac{55.56 \epsilon_r}{\ln(r_s/r_w)} \quad (\text{in } pF/m)$$

$$= \frac{1.4 \epsilon_r}{\ln(r_s/r_w)} \quad (\text{in } pF/inch)$$

Observe that

$$l_e c = \mu \epsilon \tag{4.36}$$

$$= \frac{\sqrt{\epsilon_r}}{v_o^2}$$

where the medium interior to the shield is homogeneous and characterized by $\mu = \mu_o$, $\epsilon = \epsilon_o \epsilon_r$, and $v_o = 3 \times 10^8$ m/s.

As a numerical example, consider a typical coaxial cable, RG-58U, which consists of an inner #20 gauge solid wire ($r_w = 16$ mils) and a shield (braided) having an inner radius of 58 mils. The interior dielectric is polyethylene ($\epsilon_r = 2.3$). The manufacturer specifies several nominal parameters for the cable. The first is the characteristic impedance, which is specified as 51 Ω. The next is the velocity of propagation as a percentage of the speed of light in free space, which is specified as 66 %. The third is the per-unit-length capacitance, which is specified as 30.8 pF/foot. The relative dielectric constant of the medium is related to the velocity of propagation by $v = v_o / \sqrt{\epsilon_r}$. Thus $\sqrt{\epsilon_r} = 1/0.66$, giving $\epsilon_r = 2.3$. The per-unit-length external inductance is found from (4.32) as $l_e = 0.2576 \ \mu H/m = 6.54$ nH/inch. The per-unit-length capacitance can be found from (4.35) as $c = 99.2$ pF/m $= 30.24$ pF/foot, which is reasonably close to the manufacturer's nominal specification of 30.8 pF/foot.

The line configurations shown in Fig. 4.2 are typical of land configurations on PCBs and can also be represented with a transmission-line model. The lines in Fig. 4.2 represent lines in an *inhomogeneous medium*, whereas those of Fig. 4.1 represent lines in a homogeneous medium. This observation does not affect the solution of the transmission-line equations, but does impact the derivation of the per-unit-length parameters. Let us denote the per-unit-length capacitance *with the board dielectric removed* as c_o. Clearly since all dielectrics have $\mu = \mu_o$, the fundamental result in (4.9) provides that

$$l_e c_o = \mu_o \epsilon_o \tag{4.37}$$

$$= \frac{1}{v_o^2}$$

where $v_o \cong 3 \times 10^8$ m/s is the speed of light in free space. Thus the per-unit-length external inductance for the lines of Fig. 4.2 can be found from the per-unit-length capacitance *with the dielectric removed*. Therefore we only need to find the per-unit-length capacitance with the dielectric inhomogeneity present, c, and with it replaced by free space, c_o. Although there exist some approximate, closed form solutions, numerical methods are typically used to determine the per-unit-length parameters for these structures, and these will be discussed in Chapter 6.

We now turn our attention to the derivation of the internal parameters of the conductors. If the line conductors are considered to be perfect conductors, any currents and charges must exist only on the surfaces of the conductors. Imperfect conductors are those for which the conductivities are finite but nonzero. For good conductors the current penetrates into the conductors so that a portion of its magnetic flux links a portion of the current internal to the conductor and as such produces an *per-unit-length internal inductance* l_i. The current internal to the conductors also obviously produces a per-unit-length resistance r. These parameters are derived for solid, round wires in [1] and will be stated here. Simple derivations of these parameters recognize that at dc, the current is uniformly distributed over the cross section of a solid conductor, as illustrated in Fig. 4.11. As the frequency of excitation increases, the current tends to crowd closer to the outside of the wire. This is referred to as *skin effect*, and for well-developed skin effect the current can be considered to be uniformly distributed over an annulus at the surface of the wire of depth δ, a skin depth. Thus, the resistance per unit length can be simply determined for either case as the reciprocal of the product of the wire conductivity σ and the cross-sectional area occupied by the current: πr_w^2 for dc and $2\pi r_w \delta$ for well-developed skin effect. For the case of a wire having parameters σ, μ_o, and ϵ_o the resistance and internal inductance are given in terms of the relation of

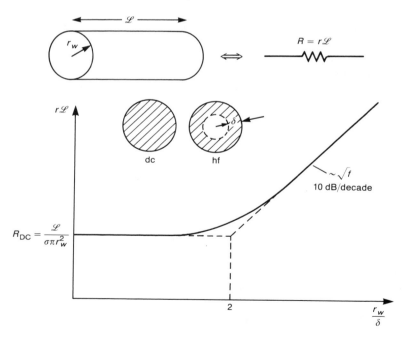

FIGURE 4.11 Illustration of the dependence of the per-unit-length resistance of a wire on frequency (skin effect).

the wire radius r_w to a skin depth δ by [1]

$$r = r_{DC} = \frac{1}{\sigma \pi r_w^2} \quad \text{(in } \Omega/\text{m)} \quad \text{for } r_w \ll \delta \tag{4.38a}$$

$$r = \frac{1}{\sigma 2\pi r_w \delta} \quad \text{(in } \Omega/\text{m)} \quad \text{for } r_w \gg \delta \tag{4.38b}$$

$$= \frac{r_w}{2\delta} r_{DC}$$

$$= \frac{1}{2r_w} \sqrt{\frac{\mu_o}{\pi \sigma}} \sqrt{f}$$

and

$$l_i = l_{i, DC} = \frac{\mu_o}{8\pi} = 50 \text{ nH/m} \quad \text{for } r_w \ll \delta \tag{4.39a}$$

$$l_i = \frac{2\delta}{r_w} l_{i, DC} \quad \text{for } r_w \gg \delta \tag{4.39b}$$

$$= \frac{1}{4\pi r_w} \sqrt{\frac{\mu_o}{\pi \sigma}} \frac{1}{\sqrt{f}}$$

and δ is the skin depth given by $\delta = 1/\sqrt{\pi f \mu \sigma}$.

It is important to discuss some important facts about these parameters. The wire resistance is equal to the dc resistance up to the frequency where the wire radius is on the order of a skin depth. Above that frequency, the resistance *increases as* $\sqrt{f}$ or 10 dB/decade. The wire internal inductance equals its dc inductance of 50 nH/m or 1.27 nH/inch, independently of the wire radius up to the frequency where the wire radius is on the order of a skin depth. Above this frequency, the wire internal inductance *decreases as* $\sqrt{f}$ or -10 dB/decade. The total per-unit-length inductance of a transmission line is the sum of the external and internal inductances as

$$l = l_i + l_e \tag{4.40}$$

$$\cong l_e$$

If the external inductance is larger than the dc internal inductance of 50 nH/m = 1.27 nH/inch then we may neglect the internal inductance since it decreases as $\sqrt{f}$ for higher frequencies but the external inductance is frequency-independent! This is typically the case, so that *in future work we will neglect the internal inductance parameter.*

Reasonable approximations for the internal parameters of rectangular–cross section conductors that typify lands of typical PCB-type lines shown in Fig. 4.2 can be made on the assumption that the current distribution over the rectangular cross section of a PCB land lies predominantly within a skin depth of the surface of the land. For a land of width w and thickness t the per-unit-length resistance becomes

$$r = r_{DC} = \frac{1}{\sigma w t} \quad \text{(in } \Omega/\text{m)} \quad \text{for } t \ll \delta \tag{4.41a}$$

$$= \frac{1}{2\sigma\delta(w + t)} \cong \frac{1}{2\sigma\delta w} \quad \text{(in } \Omega/\text{m)} \quad \text{for } t \gg \delta \tag{4.41b}$$

The land also, no doubt, possesses an internal inductance. However, this will be neglected on the reasonable assumption that it is dominated by the external inductance as for wires.

As numerical examples, we will compute typical values for wires and PCB lands. The per-unit-length resistance of a #20 gauge, solid copper wire is $r = 33.2 \text{ m}\Omega/\text{m} = 0.844 \text{ m}\Omega/\text{inch}$ at dc and $r = 0.323 \ \Omega/\text{m} = 8.2 \text{ m}\Omega/\text{inch}$ at 10 MHz. The skin depth at 10 MHz is 20.9 μm. The per-unit-length resistance of a PCB land that is 15 mils in width and etched from 1 oz copper (thickness of 1.38 mils) is $r = 32.8 \text{ m}\Omega/\text{inch}$ at dc and $r = 27.5 \text{ m}\Omega/\text{inch}$ at 10 MHz. Since the external inductance of typical parallel PCB lands is on the order of 15 nH/inch, giving an impedance at 10 MHz of $\omega l_e = 940 \text{ m}\Omega/\text{inch}$, we can begin to see why *the external inductance of wires and PCB lands is the primary contributor to the impedance of those conductors at high frequencies.* The conductor resistance is negligible compared with the external inductance impedance at high frequencies! This observation will be extremely important when we examine the voltage drop along conductors of PCBs in Chapter 13.

The internal parameters of resistance r and internal inductance l_i vary as the square root of frequency. Thus modeling them in the time domain is difficult. A useful time–domain model of this skin effect phenomenon is found in [3].

4.3   TIME-DOMAIN SOLUTION (TRANSIENTS)

The *time–domain solution* of the transmission-line equations refers to the *complete solution of those equations with no assumptions as to the time form of the line excitation.* The other solution of interest is the sinusoidal steady-state or frequency-domain solution considered in the next section, where the time form of the line excitation is restricted to being sinusoidal and furthermore the sinusoidal source is assumed to have been attached a sufficient length of time so that the transients have decayed to zero, leaving the steady-state solution. The time–domain solution is often referred to as being the transient solution, which is a misnomer since the time-domain solution gives the total solution—transient plus steady state.

4.3.1 Graphical Solutions

We will restrict our solution to *lossless lines*, since the inclusion of losses makes the solution of the transmission line equations difficult in the time domain. The solution for a lossless line also provides a reasonable approximation to the solution for the lossy line. For a *lossless line* we set $r = g = 0$ in (4.8), resulting in

$$\frac{\partial V(z, t)}{\partial z} = -l \frac{\partial}{\partial t} I(z, t) \tag{4.42a}$$

$$\frac{\partial I(z, t)}{\partial z} = -c \frac{\partial}{\partial t} V(z, t) \tag{4.42b}$$

with

$$r = g = 0 \quad \text{(lossless line)}$$

Differentiating one equation with respect to z and the other with respect to t and substituting gives uncoupled second-order equations:

$$\frac{\partial^2 V(z, t)}{\partial z^2} = lc \frac{\partial^2 V(z, t)}{\partial t^2} \tag{4.43a}$$

$$\frac{\partial^2 I(z, t)}{\partial z^2} = cl \frac{\partial^2 I(z, t)}{\partial t^2} \tag{4.43b}$$

with

$$r = g = 0 \text{ (lossless line)}$$

The solutions of the uncoupled, second-order form are [1]

$$V(z, t) = V^+ \left(t - \frac{z}{v} \right) + V^- \left(t + \frac{z}{v} \right) \tag{4.44a}$$

$$I(z, t) = \frac{1}{Z_C} V^+ \left(t - \frac{z}{v} \right) - \frac{1}{Z_C} V^- \left(t + \frac{z}{v} \right) \tag{4.44b}$$

where Z_C is the characteristic impedance of the line,

$$Z_C = \sqrt{\frac{l}{c}} \tag{4.45}$$

$$= vl$$

$$= \frac{1}{vc}$$

and v is the velocity of propagation on the line,

$$v = \frac{1}{\sqrt{lc}} \tag{4.46}$$

$$= \frac{1}{\sqrt{\mu\epsilon}}$$

where the medium surrounding the conductors is characterized by μ and ε. These results apply to lines in an inhomogeneous medium, where we would use $\epsilon = \epsilon_0 \epsilon_r'$ and ϵ_r' is the effective dielectric constant. The general forms of the solution given in (4.44) are in terms of the functions $V^+(t - z/v)$ and $V^-(t + z/v)$. The precise forms of these functions will be determined by the functional time-domain form of the excitation source, $V_S(t)$. Nevertheless, they show that time and position must be related as $t - z/v$ and $t + z/v$ in these forms. The function V^+ represents a *forward-traveling wave* traveling in the $+z$ direction. This is again clear, since as time increases, z must also increase to keep the argument of the function constant; that is, in order to track the movement of a point on the wave. Similarly, the function V^- represents a *backward-traveling wave* traveling in the $-z$ direction on the line. Thus the total solution consists of the sum of forward-traveling and backward-traveling waves. The current of each wave is related to the voltage of that wave by the characteristic impedance:

$$I^+\left(t - \frac{z}{v}\right) = \frac{1}{Z_C} V^+\left(t - \frac{z}{v}\right) \tag{4.47a}$$

$$I^-\left(t + \frac{z}{v}\right) = -\frac{1}{Z_C} V^-\left(t + \frac{z}{v}\right) \tag{4.47b}$$

We will consider lines of total length $\mathscr{L}$. The forward- and backward-traveling waves are related at the load, $z = \mathscr{L}$, by the *load reflection coefficient* as [1]

$$\Gamma_L = \frac{V^-(t + \mathscr{L}/v)}{V^+(t - \mathscr{L}/v)} \tag{4.48}$$

$$= \frac{R_L - Z_C}{R_L + Z_C}$$

Therefore the reflected waveform at the load can be found from the incident wave using the reflection coefficient as

$$V^-\left(t + \frac{\mathscr{L}}{v}\right) = \Gamma_L V^+\left(t - \frac{\mathscr{L}}{v}\right) \tag{4.49}$$

The reflection coefficient given in (4.48) applies to voltages only. A current reflection coefficient can be derived by substituting (4.48) into (4.47), so that

$$I^-\left(t + \frac{\mathscr{L}}{v}\right) = -\Gamma_L I^+\left(t - \frac{\mathscr{L}}{v}\right) \tag{4.50}$$

This reflection of waves at the load discontinuity is illustrated in Fig. 4.12. The reflection process can be viewed as a mirror that produces, as a reflected V^-, a replica of V^+ that is "flipped around," and all points on the V^- waveform are the corresponding points on the V^+ waveform multiplied by Γ_L. Note that the total voltage at the load, $V(\mathscr{L}, t)$, is the *sum* of the individual waves present at the load at a particular time as shown by (4.44).

Now let us consider the portion of the line at the source, $z = 0$, shown in Fig. 4.13. When we initially connect the source to the line, we reason that a

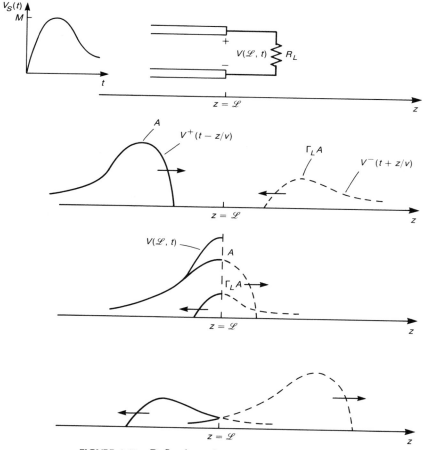

FIGURE 4.12   Reflection of waves at a termination.

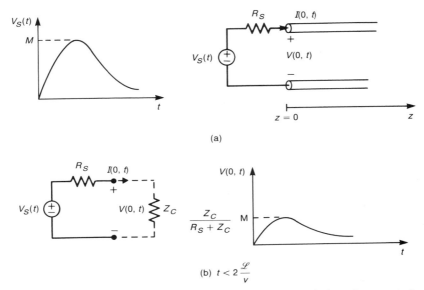

FIGURE 4.13   The equivalent circuit seen at the input to a line before the wave reflected at the load arrives.

forward-traveling wave will be propagated down the line. We would not expect a backward-traveling wave to appear on the line until this initial forward-traveling wave has reached the load, a time delay of $T = \mathcal{L}/v$, since the incident wave will not have arrived to produce this reflected wave. The portion of the incident wave that is reflected at the load will require an additional time T to move back to the source at $z = 0$. Therefore, for $0 \leq t \leq 2\mathcal{L}/v$, no backward-traveling waves will appear at $z = 0$, and for any time less than $2T$ the *total* voltage and current at $z = 0$ will consist only of forward-traveling waves, V^+ and I^+. Therefore

$$V(0, t) = V^+\left(t - \frac{0}{v}\right) \tag{4.51a}$$

$$I(0, t) = I^+\left(t - \frac{0}{v}\right) \tag{4.51b}$$

$$= \frac{V^+(t - 0/v)}{Z_C} \quad \text{for } 0 \leq t \leq \frac{2\mathcal{L}}{v}$$

Since the ratio of the *total* voltage and *total* current on the line is Z_C for $0 \leq t \leq 2\mathcal{L}/v$, as shown in (4.51), the line appears to have an input resistance of Z_C over this time interval, as shown in Fig. 4.13(b). Thus the forward-traveling voltage and current waves that are initially launched are related to the source voltage by

$$V(0, t) = \frac{Z_C}{R_S + Z_C} V_S(t) \tag{4.52a}$$

$$I(0, t) = \frac{V_S(t)}{R_S + Z_C} \tag{4.52b}$$

The initially launched waves have the same shape as the source voltage.

The initially launched wave travels toward the load, requiring a time $T = \mathcal{L}/v$ for the leading edge of the pulse to reach the load. When the pulse reaches the load, a reflected pulse is initiated, as shown in Fig. 4.12. This reflected pulse requires an additional time $T = \mathcal{L}/v$ for its leading edge to reach the source. At the source we can obtain a voltage reflection coefficient

$$\Gamma_S = \frac{R_S - Z_C}{R_S + Z_C} \tag{4.53}$$

as the ratio of the incoming incident wave (which is the reflected wave at the load) and the reflected portion of this incoming wave (which is sent back toward the load). A forward-traveling wave is therefore initiated at the source in the same fashion as at the load. This forward-traveling wave has the same shape as the incoming backward-traveling wave (which is the original pulse sent out by the source and reflected at the load), but corresponding points on the incoming wave are reduced by Γ_S. This process of repeated reflections continue as re-reflections at the source and load. At any time, the total voltage (current) at any point on the line is the sum of all the individual voltage (current) waves existing on the line at that point and time, as is shown by (4.44).

As an example consider the transmission line shown in Fig. 4.14(a). At $t = 0$ a 30 V battery with zero source resistance is attached to the line, which has a total length of $\mathcal{L} = 400$ m, a velocity of propagation of $v = 200$ m/μs and a characteristic impedance of $Z_C = 50\ \Omega$. The line is terminated in a 100 Ω resistor, so that the load reflection coefficient is

$$\Gamma_L = \frac{100 - 50}{100 + 50}$$

$$= \tfrac{1}{3}$$

and the source reflection coefficient is

$$\Gamma_S = \frac{0 - 50}{0 + 50}$$

$$= -1$$

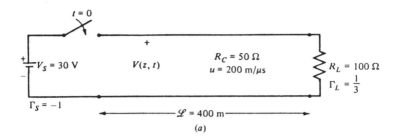

(a)

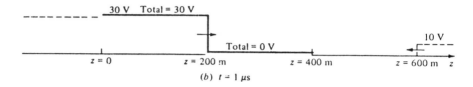

(b) $t = 1\ \mu s$

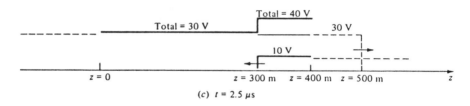

(c) $t = 2.5\ \mu s$

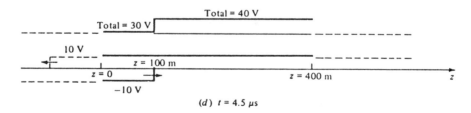

(d) $t = 4.5\ \mu s$

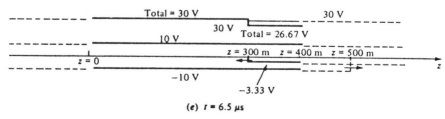

(e) $t = 6.5\ \mu s$

FIGURE 4.14 An example to illustrate sketching the voltage on a line as a function of position [1].

The one-way transit time is $T = \mathscr{L}/v = 2\,\mu\text{s}$. At $t = 0$ a 30 V pulse is sent down the line, and the line voltage is zero prior to the arrival of the pulse and 30 V after the pulse has passed. At $t = 2\,\mu\text{s}$ the pulse has arrived at the load, and a backward-traveling pulse of magnitude $30\Gamma_L = 10$ V is sent back toward the source. When this reflected pulse arrives at the source, a pulse of magnitude Γ_S of the incoming pulse or $\Gamma_S\Gamma_L 30 = -10$ V is sent back toward the load. This pulse travels to the load, at which time a reflected pulse of Γ_L of this incoming pulse or $\Gamma_L\Gamma_S\Gamma_L 30 = -3.33$V is sent back toward the source. At each point on the line the total line voltage is the sum of the waves present on the line at that point.

The previous example has illustrated the process of sketching the line voltage at various points along the line and at discrete times. Generally we are only interested in the voltage at the source and load ends of the line, $V(0, t)$ and $V(\mathscr{L}, t)$, as continuous functions of time. In order to illustrate this process, let us reconsider the previous example and sketch the voltage at the line output, $z = \mathscr{L}$, as a function of time, as is illustrated in Fig. 4.15. At $t = 0$ a 30 V pulse is sent out by the source. The leading edge of this pulse arrives at the load at $t = 2\,\mu\text{s}$. At this time a pulse of $\Gamma_L 30 = 10$ V is sent back toward the source. This 10 V pulse arrives at the source at $t = 4\,\mu\text{s}$, and a pulse of $\Gamma_S\Gamma_L 30 = -10$ V is returned to the load. This pulse arrives at the load at $t = 6\,\mu\text{s}$, and a pulse of $\Gamma_L\Gamma_S\Gamma_L 30 = -3.33$V is sent back toward the source. The contributions of these waves at $z = \mathscr{L}$ are shown in Fig. 4.15 as dashed lines, and the total voltage is shown as a solid line. Note that the load voltage oscillates during the transient time interval about 30 V, but asymptotically converges to the expected steady-state value of 30 V. If we had attached an oscilloscope across the load to display this voltage as a function of time, and the time scale were set to 1 ms per division, it would appear that the load voltage immediately assumed a value of 30 V. We would see the picture in Fig. 4.15 including the transient time interval only if the time scale of the oscilloscope were sufficiently reduced to, say, 1 μs per division. In order to sketch the load current $I(\mathscr{L}, t)$, we could divide the previously sketched load voltage by R_L. We could also sketch this directly by using current reflection coefficients $\Gamma_S = 1$ and $\Gamma_L = -\frac{1}{3}$ and an initial current pulse of 30 V$/Z_C = 0.6$ A. The current at the input to the line is sketched in this fashion in Fig. 4.15(c). Observe that this current oscillates about an expected steady-state value of 30 V$/R_L = 0.3$ A.

4.3.2 Numerical Methods

The previous section has demonstrated a graphical method of sketching the time-domain solution of the transmission-line equations. It is frequently desirable to have a numerical method that is suitable for a digital computer and will handle nonlinear as well as dynamic loads. The following method is attributed to Branin, and was originally described in [4]. It is only valid for lossless lines.

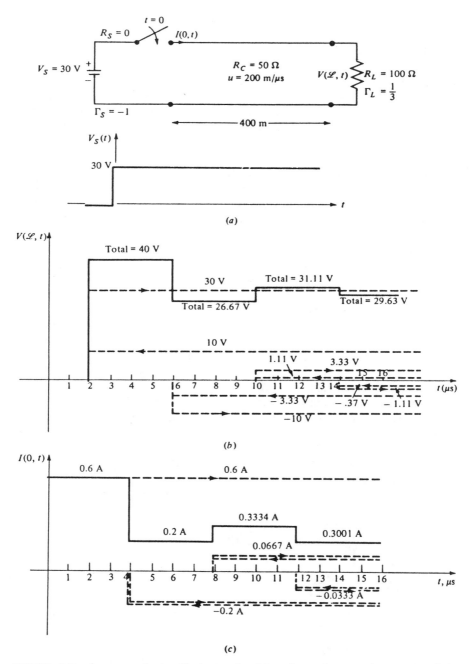

FIGURE 4.15 An example to illustrate sketching the voltage and current at the terminations as a funtion of time [1].

In order to obtain the required equations, we simply manipulate the solutions of the lossless transmission-line equations given in (4.44). Rewrite these as

$$V(z, t) = V^+\left(t - \frac{z}{v}\right) + V^-\left(t + \frac{z}{v}\right) \tag{4.54a}$$

$$Z_C I(z, t) = V^+\left(t - \frac{z}{v}\right) - V^-\left(t + \frac{z}{v}\right) \tag{4.54b}$$

Evaluating these at the source end, $z = 0$, and the load end, $z = \mathscr{L}$, gives

$$V(0, t) = V^+(t) + V^-(t) \tag{4.55a}$$

$$Z_C I(0, t) = V^+(t) - V^-(t) \tag{4.55b}$$

and

$$V(\mathscr{L}, t) = V^+(t - T) + V^-(t + T) \tag{4.56a}$$

$$Z_C I(\mathscr{L}, t) = V^+(t - T) - V^-(t + T) \tag{4.56b}$$

where the one-way time delay for the line is

$$T = \frac{\mathscr{L}}{v} \tag{4.57}$$

Adding and subtracting (4.55) and (4.56) gives

$$V(0, t) + Z_C I(0, t) = 2V^+(t) \tag{4.58a}$$

$$V(0, t) - Z_C I(0, t) = 2V^-(t) \tag{4.58b}$$

$$V(\mathscr{L}, t) + Z_C I(\mathscr{L}, t) = 2V^+(t - T) \tag{4.58c}$$

$$V(\mathscr{L}, t) - Z_C I(\mathscr{L}, t) = 2V^-(t + T) \tag{4.58d}$$

Shifting both (4.58a) and (4.58d) ahead in time by subtracting T from t along with a rearrangement of the equations gives

$$V(0, t) = Z_C I(0, t) + 2V^-(t) \tag{4.59a}$$

$$V(\mathscr{L}, t) = -Z_C I(\mathscr{L}, t) + 2V^+(t - T) \tag{4.59b}$$

$$V(0, t - T) + Z_C I(0, t - T) = 2V^+(t - T) \tag{4.59c}$$

$$V(\mathscr{L}, t - T) - Z_C I(\mathscr{L}, t - T) = 2V^-(t) \tag{4.59d}$$

Substituting (4.59d) into (4.59a) gives

$$V(0, t) = Z_C I(0, t) + E_r(\mathscr{L}, t - T) \tag{4.60a}$$

where

$$E_r(\mathscr{L}, t - T) = V(\mathscr{L}, t - T) - Z_C I(\mathscr{L}, t - T) \tag{4.60b}$$
$$= 2V^-(t)$$

Similarly, substituting (4.59c) into (4.59b) gives

$$V(\mathscr{L}, t) = -Z_C I(\mathscr{L}, t) + E_i(0, t - T) \tag{4.61a}$$

where

$$E_i(0, t - T) = V(0, t - T) + Z_C I(0, t - T) \tag{4.61b}$$
$$= 2V^+(t - T)$$

Equations (4.60) and (4.61) suggest the equivalent circuit of the total line shown in Fig. 4.16. The controlled source $E_i(0, t - T)$ is produced by the voltage and

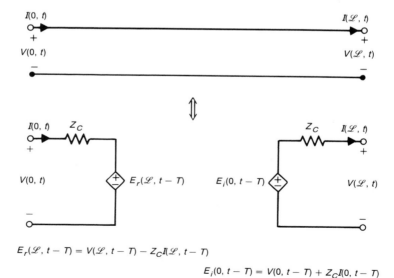

FIGURE 4.16 A time-domain equivalent circuit of a two-conductor, lossless line.

current at the input to the line at a time equal to a one-way transit delay earlier than the present time. Similarly, the controlled source $E_r(\mathscr{L}, t - T)$ is produced by the voltage and current at the line output at a time equal to a one-way transit delay earlier than the present time.

The equivalent circuit shown in Fig. 4.16 is an exact solution of the transmission-line equations for a lossless, two-conductor, uniform transmission line. The circuit analysis program SPICE contains this *exact model* among its list of available circuit element models that the user may call [5]. The model is the TXXX element, where XXX is the model number chosen by the user. SPICE uses controlled sources having time delay to construct the equivalent circuit of Fig. 4.16. The user need only input the characteristic impedance of the line Z_C (SPICE refers to this parameter as Z0) and the one-way transit delay T (SPICE refers to this as TD). Thus *SPICE will produce exact solutions of the transmission-line equations.* Furthermore, nonlinear terminations such as diodes and BJTs, as well as dynamic terminations such as capacitors and inductors, are easily handled with the SPICE code, whereas a graphical solution or the hand solution of the equivalent circuit in Fig. 4.16 for these types of loads would be quite difficult. This author highly recommends the use of SPICE for the incorporation of two-conductor transmission line effects into any analysis of an electronic circuit. It is simple and straightforward to incorporate the transmission-line effects in any time-domain analysis of an electronic circuit, and, more importantly, models of the complicated, but typical, nonlinear loads such as diodes and transistors as well as inductors and capacitors already exist in the code and can be called on by the user rather than the user needing to develop models for these loads.

As an example of the use of SPICE to model two-conductor, lossless transmission lines in the time domain, consider the time-domain analysis of the circuit of Fig. 4.15. The 30 V source is modeled with the PWL (piecewise linear) model as transitioning from 0 V to 30 V in 0.1 μs and remaining there throughout the analysis time interval of 20 μs. The SPICE program is

```
FIGURE 4.15
VS 1 0 PWL(0 0 .1U 30 20U 30)
T 1 0 2 0 Z0=50 TD=2U
RL 2 0 100
.TRAN .1U 20U
.PRINT TRAN V(2) I(VS)
.PLOT TRAN V(2) I(VS)
.END
```

The results for the load voltage are plotted using the .PROBE option of PSPICE [5] in Fig. 4.17(a) and the input current to the line is plotted in Fig. 4.17(b). Comparing these with the hand-calculated results shown in Fig. 4.15 shows good agreement.

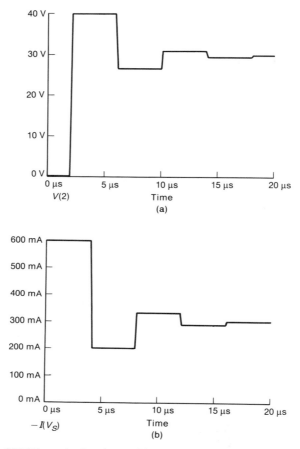

FIGURE 4.17   SPICE results for the problem: (a) output voltage; (b) input current.

4.4 FREQUENCY-DOMAIN SOLUTION (SINUSOIDAL STEADY STATE)

We now consider the case in which the source is a sinusoidal waveform; that is,

$$V_S(t) = V_S \cos \omega t \qquad (4.62)$$
$$= \mathscr{R}e\{V_S \, e^{j\omega t}\}$$

We will also assume that the source has been attached to the line for a length of time so that transients have decayed sufficiently, leaving a sinusoidal voltage and current at each point on the line. Thus we replace the source with its phasor equivalent $\hat{V}_S = V_S \, \underline{/0°}$. The resulting line voltages and currents are then given in terms of the phasor line voltages and currents as

$$V(z, t) = \mathcal{R}e\{\hat{V}(z)\,e^{j\omega t}\} \tag{4.63a}$$

$$I(z, t) = \mathcal{R}e\{\hat{I}(z)\,e^{j\omega t}\} \tag{4.63b}$$

4.4.1 Voltage and Current as Functions of Position

The phasor transmission line equations for a *lossless line* are obtained by substituting the phasor forms in (4.63) into the transmission-line equations of (4.42), where we replace time derivatives with $j\omega$ to give

$$\frac{d\hat{V}(z)}{dz} = -j\omega l\hat{I}(z) \tag{4.64a}$$

$$\frac{d\hat{I}(z)}{dz} = -j\omega c\hat{V}(z) \tag{4.64b}$$

Differentiating one with respect to z and substituting the other gives these equations as uncoupled, second-order equations:

$$\frac{d^2\hat{V}(z)}{dz^2} + \omega^2 lc\,\hat{V}(z) = 0 \tag{4.65a}$$

$$\frac{d^2\hat{I}(z)}{dz^2} + \omega^2 lc\,\hat{I}(z) = 0 \tag{4.65b}$$

The solutions to these equations are simple to obtain as [1]

$$\hat{V}(z) = \hat{V}^+\,e^{-j\beta z} + \hat{V}^-\,e^{j\beta z} \tag{4.66a}$$

$$\hat{I}(z) = \frac{\hat{V}^+}{Z_C}e^{-j\beta z} - \frac{\hat{V}^-}{Z_C}e^{j\beta z} \tag{4.66b}$$

where $\hat{V}^+$ and $\hat{V}^-$ are complex constants that are, as yet, undetermined. The other items are the familiar characteristic impedance and phase constant, given by

$$Z_C = \sqrt{\frac{l}{c}} \tag{4.67}$$

$$\beta = \frac{\omega}{v} \tag{4.68}$$

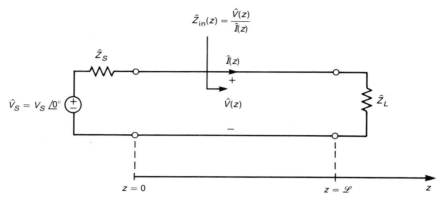

FIGURE 4.18   Definition of terms for the sinusoidal steady-state (phasor) analysis of transmission lines.

and v is the velocity of propagation on the line

$$v = \frac{1}{\sqrt{lc}} \tag{4.69}$$

$$= \frac{1}{\sqrt{\mu\epsilon}}$$

and the medium surrounding the line is assumed to be homogeneous and characterized by μ and ϵ. These results also apply to lines in an inhomogeneous medium such as in Fig. 4.2. These cases are characterized by an effective permittivity $\epsilon = \epsilon_0 \epsilon_r'$.

Consider the transmission line shown in Fig. 4.18. The line is terminated at $z = \mathscr{L}$ in a complex impedance $\hat{Z}_L$, and the source is a phasor source with source voltage $\hat{V}_S = V_S \underline{/0^\circ}$ and complex source impedance $\hat{Z}_S$. Let us define a complex voltage reflection coefficient at a particular point z on the line as the ratio of the phasor voltages of the backward- and forward-traveling waves:

$$\hat{\Gamma}(z) = \frac{\hat{V}^- \, e^{j\beta z}}{\hat{V}^+ \, e^{-j\beta z}} \tag{4.70}$$

$$= \frac{\hat{V}^-}{\hat{V}^+} e^{j2\beta z}$$

In terms of this reflection coefficient, the voltage and current expressions in (4.66) may be written as

$$\hat{V}(z) = \hat{V}^+ \, e^{-j\beta z}[1 + \hat{\Gamma}(z)] \tag{4.71a}$$

$$\hat{I}(z) = \frac{\hat{V}^+}{Z_C} e^{-j\beta z}[1 - \hat{\Gamma}(z)] \tag{4.71b}$$

We may also define an *input impedance* to the line at any point on the line as the ratio of the *total* voltage and current at that point on the line, as shown in Fig. 4.18; that is,

$$\hat{Z}_{in}(z) = \frac{\hat{V}(z)}{\hat{I}(z)} \tag{4.72}$$

$$= Z_C \frac{1 + \hat{\Gamma}(z)}{1 - \hat{\Gamma}(z)}$$

The reflection coefficient at the load is

$$\hat{\Gamma}_L = \frac{\hat{Z}_L - Z_C}{\hat{Z}_L + Z_C} \tag{4.73}$$

The reflection coefficient at some other point on the line is related to the load reflection coefficient as [1]

$$\hat{\Gamma}(z) = \hat{\Gamma}_L e^{j2\beta(z - \mathscr{L})} \tag{4.74}$$

Thus *the reflection coefficient at any point on the line can be related to the load reflection coefficient, which can be calculated directly from (4.73).* This gives a general procedure for calculation of the line voltage and current. The procedure is as follows:

1. Compute the load reflection coefficient from (4.73).
2. Compute the reflection coefficient at the line input from (4.74) as

$$\hat{\Gamma}(0) = \hat{\Gamma}_{in} \tag{4.75}$$

$$= \hat{\Gamma}_L e^{-j2\beta\mathscr{L}}$$

3. Compute the input impedance to the line from (4.72) as

$$\hat{Z}_{in}(0) = Z_C \frac{1 + \hat{\Gamma}(0)}{1 - \hat{\Gamma}(0)} \tag{4.76}$$

4. Compute the phasor input voltage to the line (by voltage division) as

$$\hat{V}(0) = \frac{\hat{Z}_{in}(0)}{\hat{Z}_{in}(0) + \hat{Z}_S} \hat{V}_S \tag{4.77}$$

5. Compute the undetermined constant $\hat{V}^+$ from (4.71) as

$$\hat{V}^+ = \frac{\hat{V}(0)}{1 + \hat{\Gamma}(0)} \tag{4.78}$$

6. Finally compute the line voltage and current at any point on the line from (4.71).

The *time-domain* line voltages and currents can then be found from these phasor voltages and currents using (4.63).

It is also possible to obtain explicit formulae for these results. The above results can be combined to obtain equations for the phasor voltage and current at any point on the line in terms of the source voltage:

$$\hat{V}(z) = \frac{1 + \hat{\Gamma}_L e^{-j2\beta\mathscr{L}} e^{j2\beta z}}{1 - \hat{\Gamma}_S \hat{\Gamma}_L e^{-j2\beta\mathscr{L}}} \frac{Z_C}{Z_C + \hat{Z}_S} \hat{V}_S e^{-j\beta z} \tag{4.79a}$$

$$\hat{I}(z) = \frac{1 - \hat{\Gamma}_L e^{-j2\beta\mathscr{L}} e^{j2\beta z}}{1 - \hat{\Gamma}_S \hat{\Gamma}_L e^{-j2\beta\mathscr{L}}} \frac{1}{Z_C + \hat{Z}_S} \hat{V}_S e^{-j\beta z} \tag{4.79b}$$

The input impedance can also be written as

$$\hat{Z}_{in}(0) = Z_C \frac{1 + \hat{\Gamma}_L e^{-j2\beta\mathscr{L}}}{1 - \hat{\Gamma}_L e^{-j2\beta\mathscr{L}}} \tag{4.80}$$

$$= Z_C \frac{\hat{Z}_L + jZ_C \tan \beta\mathscr{L}}{Z_C + j\hat{Z}_L \tan \beta\mathscr{L}}$$

A number of interesting properties can be deduced from these results. For example, from (4.73) we see that if the line is matched, $\hat{Z}_L = Z_C$, then the load reflection coefficient is zero, $\hat{\Gamma}_L = 0$. Equations (4.79) then show that for a matched line the *magnitudes* of the voltage and current are independent of position along the line:

$$\hat{V}(z) = \frac{Z_C}{Z_C + \hat{Z}_S} \hat{V}_S e^{-j\beta z} \tag{4.81a}$$

$$\hat{I}(z) = \frac{1}{Z_C + \hat{Z}_S} \hat{V}_S e^{-j\beta z} \tag{4.81b}$$

with

$$\hat{Z}_L = Z_C \quad \text{(matched line)}$$

Where possible, it is desirable to match transmission lines, $\hat{Z}_L = Z_C$, in order to eliminate reflections. This will also cause the magnitudes of the line voltage and current to be constant along the line. Only the phase of these quantities will change between two points along the line as shown by the $e^{-j\beta z}$ factor in (4.81). It is generally not possible to precisely match a line, so a quantitative criterion for the closeness of the match is desired. This is provided by the *voltage standing wave ratio or VSWR*. The VSWR is defined as *the ratio of the magnitude of the maximum voltage on the line to the minimum voltage on the line,* or [1]

$$\text{VSWR} = \frac{|\hat{V}(z)|_{\max}}{|\hat{V}(z)|_{\min}} \qquad (4.82)$$

$$= \frac{1 + |\hat{\Gamma}_L|}{1 - |\hat{\Gamma}_L|}$$

The load reflection coefficient for a short-circuit load is $\hat{\Gamma}_L = -1$, whereas that for an open-circuit load is $\hat{\Gamma}_L = 1$. For both cases the magnitude is unity making the VSWR equal to infinity. For either case the minimum voltage on the line is zero, which is why the VSWR is infinite. A large VSWR indicates an extreme variation of the voltage and current magnitude along the line, which is undesirable. On the other hand, a matched load gives a reflection coefficient of zero and a VSWR of unity. Therefore the VSWR of a line will be bounded by unity and infinity:

$$1 \le \text{VSWR} \le \infty$$

The closer the VSWR to unity, the better the match. Equation (4.80) also shows that *the input impedance to a lossless transmission line replicates for line lengths that are multiples of a half wavelength*; that is, a length of line shows the same input impedance if we add or subtract lengths that are multiples of a half wavelength. This is seen by observing that $\beta = 2\pi/\lambda$. Equation (4.80) also shows that *a quarter-wavelength line that is terminated in a short circuit (open circuit) appears at its input terminals as an open circuit (short circuit)*.

As a numerical example, consider the parallel-wire line shown in Fig. 4.19(a). The per-unit-length parameters of the line are $c = 200\,\text{pF/m}$ and $l = 0.5\,\mu\text{H/m}$, from which we compute

$$Z_C = \sqrt{\frac{l}{c}}$$

$$= 50\,\Omega$$

$$v = \frac{1}{\sqrt{lc}}$$

$$= 100\,\text{m/}\mu\text{s}$$

The frequency of the source is 30 MHz, so the line length of 1 m is 0.3λ in electrical length. The reflection coefficient at the load is

$$\hat{\Gamma}_L = \frac{\hat{Z}_L - Z_C}{\hat{Z}_L + Z_C}$$

$$= \frac{100 + j50 - 50}{100 + j50 + 50}$$

$$= 0.447 \,\underline{/26.6^\circ}$$

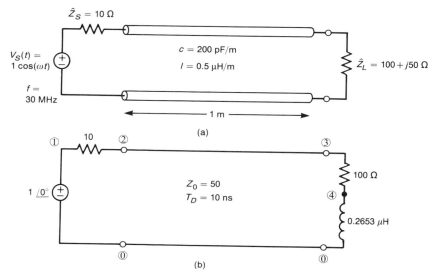

FIGURE 4.19 An example illustrating the phasor analysis of transmission lines: (a) problem definition; (b) the SPICE circuit.

and the VSWR is

$$\text{VSWR} = \frac{1 + |\hat{\Gamma}_L|}{1 - |\hat{\Gamma}_L|}$$

$$= \frac{1 + 0.447}{1 - 0.447}$$

$$= 2.62$$

The input impedance to the line is computed from the above results as

$$\hat{\Gamma}(0) = \hat{\Gamma}_L \, e^{j2\beta(0 - \mathscr{L})}$$

$$= \hat{\Gamma}_L \, e^{-j2\beta\mathscr{L}}$$

$$= 0.447 \, \underline{/26.6^\circ} \, e^{-j216^\circ}$$

$$= 0.447 \, \underline{/-189.4^\circ}$$

Thus

$$\hat{Z}_{\text{in}}(0) = Z_C \frac{1 + \hat{\Gamma}(0)}{1 - \hat{\Gamma}(0)}$$

$$= 19.54 \, \underline{/10.34^\circ} \, \Omega$$

The voltage at the input to the line becomes

$$\hat{V}(0) = \frac{\hat{Z}_{in}(0)}{\hat{Z}_{in}(0) + \hat{Z}_S}\, \hat{V}_S$$

$$= 0.664\ \underline{/3.5^\circ}\ \text{V}$$

and

$$\hat{I}(0) = \frac{\hat{V}(0)}{\hat{Z}_{in}(0)}$$

$$= 34 \times 10^{-3}\ \underline{/-6.8^\circ}\ \text{A}$$

Thus the time-domain voltage and current at the input to the line are

$$V(0, t) = 0.664 \cos(60\pi \times 10^6 t + 3.5^\circ)\ \text{V}$$
$$I(0, t) = 0.034 \cos(60\pi \times 10^6 t - 6.8^\circ)\ \text{A}$$

The phasor load voltage and current are obtained by first determining the constant $\hat{V}^+$ as

$$\hat{V}^+ = \frac{\hat{V}(0)}{1 + \hat{\Gamma}(0)}$$

$$= 1.18\ \underline{/-3.94^\circ}$$

Thus

$$\hat{V}(\mathscr{L}) = \hat{V}^+\, e^{-j\beta\mathscr{L}}(1 + \hat{\Gamma}_L)$$

$$= 1.18\ \underline{/-3.94^\circ}\, e^{-j108^\circ}(1 + 0.447\ \underline{/26.6^\circ})$$

$$= 1.668\ \underline{/-103.8^\circ}\ \text{V}$$

Therefore the time-domain load voltage is

$$V(\mathscr{L}, t) = 1.668 \cos(60\pi \times 10^6 t - 103.8^\circ)\ \text{V}$$

SPICE can be used to solve frequency-domain transmission line problems for lossless, two-conductor line using the exact transmission-line model discussed for time-domain solutions in the previous section. As an illustration, we will use SPICE to solve the previous problem. The SPICE circuit is shown in Fig.

4.19(b), and the SPICE program is

```
FIGURE 4.19
VS 1 0 AC 1
RS 1 2 10
T 2 0 3 0 ZO=50 TD=10N
RL 3 4 100
LL 4 0 .2653U
.AC DEC 1 30E6 30E6
.PRINT AC VM(2) VP(2) VM(3) VP(3)
.END
```

The SPICE results are

$$\hat{V}(0) = 0.6638 \ \underline{/3.513°}, \quad \hat{V}(\mathscr{L}) = 1.666 \ \underline{/-103.8°}$$

which were computed by hand. Note that in order to use SPICE for this frequency-domain computation, we must be able to construct a circuit to represent the complex load impedance $\hat{Z}_L = 100 + j50$. A 100 Ω resistor in series with a 0.2653 μH inductor will give this impedance at 30 MHz.

4.4.2 Power Flow

From the transverse nature of the TEM mode of propagation it is clear that power flow is in the $\pm z$ direction, since the Poynting vector is in this direction. In terms of the line voltage and current, the average power flow in the $+z$ direction is given by

$$P_{av}(z) = \tfrac{1}{2} \mathscr{R}e\{\hat{V}(z)\hat{I}^*(z)\} \tag{4.83}$$

Substituting the forms of the phasor voltages and currents given in (4.71) into (4.83) gives [1]

$$P_{av}(z) = \frac{1}{2} \frac{|\hat{V}^+|^2}{Z_C} (1 - |\hat{\Gamma}_L|^2) \tag{4.84}$$

This result could have been derived by adding the average powers of the forward- and backward-traveling waves. If we refer to the forward-traveling wave as the *incident wave* and the backward-traveling wave as the *reflected wave*, the ratio of the powers in these waves is

$$\frac{P_{av,\ reflected}}{P_{av,\ incident}} = |\hat{\Gamma}_L|^2 \tag{4.85}$$

When the load is either an open circuit or a short circuit, the magnitude of the reflection coefficient is unity, and (4.85) shows that all the incident power is reflected; a sensible result. Similarly, if the line is matched then the reflection coefficient is zero, and (4.84) shows that all the incident power is absorbed in the load.

For the example shown in Fig. 4.19 the average power delivered to the input of the line is

$$P_{\text{av, to line}} = \tfrac{1}{2} \mathscr{R}e\{\hat{V}(0)\hat{I}*(0)\}$$

$$= \tfrac{1}{2} \mathscr{R}e\{0.664 \; \underline{/3.5^\circ} \; 0.034 \; \underline{/6.8^\circ}\}$$

$$= 11.1 \text{ mW}$$

Since the line is assumed to be lossless, all of this average power is delivered to the load, which can be checked directly:

$$P_{\text{av, to load}} = \tfrac{1}{2} \mathscr{R}e\{\hat{V}(\mathscr{L})\hat{I}*(\mathscr{L})\}$$

$$= \frac{1}{2} \frac{|\hat{V}(\mathscr{L})|^2}{|\hat{Z}_L|} \cos \underline{/\hat{Z}_L}$$

$$= \frac{1}{2} \frac{(1.668)^2}{111.8} \cos 26.57^\circ$$

$$= 11.1 \text{ mW}$$

4.4.3   Inclusion of Losses

Imperfect line conductors and/or surrounding medium can easily be included in the above frequency-domain results in an *exact* manner, as opposed to the time-domain results. Furthermore, skin effect can also be accounted for in an *exact* fashion by simply computing the line resistance at the frequency of interest and including the value as a constant at that frequency. This is not the case for the time-domain results, since inclusion of the line resistance in the transmission-line equations significantly complicates their solution even if we assume that the resistance is independent of frequency (which is not true due to skin effect).

The transmission-line equations were derived from the equivalent circuit in Fig. 4.3(a) and are given in equation (4.8). Substituting the phasor voltage and current expressions gives

$$\frac{d\hat{V}(z)}{dz} = -\hat{z}\,\hat{I}(z) \tag{4.86a}$$

$$\frac{d\hat{I}(z)}{dz} = -\hat{y}\,\hat{V}(z) \tag{4.86b}$$

where the per-unit-length impedance $\hat{z}$ and admittance $\hat{y}$ are given by

$$\hat{z} = r(f) + j\omega l \qquad (4.87\text{a})$$

$$\hat{y} = g + j\omega c \qquad (4.87\text{b})$$

We have shown the per-unit-length conductor resistance as a function of frequency to emphasize its dependence on skin effect. Equations (4.86) are again a set of coupled differential equations. They may be uncoupled by differentiating one and substituting the other to give

$$\frac{d^2 \hat{V}(z)}{dz^2} - \hat{z}\hat{y}\,\hat{V}(z) = 0 \qquad (4.88\text{a})$$

$$\frac{d^2 \hat{I}(z)}{dz^2} - \hat{y}\hat{z}\,\hat{I}(z) = 0 \qquad (4.88\text{b})$$

The general solution of these equations is quite similar in form to that of the lossless case:

$$\hat{V}(z) = \hat{V}^+\, e^{-\alpha z}\, e^{-j\beta z} + \hat{V}^-\, e^{\alpha z}\, e^{j\beta z} \qquad (4.89\text{a})$$

$$\hat{I}(z) = \frac{\hat{V}^+}{\hat{Z}_C}\, e^{-\alpha z}\, e^{-j\beta z} - \frac{\hat{V}^-}{\hat{Z}_C}\, e^{\alpha z}\, e^{j\beta z} \qquad (4.89\text{b})$$

where the characteristic impedance is

$$\hat{Z}_C = \sqrt{\frac{\hat{z}}{\hat{y}}} \qquad (4.90)$$

$$= \sqrt{\frac{r(f) + j\omega l}{g + j\omega c}}$$

and the propagation constant is

$$\hat{\gamma} = \sqrt{\hat{z}\hat{y}} \qquad (4.91)$$

$$= \alpha + j\beta$$

The solution process for lossy lines is virtually unchanged from that for lossless lines [1]. There are, however, some important differences in the behavior of the line voltage and current that are worth pointing out. Observe in (4.89) that the forward- and backward-traveling voltage and current waves are attenuated as they travel alone the line, which is evidenced by the $e^{-\alpha z}$ and $e^{\alpha z}$ amplitude factors. This is virtually identical to the phenomenon observed for uniform plane waves traveling through a lossy medium. In fact, there is a direct duality between these where $\vec{E}(z)$ corresponds to $\hat{V}(z)$ and $\vec{H}(z)$ corresponds to $\hat{I}(z)$. The second important difference between the lossless and lossy cases is that average power is dissipated in a lossy line as the waves travel along the line. This was discussed briefly in Chapter 1 with regard to characterizing cable losses.

4.5 LUMPED-CIRCUIT APPROXIMATE MODELS

Typical circuit analysis computer programs such as SPICE contain an exact transmission-line model for two-conductor, lossless transmission lines that can be used for either time-domain or frequency-domain analysis of the line as we have observed. Nevertheless, an approximate method of modeling transmission lines that is frequently used is the construction of lumped-circuit models that are intended to be lumped-circuit approximations to the exact per-unit-length distributed-parameter model of Fig. 4.3. The typical such circuits are the lumped-Pi or lumped-Tee circuits shown in Fig. 4.20. These models can be used for either time- or frequency-domain analyses of the line.

In order to use the lumped-circuit models in Fig. 4.20 to model the transmission line, the line must be electrically short at the highest frequency of interest of the

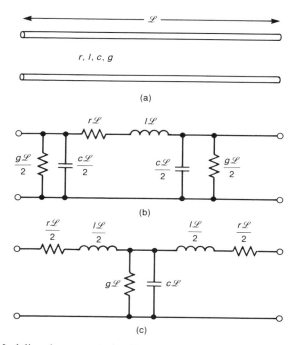

FIGURE 4.20 Modeling the transmission line with lumped circuits: (a) problem definition; (b) the lumped-Pi circuit; (c) the lumped-Tee circuit.

input source waveform. In other words, *the lumped-circuit model will correctly process only those frequency components of the input source that are below the frequency where the line becomes a significant fraction of a wavelength.* For single-frequency sinusoidal excitations this presents no problems. However, for time-domain signals such as periodic pulse trains this requires the user to determine whether incorrect processing of those spectral components of the source that are above the frequency where the line becomes electrically long will contribute any significant error. In order to increase the frequency range of validity of the lumped models, one might expect to subdivide the line into several electrically small subsections and model each with a lumped-circuit approximate model such as that in Fig. 4.20. Other work has shown that the gain in the extension of the valid frequency range is not significant, so that modeling a line with a large number of lumped-Pi sections, for example, would not give a significant enough extension of the frequency range to make the additional programming required by the large circuit structure worth the trouble [6].

As an example, consider the problem illustrated in Fig. 4.19. The line is modeled using one lumped-Pi section, as shown in Fig. 4.21(a). The line is one wavelength at 100 MHz. The results of the lumped-circuit model were obtained

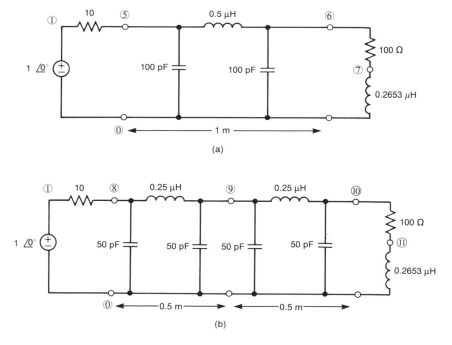

(a)

(b)

FIGURE 4.21   An example illustrating the use of lumped circuits to model a line and SPICE to perform the computations: (a) one Pi section; (b) two Pi sections.

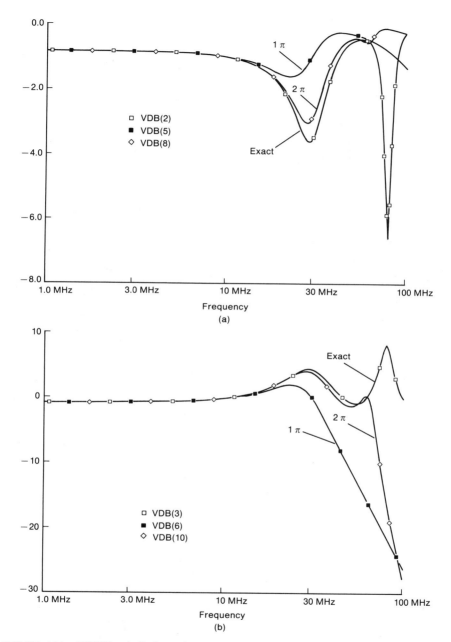

FIGURE 4.22 SPICE predictions for the circuits of Figs. 4.19 and 4.21: (a) input voltage; (b) load voltage.

from 1 MHz to 100 MHz using the SPICE program:

```
LUMPED PI MODEL FIGURE 4.21(A)
VS 1 0 AC 1
RS 1 5 10
C1 5 0 100P
L1 5 6 0.5U
C2 6 0 100P
RL 6 7 100
LL 7 0 0.2653U
.AC DEC 50 1E6 100E6
.PRINT AC VM(5) VP(5) VM(6) VP(6)
.PLOT AC VM(5) VP(5) VM(6) VP(6)
.END
```

The results are compared with those of the exact transmission-line model in Fig. 4.22. Observe that the predictions of the lumped-Pi model deviate from those of the exact model above approximately 20 MHz, where the line is $\frac{1}{5}\lambda$ in length. In order to extend the frequency range of prediction of the lumped-circuit model, we divide the line into two sections each of length 0.5 m and model each section with a lumped-Pi model as shown in Fig. 4.21(b). The predicted input and output voltages of the line are compared with the exact transmission-line model and the single-section, lumped-Pi model in Fig. 4.22. The two-section, lumped-Pi model has extended the prediction frequency range to approximately 60 MHz, where the line is 0.6λ in length.

4.6 TIME-DOMAIN TO FREQUENCY-DOMAIN TRANSFORMATIONS

A common method for obtaining the time-domain response of a transmission line is to first obtain the frequency response of the line and associated terminations by replacing the source voltage $V_S(t)$ with a unit-magnitude, variable frequency sinusoidal source. Then the actual time-domain source voltage $V_S(t)$ is decomposed into its frequency components using the Fourier series for periodic waveforms or the Fourier transform for a nonperiodic waveform (Chapter 7). If the line and associated terminations are *linear*, we may use *superposition* to pass each of the spectral components of the source through the frequency-domain transfer function, giving the spectral components (magnitude and phase) of the desired output of the line. These spectral components can then be combined using the inverse Fourier transform to give the time-domain output of the line. This process is described for linear circuits in [5] and is frequently referred to as the *time-domain to frequency-domain transformation. The method requires a linear line and linear terminations in order to be valid, since superposition was used.*

REFERENCES

[1]   C.R. Paul and S.A. Nasar, *Introduction to Electromagnetic Fields*, second edition, McGraw-Hill, NY (1987).

[2]   S. Ramo, J.R. Whinnery, and T. VanDuzer, *Fields and Waves in Communication Electronics*, John Wiley, NY (1965).

[3]   C.S. Yen, Z. Fazarinc, and R.L. Wheeler, Time-domain skin-effect model for transient analysis of lossy transmission lines, *Proc. IEEE*, **70**, 750–757 (1982).

[4]   F.H. Branin, Jr., Transient analysis of lossless transmission lines, *Proc. IEEE*, **55**, 2012–2013 (1967).

[5]   C.R. Paul, *Analysis of Linear Circuits*, McGraw-Hill, NY (1989).

[6]   C.R. Paul and W.W. Everett, III, Lumped model approximations of transmission lines: effect of load impedances on accuracy, *Technical Report, Rome Air Development Center, Griffiss AFB, NY*, RADC-TR-82-286, Vol. IV E (August 1984).

PROBLEMS

4.1   For the per-unit-length representations of a lossless transmission line shown in Fig. P4.1 derive the transmission-line equations in the limit as $\Delta z \to 0$. Note that the total per-unit-length inductance and capacitance in each circuit are l and c, respectively. This shows that the *structure* of the per-unit-length equivalent circuit is not important in the limit as $\Delta z \to 0$.

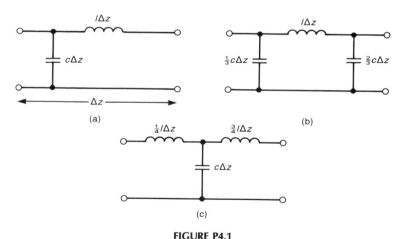

FIGURE P4.1

4.2   Compare the per-unit-length resistance (dc), inductance, capacitance and characteristic impedances for (a) two, identical #24 gauge stranded (7×32) wires $(r_w = 12 \text{ mils})$ that are separated by 100 mils, (b) a #24 gauge stranded (7×32) wire $\frac{1}{4}$ inch above an infinite, perfectly conducting ground plane, and (c) a coaxial cable composed of a #24 gauge stranded (7×32) interior wire in Teflon ($\epsilon_r = 2.1$) and a solid shield of interior radius 50 mils with wall thickness of 5 mils. [(a) 152 mΩ/m, 21.5 nH/inch, 0.333 pF/inch, 254.4 Ω; (b) 76 mΩ/m, 18.95 nH/inch, 0.378 pF/inch, 223.8 Ω; (c) 76.4 mΩ/m, 7.25 nH/inch, 2.06 pF/inch, 59.1 Ω]

4.3   One phase of a high-voltage, three-phase power transmission line consists of a stranded, steel wire of radius $\frac{1}{4}$ inch and is suspended above the earth a distance of 50 m. Determine the inductance and capacitance per mile as well as the characteristic impedance. [3.11 mH, 9.25 nF, 580 Ω]

4.4   The characteristic impedance, per-unit-length inductance, and per-unit-length capacitance vary as the natural logarithm of the radii and separation distances. This shows that increasing or decreasing the separation distances does not give a linear increase or decrease in these quantities. In order to investigate this, suppose two #20 gauge solid wires ($r_w = 16$ mils) have their separations increased from 50 mils to 250 mils. Determine the characteristic impedance, per-unit-length inductance, and per-unit-length capacitance for these two separations. [137 Ω, 330 Ω, 0.456 μH/m, 1.1 μH/m, 24.4 pF/m, 10.1 pF/m]

4.5   Determine the dc and high-frequency resistance and internal inductance of a #28 gauge solid wire ($r_w = 6.3$ mils) and the frequency where the wire radius is equal to a skin depth. [0.214 Ω/m, $2.6 \times 10^{-4}\sqrt{f}$ Ω/m, 50 nH/m, $41.3/\sqrt{f}\mu$H/m, 170.6 kHz]

4.6   Compute the dc and approximate high-frequency resistance of a 100 mil wide land of 1 ounce copper. [4.85 mΩ/inch, $1.3\sqrt{f}\mu\Omega$/inch] Determine the approximate frequency where these are equal. [13.9 MHz]

4.7   Determine the velocity of propagation and characteristic impedance for the following transmission lines:

   (a) $l = 0.25$ μH/m, $c = 100$ pF/m;
   (b) coaxial cable; $c = 50$ pF/m, $\epsilon_r = 2.1$;
   (c) two bare #28 gauge solid wires ($r_w = 6.3$ mils) separated by 100 mils;
   (d) one bare #16 gauge solid wire ($r_w = 25.4$ mils) $\frac{1}{4}$ inch above a ground plane. [50 Ω, 200 m/μs, 96.6 Ω, 207 m/μs, 331.8 Ω, 300 m/μs, 178.8 Ω, 300 m/μs]

4.8   Consider a transmission line that has $R_S = 300$ Ω, $R_L = 60$ Ω, $Z_C = 100$ Ω, $v = 200$ m/μs, $\mathscr{L} = 200$ m, and $V_S(t) = 400u(t)$ V, where $u(t)$ is the unit step function. Sketch $V(0, t), I(0, t), V(\mathscr{L}, t),$ and $I(\mathscr{L}, t)$ for $0 \le t \le 10$ μs. Do the results converge to expected steady-state values? [66.7 V, 1.11 A] Confirm your results using SPICE.

4.9   Repeat Problem 4.8 for $R_L = 0$ (short-circuit load). [0 V, 1.33 A]

4.10 Repeat Problem 4.8 for $R_L = \infty$ (open-circuit load). [400 V, 0 A]

4.11 A time-domain reflectometer (TDR) is an instrument used to determine properties of transmission lines. In particular, it can be used to detect the locations of imperfections such as breaks. The instrument launches a pulse down the line, and records the transit time for that pulse to be reflected at some discontinuity and to return to the line input. Suppose a TDR having a source impedance of 50 Ω is attached to a 50 Ω coaxial cable having some unknown length and load resistance. The dielectric of the cable is Teflon ($\epsilon_r = 2.1$). The open-circuit voltage of the TDR is a pulse of duration 10 μs. If the recorded voltage at the input of the TDR is as shown in Fig. P4.11, determine (a) the length of the cable and (b) the unknown load resistance. [621 m, 75 Ω] Confirm your results using SPICE.

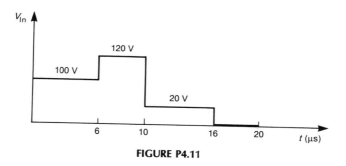

FIGURE P4.11

4.12 A 12 V battery ($R_S = 0$) is attached to an unknown length of transmission line that is terminated in a resistance. If the current to that line for 6 μs is as shown in Fig. P4.12, determine (a) the line characteristic impedance and (b) the unknown load resistance. [80 Ω, 262.7 Ω] Confirm your results using SPICE.

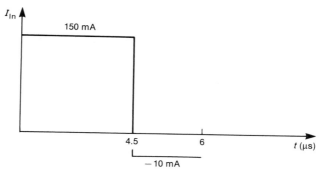

FIGURE P4.12

4.13 Digital data pulses should ideally consist of rectangular pulses. Actual data, however, have a trapezoidal shape with certain rise/fall times. Matching the data transmission line eliminates reflections and potential logic errors arising from these reflections. However, matching cannot always be accomplished. In order to investigate this problem, consider a line having $R_S = 0$ and $R_L = \infty$. Assume that the source voltage $V_S(t)$ is a ramp waveform given by

$$V_S(t) = \begin{cases} 0 & \text{for} \quad t \leq 0 \\ \dfrac{t}{\tau_r} & \text{for} \quad t \leq \tau_r \\ 1 & \text{for} \quad t \geq \tau_r \end{cases}$$

where τ_r is the rise time of the trapezoidal pulses. Sketch the load voltage for line lengths having one-way transit times T such that (a) $\tau_r = \frac{1}{10}T$, (b) $\tau_r = 2T$, (c) $\tau_r = 3T$, and (d) $\tau_r = 4T$. This example shows that in order to avoid problems resulting from mismatch, one should choose line lengths short enough that $\tau_r \gg T$ for the desired data. Confirm your results using SPICE.

4.14 A lossless transmission line is to be used to transmit digital. The data are in the form of trapezoidal pulses with rise/fall times of T as shown in Fig. P4.14, where T is the one-way transit time of the line. Sketch the load and source voltages for $0 \leq t < 8T$ for $R_S = Z_C$ and load resistances of $R_L = 2Z_C, \frac{1}{2}Z_C$. Confirm your results using SPICE.

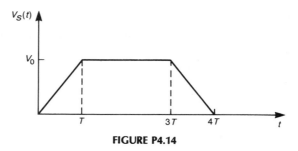

FIGURE P4.14

4.15 Sketch the line voltage as a function of position at $t = \frac{1}{2}T, \frac{3}{2}T, 2T, 3T, 4T$, and $5T$ for the line of Problem 4.14. Do these sketches verify the plots obtained for that problem?

4.16 A step voltage source having $V_S(t) = 400u(t)$ V and $R_S = 350\ \Omega$ is applied to a length of $50\ \Omega$ line that has a short-circuit load. Plot the current in the short-circuit load as a function of the line's one-way delay T for $0 \leq t \leq 12T$. Does the result converge to the expected steady-state value?

[1.143 A] At what time will the current be within 10% of the steady-state value? [15T] Confirm your results using SPICE.

4.17 Digital data are being transmitted bidirectionally on a transmission line. Ordinarily the data are transmitted such that there are no conflicts at any point. Model this line as shown in Fig. P4.17 and sketch the voltages at the terminal of each device for $0 \le t \le 7T$, where T is the line's one-way delay. Confirm your results using SPICE.

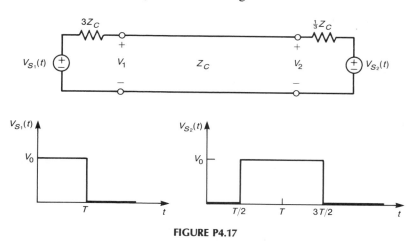

FIGURE P4.17

4.18 A transmission line having $\mathscr{L} = 200$ m, $v = 200$ m/μs, $Z_C = 50\,\Omega$, and $R_L = 20\,\Omega$ is driven by a source that has $R_S = 100\,\Omega$ and that has a source voltage that is a rectangular pulse of magnitude 6 V and duration 3 μs. Sketch the input current to the line for a total time of 5 μs. Confirm your results using SPICE.

4.19 Consider a lossless transmission line operated in the sinusoidal steady state. For the following problem specifications, determine (i) the line length as a fraction of a wavelength, (ii) the voltage reflection coefficient at the load and the input to the line, (iii) the VSWR, (iv) the input impedance to the line, (v) the time-domain voltage at the line input and at the load, and (vi) the average power delivered to the load:

   (a) $\mathscr{L} = 1$m, $f = 262.5$ MHz, $Z_C = 50\,\Omega$, $\hat{Z}_L = (30 - j200)\,\Omega$, $\hat{Z}_S = (100 + j50)\,\Omega$, $v = 300$ m/μs, and $\hat{V}_S = 10\,\underline{/30°}$ V; [0.875, 0.933 $\underline{/-27.5°}$, 0.933 $\underline{/-297.5°}$, 28.85, 82.3 $\underline{/85.5°} = 6.42 + j82.1$, 4.85 $\underline{/64.4°}$, 5.513 $\underline{/-293.9°}$, 11.2 mW]

   (b) $\mathscr{L} = 36$ m, $f = 28$ MHz, $Z_C = 150\,\Omega$, $\hat{Z}_S = (500 + j0)\,\Omega$, $\hat{Z}_L = -j30\,\Omega$, $v = 300$ m/μs, and $\hat{V}_S = 100\,\underline{/0°}$ V; [10.714, 1.0 $\underline{/-157.4°}$, 1.0 $\underline{/-56.65°}$, ∞, 278.4 $\underline{/-90°}$, 48.64 $\underline{/-119°}$, 27.62 $\underline{/241.5°}$, 0 W]

(c) $\mathscr{L} = 2$ m, $f = 175$ MHz, $Z_C = 100 \, \Omega$, $\hat{Z}_L = (200 - j30) \, \Omega$, $\hat{Z}_S = 50 \, \Omega$, $v = 200$ m/μs, and $\hat{V}_S = 10 \, \underline{/0°}$ V. [1.75, $0.346 \, \underline{/-11°}$, $0.346 \, \underline{/-191°}$, 2.058, $49.48 \, \underline{/8.52°} = 48.9 + j7.33$, $4.99 \, \underline{/4.28°}$, $10.07 \, \underline{/-274.2°}$, 248 mW]

4.20 Lines that are electrically short can be suitably represented using lumped-circuit models (π or T circuits). In order to illustrate and investgate this, use SPICE to compute and plot the frequency of the input impedance of a line using the exact model available in SPICE and using 1π and $1T$ sections to approximate the line. Plot the frequency response over the range where the line is electrically short up to the frequency where the line is one wavelength long. Show these plots for the following conditions: (a) $R_L = 100Z_C$, (b) $R_L = 10Z_C$, (c) $R_L = 0.1Z_C$, and (d) $R_L = 0.01Z_C$.

4.21 A section of lossless coaxial cable having $Z_C = 50 \, \Omega$ and $v = 200$ m/μs is terminated in a short circuit and operated at a frequency of 10 MHz. Determine the shortest length of the line such that, at the input terminals, the line appears to be a 100 pF capacitor. [9.04 m] Determine the shortest length such that the line appears to be a 1 μH inductor. [2.86 m]

4.22 Repeat Problem 4.21 for an open-circuit load. [5 m, 7.14 m]

4.23 An antenna having an input impedance at 100 MHz of $(72 + j40) \, \Omega$ is connected to a 100 MHz generator via a section of 300 Ω air-filled line of length 1.75 m. If the generator has a source voltage of 10 V and a source impedance of 50 Ω, determine the average power delivered to the antenna. [76.68 mW]

4.24 A lossy transmission line is operated at 100 MHz. Measurements on the line indicate that $\hat{Z}_C = 75 + j0 \, \Omega$, $\alpha = 0.02$ Np/m, and $\beta = 3$ rad/m. Determine the per-unit-length resistance, inductance, capacitance, and conductance of the line. If a 7 m length line is terminated in $\hat{Z}_L = (150 + j0) \, \Omega$ and driven by a source having $\hat{V}_S = 10 \, \underline{/0°}$ V and $\hat{Z}_S = (75 + j0) \, \Omega$, determine the average power delivered to the line and to the load. Repeat these calculations if the line is matched at the load. [$r = 1.5 \, \Omega$/m, $g = 2.667 \times 10^{-4}$ S/m, $l = 358$ nH/m, $c = 63.7$ pF/m, 156.1 mW and 28 mW, 166.7 mW and 126 mW]

4.25 In order to investigate the use of lumped-circuit parameter models to approximate transmission lines for the time domain, use SPICE to compute the input and load voltages using the exact model and one π section. Assume $R_S = \frac{1}{3}Z_C$, $R_L = 2Z_C$, and $V_S(t)$ as a pulse described by $V_S(t) = t$ for $0 \leq t \leq \tau_r$, $V_S(t) = 1$ for $\tau_r \leq t$. Show results for $\tau_r = 0.1T$, $1T$, and $10T$, where T is the one-way delay of the line.

CHAPTER FIVE

Antennas

Antennas are obviously a major ingredient in the discipline of EMC. *Intentional antennas* such as AM, FM, and radar antennas generate fields that couple to electronic devices and result in susceptibility problems. Intentional antennas are also used to measure the radiated emissions of a product for determining compliance to the regulatory limits. *Unintentional antennas* are responsible for producing the radiated emissions that are measured by the measurement antenna and may result in the product being out of compliance. In this chapter we will review the basics of intentional antennas. This will provide the ability to determine the electromagnetic field levels *in the vicinity of the product* that will be used to determine its susceptibility to interference. Antennas used to verify regulatory compliance will be discussed. The analysis of intentional antennas also provides understanding of the ability of unintentional antennas to radiate, which is what we are trying to minimize or prevent. A more detailed discussion is found in Chapter 9 of [1] or other similar texts on elementary electromagnetic field theory.

5.1 ELEMENTAL DIPOLE ANTENNAS

If we know the current distribution over the surface of the antenna, we can obtain the radiated electric and magnetic fields by performing an integral involving this current distribution [1]. Although this result, in principle, allows us to characterize the radiated fields of all antennas, there are two practical complications: the need to obtain the current distribution over the antenna and the need to perform a difficult integration. Typically, we make a reasonable guess as to the current *distribution*. The integration is deceptively complicated. In this section we will investigate some simple antennas that, although not practical, are easy to solve and whose radiated fields resemble those of practical antennas so long as the field point is sufficiently distant from the antenna.

5.1.1   The Electric (Hertzian) Dipole

The Hertzian dipole consists of an infinitesimal current element of length dl carrying a phasor current $\hat{I}$ that is assumed to be the same (in magnitude and phase) at all points along the element length, as illustrated in Fig. 5.1. A spherical coordinate system is commonly used to describe antennas. The location of a point in this coordinate system is described by the radial distance to the point, r, and the angular positions of a radial line to the point from the z axis, θ, and between its projection on the xy plane and the x axis, ϕ, as shown in Fig. 5.1. Orthogonal unit vectors $\vec{a}_r$, $\vec{a}_\theta$, $\vec{a}_\phi$, point in the directions of increasing values of these coordinates. The components of the magnetic field intensity fector become [1]

$$\hat{H}_r = 0 \tag{5.1a}$$

$$\hat{H}_\theta = 0 \tag{5.1b}$$

$$\hat{H}_\phi = \frac{\hat{I}\,dl}{4\pi}\beta_0^2 \sin\theta\left(j\frac{1}{\beta_0 r} + \frac{1}{\beta_0^2 r^2}\right)e^{-j\beta_0 r} \tag{5.1c}$$

Similarly, the components of the electric field intensity vector are [1]

$$\hat{E}_r = 2\frac{\hat{I}\,dl}{4\pi}\eta_0\beta_0^2 \cos\theta\left(\frac{1}{\beta_0^2 r^2} - j\frac{1}{\beta_0^3 r^3}\right)e^{-j\beta_0 r} \tag{5.2a}$$

$$\hat{E}_\theta = \frac{\hat{I}\,dl}{4\pi}\eta_0\beta_0^2 \sin\theta\left(j\frac{1}{\beta_0 r} + \frac{1}{\beta_0^2 r^2} - j\frac{1}{\beta_0^3 r^3}\right)e^{-j\beta_0 r} \tag{5.2b}$$

$$\hat{E}_\phi = 0 \tag{5.2c}$$

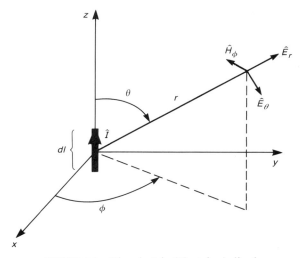

FIGURE 5.1   The electric (Hertzian) dipole.

where $\eta_0 = \sqrt{\mu_0/\epsilon_0}$ is the intrinsic impedance of free space. Note that the fields in (5.1) and (5.2) can be viewed as functions of *electrical distance from the antenna*, since $\beta_0 r = 2\pi r/\lambda_0$ and $\lambda_0 = v_0/f$ is the wavelength at the frequency of the antenna current.

The complete fields in (5.1) and (5.2) are quite complicated [2]. Our main interest is in the *far fields* where the field point is sufficiently distant from the antenna. How far is "far enough?" Note that the $1/r^3$ and $1/r^2$ terms dominate at very close distances to the antenna. As we move further from the antenna, the $1/r$ terms begin to dominate. The point where the $1/r^3$ and $1/r^2$ terms become insignificant compared with the $1/r$ terms is referred to as the boundary between the *near field* and the *far field*. This occurs where $r = \lambda_0/2\pi \cong \frac{1}{6}\lambda_0$. The reader is cautioned that *the boundary between the near and far fields for other antennas is not simply* $\lambda_0/2\pi$, as is frequently assumed. A more realistic choice for the boundary between the near and far fields will be discussed later, but can be summarized as being the larger of $3\lambda_0$ or $2D^2/\lambda_0$, where D is the largest dimension of the antenna. Typically the first criterion is used for "wire-type" antennas and the second for "surface-type" antennas such as parabolics or horns. This boundary between the near and far fields is not meant to be a precise criterion, but is only intended to indicate a general region where the fields transition from complicated to quite simple structure. In the use of antennas for communication this question of whether the receiving antenna is in the near or far field of the transmitting antenna never arises because these antennas are used for communication over large distances. However, in the area of EMC and interference caused by emissions the receiver (which may be an intentional antenna used for compliance measurement) is frequently in the near field of the transmitting antenna (which may be the product). This is particularly true for the lower frequencies of the FCC, Class B radiated emission measurement, and is investigated in detail in [2]. Nevertheless, for the moment we will assume that the field point is in the far field of the Hertzian dipole. Retaining only the $1/r$ terms in the field expressions in (5.1) and (5.2) gives the *far-field vectors*:

$$\vec{E}_{\text{far field}} = j\eta_0 \beta_0 \frac{\hat{I}\, dl}{4\pi} \sin\theta \frac{e^{-j\beta_0 r}}{r} \vec{a}_\theta \tag{5.3a}$$

$$\vec{H}_{\text{far field}} = j\beta_0 \frac{\hat{I}\, dl}{4\pi} \sin\theta \frac{e^{-j\beta_0 r}}{r} \vec{a}_\phi \tag{5.3b}$$

The time-domain fields are obtained by multiplying the phasor fields by $e^{j\omega t}$ and taking the real part of the results:

$$\vec{E}_{\text{far field}} = \mathcal{R}e\{\vec{E}_{\text{far field}} e^{j\omega t}\} \tag{5.4a}$$

$$= \frac{E_m}{r} \cos\left[\omega\left(t - \frac{r}{v_0}\right) + 90°\right]\vec{a}_\theta$$

$$= -\frac{E_m}{r} \sin\left[\omega\left(t - \frac{r}{v_0}\right)\right]\vec{a}_\theta$$

$$\vec{H}_{\text{far field}} = \mathscr{R}e\{\vec{\hat{H}}_{\text{far field}}e^{j\omega t}\} \tag{5.4b}$$

$$= \frac{E_m}{\eta_0 r}\cos\left[\omega\left(t - \frac{r}{v_0}\right) + 90°\right]\vec{a}_\phi$$

$$= -\frac{E_m}{\eta_0 r}\sin\left[\omega\left(t - \frac{r}{v_0}\right)\right]\vec{a}_\phi$$

where

$$E_m = \frac{\eta_0\beta_0 I\,dl}{4\pi}\sin\theta \tag{5.4c}$$

The far fields of the Hertzian dipole satisfy many of the properties of uniform plane waves. In fact, "locally" the fields resemble uniform plane waves, although they are more correctly classified as spherical waves. These properties are as follows:

1. The fields are proportional to $1/r$, $\hat{I}$, dl, and $\sin\theta$.
2. $|\vec{\hat{E}}_{\text{far field}}|/|\vec{\hat{H}}_{\text{far field}}| = \eta_0$.
3. $\vec{\hat{E}}_{\text{far field}}$ and $\vec{\hat{H}}_{\text{far field}}$ are *locally orthogonal*.
4. $\vec{\hat{E}}_{\text{far field}} \times \vec{\hat{H}}_{\text{far field}} = K\vec{a}_r$.
5. A phase term $e^{-j\beta_0 r}$ translates to a time delay in the time domain of $\sin[\omega(t - r/v_0)]$.

This is the origin of the technique of translating fields using the inverse-distance relationship. For example, the electric and magnetic fields at distances D_1 and D_2 are related by $|\vec{E}_{D_2}| = (D_1/D_2)|\vec{E}_{D_1}|$; that is, the fields decay inversely with increasing distance away from the radiator. It is important to remember that *the inverse-distance rule holds only if both D_1 and D_2 are in the far field of the radiating element.* This important restriction is frequently not adhered to. If either of the two distances are in the near field, the inverse-distance rule cannot be used.

We next obtain the total average power radiated by integrating the average power Poynting vector over a suitable closed surface surrounding the antenna. First we compute the Poynting vector using the *total* phasor fields in (5.1) and (5.2) as

$$\vec{S}_{\text{av}} = \tfrac{1}{2}\mathscr{R}e\{\vec{\hat{E}} \times \vec{\hat{H}}^*\} \tag{5.5}$$

$$= \tfrac{1}{2}\mathscr{R}e\{\hat{E}_\theta\hat{H}_\phi^*\vec{a}_r - \hat{E}_r\hat{H}_\phi^*\vec{a}_\theta\}$$

$$= 15\pi\left(\frac{dl}{\lambda_0}\right)^2|\hat{I}|^2\frac{\sin^2\theta}{r^2}\vec{a}_r \quad (\text{in W/m}^2)$$

This shows that average power is flowing away from the current element: our first hint of "radiation." It is instructive to note that this average power density vector could have been obtained solely from the far-field expressions given in (5.3). Integrating this result over a sphere of radius r enclosing the current element gives the total average power radiated by the current element [1]:

$$P_{rad} = \oint_S \vec{S}_{av} \cdot d\vec{s} \tag{5.6}$$

$$= 80\pi^2 \left(\frac{dl}{\lambda_0}\right)^2 \frac{|\hat{I}|^2}{2} \quad \text{(in W)}$$

Denoting $\hat{I}/\sqrt{2} = \hat{I}_{rms}$, we can compute a *radiation resistance*:

$$R_{rad} = \frac{P_{rad}}{|\hat{I}_{rms}|^2} \tag{5.7}$$

$$= 80\pi^2 \left(\frac{dl}{\lambda_0}\right)^2 \quad \text{(in } \Omega\text{)}$$

The radiation resistance represents a fictitious resistance that dissipates the same amount of power as that radiated by the Hertzian dipole when both carry the same value of rms current.

The Hertzian dipole is a very ineffective radiator. For example, for a length $dl = 1$ cm and a frequency of 300 MHz ($\lambda_0 = 1$ m), the radiation resistance is 79 mΩ. In order to radiate 1 W of power, we require a current of 3.6 A! If the frequency is changed to 3 MHz ($\lambda_0 = 100$ m), the radiation resistance is 7.9 $\mu\Omega$ and the current required to radiate 1 W is 356 A! Nevertheless the Hertzian dipole serves a useful purpose in that the *far fields of a Hertzian dipole are virtually identical to the far fields of most other practical antennas.*

5.1.2 The Magnetic Dipole (Loop)

A dual to the elemental electric dipole is the *elemental magnetic dipole or current loop* shown in Fig. 5.2. A very small loop of radius b lying in the xy plane carries a phasor current $\hat{I}$. This loop constitutes a magnetic dipole moment

$$\hat{m} = \hat{I}\pi b^2 \quad \text{(in A m}^2\text{)} \tag{5.8}$$

where πb^2 is the area enclosed by the loop. The radiated fields are [1]

$$\hat{E}_r = 0 \tag{5.9a}$$

$$\hat{E}_\theta = 0 \tag{5.9b}$$

$$\hat{E}_\phi = -j\frac{\omega\mu_0 \hat{m}\beta_0^2}{4\pi} \sin\theta \left(j\frac{1}{\beta_0 r} + \frac{1}{\beta_0^2 r^2}\right) e^{-j\beta_0 r} \tag{5.9c}$$

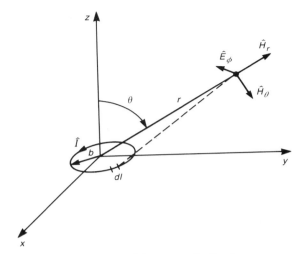

FIGURE 5.2 The magnetic dipole.

and

$$\hat{H}_r = j2\frac{\omega\mu_0\hat{m}\beta_0^2}{4\pi\eta_0}\cos\theta\left(\frac{1}{\beta_0^2 r^2} - j\frac{1}{\beta_0^3 r^3}\right)e^{-j\beta_0 r} \qquad (5.10a)$$

$$\hat{H}_\theta = j\frac{\omega\mu_0\hat{m}\beta_0^2}{4\pi\eta_0}\sin\theta\left(j\frac{1}{\beta_0 r} + \frac{1}{\beta_0^2 r^2} - j\frac{1}{\beta_0^3 r^3}\right)e^{-j\beta_0 r} \qquad (5.10b)$$

$$\hat{H}_\phi = 0 \qquad (5.10c)$$

Comparing (5.9) and (5.10) with the fields of the Hertzian dipole given in (5.1) and (5.2), we see that duality exists between the fields of these structures. Observations about the electric (magnetic) field of the electric dipole apply to the magnetic (electric) field of the magnetic dipole.

The far field of the magnetic dipole is characterized by the $1/r$-dependent terms:

$$\vec{E}_{\text{far field}} = \frac{\omega\mu_0\hat{m}\beta_0}{4\pi}\sin\theta\frac{e^{-j\beta_0 r}}{r}\vec{a}_\phi \qquad (5.11a)$$

$$\vec{H}_{\text{far field}} = -\frac{\omega\mu_0\hat{m}\beta_0}{4\pi\eta_0}\sin\theta\frac{e^{-j\beta_0 r}}{r}\vec{a}_\theta \qquad (5.11b)$$

As was the case for the Hertzian (electric) dipole, the far field of the magnetic dipole is such that the fields (1) decay as $1/r$, (2) lie in a (local) plane perpendicular to the radial direction, and (3) are related by η_0.

As was done for the Hertzian dipole, we may determine a radiation resistance of the magnetic dipole, which becomes [1]

$$R_{\text{rad}} = \frac{P_{\text{av}}}{|\hat{I}_{\text{rms}}|^2} \tag{5.12}$$

$$= 31{,}170 \left(\frac{A}{\lambda_0^2} \right)^2$$

where $A = \pi b^2$ is the area of the loop. Like the Hertzian dipole, the magnetic dipole is not an efficient radiator. Consider a loop of radius 1 cm. At 300 MHz the radiation resistance is 3.08 mΩ. In order to radiate 1 W, the loop requires a current of 18 A. At 3 MHz the radiation resistance is 3.08×10^{-11} Ω, and the current required to radiate 1 W is 1.8×10^5 A!

If a loop is electrically small, its shape is not important with respect to the far fields that it generates [1]. In order to illustrate the application of this result to radiated emissions, consider a 1 cm × 1 cm current loop on a PCB (an equivalent loop radius of 5.64 mm). Suppose the loop carries a 100 mA current at a frequency of 50 MHz. At a measurement distance of 3 m (FCC Class B) the electric field is a maximum in the plane of the loop and is $|\hat{E}| = 109.6 \ \mu\text{V/m} = 40.8 \ \text{dB}\mu\text{V/m}$. Recall that the FCC Class B limit from 30 MHz to 88 MHz is $40 \ \text{dB}\mu\text{V/m}$. Therefore a 1 cm × 1 cm loop carrying a 50 MHz, 100 mA current will cause a radiated emission that will fail to comply with the FCC Class B regulatory limit! This should serve to illustrate to the reader that passing the regulatory requirements on radiated emissions is not a simple matter, since the above dimensions and current levels are quite representative of those found on PCBs of electronic products.

5.2   THE HALF-WAVE DIPOLE AND QUARTER-WAVE MONOPOLE

The Hertzian dipole considered in Section 5.1.1 is an obviously impractical antenna for several reasons. Primarily, the length of the dipole was assumed to be infinitesimal in order to simplify the computation of the fields. Also, the current along the Hertzian dipole was assumed to be constant along the dipole. This latter assumption required the current to be nonzero at the endpoints of the dipole—an unrealistic and, moreover, physically impossible situation since the surrounding medium, free space, is nonconductive. Also, the Hertzian dipole is a very inefficient radiator since the radiation resistance is quite small, requiring large currents in order to radiate significant power. The magnetic dipole suffers from similar problems. In this section we will consider two practical and more frequently used antennas: the long-dipole and monopole antennas.

The long-dipole antenna (or, simply, the dipole antenna) consists of a thin wire that is fed or excited via a voltage source inserted at the midpoint, as shown in Fig. 5.3(a). Each leg is of length $\frac{1}{2}l$.

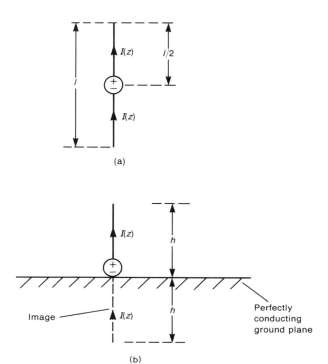

FIGURE 5.3 Illustration of (a) the dipole antenna and (b) the monopole antenna.

The monopole antenna shown in Fig. 5.3(b) consists of a single leg perpendicular to a ground plane. The monopole is fed at its base with respect to the ground plane. For purposes of analysis, the ground plane is considered to be infinite and perfectly conducting. In practice, this ideal ground plane is approximated. On aircraft, for example, the metallic fuselage simulates this ground plane. For ground-based stations the earth simulates, to some degree, this ground plane. Since the earth is much less of an approximation to a perfectly conducting plane that is metal, ground-based stations are usually augmented by a grid of wires lying on the ground to simulate the ground plane. The monopole antenna can be analyzed by replacing the ground plane with the image of the current element that is above the ground plane, as indicated in Fig. 5.3(b). Images are discussed in Section 5.6.1. Once the ground plane is replaced with the image, the problem reduces to the dipole problem so that a separate analysis of the monopole is not needed.

We observed previously that if we know the current distribution over the surface of an antenna, we may compute the resulting radiated fields. In practice, one often makes a reasonable guess as to the *current distribution* over the antenna surface. It can be shown that the current distribution of the long-dipole antenna follows (approximately) the same distribution as on a transmission line; that is, $\hat{I}(z)$ is proportional to $\sin \beta_0 z$ [1]. Placing the center of the dipole at the origin of a spherical coordinate system as shown in Fig. 5.4(a), with the dipole directed along the z axis, we may therefore write an expression for the current distribution along the wire is

$$\hat{I}(z) = \begin{cases} \hat{I}_m \sin[\beta_0(\tfrac{1}{2}l - z)] & \text{for } 0 < z < \tfrac{1}{2}l \\ \hat{I}_m \sin[\beta_0(\tfrac{1}{2}l + z)] & \text{for } -\tfrac{1}{2}l < z < 0 \end{cases} \tag{5.13}$$

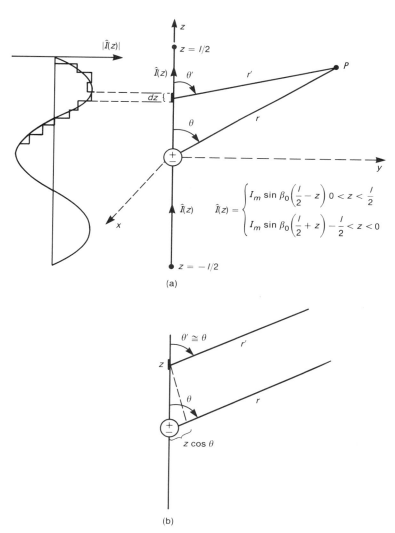

(a)

(b)

FIGURE 5.4 Computation of the radiated fields of the dipole antenna: (a) superposition of the Hertzian dipole fields; (b) the far-field approximation.

Note that this current distribution satisfies the necessary criteria: (1) the variation with z is proportional to $\sin \beta_0 z$ and (2) the current goes to zero at the endpoints, $z = -\frac{1}{2}l$ and $z = \frac{1}{2}l$.

Now that we have assumed the current distribution along the dipole, we may compute the fields of the long dipole as being the superpositions of the fields due to many small Hertzian dipoles of length dz having a current that is constant and equal to the value of $\hat{I}(z)$ at that point along the dipole, as shown in Fig. 5.4(a). We also assume that the desired field point is in the far field of these current elements, so that we only need the far-field expressions for the Hertzian dipole given in (5.3). For example, the field at point P in Fig. 5.4(a) due to the segment dz is

$$d\hat{E}_\theta = j\eta_0 \beta_0 \frac{\hat{I}(z) \sin \theta'}{4\pi r'} e^{-j\beta_0 r'} \, dz \qquad (5.14)$$

Our ultimate objective is to obtain the fields at point P in terms of the radial distance r from the midpoint of the dipole and angle θ. Since we are considering only the far field, the radial distance r from the center of the dipole at point P and the distance r' from the current element to point P will be approximately equal ($r \cong r'$) and the angles θ and θ' will be approximately equal ($\theta \cong \theta'$), as shown in Fig. 5.4(b).

We may substitute $r' = r$ into the denominator of (5.14), but we should not make this substitution into the $e^{-j\beta_0 r'}$ term for the following reason. This term may be written as $e^{-j\beta_0 r'} = \underline{/2\pi r'/\lambda_0}$, and its value depends not on physical distance r' but on electrical distance r'/λ_0. Therefore, even though r' and r may be approximately equal, the term may depend significantly on the difference in electrical distances. For example, suppose $r = 1000$ m and $r' = 1000.5$ m and the frequency is $f = 300$ MHz. We have ($\lambda_0 = 1$ m) $\beta_0 r = 2\pi(1000) = 360{,}000°$ and $\beta_0 r' = 2\pi(1000.5) = 360{,}180°$. Note that the fields at 1000 m and those only 0.5 m farther away are 180° out of phase! We will see a more striking example of this in the next section, where the far fields from two antennas that are widely spaced physically but separated on the order of a wavelength may actually be completely out of phase and add destructively to yield a result of zero.

Thus it is not a reasonable approximation to substitute r for r' in the phase term in (5.14). However, we may still write the result in terms of r. Consider Fig. 5.4(b), which shows the two radial distances r and r' as being approximately parallel. Thus we are assuming that the field point is sufficiently far, physically, from the antenna. From Fig. 5.4(b) we may obtain

$$r' \cong r - z \cos \theta \qquad (5.15)$$

Substituting (5.15) into the phase term in (5.14) and r into the denominator gives

$$d\hat{E}_\theta = j\eta_0 \beta_0 \frac{\hat{I}(z) \sin \theta}{4\pi r} e^{-j\beta_0(r - z \cos \theta)} \, dz \qquad (5.16)$$

The total electric field is the sum of these contributions:

$$\hat{E}_\theta = \int_{z=-l/2}^{z=l/2} j\eta_0 \beta_0 \frac{\hat{I}(z)\sin\theta}{4\pi r} e^{-j\beta_0 r} e^{j\beta_0 z\cos\theta}\, dz \tag{5.17}$$

Substituting the expressions for $\hat{I}(z)$ given in (5.13), we obtain [1]

$$\hat{E}_\theta = j\frac{\eta_0 \hat{I}_m e^{-j\beta_0 r}}{2\pi r} F(\theta) \tag{5.18}$$

$$= j\frac{60\hat{I}_m e^{-j\beta_0 \tau}}{r} F(\theta)$$

where the θ-variation term in this result is denoted by

$$F(\theta) = \frac{\cos[\beta_0(\tfrac{1}{2}l)\cos\theta] - \cos\beta_0(\tfrac{1}{2}l)}{\sin\theta} \tag{5.19}$$

$$= \frac{\cos[(\pi l/\lambda_0)\cos\theta] - \cos(\pi l/\lambda_0)}{\sin\theta}$$

since $\beta_0 = 2\pi/\lambda_0$. The magnetic field in the far-field region of the Hertzian dipole is orthogonal to the electric field and related by η_0. If we carry through the above development for the magnetic field, we obtain

$$\hat{H}_\phi = \frac{\hat{E}_\theta}{\eta_0} \tag{5.20}$$

where $\hat{E}_\theta$ is given by (5.18).

The most frequently encountered case is the half-wave dipole, in which the total dipole length is $l = \tfrac{1}{2}\lambda_0$. Substituting into (5.19), we obtain

$$F(\theta) = \frac{\cos(\tfrac{1}{2}\pi\cos\theta)}{\sin\theta} \quad \text{(half-wave dipole, } l = \tfrac{1}{2}\lambda_0) \tag{5.21}$$

The electric field will be a maximum for $\theta = 90°$ (broadside to the antenna). For this case, $F(90°) = 1$, and the maximum electric field for the half-wave dipole becomes

$$|\hat{E}|_{\max} = 60\frac{|\hat{I}_m|}{r} \quad (\theta = 90°) \tag{5.22}$$

The field is directed in the θ direction and is independent of ϕ, which makes sense from symmetry considerations.

Now let us compute the average power radiated by the dipole. Once again, integration of the Poynting vector over a sphere of radius r gives the total radiated power [1]:

$$P_{\text{rad}} = 73|\hat{I}_{\text{in,rms}}| \quad (\text{in W}) \quad (\text{half-wave dipole}) \qquad (5.23)$$

Thus, if we know the rms value of the *input current* at the terminals of a half-wave dipole, we may find the total average power radiated by multiplying the square of the rms current by 73 Ω. This suggests that we define a radiation resistance of the half-wave dipole as

$$R_{\text{rad}} = 73 \, \Omega \quad (\text{half-wave dipole}) \qquad (5.24)$$

There is one important difference between the dipole and the monopole. Although the field patterns are the same, the monopole radiates only half the power of the dipole: the power radiated out of the half-sphere above the ground plane. Thus the radiation resistance for the monopole is half that of the corresponding dipole. In particular, for a quarter-wave monopole of length $h = \frac{1}{4}\lambda_0$ (which corresponds to a half-wave dipole), we have

$$R_{\text{rad}} = 36.5 \, \Omega \quad (\text{quarter-wave monopole}) \qquad (5.25)$$

Up to this point, we have not considered the total input impedance $\hat{Z}_{\text{in}}$ seen at the terminals of the dipole or monopole antenna. The input impedance will, in general, have a real and an imaginary part as

$$\hat{Z}_{\text{in}} = R_{\text{in}} + jX_{\text{in}} \qquad (5.26)$$

The input resistance will consist of the sum of the radiation resistance and the resistance of the imperfect wires used to construct the dipole, so that

$$\hat{Z}_{\text{in}} = R_{\text{loss}} + R_{\text{rad}} + jX_{\text{in}} \qquad (5.27)$$

Figure 5.5 shows the radiation resistance and reactance referred to the base of a monopole antenna for various lengths of the antenna [3]. Figure 5.5 can be used to give the input impedance for a dipole by doubling the values given in the figure. The input reactance for a half-wave dipole (quarter-wave monopole) is $X_{\text{in}} = 42.5 \, \Omega$ ($X_{\text{in}} = 21.25 \, \Omega$). Note that for monopoles that are shorter than one-quarter wavelength (or dipoles shorter than one-half wavelength) the radiation resistance becomes much smaller and the reactive part becomes negative, symbolizing a capacitive reactance. Thus monopoles that are shorter than one-quarter wavelength appear at their input as a small resistance in series with a capacitance, which is, intuitively, a sensible result. Also observe in Fig. 5.5 that the reactive part of the input impedance is zero for a monopole that is slightly shorter than a quarter wavelength. Having a zero reactive part is

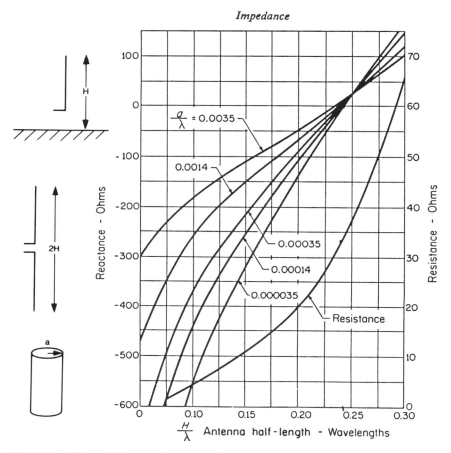

FIGURE 5.5 The radiation resistance and reactance of a dipole antenna as a function of length and wire radius [3].

obviously desirable from the standpoint of maximizing the power that is delivered from a source that has a real source resistance (such as 50 Ω) to the antenna and subsequently radiated. This is why monopoles are cut to lengths of slightly shorter than a quarter wavelength. If the physical length of a quarter-wave monopole is excessive for the intended installation, it can be shortened, but that introduces a large capacitive reactance to the input impedance, which necessitates a larger value of source excitation voltage to produce the same level of radiated power (dissipated in R_{rad}). In order to overcome this problem, short antennas have "loading coils" or inductors inserted in series with their input to cancel this capacitive reactance and increase the radiated power. This is sometimes referred to as "tuning the antenna."

The important point to realize here is that, knowing the input impedance of the antenna, we can compute the total average power radiated by the antenna by computing the average power dissipated in R_{rad}. For example, consider the half-wave dipole driven by a 100 V rms, 150 MHz, 50 Ω source as shown in Fig. 5.6(a). Replacing the antenna with its equivalent circuit at its input terminals gives the circuit shown in Fig. 5.6(b). The input current to the antenna is

$$\hat{I}_{ant} = \frac{\hat{V}_S}{R_{loss} + R_{rad} + jX}$$

Assume the wires are #20 AWG solid copper. The wires have radii much larger than a skin depth at the operating frequency of 150 MHz ($\delta = 5.4 \times 10^{-6}$ m),

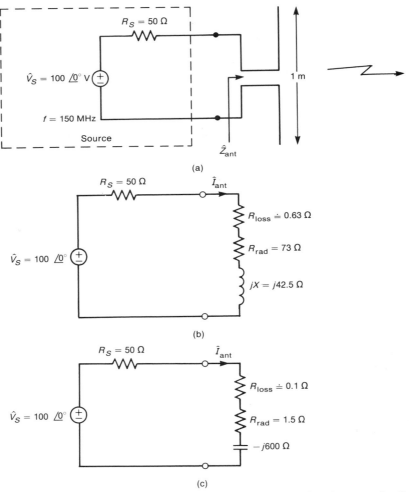

FIGURE 5.6   An example illustrating the computation of the radiated power of a dipole antenna: (b) equivalent circuit of a half-wave dipole; (c) equivalent circuit of a $\frac{1}{8}\lambda_o$ dipole.

so the high-frequency approximation for wire resistance developed in Chapter 4 can be used to compute R_{loss} as

$$r_{\text{wire}} = \frac{1}{2\pi r_w \sigma \delta}$$

$$= 1.25 \ \Omega/\text{m}$$

Using this result, the net ohmic resistance of the wires used to construct the dipole can be obtained as [1]

$$R_{\text{loss}} = r_{\text{wire}} \tfrac{1}{2} l$$

$$= 0.63 \ \Omega$$

Since the dipole is a half-wave dipole, its impedance is $(73 + j42.5) \ \Omega$ (see Fig. 5.5), so that

$$\hat{Z}_{\text{ant}} = R_{\text{loss}} + R_{\text{rad}} + jX_{\text{in}}$$

$$= (0.63 + 73 + j42.5) \ \Omega$$

Thus the current at the antenna input terminals is

$$\hat{I}_{\text{ant}} = \frac{100 \ \underline{/0^\circ}}{50 + 73.63 + j42.5}$$

$$= 0.765 \ \underline{/-18.97^\circ} \ \text{A}$$

The total average power dissipated in antenna losses is

$$P_{\text{loss}} = \tfrac{1}{2} |\hat{I}_{\text{ant}}|^2 R_{\text{loss}}$$

$$= 184 \ \text{mW}$$

The total average power radiated is

$$P_{\text{rad}} = \tfrac{1}{2} |\hat{I}_{\text{ant}}|^2 R_{\text{rad}}$$

$$= 21.36 \ \text{W}$$

In order to illustrate the effect of short antennas on radiated power, consider the previous problem where the dipole antenna is shortened to $\tfrac{1}{8}\lambda_0$ in total length. From Fig. 5.5, $R_{\text{rad}} \cong 1.5 \ \Omega$ and $X_{\text{in}} \cong -600 \ \Omega$. The equivalent circuit is shown in Fig. 5.6(c). The current at the input to the antenna is

$$\hat{I}_{\text{ant}} = \frac{100 \ \underline{/0^\circ}}{50 + 0.16 + 1.5 - j600}$$

$$= 0.166 \ \underline{/85.1^\circ} \ \text{A}$$

Thus the radiated power is

$$P_{\text{av,rad}} = \tfrac{1}{2}|\hat{I}_{\text{ant}}|^2 R_{\text{rad}}$$

$$= 20.7 \text{ mW}$$

Thus the reduced radiation resistance along with the large increase in reactive part in the input impedance caused by shortening the length of the dipole has significantly reduced the radiated power of the antenna. If an inductor having an inductance of 0.637 μH giving a reactance of $+j600$ is inserted in series with this antenna, the radiated power is increased to 2.81 W! This illustrates the extreme effect of the reactive part of the input impedance.

5.3 ANTENNA ARRAYS

The radiation characteristics of the Hertzian dipole, the magnetic dipole, the long dipole, and the monopole are evidently omnidirectional in any plane perpendicular to the antenna axis, since all fields are independent of ϕ. This characteristic follows from the symmetry of these structures about the z axis. From the standpoint of communication, we may wish to focus the radiated signal since any of the radiated power that is not transmitted in the direction of the receiver is wasted. On the other hand, from the standpoint of EMC, we may be interested in directing the radiated signal away from another receiver in order to prevent interference with that receiver. If the transmitting antenna has an omnidirectional pattern, we do not have these options. In this section we will investigate how to use two or more omnidirectional antennas to produce maxima and/or nulls in the resulting pattern. This results from phasing the currents to the antennas and separating them sufficiently such that the combined fields will add constructively or destructively to produce these resulting maxima or nulls. This result, although applied to the emission patterns of communications antennas, has direct application in the radiated emissions of products, since it illustrates how multiple emissions may combine. In addition, we will use the simple results obtained here to obtain simple models for predicting the radiated emissions from wires and PCB lands in Chapter 8.

Consider two omnidirectional antennas such as half-wave dipoles in free space or quarter-wave monopoles above ground, as shown in Fig. 5.7(a). The current elements lie on the y axis and are directed in the z direction. They are separated by a distance d and are equally spaced about the origin. Assuming the field point P is in the far field of the antennas, $\theta_1 \cong \theta_2 \cong \theta$, the far fields at point P due to each antenna are of the form

$$\hat{E}_{\theta 1} = \frac{\hat{M}I \,\underline{/\alpha}}{r_1} e^{-j\beta_0 r_1} \tag{5.28a}$$

$$\hat{E}_{\theta 2} = \frac{\hat{M}I \,\underline{/0}}{r_2} e^{-j\beta_0 r_2} \tag{5.28b}$$

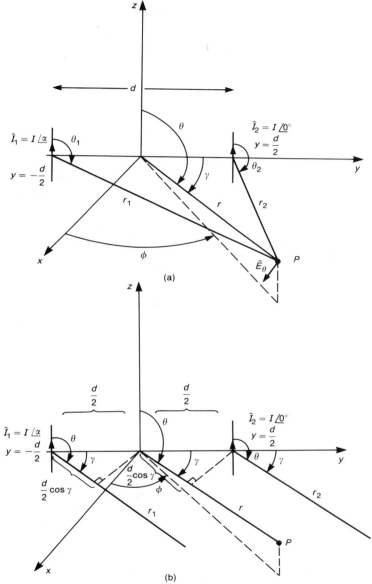

FIGURE 5.7 Computation of the radiated fields of an array of two dipoles: (a) definitions; (b) the far-field approximation.

where $\hat{I}_1 = I \underline{/\alpha}$ and $\hat{I}_2 = I \underline{/0}$, and we assume the currents of the two antennas are equal in magnitude but the current in antenna #1 leads that of antenna #2 by α. The factor $\hat{M}$ depends on the type of antennas used. For Hertzian dipoles $\hat{M} = j\eta_0\beta_0(dl/4\pi)\sin\theta$ (see (5.3a)). For long dipoles $\hat{M} = j60F(\theta)$ (see

(5.18)). The total field at point P is the sum of the fields of the two antennas:

$$\hat{E}_\theta = \hat{E}_{\theta 1} + \hat{E}_{\theta 2} \tag{5.29}$$

$$= \hat{M}I\left(\frac{e^{-j\beta_0 r_1}}{r_1}e^{j\alpha} + \frac{e^{-j\beta_0 r_2}}{r_2}\right)$$

$$= \hat{M}Ie^{j\alpha/2}\left(\frac{e^{-j\beta_0 r_1}e^{j\alpha/2}}{r_1} + \frac{e^{-j\beta_0 r_2}e^{-j\alpha/2}}{r_2}\right)$$

In order to simplify this equation, we observe that, since P is sufficiently far from the origin, $r_1 \cong r_2 \cong r$, where r is the distance from the midpoint of the array to point P. This approximation can be used in the denominator terms of (5.29) but cannot be used in the phase terms for reasons discussed in the previous section. If we draw the radius vectors r_1 and r_2 parallel to the radius vector r as shown in Fig. 5.7(b) then we can obtain a reasonable approximation for the phase terms. To do this, we observe that the path lengths can be written in terms of the angle γ between the radius vector and the y axis as

$$r_1 \cong r + \tfrac{1}{2}d \cos \gamma \tag{5.30a}$$

$$r_2 \cong r - \tfrac{1}{2}d \cos \gamma \tag{5.30b}$$

The angle γ can be obtained as the dot product of the unit vector in the radial direction and the unit vector in the y direction as [1]

$$\cos \gamma = \vec{a}_r \cdot \vec{a}_y \tag{5.31}$$

$$= \sin \theta \sin \phi$$

Substituting gives

$$r_1 \cong r + \tfrac{1}{2}d \sin \theta \sin \phi \tag{5.32a}$$

$$r_2 \cong r - \tfrac{1}{2}d \sin \theta \sin \phi \tag{5.32b}$$

where d is the separation between the two antennas. Substituting (5.32) into (5.29) gives

$$\hat{E}_\theta = \frac{\hat{M}I}{r}e^{j\alpha/2}e^{-j\beta_0 r}\left(e^{j(\beta_0(d/2)\sin\theta\sin\phi - \alpha/2)}\right. \tag{5.33}$$

$$\left. + e^{-j(\beta_0(d/2)\sin\theta\sin\phi - \alpha/2)}\right)$$

$$= 2e^{j\alpha/2}\underbrace{\left[\hat{M}\frac{Ie^{-j\beta_0 r}}{r}\right]}_{\substack{\text{pattern of}\\\text{individual}\\\text{elements}}} \times \underbrace{\cos\left(\frac{\pi d}{\lambda_0}\sin\theta\sin\phi - \frac{\alpha}{2}\right)}_{F_{\text{array}}(\theta, \phi)}$$

We have substituted $\beta_0 = 2\pi/\lambda_0$ and $\cos\psi = \frac{1}{2}(e^{j\psi} + e^{-j\psi})$. Observe that the resultant field is the *product* of the pattern of the individual (identical) antenna elements and the array factor $F_{\text{array}}(\theta, \phi)$, which depends only on the antenna spacing in electrical dimensions, d/λ_0, and phasing of the currents, α. This is referred to as *the principle of pattern multiplication.*

The antenna spacing and/or the relative phase of the currents can be chosen to give maxima/minima in the radiation pattern of the array. For example,

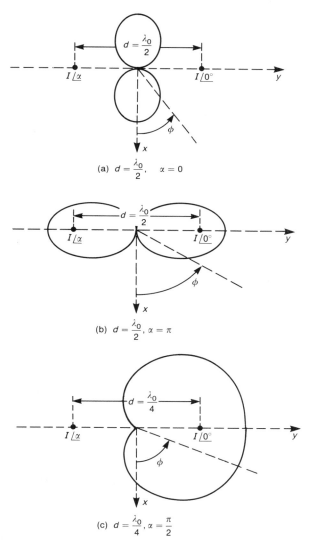

(a) $d = \dfrac{\lambda_0}{2}$, $\alpha = 0$

(b) $d = \dfrac{\lambda_0}{2}$, $\alpha = \pi$

(c) $d = \dfrac{\lambda_0}{4}$, $\alpha = \dfrac{\pi}{2}$

FIGURE 5.8 Patterns of a two-element array: (a) $d = \frac{1}{2}\lambda_0$, $\alpha = 0°$; (b) $d = \frac{1}{2}\lambda_0$, $\alpha = 180°$; (c) $d = \frac{1}{4}\lambda_0$, $\alpha = 90°$.

suppose the spacing between the antennas is one-half wavelength and $\alpha = 0°$. The array factor in the plane perpendicular to the array, $\theta = 90°$, is

$$F_{\text{array}}(\theta = 90°, \phi) = \cos(\tfrac{1}{2}\pi \sin \phi)$$

This pattern is plotted in Fig. 5.8(a). We have produced a pattern that has nulls at $\phi = 90°$ and $\phi = 270°$. As another example, suppose the antenna spacing is again one-half wavelength but $\alpha = 180°$. The array factor is

$$F_{\text{array}}(\theta = 90°, \phi) = \cos(\tfrac{1}{2}\pi \sin \phi - \tfrac{1}{2}\pi)$$

The pattern is plotted in Fig. 5.8(b). This has produced nulls at $\phi = 0°$ and $\phi = 180°$. As a final example, suppose the antennas are spaced one-quarter wavelength and $\alpha = 90°$. The array factor is

$$F_{\text{array}}(\theta = 90°, \phi) = \cos(\tfrac{1}{4}\pi \sin \phi - \tfrac{1}{4}\pi)$$

The pattern is plotted in Fig. 5.8(c), and we have produced a null at $\phi = 270°$ and a maximum at $\phi = 90°$.

5.4   CHARACTERIZATION OF ANTENNAS

The antennas we have considered previously are quite simple to analyze. The analysis of other antennas to determine their total radiated power and, more importantly, the shapes of their radiated emission patterns is not so simple. These more complicated antennas are more commonly characterized by measured parameters such as directivity and gain, effective aperture, and/or antenna factor. The purpose of this section is to investigate these criteria.

5.4.1   Directivity and Gain

The *directive gain* of an antenna, $D(\theta, \phi)$, is a measure of the concentration of the radiated power in a particular θ, ϕ direction at a fixed distance r away from the antenna. For the elemental dipoles, the long dipole, and the monopole, we noted that the radiated power is a maximum for $\theta = 90°$ and is zero for $\theta = 0°$ and $\theta = 180°$. To obtain a more quantitative measure of this concentration of radiated power, we will define the radiation intensity $U(\theta, \phi)$.

We found that the far-field, radiated average-power densities for the Hertzian dipole, the magnetic dipole, the long dipole, and the monopole are of the form

$$\vec{S}_{\text{av}} = \frac{|\vec{E}_{\text{far field}}|^2}{2\eta_0} \vec{a}_r \tag{5.34}$$

$$= \frac{E_0^2}{2\eta_0 r^2} \vec{a}_r$$

where E_0 depends on θ, the antenna type and the antenna current. To obtain a power pattern relationship that is independent of distance from the antenna, we multiply (5.34) by r^2 and define the resulting quantity to be the *radiation intensity*, that is,

$$U(\theta, \phi) = r^2 S_{av} \tag{5.35}$$

The radiation intensity will be a function of θ and ϕ but will be independent of distance from the antenna. The total average power radiated will be

$$P_{rad} = \oint \vec{S}_{av} \cdot \vec{ds} \tag{5.36}$$

$$= \int_{\theta=0}^{\pi} \int_{\phi=0}^{2\pi} U(\theta, \phi) \sin \theta \, d\phi \, d\theta$$

$$= \oint_{S} U(\theta, \phi) \, d\Omega$$

The quantity $d\Omega = \sin \theta \, d\phi \, d\theta$ is an element of *solid angle* Ω, and the unit of solid angle is the steradian (sr). The units of U are therefore watts per steradian (W/sr). Note that for $U = 1$, (5.36) integrates to 4π. *The total radiated power is therefore the integral of the radiation intensity over a solid angle of 4π sr.* Note also that the average radiation intensity is the total radiated power divided by 4π sr:

$$U_{av} = \frac{P_{rad}}{4\pi} \tag{5.37}$$

The radiation intensity for more complicated antennas is similarly defined. *The directive gain of an antenna in a particular direction, $D(\theta, \phi)$, is the ratio of the radiation intensity in that direction to the average radiation intensity:*

$$D(\theta, \phi) = \frac{U(\theta, \phi)}{U_{av}} \tag{5.38}$$

$$= \frac{4\pi U(\theta, \phi)}{P_{rad}}$$

The directivity of the antenna is the directive gain in the direction that yields a maximum:

$$D_{max} = \frac{U_{max}}{U_{av}} \tag{5.39}$$

As an example, let us compute the directive gain and directivity of a Hertzian dipole. The radiation intensity is found from (5.5) and (5.35),

$$U(\theta, \phi) = r^2 S_{av}$$

$$= 15\pi \left(\frac{dl}{\lambda_0}\right)^2 |\hat{I}|^2 \sin^2 \theta$$

and the radiated power is given by (5.6),

$$P_{rad} = 40\pi^2 \left(\frac{dl}{\lambda_0}\right)^2 |\hat{I}|^2$$

Thus the directive gain is

$$D(\theta, \phi) = \frac{4\pi U(\theta, \phi)}{P_{rad}}$$

$$= 1.5 \sin^2 \theta$$

The directivity is therefore the directive gain at $\theta = \frac{1}{2}\pi$:

$$D_{max} = 1.5$$

For the half-wave dipole we obtain

$$D(\theta, \phi) = \frac{\eta_0}{\pi R_{rad}} F^2(\theta)$$

$$= 1.64 \, F^2(\theta)$$

where $F(\theta)$ is given by (5.21) and $R_{rad} = 73 \, \Omega$ and

$$D_{max} = 1.64$$

which occurs for $\theta = \frac{1}{2}\pi$.

The directive gain $D(\theta, \phi)$ of an antenna is simply a function of the shape of the antenna pattern. The *power gain* $G(\theta, \phi)$, on the other hand, takes into account the losses of the antenna. Suppose that a total power P_{app} is applied to the antenna and only P_{rad} is radiated. The difference is consumed in ohmic losses of the antenna as well as in other inherent losses such as those in an imperfect ground for monopole antennas. If we define an *efficiency factor e* as

$$e = \frac{P_{rad}}{P_{app}} \qquad (5.40)$$

then the power gain is related to the directive gain as

$$G(\theta, \phi) = eD(\theta, \phi) \tag{5.41}$$

where we have defined the power gain as

$$G(\theta, \phi) = \frac{4\pi U(\theta, \phi)}{P_{app}} \tag{5.42}$$

For most antennas, the efficiency is nearly 100%, and thus the power gain and directive gain are nearly equal.

We also need to discuss the concept of an isotropic point source. An *isotropic point source* is a fictitious lossless antenna that radiates power equally in all directions. Since this antenna is lossless, its directive and power gains are equal. For an isotropic point source radiating or transmitting a total power P_T, the power density at some distance d away is the total radiated power divided by the area of a sphere of radius d:

$$\vec{S}_{av} = \frac{P_T}{4\pi d^2} \vec{a}_r \tag{5.43}$$

We can calculate the electric and magnetic fields for the isotropic point source from the realization that the waves resemble (locally) uniform plane waves so that

$$\vec{S}_{av} = \frac{|\vec{E}|^2}{2\eta_0} \quad (\text{in W/m}^2) \tag{5.44}$$

Combining (5.43) and (5.44) gives

$$|\vec{E}| = \frac{\sqrt{60 P_T}}{d} \vec{a}_\theta \quad (\text{in V/m}) \tag{5.45}$$

where we have substituted $\eta_0 = 120\pi \ \Omega$.

The isotropic point source, although quite idealistic, is useful as a standard or reference antenna to which we refer many of our calculations. For example, since the isotropic point source is lossless, the directive and power gains are equal, and both will be designated by G_0. The gain becomes

$$G_0(\theta, \phi) = \frac{4\pi U_0(\theta, \phi)}{P_T} \tag{5.46}$$

$$= 1$$

Therefore the directive gain and power gain of other antennas may be thought of as being determined with respect to an isotropic point source. In certain other cases, the gain of an antenna may be referred to the gain of a half-wave dipole. When discussing gain, one must be careful to determine the reference antenna.

Quite often the gain (directive or power) of an antenna is given in decibels, where

$$G_{dB} = 10 \log_{10} G \qquad (5.47)$$

For example, the Hertzian dipole has a directivity of 1.76 dB and the isotropic point source has a directivity of 0 dB. The half-wave dipole has a maximum gain of 2.15 dB. Equivalently, we say that the gain of an antenna is the gain over (or with respect to) an isotropic antenna:

$$G_{dB} = 10 \log_{10}\left(\frac{G}{G_0}\right) \qquad (5.48)$$

We are frequently interested in the coupling between two antennas, one of which is used as a transmitter and the other as a receiver. An important principle in this problem is that of *reciprocity* [1, 3–6]. Reciprocity provides that the source and receiver can be interchanged without affecting the results so long as the impedances of the source and receiver are the same. Several additional properties can be proven. The impedance seen looking into an antenna terminals when it is used for transmission is the same as the Thévenin source impedance seen looking back into its terminals when it is used for reception. In addition, the transmission pattern of the antenna is the same as its reception pattern.

5.4.2 Effective Aperture

An additional useful concept is that of an antenna's effective aperture. The *effective aperture* of an antenna is related to the ability of the antenna to extract energy from a passing wave. *The effective aperture of an antenna, A_e, is the ratio of the power received (in its load impedance), P_R, to the power density of the incident wave, S_{av}, when the polarization of the incident wave and the polarization of the receiving antenna are matched:*

$$A_e = \frac{P_R}{S_{av}} \quad (\text{in } m^2) \qquad (5.49)$$

The *maximum effective aperture* A_{em} is the ratio in (5.49) when the load impedance is the conjugate of the antenna impedance, which means that maximum power transfer to the load takes place. For a linearly polarized incident wave and a receiving antenna such as a dipole or monopole that produces linearly polarized waves when it is used for transmission, the requirement for matched polarization essentially means that the antenna is oriented with respect to the incident wave to produce the maximum response; that is, the electric field vector of the incident wave is parallel to the electric field vector that would be produced by this antenna when it is used for transmission.

For example, let us compute the maximum effective aperture of a Hertzian dipole antenna. If the dipole is terminated in an impedanze $\hat{Z}_L$, we assume that $\hat{Z}_L = R_{rad} - jX$ where the input impedance to the dipole is $\hat{Z}_{in} = R_{rad} + jX$ and the dipole is assumed to be lossless. Suppose that the incident wave is arriving at an angle θ, with the electric field vector in the θ direction as shown in Fig. 5.9. The open-circuit voltage produced at the terminals of the antenna is

$$|\hat{V}_{oc}| = |\hat{E}_\theta|\, dl \sin \theta \qquad (5.50)$$

The power density in the incident wave is

$$S_{av} = \frac{1}{2} \frac{|\hat{E}_\theta|^2}{\eta_0} \qquad (5.51)$$

Since the load is matched for maximum power transfer, the power received is

$$P_R = \frac{|\hat{V}_{oc}|^2}{8R_{rad}} \qquad (5.52)$$

$$= \frac{|\hat{E}_\theta|^2\, dl^2 \sin^2 \theta}{8R_{rad}}$$

Substituting the value for R_{rad} given in (5.7) gives

$$P_R = \frac{|\hat{E}_\theta|^2 \lambda_0^2}{640\pi^2} \sin^2 \theta \qquad (5.53)$$

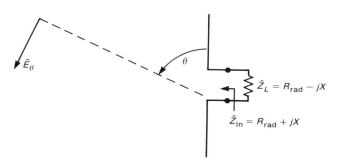

FIGURE 5.9 Illustration of the computation of the maximum effective aperture A_{em} of a linear antenna.

Thus the maximum effective aperture is

$$A_{em}(\theta, \phi) = \frac{P_R}{S_{av}} \tag{5.54}$$

$$= 1.5 \sin^2 \theta \frac{\lambda_0^2}{4\pi}$$

$$= \frac{\lambda_0^2}{4\pi} D(\theta, \phi)$$

where we have substituted the directive gain of the Hertzian dipole

$$D(\theta, \phi) = 1.5 \sin^2 \theta \tag{5.55}$$

and θ is the direction of the incoming wave. Observe that the maximum effective aperature of an antenna is not necessarily related to its "physical aperture."

It can be shown that the result in (5.54) is a general result valid for more general antennas; that is, the maximum effective aperture of an antenna used for reception is related to the directive gain *in the direction of the incoming wave* of that antenna when it is used for transmission as [1, 3–6]:

$$G(\theta, \phi) = \frac{4\pi}{\lambda_0^2} A_{em}(\theta, \phi) \tag{5.56}$$

The direction for A_{em} (the direction of the incoming incident wave with respect to the receiving antenna) is the direction of the gain G (the gain of the antenna in this direction when it is used for transmission). We have interchanged directive gain D and power gain G on the assumption that the antennas are lossless.

5.4.3 Antenna Factor

The above properties of antennas are more commonly used in the area of the use of antennas for communication such as signal transmission and radar applications. In the area of their use in EMC a more common way of characterizing the reception properties of an antenna is with the notion of its *antenna factor*. Consider a dipole antenna that is used to measure the electric field of an incident, linearly polarized uniform plane wave as shown in Fig. 5.10(a). A receiver such as a spectrum analyzer is attached to the terminals of this measurement antenna. The voltage measured by this instrument is denoted as $\hat{V}_{rec}$. It is desired to relate this received voltage to the incident electric field. This is done with the antenna's *antenna factor*, which is defined as *the ratio of the incident electric field at the surface of the measurement antenna to the received*

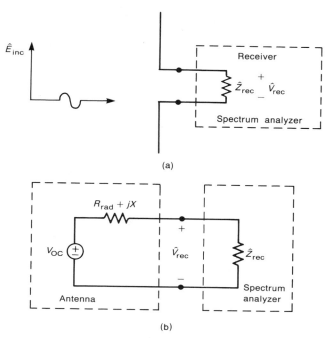

FIGURE 5.10   The antenna factor AF: (a) general circuit; (b) equivalent circuit.

voltage at the antenna terminals:

$$\text{AF} = \frac{\text{V/m in incident wave}}{\text{V received}} \quad (\text{in } 1/\text{m}) \qquad (5.57)$$

$$= \frac{|\hat{E}_{\text{inc}}|}{|\hat{V}_{\text{rec}}|}$$

This is frequently expressed in dB as

$$\text{AF}_{\text{dB}} = \text{dB}\mu\text{V/m (incident field)} - \text{dB}\mu\text{V (received voltage)} \quad (5.58a)$$

or

$$\text{dB}\mu\text{V/m (incident field)} = \text{dB}\mu\text{V (received voltage)} + \text{AF}_{\text{dB}} \quad (5.58b)$$

Note that the units of the antenna factor are $1/\text{m}$. The units are frequently ignored, and the antenna factor is stated in dB. The antenna factor is usually furnished by the manufacturer of the antenna as measured data at various frequencies in the range of intended use of the measurement antenna. A typical such plot provided by the manufacturer of a biconical measurement antenna

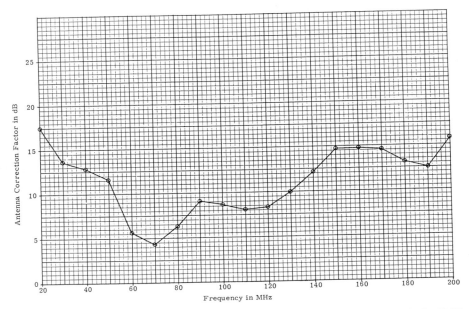

FIGURE 5.11 The antenna factor versus frequency for a typical biconical EMC measurement antenna (courtesy the Eaton Corporation).

is shown in Fig. 5.11. A known field is provided by some standard antenna at a calibrated test site such as the National Institute of Standards and Technology (NIST) in Boulder, Colorado in the US (formerly known as the National Bureau of Standards or NBS). The ratio of the known value of the incident field to the measured voltage at the terminals of the antenna in dB according to (5.57) or (5.58) is plotted for the antenna versus frequency. The reciprocal of the antenna factor is referred to as the *antenna effective height* h_e [3–6].

There are several important implicit assumptions in these measured antenna factor data. *If any of these implicit assumptions are not adhered to in the course of using this antenna for measurement then the measured data are invalid.* The first important assumption is that *the incident field is polarized for maximum response of the antenna.* For a dipole or other wire-type antenna this means that *the response will be the component of the incident field that is parallel to the antenna axis.* Ordinarily this is what is desired, since the antenna will be typically used to measure vertical and horizontal fields in testing for compliance to the radiated emission regulatory limits. The second important implicit assumption has to do with *the input impedance of the receiver that is used not only to make the measurement but also to calibrate the antenna.* The most common impedance is the typical input impedance to virtually all spectrum analyzers, and that is 50 Ω. Nevertheless, the antenna manufacturer should explicitly state what termination impedance was used in the calibration. Note that *this does not assume that the receiver is matched to the antenna,* and usually it will not be.

However, from the standpoint of using the antenna factor calibration chart for that antenna it is only important to use a termination impedance that is the same as was used to calibrate the antenna.

On the other hand, suppose we wish to *calculate* the antenna factor of an ideal antenna such as a dipole from the field equations, maximum effective aperture, etc. for that antenna. Since the spectrum analyzer input impedance is $\hat{Z}_{rec} = (50 + j0)\,\Omega$ and is therefore not matched, we must use the equivalent circuit of Fig. 5.10(b) to obtain this. First compute $\hat{V}_{rec,matched}$ *assuming a matched load*, $\hat{Z}_{rec} = R_{rad} - jX$, using the results in the previous sections. Then use this result to obtain the open-circuit voltage $\hat{V}_{OC} = 2\hat{V}_{rec,matched}$. Then use the equivalent circuit in Fig. 5.10(b) to compute the actual received voltage $\hat{V}_{rec}$, and from that the antenna factor.

As an example of the use of measured data to determine the antenna factor, consider the calibration of a measurement antenna shown in Fig. 5.12. A known, incident, linearly polarized, uniform plane wave is incident on the antenna, and the electric field *at the position of the antenna in the absence of the antenna* is 60 dBμV/m. A 30 foot length of RG-58U coaxial cable is used to connect the antenna to a 50 Ω spectrum analyzer. The spectrum analyzer measures 40 dBμV. Since the antenna factor relates the incident electric field to the *voltage at the base of the antenna*, we must relate the spectrum analyzer reading to the voltage at the base of the antenna. The coaxial cable has 4.5 dB/100 feet loss at the frequency of the incident wave, 100 MHz. Thus the cable loss of 1.35 dB must be *added* to the spectrum analyzer reading to give the voltage at the antenna terminals of 41.35 dBμV. Therefore the antenna factor is

$$AF_{dB} = 60\ dB\mu V/m - 41.35\ dB\mu V$$

$$= 18.65\ dB$$

It is a simple matter to convert the spectrum analyzer readings to the value of incident field; *add the antenna factor in dB to the spectrum analyzer reading*

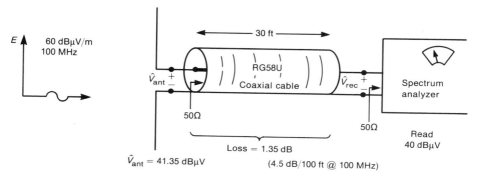

FIGURE 5.12 An example illustrating the use of the antenna factor to compute the received voltage.

in dBμV and add the connection loss in dB to give the incident electric field in dBμV/m:

$$E\,(\text{dB}\mu\text{V/m}) = \text{AF}\,(\text{dB}) + V_{SA}\,(\text{dB}\mu\text{V}) + \text{Cable Loss}\,(\text{dB}) \qquad (5.59)$$

Observe that the connection cable loss must be added and not subtracted, since the antenna factor is with respect to the base of the antenna and does not include any connection cable loss (unless explicitly stated by the antenna manufacturer).

5.4.4   Effects of Balancing and Baluns

The ideal antennas that we are considering are inherently *balanced structures*. There are numerous definitions of this concept of *balanced structure*. Generally, but not always, these seemingly different definitions lead to the same result. For example, consider the long dipole antenna shown in Fig. 5.3(a). In the analysis of this antenna we assumed that the current $\hat{I}(z_1)$ at a point z_1 on the upper arm is the same in magnitude as the current at the corresponding position on the lower arm, $-z_1$ (a point that is the same distance from the feed point as the point on the upper arm). From this standpoint of *symmetry of the antenna currents*, the antenna is *inherently* a balanced structure. This also inherently assumes that the current entering one terminal of the antenna is equal but opposite to the current entering the other terminal. Nearby metallic obstacles such as ground planes can upset this balance, causing the pattern to deviate substantially from the ideal pattern that was obtained from the assumption of balanced currents on the arms of the antenna [7].

Other factors can upset the balance of the currents on the antenna structure. The most common type of feedline that is used to supply signals to antennas is the coaxial cable. Under ideal conditions, the current returns to its source on the *interior* of the overall shield. If this type of cable is attached to an inherently balanced structure such as a dipole antenna, some of the current may flow *on the outside of the shield*. This current will radiate, whereas the current going down the interior wire and returning on the inside of the shield will not. The amount of current that flows on the outside of the shield depends on "the impedance to ground" between the shield exterior and the ground, $\hat{Z}_G$, along with the excitation of the shield exterior (unintentional excitation).

The common way of preventing unbalance due to a coaxial feed cable is the use of a *balun*, which is an acronym for BALanced to UNbalanced, referring to the transition from an unbalanced coaxial cable to a balanced antenna. The balun is inserted at the input to the antenna, as shown in Fig. 5.13(a). In the case of the coaxial feed cable, the intent of the balun is to increase the impedance between the outside of the shield and ground. A common form is the "bazooka balun" shown in Fig. 5.13(b). A quarter-wavelength section of shield is inserted over the shield of the original cable, and these are shorted together a quarter-wavelength from the feed point. A quarter-wavelength, short-circuited transmission line is formed between the outer coax and the inner coax. We found in the previous chapter that a short-circuited, quarter-wavelength

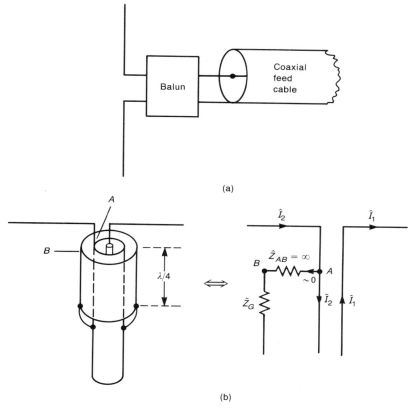

(a)

(b)

FIGURE 5.13 Use of baluns to reduce common-mode currents on antennas: (a) use of a discrete balun to connect an unbalanced coaxial cable to a balanced antenna; (b) the bazooka balun.

transmission line appears at its input as an open circuit. Therefore the impedance between points A and B is very large (theoretically infinite), so that the impedance between the inner shield and ground is infinite also.

There are other methods of producing balanced feeds. They also seek to block this current on the outside of the shield. One obvious method is to add ferrite beads (discussed in Chapter 6) around the feed coax as shown in Fig. 5.14(a). The beads act as common-mode chokes [8]. Another way of accomplishing this is to custom wind a ferrite toroid as shown in Fig. 5.14(b) [5]. The equivalent circuit is also shown. Ferrite baluns tend to provide "wideband balancing" over bandwidths of as much as 3:1, that is, the ratio of the upper useable frequency to the lowest one is a factor of 3. The bazooka balun works only at the frequency where its length is one-quarter wavelength, so that its bandwidth is limited.

Further discussions of baluns are given in [3–7]. Balancing is a critical factor in the accurate measurement of radiated emissions. If the antenna-feed

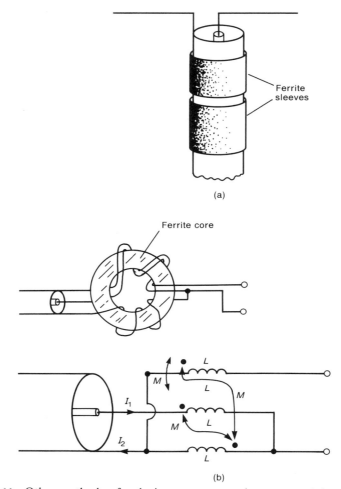

(a)

(b)

FIGURE 5.14   Other methods of reducing common-mode currents: (a) use of ferrite sleeves; (b) use of a ferrite toroid.

line combination is not balanced, the measured data may appear to comply with the regulatory limits when in fact they may not because of the pattern distortion caused by the unbalance. Broadband baluns are obviously very desirable in measurements for compliance verification, since we do not wish to have to "retune" the balun for each frequency in the test.

5.4.5   Impedance Matching and the Use of Pads

The ability to make *swept-frequency measurements* in the course of measuring the radiated emissions of a product for the verification of compliance with the regulatory limits is obviously important from the standpoint of rapid gathering of the test data. The FCC prefers the use of the halfwave dipole. In order to use this antenna to measure radiated emissions over the frequency range of the limit, 30 MHz–1 GHz, we must change the length of the dipole at each measurement frequency so that the dipole length is one-half wavelength at that frequency. A more efficient method is to use *broadband measurement antennas* such as the biconical and log-periodic antennas discussed in Section 5.7. These antennas are calibrated as discussed above, and the calibration data usually consist of a plot of the antenna factor for the antenna versus frequency. As pointed out above, the antenna factor data assume that not only is the antenna balanced (measurement antennas are usually provided with built-in baluns) but the *termination impedance seen at the antenna terminals is 50 Ω.* Although the spectrum analyzer or receiver used to measure the antenna terminal voltage usually has an input impedance of 50 Ω, it is almost always necessary to connect it to the antenna with a connection cable such as a coaxial cable, as shown in Fig. 5.12. If the characteristic impedance of the coaxial cable is also 50 Ω (as is usually the case) then the impedance seen looking into the coaxial cable with the receiver attached is also 50 Ω, since this cable is matched at the receiver. Thus the antenna sees an impedance of 50 Ω for all frequencies, as was assumed in the course of its calibration. If for some reason the termination impedance at the receiver end of the cable is not 50 Ω, then the impedance that the antenna terminals sees looking into the cable is not 50 Ω, because the cable is not matched, and, moreover, the impedance seen looking into the cable will vary with frequency. *The input impedance to a cable will be independent of frequency and equal to its characteristic resistance Z_C (normally 50 Ω) only if the cable is terminated in a matched load impedance, $\hat{Z}_L = Z_C$.* A way of providing a matched termination for other values of termination impedances is with the use of a *pad.*

A pad is simply a resistive network whose input impedance remains fairly constant regardless of its termination impedance. A typical topology of a pad is the "pi" structure shown in Fig. 5.15(a) with reference to its resemblance to the symbol π. It is also possible to choose other structures such as the "tee" structure. Being resistive circuits, these pads provide matching over wide frequency ranges (they are said to be broadband devices) but they also give an insertion loss. The resistor values and schematic of a 50 Ω, 6 dB pad are shown in Fig. 5.15(b), and a photograph of a commercially available pad is shown in Fig. 5.15(c).

The insertion loss (IL) is defined as the ratio of the power delivered to the load with and without the pad:

$$IL_{dB} = 10 \log_{10}\left(\frac{P_{L,\text{without pad}}}{P_{L,\text{with pad}}}\right) \tag{5.60}$$

$$= 20 \log_{10}\left(\frac{|\hat{V}_L|_{\text{without pad}}}{|\hat{V}_L|_{\text{with pad}}}\right)$$

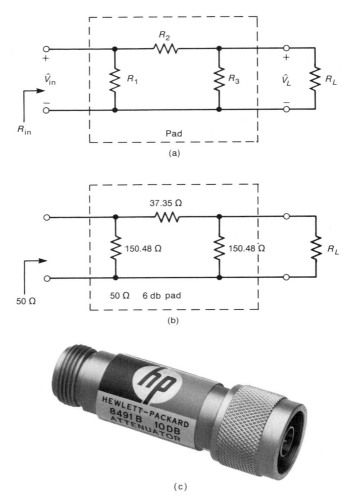

FIGURE 5.15 Use of pads to match transmission lines: (a) a pi pad structure; (b) schematic of a 50 Ω, 6 dB pad; (c) a commercially available pad (courtesy of the Hewlett Packard Corporation).

A rudimentary idea of how the pad may accomplish matching for a wide range of load impedances is the following. If we choose R_3 in Fig. 5.15(a) to be much less than the value of all possible values of load impedance used and choose R_2 to be much greater than R_3 then the input impedance seen looking into the pad is

$$R_{\text{in}} = R_1 \parallel (R_2 + R_3 \parallel R_L) \tag{5.61}$$

$$\cong R_1 \parallel R_2$$

Of course the smaller R_3, the wider the range of load resistances that may be accommodated. The larger R_2, the less important R_3 becomes. However, both options result in increased insertion loss. Typically, *the larger the insertion loss that one can tolerate, the better the matching ability of the pad, i.e., the wider the range of load impedances such that* $R_{in} \cong Z_C$. As an illustration of this point, consider the 50 Ω, 6 dB pad shown in Fig. 5.15(b). Consider the extreme values of termination impedances: open circuit and short circuit. The open-circuit impedance can be calculated to be

$$R_{in,oc} = 150.48 \parallel (37.35 + 150.48)$$
$$= 83.55 \ \Omega$$

The short-circuit input impedance is

$$R_{in,sc} = 150.48 \parallel 37.35$$
$$= 29.92 \ \Omega$$

If this pad is inserted between a 50 Ω coaxial cable and the load, the voltage standing wave ratio (VSWR) on the cable will be less than 1.67 for any load between an open circuit and a short circuit. Usually, an acceptable VSWR is less than 1.2, which will be more closely achieved for realistic loads. A 20 dB, 50 Ω paid will give a smaller range of VSWR at the expense of more loss. For example, the values of resistors for a 20 dB, 50 Ω pad are $R_1 = R_3 = 61.11 \ \Omega$ and $R_2 = 247.50 \ \Omega$. For the short-circuit load the input impedance is 49.01 Ω, with a VSWR of 1.02. For the open-circuit load the input impedance is 51.01 Ω, with a VSWR of 1.02. The insertion loss is obtained from Fig. 5.15(a) [1]:

$$\text{IL} = 20 \log_{10}\left[\left(\frac{R_3 \parallel R_L}{R_2 + R_3 \parallel R_L}\right)^{-1}\right] \tag{5.62}$$

Solving this gives

$$R_1 = R_3 \tag{5.63a}$$
$$= \frac{R_L(1 + X)}{(R_L/Z_C)X - 1}$$
$$R_2 = (R_3 \parallel R_L)(X - 1) \tag{5.63b}$$

where

$$X = 10^{\text{IL}/20} \tag{5.63c}$$

and we substitute $R_L = Z_C$.

5.5   THE FRIIS TRANSMISSION EQUATION

Exact calculation of the coupling between two antennas is usually a formidable problem. For this reason, many practical calculations of antenna coupling are carried out, approximately, with the Friis transmission equation. Consider two antennas in free space, shown in Fig. 5.16. One antenna is transmitting a total power P_T and the other is receiving power P_R in its terminal impedance. The transmitting antenna has a gain of $G_T(\theta_T, \phi_T)$ and an effective aperture $A_{eT}(\theta_T, \phi_T)$ in the direction of transmission, θ_T, ϕ_T. The receiving antenna has a gain and effective aperture of $G_R(\theta_R, \phi_R)$ and $A_{eR}(\theta_R, \phi_R)$ in this direction of transmission, θ_R, ϕ_R. The power density at the receiving antenna is the power density of an isotropic point source multiplied by the gain of the transmitting antenna *in the direction of transmission*:

$$S_{av} = \frac{P_T}{4\pi d^2} G_T(\theta_T, \phi_T) \tag{5.64}$$

The received power is the product of this power density and the effective aperture of the receiving antenna *in the direction of transmission*:

$$P_R = S_{av} A_{eR}(\theta_R, \phi_R) \tag{5.65}$$

Substituting (5.64) into (5.65) gives

$$\frac{P_R}{P_T} = \frac{G_T(\theta_T, \phi_T) A_{eR}(\theta_R, \phi_R)}{4\pi d^2} \tag{5.66}$$

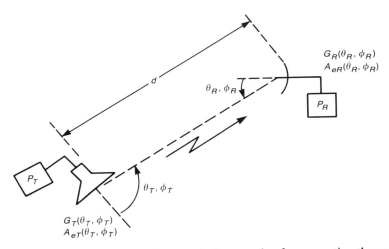

FIGURE 5.16   Illustration of the Friis transmission equation for computing the coupling between two antennas in terms of their gains.

Replacing the effective aperture of the receiving antenna with its gain via (5.56) (assuming a matched load and matched polarization so that the effective apertures are the maximum effective apertures), we obtain the most common version of the *Friis transmission equation*:

$$\frac{P_R}{P_T} = G_T(\theta_T, \phi_T) G_R(\theta_R, \phi_R) \left(\frac{\lambda_0}{4\pi d} \right)^2 \tag{5.67}$$

The electric field intensity of the transmitted field at a distance d from the transmitting antenna can also be computed. The power density in the transmitted wave is that of a uniform plane wave (locally):

$$S_{\text{av}} = \frac{1}{2} \frac{|\hat{E}|^2}{\eta_0} \tag{5.68}$$

Combining this with (5.64) gives

$$|\hat{E}| = \frac{\sqrt{60 P_T G_T(\theta_T, \phi_T)}}{d} \tag{5.69}$$

since $\eta_0 = 120\pi$.

Quite often, in practice, the antenna gains are stated in decibels as is the transmitted power. In decibels, the Friis transmission equation becomes

$$10 \log_{10}\left(\frac{P_R}{P_T}\right) = G_{T,\text{dB}} + G_{R,\text{dB}} - 20 \log_{10} f - 20 \log_{10} d + 147.6 \tag{5.70}$$

There are a number of assumptions inherent in the Friis transmission equation. In order that the relationship between gain and effective aperture given in (5.56) be valid, the receiving antenna must be matched to its load impedance and the polarization of the incoming wave; otherwise, the Friis equation will result in an upper limit on the coupling ("worst case"). We also require that the two antennas be in the *far field* of each other. The far-field criterion is usually taken to be the larger of

$$d_{\text{far field}} > 2D^2/\lambda_0 \quad \text{(surface antennas)}$$

or

$$d_{\text{far field}} > 3\lambda_0 \quad \text{(wire antennas)}$$

where D is the maximum dimension of the antenna [1]. The first criterion is used for "surface-type antennas," whereas the second criterion is used for "wire-type antennas." Inherent in the effective-aperture concept is the assumption that the incoming field resembles a uniform plane wave in the vicinity of the receiving antenna. The transmitted wave in the far field of the transmitting antenna resembles a spherical wave from a point source that only locally resembles a uniform plane wave, as is assumed in all of the previous derivations. The criterion that the two antennas be separated by $2D^2/\lambda_0$ insures that the spherical incoming wave differs in phase from a plane wave at the extremities of the antenna surface by at most $\frac{1}{16}\lambda_0$ [1]. The separation criterion of $3\lambda_0$ insures that "wave impedance" of the incoming wave is approximately that of free space.

As an example, let us calculate the coupling between two half-wave dipole antennas. Assume that the dipoles are separated by a distance of 1000 m, are operated at a frequency of 150 MHz, and are oriented parallel to each other for maximum reception. The transmitting dipole is driven by a 100 V, 50 Ω, source, as shown in Fig. 5.6. The radiated power was calculated in Section 5.2 from the equivalent circuit of Fig. 5.6(b) to be 21.36 W. The gain of a half-wave dipole in the main beam, broadside to the antenna, is 2.15 dB (1.64 absolute). The electric field at the receiving antenna is calculated from the Friis transmission equation given in (5.69) as

$$|\hat{E}| = \frac{\sqrt{60 \times 21.36 \times 1.64}}{1000}$$

$$= 45.85 \text{ mV/m}$$

The value calculated from (5.22) is

$$|\hat{E}|_{\max} = 60\frac{\hat{I}_m}{r} \quad (\theta = 90°)$$

$$= 60\frac{0.765\text{A}}{1000}$$

$$= 45.90 \text{ mV/m}$$

The average power density at the receiving antenna is

$$S_{av} = \frac{1}{2}\frac{|\hat{E}|^2}{\eta_0}$$

$$= 2.794 \ \mu\text{W/m}^2$$

Thus the average power received in a matched load is

$$P_R = S_{av} A_{eR}$$

$$= S_{av}\frac{\lambda_0^2}{4\pi} G_R$$

$$= 2.794 \times 10^{-6} \times \frac{2^2}{4\pi} \times 1.64$$

$$= 1.459 \ \mu\text{W}$$

$$= -28.36 \text{ dBm}$$

The radiated power of 21.36 W is 43.3 dBm. Therefore the ratio of the received and transmitted powers is

$$\frac{P_R}{P_T} = -28.36 \text{ dBm} - 43.3 \text{ dBm}$$

$$= -71.66 \text{ dB}$$

As an alternative, using the Friis transmition equation in (5.70) gives

$$\frac{P_R}{P_T} = 2.15 + 2.15 - 20 \log_{10}(150 \times 10^6) - 20 \log_{10}(1000) + 147.6$$

$$= -71.66 \text{ dB}$$

5.6 EFFECTS OF REFLECTIONS

Radiated electromagnetic fields will be reflected at a conducting plane in order to satisfy the boundary conditions. In this section we will examine this important aspect.

5.6.1 The Method of Images

Consider the point charge Q located a distance h above a perfectly conducting ground plane shown in Fig. 5.17(a). Solving for the fields generated by this charge above the ground plane is a difficult problem unless we replace the ground plane with the *image* of the charge. This image must be such that the electric field distribution in the space above the previous position of the ground plane remains unchanged. Replacing the ground plane with a negative point charge of value Q at a distance h below the ground plane will yield a field distribution above the ground plane that is identical to that before the replacement of the ground plane with the image [1]. This is intuitively clear from the sketch.

Next consider a current element I parallel to and at a distance h above a perfectly conducting ground plane, as shown in Fig. 5.17(b). Since current represents the flow of charge, we can visualize positive charge accumulating at the right end of the current element (in the direction of current flow) and negative charge of equal amount at the left end. This observation allows us to generate the image of the current element by analogy to static charge distributions, as shown in Fig. 5.17(b): the current image is located parallel to and at a distance h below the ground plane and is directed opposite to the original current direction. Similarly, a current element vertical to the ground plane, as shown in Fig. 5.17(c), is replaced with a vertical current image at the same distance below the ground plane and directed in the same direction as the original current element as the static charge analogy shows. A current that

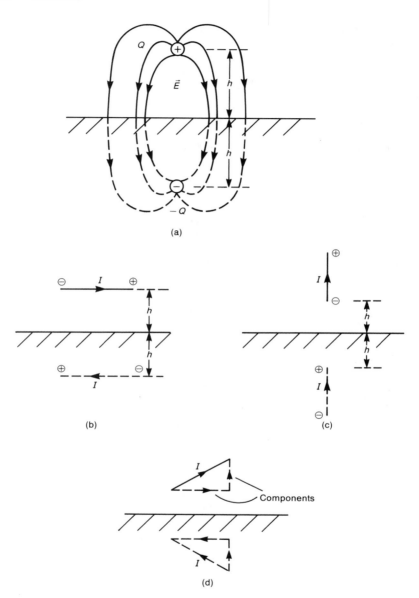

(a)

(b)   (c)

(d)

FIGURE 5.17 Illustration of the technique of images for charges and currents above infinite, perfectly conducting ground planes: (a) the image of a static charge; (b) the image of a current parallel to the plane; (c) the image of a current perpendicular to the plane; (d) decomposition of a current into its horizontal and vertical components in order to determine its image.

is neither horizontally nor vertically directed can be decomposed into its components as shown in Fig. 5.17(d), from which the image components can be constructed as before, resulting in the total image.

5.6.2 Normal Incidence of Uniform Plane Waves on Plane, Material Boundaries

Consider the incidence of a uniform plane wave normal to the plane boundary between two media shown in Fig. 5.18. Intuitively, it is clear that a wave will be reflected at the boundary and a wave will be transmitted across the boundary. We denote the incident wave in phasor form as

$$\vec{E}_i = \hat{E}_i e^{-\hat{\gamma}_1 z} \vec{a}_x = \hat{E}_i e^{-\alpha_1 z} e^{-j\beta_1 z} \vec{a}_x \tag{5.71a}$$

$$\vec{H}_i = \frac{\hat{E}_i}{\hat{\eta}_1} e^{-\hat{\gamma}_1 z} \vec{a}_y = \frac{\hat{E}_i}{\eta_1} e^{-\alpha_1 z} e^{-j\beta_1 z} e^{-j\theta_{\eta_1}} \vec{a}_y \tag{5.71b}$$

where

$$\hat{\gamma}_1 = \sqrt{j\omega\mu_1(\sigma_1 + j\omega\epsilon_1)} = \alpha_1 + j\beta_1 \tag{5.71c}$$

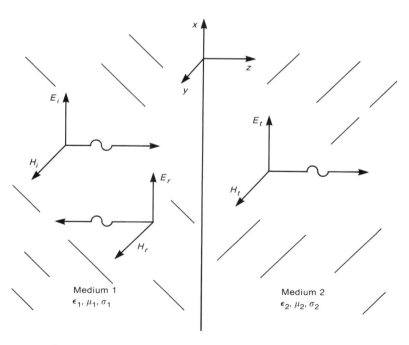

FIGURE 5.18 A uniform plane wave with normal incidence to the interface between two media.

and

$$\hat{\eta}_1 = \sqrt{\frac{j\omega\mu_1}{\sigma_1 + j\omega\epsilon_1}} = \eta_1 \angle\theta_{\eta 1} \tag{5.71d}$$

The reflected wave is represented as

$$\vec{E}_r = \hat{E}_r e^{\hat{\gamma}_1 z}\vec{a}_x = \hat{E}_r e^{\alpha_1 z} e^{j\beta_1 z}\vec{a}_x \tag{5.72a}$$

$$\vec{H}_r = -\frac{\hat{E}_r}{\hat{\eta}_1} e^{\hat{\gamma}_1 z}\vec{a}_y = -\frac{\hat{E}_r}{\eta_1} e^{\alpha_1 z} e^{j\beta_1 z} e^{-j\theta_{\eta 1}}\vec{a}_y \tag{5.72b}$$

in accordance with the general solution developed in Section 3.6. Note that the directions of the electric and magnetic field vectors are such that power flow in the reflected wave is in the negative z direction according to Poynting vector for that wave, $\vec{S} = \vec{E} \times \vec{H}$, as it must be. The wave transmitted into medium #2 is represented as

$$\vec{E}_t = \hat{E}_t e^{-\hat{\gamma}_2 z}\vec{a}_x = \hat{E}_t e^{-\alpha_2 z} e^{-j\beta_2 z}\vec{a}_x \tag{5.73a}$$

$$\vec{H}_t = \frac{\hat{E}_t}{\hat{\eta}_2} e^{-\hat{\gamma}_2 z}\vec{a}_y = \frac{\hat{E}_t}{\eta_2} e^{-\alpha_2 z} e^{-j\beta_2 z} e^{-j\theta_{\eta 2}}\vec{a}_y \tag{5.73b}$$

where

$$\hat{\gamma}_2 = \sqrt{j\omega\mu_2(\sigma_2 + j\omega\epsilon_2)} = \alpha_2 + j\beta_2 \tag{5.73c}$$

and

$$\hat{\eta}_2 = \sqrt{\frac{j\omega\mu_2}{\sigma_2 + j\omega\epsilon_2}} = \eta_2 \angle\theta_{\eta 2} \tag{5.73d}$$

At the boundary, $z = 0$, the boundary conditions require that the total electric field tangent to the boundary be continuous across the boundary. Since the electric fields are defined in the x direction, we obtain

$$\vec{E}_i + \vec{E}_r = \vec{E}_t \quad \text{at } z = 0 \tag{5.74a}$$

Similarly, since neither medium is a perfect conductor, the tangential components of the total magnetic field intensity vector must be continuous across the boundary, resulting in

$$\vec{H}_i + \vec{H}_r = \vec{H}_t \quad \text{at } z = 0 \tag{5.74b}$$

Substituting the forms of the field vectors given above and evaluating them at $z = 0$ gives [1]

$$\hat{\Gamma} = \frac{\hat{E}_r}{\hat{E}_i} = \frac{\hat{\eta}_2 - \hat{\eta}_1}{\hat{\eta}_2 + \hat{\eta}_1} = \Gamma \ \underline{/\theta_\Gamma} \qquad (5.75)$$

and

$$\hat{T} = \frac{\hat{E}_t}{\hat{E}_i} = \frac{2\hat{\eta}_2}{\hat{\eta}_2 + \hat{\eta}_1} = T \ \underline{/\theta_T} \qquad (5.76)$$

It is also possible to show that $1 + \hat{\Gamma} = \hat{T}$. The quantities $\hat{\Gamma}$ and $\hat{T}$ are the *reflection* and *transmission coefficients*, respectively, of the boundary. It is a simple matter to also show that $|\hat{\Gamma}| \leq 1$. The magnitude of $\hat{T}$ may exceed unity. Note that $\hat{\Gamma}$ and $\hat{T}$ will be real only if both regions are lossless, i.e., $\sigma_1 = \sigma_2 = 0$; otherwise, $\hat{\Gamma}$ and $\hat{T}$ will in general be complex.

We now assume that the form of the incident wave is $\vec{\hat{E}} = E_m e^{-\hat{\gamma}_1 z} \vec{a}_x$, where the magnitude of the incident wave E_m is assumed known. For example, the incident wave may be produced by some distant antenna, and we know how to calculate the value of the electric field at the boundary *in the absence of the boundary* using the Friis transmission equation from the results of Section 5.5. Thus the phasor forms of the field vectors become, in terms of the incident field magnitude,

$$\vec{\hat{E}}_i = E_m e^{-\hat{\gamma}_1 z} \vec{a}_x \qquad (5.77a)$$

$$\vec{\hat{H}}_i = \frac{E_m}{\hat{\eta}_1} e^{-\hat{\gamma}_1 z} \vec{a}_y \qquad (5.77b)$$

$$\vec{\hat{E}}_r = \hat{\Gamma} E_m e^{\hat{\gamma}_1 z} \vec{a}_z \qquad (5.77c)$$

$$\vec{\hat{H}}_r = -\frac{\hat{\Gamma} E_m}{\hat{\eta}_1} e^{\hat{\gamma}_1 z} \vec{a}_y \qquad (5.77d)$$

$$\vec{\hat{E}}_t = \hat{T} E_m e^{-\hat{\gamma}_2 z} \vec{a}_x \qquad (5.77e)$$

$$\vec{\hat{H}}_t = \frac{\hat{T} E_m}{\hat{\eta}_2} e^{-\hat{\gamma}_2 z} \vec{a}_y \qquad (5.77f)$$

Multiplying these phasor forms by $e^{j\omega t}$ and taking the real part of the result gives the time-domain forms of the field vectors:

$$\vec{E}_i = E_m e^{-\alpha_1 z} \cos(\omega t - \beta_1 z) \vec{a}_x \qquad (5.78a)$$

$$\vec{H}_i = \frac{E_m}{\eta_1} e^{-\alpha_1 z} \cos(\omega t - \beta_1 z - \theta_{\eta_1}) \vec{a}_y \qquad (5.78b)$$

$$\vec{E}_r = \Gamma E_m e^{\alpha_1 z} \cos(\omega t + \beta_1 z + \theta_\Gamma) \vec{a}_x \qquad (5.78c)$$

$$\vec{H}_r = -\frac{\Gamma E_m}{\eta_1} e^{\alpha_1 z} \cos(\omega t + \beta_1 z + \theta_\Gamma - \theta_{\eta 1}) \vec{a}_y \qquad (5.78\text{d})$$

$$\vec{E}_t = T E_m e^{-\alpha_2 z} \cos(\omega t - \beta_2 z + \theta_T) \vec{a}_x \qquad (5.78\text{e})$$

$$\vec{H}_t = \frac{T E_m}{\eta_2} e^{-\alpha_2 z} \cos(\omega t - \beta_2 z + \theta_T - \theta_{\eta 2}) \vec{a}_y \qquad (5.78\text{f})$$

The average power density vector of the wave transmitted into the second medium is

$$\vec{S}_{\text{av},t} = \tfrac{1}{2} \mathscr{R}e\{\vec{E}_t \times \vec{H}_t^*\} \qquad (5.79)$$

$$= \tfrac{1}{2} \mathscr{R}e\left\{ \hat{T} E_m e^{-\hat{\gamma}_2 z} \frac{\hat{T}^* E_m e^{-\hat{\gamma}_2^* z}}{\hat{\eta}_2^*} \right\} \vec{a}_z$$

$$= \frac{1}{2} \frac{E_m^2 T^2}{\eta_2} e^{-2\alpha_2 z} \cos\theta_{\eta 2} \, \vec{a}_z$$

where we denote $|\hat{T}| = T$ and $|\hat{\eta}_2| = \eta_2$. Note that this is a simple calculation in the second medium because there is only one wave in this medium. In medium #1 there are two waves: a forward-traveling wave and a backward-traveling wave. In this medium the Poynting vector contains a cross-coupling term unless the medium is lossless.

To close this section, we consider incidence on the surface of a perfect conductor, $\sigma_2 = \infty$, from a lossless region, $\sigma_1 = 0$. The intrinsic impedance of the perfect conductor is zero, $\eta_2 = 0$, so that the reflection coefficient is $\hat{\Gamma} = -1$ (and the transmission coefficient is zero, which is a sensible result). Thus the reflected electric field is equal but opposite to the incident electric field: $\vec{E}_r = -\vec{E}_i$. This is a sensible result because the reflected electric field must cancel the incident electric field at the boundary in order to produce zero total tangential electric field at the surface of the perfect conductor, as required by the boundary conditions. Recall that for a perfect conductor the total tangential magnetic field intensity vector is not zero but equals the linear surface current density on the perfect conductor. Thus the total fields in region #1 become

$$\vec{E}_1 = \vec{E}_i + \vec{E}_r \qquad (5.80\text{a})$$

$$= E_m(e^{-j\beta_1 z} - e^{j\beta_1 z}) \vec{a}_x$$

$$= -2j E_m \sin(\beta_1 z) \vec{a}_x$$

$$\vec{H}_i = \vec{H}_i + \vec{H}_r \qquad (5.80\text{b})$$

$$= \frac{E_m}{\eta_1}(e^{-j\beta_1 z} + e^{j\beta_1 z}) \vec{a}_y$$

$$= \frac{2 E_m}{\eta_1} \cos(\beta_1 z) \vec{a}_y$$

The time-domain expressions become

$$\vec{E}_1 = \mathscr{R}e\{\hat{\vec{E}}_1 e^{j\omega t}\} \tag{5.81a}$$

$$= 2E_m \sin(\beta_1 z) \sin(\omega t)\, \vec{a}_x$$

$$\vec{H}_1 = \mathscr{R}e\{\hat{\vec{H}}_1 e^{j\omega t}\} \tag{5.81b}$$

$$= \frac{2E_m}{\eta_1} \cos(\beta_1 z) \cos(\omega t)\, \vec{a}_y$$

These total fields represent *standing waves*. The magnitudes of the fields are

$$|\hat{E}_1| = 2E_m|\sin(\beta_1 z)| \tag{5.82a}$$

$$= 2E_m \left|\sin\left(\frac{2\pi z}{\lambda_1}\right)\right|$$

$$|\hat{H}_1| = 2\frac{E_m}{\eta_1}|\cos(\beta_1 z)| \tag{5.82b}$$

$$= 2\frac{E_m}{\eta_1}\left|\cos\left(\frac{2\pi z}{\lambda_1}\right)\right|$$

These are plotted in Fig. 5.19. Note that the electric field achieves a maximum at distances of $\frac{1}{4}\lambda_1$, $\frac{3}{4}\lambda_1$, ... and achieves minima (zero) at distances of

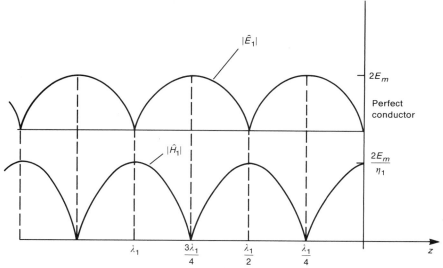

FIGURE 5.19 The total (incident plus reflected) fields for a uniform plane wave of normal incidence on a perfect conductor. The electric field goes to zero at multiples of $\frac{1}{2}\lambda$ from the boundary, and the magnetic field goes to zero at odd multiples of $\frac{1}{4}\lambda$ from the boundary.

$\frac{1}{2}\lambda_1$, λ_1, ... The magnetic field minima and maxima are separated from the corresponding points of the electric field by $\frac{1}{4}\lambda_1$. Also observe that corresponding points on the waveform *replicate for distances that are multiples of $\frac{1}{2}\lambda_1$*.

5.6.3 Multipath Effects

Radiated fields may travel to the receiving antenna via many different paths. Depending on the electrical length of these paths, the signals may arrive at the receiving antenna in phase, out of phase or some gradation in between. The total signal at the receiving antenna will be the phasor/vector sum of all the waves incident on the antenna. Because the electrical lengths of the various paths may be significant, the signals may add constructively or destructively, as with the case of antenna arrays discussed in Section 5.3. Consider the case of a transmitting antenna and a receiving antenna situated above a perfectly conducting ground plane, as illustrated in Fig. 5.20. The received signal at the measurement antenna is a contribution of two signals: a *direct wave* that travels a line-of-sight path between the product emission point and the measurement antenna and a *reflected wave* that is reflected at the ground plane. In the case

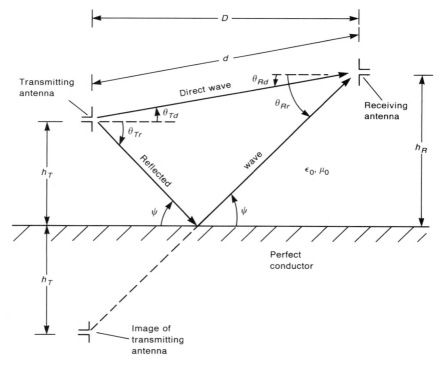

FIGURE 5.20 Illustration of the problem of communication between two antennas above a ground plane. The received field is the sum of the contributions due to the direct and the reflected waves.

of measurement of radiated emissions from a product for verification of compliance to the regulatory limits, the test setup is also situated above a ground plane so that the received emission will be the combination of a direct and a reflected wave. We will now investigate this phenomenon quantitatively.

Suppose the electric field intensity *patterns* of the transmitting and receiving antennas are described by $E_T(\theta_T, \phi_T)$ and $E_R(\theta_R, \phi_R)$, respectively. The antennas are at heights h_T and h_R and are separated a horizontal distance D. The angle of incidence of the reflected wave at the ground plane, ψ, and the angle of reflection of the reflected wave can be shown to be equal [1]. This is referred to as *Snell's law*. The length of the direct wave path is

$$d = \sqrt{D^2 + (h_R - h_T)^2} \tag{5.83}$$

The reflected wave can be thought of as being due to the image of the transmitting antenna, as shown in Fig. 5.20. Replacing the perfectly conducting ground plane with the image of the transmitting antenna shows that the length of the reflected wave path is

$$d_r = \sqrt{D^2 + (h_R + h_T)^2} \tag{5.84}$$

The received voltage at the base of the receiving antenna and due to the direct wave is proportional to

$$\hat{V}_d = \hat{V}_0 E_T(\theta_{Td}, \phi_{Td}) E_R(\theta_{Rd}, \phi_{Rd}) \frac{e^{-j\beta_0 d}}{d} \tag{5.85}$$

We have assumed that the antennas are in the far fields of each other so that the electric field at the receiving antenna resembles, locally, a uniform plane wave. This gives the $1/d$ and $e^{-j\beta_0 d}$ dependence. The reflected wave will be the transmitted wave (in the direction θ_{Tr}, ϕ_{Tr}) that is multiplied by the reflection coefficient at the point of reflection, $\hat{\Gamma}$. The form of the received voltage at the base of the receiving antenna is

$$\hat{V}_r = \hat{V}_0 E_T(\theta_{Tr}, \phi_{Tr}) E_R(\theta_{Rr}, \phi_{Rr}) \hat{\Gamma} \frac{e^{-j\beta_0 d_r}}{d_r} \tag{5.86}$$

The total received voltage is the sum of (5.85) and (5.86):

$$\hat{V} = \hat{V}_d + \hat{V}_r \tag{5.87a}$$

$$= \hat{V}_0 E_T(\theta_{Td}, \phi_{Td}) E_R(\theta_{Rd}, \phi_{Rd}) \frac{e^{-j\beta_0 d}}{d}$$

$$+ \hat{V}_0 E_T(\theta_{Tr}, \phi_{Tr}) E_R(\theta_{Rr}, \phi_{Rr}) \hat{\Gamma} \frac{e^{-j\beta_0 d_r}}{d_r}$$

$$= \hat{V}_0 E_T(\theta_{Td}, \phi_{Td}) E_R(\theta_{Rd}, \phi_{Rd}) \frac{e^{-j\beta_0 d}}{d} \hat{F}$$

where

$$\hat{F} = 1 + \frac{E_T(\theta_{Tr}, \phi_{Tr})E_R(\theta_{Rr}, \phi_{Rr})}{E_T(\theta_{Td}, \phi_{Td})E_R(\theta_{Rd}, \phi_{Rd})} \hat{\Gamma} \frac{d}{d_r} e^{-j\beta_0(d_r - d)} \qquad (5.87b)$$

Thus the ground reflection modifies the free space direct wave propagation (the coupling without the ground plane present) by the multiplicative factor $\hat{F}$. Consequently the Friis transmission equation can be modified to account for the ground reflection by multiplying it by the square of the magnitude of $\hat{F}$ (since the Friis transmission equation involves power).

Let us now consider the reflection coefficient $\hat{\Gamma}$. Two cases arise: *parallel polarization* and *perpendicular polarization*. These cases correspond to the two required measurement antenna polarizations in compliance measurements: *vertical* and *horizontal*. The case of perpendicular (horizontal) polarization is shown in Fig. 5.21(a). The term "perpendicular" refers to the fact that the incident electric field is perpendicular to the *plane of incidence*. The plane of incidence contains the propagation vector of the wave and the normal to the surface. The incident and reflected electric field vectors are parallel to the ground plane for this polarization. The reflection coefficient at the ground plane becomes [1]

$$\hat{\Gamma}_H = \frac{\hat{E}_r}{\hat{E}_i} \qquad (5.88)$$

$$= -1$$

and we have used the H subscript to indicate horizontal polarization of the antennas. This result is intuitively obvious if we recall the boundary condition that the total tangential electric field at the surface of a perfect conductor must be zero. The electric fields of the incident and reflected fields are both tangent to the conductor. Thus the reflected electric field must be opposite to that of the incident wave.

In the case of parallel polarization shown in Fig. 5.21(b) the electric field vectors are parallel to the plane of incidence. This corresponds to the case of vertical antenna orientations. The reflection coefficient for this polarization is [1]

$$\hat{\Gamma}_V = \frac{\hat{E}_r}{\hat{E}_i} \qquad (5.89)$$

$$= +1$$

where the V subscript refers to the vertical polarization of the antennas. Again, this result is intuitively obvious if we consider the boundary condition that the total tangential electric field at the surface of a product conductor must be zero. In this case the tangential components are the z components, which must be

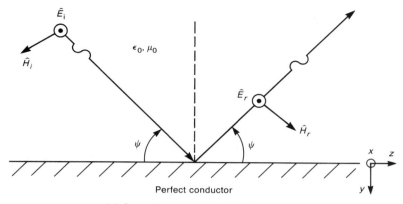

(a) Perpendicular (horizontal) polarization

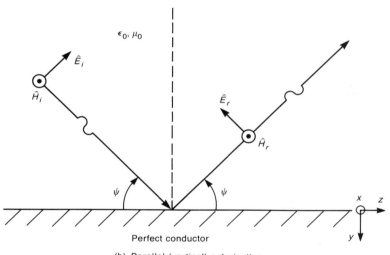

(b) Parallel (vertical) polarization

FIGURE 5.21   Illustration of the determination of the reflection coefficients for a uniform plane wave with oblique incidence on a perfect conductor for (a) perpendicular (horizontal) polarization and (b) parallel (vertical) polarization of the incident wave.

equal and opposite at the surface. Therefore the incident and reflected electric fields must remain unchanged with respect to their propagation vectors, as shown in Fig. 5.21(b).

As a practical example, let us compute the ground plane reflection factor for a current element (a Hertzian dipole) and a linear receiving antenna such as a dipole for a typical FCC Class B measurement. The separation between the current element and the receiving antenna is 3 m. The current element (the product) is placed at a height of 1 m above the ground plane. We will consider

the current element and the measurement antenna to be oriented parallel to
each other, and the height of the measurement antenna above the ground plane
must be scanned from 1 m to 4 m. First consider the case of *horizontal
polarization* shown in Fig. 5.22(a). The factor in (5.87b) becomes

$$\hat{F}_H = 1 - \frac{d}{d_r} e^{-j(2\pi/\lambda_0)(d_r - d)}$$

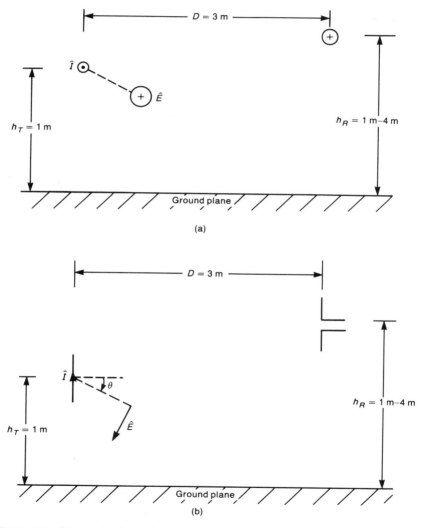

FIGURE 5.22 Determination of the reflected wave correction factor for the FCC
Class B test: (a) horizontal polarization; (b) vertical polarization.

where we have substituted $\beta_0 = 2\pi/\lambda_0$. The reflection coefficient is $\hat{\Gamma}_H = -1$ for this case, and the pattern is omnidirectional in a plane perpendicular to the current element and the horizontally-positioned measurement antenna. Thus the electric field patterns in (5.87b) are unity. The minimum scan height of the measurement antenna of 1 m gives $d = 3$ m and $d_r = \sqrt{13}$ m. The maximum scan height of 4 m gives $d = \sqrt{18}$ m and $d_r = \sqrt{34}$ m. Thus the factor is

$$\hat{F}_{1m} = 1 - 3\sqrt{\tfrac{1}{13}}e^{-j(2\pi/\lambda_0)(\sqrt{13}-3)}$$

and

$$\hat{F}_{4m} = 1 - \sqrt{\tfrac{18}{34}}e^{-j(2\pi/\lambda_0)(\sqrt{13}-\sqrt{18})}$$

Table 5.1 shows this factor for selected frequencies in the radiated emission range of 30 MHz–1 GHz. The factor for *vertical polarization* requires that we consider the pattern for the antenna. Recall that the pattern is a maximum

TABLE 5.1   Correction Factor for Horizontal Polarization.

Frequency (MHz)	F_{1m} (dB)	F_{4m} (dB)
30	−8.32	−1.30
40	−6.24	+0.73
50	−4.54	+2.21
60	−3.12	+3.28
70	−1.92	+4.03
80	−0.90	+4.50
90	+0.00	+4.73
100	+0.78	+4.71
110	+1.47	+4.46
120	+2.07	+3.96
130	+2.61	+3.18
140	+3.08	+2.07
150	+3.49	+0.53
160	+3.85	−1.58
170	+4.17	−4.54
180	+4.44	−8.63
190	+4.67	−11.2
200	+4.86	−7.63
300	+4.78	+4.42
400	+0.43	−3.42
500	−15.1	+3.81
600	+1.10	−0.55
700	+4.93	+2.85
800	+4.69	+1.45
900	+0.07	+1.44
1000	−14.0	+2.86

broadside to the antenna and is zero off the ends of the antenna. The pattern varies as the *cosine* of the various angles in Fig. 5.20: θ_{Tr}, θ_{Td}, θ_{Rr}, and θ_{Rd}. Thus the factor becomes

$$\hat{F}_V = 1 + \frac{\cos \theta_{Tr} \cos \theta_{Rr}}{\cos \theta_{Td} \cos \theta_{Rd}} \hat{\Gamma}_V \frac{d}{d_r} e^{-j(2\pi/\lambda_0)(d_r - d)}$$

The various cosines are

$$\cos \theta_{Tr} = \frac{D}{d_r}$$

$$\cos \theta_{Rr} = \frac{D}{d_r}$$

$$\cos \theta_{Td} = \frac{D}{d}$$

$$\cos \theta_{Td} = \frac{D}{d}$$

TABLE 5.2   Correction Factor for Vertical Polarization.

Frequency (MHz)	F_{1m} (dB)	F_{4m} (dB)
30	+3.80	+1.95
40	+3.69	+1.24
50	+3.54	+0.32
60	+3.36	−0.81
70	+3.14	−2.08
80	+2.88	−3.32
90	+2.58	−4.13
100	+2.24	−4.08
110	+1.86	−3.20
120	+1.42	−1.94
130	+0.94	−0.67
140	+0.40	+0.44
150	−0.19	+1.33
160	−0.85	+2.01
170	−1.58	+2.48
180	−2.37	+2.75
190	−3.24	+2.83
200	−4.15	+2.71
300	−3.73	−3.07
400	+2.43	+2.35
500	+3.95	−1.65
600	+2.07	+1.74
700	−4.95	−0.28
800	−3.31	+0.86
900	+2.25	+0.87
1000	+3.94	−0.29

Substituting these along with $\hat{\Gamma}_V = +1$ gives

$$\hat{F}_V = 1 + \left(\frac{d}{d_r}\right)^3 e^{-j(2\pi/\lambda_0)(d_r - d)}$$

Table 5.2 gives these values for selected frequencies.

The above results assume that the receiving antenna is in the far field of the transmitting antenna, since the fields were assumed to vary with distance as $e^{-j\beta r}/r$. At the lower radiated emission measurement frequencies the receiving (measurement) antenna is probably in the near field of the radiator (the product). Thus the above results are not applicable! This important consideration is investigated in [2].

5.7   BROADBAND MEASUREMENT ANTENNAS

As has been pointed out, the FCC prefers the use of tuned, half-wave dipoles for measurement of radiated emissions. From the standpoint of rapid and efficient gathering of data over the frequency range of the radiated emission limits of 30 MHz–1 GHz, the tuned half-wave dipole is not an attractive measurement antenna. Its length must be physically adjusted to provide a total length of $\frac{1}{2}\lambda_0$ at each measurement frequency. Also, in the measurement of the vertically polarized emissions at the lowest frequency of the limit, 30 MHz, the dipole length is 5 m or approximately 15 feet. Thus the antenna cannot be scanned from 1 m to 4 m at these lower frequencies in the vertical polarization mode.

A more practical measurement technique is the use of *broadband measurement antennas* such as the *biconical* and *log-periodic* antennas. A broadband antenna is an antenna that has the following two characteristics over the band of frequencies of its intended use:

1. *The input/output impedance is fairly constant over the frequency band.*
2. *The pattern is fairly constant over the frequency band.*

In the measurement of radiated emissions for compliance verification, the biconical antenna is typically used in the frequency range of 30 MHz–200 MHz, whereas the log-periodic antenna is typically used in the remainder of the band from 200 MHz to 1 GHz. In this section we will characterize these broadband measurement antennas.

5.7.1   The Biconical Antenna

The *infinite* biconical antenna is constructed of two cones of half angle θ_h with a small gap at the feed point, as shown in Fig. 5.23. A voltage source feeds the antenna at this gap. A spherical coordinate system is appropriate to use for the

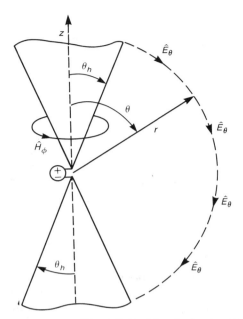

FIGURE 5.23   The infinite, biconical antenna.

analysis here. In the space surrounding the cones (assumed to be free space), $\vec{J} = 0$, and symmetry suggests that the fields are $\vec{H} = \hat{H}_\phi \vec{a}_\phi$ and $\vec{E} = \hat{E}_\theta \vec{a}_\theta$. Faraday's and Ampère's laws can be solved to give the forms of the fields as [3–6]

$$\hat{H}_\phi = \frac{H_0}{\sin \theta} \frac{e^{-j\beta_o r}}{r} \tag{5.90}$$

and

$$\hat{E}_\theta = \frac{\beta_0}{\omega \epsilon_0} \frac{H_0}{\sin \theta} \frac{e^{-j\beta_o r}}{r} \tag{5.91}$$

$$= \eta_0 \hat{H}_\phi$$

where H_0 is a constant.

Note that the radiated fields are of the *transverse electromagnetic (TEM) mode* type in that the electric field and the magnetic field are orthogonal and transverse to the direction of propagation, the r direction, as shown by the presence of the $e^{-j\beta_o r}$ term. Therefore we may uniquely define voltage between two points on the cones as was the case for transmission lines. The voltage produced between two points on the two cones that are a distance r from the feed point is

$$\hat{V}(r) = -\int_{\theta = \pi - \theta_h}^{\theta_h} \vec{E} \cdot d\vec{l} \tag{5.92}$$

$$= 2\eta_0 H_0 e^{-j\beta_0 r} \ln(\cot \tfrac{1}{2}\theta_h)$$

The current on the surface of the cones is found by using Ampère's law in integral form:

$$\hat{I}(r) = \int_{\phi = 0}^{2\pi} \hat{H}_\phi r \sin \theta \, d\phi \tag{5.93}$$

$$= 2\pi H_0 e^{-j\beta_0 r}$$

The input impedance at the feed terminals can be obtained as the ratio of the voltage and current for $r = 0$:

$$\hat{Z}_{\text{in}} = \frac{\hat{V}(r)}{\hat{I}(r)}\bigg|_{r = 0} \tag{5.94}$$

$$= \frac{\eta_0}{\pi} \ln(\cot \tfrac{1}{2}\theta_h)$$

$$= 120 \ln(\cot \tfrac{1}{2}\theta_h)$$

which is purely resistive. Usually the cone half angle is chosen to provide a match to the feed line characteristic resistance Z_C. A balun is also normally included at the antenna input. The radiation resistance R_{rad} can be shown to also be equal to $\hat{Z}_{\text{in}}$ given in (5.94), which is a sensible result. To do this, we compute the total radiated average power as

$$P_{\text{rad}} = \oint_S \vec{S}_{\text{av}} \cdot d\vec{s} \tag{5.95}$$

$$= \int_{\phi = 0}^{2\pi} \int_{\theta = \theta_h}^{\pi = \theta_h} \frac{|\hat{E}_\theta|^2}{2\eta_0} r^2 \sin \theta \, d\theta \, d\phi$$

$$= \pi \eta_0 H_0^2 \int_{\theta = 0}^{\theta_h} \frac{d\theta}{\sin \theta}$$

$$= 2\pi \eta_0 |H_0|^2 \ln(\cot \tfrac{1}{2}\theta_h)$$

The radiation resistance is defined by

$$P_{\text{rad}} = \tfrac{1}{2}|\hat{I}(0)|^2 R_{\text{rad}} \tag{5.96}$$

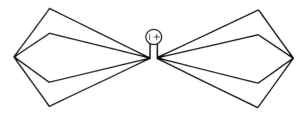

FIGURE 5.24   The truncated biconical antenna composed of wire elements.

Substituting (5.93) evaluated at $r = 0$ into (5.96) shows that

$$R_{\text{rad}} = \hat{Z}_{\text{in}} \qquad (5.97)$$

Observe that the radiated fields are spherical waves with $\vec{E}$ in the θ direction and $\vec{H}$ in the ϕ direction. For linearly polarized waves incident on the antenna from the broadside direction, $\theta = 90°$, the antenna responds to the component that is parallel to its axis. Thus this antenna can be used to perform vertical and horizontal field measurements for regulatory compliance verification. Also observe that the input impedance and pattern are, theoretically, constant over an infinite range of frequencies. Unfortunately, infinite-length cones are obviously not practical, so that practical biconical antennas are constructed of *truncated cones*. Finite-length cones cause discontinuities at the ends, which result in reflections as the waves travel outward along the cones. This produces standing waves on the cones that result in the input impedance having an imaginary part rather than being purely real and independent of frequency. References [3–5] show variations in the input impedance for various cone lengths and half angles. Another practical method of construction for biconicals is to use wires to approximate the cone surfaces, as illustrated in Fig. 5.24. Other variations are the *discone* antenna shown in Fig. 5.25(a) consisting of one cone

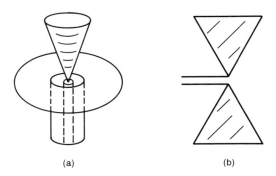

(a)   (b)

FIGURE 5.25   Other implementations of truncated biconical antennas: (a) the discone antenna; (b) the bow-tie antenna.

(truncated) above a circular ground plane. This provides a convenient balanced feed with a coaxial cable. The method of images shows that the fields of the discone antenna above the ground plane are the same as those of the biconical antenna. Also, the radiation resistance is one-half that of the biconical, since only one-half of the biconical radiated power is radiated by the discone antenna. Another vision is the *bow-tie* antenna shown in Fig. 5.25(b). This consists of flat, triangular plates, which can be constructed of metal sheets or by using a wire to outline the area of each plate in order to reduce weight and wind loading. The bow-tie antenna is frequently used for reception of UHF television signals. Use of wires to outline the triangular sheet reduces the bandwidth over solid metal triangles or cones.

5.7.2 The Log-Periodic Antenna

The log-periodic antenna is a member of a general class of frequency-independent antennas that rely on repetitive dimensions of their structures. The structural dimensions increase in proportion to the distance from the origin of the structure. This results in the input impedance and radiation properties repeating periodically as the logarithm of frequency. Thus they are said to be broadband antennas.

The most common form of log-periodic antenna used for the measurement of radiated emissions from 200 MHz to 1 GHz is the *log-periodic dipole array* shown in Fig. 5.26(a). This antenna shares the properties of all other log-periodic structures in that the element distances, lengths and separations are related by a constant such as

$$\tau = \frac{l_n}{l_{n-1}} = \frac{d_n}{d_{n-1}} = \frac{R_n}{R_{n-1}} \tag{5.98}$$

There are two ways of feeding the elements, one of which does not work well. If all the elements are connected in parallel and the antenna fed at the apex as shown in Fig. 5.26(b), the currents in the adjacent elements are in the same direction. This can be viewed as a dipole array using the techniques of Section 5.3. The elements will be spaced closely in terms of wavelengths so that currents in adjacent elements will interact. Because the current progression is to the right, the pattern will be a beam directed to the right. However, elements to the right will interfere with the radiation from elements to the left, causing interference effects. If the currents to the adjacent elements are reversed in phase by criss-crossing the feed wires as shown in Fig. 5.26(c), the beam will be directed to the left and negligible interference will be caused by the shorter elements, whose currents have alternate phasing. A practical method of feeding the log-periodic antenna with a coaxial cable and at the same time producing the 180° phase shift between adjacent elements as well as *balanced* operation is illustrated in Fig. 5.27. A coaxial cable is passed through a hollow pipe, to which half of the set of elements are attached. The coaxial cable shield is attached to this pipe at point *A*, whereas its center conductor is connected to the other

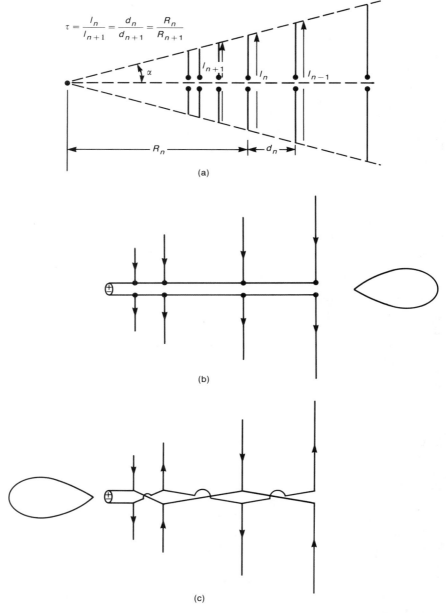

FIGURE 5.26 Log-periodic antennas: (a) periodicity of the structure; (b) nonpreferred excitation method; (c) preferred excitation method.

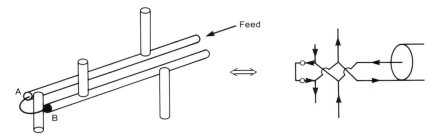

FIGURE 5.27   Practical feed of a log-periodic antenna.

pipe at the same point, *B*. This provides the ability to feed the antenna from its rear so that the feedline does not interfere with the pattern and at the same time produces the 180° phase shift between adjacent elements.

The cutoff frequencies of the log-periodic dipole array (its bandwidth) can be approximately computed by determining the frequency where the shortest elements are one-half wavelength (the highest frequency of operation) and the frequency where the longest elements are one-half wavelength (the lowest frequency of operation). For a particular operating frequency only a few elements will be active, which will be in the vicinity of where the elements are approximately one-half wavelength. So the active region dynamically adjusts to using the elements that are efficient radiators at the operating frequency. As is somewhat obvious, a linearly polarized uniform plane wave incident along the antenna axis will produce the maximum response when its electric field vector is oriented parallel to the elements of the array. Thus the antenna can be used to measure vertical and horizontal radiated emissions in the verification of regulatory compliance.

The input impedance to the log-periodic dipole array tends to be resistive, frequency-independent, and between 50 and 100 Ω. A direct analysis is more difficult than for the ideal biconical antenna. Consequently, various design equations are usually employed. For a discussion of these see [5]. The VSWR can be kept to below 2.0 over the typical frequency range of its use in radiated emission measurements: 200 MHz–1 GHz.

REFERENCES

[8]   C.R. Paul and S.A. Nasar, *Introduction to Electromagnetic Fields*, second edition, McGraw-Hill, NY (1987).

[2]   T.J. Dvorak, Fields at a measuring site, *Proceedings of the 1988 IEEE International Symposium on Electromagnetic Compatibility, Seattle, WA, August 1988.*

[3]   E.C. Jordan and K.G. Balmain, *Electromagnetic Waves and Radiating Systems*, second edition, Prentice-Hall, Englewood Cliffs, NJ (1968).

[4]   W.L. Stutzman and G.A. Thiele, *Antenna Theory and Design*, John Wiley, NY (1981).

[5] C.A. Balanis, *Antenna Theory Analysis and Design*, Harper & Row, NY (1982).

[6] J.D. Kraus, *Antennas*, McGraw-Hill, NY (1950).

[7] *ARRL Antenna Book*, 14th edition, American Radio Relay League, Newington, Connecticut (1984).

[8] J. DeMarinis, The antenna cable as a source of error in EMI measurements, *Proceedings of the 1988 IEEE International Symposium on Electromagnetic Compatibility, Seattle, WA, August 1988*.

PROBLEMS

5.1 Consider a Hertzian dipole of length 1 cm carrying a phasor current of $\hat{I} = 10 \underline{/30°}$ A. If the frequency is 100 MHz, determine the electric and magnetic fields at a distance of 10 cm away from the dipole and $\theta = 45°$. [$\hat{E}_r = 2069.67 \underline{/-60.17°}$, $\hat{E}_\theta = 991.4 \underline{/-59.64°}$, $\hat{H}_\phi = 0.57484 \underline{/29.83°}$] Compute the ratios $|\hat{E}_\theta|/|\hat{E}_r|$ and $|\hat{E}_\theta|/|\hat{H}_\phi|$ at this distance. [0.479, 1724.6] Repeat for distances of 1 m and 10 m, and $\theta = 45°$. [$\hat{E}_r = 4.701 \underline{/-115.5°}$, $\hat{E}_\theta = 4.033 \underline{/-31.73°}$, $\hat{H}_\phi = 1.306 \times 10^{-2} \underline{/-25.52°}$, 0.8579, 308.8, $\hat{E}_r = 4.247 \times 10^{-2} \underline{/-92.73°}$, $\hat{E}_\theta = 0.444 \underline{/-2.74°}$, $\hat{H}_\phi = 1.18 \times 10^{-3} \underline{/-2.7°}$, 10.45, 376.08] Determine these distances in wavelengths [0.033, 0.33, 3.33]

5.2 Compute the radiation resistance of and the total average power radiated by the dipole of Problem 5.1. [8.77×10^{-3} Ω, 438.65 mW]

5.3 Consider a magnetic dipole (loop) antenna of radius 1 cm carrying a current of $\hat{I} = 10 \underline{/30°}$. If the frequency is 100 MHz, compute the electric and magnetic field intensities at $r = 10$ cm, 1 m, and 10 m, and $\theta = 45°$. [$\hat{E}_\phi = 14.26 \underline{/-90°}$, $0.0324 \underline{/-146°}$, $2.93 \times 10^{-3} \underline{/-123°}$, $\hat{H}_\theta = 0.173 \underline{/0.355°}$, $7.058 \times 10^{-4} \underline{/28.3°}$, $7.82 \times 10^{-5} \underline{/57.3°}$, $\hat{H}_r = 0.3612 \underline{/-0.17°}$, $8.223 \times 10^{-4} \underline{/-56°}$, $7.42 \times 10^{-6} \underline{/-32.7°}$] Determine the ratios $|\hat{H}_\theta|/|\hat{H}_r|$ and $|\hat{E}_\phi|/|\hat{H}_\theta|$ at these distances. [0.472, 82.4, 0.858, 459.1, 10.55, 375.64]

5.4 Compute the radiation resistance of and the total average power radiated by the magnetic dipole of Problem 5.3. [3.8×10^{-5} Ω, 1.9 mW]

5.5 Determine the magnitudes of the electric and magnetic fields of a half-wave dipole operated at a frequency of 300 MHz at a distance of 100 m in the broadside plane, i.e., $\theta = 90°$. The input current to the terminals is $100 \underline{/0°}$ mA. [60 mV/m, 159.15 μA/m] Determine the total average power radiated. [365 mW]

5.6 A lossless quarter-wave monopole antenna is situated above a perfectly conducting ground plane and is driven by a 100 V, 300 MHz source that has an internal impedance of 50 Ω. Compute the total average power radiated. [23 W] Repeat this calculation if a $\frac{1}{5}\lambda$ lossless monopole having an input impedance $(20 - j50)$ Ω is substituted. [13.5 W] Repeat for a short, $\frac{1}{10}\lambda$ lossless monopole that has an input impedance of $(4 - j180)\,\Omega$. [0.566 W]

5.7   A lossless dipole antenna is attached to a source with a length of lossless 50 Ω coaxial cable. The source has an open-circuit voltage of 100 V rms and a source impedance of 50 Ω. If the frequency of the source is such that the dipole length is one-half wavelength and the transmission-line length is 1.3λ, determine the total average power radiated by the antenna and the VSWR on the cable. [43.1 W, 2.18]

5.8   Two identical monopole antennas are perpendicular to earth. The antennas are separated by distance d and fed with currents of equal magnitude, as shown in Fig. 5.7(a). Sketch the patterns of the array in a plane parallel to the earth for the following conditions: (a) $d = \frac{1}{2}\lambda_0$, $\alpha = 90°$; (b) $d = \frac{5}{8}\lambda_0, \alpha = 45°$; (c) $d = \lambda_0, \alpha = 180°$; (d) $d = \frac{1}{4}\lambda_0, \alpha = 180°$.

5.9   An AM broadcast-band transmitting station consists of two vertical monopoles above earth. The two antennas are separated by 164 feet and the transmitting frequency is 1500 kHz. The antennas are fed with signals of equal amplitude and a phase difference of 135°. Sketch the electric field pattern at the surface of the earth.

5.10 Consider two lossless, widely separated half-wave dipoles. If 10 W is delivered to the transmitting antenna and 1 mW is received in the other antenna in a matched load, determine the received power if the receiving antenna's load is changed to $(10 + j0)\,\Omega$. [0.336 mW]

5.11 Determine the maximum effective aperture of a half-wave dipole operated at 300 MHz. [0.1305 m²]

5.12 Determine an expression for the antenna factor of an antenna that is terminated in a matched load. Assume that the antenna input impedance is $\hat{Z}_a = R_a + jX_a$ and its maximum directive gain is G. [$\mathrm{AF} = (2f/v_0|\hat{Z}_a|)\sqrt{\pi\eta_0 R_a/G}$] Write this result in dB. [$\mathrm{AF}_{\mathrm{dB}} = 20\log f(\mathrm{MHz}) - G(\mathrm{dB}) - 20\log|\hat{Z}_a| + 10\log R_a - 12.79$] Repeat this if the antenna is terminated not in a matched impedance but in a general impedance $\hat{Z}_L$. [$\mathrm{AF} = (|\hat{Z}_a + \hat{Z}_L|/|\hat{Z}_L|)(f/v_0|\hat{Z}_a|)\sqrt{\pi\eta_0 R_a/G}$, $\mathrm{AF}_{\mathrm{dB}} = 20\log f(\mathrm{MHz}) - G(\mathrm{dB}) - 20\log|\hat{Z}_L| + 20\log|\hat{Z}_a + \hat{Z}_L| - 10\log R_a - 18.81$] Compute the antenna factor for a half-wave dipole operated at 300 MHz and terminated in (1) a matched load and (2) 50 Ω. [(1) $\mathrm{AF}_{\mathrm{dB}} = 14.7$, (2) $\mathrm{AF}_{\mathrm{dB}} = 18.26$]. An FM antenna has an impedance of $\hat{Z}_a = 300\,\Omega$ and a gain of 2.15 dB. The matched receiver ($\hat{Z}_L = 300\,\Omega$) requires a minimum signal of 1 μV or 0 dBμV for adequate reception. Determine the minimum electric field intensity at 100 MHz. [$E_{\mathrm{inc,min}} = 0.29\ \mathrm{dB\mu V/m} = 1.03\ \mu\mathrm{V/m}$]

5.13 A 1.5 m dipole is connected to a 50 Ω spectrum analyzer with 200 feet of RG-58U coaxial cable. A 100 MHz uniform plane wave is incident on the antenna. Determine the relation between the incident electric field and the voltage received at the input terminals of the spectrum analyzer. $[\,|\hat{V}_{\text{rec}}|_{\text{dB}\mu\text{V}} = -14 + |\hat{E}_{\text{inc}}|_{\text{dB}\mu\text{V/m}}]$ Determine the received voltage level corresponding to the FCC Class B limit. [29.36 dBμV]

5.14 Design a 20 dB pad to be used in a 300 Ω system. [$R_1 = R_3 = 366.67$ Ω, $R_2 = 1485$ Ω]

5.15 A quarter-wave transformer illustrated in Fig. P5.15. is used to match an antenna to a transmission line. If the input impedance to the terminals of the antenna is purely real and given by R_{in}, show that a $\frac{1}{4}\lambda$ length of transmission line with characteristic impedance Z_T will give an input impedance of Z_C if $Z_T = \sqrt{R_{\text{in}}Z_C}$.

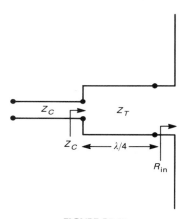

FIGURE P5.15

5.16 An aircraft transmitter is designed to communicate with a ground station. The ground receiver must receive at least 1 μW for proper reception. Assume that both antennas are omnidirectional. After takeoff, the airplane flies over the station at an altitude of 5000 ft. When the aircraft is directly over the ground station, a signal of 500 mW is received by the station. Determine the maximum communication range of the airplane. [670 miles]

5.17 A telemetry transmitter placed on the moon is to transmit data to the earth. The transmitter power is 100 mW and the gain of the transmitting antenna in the direction of transmission is 12 dB. Determine the minimum gain of the receiving antenna in order to receive 1 nW. The distance from the moon to earth is 238,857 miles and the transmitter frequency is 100 MHz. [92.1 dB]

5.18 A microwave relay link is to be designed. The transmitting and receiving antennas are separated by 30 miles, and the power gain in the direction of transmission for both antennas is 45 dB. If both antennas are lossless and matched and the frequency is 3 GHz, determine the minimum transmitter power if the received power is to be 1 mW. [36.5 W]

5.19 An antenna on an aircraft is being used to jam an enemy radar. If the antenna has a gain of 12 dB in the direction of transmission and the transmitted power is 5 kW, determine the electric field intensity in the vicinity of the enemy radar, which is 2 miles away. The frequency of transmission is 7 GHz. [0.68 V/m]

5.20 A uniform plane wave traveling in free space is incident normally to the surface of a perfect conductor. If the total electric field is zero at a distance of 1 m away from the surface of the perfect conductor, determine the lowest possible frequency of the incident wave. [150 MHz]

5.21 A uniform plane wave whose electric field is given by $\vec{E}_i = 100 \cos(\omega t - 6\pi x) \vec{a}_z$ V/m is incident from a region having $\epsilon_r = 4$, $\mu_r = 1$, and $\sigma = 0$ normally to the plane surface of a material having $\epsilon_r = 9$, $\mu_r = 4$, and $\sigma = 0$. Write complete time-domain expressions for the incident, reflected, and transmitted electric and magnetic fields, and determine the average power transmitted through a 2 m^2 area of the surface. $[\vec{H}_i = -(100/60\pi)\cos(\omega t - 6\pi x)\vec{a}_y, \vec{E}_r = 14.29 \cos(\omega t + 6\pi x) \vec{a}_z, \vec{H}_r = (14.29/60\pi)\cos(\omega t + 6\pi x) \vec{a}_y, \vec{E}_t = 114.29 \cos(\omega t - 18\pi x) \vec{a}_z, \vec{H}_t = -(114.29/80\pi)\cos(\omega t - 18\pi x)\vec{a}_y, 51.97$ W]

5.22 A 200 MHz uniform plane wave traveling in free space strikes a large block of material having $\epsilon_r = 4$, $\mu_r = 9$, and $\sigma = 0$ normal to the surface. If the incident magnetic field intensity vector is given by $\vec{H}_i = 1 \cos(\omega t - \beta y) \vec{a}_z$ A/m, write complete time-domain expressions for the incident, reflected, and transmitted field vectors. Determine the average power crossing a 5 m^2 area of the surface. $[\vec{H}_i = 1 \cos(\omega t - 4\pi/3y) \vec{a}_z, \vec{E}_i = -120 \pi \cos(\omega t - 4\pi/3y) \vec{a}_x, \vec{E}_r = -\frac{120}{5}\pi \cos(\omega t + 4\pi/3y) \vec{a}_x, \vec{H}_r = -\frac{1}{5} \cos(\omega t + 4\pi/3y) \vec{a}_z, \vec{E}_t = -\frac{720}{5}\pi \cos(\omega t - 8\pi y) \vec{a}_x, \vec{H}_t = 0.8 \cos(\omega t - 8\pi y) \vec{a}_z, 904.78$ W]

5.23 A 10 V/m uniform plane wave of frequency 3 MHz is incident from the free space normal to the surface of a material having $\epsilon_r = 4$, $\mu_r = 1$, and $\sigma = 10^3$ S/m. Determine the average power dissipated in a volume of the material that is 1 mm deep and has a surface area of 2 m^2. [59.8 μW].

5.24 Modify the ground reflection factors that were derived for Hertzian dipole antennas and given in Tables 5.1 and 5.2 for half-wave dipole antennas. [Horizontal—no change, vertical—replace $\cos \theta_i$ with $\cos(\frac{1}{2}\pi \sin \theta_i)/\cos \theta_i$ in $\hat{F}_V$]

5.25 A 300 Ω twin-lead transmission line is attached to an infinite biconical antenna. Determine the cone angle that will match the line to the antenna. [9.38°] Repeat this for a 50 Ω line. [66.79°] Repeat for the discone antenna shown in Fig. 5.25(a). [0.77°, 47°]

5.26 Determine the directive gain of the infinite biconical antenna. $[D(\theta, \phi) = [\sin^2 \theta \ln(\cot \frac{1}{2}\theta_h)]^{-1}]$ Determine the maximum gain for a cone angle that will match to a 50 Ω line. [2.84]

5.27 A log-periodic dipole array is to operate over the frequency band of 200 MHz–1 GHz. Determine the shortest and longest member lengths. [75 cm, 15 cm]

PART II

APPLICATIONS TO EMC DESIGN

CHAPTER SIX

Nonideal Behavior of Components

In this chapter we will discuss the typical circuit components used in the design of electronic systems and particularly in digital electronics. Our concentration will be on their role in suppression of radiated and conducted emissions and on their *nonideal* behavior. The latter is critical to their ability to provide adequate suppression. The reader must begin to think in terms not only of the component's *ideal* behavior but also of its *nonideal* behavior. An example is the frequency response of a capacitor's impedance. These components are often used to bypass or divert a high-frequency signal from, for example, a cable where the signal may radiate very efficiently. If the desired frequency of the emission is above the *self-resonant frequency* of the capacitor, the behavior of the capacitor will resemble that of an inductor, and the low impedance desired will not be realized.

Throughout this and later chapters it is critical for the reader to remember: *the primary frequencies of interest are those of the applicable governmental regulations.* For example, if the product is intended to be marketed only in the US then the emissions in the frequency range of the FCC limits (450 kHz–30 MHz for conducted emissions and 30 MHz–40 GHz for radiated emissions) are the *primary* frequencies of interest. A radiated emission occurring at 29 MHz is of no consequence in meeting the FCC regulatory limits! However, we cannot be totally unconcerned about the levels of emissions that are outside the frequency ranges of the regulatory limits, since these emissions may cause interference with other products, which will result in field problems and customer complaints. *Merely satisfying the applicable regulatory requirements does not represent a complete system design from the standpoint of EMC.*

In this chapter we will develop *mathematical models* that yield considerable insight into the *nonideal behavior* of components. Certain approximations will need to be made in developing a relatively simple model. Throughout this text we will frequently show experimentally measured data that serve to illustrate the prediction accuracy of the models that are developed. It is important to keep in mind that *if a postulated model fails to predict experimentally observed phenomena, it is useless!*

Our interest in a component's behavior will focus on the high frequencies of the regulations where it is to be used, to reduce conducted and/or radiated emissions. The ultimate test of whether a component will provide the anticipated performance at the desired frequency is to *experimentally measure the desired behavior* (*e.g., impedance*) *of the component at the desired frequency!* There exist a large number of commercially available test instruments that measure the high-frequency behavior of components. Most of these devices are computer-controllable and quite simple to use. One can therefore quickly and accurately determine whether a component will provide the desired EMI suppression through measurement.

6.1 WIRES

The conductors of a system (wires and printed circuit board, PCB, lands) are frequently overlooked as being important *components* of the system. Their behavior at the regulatory frequencies will be our primary concern here. In the radiated emission range (30 MHz–40 GHz) and to a lesser degree in the conducted emission range (150 kHz–30 MHz) the behavior of these elements is far from the ideal. Perhaps the most important effect, at least in digital circuits, is the conductor *inductance.* The resistance of the conductors is generally more important in the functional design as in determining the required land size and/or wire gauge to insure minimum voltage drop along them in a power distribution circuit. However, *at the frequencies of the regulatory limits and particularly in the radiated emission range the inductance of the conductors is considerably more important than the resistance.* We examine these topics in this section.

The term *wire* will be used in this text to refer to conductors that consist of one or more *solid, cylindrical conductors.* A single conductor is referred to as a *solid wire.* The wire has radius r_w and conductivity σ. The vast majority of conductor materials are copper (Cu), which has a dc conductivity $\sigma_{Cu} = 5.8 \times 10^7$ S/m. Normally the conductor is not ferromagnetic, and as such its permeability μ is that of free space: $\mu = \mu_o = 4\pi \times 10^{-7}$ H/m. Also, the permittivity of virtually all conductors is that of free space: $\epsilon = \epsilon_o \cong (1/36\pi) \times 10^{-9}$ F/m. Table 6.1 gives the relative conductivities (relative to Cu) σ_r and relative permeabilities (relative to free space) μ_r for various conducting materials.

Stranded wires are composed of several strands of solid wires of radii r_{ws} that are placed parallel to each other to give flexibility. *As a reasonable approximation, the resistance and internal inductance of a stranded wire consisting of S strands can be computed by dividing the resistance or internal inductance of a single strand of radius r_{ws} by the number of strands, S.* Thus we are essentially treating the stranded wire as being S identical wires that are connected, electrically, in *parallel. The external parameters of inductance and capacitance*

TABLE 6.1 Conductivities (Relative to Copper) and Permeabilities (Relative to Free Space) of Conductors.

Conductor	σ_r	μ_r
Silver	1.05	1
Copper–annealed	1.00	1
Gold	0.70	1
Aluminum	0.61	1
Brass	0.26	1
Nickel	0.20	1
Bronze	0.18	1
Tin	0.15	1
Steel (SAE 1045)	0.10	1000
Lead	0.08	1
Monel	0.04	1
Stainless steel (430)	0.02	500
Zinc	0.29	1
Iron	0.17	1000
Beryllium	0.10	1
Mu-metal (at 1 kHz)	0.03	20,000
Permalloy (at 1 kHz)	0.03	80,000

can be approximately computed by replacing the stranded wire with a solid wire of equivalent radius. Consequently, we can obtain the parameters we will need by considering only solid wires.

Numerous handbooks from wire manufacturers list not only the radius and number of strands of stranded wires but also list an equivalent *gauge* that roughly represents the overall radius of the bundle of strands. Wires are referred to by *gauge*, which represents a solid wire of a certain radius. Although there are several gauge definitions, the most common is the *American Wire Gauge* (*AWG*). Manufacturer handbooks also list the wire radius corresponding to the various wire gauges. Wire radii are typically given in the English unit system in terms of *mils*, where 1 mil = $\frac{1}{1000}$ inch = 0.001 inch. Table 6.2 gives the wire *diameters* for typical wire gauges.

TABLE 6.2 Wire Gauges (AWG) and Wire Diameters.

Wire Gauge	Wire Diameter (mils)	
	Solid	Stranded
4/0	460.1	522.0 (427 × 23)
		522.0 (259 × 21)
3/0	409.6	464.0 (427 × 24)
		464.0 (259 × 23)
2/0	364.8	414.0 (259 × 23)
		414.0 (133 × 20)

TABLE 6.2   Continued.

Wire Gauge	Wire Diameter (mils)	
	Solid	Stranded
1/0	324.9	368.0 (259 × 24)
		368.0 (133 × 21)
1	289.3	328.0 (2109 × 34)
		328.0 (817 × 30)
2	257.6	292.0 (2646 × 36)
		292.0 (665 × 30)
4	204.3	232.0 (1666 × 36)
6	162.0	184.0 (1050 × 36)
		184.0 (259 × 30)
8	128.5	147.0 (655 × 36)
10	101.9	116.0 (105 × 30)
		115.0 (37 × 26)
12	80.0	95.0 (165 × 34)
		96.0 (7 × 20)
14	64.1	73.0 (105 × 34)
		73.0 (41 × 30)
		73.0 (7 × 22)
16	50.8	59.0 (105 × 36)
		59.0 (26 × 30)
		60.0 (7 × 24)
18	40.3	47.0 (65 × 36)
		49.0 (19 × 30)
		47.0 (16 × 30)
		48.0 (7 × 26)
20	32.0	36.0 (41 × 36)
		36.0 (26 × 34)
		37.0 (19 × 32)
		35.0 (10 × 30)
22	25.3	30.0 (26 × 36)
		31.0 (19 × 34)
		30.0 (7 × 30)
24	20.1	23.0 (41 × 40)
		24.0 (19 × 36)
		23.0 (10 × 34)
		24.0 (7 × 32)
26	15.9	19.0 (7 × 34)
		20.0 (19 × 38)
		21.0 (10 × 36)
28	12.6	16.0 (19 × 40)
		15.0 (7 × 36)
30	10.0	12.0 (7 × 38)
32	8.0	8.0 (7 × 40)
34	6.3	7.5 (7 × 42)
36	5.0	6.0 (7 × 44)
38	4.0	

Stranded wires are specified in terms of a diameter equivalent to a corresponding solid wire. Stranded wires are also specified in terms of the number and gauge of the solid wires that make up the stranded wire as (number × gauge). It is a simple matter to convert wire radii in mils to wire radii in meters. For example, the radius of a #20 AWG wire is $r_w = 16$ mils. To convert this to meters, we multiply by unit ratios as described in Chapter 1:

$$r_w = 16 \text{ mils} \times \frac{1 \text{ inch}}{1000 \text{ mils}} \times \frac{2.54 \text{ cm}}{1 \text{ inch}} \times \frac{1 \text{ m}}{100 \text{ cm}}$$

$$= 16 \text{ mils} \times (2.54 \times 10^{-5} \text{ m/mil})$$

$$= 0.4064 \text{ mm}$$

Therefore, *to convert wire radii from mils to meters, multiply by* 2.54×10^{-5}.

Wires are normally covered with a cylindrical dielectric insulation for obvious reasons. The thickness of the dielectric insulation is typically of the order of the wire radius. There are various types of dielectric insulations used by wire manufacturers. Their handbooks list the dc (or low-frequency) values of relative permittivity ϵ_r for the various insulation materials. Table 6.3 lists ϵ_r for various insulation materials.

TABLE 6.3   Relative Permittivities of Insulation Dielectrics.

Material	ϵ_r
Air	1.0
Polyethylene foam	1.6
Cellular polyethylene	1.8
Teflon	2.1
Polyethylene	2.3
Polystyrene	2.5
Nylon	3.0
Silicone rubber	3.1
Polyvinylchloride (PVC)	3.5
Epoxy resin	3.6
Quartz (fused)	3.8
Epoxy glass	4.7
Bakelite	4.9
Glass (pyrex)	5.0
Mylar	5.0
Porcelain	6.0
Neoprene	6.7
Polyurethane	7.0

It is important to remember that *dielectric materials are not ferromagnetic and thus have relative permeabilities of free space,* $\mu_r = 1$. Therefore *wire insulations do not affect magnetic field properties caused by currents of the wires.*

6.1.1   Resistance and Internal Inductance of Wires

The dc resistance of a round wire of radius r_w, conductivity σ, and total length $\mathscr{L}$ was obtained in Chapter 4 and is given by

$$R = \frac{\mathscr{L}}{\sigma \pi r_w^2} \quad \Omega \tag{6.1}$$

As the frequency is increased, the current over the wire cross section tends to crowd closer to the outer periphery due to a phenomenon known as *skin effect*, as discussed in Chapter 4. Essentially, the current can be assumed to be concentrated in an annulus at the wire surface of thickness equal to the skin depth

$$\delta = \frac{1}{\sqrt{\pi f \mu_o \sigma}} \tag{6.2}$$

when the skin depth is less than the wire radius. Table 6.4 gives the skin depth of copper ($\sigma = 5.8 \times 10^7\ \text{S/m}$, $\epsilon_r = 1$, $\mu_r = 1$) at various frequencies.

Note that the skin depth becomes extremely small at frequencies in the range of the radiated emission regulatory limits. At roughly the middle of this band, 100 MHz, the skin depth is 0.26 mils. Current tends to be predominantly concentrated in a strip near the surface of a conductor of depth δ. Therefore a conductor carrying a high-frequency current utilizes only a very small fraction of the metal of that conductor.

Figure 6.1 illustrates the fact that the current in a round wire is uniformly distributed over the cross section at dc, but is increasingly concentrated in a narrow thickness of approximately one skin depth near the outer surface for

TABLE 6.4   Skin Depth of Copper.

f	δ
60 Hz	8.5 mm
1 kHz	2.09 mm
10 kHz	0.66 mm
100 kHz	0.21 mm
1 MHz	2.6 mils
10 MHz	0.82 mils
100 MHz	0.26 mils
1 GHz	0.0823 mils

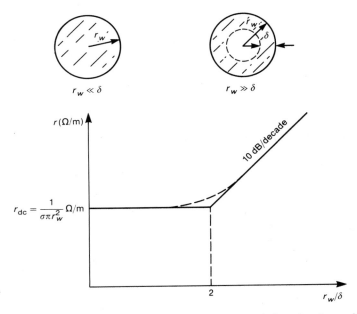

FIGURE 6.1 Illustration of the dependence of the per-unit-length wire resistance on frequency (skin effect).

higher frequencies. Since the resistance is proportional to cross-sectional area occupied by the current, the *per-unit-length* resistance becomes

$$r_{LF} = r_{DC} \quad \text{for } r_w \ll \delta \qquad (6.3a)$$

$$= \frac{1}{\sigma \pi r_w^2} \quad (\text{in } \Omega/\text{m})$$

$$r_{HF} = \frac{1}{\sigma [\pi r_w^2 - \pi(r_w - \delta)^2]} \quad \text{for } r_w \gg \delta \qquad (6.3b)$$

$$\cong \frac{1}{\sigma 2\pi r_w \delta}$$

$$= \frac{r_w}{2\delta} r_{DC}$$

$$= \frac{1}{2r_w} \sqrt{\frac{\mu_o}{\pi \sigma}} \sqrt{f} \quad (\text{in } \Omega/\text{m})$$

This is plotted in Fig. 6.1. Observe from (6.2) that the skin depth decreases with increasing frequency as the inverse square root of the frequency, $\sqrt{f}$. Thus the high-frequency resistance r_{HF} increases at a rate of 10 dB/decade. The resistance remains at the dc value up to the frequency where these two asymptotes meet, or $r_w = 2\delta$. The resistance in (6.3) is a *per-unit-length* resistance, with units of Ω/m. A length $\mathscr{L}$ of wire would have a total resistance $R = r\mathscr{L}$.

The isolated wire also has an inductance that is frequency-dependent. This is referred to as the *internal inductance*, since it is due to magnetic flux internal to the wire. The *dc internal inductance* was given in Chapter 4 and derived in [1] as

$$l_{i,DC} = \frac{\mu_o}{8\pi} \quad \text{for } r_w \ll \delta \tag{6.4a}$$

$$= 0.5 \times 10^{-7} \text{ H/m}$$

$$= 50 \text{ nH/m}$$

$$= 1.27 \text{ nH/inch}$$

This is also a *per-unit-length* parameter. A length $\mathscr{L}$ of conductor would have a total internal inductance $L_i = l_i\mathscr{L}$. For high-frequency excitation the current again tends to crowd toward the wire surface, and tends to be concentrated in a thickness δ. The per-unit-length internal inductance for these higher frequencies was also given in Chapter 4 and derived in [1], and becomes

$$l_{i,HF} = \frac{2\delta}{r_w} l_{i,DC} \quad \text{for } r_w \gg \delta \tag{6.4b}$$

$$= \frac{1}{4\pi r_w} \sqrt{\frac{\mu_o}{\pi\sigma}} \frac{1}{\sqrt{f}}$$

Since the skin depth δ decreases with increasing frequency as the inverse square root of the frequency, (6.4b) shows that the high-frequency, per-unit-length internal inductance *decreases* at a rate of -10 dB/decade for $r_w \gg \delta$. This frequency behavior is plotted in Fig. 6.2.

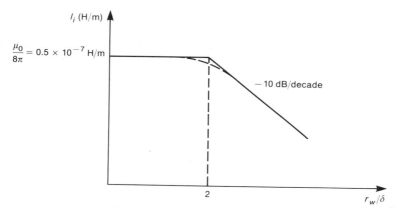

FIGURE 6.2 Illustration of the dependence of the per-unit-length internal inductance of wires on frequency (skin effect).

6.1.2 External Inductance and Capacitance of Parallel Wires

The resistance and internal inductance derived previously are uniquely attributable to or associated with a wire. Currents require a return path. The most common configuration is a pair of parallel wires of equal radii r_w, length $\mathscr{L}$, and separation s, as shown in Fig. 6.3. The magnetic flux external to each wire contributes to the total flux penetrating the area between the two wires. The per-unit-length *external inductance* l_e of a pair of wires is the ratio of the flux between the two wires, ψ_e, per unit of line length to the current producing that flux. This was derived in Chapter 4, and, assuming that the wires are separated sufficiently ($s/r_w > 5$) such that the current is uniformly distributed around the wire peripheries so that *proximity effect* is not a factor, is given as

$$l_e = \frac{\psi_e / \mathscr{L}}{I} \tag{6.5}$$

$$= \frac{\mu_o}{\pi} \ln\left(\frac{s}{r_w}\right) \quad \text{(in H/m)}$$

$$= 0.4 \ln\left(\frac{s}{r_w}\right) \quad \text{(in } \mu\text{H/m)}$$

$$= 10.16 \ln\left(\frac{s}{r_w}\right) \quad \text{(in nH/inch)}$$

The total *loop inductance* is the sum of the product of the line length and the internal inductances of the two wires and the product of the per-unit-length external inductance and the line length, i.e., $L_{\text{loop}} = 2l_i\mathscr{L} + l_e\mathscr{L}$. Note that $l_e\mathscr{L}$ is the inductance of the loop bounded by the two wires. A way of uniquely attributing portions of this external inductance to each wire is described in Section 6.1.4 by the use of *partial inductances* [2]. For our present purposes it is sufficient to observe that the external inductance is a loop inductance and may be assigned to either wire in the loop.

Charge on the wires contributes to a *per-unit-length capacitance c* between the two wires that depend on the wire separation and radii, as did the external inductance. This *per-unit-length capacitance* was derived in Chapter 4, and is

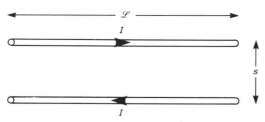

FIGURE 6.3 A pair of parallel wires to be modeled with an equivalent circuit.

the ratio of the per-unit-length charge on the wires, $Q/\mathscr{L}$, to the voltage between them:

$$c = \frac{Q/\mathscr{L}}{V} \qquad (6.6)$$

$$= \frac{\pi\epsilon_0}{\ln(s/r_w)} \quad \text{(in F/m)}$$

$$= \frac{27.78}{\ln(s/r_w)} \quad \text{(in pF/m)}$$

$$= \frac{0.706}{\ln(s/r_w)} \quad \text{(in pF/inch)}$$

This result assumes that the wires are separated sufficiently ($s/r_w > 5$) such that the charge is uniformly distributed around the wire peripheries and *proximity effect* is not a factor. The total capacitance between a pair of parallel wires of total length $\mathscr{L}$ is the product of the per-unit-length capacitance and the line length: $C = c\mathscr{L}$.

6.1.3   Lumped Equivalent Circuits of Parallel Wires

Each of these per-unit-length parameters when multiplied by the line length gives the total parameter for that length of line. If the total line length $\mathscr{L}$ is *electrically short*, i.e., $\mathscr{L} \ll \lambda$, at the frequency of excitation, we may lump these distributed parameters and obtain *lumped equivalent circuits* of the pair of wires. Combining the above elements gives several possible lumped-circuit models of the pair of parallel wires shown in Fig. 6.4. The *lumped-backward gamma* model of Fig. 6.4(a) is so named because of its resemblance to the Greek letter Γ. The remaining lumped-circuit models, *lumped-Pi*, *lumped-T*, and *lumped-Γ models* in the remaining parts of the figure are similarly named. Either of these models would constitute acceptable approximations of the line so long as the line is electrically short. However, depending on the impedance level of the load attached to the endpoint of the wires, one model will extend the prediction accuracy of the model further in frequency than another model. This is discussed and investigated in [3]. For example, if the load impedance Z_L is a "low impedance," i.e., much less than the *characteristic impedance* of the line,

$$Z_C = \sqrt{\frac{l_e}{c}} \qquad (6.7)$$

$$= 120 \ln\left(\frac{s}{r_w}\right) \quad \Omega$$

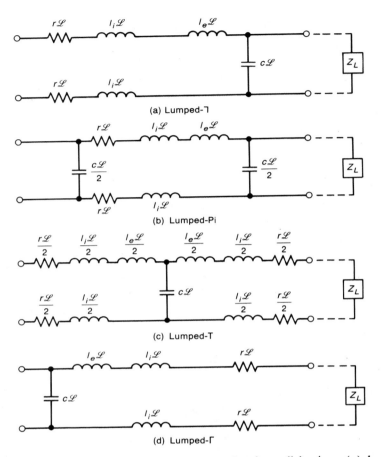

FIGURE 6.4 Lumped equivalent circuits for a pair of parallel wires: (a) lumped-backward Γ; (b) lumped-Pi; (c) lumped-T; (d) lumped-Γ.

the lumped-Γ model of Fig. 6.4(d) and the lumped-T model of Fig. 6.4(c) would extend the frequency range of adequate prediction slightly higher than the lumped-backward gamma and lumped-Pi models of Figs. 6.4(a) and 6.4(b). This is intuitively reasonable because a low impedance load would be in parallel with the rightmost parallel capacitance element of the lumped-backward gamma and lumped-Pi models. Thus these elements will be rendered ineffective by the low impedance load. The converse applies to "high-impedance" loads: the rightmost resistance and inductance elements of the lumped-Γ and lumped-T models are in series with this high-impedance load, and are therefore rendered ineffectual.

Once the per-unit-length parameters of the wires are determined and these lumped circuits are constructed, any lumped-circuit analysis program such as SPICE can be used to analyze the resulting circuit with the loads (gates, etc.)

attached. Important time-domain parameters such as rise time, wave shape, and time delay can be easily determined from that simulation.

An important point concerning these models that is frequently misunderstood needs to be discussed. Note that in either of the lumped circuits in Fig. 6.4 the external inductance l_e is in series with the internal inductance l_i. The impedance of the external inductance is $\omega L_e = 2\pi f l_e \mathscr{L}$, and therefore increases directly with frequency. (The external inductance is frequency-independent.) The impedance of the internal inductance is $\omega L_i = 2\pi f l_i \mathscr{L}$ and also appears to increase directly with frequency. However, upon closer examination, we recall that the per-unit-length internal inductance decreases with increasing frequency as the inverse square root of the frequency. Thus *the impedance of the internal inductance increases only as the square root of the frequency.* Therefore the impedance of the external inductance increases with frequency at a rate faster than that of the impedance of the internal inductance! Also, the external inductance for typical wire sizes and separations is usually much larger than the internal inductance. For example, consider a pair of #20 gauge solid copper wires that are separated by a distance of 50 mils (typical separation between adjacent conductors in a ribbon cable). The per-unit-length internal inductance is $l_{i,DC} = 0.05\ \mu\mathrm{F/m} = 1.27\ \mathrm{nH/inch}$, whereas the per-unit-length external inductance is $l_e = 0.456\ \mu\mathrm{H/m} = 11.58\ \mathrm{nH/inch}$, which is larger than the internal inductance by a factor of 10! Consider higher frequencies where $r_w > 2\delta$. The per-unit-length external inductance is larger than the per-unit-length internal inductance by a factor of 10, and above this frequency the difference increases since the external inductance remains constant with increasing frequency but the internal inductance decreases as $1/\sqrt{f}$. Consequently, *the impedance of the internal inductance is usually much smaller than the impedance of the external inductance, and we may therefore neglect the internal inductance in the model.* It is important to make these types of observations where possible based on typical dimensions in order to obtain the simplest model so that qualitative behavior can be more easily extracted from the model.

Several final points need to be discussed. In the above derivations of the per-unit-length external parameters l_e and c we assumed that the medium surrounding the wires was *homogeneous* and was that of free space with permittivity ϵ_o and permeability μ_o. Thus we assumed *bare wires in free space.* Wires often have circular dielectric insulation surrounding them to prevent contact with other wires. This type of medium is said to be *inhomogeneous*, since the electric and magnetic fields exist partly in the dielectric insulations (ϵ_r) and partly in air. Dielectrics are not ferromagnetic, and thus have $\mu = \mu_o$. Thus *the presence of inhomogeneous dielectric media does not affect the external inductance parameter.* However, since the surrounding medium is inhomogeneous in permittivity ϵ, equation (6.6) for the per-unit-length capacitance of the line *does not apply.* We cannot simply replace ϵ_o in that equation with the permittivity of the dielectric insulation, since the electric fields are not confined to the dielectric. Derivation of the per-unit-length capacitance for an inhomogeneous medium is a difficult problem, and closed-form expressions do not exist for this case—contrary to what some handbooks imply. Numerical methods must be applied in this case [4]. In spite of these technicalities, *we can obtain reasonable approximations for the inhomogeneous medium case by ignoring the dielectric insulations* and using the above expressions for l_e, (6.5), c, (6.6), for typical wire sizes, dielectric insulation thicknesses, and wire separations.

6.1.4   The Concept of Partial Inductance

The notion of the inductance of a subsegment of an isolated wire or PCB land is frequently misunderstood. Although inductance is a property of a *closed loop*, it is possible to uniquely ascribe portions of this loop inductance to segments of the loop. This is often mistakenly thought to be the internal inductance of the conductor due to magnetic flux internal to the conductor. This internal inductance decreases as the inverse square root of the frequency due to skin effect. However, this internal inductance is not the dominant inductance of the conductor segment. It is dominated by an external inductance (the *partial inductance* of the segment) that is frequency-independent.

In order to demonstrate this dominant, external inductance, consider the rectangular loop that supports a current I, as shown in Fig. 6.5(a). The current gives rise to a magnetic flux density $\vec{B}$. The inductance of this closed loop, L, is defined as the total magnetic flux ψ_m penetrating the surface s bounded by the current,

$$\psi_m = \int_s \vec{B} \cdot d\vec{s} \tag{6.8a}$$

per unit of that current [1, 2]:

$$L = \frac{\psi_m}{I} \tag{6.8b}$$

For later purposes we will identify currents associated with individual sides of the rectangle, I_i, but $I = I_1 = I_2 = I_3 = I_4$. We now wish to construct the equivalent circuit of the loop shown in Fig. 6.5(b). Its inductances L_{pii} are referred to as *self partial inductances* and the inductances L_{pij} with $i \neq j$ are referred to as *mutual partial inductances* [5, 6]. These may be defined in a unique and meaningful way by using an alternate form of (6.8). Since $\nabla \cdot \vec{B} = 0$, we may write $\vec{B}$ in terms of the magnetic vector potential $\vec{A}$ as [1]

$$\vec{B} = \nabla \times \vec{A} \tag{6.9}$$

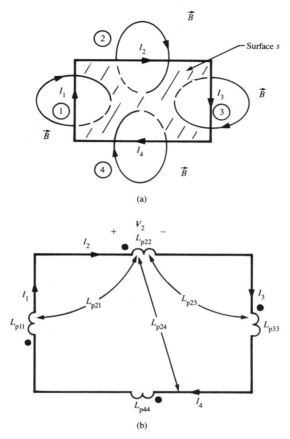

(a)

(b)

FIGURE 6.5 Illustration of the decomposition of loop inductance into partial inductances: (a) the physical circuit; (b) the equivalent circuit in terms of partial inductances.

Substituting this into (6.8) and using Stokes' theorem gives [1]

$$L = \frac{\oint_c \vec{A} \cdot d\vec{l}}{I} \qquad (6.10a)$$

where c is the contour of the loop. The curl of the magnetic vector potential $\vec{A}$ is defined in (6.9). To completely define a vector field, we must define its curl *and* its divergence [1]. The definition of the divergence of $\vec{A}$ is usually accomplished with the Lorentz choice of gauge [1]. Although there are other choices of gauge, so long as we are consistent the contour integral in (6.10a) can be broken into contributions uniquely associated with each segment of the loop as

$$
L = \frac{\int_{c_1} \vec{A} \cdot d\vec{l}}{I} + \frac{\int_{c_2} \vec{A} \cdot d\vec{l}}{I} + \frac{\int_{c_3} \vec{A} \cdot d\vec{l}}{I} + \frac{\int_{c_4} \vec{A} \cdot d\vec{l}}{I} \tag{6.10b}
$$

$$
= L_1 + L_2 + L_3 + L_4
$$

where c_i are identified with the four individual segments of the loop in Fig. 6.5(a) and

$$
L_i = \frac{\int_{c_i} \vec{A} \cdot d\vec{l}}{I} \tag{6.10c}
$$

This observation suggests that we may uniquely attribute portions of the loop inductance to segments of the loop.

The alternate result in (6.10) suggests that the partial inductances of the equivalent circuit in Fig. 6.5(b) be defined as

$$
L_{pij} = \frac{\int_{c_i} \vec{A}_{ij} \cdot d\vec{l_i}}{I_j} \tag{6.11}
$$

where $\vec{A}_{ij}$ is the magnetic vector potential along segment l_i due to current I_j on segment l_j. If $i = j$, these are referred to as self partial inductances, and, if $i \neq j$, as mutual partial inductances. With this definition, the voltage developed across a segment of a conductor can be uniquely and meaningfully obtained. For example, the voltage developed across segment 2 is

$$
V_2 = L_{p22} \frac{dI_2}{dt} + L_{p21} \frac{dI_1}{dt} + L_{p23} \frac{dI_3}{dt} + L_{p24} \frac{dI_4}{dt} \tag{6.12}
$$

We now turn to the important calculation and interpretation of these partial inductances. Ruehli [5] has shown that an alternative to (6.11) is

$$
L_{pij} = \frac{\int_{s_i} \vec{B}_{ij} \cdot d\vec{s_i}}{I_j} \tag{6.13}
$$

where s_i is the area bounded by the conductor i and infinity and by straight lines that are located at the ends of segment j and are perpendicular to segment j. This is illustrated in Fig. 6.6 for parallel segments. The extension to nonparallel

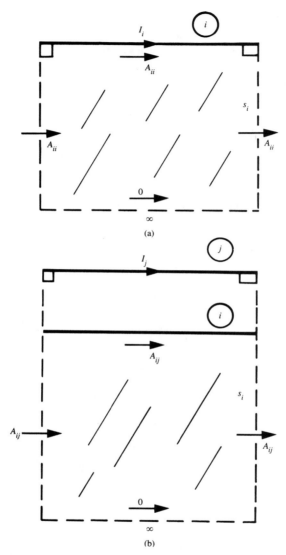

FIGURE 6.6   Illustration of the definition of partial inductances: (a) self partial inductance; (b) mutual partial inductance.

segments is straightforward and is given in [5]. The equivalence between (6.11) and (6.13) is important and simple to prove. Utilizing $\vec{B} = \nabla \times \vec{A}$ and Stokes' theorem, the numerator of (6.13) can be written as

$$\int_{s_i} \vec{B}_{ij} \cdot d\vec{s}_i = \oint_{c_i} \vec{A}_{ij} \cdot d\vec{l}_i \tag{6.14}$$

where $\vec{B}_{ij}$ is the total magnetic flux density penetrating s_i (which extends from segment i to infinity) and $\vec{A}_{ij}$ is the corresponding magnetic vector potential associated with $\vec{B}_{ij}$ along the closed contour c_i that bounds s_i. The magnetic vector potential $\vec{A}$ has two important properties that are crucial to this proof. These are that (1) $\vec{A}$ is parallel to the current producing it, and (2) $\vec{A}$ goes to zero as the distance away from the current increases [1]. By construction, the sides of s_i are perpendicular to segment j, whose current I_j produces $\vec{A}_{ij}$. Since $\vec{A}_{ij}$ is aligned with the current producing it, $\vec{A}_{ij}$ is perpendicular to the sides of s_i, and as such contributes nothing to the right-hand side of (6.14) along this portion of c_i. Also, $\vec{A}_{ij}$ at infinity is zero, and no contribution is obtained along the part of the contour. Consequently, the only contribution to the right-hand side of (6.14) is along segment i, as was to be proven. The important result is that *the partial inductance L_{pij} is the ratio of the magnetic flux penetrating the surface between segment i and infinity and the current I_j that produces it.* Equivalently, L_{pij} can also be determined in terms of the magnetic vector potential along segment i, as in (6.11). Either concept may be used in computing L_{pij}, but the notion of the relation to magnetic flux through the surface bounded by the segment and infinity is more useful in visualizing qualitative results.

The net inductance of a segment of a loop is the sum of the self and mutual partial inductances of that segment:

$$L_i = \sum_{j=1}^{N} \pm L_{pij} \qquad (6.15)$$

where the loop contains a total of N current segments, each supporting a current I_j, and the sign of each term is related to the relative orientation of the currents assigned to segments i and j. Note that for the rectangular loop shown in Fig. 6.5(a) mutual partial inductances exist only between segments that are not perpendicular to each other, since $\vec{A}_{ij}$ is perpendicular to segment i for segments j that are perpendicular to segment i. For example, $L_{p23} = L_{p21} = 0$ and $L_{p34} = L_{p14} = 0$. The total loop inductance can then be obtained as

$$L = \sum_{i=1}^{S} L_i \qquad (6.16)$$

where the loop is broken into S segments. Note that from knowledge of the self and mutual partial inductances of the loop the total loop inductance can be determined via (6.15) and (6.16), but the reverse is not true: each partial inductance must be computed directly and cannot be determined from a knowledge of the total loop inductance L alone. *In order for this concept of partial inductance to be meaningful, the segments must form a closed loop.* These notions can also be extended to systems of more than one loop. Each loop can be divided into segments, and the partial inductances due to currents on all segments of all loops can be obtained. Implicit in these results is the requirement that the electrical dimensions of a portion of the physical system that is represented with these lumped partial inductances be small.

We now give the results for the partial inductances of parallel current filaments of length l and separation d as illustrated in Fig. 6.7. The currents are assumed to be constant along the wire length and the segments are not offset. The result for offset segments is given in [5, 6, 8]. We must simply determine the magnetic flux between a conductor and infinity due to a current on that conductor L_{pii}, or on another parallel conductor, L_{pij}, according to (6.13). These basic results give the self and mutual partial inductances of *filamentary wires*; that is, we approximate wires of radii r_w with filaments of current. For round wires this is normally a reasonable approximation for the mutual partial inductance so long as the wires are separated sufficiently in terms of the ratio of their separation d to their radii r_w. Typically, a ratio $d/r_w > 5$ is sufficient in order to replace the wires with filaments. The self partial inductances will contain an internal self partial inductance due to flux internal to the conductor. Typically, this internal inductance is insignificant compared with the external partial self inductance due to flux external to the conductor. We showed previously that the dc internal inductance of a round wire of any radius is $\mu_o/8\pi = 0.5 \times 10^{-7}$ H/m $= 50$ nH/m $= 1.27$ nH/inch. In the following we will ignore the internal inductance.

We first give the mutual partial inductance for Fig. 6.7(b) according to (6.13) and set $d = r_w$ in that result to obtain the self partial inductance for Fig. 6.7(a). The magnetic flux density due to a dc current on a finite-length filament can be obtained with the Biot–Savart law [1] or by evaluating the field of a Hertzian dipole for $\omega = 0$ in equation (5.1c). In either case we obtain the magnetic flux density penetrating the desired surface in Fig. 6.7(b) due to a differential current

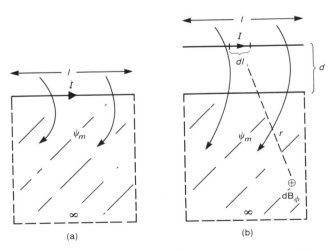

FIGURE 6.7   Illustration of the equivalence between vector magnetic potential and magnetic flux definitions of (a) self partial inductance; (b) mutual partial inductance.

element of length dl as $dB_\phi = (\mu_o I/4\pi r^2)\sin\theta\, dl$. Observe that the flux is in the ϕ direction, and is therefore perpendicular to the desired surface. Substituting this into (6.13) and performing the integration, the mutual partial inductance becomes [8]

$$L_{pij} = \frac{\mu_o}{2\pi}l\left\{\ln\left[\frac{l}{d}+\sqrt{\left(\frac{l}{d}\right)^2+1}\right]+\frac{d}{l}-\sqrt{\left(\frac{d}{l}\right)^2+1}\right\} \qquad (6.17)$$

The self partial inductance is obtained by substituting the wire radius r_w for d in (6.17), giving

$$L_{pii} = \frac{\mu_o}{2\pi}l\left\{\ln\left[\frac{l}{r_w}+\sqrt{\left(\frac{l}{r_w}\right)^2+1}\right]+\frac{r_w}{l}-\sqrt{\left(\frac{r_w}{l}\right)^2+1}\right\} \qquad (6.18)$$

Approximate values for these self and mutual partial inductances can be obtained for wire lengths that are much longer than the separation or wire radii, i.e., $l/d \gg 1$, $l/r_w \gg 1$. To obtain these, we use the series expansion for the inverse hyperbolic sine [7],

$$\sinh^{-1}x = \ln[x+\sqrt{x^2+1}] = \ln 2x + \frac{1}{4x^2}-\frac{3}{32x^4}+\frac{5}{96x^6}\cdots \qquad (6.19a)$$

and the binomial expansion to evaluate the square root,

$$(x+1)^n = 1 + \frac{nx}{1!}+\frac{n(n-1)x^2}{2!}+\frac{n(n-1)(n-2)x^3}{3!}\cdots \quad \text{for } |x|<1 \qquad (6.19b)$$

Application of the binomial expansion to the last square root in (6.17) gives, for $x = d/l \ll 1$ and $n = \frac{1}{2}$,

$$\sqrt{\left(\frac{d}{l}\right)^2+1} = 1 + \frac{1}{2}\frac{d}{l}-\frac{1}{8}\frac{d^2}{l^2}\cdots \qquad (6.19c)$$

Equation (6.17) for the mutual partial inductance is approximated for $d/l \ll 1$ by substituting the results in (6.19):

$$L_{pij} = \frac{\mu_o}{2\pi}l\left[\ln\left(\frac{2l}{d}\right)-1+\frac{d}{2l}+\frac{3d^2}{8l^2}\cdots\right] \qquad (6.20)$$

$$\cong 2\times 10^{-7}l\left[\ln\left(\frac{2l}{d}\right)-1+\frac{d}{2l}\right] \quad \text{for } \frac{d}{l}\ll 1$$

Frequently the last term, $d/2l$, can be removed from (6.20). Similarly, equation (6.18) for the self partial inductance is approximated for $r_w/l \ll 1$ by substituting the results in (6.19):

$$L_{pii} = \frac{\mu_0}{2\pi} l \left[\ln\left(\frac{2l}{r_w}\right) - 1 + \frac{r_w}{2l} + \frac{3r_w^2}{8l^2} \cdots \right] \tag{6.21}$$

$$\cong 2 \times 10^{-7} l \left[\ln\left(\frac{2l}{r_w}\right) - 1 \right]$$

It is worthwhile examining some numerical results. First consider a 1 inch length of #20 gauge wire ($r_w = 16$ mils). The exact self partial inductance from (6.18) is $L_{pii} = 19.53$ nH, whereas the approximate expression in (6.21) gives $L_{pii} = 19.49$ nH. Since the length is 1 inch, this shows that the often-quoted estimated value of 15 nH/inch for conductors is reasonable. A total length of 10 inches gives an exact value according to (6.18) of $L_{pii} = 311.5$ nH, whereas (6.21) gives an approximate value of $L_{pii} = 311.4$ nH. If we divide this by the total length of 10 inches, we obtain a per-unit-length self partial inductance of 31.15 nH/inch. The self partial inductance for a length of 20 inches is 693.3 nH, a factor of 2.2 over the value for a length of 10 inches. Dividing by the length gives a per-unit-length self partial inductance for the length of 20 inches of some 35 nH/inch. Observe that (6.21) can be written as

$$\frac{L_{pii}}{l} = \frac{\mu_0}{2\pi} \left[\ln\left(\frac{2l}{r_w}\right) - 1 \right] \tag{6.22}$$

and a per-unit-length self partial inductance cannot be discussed, since the right-hand side still contains the length l of the wire. The same remarks apply to the mutual partial inductance given in (6.17) or in (6.20). However, as a rough estimate, we can speak of a per-unit-length self partial inductance of a wire as being "around 15 nH/inch to 30 nH/inch." As a numerical example of mutual partial inductances, consider two #20 gauge wires of length 1 inch and separated a distance of $\frac{1}{4}$ inch. The result is obtained from (6.17) as $L_{pij} = 6.675$ nH. The approximate relation in (6.20) gives $L_{pij} = 6.119$ nH. Increasing these common lengths to 10 inches but preserving the separation of $\frac{1}{4}$ inch gives $L_{pij} = 173.1$ nH according to the exact expression in (6.17) or $L_{pij} = 172.4$ nH according to the approximate expression in (6.20). Increasing the lengths to 20 inches increases the mutual partial inductance to 414.7 nH, an increase of a factor of 2.5 over the value for 10 inch lengths. This illustrates again that it is not possible to give a unique value of per-unit-length partial inductance, self or mutual.

It is also important to recognize that increasing the wire radius does not substantially lower the wire self partial inductance. This is because the wire radius in (6.21) is involved in a natural logarithm term. For example, the self partial inductance of a 1 inch length of #28 gauge solid wire ($r_w = 6.3$ mils) is 24.18 nH, whereas that of a 1 inch length of #20 gauge solid wire ($r_w = 16$ mils) is 19.49 nH.

The decrease (increase) in the mutual partial inductance with an increase (decrease) in wire spacing also does not vary directly with the change in wire spacing, because the separation distance d is involved in a natural logarithm term. Table 6.5 shows this using the expression in (6.20) for three lengths (1 inch, 10 inches, and 20 inches) and for four separations ($\frac{1}{2}$ inch, $\frac{1}{4}$ inch, $\frac{1}{8}$ inch, and $\frac{1}{16}$ inch).

The self partial inductance is, of course, theoretically independent of the proximity to adjacent conductors. In fact, closely spaced conductors can alter the self partial inductance of one or both of the conductors. This is because closely spaced conductors will interact and cause the current distributions over the conductor cross sections to deviate from a uniform one which was implicitly assumed in the above derivations. Typically, this occurs only to a minor degree unless the ratio of wire separation to radius is less than approximately 5 : 1 [1]. A separation to radius ratio of 4 : 1 for two identical wires would mean that another wire of the same radius would exactly fit the space between the wires. Thus the above "wide-separation" approximations are reasonable from a practical standpoint.

The inductance of a closed loop can be computed from the equivalent circuit of partial inductances, but the reverse is not true. As an illustration, we will compute the inductance of the rectangular loop of side lengths l and d shown in Fig. 6.8(a). The equivalent circuit in terms of partial inductances is shown in Fig. 6.8(b). The voltage across a gap in the loop is computed as

$$V = sL_{pd}I - sM_{pd}I + sL_{pl}I - sM_{pl}I + sL_{pd}I - sM_{pd}I + sL_{pl}I - sM_{pl}I$$
$$= s[2(L_{pd} - M_{pd}) + 2(L_{pl} - M_{pl})]I \qquad (6.23)$$

from which we identify the inductance of the loop as

$$L_{\text{loop}} = 2(L_{pd} - M_{pd}) + 2(L_{pl} - M_{pl}) \qquad (6.24)$$

TABLE 6.5   Mutual Partial Inductance Between Two Parallel Wires For Various Spacings.

Separation	Common Length		
	1 inch	10 inches	20 inches
$\frac{1}{2}$ inch	3.23 nH	137.9 nH	344.9 nH
$\frac{1}{4}$ inch	6.12 nH	172.4 nH	414.7 nH
$\frac{1}{8}$ inch	9.32 nH	207.3 nH	484.8 nH
$\frac{1}{16}$ inch	12.7 nH	242.4 nH	555.0 nH

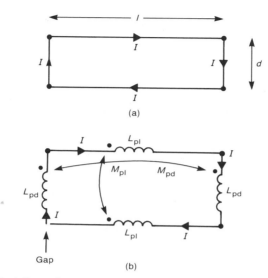

FIGURE 6.8 Modeling of a rectangular loop with partial inductances: (a) physical dimensions; (b) equivalent circuit.

It is worthwhile comparing the loop inductance of a pair of parallel, closely spaced wires computed using the per-unit-length loop inductance found earlier for a transmission line with the value computed using partial inductances in (6.24). For parallel wires whose length l is much larger than the separation d the first term in (6.24) is negligible compared with the second term. For example, consider a pair of #20 gauge solid wires of length 1 inch and separated by $\frac{1}{4}$ inch ($l = 1$ inch = 0.0254 m, $d = 0.25$ inch = 0.00635 m). Equation (6.24), using the partial inductance relations in (6.17) and (6.18), gives

$$L_{pd} = 3.181 \text{ nH}$$

$$M_{pd} = 0.1579 \text{ nH}$$

$$L_{pl} = 19.53 \text{ nH}$$

$$M_{pl} = 6.675 \text{ nH}$$

$$L_{\text{loop}} = 2(3.181 \text{ nH} - 0.1579 \text{ nH}) + 2(19.53 \text{ nH} - 6.675 \text{ nH})$$

$$= 6.046 \text{ nH} + 25.71 \text{ nH}$$

$$= 31.75 \text{ nH}$$

Computing these using the per-unit-length inductance in (6.5) gives

$$L = l_e \mathscr{L} = 27.93 \text{ nH}$$

If we increase the wire length to 10 inches and use the same separation of $\frac{1}{4}$ inch, the above values become

$$L_{pd} = 3.181 \text{ nH}$$

$$M_{pd} = 0.01579 \text{ nH}$$

$$L_{pl} = 311.5 \text{ nH}$$

$$M_{pl} = 173.1 \text{ nH}$$

$$L_{\text{loop}} = 2(3.181 \text{ nH} - 0.01579 \text{ nH}) + 2(311.5 \text{ nH} - 173.1 \text{ nH})$$

$$= 6.33 \text{ nH} + 276.9 \text{ nH}$$

$$= 283.2 \text{ nH}$$

Computing these using the per-unit-length inductance in (6.5) gives

$$L = 279.3 \text{ nH}$$

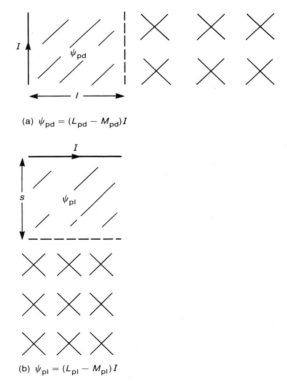

FIGURE 6.9 Physical understanding of partial inductances in terms of net magnetic flux: (a) end segments; (b) side segments.

Observe that for parallel wires with large length-to-separation ratios, the per-unit-length inductance derived assuming infinite-length wires (which does not include the inductances of and between the end sections) gives a close approximation to the correct value obtained with partial inductances.

Figure 6.9 explains the result for the loop inductance obtained using partial inductances and given in (6.24). The first term in (6.24) is the net magnetic flux threading the loop due to the currents in the two vertical end segments. In the context of partial inductances, $L_{pd} - M_{pd}$ is the ratio of the flux penetrating the loop divided by the current I in the left vertical segment. The self partial inductance L_{pd} multiplied by the current I is the magnetic flux due to the current in the left vertical segment that passes between the left segment and infinity. The mutual partial inductance M_{pd} multiplied by the current I is the magnetic flux due to the current in the left vertical segment that passes between the right segment and infinity. The difference in these two quantities is the net flux penetrating the loop divided by the current in the left vertical segment. This is multiplied by two to account for the flux penetrating the loop due to the vertical right segment. Similar interpretations apply to the horizontal segments, giving the second term. The concept of partial inductances as the ratio of the magnetic flux between the conductor and infinity for self partial inductances and between the other condutor and infinity for mutual partial inductances provides considerable insight into the meaning of inductance. A great deal of seemingly "mysterious" consequences of inductance can be easily explained using the concepts of partial inductances.

6.2 PRINTED CIRCUIT BOARD (PCB) LANDS

Wires are generally found in cables that interconnect subsystems and PCBs within systems. The conductors on PCBs have rectangular cross sections, as opposed to wires, whose cross sections ae circular. PCBs are composed of a dielectric substrate (typically glass–epoxy with $\epsilon_r \cong 4.7$) on which rectangular cross-section conductors (*lands*) are etched. Typical board thicknesses are of order 50 mils. Land thicknesses are specified in terms of the thickness of the board cladding that was etched away to form the lands. Typical cladding thicknesses are 1 ounce Cu and 2 ounce Cu. This refers to the weight of that thickness of the copper material that occupies an area of 1 square foot. For example, the thickness of 1 ounce Cu cladding is 1.38 mils, and a 1 square foot area would weigh 1 ounce. The thickness of 2 ounce Cu is double this, or 2.76 mils. Throughout this text we will assume the most common thickness of 1 ounce Cu or 1.38 mils.

The current distribution over the land cross section behaves in a manner that is quite similar to that of wires. For dc or low-frequency excitation the current is approximately uniformly distributed over the land cross section. Thus the per-unit-length low-frequency resistance of the land is

$$r_{LF} = r_{DC} \tag{6.25a}$$

$$= \frac{1}{\sigma w t} \quad (\text{in } \Omega/\text{m})$$

where w is the land width and t is the land thickness (1.38 mils). For high-frequency excitation the current tends to crowd to the outer edges of the land. Calculation of the high-frequency resistance is a difficult problem, but can be reasonably approximated by assuming that the current is uniformly distributed over a skin depth δ to give

$$r_{HF} = \frac{1}{\sigma(2\delta w + 2\delta t)} \tag{6.25b}$$

$$\cong \frac{1}{2\sigma\delta w} \quad (\text{in } \Omega/\text{m})$$

The land also possesses an internal inductance due to magnetic flux internal to the land in a fashion similar to that of a wire. However, in the case of a land computation of this internal inductance is a difficult problem. We will ignore this internal inductance parameter on the assumption that it will be neglible in comparison with the external inductance parameter in any lumped-circuit model.

We now consider two-conductor transmission lines formed from these rectangularly shaped lands. In addition to the resistance and internal inductance discussed previously, pairs of these lands will be characterized by external parameters of inductance l_e and capacitance c. These parameters are due to magnetic and electric flux, respectively, that are external to the lands in a fashion similar to pairs of wires. There are a number of ways to arrange these lands on the surface(s) of the dielectric board. The common ones are shown in cross section in Fig. 6.10. Figure 6.10(a) shows a configuration having a land on one side and a ground plane on the other, and is referred to in the microwave literature as the *microstrip* configuration. This configuration is common to microwave integrated circuits and is less-commonly found in PCB applications. The case of two parallel strips on the same side of the board shown in Fig. 6.10(b) is the most common form of transmssion line that will be encountered, and is referred to in the microwave literature as the *coplanar strips* configuration. This represents clock lands, address/data bus lands, etc. This configuration is also commonly used in microwave integrated circuits. The third case of two parallel strips on the opposite side of the board, shown in Fig. 6.10(c), is a less commonly encountered configuration for an important practical reason. Lands are normally etched on the top side of the board in one direction and are etched on the bottom side in a direction perpendicular to those of the top side on

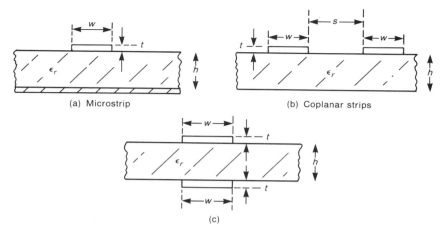

FIGURE 6.10 Cross-sectional dimensions of (a) microstrip, (b) coplanar strips, and (c) strips on opposite sides of the PCB.

double-sided boards. This is done to facilitate *wiring of the board*; that is, making it possible to interconnect the lands to provide the proper functionality of these signal paths. A signal path may follow a land on the top side of the board and then move to the bottom side using a *via* (a metallic hole through the board). The signal can then proceed in an orthogonal direction and move back to the top side with another via. This allows signal paths in a larger number of signal directions than would be possible on a *one-sided board*. Therefore signal lands on one side of a board are usually orthogonal to those on the other side. Thus the case of two parallel lands on opposite sides of the board shown in Fig. 6.10(c) is rarely encountered. Arranging lands on opposite sides of the board in orthogonal directions also has the important effect of minimizing *crosstalk* between these lands, which is discussed in Chapter 10.

Exact closed-form equations for the per-unit-length parameters of external inductance and capacitance for the configurations of Fig. 6.10 generally cannot be obtained. However, approximate, closed-form expressions can be obtained. These have generally been developed for microwave integrated circuit applications, but are applicable here since they are derived from static field considerations also. The methods of computation fall into two broad categories:

1. The method of conformal transformations.
2. Numerical methods such as
   (a) the finite-difference method [1],
   (b) the finite-element method, and
   (c) integral equation methods (method of moments) [1].

For these applications the reader is referred to [9–14]. All of these methods seek to determine the static (dc) per-unit-length capacitance for an infinitely long section of line (a) with the dielectric present, c, and (b) with the dielectric removed (replaced with air), c_0. Using the relation $l_e c = \mu \epsilon$ obtained for a *homogeneous medium* (μ, ϵ) in Chapter 4 and the fact that the inductance and magnetic field are not affected by an inhomogeneity in ϵ gives the per-unit-length inductance $l_e = (\mu_o \epsilon_o)/c_o = 1/v_o^2 c_o$.

These methods normally specify the per-unit-length parameters in terms of the characteristic impedance of the line, Z_C, which is

$$Z_C = \sqrt{\frac{l_e}{c}} = v l_e = \frac{1}{vc} \qquad (6.26)$$

The velocity of the waves on this line, v, is not the same as in free space due to the influence of the intervening dielectric board. This velocity is commonly expressed in terms of the *effective dielectric of the surrounding medium*, ϵ_r', as

$$v = \frac{1}{\sqrt{l_e c}} = \frac{1}{\sqrt{\mu_o \epsilon_o \epsilon_r'}} = \frac{v_o}{\sqrt{\epsilon_r'}} \qquad (6.27)$$

Thus, with knowledge of the characteristic impedance and effective dielectric constant, one can compute the per-unit-length external inductance and capacitance as

$$l_e = \frac{Z_C}{v} \qquad (6.28a)$$

$$c = \frac{1}{vZ_C} \qquad (6.28b)$$

Denoting the characteristic impedance, velocity of propagation, and per-unit-length capacitance *with the dielectric board replaced by free space* (*removed*) as Z_{C0}, v_o, and c_o, we can obtain alternate forms for the per-unit-length inductance as

$$l_e = \frac{Z_{C0}}{v_o} = \frac{1}{v_o^2 c_o} \qquad (6.29)$$

This result is obtained by realizing that *the dielectric of the board does not affect magnetic field parameters such as inductance, since the board permeability in the same as that of free space*, μ_0.

The notion of an effective dielectric constant for the PCB is important to understand. The medium surrounding the conductors is referred to as an *inhomogeneous medium* since the electric field exists partly in the board substrate and partly in the air surrounding the board, as illustrated in Fig. 6.11. These sketches of the electric field lines are easily seen if we apply a dc voltage between

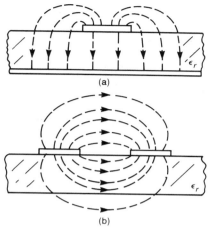

FIGURE 6.11 Cross-sectional field distributions for (a) microstrip and (b) coplanar strips.

the two conductors. The effective dielectric constant ϵ_r' is that of a fictitious dielectric material such that if the original conductors are immersed in a homogeneous material having this dielectric constant as shown in Fig. 6.12, *the corresponding lines in Figs. 6.11 and 6.12 will have the same characteristic impedance Z_C and velocity of propagation $v = 1/\sqrt{\mu_0 \epsilon_0 \epsilon_r'}$.* Evidently this effective dielectric constant will be a function of the dielectric constant of the board, ϵ_r, as well as the board thickness h, the land widths w, the land separations s, and,

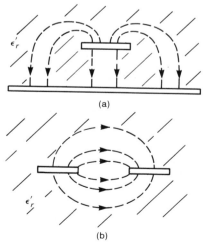

FIGURE 6.12 Replacement of an inhomogeneous dielectric with a homogeneous one having an effective dielectric constant of ϵ_r' for (a) microstrip and (b) coplanar strips.

to a lesser degree, the land thickness t. An approximate equation for this effective dielectric constant can be obtained by observing that, under certain conditions, roughly half the electric field lines will be in the air and the other half will be confined to the dielectric board. This would seem to be reasonable for the microstrip configuration if the board thickness is very large compared with the land width, $w/h \ll 1$, and for the case of coplanar strips if the strips are narrow and closely separated compared with the board thickness. For these cases we can reason that an effective dielectric constant would be the average of the dielectric constant of air and that of the dielectric board:

$$\epsilon_r' \cong \tfrac{1}{2}(\epsilon_r + 1) \tag{6.30}$$

More exact expressions are derived for microwave circuit applications in [10–14]. These are summarized for strips of zero thickness, $t = 0$, as

$$\epsilon_r' = \frac{\epsilon_r + 1}{2} + \frac{\epsilon_r - 1}{2}\left(1 + 10\frac{h}{w}\right)^{-\frac{1}{2}} \quad \text{(microstrip)} \tag{6.31}$$

$$\epsilon_r' = \frac{\epsilon_r + 1}{2}\left\{\tanh\left[0.775\ln\left(\frac{h}{w}\right) + 1.75\right]\right.$$

$$+ \frac{kw}{h}\left[0.04 - 0.7k + 0.01(1 - 0.1\epsilon_r)(0.25 + k)\right]\right\} \quad \text{(coplanar strips)} \tag{6.32a}$$

where k in (6.32a) is given by

$$k = \frac{s}{s + 2w} \tag{6.32b}$$

An empirical equation for the effective dielectric constant for the microstrip configuration is given in [9] as $\epsilon_r' = 0.475\epsilon_r + 0.67$, which was obtained from measured propagation delays of several microstrip boards. Observe that this empirical relation is a close approximation to (6.30): $\epsilon_r' = 0.5\epsilon_r + 0.5$. More exact expressions are rarely needed for our purposes, since dielectric constants of typical PCB board material have values of about 4.7. For example, consider typical dimensions of board thickness of 62 mils, land widths of 15 mils, and edge-to-edge spacing of 15 mils. Equation (6.30) gives $\epsilon_r' = 2.85$, equation (6.31) gives $\epsilon_r' = 3.134$, equation (6.32) gives $\epsilon_r' = 2.787$. Further realizing that ϵ_r' will enter into Z_C and v as the square root makes the approximation in (6.30) more adequate ($\sqrt{\epsilon_r'} = 1.688, 1.77, 1.669$). Therefore *it would be reasonable to use the approximate expression for the effective dielectric constant given in (6.30).* The velocity of propagation is then given by (6.27) in terms of this effective dielectric constant.

The value of the characteristic impedance for the microstrip configuration is [11–14]

$$
Z_C = \begin{cases} \dfrac{60}{\sqrt{\epsilon_r}} \ln\left(\dfrac{8h}{w} + 0.25 \dfrac{w}{h} \right) & \text{for } \dfrac{w}{h} \le 1 \\[4mm] \dfrac{377}{\sqrt{\epsilon_r}} \left[\dfrac{w}{h} + 1.393 + 0.667 \ln\left(\dfrac{w}{h} + 1.444 \right) \right]^{-1} & \text{for } \dfrac{w}{h} \ge 1 \end{cases}
\tag{6.33}
$$

The characteristic impedance for the coplanar strip configuration is [11–14]

$$
Z_C = \frac{377}{\sqrt{\epsilon_r}} \frac{K(k)}{K'(k)}
\tag{6.34a}
$$

where k is given by (6.32b), $K(\ )$ is the complete elliptic function of the first kind, and $K'(\ )$ is its complementary function given by $K'(k) = K(k')$, with $k' = \sqrt{1 - k^2}$. An accurate and simple expression for $K(k)/K'(k)$ is [13]

$$
\frac{K(k)}{K'(k)} = \begin{cases} \dfrac{1}{\pi} \ln\left(2 \dfrac{1 + \sqrt{k}}{1 - \sqrt{k}} \right) & \text{for } \dfrac{1}{\sqrt{2}} \le k \le 1 \\[4mm] \pi \left/ \ln\left(2 \dfrac{1 + \sqrt{k'}}{1 - \sqrt{k'}} \right) \right. & \text{for } 0 \le k \le \dfrac{1}{\sqrt{2}} \end{cases}
\tag{6.34b}
$$

For the microstrip line with $w = 15$ mils, $h = 62$ mils, and $\epsilon_r = 4.7$, we compute from (6.33) $Z_C = 118.635 \ \Omega$. For coplanar strips with $w = 15$ mils, $s = 15$ mils, $h = 62$ mils, and $\epsilon_r = 4.7$ we compute from (6.34) $Z_C = 144.445 \ \Omega$. A simpler but more approximate relation for the characteristic impedance of the microstrip configuration is given in [9] as

$$
Z_C = \frac{87}{\sqrt{\epsilon_r + 1.41}} \ln\left(\frac{5.98h}{0.8w + t} \right).
$$

This relation is valid for land thickness-to-width ratios of $0.1 \le t/w \le 0.8$. For 1 ounce Cu land, where $t = 1.38$ mils, this restriction is satisfied so long as 1.725 mils $\le w \le 13.8$ mils. For the microstrip with $w = 15$ mils, $h = 62$ mils, and $t = 0$, we obtain from this relationship $Z_C = 120.747 \ \Omega$, as compared with the value of $118.635 \ \Omega$ computed from (6.33).

For the case of two equal-width strips that are on opposite sides of the board, as shown in Fig. 6.10(c), results are obtained in [10]. These are

$$Z_C = \frac{377}{\sqrt{\epsilon_r}\left\{\frac{w}{h} + 0.441 + \frac{\epsilon_r + 1}{2\pi\epsilon_r}\left[\ln\left(\frac{w}{h} + 0.94\right) + 1.451\right] + 0.082\frac{\epsilon_r - 1}{(\epsilon_r)^2}\right\}} \quad (6.35a)$$

$$\text{for } \frac{w}{h} > 1$$

and

$$Z_C = \frac{377\sqrt{2}}{\pi\sqrt{\epsilon_r + 1}}\left[\ln\left(\frac{4h}{w}\right) + \frac{1}{8}\left(\frac{w}{h}\right)^2 - \frac{1}{2}\frac{\epsilon_r - 1}{\epsilon_r + 1}\left(0.452 + \frac{0.242}{\epsilon_r}\right)\right] \quad (6.35b)$$

$$\text{for } \frac{w}{h} < 1$$

For a glass–epoxy board with $w = 200$ mils and $h = 62$ mils we obtain from (6.35a) $Z_C = 41.05\ \Omega$. For coplanar strips with $w = 200$ mils, $s = 62$ mils, and $h = 62$ mils we compute from (6.34) $Z_C = 155.7\ \Omega$. This illustrates the point that *pairs of wide lands placed on opposite sides of the board will give characteristic impedances lower than placing the lands on the same side of the board* (and using a spacing equal to the board width). Low-impedance power distribution circuits can be obtained in this fashion. A rough approximation for strips on the opposite of the board that are wide compared with the board thickness can be obtained by making the analogy to a parallel-plate capacitor (the strips must be wide compared to the board thickness in order to neglect the fringing of the fields) to give

$$Z_C \cong \frac{377}{\sqrt{\epsilon_r}}\frac{h}{w} \quad \text{for } \frac{w}{h} \gg 10 \quad (6.36)$$

The effective dielectric constant can be approximated as the board dielectric constant if the strips are wide compared with the board thickness, so that fringing can be neglected. This restriction of strips much wider than their separation (board thickness) is not generally fulfilled for typical printed circuit boards. For example, for $w = 15$ mils and $h = 62$ mils $w/h = 0.242$, and (6.36) would not be applicable. However, for $w = 200$ mils and $h = 62$ mils equation (6.36) gives $53.9\ \Omega$ (using $\epsilon_r = 4.7$), whereas (6.35a) gives $41.05\ \Omega$. For this latter case $w/h = 3.2$.

The concepts of partial inductance developed for wires in Section 6.1.4 apply to PCB lands in like fashion. Consider a PCB land of length l, width w, and thickness t. The self partial inductance is given in [8]. Typical PCB lands have thicknesses t that are much smaller than the land widths w. For example, 1 ounce copper lands have thicknesses of about 1.4 mils. Typical widths are of order 10–20 mils. Thus we will assume that the PCB land is a "tape" with zero thickness. Assuming the current to be uniformly distributed over the tape cross section, the result for the self partial inductance given in [8] is

$$L_{p\,\text{tape}} = \frac{\mu_0}{2\pi} l \left\{ \ln(u + \sqrt{u^2 + 1}) + u \ln\left[\frac{1}{u} + \sqrt{\left(\frac{1}{u}\right)^2 + 1}\right] \right.$$
$$\left. + \frac{u^2}{3} + \frac{1}{3u} - \frac{1}{3u}(u^2 + 1)^{3/2} \right\} \qquad (6.37)$$

where $u = l/w$. This result can be simplified for land lengths l that are long compared with their width, i.e., $u = l/w \gg 1$. Using the binomial expansion given in (6.19b) to expand the last term in (6.37) as $(1 + u^2)^{3/2} \cong u^3 + \frac{3}{2}u$ and approximating the two logarithmic terms gives

$$L_{p\,\text{tape}} \cong \frac{\mu_0}{2\pi} l \left(\ln 2u + 1 + \frac{1}{3u} - \frac{1}{2} \right) \qquad (6.38)$$
$$= 2 \times 10^{-7} l \left[\ln\left(\frac{2l}{w}\right) + \frac{1}{2} \right]$$

which agrees with Grover's approximate result [6, p. 35].

The partial mutual inductance between two parallel lands can be reliably approximated as being that between two filamentary wires if the separation between the two lands is large [5]. The expression given in (6.20) can be used for this case. If the lands are close together, the exact expression is given in [8]. It is quite complicated and will not be reproduced here.

As a numerical example, consider a PCB land of width 10 mils and length 1 inch. The ratio of land width to land thickness is $w/t = 7.14$. The expression in (6.37), which assumes a zero-thickness land, should be adequate for this case and gives ($u = 100$) $L_p = 29.47$ nH. The alternative expression in (6.38) gives $L_p = 29.46$ nH. For a PCB land of width 10 mils and length 10 inches ($u = 1000$) equation (6.37) gives $L_p = 411.5$ nH, whereas (6.38) gives $L_p = 411.5$ nH. This shows that, although we cannot speak of a per-unit-length inductance, for the 10 mil wide land a rough idea of the per-unit length inductance is of order 30–40 nH/inch.

It is instructive to compare the loop inductance between a pair of parallel PCB lands computed using partial inductances and computed using the characteristic impedance. For example, consider the case of coplanar strips illustrated in Fig. 6.10(b). Assume the lands are 1 inch in length, are both of width 15 mils, and are separated center-to-center by 30 mils ($s = 15$ mils). Assume that the board is glass–epoxy ($\epsilon_r = 4.7$) and of thickness 62 mils. The characteristic impedance was computed earlier as $Z_C = 144.445\ \Omega$ using (6.34), and the effective dielectric constant was computed as $\epsilon_r' = 2.787$ using (6.32). The per-unit-length external inductance is computed using the relations in (6.27) and (6.28a) to be $l_e = Z_C/v = (Z_C/v_o)\sqrt{\epsilon_r'} = 803.7$ nH/m. Multiplying this by the line length of $\mathscr{L} = 1$ inch $= 0.0254$ m gives a total loop inductance $L = 20.42$ nH. Alternatively, we may compute this using partial inductances. We will use only the second term of (6.24), since we showed that the first term due to the vertical end segments is usually negligible in comparison. Equation (6.38) gives the self partial inductance as $L_{pl} = 27.40$ nH. The result for mutual partial inductance between two filaments given in (6.20) gives $M_{pl} = 16.41$ nH. Thus the net loop inductance is $L = 2(L_{pl} - M_{pl}) = 21.98$ nH, which is quite close to the value of 20.42 nH computed using the characteristic impedance.

6.3 EFFECT OF COMPONENT LEADS

We now embark on an examination of the various discrete components, resistors, capacitors, inductors, etc., that are employed in electronic systems. Our emphasis will be on their *nonideal behavior* in the high-frequency range of the regulatory limits. A component must inevitably be connected to the circuit via *leads*. These connection leads usually take the form of bare wires such as the attachment leads of resistors, capacitors, etc. This is referred to as *discrete–lead attachment*. There is an increasing use of other attachment techniques that speed automated assembly of the components on the printed circuit boards (PCBs). Perhaps the most common alternative is the *surface–mount technology* or SMT method. With this method, flat, rectangular–cross-section "tabs" attached to the component package are soldered directly to the PCB. Not only does this reduce the length of the attachment leads (an important factor in achieving the desired behavior of the component), but it also speeds the automated attachment of the component to the PCB. It also allows an increased number of components to be placed on the PCB over the discrete–lead attachment method. Components are normally placed on only one side of a PCB. With the use of SMT components, many of the smaller components such as resistors and capacitors can be placed on the other side of the PCB, thereby increasing the component density. Most PCBs in today's electronic systems could not be "populated" in the allowable board space without the use of SMT components. We will concentrate on the discrete–lead components, although many of our results will also be applicable to SMT components.

One of the most important factors that affect the high-frequency behavior of components is the *length of the component attachment leads*. Unnecesarily long attachment leads cause the component behavior to deviate from the ideal at high frequencies, which often fall in the frequency range of the regulatory limits where we want the component to behave as expected. The *length and separation of the component leads* cause the component to have, in addition to the ideal behavior, an *inductive* element and a *capacitive* element. These elements in combination with the component can give an overall behavior that is far from the desired ideal behavior.

In order to model the inductance of the attachment leads, consider the discrete lead attachment shown in Fig. 6.13(a). The inductance of this loop can be obtained using the previous results by either multiplying the per-unit-length inductance of the pair of parallel wires given in (6.5) by the lead length or by using the concept of partial inductance given in (6.24). As an example, consider typical component leads that are #20 gauge solid wires ($r_w = 16$ mils). Suppose the leads are 0.5 inch long and are separated by 0.25 inch. The expression in (6.5) gives 14 nH. The equivalent circuit becomes as shown in Fig. 6.13(b). The inductance we have computed is the inductance of the loop, and as such we may lump it and place it in either lead. By way of comparison using partial inductances, (6.24) gives a loop inductance of 17.63 nH. Using only the second term of (6.24) gives 12 nH. Both methods give about the same value of loop inductance, but to simplify calculation, we will compute the lead loop inductance using the per-unit-length result in (6.5).

The next effect that we will consider is the capacitance between the leads. This may be computed by multiplying the per-unit-length capacitance given in (6.6) by the lead length. As an example, consider two #20 gauge leads of length 0.5 inch separated by 0.25 inch. The capacitance is 0.128 pF.

The lumped–circuit model of lead inductance is shown in Fig. 6.13(b), and the model of lead capacitance is shown in Fig. 6.14(b). How shall we combine these two effects into a single model? There is no unique answer to this question, since these are *distributed parameter elements*. In other words, their effects are distributed along the length of the leads, as illustrated in Fig. 6.15(a).

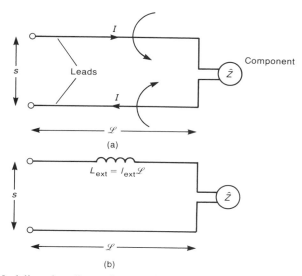

FIGURE 6.13 Modeling the effects of magnetic fields of component leads: (a) physical configuration; (b) the equivalent circuit.

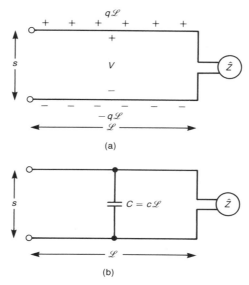

FIGURE 6.14 Modeling the effects of electric fields of component leads: (a) physical configuration; (b) the equivalent circuit.

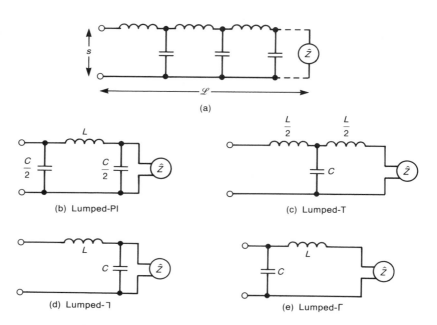

FIGURE 6.15 Equivalent circuits of component leads: (a) distributed parameter; (b) lumped-Pi; (c) lumped-T; (d) lumped-backward Γ; (e) lumped-Γ.

Nevertheless, if the lead length $\mathscr{L}$ and separation s are *electrically short* at the frequencies of interest, we may lump L and C and produce several *lumped–circuit models* that are identical in structure to those derived for the pair of parallel wires in Section 6.1.3. Figures 6.15(b)–(e) again show these four possible equivalent circuits. Again, although either circuit would be an approximate representation of this distributed-parameter phenomenon for electrically short lead lengths, one structure may be a better approximation than the other, depending on the impedance $\hat{Z}$ of the component, as was discussed in Section 6.1.3. In fact, we are only interested in estimates of the effect of the leads, and for this purpose either model would be adequate.

6.4   RESISTORS

Resistors are perhaps the most common component in electronic systems. These components are constructed in basically three forms: (1) carbon composition, (2) wire wound, and (3) thin film. Carbon-composition resistors are the most common. They are constructed by forming a cylindrical block of carbon and attaching two wires to the ends. Wire-wound resistors are formed by winding a length of wire that has the desired *dc resistance* on a cylindrical form to conserve space. Wire-wound resistors have a significant amount of inductance due to the construction technique. It is usually difficult to determine whether a resistor is carbon-composition or wire-wound by simply looking at it. The desired length of the wire used to construct a wire-wound resistor can be computed from (6.1). Thin-film resistors are constructed by depositing a thin, metallic film on an insulating substrate. Leads are attached to the ends of the metallic film, and the package resembles an axial–lead resistor. Because of the construction technique, this resistor has more precise values of resistance than the carbon-composition type but less inductance than the wire-wound type.

The ideal frequency response of a resistor has a magnitude equal to the value of the resistor and a phase angle of 0° for all frequencies as shown in Fig. 6.16.

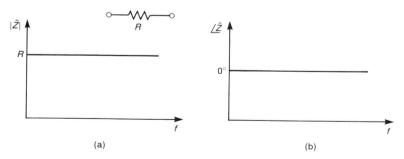

FIGURE 6.16   Frequency behavior of the impedance of an ideal resistor: (a) magnitude; (b) phase.

We denote this as

$$\hat{Z} = R \, \underline{/0^\circ} \tag{6.39}$$

Actual resistors behave somewhat differently than this ideal at higher frequencies, with the degree to which they differ depending on the construction technique used. For example, since a wire-wound resistor is constructed of turns of wire, we would expect this resistor to have a significant inductive behavior at higher frequencies. Carbon-composition resistors would not be expected o exhibit this to the same degree. Consequently, if the current passing through the resistor has a large di/dt factor, we would be well advised to use a carbon-composition resistor here instead of a wire-wound resistor. Otherwise, the wire-wound resistor would have a terminal behavior represented by $v(t) = Ri(t) + L \, di(t)/dt$. An example of where this nonideal behavior would be very undesirable would be in the use of this resistor as a "sense resistor" in the source lead of a field-effect transistor that is used as the switch element in a switching power supply. The voltage developed across this resistor is *intended* to be a replica of the current passing out this lead of the transistor, and is used as a control to affect the duty cycle of the transistor switch. However, since the current is rapidly changing with time, the inductive nature of a wire-wound resistor may cause the voltage developed across it to resemble the derivative of this current, which would not be desirable.

The advantage of wire-wound resistors over carbon-composition ones is that much tighter tolerances on element value can be obtained. For example, carbon resistors typically have tolerances of 5–10%. This means that the manufacturer guarantees only that, for example, a 1 kΩ resistor would have a value between 1.1 kΩ and 900 Ω for a 10% tolerance. For the *sense* resistor in the above switching power supply it is important to use a small value of resistance so that the functional performance of the transistor switch will not be impaired. Typical values are of order 1 Ω. The proper operation of the switcher depends on obtaining accurate values of the sampled current, which a 10% tolerance carbon resistor may not give. Consequently a wire-wound resistor might be used in this application. From a functional standpoint, this inductive behavior of the wire-wound resistor can be tolerated. From an EMC standpoint, however, the differentiation of this switch waveform causes pulses of voltage to be developed across the resistor that have a repetition rate of the basic switch frequency *and* very fast rise/fall times. We will see in the next chapter that the spectral content of such signals extends well above the repetition rate of the signal, so that this could cause radiated and/or conducted emission problems.

Both carbon-composition and wire-wound resistors exhibit other nonideal effects. For example, there is a certain "bridging capacitance" from end-to-end due to charge leakage around the resistor body. Usually this is a minor effect. A more significant effect is represented by the *inductance and capacitance of the leads attached to the element*, as was discussed in the previous section. Replacing the leads with a lumped-backward Γ equivalent circuit gives the model shown in Fig. 6.17(a). We could have also chosen to use any of the other models of Fig. 6.15, but will choose the lumped-backward Γ model for simplicity. Thus the equivalent circuit of the resistor is as shown in Fig. 6.17(b). The lead inductance L_{lead} in this model refers to the inductance of the loop area bounded by the two leads. Values calculated for typical lead lengths of 0.5 inch, lead separations of 0.25 inch, and lead wires ($\#20$ gauge with $r_w = 16$ mils) using (6.5) give L_{lead} of some 14 nH. (The separation is largely determined by the length of the resistor body when the leads are bent at right angles to the body.) The *parasitic* capacitance in this model refers to the parallel combination of the lead and leakage capacitances, $C_{par} = C_{lead} + C_{leakage}$. Typical values are $C_{par} \cong 1$–2 pF. Values of C_{lead} calculated for typical lead lengths of 0.5 inch, lead separations of 0.25 inch, and lead wires ($\#20$ gauge with $r_w = 16$ mils) using (6.6) give C_{lead} of some 0.128 pF. This is probably smaller than the leakage

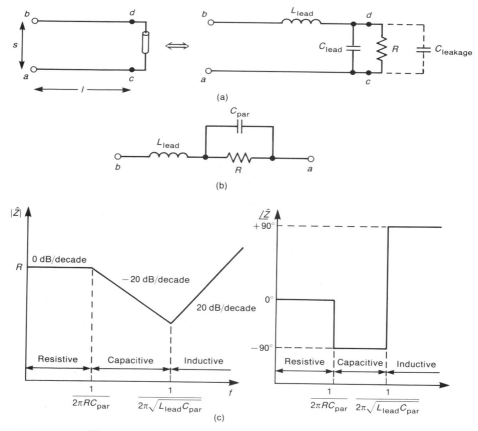

(a)

(b)

(c)

FIGURE 6.17   The nonideal resistor including the effects of the leads: (a) equivalent circuit; (b) simplified equivalent circuit; (c) Bode plots of the impedance variation with frequency.

capacitance of the resistor body. It is instructive to calculate the effect of these elements. Consider a 1 kΩ resistor. If we assume a value of the parasitic capacitance of 1 pF, the impedance of C_{par} is 1 kΩ at a frequency of approximately 159 MHz. The inductance and capacitance resonate at a frequency of approximately 1.3 GHz. This illustrates that for high-impedance resistors the parasitic capacitance is the dominant element effect.

In order to examine the frequency response of this model, we first derive the equation for the impedance of the model. A simple way of doing this is described in [15]. First replace the inductors with their impedances in terms of the complex variable $p = \sigma + j\omega$ to give $\hat{Z}_L = pL$, and replace the capacitors with their impedance $\hat{Z}_C = 1/pC$. Derive the expression for the impedance of the element, $\hat{Z}(p)$, and then substitute $p = j\omega$ in that expression. The impedance $\hat{Z}(p)$ here is a general form of a "transfer function" giving the *ratio* of two circuit quantities (current and/or voltage) [15]. The transfer function here is the ratio of the terminal voltage $\hat{V}(p)$ and the terminal current $\hat{I}(p)$ as $\hat{Z}(p) = \hat{V}(p)/\hat{I}(p)$, where the voltage and current are defined with the passive sign convention [15]. For the resistor model of Fig. 6.17(b) one can derive

$$\hat{Z}(p) = L_{lead} \frac{p^2 + p/RC_{par} + 1/L_{lead}C_{par}}{p + 1/RC_{par}} \tag{6.40}$$

Substituting $p = j\omega$ into this expression gives

$$\hat{Z}(j\omega) = L_{lead} \frac{1/L_{lead}C_{par} - \omega^2 + j\omega/RC_{par}}{j\omega + 1/RC_{par}} \tag{6.41}$$

The corresponding Bode or asymptotic plot [15] of the magnitude and phase angle of this impedance is given in Fig. 6.17(c). We will frequently employ the logarithmic or Bode plot method of displaying the frequency response of elements. The reader should review this method, which is described in any typical circuit analysis text. A complete discussion is given in [15]. The basic method is to plot not the magnitude $|\hat{Z}(j\omega)|$ but the logarithm of the magnitude, $|\hat{Z}(j\omega)|_{dB} = 20 \log_{10} |\hat{Z}(j\omega)|$, in decibels (above or relative to a reference level of 1 Ω). In order that straight lines on the "unlogged" plot translate to straight lines on the log plot, the frequency axis must be plotted as $\log_{10} f$. This is usually more easily accomplished using semilog graph paper where the vertical axis has linear tick mark spacing for plotting $|\hat{Z}(j\omega)|_{dB} = 20 \log_{10} |\hat{Z}(j\omega)|$ and logarithmic spacing of the tick marks on the horizontal axis for plotting the frequency on a logarithmic basis. This is implied when we label the horizontal axes of these plots as simply f. We could also use log–log graph paper with logarithmically spaced tick marks on the vertical axis for plotting the absolute magnitude instead of the magnitude in dB (relative to 1 Ω).

It is important at this point to consider another computational technique. The reader should be able to not only compute the "transfer function" of a circuit but also quickly check the accuracy of the result and determine the gross behavior of the frequency response. In order to do this we simply check, directly from the circuit, the behavior at two frequencies: dc and infinite frequency. In order to check the behavior at dc, we simply substitute $p = 0$ into any impedance expression. For example, substituting $p = 0$ into $\hat{Z}_L = pL$ and $\hat{Z}_C = 1/pC$ gives

$$\hat{Z}_L = 0 \big|_{f=0} \tag{6.42a}$$

$$\hat{Z}_C = \infty \big|_{f=0} \tag{6.42b}$$

In other words, an inductor (an ideal one) is a *short circuit* at dc and a capacitor is an *open circuit* at dc. This can be checked directly from the circuit by replacing the inductor with a short circuit and replacing the capacitor with an open circuit. Once this is done, we see that the behavior of the model at dc is the same as an ideal resistor. As we increase the frequency, the impedance of the capacitor decreases and tends to "short out" the resistor of the model. This begins to occur at a frequency where the impedance of the capacitor equals the resistance, or $\omega_1 = 1/RC_{\text{par}}$. Thus the net impedance decreases at -20 dB/decade and the phase angle approaches $-90°$ above this frequency. At a point where the inductor and capacitor of the model *resonate*, $\omega_0 = 1/\sqrt{L_{\text{lead}}C_{\text{par}}}$, the impedance of the model is at a minimum. (Actually, this minimum occurs at a frequency that is slightly above this resonant frequency, with it approaching this frequency the smaller the value of R.) Above this resonant frequency, the impedance of the inductor becomes dominant and the magnitude of the impedance increases at 20 dB/decade and the phase angle approaches $+90°$. Finally, as the frequency approaches infinity, the inductor behaves as an open circuit and the capacitor behaves as a short circuit, so that the net impedance of the model approaches that of an open circuit (due primarily to the inductor):

$$\hat{Z}_L = \infty \big|_{f=\infty} \tag{6.43a}$$

$$\hat{Z}_C = 0 \big|_{f=\infty} \tag{6.43b}$$

Since the inductance was dominant for higher frequencies, the phase angle approaches $90°$. All of this behavior is borne out by the transfer function that we derived. However, it is always a good idea to perform these simple checks. Also, an understandng of the above simple principles can be an aid, along with the understanding of the physical construction of the element, in the construction of a suitable model that will represent this nonideal behavior. This examination of the model over distinct frequency ranges is represented in Fig. 6.18. The reader should study this method since it will be used on numerous occasions to examine and construct models of elements and devices.

We will frequently present and examine experimentally obtained data. The purpose in doing so is twofold. First, it is not possible to construct one model that will apply to all frequencies, so we will need to accept some approximate

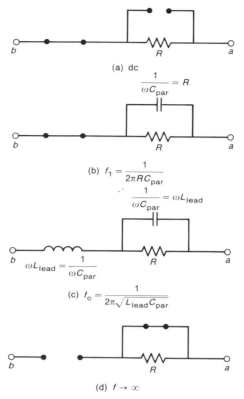

(a) dc

$$\frac{1}{\omega C_{par}} = R$$

(b) $f_1 = \dfrac{1}{2\pi R C_{par}}$

$$\frac{1}{\omega C_{par}} = \omega L_{lead}$$

$\omega L_{lead} = \dfrac{1}{\omega C_{par}}$

(c) $f_o = \dfrac{1}{2\pi \sqrt{L_{lead} C_{par}}}$

(d) $f \rightarrow \infty$

FIGURE 6.18 Simplification of the equivalent circuit of a resistor for various frequencies: (a) dc; (b) $f_1 = 1/2\pi R C_{par}$; (c) $f_o = 1/2\pi \sqrt{L_{lead} C_{par}}$; (d) as $f \rightarrow \infty$.

behavior in exchange for model simplicity. A model that will predict the behavior of an element for a very wide frequency range can always be constructed. However, that model will of necessity be very complex, and consequently will yield very little insight into the device behavior. Experimental data will reveal the adequacy of the simpler model. Secondly, it is important for the reader to obtain some appreciation for the typical range of numerical results. Examining experimental data obtained from actual devices will serve this latter purpose. An example of such data is shown in Fig. 6.19, where the measured impedance of a 1 kΩ, $\frac{1}{8}$ W carbon resistor having 0.5 inch lead lengths and 0.25 inch lead separation is shown over a frequency range of 1–500 MHz. Comparing Fig. 6.19 with Fig. 6.17(c), we see that the first breakpoint f_1 occurs at approximately 90 MHz, but the resonant frequency of the model, f_o, is somewhat above the highest measured frequency of 500 MHz. Nevertheless, the model of Fig. 6.17(b) gives an adequate description of the resistor if we choose $R = 1.05$ kΩ, $C_{par} = 1.2$ pF, and $L_{lead} = 14$ nH. A SPICE simulation of this model is shown in Fig. 6.20 using these model element values (see [15] for a discussion of the

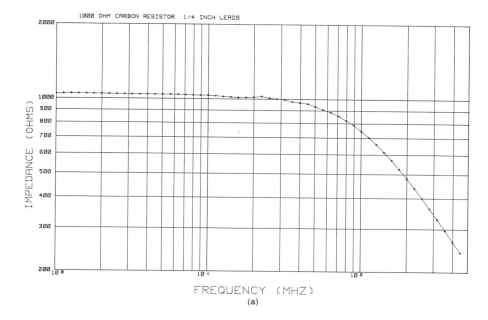

FREQUENCY (MHZ)
(a)

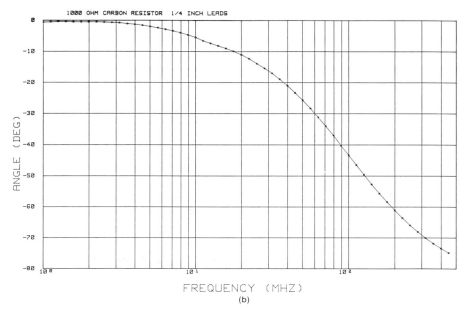

FREQUENCY (MHZ)
(b)

FIGURE 6.19   Measured impedance of a 1000 Ω, carbon resistor having $\frac{1}{4}$ inch lead lengths: (a) magnitude; (b) phase.

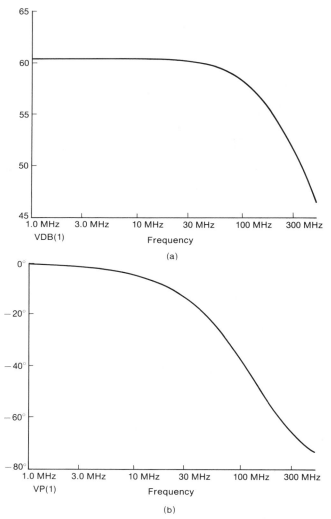

FIGURE 6.20 SPICE predictions of the impedance of a 1000 Ω resistor having $\frac{1}{4}$ inch lead lengths: (a) magnitude; (b) phase.

SPICE circuit analysis program). The SPICE program is obtained from the circuit in Fig. 6.21 as

```
SIMULATION OF 1K OHM CARBON RESISTOR
IS 0 1 AC 1
L 1 2 14NH
C 2 0 1.2PF
R 2 0 1.05K
.AC DEC 50 1MEG 500MEG
.PLOT AC VDB(1) VP(1)
END
```

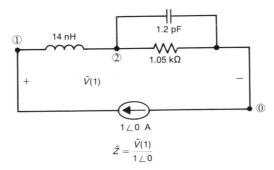

$$\hat{Z} = \frac{\hat{V}(1)}{1\angle 0}$$

FIGURE 6.21   The SPICE simulation circuit for the 1000 Ω resistor having $\frac{1}{4}$ inch lead lengths.

6.5   CAPACITORS

The ideal behavior of a capacitor is shown in Fig. 6.22. The impedance is $\hat{Z}(p) = 1/pC$, or, by substituting $p = j\omega$,

$$\hat{Z}(j\omega) = \frac{1}{j\omega C} \tag{6.44}$$

$$= -j\frac{1}{\omega C}$$

$$= \frac{1}{\omega C} \angle -90°$$

The magnitude of the impedance decreases linearly with frequency, or -20 dB/decade, and the phase angle is constant at $-90°$.

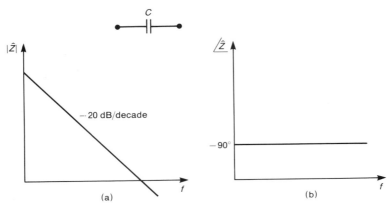

FIGURE 6.22   Frequency response of the impedance of an ideal capacitor: (a) magnitude; (b) phase.

There are numerous types of capacitors. For the purposes of EMC suppression the typical types are ceramic and tantalum electrolytic. Large values of capacitance ($1-1000\ \mu F$) can be obtained in a small package with the tantalum electrolytic capacitor. Ceramic capacitors give smaller values of capacitance ($1\ \mu F-5$ pF) than electrolytic capacitors, yet they tend to maintain their ideal behavior up to a much higher frequency than the latter. Thus ceramic capacitors are typically used for suppression in the radiated emission frequency range, whereas electrolytic capacitors, by virtue of their much larger values, are typically used for suppression in the conducted emission band and also for providing bulk charge storage on printed-circuit boards, as we will see. For a more complete discussion of capacitor types see [16].

Both types of capacitors have similar equivalent circuits, but the model element values differ substantially. This accounts for their different behavior over different frequency waves. Both types of capacitor can be viewed as a pair of parallel plates separated by a dielectric, as illustrated in Fig. 6.23. The loss (polarization and ohmic) in the dielectric is represented as a parallel resistance R_{diel}[1]. Usually this is a large value, as one would expect (hope). The resistance of the plates is represented by R_{plate}. For small ceramic capacitors, this is usually small enough in relation to the other elements to be neglected. Once again, the leads attached to the capacitor have a certain inductance represented by L_{lead} and capacitance C_{lead}. Again, these parasitic element values depend on the configuration of the two leads. If the leads are formed in the shape of a U or bent $90°$ to the body of the capacitor as is the usual custom then these parasitic lead components are as calculated previously. Usually R_{diel} is so large that it can be neglected. Similarly C_{lead} is usually much less than the ideal capacitance C, and thus may be neglected. Thus the equivalent circuit of the capacitor alone consists of the series combination of C and R_{plate}. The resistance R_{plate} is referred to as the equivalent series resistance or ESR and denoted as R_s. Thus the model consists of the series combination of C, L_{lead}, and R_s, as shown in Fig. 6.24.

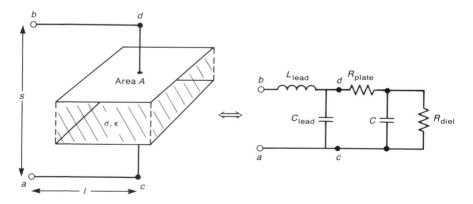

FIGURE 6.23   Modeling of a physical capacitor with an equivalent circuit.

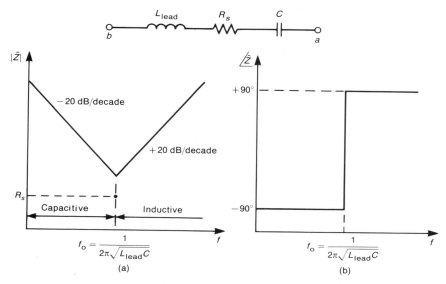

FIGURE 6.24 A simplified equivalent circuit of a capacitor including the effects of lead length showing Bode plots of the impedance: (a) magnitude; (b) phase.

The ESR is typically several ohms for electrolytic capacitors and varies with frequency. For ceramic capacitors over the regulatory limit frequency range the series resistance is usually negligible. The impedance of this model is

$$\hat{Z}(p) = L_{\text{lead}} \frac{p^2 + R_s p/L_{\text{lead}} + 1/L_{\text{lead}}C}{p} \tag{6.45}$$

Substituting $p = j\omega$ gives

$$\hat{Z}(j\omega) = L_{\text{lead}} \frac{1/L_{\text{lead}}C - \omega^2 + j\omega R_s/L_{\text{lead}}}{j\omega} \tag{6.46}$$

The Bode plots of this impedance are shown in Fig. 6.24. At dc the circuit appears as an open circuit (replace the inductor with a short circuit and the capacitor with an open circuit). As frequency increases, the impedance of the capacitor dominates and decreases linearly with frequency at a rate of -20 dB/decade. The impedance of the inductor increases until it equals that of the capacitor at $f_o = 1/2\pi\sqrt{L_{\text{lead}}C}$. At this frequency the series combination appears as a short circuit (although the magnitudes of the impedances are equal they are of opposite sign) and the net impedance of the branch is R_s. The frequency f_o is referred to as the *self-resonant frequency of the capacitor*. For higher frequencies the magnitude of the impedance of the inductor dominates and the impedance increases at a rate of $+20$ dB/decade, while the phase angle approaches $+90°$. If one is relying on this element to provide a low impedance such as for shunting noise currents to ground then the frequency of the current to be suppressed must be lower than the *self-resonant frequency* f_o of the capacitor or else the impedance will be larger than anticipated on the basis of the ideal behavior of the capacitor.

As an example, suppose the leads of a capacitor are formed into a U shape with a separation of 0.25 inch and length 0.5 inch. We calculated previously that the inductance of the loop formed by these leads is $L_{lead} \cong 14$ nH. Therefore a 470 pF capacitor will resonate at a frequency of 62 MHz and a 0.1 μF capacitor will resonate at a frequency of 4.25 MHz. This points out the important fact that, *for a fixed lead length and spacing, the larger the capacitance value the lower the self-resonant frequency* (by the square root of the capacitance ratios). Figures 6.25 and 6.26 show measured impedances of a 470 pF ceramic capacitor from 1 MHz to 500 MHz. Two lead lengths are shown: essentially no lead lengths (Fig. 6.25) and $\frac{1}{2}$ inch lead lengths (Fig. 6.26). Note that the self-resonant frequency for the $\frac{1}{2}$ inch lead length is about 62 MHz, as calculated above. Thus if one is interested in providing a low impedance to shunt a 200 MHz signal, the 470 pF capacitor with either lead length will give an impedance larger than expected. A 0.15 μF tantalum capacitor was measured, and results are shown from 1 MHz to 500 MHz for essentially no lead lengths and for $\frac{1}{2}$ inch lead lengths in Fig. 6.27 and 6.28, respectively. Note that the frequency response of the tantalum capacitor is not as ideal as that of the ceramic capacitor. This is due to the more significant ESR of the tantalum capacitor.

A frequent mistake made in suppression is in the choice and effectiveness of capacitors [17]. Capacitors are generally the common choice for suppression element since they are easily installable after the product is constructed—simply solder them across the two terminals in a connector or on a PCB to provide a low-impedance path to divert the noise current. Suppose it is desired to reduce the radiated emission at, say, 100 MHz. Also suppose it is found that the noise current present on a particular peripheral cable is the primary radiation source point. One might place a capacitor between the signal and return wires of the cable where it exits the product in order to divert the high-frequency noise current and prevent it from being present on the pheripheral cable, where its radiation efficiency will be considerably greater. One might choose a ceramic capacitor of value 100 pF. Suppose the product's radiated emissions are remeasured and found to be reduced, yet this reduction is still not sufficient for the product to comply with the regulatory limit at this frequency. In order to reduce the emission still further, one might be tempted to increase the value of capacitance to, say, 10,000 pF (0.01 μF). When this capacitor is substituted, it will be found that, instead of the radiated emission being reduced (by the expected 40 dB), they are actually increased! What has happened is that the self-resonant frequency of the larger capacitor has been reduced from that of the smaller capacitor not because of any change in L_{lead} but simply because of the larger value of C. Since the self-resonant frequency of the 10,000 pF capacitor is now below the frequency of interest (100 MHz), the capacitor appears

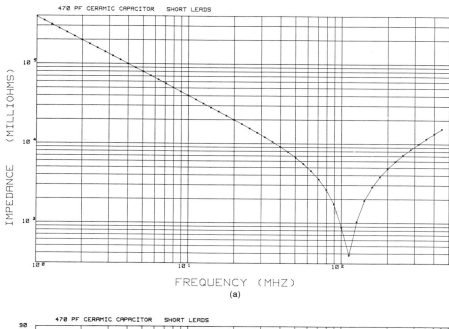

FREQUENCY (MHZ)
(a)

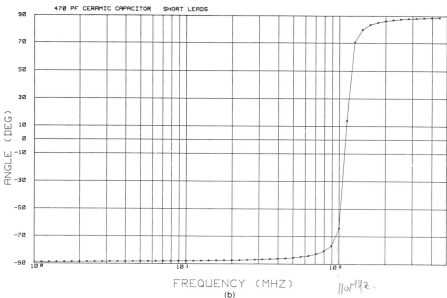

FREQUENCY (MHZ)
(b)

FIGURE 6.25 Measured impedance of a 470 pF ceramic capacitor with short lead lengths: (a) magnitude; (b) phase.

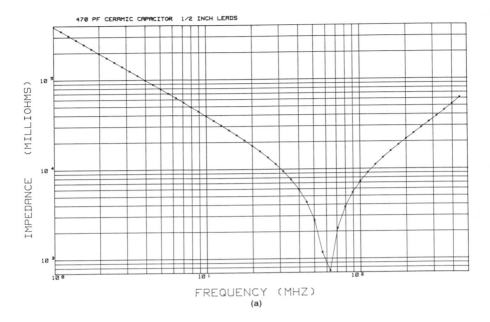

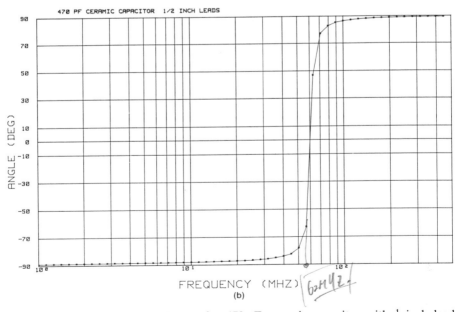

FIGURE 6.26   Measured impedance of a 470 pF ceramic capacitor with $\frac{1}{2}$ inch lead lengths: (a) magnitude; (b) phase.

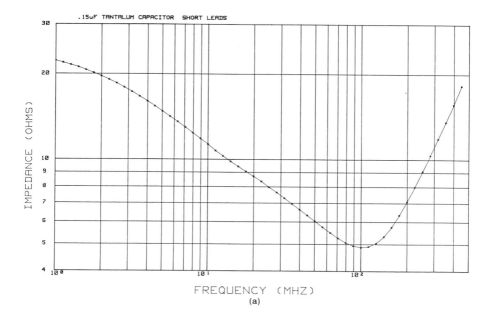

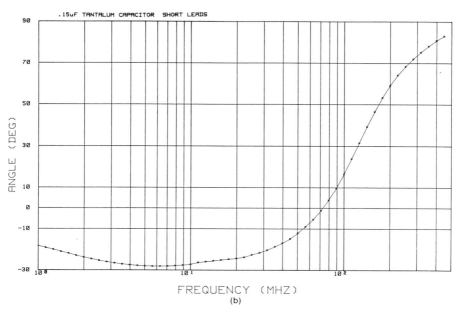

FIGURE 6.27 Measured impedance of a 0.15 μF tantalum capacitor with short lead lengths: (a) magnitude; (b) phase.

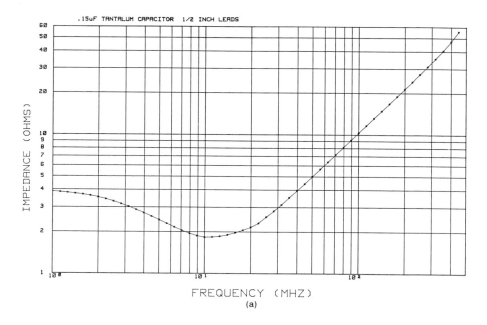

(a)

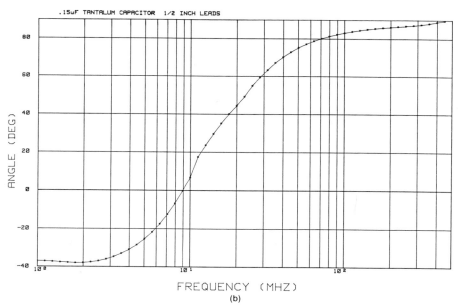

(b)

FIGURE 6.28   Measured impedance of a $0.15\ \mu$F tantalum capacitor with $\frac{1}{2}$ inch lead lengths: (a) magnitude; (b) phase.

inductive, giving an impedance larger than expected. Measured data for 100 pF and 10,000 pF ceramic capacitors both having 0.5 inch lead lengths show that the impedance of the 100 pF capacitor at 100 MHz is 3 Ω, whereas the 10,000 pF capacitor has an impedance of 12 Ω at this frequency!

Another caution that should be observed is the effect of the added suppression element on the *functional signals*. Placing a capacitor across the signal and return leads of a cable in order to divert high-frequency signal components from the cable can produce *ringing* by virtue of the resonance created by the capacitor in parallel with the inductance of the cable. Resistors are also frequently inserted in series with the cable in order to block these high-frequency signals. This is frequently implemented by inserting "*RC* packs" in the PCB where the off-board cable connector exits the PCB. The values of *R* and *C* in the above implementation should be chosen carefully. Suppose the input to this circuit is a trapezoidal pulse train representing a typical digital signal such as digital data that are being transmitted over the peripheral cable into which the *RC* circuit has been inserted. For the present we will ignore the effect of the peripheral cable. The transfer function of the *RC* circuit so formed is flat out to the break frequency of $1/2\pi RC$ and decreases at a rate of -20 dB/decade above that. Thus we have formed *a low-pass filter*. If the break frequency occurs low enough in frequency in comparison with the spectrum of the signal to be passed by the cable due to large values of *R* or *C*, the waveform of the signal can be adversely affected, resulting in functional performance problems. On the other hand, if these values are too small, very little filtering of the high-frequency noise on the cable may occur. This indicates two important points: (1) one must be careful to not adversely affect the functional signal with a suppression scheme, or else passing the regulatory limits will be a moot point; and (2) if the added suppression scheme does not produce a sufficient reduction, one should not be confused, since there is a reason why it does not.

It is also important to understand the concept illustrated in Fig. 6.29. Suppose a capacitor is to be placed in parallel with a cable or a pair of lands on a PCB in order to divert a noise current $\hat{I}_{\text{NOISE}}$. The impedance of the capacitor is represented by $\hat{Z}_{\text{CAP}}$, and the impedance seen looking into the pair of conductors that we wish to divert the noise current from is designated by $\hat{Z}_{\text{LOAD}}$. By current

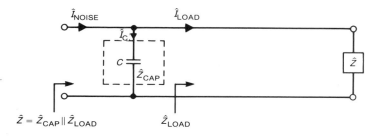

FIGURE 6.29 An important consideration in the diversion of currents with a parallel element: current division and the impedance of the load.

division, the portion of the noise current that is diverted through the capacitor is given by [15]

$$\hat{I}_C = \frac{\hat{Z}_{LOAD}}{\hat{Z}_{CAP} + \hat{Z}_{LOAD}} \hat{I}_{NOISE} \tag{6.47}$$

If $\hat{Z}_{LOAD}$ is *large* compared with $\hat{Z}_{CAP}$ then the capacitor will be effective in keeping $\hat{I}_{NOISE}$ off the cable. On the other hand, if $\hat{Z}_{LOAD}$ is *small* compared with $\hat{Z}_{CAP}$ then the capacitor will be ineffective in diverting the noise current! This is why the use of parallel capacitors in *low-impedance* circuits is usually *ineffectual.* They are most effective with *high-impedance* loads. *Whenever a parallel suppression component is to be used, the impedance levels of not only the element but also the parallel path should be computed or estimated at the desired frequency. If $\hat{Z}_{LOAD} \ll \hat{Z}_{CAP}$ then the suppression component will be ineffectual.* Therefore it is important to remember that *parallel capacitors work best in high-impedance circuits* with regard to diverting noise currents.

6.6   INDUCTORS

The impedance of an ideal inductor is shown versus frequency in Fig. 6.30, and is given by

$$\hat{Z}_L = j\omega L \tag{6.48}$$

$$= \omega L \ \underline{/90^\circ}$$

The magnitude increases linearly with frequency at a rate of $+20\,\text{dB/decade}$ and the angle is $+90^\circ$ for all frequencies. There are numerous variations of the basic construction technique of winding turns of wire on a cylindrical form.

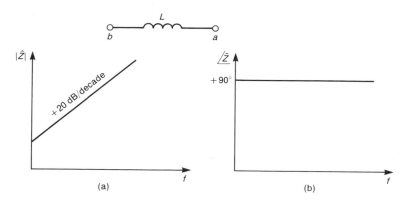

(a)   (b)

FIGURE 6.30   The frequency response of an ideal inductor.

The specific construction technique will determine the values of the *parasitic* elements in the model of the nonideal inductor that is shown in Fig. 6.31. The process of winding turns of wire on a cylindrical form introduces resistance of the wire as well as capacitance between neighboring turns. This produces the parasitic capacitance elements R_{par} and C_{par} in the nonideal model. Some construction techniques wind the turns of wire in *layers* to shorten the length of the inductor body. But this adds capacitance between layers, which *substantially increases* C_{par}. The nonideal inductor should also include the inductance of the attachment leads L_{lead}, as with all other elements. However since the *intentional element* is an inductance and its value is usually much larger than the lead inductance, we may generally neglect L_{lead} in this nonideal model. Similarly, the lead capacitance C_{lead} is frequently less than the parasitic capacitance C_{par}, so that we may neglect the lead capacitance. Thus the model consists of the series combination of R_{par} and L in parallel with C_{par}. The impedance of this model becomes

$$\hat{Z}_L(p) = R_{par} \frac{1 + pL/R_{par}}{p^2 L C_{par} + p R_{par} C_{par} + 1} \tag{6.49}$$

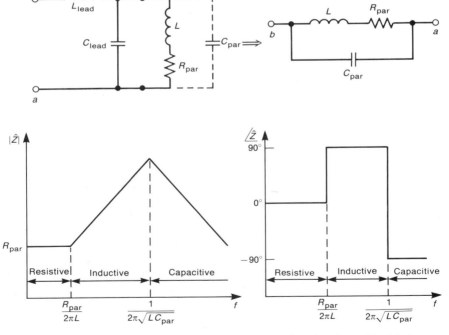

FIGURE 6.31 An equivalent circuit of an inductor, including the effect of lead inductance and parasitic capacitance. Bode plots of the frequency response are also shown.

Subsituting $p = j\omega$ gives

$$\hat{Z}_L(j\omega) = R_{par} \frac{1 + j\omega L/R_{par}}{1 - \omega^2 L C_{par} + j\omega R_{par} C_{par}} \tag{6.50}$$

At low frequencies the resistance dominates, and the impedance is R_{par}. As frequency is increased, the inductance of the model begins to dominate at a frequency of $\omega = R_{par}/L$, and the impedance increases at 20 dB/decade while the angle is $+90°$. As frequency is further increased, the impedance of the parasitic capacitance decreases until its magnitude equals that of the inductor impedance. This occurs at the *self-resonant frequency* of the inductor, $f_o = 1/2\pi\sqrt{LC_{par}}$. The Bode plot of the model is also shown in Fig. 6.31.

The measured impedance of a 1.2 μH inductor is shown in Fig. 6.32 from 1 MHz to 500 MHz. The self-resonant frequency of this inductor is of order 110 MHz. This gives a value of C_{par} of 1.7 pF. The measured impedance of a 10 μH inductor gives a value of parasitic capacitance 1.6 pF and a self-resonant frequency of around 40 MHz. This result is reasonable to expect, since the resonant frequency should be reduced by the square root of the ratio of the inductances if the lead lengths and parasitic capacitances are the same. Therefore it appears that the dominant capacitance here is due to the lead capacitance. Once again, it is important to remember that *increasing the value of an inductor will not necessarily give a lower impedance at high frequencies, since the larger value of inductance will serve to lower the self-resonant frequency, even though the lead lengths remain identical.*

Capacitors are used to *divert* noise currents, whereas inductors are placed in series with wires or lands to *block* noise currents. This will be effective *if the impedance of the inductor at the frequency of the noise current is larger than the original series impedance seen looking into the wires or lands*, $\hat{Z}_{LOAD}$, as shown in Fig. 6.33. The choice of whether to use a parallel capacitor to divert noise currents or a series inductor to block noise currents depends strongly on the impedance that it is placed in series or parallel with. *If $\hat{Z}_{LOAD}$ is large then a rather large value of inductance will be required in order to increase the net impedance of the circuit and provide any blockage of the noise current!* This is why *series inductors are most effective in low-impedance circuits.* Conversely, parallel capacitors must present a much smaller impedance than $\hat{Z}_{LOAD}$ in order to divert noise currents, so that *parallel capacitors are most effective in high-impedance circuits.*

As with parallel suppression capacitors, one must be concerned with the effect of the suppression element on the functional signal. Addition of series inductors can cause ringing, which can affect the desired performance of the system. However, they are quite effective in lines that do not carry high-speed signals and operate infrequently, such as reset lines of digital devices.

One final, and very important, point concerning inductors should be discussed. In order to increase the value of inductance having a certain number of turns and cross-sectional area, inductors are frequently wound on *ferromagnetic*

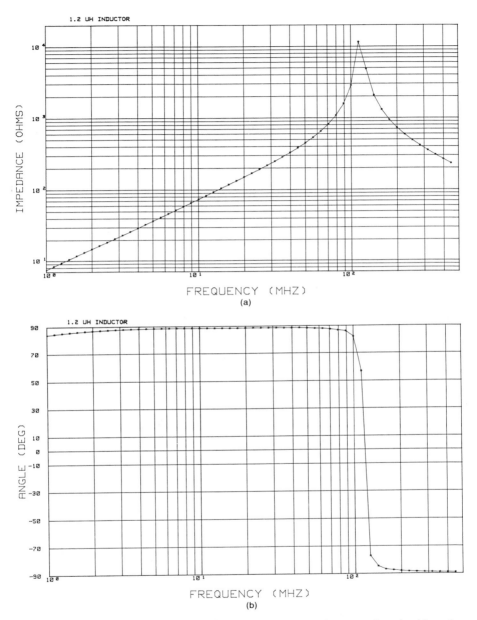

FIGURE 6.32   Measured impedance of a 1.2 μH inductor having very short lead lengths: (a) magnitude; (b) phase.

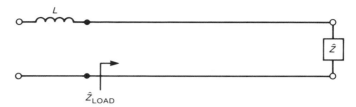

FIGURE 6.33 An important consideration in the blocking of currents with a series element: the impedance of the load.

cores. There are numerous types of these ranging from iron to powdered ferrite material. All types of ferromagnetic materials have large relative permeabilities μ_r, where the permeability is $\mu = \mu_r \mu_0$. For example, steel (SAE 1045) has a relative permeability of $\mu_r = 1000$ and mumetal has $\mu_r = 20,000$. Non-ferromagnetic metals such as copper and aluminum have relative permeabilities of free space, $\mu_r = 1$. The values of relative permeability cited for these materials are values measured at *low currents* and at *low frequencies*, typically 1 kHz or lower. Ferromagnetic materials suffer from the property of *saturation*, illustrated in Fig. 6.34. Consider a ferromagnetic *toroid* that has N turns of wire wound on it. An approximate value for the inductance of this toroid (assuming all the magnetic flux is confined to the core) is $L = \mu_r \mu_0 N^2 A / l$, where A is the core cross-sectional area and l is the mean path length of the core [1]. Suppose a current I is passed through the turns. This current creates a *magnetic field intensity H* that is proportional to the product of the number of turns and the current, NI. A *magnetic flux density B* is produced in the core. The product of B and the cross-sectional area of the core, A, gives the magnetic flux $\psi = BA$, whose units are webers. The relationship between H and B is also shown in Fig. 6.34. The permeability is the *slope* of this $B-H$ curve.

$$\mu = \frac{\Delta B}{\Delta H} \tag{6.51}$$

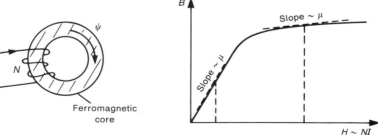

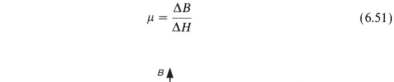

FIGURE 6.34 The nonlinear relationship between magnetic flux density and magnetic field intensity for a ferromagnetic core inductor.

At low values of current I the slope of the $B-H$ curve is large, as is the permeability. As current is increased, the *operating point* moves up the curve and the slope decreases. Thus the *permeability decreases with increasing current.* Since the inductance is a direct function of the permeability of the core, the *inductance decreases with increasing current.* We will have numerous occasions to see this phenomenon in the future.

6.7 FERRITES AND COMMON-MODE CHOKES

Ferromagnetic materials have considerable effect on magnetic fields, as was discussed in the previous section. *Magnetic fields tend to concentrate in high-permeability materials.* For example, in the ferromagnetic-core inductor shown in Fig. 6.34 we indicated that the magnetic flux ψ was confined to the ferromagnetic core. This is correct to a reasonable approximation. Some of the flux *leaks out* and completes the magnetic path through the surrounding air. The division between how much of the total flux remains in the core and how much leaks out depends on the *reluctance* of the core [15]. The quantity of reluctance $\mathscr{R}$ depends on the permeability μ of the magnetic path, its cross-sectional area A, and its length l as [1]

$$\mathscr{R} = \frac{l}{\mu A} \qquad (6.52)$$

An important analogy to ordinary lumped circuits can be used to analyze magnetic circuits. It consists of making the analogy of voltage to *magnetomotive force (mmf),* which is given in ampere turns, NI and current to magnetic flux ψ as

$$\mathscr{R} = \frac{NI}{\psi} \qquad (6.53)$$

The equivalent circuit for the toroidal inductor of Fig. 6.34 is given in Fig. 6.35. By current division, the portion of the total flux ψ that remains in the core is

$$\psi_{\text{core}} = \frac{\mathscr{R}_{\text{air}}}{\mathscr{R}_{\text{air}} + \mathscr{R}_{\text{core}}} \psi \qquad (6.54)$$

For high-permeability cores $\mathscr{R}_{\text{core}} \ll \mathscr{R}_{\text{air}}$, so that the majority of the flux is confined to the core. The reluctances of the paths are proportional to the permeabilities of the paths, so that the portion of the total flux that remains in the core is proportional to the ratios of the *relative permeabilities* of the two paths. Cores constructed from ferromagnetic materials such as steel, which has $\mu_r = 1000$, tend to have small leakage flux. We will use this notion of *lowering the reluctance of a magnetic path in order to concentrate magnetic flux in that path* on numerous occasions in the future.

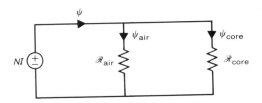

FIGURE 6.35 An equivalent circuit relating core and air fluxes for a ferromagnetic core inductor.

We now embark on a discussion of one of the most important topics affecting the radiated emissions of products, *common-mode* and *differential-mode currents*. Consider the pair of parallel conductors carying currents $\hat{I}_1$ and $\hat{I}_2$, as shown in Fig. 6.36. We may *decompose* these two currents into two auxiliary currents, which we refer to as *differential-mode*, $\hat{I}_D$, and *common-mode*, $\hat{I}_C$:

$$\hat{I}_1 = \hat{I}_C + \hat{I}_D \qquad (6.55a)$$

$$\hat{I}_2 = \hat{I}_C - \hat{I}_D \qquad (6.55b)$$

Solving these two equations gives

$$\hat{I}_D = \tfrac{1}{2}(\hat{I}_1 - \hat{I}_2) \qquad (6.56a)$$

$$\hat{I}_C = \tfrac{1}{2}(\hat{I}_1 + \hat{I}_2) \qquad (6.56b)$$

The differential-mode currents $\hat{I}_D$ are equal in magnitude but oppositely directed in the two wires. These are the desired or functional currents. The common-mode currents $\hat{I}_C$ are equal in magnitude but are directed in the same direction. These are not intended to be present, but will be present in practical systems. Standard lumped-circuit theory does not predict these common-mode currents. They are frequently referred to as *antenna-mode currents*.

Let us now investigate the significance of each current on the radiated emissions from this pair of conductors, which may be wires or lands on a PCB. This will be investigated in more detail in Chapter 8. For the present it suffices to give a general discussion. The radiated electric fields $\hat{E}$ due to each current can be superimposed to give the total radiated electric field. First consider the radiated fields due to differential-mode currents, as illustrated in Fig. 6.37(a).

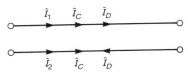

FIGURE 6.36 Decomposition of the currents on a two-wire transmission line into common-mode, $\hat{I}_C$, and differential-mode, $\hat{I}_D$, components.

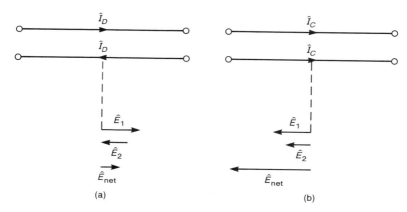

FIGURE 6.37 Illustration of the relative radiated emission potential of (a) differential-mode currents and (b) common-mode currents.

The differential-mode currents are oppositely directed. Thus the resulting electric fields will also be oppositely directed. However, since the two conductors are not collocated, the fields will not exactly cancel, but will subtract to give a small net radiated electric field. On the other hand, since the common-mode currents are directed in the same direction, their radiated fields will add giving a much larger contribution to the total radiated field than will the differential-mode currents, as is illustrated in Fig. 6.37(b). Thus *a small common-mode current can produce the same level of radiated electric field as a much larger value of differential-mode current*. In short, *common-mode currents have a much higher potential for producing radiated emissions than do differential-mode currents!* We will find in Chapter 8 that *the predominant mechanisms for producing radiated electric fields in practical products are the common-mode currents on the conductors!* For example, we will find that microamps of common-mode current will produce the same level of radiated electric field as tens of milliamps of differential-mode current! Common-mode currents are not intended to be present on the conductors of an electronic system, but nevertheless are present in all practical systems. Because of their considerable potential for producing radiated electric fields, we must determine a method for reducing them.

One of the most effective methods for reducing common-mode currents is with *common-mode chokes*. A pair of wires carrying currents $\hat{I}_1$ and $\hat{I}_2$ are wound around a ferromagnetic core as shown in Fig. 6.38(a). Note the directions of the windings. The equivalent circuit is also shown. Here we assume that the windings are identical, such that $L_1 = L_2 = L$. In order to investigate the effect of the core on blocking the common-mode current, we calculate the impedance of one winding:

$$\hat{Z}_1 = \frac{\hat{V}_1}{\hat{I}_1} = \frac{pL\,\hat{I}_1 + pM\,\hat{I}_2}{\hat{I}_1} \tag{6.57}$$

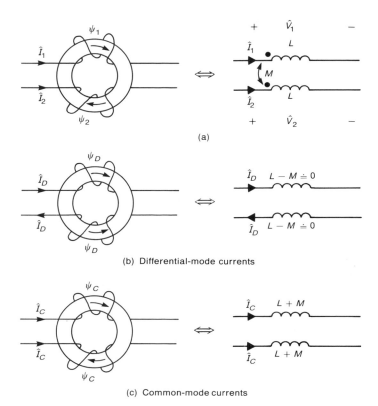

FIGURE 6.38 Modeling the effect of a common-mode choke on (a) the currents of a two-wire line, (b) the differential-mode components, and (c) the common-mode components.

Now let us investigate the contribution to the series impedance due to each component of the current. First let us consider common-mode currents in which $\hat{I}_1 = \hat{I}_C$ and $\hat{I}_2 = \hat{I}_C$. Substituting into (6.57) gives

$$\hat{Z}_{CM} = p(L + M) \qquad (6.58)$$

The contribution to the series impedance due to differential-mode currents where $\hat{I}_1 = \hat{I}_D$ and $\hat{I}_2 = -\hat{I}_D$ is

$$\hat{Z}_{DM} = p(L - M) \qquad (6.59)$$

If the windings are symmetric and all the flux remains in the core, i.e., the flux of one winding completely links the other winding, then $L = M$ and $\hat{Z}_{DM} = 0$! Thus in the ideal case where $L = M$ *a common-mode choke has no effect on differential-mode currents, but selectively places an inductance (impedance) 2L in series with the two conductors to common-mode currents.* These notions are illustrated in Fig. 6.38.

Thus common-mode chokes can be effective in blocking common-mode currents. In order to provide this impedance to common-mode currents, the wires must be wound around the core such that the fluxes due to the two common-mode currents *add* in the core whereas the fluxes due to the two differential-mode currents *subtract* in the core. Whether the wires have been wound properly can be checked with the *right-hand rule*, where, if one places the thumb of one's right hand in the direction of the current, the fingers will point in the resulting direction of the flux produced by that current. A foolproof way of winding two wires (or any number of wires) on a core to produce this effect is to wind the entire group around the core as illustrated in Fig. 6.39(a). In either case one should ensure that the wires entering the winding and those exiting the winding are separated from each other on the core, or else the parasitic capacitance between the input and output will shunt the core and reduce its effectiveness, as illustrated in Fig. 6.39(b).

The effectiveness of the common-mode choke relied on the assumption that the self and mutual inductances are equal, $L = M$. High-permeability cores tend to concentrate the flux in the core and reduce any leakage flux. Symmetrical windings also aid in producing this. Unfortunately, ferromagnetic materials suffer from *saturation* effects at high currents, as discussed earlier, and their permeabilities tend to deteriorate with increasing frequency more than low-permeability cores. One of the most important advantages of the common-mode choke is that *fluxes due to high differential-mode currents cancel in the core and do not staurate it.* The functional or differential-mode currents $\hat{I}_D$ are the desired

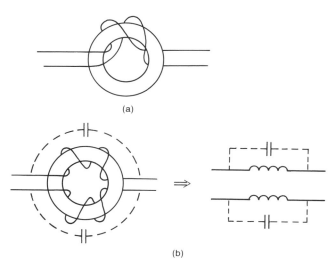

(a)

(b)

FIGURE 6.39 (a) A simple way of winding a common-mode choke. (b) Parasitic capacitance.

currents, and are usually large in magnitude. If the flux due to these high-level currents did not cancel in the core, the core would tend to saturate, and the high permeability would be lost. Thus more of the flux would leak out into the surrounding air, and the self and mutual inductances would not be approximately equal. Furthermore, since the differential-mode fluxes cancel in the core, the choke does not (ideally) affect the functional signal as do the other suppression components discussed previously. So *the functional signals are not (ideally) affected by the presence of the choke, and also do not affect the performance of the choke.*

It seems that we should select a ferrite core material that has the highest *initial permeability* possible in order to concentrate the flux in the core and give the highest possible common-mode inductance $L + M$. Ferrite core materials have different frequency responses of their permeability. A core having an initial relative permeability of 2000 at 1 kHz and low current might have that relative permeability reduced to under 10 at frequencies in the frequency range of the regulatory limit *where it is to have an effect.* Figure 6.40 illustrates this point. Manufacturers of ferrite core materials have their individual mix of materials they use to fabricate the ferrite material. However, ferrites typically are predominantly either of manganese zinc (MnZn) or nickel zinc (NiZn). Manganeze zinc ferrites tend to have the higher initial permeabilities, *but* their permeabilities deteriorate more rapidly with increasing frequency than do nickel zinc ferrites. Therefore, although a ferrite core having a large initial permeability may seem more attractive than one with a lower value, it should be remembered that in the range of the radiated emission limit (30 MHz–40 GHz) the core having the lower initial permeability may well have the higher permeability of the two, and is therefore preferred for use in suppressing the spectral components

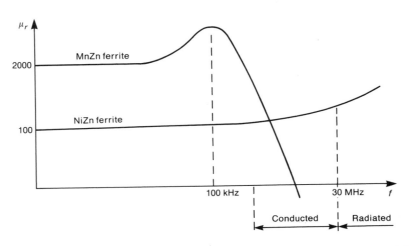

FIGURE 6.40 Frequency response of the relative permeabilities of MnZn and NiZn ferrites.

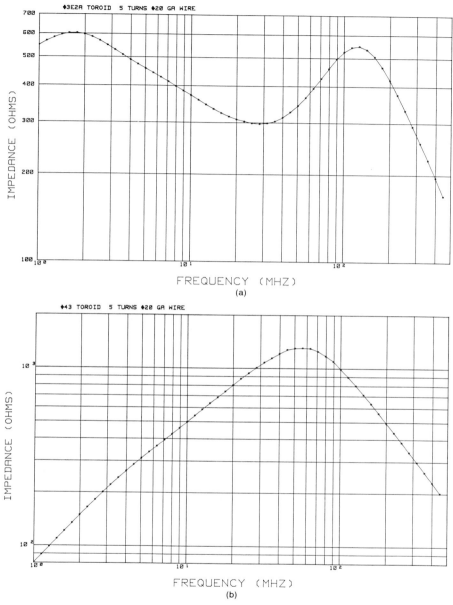

FIGURE 6.41   Measured impedance of inductors formed by winding five turns of #28 gauge wire on (a) MnZn and (b) NiZn cores.

of common-mode currents in this frequency range. Typical EMC laboratories have specific cores to be used for conducted emission suppression and others to be used for radiated emission suppression because of these considerations.

In order to illustrate this frequency dependence, we have shown the frequency response of the impedance of an inductor formed by winding five turns of #20 gauge wire on two toroids in Fig. 6.41. Figure 6.41(a) shows the impedance for a typical MnZn core, while Fig. 6.41(b) shows the impedance for a typical NiZn core. Note that the MnZn core shows an impedance of some 500 Ω at 1 MHz, whereas the NiZn core shows an impedance of some 80 Ω at 1 MHz. However, at a frequency of 60 MHz, the MnZn core shows an impedance of 380 Ω, whereas the NiZn core shows an impedance of 1200 Ω! This illustrates that the type of core to be used depends on the frequency of application (suppression of conducted emissions or radiated emissions). Unless one is careful to catalogue the cores in the inventory such as by painting them with different colors, the proper selection can be difficult.

6.8   FERRITE BEADS

Ferrite materials are basically nonconductive ceramic materials that differ from other ferromagnetic materials such as iron in that they have low eddy-current losses at frequencies up to hundreds of megahertz. Thus they can be used to provide selective attenuation of high-frequency signals that we may wish to suppress from the standpoint of EMC and not effect the more important lower-frequency components of the functional signal. These materials are available in various forms. The most common form is a *bead* as shown in Fig. 6.42. The ferrite material is formed around a wire, so that the device resembles an ordinary resistor (a black one without bands). It can be inserted in series with a wire or land and provide a high-frequency impedance in that conductor. However, unlike the common-mode choke, *the ferrite bead affects both differential-mode and common-mode currents equally.* If the high-frequency components of the differential-mode current are important from a functional standpoint then the ferrite bead may affect functional performance of the system.

The current passing along the wire produces magnetic flux in the circumferential direction, as we observed previously. This flux passes through the bead material, producing an internal inductance in much the same way as for a wire considered in Section 6.1.1. Thus the inductance is proportional to the permeability of the bead material: $L_{bead} = \mu_0 \mu_r K$, where K is some constant

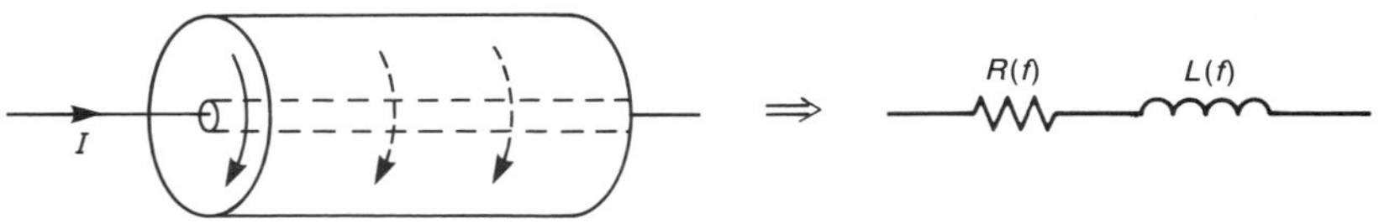

FIGURE 6.42 Ferrite beads.

depending on the bead dimensions. The bead material is characterized by a complex relative permeability as

$$\mu_r = \mu_r'(f) - j\mu_r''(f) \qquad (6.60)$$

The real part μ_r' is related to the stored magnetic energy in the bead material, while the imaginary part μ_r'' is related to the losses in the bead material. Both are shown as being functions of frequency. Substituting this into the general equation for the impedance of the bead inductance gives

$$
\begin{aligned}
j\omega L_{\text{bead}} &= j\omega\mu_o\mu_r K \qquad (6.61)\\
&= j\omega\mu_o(\mu_r' - j\mu_r'')K\\
&= \underbrace{\omega\mu_r''(f)\mu_o K}_{R(f)} + \underbrace{j\omega\mu_r'(f)\mu_o K}_{L(f)}
\end{aligned}
$$

From this result we see that the equivalent circuit consists of a resistance that is dependent on frequency in series with an inductance that is also dependent on frequency. Typical ferrite beads can be expected to give impedances of order 100 Ω above approximately 100 MHz. *Multiple-hole* ferrite beads as illustrated in Fig. 6.43 can be used to increase this high-frequency impedance. The measured impedances of a $\frac{1}{2}$-turn (a bead surrounding a wire) ferrite bead and a $2\frac{1}{2}$-turn ferrite bead from 1 to 500 MHz are shown in Fig. 6.44.

Because the impedance of ferrite beads is limited to several hundred ohms over the frequency range of their effectiveness, they are typically used in *low-impedance* circuits such as power supplies. They are also used to construct lossy filters. For example, placing a ferrite bead in series with a wire and placing a capacitor between the two wires will constitute a two-pole, low-pass filter. A series ferrite bead can also act to damp ringing in fast-risetime circuits. Ferrites are available in other forms. A recent use has been in the placement of ferrite slabs under dual-in-line (DIP) packages. The ferrite slab has holes drilled along its edges that correspond to the pin spacings of the DIP package. The pins of the DIP are placed through these holes and the combination inserted into a carrier or soldered directly to the PCB. An example of this use to damp very high-frequency oscillations is described in [18].

Ferrite beads are no different than other uses of ferrites in that they are susceptible to saturation when used in circuits that pass high-level, low-frequency currents. A ferrite bead placed in series with the 60 Hz power lead would probably be saturated by this high-level (1–10 A) current.

FIGURE 6.43   A multi-turn ferrite bead.

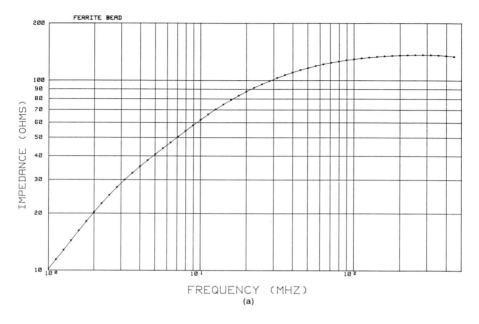

(a)

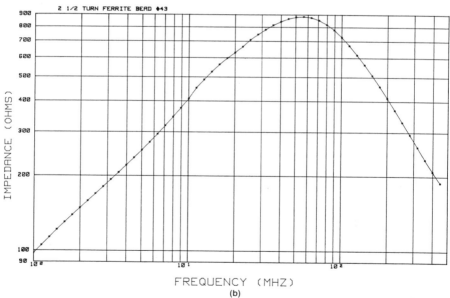

(b)

FIGURE 6.44 Measured impedances of (a) a $\frac{1}{2}$-turn ferrite bead and (b) a $2\frac{1}{2}$-turn ferrite bead.

6.9   ELECTROMECHANICAL DEVICES

A number of electronic products such as typewriters, printers, and robotic devices use small electromechanical devices such as dc motors, stepper motors, ac motors, and solenoids to translate electrical energy into mechanical motion. These seemingly innocuous (from an EMC standpoint) devices can create significant EMC problems. DC motors create high-frequency spectra due to arcing at the brushes as well as providing paths for common-mode currents through their frames. The purpose of this section is to highlight these problem areas and increase the awareness of the reader for their potential to create EMC problems.

6.9.1   DC Motors

DC motors are used to produce rotational motion, which can be used to produce translational motion using gears or belts. They rely on the property of magnetic north and south poles to attract and like poles to repel. A dc motor consists of stationary windings or coils on the *stator*, along with coils attached to the rotating member or *rotor*, as illustrated in Fig. 6.45(a). The coils are wound

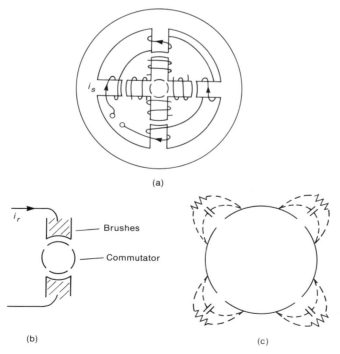

FIGURE 6.45   A dc motor illustrating (a) physical construction, (b) brushes and commutator, and (c) arc suppression elements.

on metallic protrusions, and a dc current is passed through the windings, creating magnetic poles. A commutator consists of metallic segments that are segmented such that the dc current to the rotor windings can be applied to the appropriate coils to cause the rotor to align with or repel the stator poles as the rotor rotates. Carbon brushes make contact with the rotor segments and provide a means of alternating the current and magnetic fields of the rotor poles using a dc current from a source, as illustrated in Fig. 6.45(b). As the current to the rotor coils is connected and disconnected to the dc source through the commutator segments, arcing at the brushes is created due to the periodic interruption of the current in the rotor coils (inductors). This arcing has a very high-frequency spectral content, as we will see in the next chapter. This spectral content tends to create radiated emission problems in the radiated emission regulatory limit frequency range between 200 MHz and 1 GHz, depending on the motor type. In order to suppress this arcing, resistors or capacitors may be placed across the commutator segments as illustrated in Fig. 6.45(c). These can be implemented in the form of capacitor or resistor disks that are segmented disks of capacitors or resistors attached directly to the commutator or in resistive rings placed around the commutator. In some cases it may be necessary to insert small inductors in the dc leads to block those noise currents that are not completely suppressed by the capacitor or resistor disks.

An additional source of high-frequency noise and associated radiated and conducted emissions comes not from the motors themselves but from the *driver circuits* that are used to change the direction of rotation to provide precise position control of the motor. A typical "H-drive" circuit for a small dc motor is illustrated in Fig. 6.46(a). When transistors T_1 and T'_1 are turned on, current flows through the commutator and the rotor windings, causing the rotor to turn in one direction. When these are turned off and transistors T_2 and T'_2 are turned on, the rotor turns in the opposite direction. This driver circuit is usually connected to the motor via a long pair of wires as shown in Fig. 6.46(b). For reasons of thermal cooling of the motor, its housing is usually attached to the metallic frame of the product, which acts as a heat sink. This produces a large capacitance C_{par} between the motor housing and the product frame. This provides a path for *common-mode currents* to pass through the connection wires from the rotor to the stator via capacitance between these windings, and eventually to the frame via C_{par}. The current provided to the motor by the driver, although ideally intended to be dc, typically has fast- risetime spikes present on it due to the constant interruption of the current in the driver circuit and in the rotor coils by the commutator. These spikes have very high-frequency spectral content, which is then placed on the product frame and is coupled to other parts of the product radiating in the process. The loop area formed by the leads and their return path (the product frame) also tends to be quite large. We will find in Chapter 8 that the radiation potential tends to be a direct function of the loop area occupied by that current; the larger the loop area, the larger the radiated emission. In order to block this common-mode current, a common-mode choke may be needed to be placed in the driver leads, as is

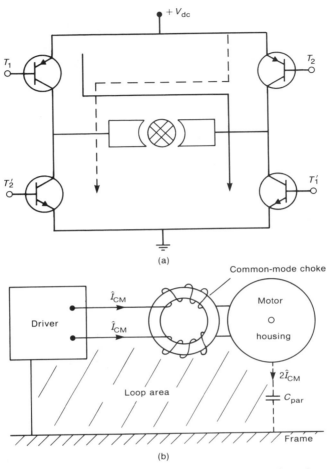

FIGURE 6.46 Illustration of (a) an H-drive circuit and (b) conversion of common-mode driver currents into differential-mode currents with a large loop area because of parasitic capacitance of the motor frame.

illustrated in Fig. 6.46(b). This shows a case where common-mode current (in the driver leads) becomes essentially a differential-mode current flowing around a large loop area. Measured common-mode impedances between the input wires (tied together) and the motor frame for a small dc motor give an impedance null around 100 MHz of some 20 Ω!

6.9.2 Stepper Motors

An alternative to the dc motor for electromechanical positioning is the *stepper motor*. There are basically two types of stepper motors: *permanent magnet* (*PM*) and *variable reluctance* (*VR*). Both types have dc current applied to the stationary windings of the stator to produce magnetic poles. The stator and the rotor are segmented into a large number of poles around their peripheries in order to provide fine positioning. The rotor of the PM stepper is a permanent magnet made of rare earth materials. The rotor of the VR stepper consists of shorted turns of wire. Flux from the stator induces currents in these shorted turns, which induces magnetic poles on the rotor. The windings of the stator are arranged in phases to provide various degrees of magnetic pole segmentation. The rotor poles tend to align with those of the stator that are energized in order to reduce the *reluctance* of the magnetic path.

Although there is no arcing to generate high-frequency signals as with the commutator of the dc motor, there remains the problem of common-mode currents between the driver circuit wires and the frame of the motor, which is again attached to the frame of the product for cooling. A typical driver circuit is shown in Fig. 6.47. Turning on transistors T_1 and T_4, for example, causes current to flow through the windings of phase A and phase D in the indicated direction, causing the motor to rotate to one desired position. Constant energization and de-energization of these stator windings again causes high-frequency noise to be passed down the connection wiring. As with other motors, parasitic capacitance exists between the input wires and the motor frame, which is attached to the product frame for thermal reasons. Thus the noise currents on the input wires are placed on the frame of the product and return to the driver via that path. This creates the same problem that was observed for dc motors, and may require that a common-mode choke be placed in the driver wires in order to block this path and reduce the radiated emissions of this

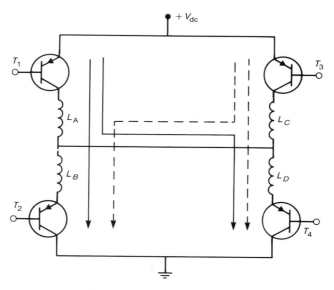

FIGURE 6.47   A typical driver circuit for a stepper motor.

common-mode current (which becomes a differential-mode current by passing through the large loop created by the driver wires–ground plane circuit). Measurements of the common-mode impedance between the input wires (tied together) and the motor frame for a typical small stepper motor show a null around 70 MHz of some 3 Ω!

6.9.3   AC Motors

AC motors are not usually used to provide positioning of mechanical elements as are dc and stepper motors, but rather are used to provide constant speed and drive small cooling fans. The absence of brushes and the attendant arcing as with dc motors and the noise currents provided by the drivers of both dc and stepper motors tends to diminish the potential of these motors for creating noise problems as opposed to dc or stepper motors. However, because the rotor and stator of these motors consist of closely spaced inductors, there remains the problem of large parasitic capacitance between the rotor and stator. If the motor frame is mechanically attached to other mechanical parts of the product then the potential exists for coupling common-mode currents from the ac power source to the product frame and vice versa. If high-frequency noise is present on the ac waveform feeding these motors or the frame to which the stator is attached then it is likely that this noise will be coupled to the product frame or to the ac power cord, where its potential for radiated or conducted emissions will usually be enhanced. Chopper-drivers are frequently used to control the power to the ac motors. Thus the input current may have a high-frequency spectral content which is transferred via common mode to the product frame. Common-mode chokes in the attachment leads of these types of motors may be needed in order to block this path.

6.9.4   Solenoids

A *solenoid* is essentially a coil of wire with a ferromagnetic slug at its center. Energizing the coil with a dc current causes a magnetic flux to be generated. The ferromagnetic slug tends to move to the center of the coil in a translational fashion in order to minimize the reluctance of the magnetic path. These types of electromechanical devices suffer from many of the problems of the above motors with the exception of commutation. Sudden energization and de-energization of the winding inductance creates high-frequency noise. Once again, the parasitic capacitance between the windings and the metallic housing of the solenoide creates a potential path for common-mode currents to be placed on the frame of the product, which may require a common-mode choke in the input leads. Measurement of the common-mode impedance between the input wires (tied together) and the solenoid frame for a small solenoid indicate a null at around 150 MHz of 8 Ω!

6.10 DIGITAL CIRCUIT DEVICES

Digital products are increasing in popularity because of their increasing ability to rapidly process data and their inherent noise immunity. However, these attributes pose problems for an EMC standpoint. Data are transmitted and processed in the form of pulses. The transitions between each state (the pulse rise/fall times) tend to be extraordinarily fast (of order 1–20 ns). We will see in the next chapter that these fast transition times tend to generate high-frequency spectral content in the frequency-domain representation of these signals, which contributes to the high-frequency radiated and conducted emissions of the product. The requirement for increased speeds of data transmission and processing will no doubt cause these EMC concerns to increase in digital products in the future.

Digital products are relatively simple in architecture and typically consist of a central processor in the form of a microprocessor, which performs computation, stores and retrieves data, and instructions, and provides sequencing of the entire process. Various read-only-memory (ROM) modules provide nonvolatile storage of program instructions. Random-access-memory (RAM) modules provide for storage of data, and various drivers or buffers provide the ability to drive peripheral devices or communicate data. One or more *system clocks* provide synchronization of the occurrence of each task within well-defined windows of time. The primary task is to input data and instructions either from external devices such as tape or disk drives or from keyboards and to process this and provide the results as output to displays or as signals to drive external processes such as motors or other actuators.

The process seems relatively straightforward from the standpoint of its affect on the EMC profile of the product. However, the subtle aspects of the process have considerable impact. For example, the particular technology and requirements of the product affect the rise/fall time of the clock and data pulses, which affect their high-frequency spectal content. Buffer gates are frequently provided to interface between low-current logic signals and high-current outputs. These have the effect of "squaring up" the signals. Suppose the rise/fall times of a clock signal have been slowed by the insertion of a low-pass filter such as a shunt capacitor. If a buffer gate has been inserted at some point further down the line, the signal may be "squared up" and have current drive added, thus increasing its high-frequency spectral content.

Conductors that are intended to carry only "rare-event" signals that only occur infrequently should not be overlooked since, *although they are not intended to carry high-frequency signals, they may have these present due to inadvertent coupling to these lines.* For example, the reset line of a microprocessor may be only active infrequently during machine operation. However inadvertent coupling of other high-frequency signals to this line can cause very high-frequency signals to be present on this line. If the reset line is routed a long distance around the PCB, it may cause significant radiated emissions, which the EMC engineer may not suspect as being the source of the emissions. It is particularly instructive to probe points on a PCB of a digital product and observe the spectral content of the signals. Virtually all signal lines in a digital product should be suspect with regard to carrying high-frequency signals, although some (such as clock lands) are clearly of more importance. It is this author's experience that the most effective method of reducing radiated and conducted emissions is to affect the source of these emissions. Although this is a seemingly obvious point, it is nevertheless important to keep in mind. Once noise signals are allowed to propagate away from their source, their suppression becomes a problem of suppressing the emanation of the same signal from different points of emission resulting in the need for many more suppression elements.

The active digital components are composed of large numbers of semiconductor diodes, bipolar junction transistors (BJTs), and field effect transistors (FETs). These are implemented in integrated circuit form on minute chips. One of the primary parasitic components of these elements that is of concern in EMC is the *parasitic capacitances* formed at the semiconductor junctions [19]. Each of these elements is formed from two types of semiconductor, *n*-type and *p*-type. This junction causes a separation of charge which acts like a capacitance. Once these parasitic capacitances are added to the ideal model of the device, it becomes clear that rise/fall times of signals will be affected. Of more importance is the effect of these parasitic capacitances in *routing signals around the element*, in effect providing a direct connection at high frequencies from the input of the device to its output.

6.11   EFFECT OF COMPONENT VARIABILITY

It is very important to remember that it will always be required to produce a large number of supposedly identical copies of a product for sale. It is important and relatively simple to produce identical products from a functional standpoint; that is, all products are able to meet the functional performance design goals. This has always been and will continue to be an important criterion in the design process. However, consistency in achieving the EMC design goals among supposedly identical copies of a product is another matter that is not generally assured by achieving consistency in meeting the functional performance design goals. For example, suppose a product prototype is "fine-tuned" to meet the EMC regulatory limits on radiated and conducted emissions. Once the product is placed in production and a large number of copies are made, it is not assured that all of these "copies" will also meet EMC regulatory limits, which require that *all units that are sold must comply*. Changes in parts vendors to reduce product cost can cause a product that previously was in compliance to suddenly be out of compliance, even though it continues to meet the functional performance objectives. One also must realize that the functional performance goals and the EMC performance goals are often in conflict. For example, functional designers are generally interested in the *maximum rise/fall times of a digital component*, whereas EMC designers are more interested in the *minimum rise/fall times of the digital component* since the shorter the rise/fall time the larger the high-frequency spectral content of the signal. Manufacturers of components cannot guarantee absolute conformance to specifications of their components, but instead specify bounds. A digital component manufacturer may guarantee maximum rise/fall times of his/her component for functional performance reasons. A large quantity of these parts used to produce "identical copies" of the producct may (and usually do) exhibit variations that, although within the bounds specified by the part manufacturer, may exceed the bounds that are being relied upon by the EMC engineer and cause one or more of the "copies" to be out of compliance.

Changing parts vendors to reduce cost at some point in the production cycle of the product can create compliance problems. An example is illustrated in [20]. An RS-232 line driver was tested for this type of variability. Several "equivalent" line drivers from different vendors were tested, and the spectrum present on the -12 V dc lead to each component was measured over the frequency range of 10–210 MHz. The reader is referred to that publication, which shows extreme variability in the emission present on the -12 V lead from vendor to vendor and within parts of the same vendor. All parts would no doubt meet the functional performance goals. Also observe that the -12 V lead of the line driver "is not supposed to carry these high-frequency signals," but does in fact have these present. This again illustrates that *just because a conductor is not intended to carry high-frequency signals does not rule out the presence of high-frequency signals on that conductor.*

6.12 MECHANICAL SWITCHES

Mechanical switches are often used in electronic products to provide the operator with a quick and easy way of changing the product behavior. On–off switches connect commercial power to the product. Other switches may simply provide a change in the status of the product, for example a reset switch on a personal computer. The EMC problems that may result from the activation of mechanical switches are quite varied, and depend strongly on the load that is switched. Arcing at the contacts is the primary interference problem since the arc waveform may contain very high-frequency spectral components of large magnitude, as we will see. Early investigations concerned the interruption of large currents by circuit breakers in power systems [21]. In the early 1940s work concentrated on the behavior of switches in telephone circuits with regard to erosion of the switch contacts as well as the interference produced by these operations [22–26]. In order to subdue this potential interference problem and to insure longer life for the switch contacts, various protection networks are often used.

In this final section we will discuss the EMC aspects of mechanical switches. The discussion will be brief but will cover the essential points. For a more complete and thorough discussion of the subject the author recommends [16] and [21–26]. These references contain virtually all the information one needs to know about mechanical switches from the standpoint of EMC. The following is a condensed summary of that information.

6.12.1   Arcing at Switch Contacts

It has been known from before the days of the Marconi spark-gap transmitter that current in the form of an arc can be conducted between two electrodes that are immersed in air. For example, consider two contacts separated a distance d in air shown in Fig. 6.48. The typical voltage–current characteristic is shown. There are three regions shown: the Townsend discharge region named for its discoverer, the glow discharge region, and the arc discharge region. The various voltage levels ae denoted as V_B, V_G, and V_A. Typical values for these variables for contacts in air are $V_B \cong 320$ V, $V_G \cong 280$ V, and $V_A \cong 12$ V. The value of V_B depends on contact separation, while that of V_A depends on the contact material. The currents at the transition regions are denoted as I_G and I_A. These are quite variable, but are of order $I_G \cong 1-100$ mA and $I_A \cong 0.1-1$ A.

There are always a few free electrons in the space between the contacts due to cosmic radiation, photon collisions with the gas molecules, etc. As the voltage between the two contacts is increased, the resulting electric field between the contacts accelerates these free electrons, causing them to strike neutral gas molecules. If the free electrons have sufficient kinetic energy imparted by the

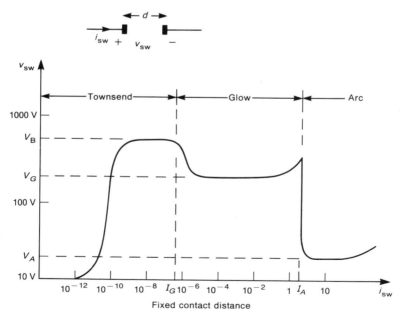

FIGURE 6.48 Illustration of the voltage–current characteristic of breakdown at mechanical contacts.

electric field, they strike the gas molecules, creating additional free electrons as electron-ion pairs. The field accelerates these newly formed electrons, causing them to strike other gas molecules and thereby releasing more free electrons. This produces a multiplicative production of free electrons and positive ions. The positive ions move toward the cathode (the negative terminal of the contact) and the electrons move toward the anode (the positive terminal of the contact). As the positive ions move toward the cathode, they create a space charge around it that increases the local field and the production of free electrons. The positive ions also strike the cathode, liberating more free electrons by secondary emission. This mechanism characterizes the Townsend discharge region of Fig. 6.48. Thermionic heating of the cathode can also liberate electrons, but this mechanism tends to predominate at the higher currents of the arc discharge region. In the early, low-current region all of the electrons emitted at the cathode are collected by the anode. Above a certain voltage level all electrons emitted by the cathode are collected, independently of further increases in voltage, and the curve rises vertically. At still higher voltages the electrons acquire sufficient kinetic energy to create electron-ion pairs in their collisions with the gas molecules, thereby increasing the free electrons and resulting in an avalanche effect. The peak or *breakdown* voltage V_B is dependent on the gas, contact separation and pressure. Paschen found that the breakdown voltage was dependent on the product of the pressure and the contact separation distance as [21]

$$V_B = \frac{K_1 pd}{K_2 + \ln pd} \tag{6.62}$$

where K_1 and K_2 are constants that depend on the gas. For air at standard atmospheric pressure the minimum breakdown voltage is approximately $V_{B, \min} \cong 320$ V and occurs at a contact separation $d_{\min} = 0.3$ mils $= 0.00762$ mm.

At the peak of the Townsend discharge region the production of free electrons and positive ions reaches a self-sustaining, *avalanche* stage wherein the current is sustained by this avalanching process but the voltage across the contacts drops to the lower *glow voltage* designated as V_G. For contacts in air at atmospheric pressure $V_G \cong 280$ V. A region near the cathode develops a faint glow, which is the origin of the term. The voltage drop across the switch remains constant at V_G for a large range of current and is primarily determined by the region between the cathode and the beginning of the glow region: the *cathode fall region*. As the current increases, the dimension of the glow region increases toward the anode, but the voltage drop across the gap remains at V_G.

When the current increases sufficiently such that cathode fall region encompasses the entire cathode area, further current increases result in higher current density. This leads to heating of the cathode and a slight increase in gap voltage. A point is reached rather quickly where the heating causes vaporization of the contact metal, resulting in a rapid drop in contact voltage, which marks the beginning of the *arc discharge region* where an arc forms between the contacts. The contact voltage drops to a very low voltage of about

$V_A \cong 12$ V. The value of V_A is determined by the contact material (since vaporization of the metal is the important process here) but is of about $11-16$ V. Once the arc is initiated, a very luminous discharge results where further increases in current do not result in any appreciable change in the contact voltage from V_A. This is the usual visual effect one sees when contacts are opened and an arc forms momentarily.

The formation of an arc discussed above was initiated by voltages large enough to cause breakdown of the intervening gas. The voltage across the contacts divided by the contact spacing exceeds the breakdown field strength of the gas. This is referred to as a *long arc*. For smaller contact spacings in a vacuum the arc can be initiated by a *field-induced emission* wherein the electric fields at the highest and sharpest points on the cathode liberate electrons. This electron stream fans out as it crosses the gap. Bombardment of the anode by this electron stream causes it to heat to several thousand kelvin, which is sufficient to vaporize the electrode. The cathode may vaporize first, depending on the contact sizes, rate of heating, etc. As the positive ions move toward the cathode, a space charge forms that further promotes the emission, resulting in an avalanche effect. Thus an arc may be formed where the voltage and contact spacing are not sufficient for a gas breakdown. This is referred to as a *short arc* or *metal-vapor arc discharge*. The required field strength is of order $E_B = 10^9$ V/m, although this varies, depending on the cleanliness of the contact surface and surface contaminants.

Figure 6.49 sumarizes the breakdown voltage of a switch with air as the intervening medium. The plot is shown as a function of the contact separation

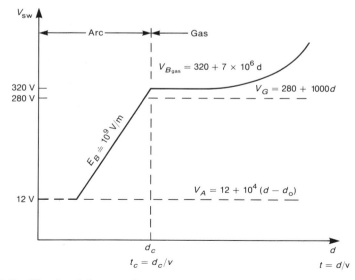

FIGURE 6.49 The breakdown voltage versus contact separation for a mechanical switch.

distance d. Dividing the separation distance by the velocity of contact closure or opening, v, gives the axis as a function of time t. For small contact separations less than d_c a short arc may form if the contact voltage divided by the contact separation equals $E_B \cong 10^9$ V/m. The contact voltage drops to $V_A \cong 12$ V. The current through the switch is determined solely by the circuit voltage and impedanace. However, a minimum current I_A is required to sustain the arc. This *minimum arc current* is quite variable, and ranges from tens of milliamps to one amp. *If the voltage across the contact available from the external circuit drops below V_A and/or the current through the contact available from the external circuit drops below I_A, the arc is extinguished.* For contact separations larger than d_c a glow discharge will form if the contact voltage exceeds the breakdown voltage, which is given by [22]

$$V_{B,\,gas} = 320 + 7 \times 10^6\, d \tag{6.63}$$

This is essentially the Paschen voltage curve for $d > d_{min}$. If the current available from the external circuit exceeds the *minimum glow discharge sustaining current* I_G, a glow discharge will form, and the contact voltage will drop to [22]

$$V_G = 280 + 1000d \tag{6.64}$$

If the current available from the circuit exceeds the minimum arc-sustaining current I_A, the glow discharge will transition to a *long arc*, and the contact voltage will again drop to $V_A \cong 12$ V. It must be re-emphasized that, *in order to sustain a glow (arc) discharge, the voltage across and current through the contact that are available from the external circuit must exceed V_G and $I_G(V_A$ and $I_A)$*. Again, the minimum sustaining voltages are rather predictable as $V_G \cong 280$ V and $V_A \cong 12$ V, whereas the minimum sustaining currents I_G and I_A are quite variable. Some representative ranges are $I_G \cong 1$–100 mA and $I_A \cong 100$ mA–1 A. *The glow discharge is characterized by large voltage and small current, whereas the arc discharge (long or short arc) is characterized by low voltage and large current.*

It is interesting to observe that, although the physics of the two processes is quite different, the arc discharge of a mechanical switch has characteristics very similar to the silicon-controlled rectifier (SCR). Consider the voltage-current characteristic of the switch shown in Fig. 6.48. If we plot this only for large currents above a few milliamps, the characteristic resembles that of a SCR. In fact, the operation of the two are quite similar. In order to "fire" an SCR, the voltage must be increased to the breakover point. Once the SCR fires, its voltage drops to a low value and the current increases substantially. The SCR can only be turned off by reducing its current below the "hold on current." The arc discharge of a mechanical switch is similar. In order to create an arc, the voltage across the switch must exceed the breakdown curve shown in Fig. 6.49. Once the arc forms, reducing the switch voltage cannot extinguish the arc (unless it is reduced below the arc voltage V_A). If the current is reduced below the minimum arcing current I_A, the arc will be extinguished!

6.12.2 Circuit Effects

The external circuit has a pronounced effect on the type of discharge or even whether a discharge occurs. For example, consider the circuit of Fig. 6.50, wherein a switch opens or closes to disconnect or connect a resistive load to a dc voltage source. The switch voltage and current waveforms for the switch closing are shown in Fig. 6.50(b). As the contacts approach, we move from right to left on the breakdown curve of Fig. 6.49. Assuming that the battery voltage (which appears across the open switch) is sufficiently large (greater than 320 V) that it will exceed the Paschen voltage at some contact separation, the voltage across the switch drops to V_G at time t_1. The current through the switch is

$$I_{sw} = \frac{V_{dc} - V_G}{R} \tag{6.65}$$

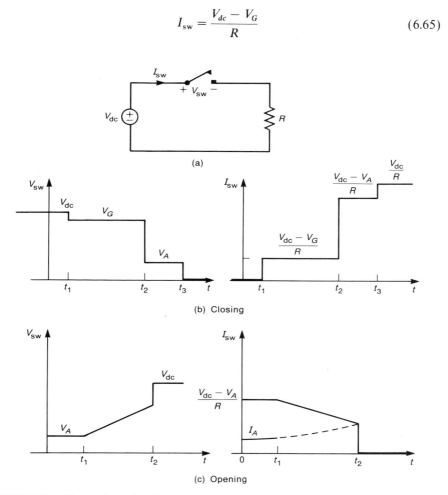

(a)

(b) Closing

(c) Opening

FIGURE 6.50 Illustration of arcing at switch contacts for a resistive load: (a) circuit; (b) closing contacts; (c) opening contacts.

If this exceeds the minimum glow discharge current I_G, the glow discharge will be sustained. If not, it will be extinguished. As the contacts move closer, the electric field will exceed $E_B \cong 10^9$ V/m, a short arc develops, and the contact voltage drops to V_A at time t_2. This short arc will be sustained *if* the switch current:

$$I_{sw} = \frac{V_{dc} - V_A}{R} \tag{6.66}$$

exceeds the minimum arc-sustaining current I_A. If not, the arc will be extinguished. The glow discharge may also transition to a long arc if the switch current in (6.65) exceeds the minimum arc-sustaining current. And finally the contacts come in contact at t_3 and the contact voltage drops to zero while the current rises to

$$I_{sw} = \frac{V_{dc}}{R} \tag{6.67}$$

The waveforms for the switch opening are shown in Fig. 6.50(c). When the contacts begin to separate, a molten bridge is formed, which soon ruptures. At these close contact spacings, the electric field will surely exceed $E_B \cong 10^9$ V/m, and a short arc will develop. The contact voltage drops to V_A almost immediately. The switch current is given by (6.66), and if it exceeds the minimum arc-sustaining current, the arc will be sustained, otherwise it will be extinguished. At this point for a resistive circuit it would seem that the arc, if sustainable at these close spacings, would continue regardless of the contact separation—which is obviously absurd. The arc will eventually be extinguished at some larger contact separation. How does this occur? The answer lies in the fact that we have oversimplified the constancy of the minimum arc-sustaining voltage V_A and minimum arc-sustaining current I_A. Figure 6.50(c) shows that, as the contacts separate, the voltage across them rises from V_A, and the minimum sustaining current rises from the value of I_A. Thus, as the contacts separate, the current through the contact is

$$I_{sw} = \frac{V_{dc} - V_A(d)}{R} \tag{6.68}$$

Since V_A increases, I_{sw} decreases, eventually meeting the increasing arc-sustaining current at time t_2, at which point the arc extinguishes.

This simple resistive circuit example yields considerable insight into the basic arc phenomenon, but is somewhat unrealistic. Switches are frequently used to interrupt inductive loads such as solenoids or motors. Interruption of these types of circuits leads to an interesting phenomenon called the *showering arc*, illustrated in Fig. 6.51. The inevitable parasitic capacitance is shown in parallel

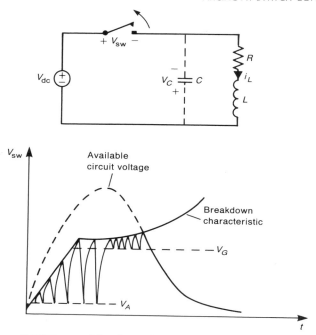

FIGURE 6.51   The showering arc for an inductive load.

with the inductive load. When the switch is closed, a steady-state current $I = V_{dc}/R$ is established in the inductor. When the switch opens, the inductor attempts to maintain this current. It is therefore diverted through the capacitance, charging the latter. The switch voltage $v_{sw}(t) = v_C(t) + V_{dc}$, and therefore increases. As this switch voltage increases, it may exceed the switch breakdown voltage, whereby a short arc forms and the switch voltage drops to V_A. The capacitor discharges through the switch, with the current being primarily limited by the local resistance and inductance of the switch wiring. If the switch current exceeds the minimum arc-sustaining current, the arc is sustained. If not, the arc is extinguished, and the capacitor begins to recharge. The switch voltage once again exceeds the switch breakdown voltage, and the switch voltage drops to V_A. If the arc is not sustainable, the capacitor begins to recharge once again. Eventually the energy stored initially is dissipated, and the capacitor voltage decays to zero, leaving $v_{sw} = V_{dc}$. This leads to a sequence of rising (as the capacitor charges) and rapidly falling (as the switch breaks down) voltages across the contacts, which has been referred to as the *showering arc* [25, 26]. As the contact separation increases, a glow discharge may develop and may or may not be sustainable, resulting in miniature showering arcs, as illustrated in Fig. 6.51. The number and duration of each showering arc depends on the circuit element values and any delays associated by interconnection transmission lines. A SPICE model useful for predicting the arcing at switches and associated crosstalk is described in [27].

6.12.3 Arc Suppression

Showering arcs clearly have significant spectral content, and may therefore create EMC problems. The wiring carrying these currents may cause significant radiated emissions, thereby creating interference problems. These signals may also be directly conducted along interconnected wiring paths, creating a potentially more troublesome effect, since the signal levels that are directly conducted to other points will be of the order of the switch voltages, which may be several hundred volts. Since these potential effects are recognized, various suppression measures are usually employed in conjunction with a mechanical switch.

There are so many unknowns involved that it is difficult to make precise calculations. For example, velocity of switch closure or opening has a significent effect on the levels and duration of the showering arc. When the capacitor in Fig. 6.51 discharges through the switch, the discharge current is limited only by the impedance of the local wiring, which is quite small and substantially unknown and variable. Thus contact protection is usually based on simple calculations that reveal starting values to be used and then using an experimental test. In either event, the goal of contact protection is to *prevent the formation of an arc (sustained or showering)*. There are generally two methods that may be employed:

1. *Prevent the switch voltage from exceeding the breakdown voltage of the switch.* Or
2. *Insure that the arc current is below the minimum arc-sustaining current.*

Technique 1 prevents the arc from forming while 2 prevents it from being sustained. There are two methods for implementing 1, as illustrated in Fig. 6.52.

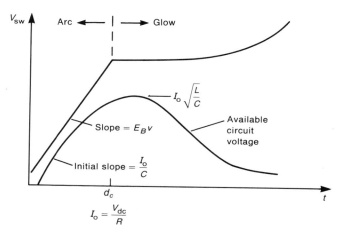

FIGURE 6.52 Contact protection by reducing the circuit available voltage.

The contact breakdown voltage profile is plotted against the available circuit voltage (in the absence of breakdown). The slope of the arc breakdown characteristic, $d < d_c$, is obtained as the product of $E_B v$. Choosing $E_B = 10^8$ V/m and a typical switch velocity $v = 0.01$ m/s gives a slope of 1 V/μs. The initial slope of the available circuit voltage for the circuit of Fig. 6.51 can be shown to be I_o/C, where $I_o = V_{dc}/R$ is the initial current through the inductor. Thus the initial rise of contact voltage should be kept below 1 V/μs, although this number is quite variable, depending on the contact surface (which affects E_B) and the contact approach velocity. The peak value of the available circuit voltage can be shown to be $I_o\sqrt{L/C}$ by neglecting the resistance R and assuming that all the energy stored in the inductor, $\frac{1}{2}LI_o^2$, is transferred to the capacitor, $\frac{1}{2}CV_{peak}^2$. In addition, the discharge waveform will be nonoscillatory (overdamped) if $\sqrt{L/C} < \frac{1}{2}R$ [15]. Even though arc discharge can be avoided by slowing the initial rise of the available circuit voltage, a glow discharge (which may transition to a long arc discharge) may develop if the peak available circuit voltage exceeds the gas discharge breakdown threshold. Therefore, in order to prevent initiation of an arc, the following two criteria should be satisfied:

(a)
$$E_B v > \frac{V_{dc}}{RC}$$

(b)
$$\frac{V_{dc}}{R}\sqrt{\frac{L}{C}} < V_{B,\,gas} \cong 320 \text{ V}$$

This can be implemented by placing a sufficiently large capacitor in parallel with the inductor or the switch to increase the net capacitance, thereby reducing the peak available circuit voltage and also reducing the initial rise of the available circuit voltage as shown in Fig. 6.53(a). This scheme has a significant drawback in that contact damage during switch closure may be significant due to the large capacitor charging current. When the switch is open, the capacitor charges to the supply voltage V_{dc}. When the switch closes, this initial voltage discharges through the switch, which results in a large current surge because this current is limited only by the resistance and inductance of the local switch wiring.

Figure 6.53(b) shows how to remedy this problem caused by a single capacitor across the switch contact: place a resistor in series with the capacitor to limit the discharge current that occurs on contact closure. Limiting this discharge current on switch closure to below I_A gives the minimum value of the resistance. On contact opening, it is desirable to have the resistance as small as possible so as to not limit the arc suppression of the capacitor. The minimum value of R is chosen to limit the discharge current during switch closure to below the minimum arcing current: $V_{dc}/R < I_{A,\,min}$. The maximum value is determined by the opening of the switch. When the switch opens, the current is diverted through the resistor, and the switch voltage is $I_o R$, where $I_o = V_{dc}/R_L$ is the initial current through the inductor. Usually the maximum value of R is chosen to be equal to R_L in order to limit the contact voltage to at most the supply

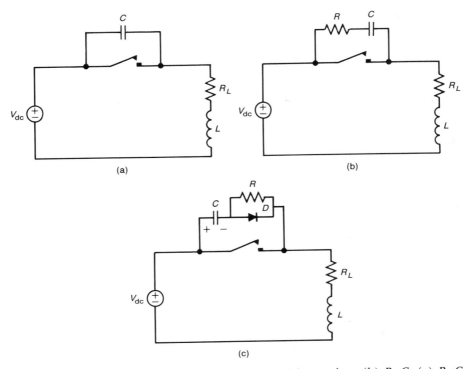

FIGURE 6.53 Various contact protection schemes: (a) capacitor; (b) R–C; (c) R–C with a diode.

voltage. Therefore the limits on choice of R are

$$\frac{V_{dc}}{I_{A,\,min}} < R < R_L \tag{6.69}$$

The capacitor is chosen to satisfy the above two criteria: (1) the initial rate of voltage rise of the available circuit voltage, I_o/C, is less than 1 V/μs to avoid an arc forming; and (2) the peak available voltage, $I_o\sqrt{L/C}$, is less than 320 V to avoid a gas breakdown, which may transition to an arc. This leads to values of C that must satisfy

$$(1) \qquad C \ge (\tfrac{1}{320}I_o)^2 L \tag{6.70a}$$

$$(2) \qquad C \ge I_o \times 10^{-6} \tag{6.70b}$$

A better but slightly more expensive network is shown in Fig. 6.53(c). A diode is placed across the resistor. While the switch is open, the capacitor charges up with polarity shown. When the switch closes, the resistor R limits the discharge current. When the switch opens, the diode shorts out the resistor, and the capacitor momentarily diverts the load current as described above. The capacitor value is chosen as above, but the resistor value is chosen to limit the current on closure to be less than the minimum arcing current:

$$R \geq V_{dc}/I_{A,\,min}.$$

Contact suppression can be employed across the switch as described previously or across the inductive load, or both. An example of applying a diode across an inductive load is shown in Fig. 6.54(a). When the switch opens, the inductor current is diverted through the diode rather than the switch. Contact arcing during switch closure is not affected. A common example of protection of inductive loads with a diode is in switching transistors. A diode ("freewheeling" diode) is placed across the inductive load, which may represent the inductance of a dc motor as shown in Fig. 6.54(b). When the transistor switch interrupts the current through the inductance I_L, the inductance kick or Faraday's law voltage across the inductor causes the diode to short out. Thus the diode clamps the collector of the transistor to $+V_{CC}$, preventing large collector–emitter voltages, which may destroy the transistor. Currents of large magnitude and fast risetimes will circulate around the inductor–diode loop. Therefore the diode must be placed very close to the inductor in order to minimize the loop radiation of this current loop.

Contact protection may or may not be required for resistive loads. If the load draws less than the minimum arcing current, no arc can be sustained, and no contact protection is generally required. If the resistive load draws a current greater than the minimum arcing current, a contact protection circuit similar to that of Fig. 6.53 may be required.

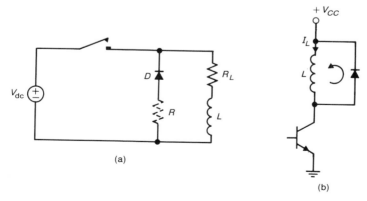

(a)

(b)

FIGURE 6.54   Diode protection for an inductive load.

REFERENCES

[1] C.R. Paul and S.A. Nasar, *Introduction to Electromagnetic Fields*, second edition, McGraw-Hill, NY (1987).

[2] C.R. Paul, Modeling electromagnetic interference properties of printed circuit boards, *IBM J. Research and Development*, **33**, 33–50 (1989).

[3] C.R. Paul and W.W. Everett, III, Lumped model approximations of transmission lines: effect of load impedances on accuracy, *Technical Report, Rome Air Development Center, Griffiss AFB, NY*, RADC-TR-82-286, Vol. IV E (August 1984).

[4] J.C. Clements, C.R. Paul, and A.T. Adams, Two-dimensional systems of dielectric-coated, cylindrical conductors, *IEEE Trans. on Electromagnetic Compatibility*, **EMC-17**, 238–248 (1975).

[5] A.E. Ruehli, Inductance calculations in a complex integrated circuit environment, *IBM J. Research and Development*, **16**, 470–481 (1972).

[6] F.W. Grover, *Inductance Calculations*, Dover, NY (1946).

[7] H.B. Dwight, *Tables of Integrals and Other Mathematical Data*, fourth edition, Macmillan, New York (1961).

[8] C. Hoer and C. Love, Exact inductance equations for rectangular conductors with applications to more complicated geometries, *J. Research National Bureau of Standards–C. Engineering Instrumentation*, **69C**, 127–137 (1965).

[9] H.R. Kaupp, Characteristics of microstrip transmission lines, *IEEE Trans. on Computers*, **EC-16**, 185–193 (1967).

[10] H.A. Wheeler, Transmission line properties of parallel strips separated by a dielectric sheet, *IEEE Trans. on Microwave Theory and Techniques*, **MTT-13**, 172–185 (1965).

[11] R. Garg and I.J. Bahl, Characteristics of coupled microstrip, *IEEE Trans. on Microwave Theory and Techniques*, **MTT-23**, 700–705 (1979) [Corrections: **MTT-28**, 272 (1980).]

[12] T.C. Edwards, *Foundations for Microstrip Engineering*, John Wiley, New York (1981).

[13] K.C. Gupta, R. Garg, and I.J. Bahl, *Microstrip Lines and Slotlines*, Artech House, Dedham, MA (1979).

[14] K.C. Gupta, R. Garg, and R. Chadha, *Computer-Aided Design of Microwave Circuits*, Artech House, Dedham, MA (1981).

[15] C.R. Paul, *Analysis of Linear Circuits*, McGraw-Hill, NY (1989).

[16] H.W. Ott, *Noise Reduction Techniques in Electronic Systems*, second edition, John Wiley Interscience, NY (1988).

[17] J.C. Engelbrecht and K. Hermes, A study of decoupling capacitors for EMI reduction, *Technical Report, International Business Machines*, TR-51.0152 (May 1984).

[18] D.M. Hanttula and S.W. Wong, Case study—effect of analog buffer amplifier on radiated emissions, *Proceedings 1985 IEEE International Symposium on Electromagnetic Compatibility, Wakefield, MA, August 1985*.

[19] C.R. Paul, S.A. Nasar, and L.E. Unnewehr, *Introduction to Electrical Engineering*, McGraw-Hill, NY (1986).

[20] D.R. Kerns, Integrated circuit construction and its effect on EMC performance, *Proceedings of the 1984 IEEE International Symposium on Electromagnetic Compatibility, San Antonio, TX, April 1984.*

[21] R. Holm, *Electric Contacts, Theory and Application,* fourth edition, Springer-Verlag, Berlin (1967).

[22] H.N. Wagar, Performance principles of switching contacts, *Physical Design of Electronic Systems,* Vol. 3, *Integrated Device and Connection Technology* (Bell Laboratories), Chap. 9, Prentice-Hall, Englewood Cliffs, N.J. (1971).

[23] J.D. Cobine, *Gaseous Conductors,* Dover, New York (1958).

[24] A.M. Curtis, Contact phenomena in telephone switching circuits, *Bell System Technical J.* **19**, 40–62 (1940).

[25] G.W. Mills, The mechanism of the showering arc, *IEEE Trans. on Parts, Materials and Packaging,* **PMP-5**, 47–55 (1969).

[26] E.K. Howell, How switches produce electrical noise, *IEEE Trans. on Electromagnetic Compatibility,* **EMC-21**, 162–170 (1979).

[27] S.A. Hall, C.R. Paul, K.B. Hardin, and A.D. Nielsen, Prediction of crosstalk due to showering arcs at switch contacts, *Proceedings of the 1991 IEEE International Symposium on Electromagnetic Compatability, Cherry Hill, NJ, August, 1991.*

PROBLEMS

6.1 Calculate the per-unit-length dc resistance of the following wires: #6 AWG (solid and 259 × 30), #20 AWG (solid and 19 × 32), #28 AWG (solid and 7 × 36), #30 AWG (solid and 7 × 38). [1.3 mΩ/m, 1.31 mΩ/m, 33.2 mΩ/m, 28 mΩ/m, 214.3 mΩ/m, 194.4 mΩ/m, 340.3 mΩ/m, 303.8 mΩ/m]

6.2 Determine the skin depth of steel (SAE 1045) at 1 MHz, 100 MHz, and 1 GHz. [0.26 mils, 0.026 mils, 0.00823 mils]

6.3 Determine the frequency where the resistance of a #20 AWG solid wire begins to increase due to skin effect. [105.8 kHz] Determine the resistance of this wire at 100 MHz. [1.022 Ω/m]

6.4 Determine the frequency where the internal inductance of a #32 AWG solid wire begins to decrease due to skin effect. [1.7 MHz] Determine the internal inductance of this wire at 100 MHz. [6.5 nH/m or 0.165 nH/inch]

6.5 Determine the resistance, internal inductance, external inductance and capacitance of a typical ribbon cable consisting of two #28 AWG (7 × 36) wires 2 m in length and separated by 50 mils at 100 MHz. [1.87 Ω, 2.97 nH, 1.518 μH, 29.28 pF] Determine the characteristic impedance of the cable. [227.7 Ω]

6.6 Determine the external inductance of the ribbon cable of Problem 6.5 using partial inductance. [1.517 μH]

6.7   Determine the dc resistance of a PCB land of width 15 mils constructed from 1 oz Cu. [1.29 Ω/m = 32.8 mΩ/inch]

6.8   Determine the effective dielectric constant and characteristic impedance of a microstrip line constructed of a glass–epoxy board of thickness of 47 mils supporting a 1 oz Cu land 100 mils in width. [3.625, 45.3 Ω] Determine the per-unit-length inductance and capacitance. [7.3 nH/inch, 3.56 pF/inch]

6.9   Determine the effective dielectric constant and characteristic impedance of a coplanar strip line constructed of a glass–epoxy board of thickness 47 mils supporting two 1 oz Cu lands 100 mils in width and separated (edge to edge) by 100 mils. [1.96, 172.2 Ω] Determine the per-unit-length inductance and capacitance. [20.4 nH/inch, 0.688 pF/inch]

6.10   Determine the characteristic impedance of two 1 oz Cu lands 100 mils in width that are located on opposite sides of a 47 mil glass–epoxy board. [56.48 Ω]

6.11   Sketch Bode plots for the following transfer functions:

(1) $\hat{H}(p) = \dfrac{2p + 20}{(p + 1)(5p^2 + 275p + 1250)}$

(2) $\hat{H}(p) = \dfrac{p^3 + p^2}{(p^2 + 2p + 1)(p + 5)}$

(3) $\hat{H}(p) = 500\,\dfrac{p^2 + 2p + 1}{p(p + 1)(p^2 + 20p + 100)}$

(4) $\hat{H}(p) = \dfrac{10p + 20}{2p^2 + 4p + 202}$

(5) $\hat{H}(p) = \dfrac{50p + 100}{2p^2 + 20p}$

6.12   The magnitude of an impedance is sketched as a Bode plot in Fig. P6.12. Determine one possible impedance expression for this. [$\hat{Z} = 0.279(p^2 + 6p + 9)/(p^3 + 10p^2)$)

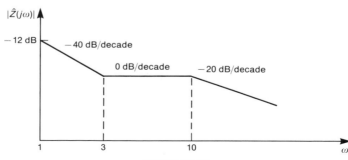

FIGURE P6.12

6.13 Determine the self-resonant frequency of the $1\,k\Omega$ carbon resistor consisting of 0.5 inch leads separated by 0.25 inch whose measured characteristic is shown in Fig. 6.19. [1.228 GHz]

6.14 Determine the lead inductance of the capacitors whose measured impedances are shown in Figs. 6.25 and 6.26. [4.45 nH, 14 nH] Use SPICE to predict the impedance of these capacitors.

6.15 A 100 MHz noise signal is being fed to the input of a transistor amplifier whose input impedance is $60\ \underline{/0°}\ \Omega$ at 100 MHz. Determine the value of a capacitor which when placed across the input will divert 60% of this current. [20 pF] Confirm this using SPICE.

6.16 A $50\ \Omega$, 100 MHz sinusoidal source is terminated in a $50\ \Omega$ load resistor. Determine the value of a capacitor that when placed across the output will reduce the output voltage by 20 dB. [633.4 pF]

6.17 A common way of attempting to overcome the low self-resonant frequency of large-value capacitors is to place a small-value (high self-resonant frequency) capacitor in parallel with the large-value (low self-resonant frequency) capacitor. The intent is for the smaller capacitor to "take over" for high frequencies that are above the self-resonant frequency of the larger capacitor. However, this ideal performance fails to consider the inevitable lead inductances of the capacitors. The equivalent circuit for this scheme is shown in Fig. P6.17, where L_1 and L_2 denote the lead inductances. Derive the input impedance of this arrangement. If the lead inductances are equal, $L_1 = L_2 = L$, and $C_2 \ll C_1$ determine the break frequencies in this impedance. $[f_1 = 1/2\pi\sqrt{LC_1} < f_2 = 1/2\pi\sqrt{2LC_2} = f_3/\sqrt{2} < f_3 = 1/2\pi\sqrt{LC_2}]$ Show the fallacy in this method by plotting the Bode plot of this impedance to show that f_2 and f_3 are separated by only $\sqrt{2}$! Measure the individual and combined impedances of a 0.022 μF ceramic capacitor and a 220 pF ceramic capacitor from 10 to 500 MHz.

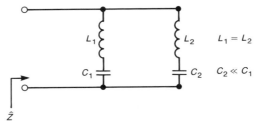

FIGURE P6.17

6.18 Determine the parasitic capacitance for the 1.2 μH inductor whose measured impedance is shown in Fig. 6.32. [1.75 pF]

6.19 An inductor is to be placed in series with a $50\ \Omega$ load to block a 100 MHz noise current. Determine a value for the inductance that will reduce the 100 MHz noise signal across the load by 20 dB. [0.8 μH]

6.20 A common-mode choke is constructed as shown in Fig. P6.20. With terminals AB connected, the impedance seen looking into terminals ab is 300,000 $\underline{/90°}$ Ω at 50 MHz. With terminals Ab connected, the impedance at 50 MHz seen looking into terminals aB is 10^6 $\underline{/90°}$ Ω. Determine the self and mutual inductances of the choke. [$L = 1.03$ mH, $M = 0.557$ mH]

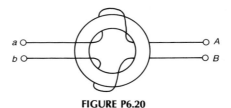

FIGURE P6.20

6.21 A common-mode choke is connected between a source and load as shown in Fig. P6.21. Determine the amplitude of the load voltage, assuming all common-mode currents have been eliminated by the common-mode choke. [2.56 V]

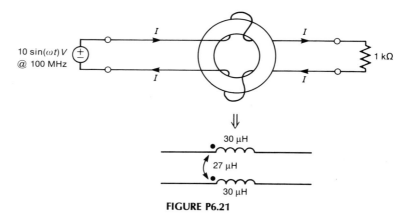

FIGURE P6.21

6.22 A component is measured and found to have an impedance whose magnitude is shown in Fig. P6.22. Synthesize an equivalent circuit to represent this impedance. [100 Ω in parallel with 15.92 μH]

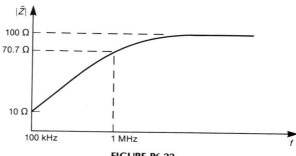

FIGURE P6.22

6.23 A component is measured and found to have an impedance whose (asymptotic) frequency response is as shown in Fig. P6.23. Synthesize an equivalent circuit to represent this impedance. [1 nH in series with the parallel combination of 100 Ω and 1 μF]

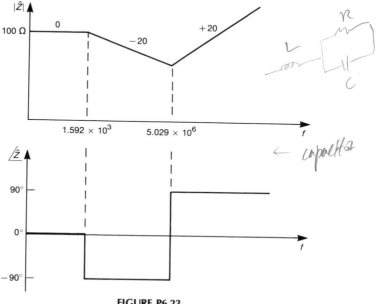

FIGURE P6.23

6.24 A small dc motor has the input impedance frequency response shown in asymptotic form in Fig. P6.24. Synthesize an equivalent circuit to represent this motor at its input terminals. [159 pF in parallel with the series combination of 100 Ω and 159 μH]

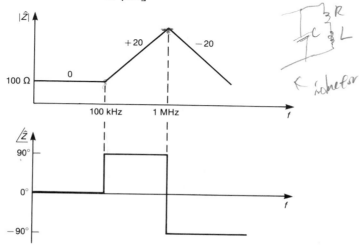

FIGURE P6.24

6.25 A component has the measured input impedance shown in asymptotic form in Fig. P6.25. Synthesize an equivalent circuit to represent this component at its input terminals. [$100\ \Omega$ in series with $0.01\ \mu F$]

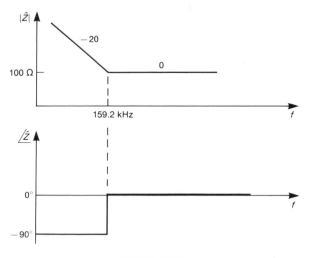

FIGURE P6.25

6.26 Determine the critical distance in Fig. 6.49 where the arc breakdown characteristic intersects the gas breakdown characteristic for $E_B = 10^9$ V/m. [$0.322\ \mu m$ or 0.0127 mils]

6.27 Consider the resistive circuit of Fig. 6.50 with the switch closing. If $V_{dc} = 400$ V, $V_G = 280$ V, $V_A = 12$ V, $I_G = 10$ mA, $I_A = 0.25$ A, and $R = 1000\ \Omega$, sketch the voltage and current across the contacts. [Since $(V_{dc} - V_A)/R = 388$ mA, the breakdown will transition immediately from glow to arc, and remain until switch contact]

6.28 For the $R-C$ switch protection network of Fig. 6.53 suppose $V_{dc} = 50$ V, $R_L = 500\ \Omega$, $L = 10$ mH, $I_A = 0.25$ A, and the switch closure/opening velocity is 0.1 m/s. Determine R and C such that the contacts will be protected. [$C > 0.1\ \mu F$, $200\ \Omega < R < 500\ \Omega$]

6.29 Measure the frequency response over portions of the conducted and radiated emission frequency range (1–500 MHz) of some of the following elements: (1) ceramic capacitors; (2) tantalum electrolytic capacitors; (3) inductors; (4) ferrite cores; (5) ferrite beads; and (6) the common-mode impedance between the input wires (tied together) and the frame for small dc, stepper, and ac motors and solenoids.

CHAPTER SEVEN

Signal Spectra

The frequency content or spectrum of the signals present in an electronic system is perhaps the most important aspect of the ability of that system to not only meet any regulatory limits but also perform compatibly with other electronic systems. In this chapter we will investigate this important aspect of EMC. We will begin the discussion with a general overview of the spectral composition of periodic signals. Once these important concepts are firmly understood, we will specialize these notions to signals that are representative of typical digital products. Bounds will be developed for these spectra that will facilitate the analysis of the effects of these signals. The use of a spectrum analyzer to measure the signal's spectral content will also be discussed, since the ability to use this important instrument correctly is critical to the correct evaluation of the product's compliance (or noncompliance) with governmental regulatory requirements. Finally, we will extend these notions to nonperiodic signals and then to random signals that represent data signals.

7.1 PERIODIC SIGNALS

Time-domain signals or waveforms that occur repetitively in time are referred to as being *periodic*. The more important signals that contribute directly to the radiated and conducted emissions of digital electronic systems are periodic signals. These types of waveforms are representative of clock and data signals that are necessary for the proper function of the system. Data streams in digital products are examples of random signals. The waveform takes on one of two levels during periodic intervals of the clock signal. However, the value in each interval (0 or 1) is a random variable. Signals whose time behavior is precisely known are referred to as *deterministic* signals. Signals whose time behavior is not known but can only be described statistically are referred to as *nondeterministic* or *random* signals. We will examine the frequency-domain description of periodic, deterministic waveforms in this section with the intent of representing clock waveforms of digital products. To some degree, this gives insight into the spectral composition of data signals. However, data signals are nondeterministic, since otherwise no information would be conveyed. The spectra of these types of signals will be determined in Section 7.6.

7.1.1   Orthogonal Basis Functions and Series Expansions

A *periodic* function of time t represented by $x(t)$ is a function (waveform or signal) that has the property [1, 2]

$$x(t \pm kT) = x(t), \quad k = 1, 2, 3,\ldots \tag{7.1}$$

That is, the function repeats itself over intervals of length T, where T is referred to as the *period* of the waveform. An example of a periodic signal is shown in Fig. 7.1. The reciprocal of the period is referred to as the *fundamental frequency* of the waveform, with units of hertz:

$$f_o = \frac{1}{T} \tag{7.2a}$$

It can be expressed in radians per second as

$$\omega_o = 2\pi f_o = \frac{2\pi}{T} \tag{7.2b}$$

The *average power* in a periodic waveform is defined by

$$P_{AV} = \lim_{T \to \infty} \frac{1}{T} \int_{t_1}^{t_1 + T} x^2(t)\, dt \tag{7.3}$$

where t_1 is some arbitrary time. That is, we only need to integrate over a time interval of length equal to the period of the signal. The *energy* in a signal is defined by

$$E = \int_{-\infty}^{\infty} x^2(t)\, dt \tag{7.4}$$

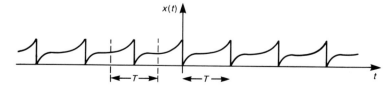

FIGURE 7.1   A periodic signal with period T.

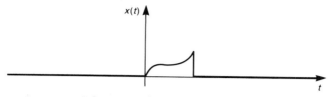

FIGURE 7.2 A nonperiodic signal.

Note that a periodic signal has infinite energy, since it must repeat indefinitely, yet the average power is finite. Hence periodic signals are referred to as *power signals.*

Signals that are not periodic are referred to as *nonperiodic.* An example is shown in Fig. 7.2. Nonperiodic signals have zero average power but finite energy. Hence they are referred to as *energy signals.*

Periodic signals can be represented as *linear combinations* of more basic signals that are referred to as *basis functions* and denoted as $\phi_n(t)$:

$$x(t) = \sum_{n=0}^{\infty} c_n \phi_n(t) \tag{7.5}$$

$$= c_0 \phi_0(t) + c_1 \phi_1(t) + c_2 \phi_2(t) + \cdots$$

The basis functions are also periodic, with the same period as $x(t)$. The coefficients c_n are referred to as the *expansion coefficients.* The advantage of such a representation is illustrated in Fig. 7.3. Consider a *linear system* having an input $x(t)$ and an output $y(t)$. This is referred to as a single-input, single-output system. The system is *linear* if it possesses the following two properties:

1. *If $x_1(t)$ produces $y_1(t)$ and $x_2(t)$ produces $y_2(t)$ then $x_1(t) + x_2(t)$ produces $y_1(t) + y_2(t)$.*
2. *If $x(t)$ produces $y(t)$ then $kx(t)$ produces $kt(t)$.*

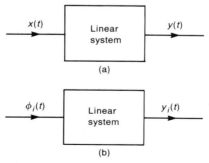

FIGURE 7.3 Processing of a signal by a linear system: (a) input $x(t)$ produces output $y(t)$; (b) a basis function in the expansion of $x(t)$, $\phi_i(t)$, produces the corresponding component of the output, $y_i(t)$.

These two properties are often collectively referred to as the property of *superposition*. Thus if we know the response of a linear system to each of the basis functions,

$$\phi_n(t) \rightarrow y_n(t) \tag{7.6a}$$

then the response to $x(t)$, which is decomposed into basis functions as in (7.5), is

$$y(t) = \sum_{n=0}^{\infty} c_n y_n(t) \tag{7.6b}$$

$$= c_0 y_0(t) + c_1 y_1(t) + c_2 y_2(t) + \cdots$$

That is, *the response to the original input signal can be found, by superposition, as the sum of the weighted responses to the individual components or basis functions that are used to represent the original input signal.* Quite often it is easier to determine the responses to the simpler basis functions than to the original signal. This then simplifies the determination of the response of the system to the more general input. This not only simplifies the calculation of the response to the original signal $y(t)$, but also yields considerable insight into how the linear system processes more general inputs. Note that the utility of this result relies on the system being *linear*, since we have used the property of superposition.

The question now becomes the appropriate choice of the basis functions that are to be used to represent a signal via (7.5). This is an important point, since a judicious choice of the basis set will determine the number of expansion terms required to give a good approximation to the function as in (7.5). For example, if $x(t)$ is a single-frequency sinusoid, the appropriate choice of basis functions would also be sinusoids; only one expansion function would be required to exactly match the function. On the other hand, if $x(t)$ is a square wave, the appropriate choice for the basis functions would also be square waves. There are obviously a large class of $x(t)$ where the "best" choice of basis functions is not so evident. Regardless of the specific choice, it is important to choose *orthogonal* basis functions. It is assumed that these basis functions are periodic, with period T. The set of functions $\phi_n(t)$ are said to be an *orthogonal* set if they possess the following property:

$$\int_{t_1}^{t_1 + T} \phi_n(t)\phi_m^*(t)\, dt = \begin{cases} \alpha_m & \text{for } m = n \\ 0 & \text{for } m \neq n \end{cases} \tag{7.7}$$

where * denotes the complex conjugate of the function. If the basis functions are real-valued then the conjugate operation is unnecessary. The orthogonality definition in (7.7) means that the functions of a basis set are orthogonal if the integral of the product of any basis function and the complex conjugate of another basis function over a time interval of length equal to the period T yields zero if the two basis functions are not identical, but yields a nonzero constant if the two basis functions are the same. The basis functions are said to be *orthonormal* if the constants α_m are unity, $\alpha_m = 1$, for all m. Although the basis functions and the expansion coefficients in (7.5) may be complex-valued, we will not use the caret ^ notation to differentiate between real and complex quantities in this chapter in order to simplify the notation.

Choosing orthogonal basis functions to expand some general function will simplify the determination of the expansion coefficients c_n. To show this property, multiply the expansion in (7.5) by $\phi_m^*(t)$ and integrate over a time interval of length T:

$$\int_{t_1}^{t_1+T} \phi_m^*(t)x(t)\,dt = \sum_{n=0}^{\infty} c_n \int_{t_1}^{t_1+T} \phi_m^*(t)\phi_n(t)\,dt \qquad (7.8)$$

$$= c_m \alpha_m$$

by the properties of orthogonality given in (7.7). Therefore the expansion coefficients are

$$c_n = \frac{1}{\alpha_n} \int_{t_1}^{t_1+T} \phi_n^*(t)x(t)\,dt \qquad (7.9)$$

Note that the computation of each expansion coefficient does not depend on the values of any of the other expansion coefficients.

Orthogonal basis functions also have the important property of minimizing the approximation error when we truncate the expansion, i.e., use a finite number of expansion terms in (7.5). This is an important consideration, since, although an exact representation of $x(t)$ theoretically requires an infinite number of terms in the expansion, any practical representation will utilize only a finite number of terms. *Orthogonal basis functions minimize the integral-square approximation error (ISE) in any finite-term expansion* [2]. The ISE is defined as

$$\text{ISE} = \int_{t_1}^{t_1+T} [x(t) - \bar{x}(t)]^2\,dt \qquad (7.10)$$

where a finite N-term expansion is denoted as

$$\bar{x}(t) = \sum_{n=0}^{N} c_n \phi_n(t) \qquad (7.11)$$

The ISE reflects the fact that "positive error" at a particular time t_i, i.e., $\bar{x}(t) > x(t_i)$, is as bad as "negative error," i.e., $\bar{x}(t_i) < x(t_i)$. In other words, the choice of orthogonal basis functions over other nonorthogonal basis sets gives a lower ISE in any finite-term expansion [2].

Although there are a number of choices for the orthogonal basis set that we could use to expand a general periodic signal, we will concentrate on the sinusoidal basis functions. The sinusoidal basis functions lead to the Fourier series representation that we will discuss next.

7.1.2   The Fourier Series Representation

The *trigonometric Fourier series* uses the sinusoidal basis functions [1]:

$$\phi_n = \begin{cases} 1 & \text{for } n = 0 \\ \cos\left(2\pi n \dfrac{t}{T}\right) \\ \sin\left(2\pi n \dfrac{t}{T}\right) \end{cases} \quad \text{for } n = 1, 2, 3, \ldots \tag{7.12}$$

It is a simple matter to show that these basis functions are orthogonal [1]. The series expansion of $x(t)$ in (7.5) becomes

$$x(t) = a_0 + \sum_{n=1}^{\infty} a_n \cos\left(2\pi n \frac{t}{T}\right) + \sum_{n=1}^{\infty} b_n \sin\left(2\pi n \frac{t}{T}\right) \tag{7.13a}$$

where the expansion coefficients are [1]

$$a_0 = \frac{1}{T} \int_{t_1}^{t_1 + T} x(t)\, dt \tag{7.13b}$$

$$a_n = \frac{2}{T} \int_{t_1}^{t_1 + T} x(t) \cos\left(2\pi n \frac{t}{T}\right) dt \tag{7.13c}$$

$$b_n = \frac{2}{T} \int_{t_1}^{t_1 + T} x(t) \sin\left(2\pi n \frac{t}{T}\right) dt \tag{7.13d}$$

and a_0 is the *average value* of the signal. The terms for $n = 1$ are referred to as the *fundamental frequency* terms, with fundamental radian frequency $\omega_o = 2\pi/T$ or cyclic frequency $f_o = 1/T$. The $n = 2$ terms are referred to as the *second-harmonic* terms with frequency $2f_o = 2/T$, the $n = 3$ terms are referred to as the *third-harmonic* terms with frequency $3f_o = 3/T$, and so forth.

As an example, consider the periodic rectangular waveform shown in Fig. 7.4. This type of waveform is often referred to as a "square wave." The

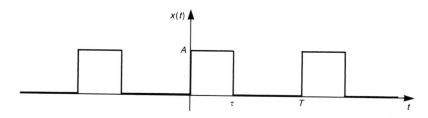

FIGURE 7.4   A periodic "square-wave" pulse train.

pulse amplitude is A and the pulse duration is τ. The expansion coefficient a_0 is obtained from (7.13b) as

$$a_0 = \frac{1}{T} \int_{t_1}^{t_1+T} x(t) \times 1 \, dt \tag{7.14a}$$

$$= \frac{1}{T} \left[\int_0^\tau A \times 1 \, dt + \int_\tau^T 0 \times 1 \, dt \right]$$

$$= \frac{A\tau}{T}$$

which is the *average value of the signal*. The other expansion coefficients are

$$a_n = \frac{2}{T} \int_0^\tau A \cos\left(2\pi n \frac{t}{T} \right) dt \tag{7.14b}$$

$$= \frac{A}{n\pi} \sin\left(2\pi n \frac{\tau}{T} \right)$$

$$b_n = \frac{2}{T} \int_0^\tau A \sin\left(2\pi n \frac{t}{T} \right) dt$$

$$= \frac{A}{n\pi} \left[1 - \cos\left(\frac{2\pi n\tau}{T} \right) \right]$$

This type of waveform is a first-order or rough approximation of clock waveforms found in digital systems. We will obtain a better representation for clock waveforms later. For the present it is interesting to note that for a 50% duty cycle, i.e., $\tau = \frac{1}{2}T$, $a_0 = \frac{1}{2}A$, $a_n = 0$, $b_n = 2A/n\pi$ for n odd and $b_n = 0$ for n even. The Fourier series representation for a 50% duty cycle becomes

$$x(t) = \frac{A}{2} + \frac{2A}{\pi} \sin(\omega_0 t) + \frac{2A}{3\pi} \sin(3\omega_0 t) + \cdots \tag{7.15a}$$

Thus for a 50% duty cycle there are no cosine terms (although a_0 is present) and only odd harmonics associated with the sine terms. In order to illustrate how the various components in the expansion of (7.15a) contribute to the overall makeup of the signal, we have shown the summation of the first seven components of a square wave of Fig. 7.4 (using (7.15a)) having $A = 1$, $T = 1$ s, and $\tau = 0.5$ s in Fig. 7.5. The trigonometric identities

$$\sin(A \pm B) = \sin A \cos B \pm \cos A \sin B$$

$$\cos(A \pm B) = \cos A \cos B \mp \sin A \sin B$$

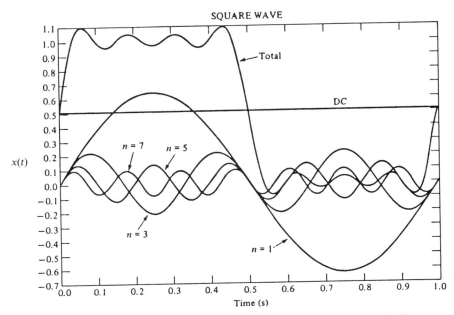

FIGURE 7.5 Illustration of the decomposition of a square wave into its frequency components [1].

can be used to place (7.15a) in the form

$$x(t) = \frac{A}{2} + \frac{2A}{\pi} \cos(\omega_o t - 90°) + \frac{2A}{3\pi} \cos(3\omega_o t - 90°) + \cdots \quad (7.15b)$$

The trigonometric Fourier series can be written in an equivalent but more useful form, referred to as the *complex exponential form*, as follows. Recall *Euler's identity*

$$e^{j\omega t} = \cos \omega t + j \sin \omega t \quad (7.16a)$$

which yields

$$\cos \omega t = \frac{e^{j\omega t} + e^{-j\omega t}}{2} \quad (7.16b)$$

$$\sin \omega t = \frac{e^{j\omega t} - e^{-j\omega t}}{2j} \quad (7.16c)$$

The complex-exponential form of the Fourier series can be derived by substituting (7.16b) and (7.16c) into (7.13a) [1]. It can also be derived directly using the properties of orthogonality of the basis functions. The basis functions in the complex-exponential form are

$$\phi_n(t) = e^{jn\omega_o t}, \quad n = -\infty, \ldots, -1, 0, 1, \ldots, \infty \tag{7.17}$$

It is simple matter to show, using (7.7), that these basis functions are orthogonal:

$$\int_{t_1}^{t_1+T} e^{-jm\omega_o t} e^{jn\omega_o t} \, dt = \begin{cases} 0 & \text{for } n \neq m \\ T & \text{for } n = m \end{cases} \tag{7.18}$$

The complex-exponential form of the expansion for a periodic $x(t)$ becomes

$$x(t) = \sum_{n=-\infty}^{\infty} c_n e^{jn\omega_o t} \tag{7.19}$$

Multiplying (7.19) by $\phi_m^*(t) = e^{-jm\omega_o t}$ and integrating over a time interval of length T gives

$$\int_{t_1}^{t_1+T} e^{-jm\omega_o t} x(t) \, dt = \sum_{n=-\infty}^{\infty} c_n \int_{t_1}^{t_1+T} e^{-jm\omega_o t} e^{jn\omega_o t} \, dt \tag{7.20}$$

$$= c_m T$$

Thus the expansion coefficients are given by

$$c_n = \frac{1}{T} \int_{t_1}^{t_1+T} x(t) e^{-jn\omega_o t} \, dt \tag{7.21}$$

Note that the complex form of the Fourier series contains, in addition to positive-valued harmonic frequencies ω_o, $2\omega_o$, $3\omega_o, \ldots$, negative-valued harmonics $-\omega_o$, $-2\omega_o$, $-3\omega_o, \ldots$. In addition, the expansion coefficients c_n may be complex-valued, whereas the expansion coefficients in the trigonometric Fourier series are real-valued. At first glance it may seem that the physical intuition present in the trigonometric Fourier series has been lost in the complex form. This is not the case. We should realize that for each positive value of n (and harmonic frequency) there is a corresponding negative value of n (and harmonic frequency). The coefficients of these, c_n and c_{-n}, are the conjugates of each other:

$$c_{-n} = \frac{1}{T} \int_{t_1}^{t_1+T} x(t) e^{jn\omega_o t} \, dt \tag{7.22}$$

$$= c_n^*$$

Since c_n may be complex-valued, let us denote it by

$$c_n = |c_n| \underline{/c_n} \tag{7.23}$$

$$= |c_n| e^{j \underline{/c_n}}$$

Thus

$$c_n^* = |c_n| e^{-j \underline{/c_n}} \tag{7.24}$$

The complex-exponential form in (7.19) may be written as

$$x(t) = c_0 + \sum_{n=1}^{\infty} c_n e^{jn\omega_o t} + \sum_{n=-1}^{-\infty} c_n e^{jn\omega_o t} \tag{7.25}$$

Changing the second summation to positive n gives

$$x(t) = c_0 + \sum_{n=1}^{\infty} c_n e^{jn\omega_o t} + \sum_{n=1}^{\infty} c_n^* e^{-jn\omega_o t} \tag{7.26}$$

Substituting (7.23) and (7.24) gives

$$x(t) = c_0 + \sum_{n=1}^{\infty} |c_n| e^{j(n\omega_o t + \underline{/c_n})} + \sum_{n=1}^{\infty} |c_n| e^{-j(n\omega_o t + \underline{/c_n})} \tag{7.27}$$

$$= c_0 + \sum_{n=1}^{\infty} |c_n| (e^{j(n\omega_o t + \underline{/c_n})} + e^{-j(n\omega_o t + \underline{/c_n})})$$

$$= c_0 + \sum_{n=1}^{\infty} 2|c_n| \cos(n\omega_o t + \underline{/c_n})$$

Therefore, *in order to obtain the expansion coefficients for the one-sided spectrum (positive frequencies only), we double the c_n for the double-sided spectrum, and the dc component c_0 remains unchanged.* Observe that the trigonometric Fourier series in (7.13a) can be placed in the form of (7.27) by combining the corresponding sin and cos terms and using trigonometric identities.

The complex-exponential expansion coefficients are usually more easily computed than are the coefficients in the trigonometric form. For example, consider the square wave of Fig. 7.4. We obtain, using (7.21),

$$c_n = \frac{1}{T} \int_{t_1}^{t_1 + T} e^{-jn\omega_o t} x(t) \, dt \tag{7.28}$$

$$= \frac{1}{T} \left[\int_0^\tau e^{-jn\omega_o t} A \, dt + \int_\tau^T e^{-jn\omega_o t} \times 0 \, dt \right]$$

$$= \frac{A}{jn\omega_o T} (1 - e^{-jn\omega_o \tau})$$

In calculations of this type it is often desirable to put the result into a sine or cosine form of function. To do this we extract $e^{-jn\omega_o\tau/2}$ from the term in parentheses to give

$$c_n = \frac{A}{jn\omega_o T} e^{-jn\omega_o\tau/2} \left(e^{jn\omega_o\tau/2} - e^{-jn\omega_o\tau/2} \right) \tag{7.29}$$

$$= \frac{A}{jn\omega_o T} e^{-jn\omega_o\tau/2} 2j \sin\left(\tfrac{1}{2} n\omega_o \tau\right)$$

$$= \frac{A\tau}{T} e^{-jn\omega_o\tau/2} \frac{\sin\left(\tfrac{1}{2} n\omega_o \tau\right)}{\tfrac{1}{2} n\omega_o \tau}$$

From this result we see that

$$|c_n| = \frac{A\tau}{T} \left| \frac{\sin\left(\tfrac{1}{2} n\omega_o \tau\right)}{\tfrac{1}{2} n\omega_o \tau} \right| \tag{7.30a}$$

$$\underline{/c_n} = \pm \tfrac{1}{2} n\omega_o \tau \tag{7.30b}$$

The $\pm$ sign of the angle comes about because the $\sin(\tfrac{1}{2} n\omega_o \tau)$ term may be positive or negative (an angle of 180°). This is added to the angle of $e^{-jn\omega_o\tau/2}$. Substituting $\omega_o = 2\pi/T$ gives

$$|c_n| = \frac{A\tau}{T} \left| \frac{\sin(n\pi\tau/T)}{n\pi\tau/T} \right| \tag{7.31a}$$

$$\underline{/c_n} = \pm \frac{n\pi\tau}{T} \tag{7.31b}$$

The amplitude spectrum is given in Fig. 7.6(b) and the phase spectrum in 7.6(c). The horizontal axes are in terms of cyclic frequency f. The amplitudes of the spectral components lie on an envelope that is

$$\frac{A\tau}{T} \left| \frac{\sin(\pi f\tau)}{\pi f\tau} \right|$$

This goes to zero where $\pi f\tau = m\pi$ or at multiples of $1/\tau$. This is a useful and commonly occurring function, usually denoted by

$$\frac{\sin x}{x}$$

At $x = 0$ the function evaluates to unity, and is zero when $x = m\pi$ for $m = 1, 2, 3, \ldots$. Although the continuous envelope bounds the spectral amplitudes,

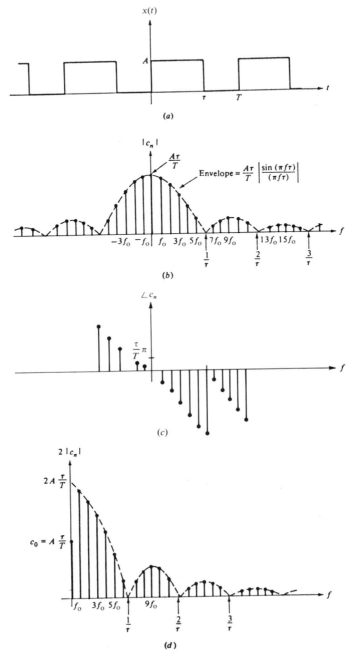

FIGURE 7.6   Frequency-domain representation of a square wave: (a) the signal; (b) the two-sided magnitude spectrum; (c) the phase spectrum; (d) the one-sided magnitude spectrum [1].

the spectral components only exist at multiples (harmonics) of the fundamental frequency $f_o = 1/T$. The phase is similarly plotted in Fig. 7.6(c). The magnitude and phase spectra given in Figs. 7.6(b) and (c) are said to be *two-sided spectra*, since both positive and negative frequency components are shown. The *one-sided spectrum* for positive frequencies only is obtained using the result in (7.27), and is shown in Fig. 7.6(d). Usually the one-sided spectra are preferred. Note that all positive frequency components *except the dc component* in the two-sided magnitude spectrum are *doubled* to give the one-sided magnitude spectrum. The one-sided phase spectrum is simply the two-sided phase spectrum for positive frequencies. Assuming a square wave with 50% duty cycle gives $|c_n| = A/n\pi$ for odd n and $|c_n| = 0$ for even n. The phase angle is $\angle c_n = \pm \frac{1}{2}n\pi$. Substituting these results into (7.27) gives

$$x(t) = \frac{A}{2} + \frac{2A}{\pi}\cos(\omega_o t - 90°) + \frac{2A}{3\pi}\cos(3\omega_o t - 90°) + \cdots \qquad (7.32)$$

which is identical to (7.15b) obtained for the trigonometric form.

7.1.3   Response of Linear Systems to Periodic Inputs

Consider the single-input, single-output linear system in Fig. 7.3(a). Suppose the input is a sinusoidal signal:

$$x(t) = X\cos(\omega t + \phi_x) \qquad (7.33a)$$

The output will also be a sinusoid at the same frequency as the input:

$$y(t) = Y\cos(\omega t + \theta_y) \qquad (7.33b)$$

Replacing the time-domain forms with their *phasor* equivalents allows a simple determination of this response [1]. The unit impulse response, denoted by $h(t)$, is the response for the system with a unit impulse function as the input, $x(t) = \delta(t)$, for zero initial conditions [1]. The phasor impulse response is denoted as $H(j\omega) = |H(j\omega)| \angle H(j\omega)$ and can be easily derived as the *transfer function* of the system [1]. In the case the phasor output becomes

$$Y \angle \theta_y = H(j\omega)X \angle \phi_x \qquad (7.34a)$$

Thus the magnitude of the output becomes

$$Y = |H(j\omega)|X \qquad (7.34b)$$

and the phase of the output is

$$\theta_y = \underline{/H(j\omega)} + \phi_x \qquad (7.34c)$$

Now suppose that $x(t)$ is periodic and its Fourier series has been obtained in the form

$$x(t) = c_0 + \sum_{n=1}^{\infty} 2|c_n| \cos(n\omega_0 t + \underline{/c_n}) \qquad (7.35)$$

We may pass each of these components through the system, determine the *sinusoidal steady-state response* to each, $Y(jn\omega_0)$, and add those responses to give the complete sinusoidal steady-state response to $x(t)$:

$$Y\underline{/\theta_y} = H(jn\omega_0)X\underline{/\phi_x} \qquad (7.36)$$

So the time-domain output $y(t)$ becomes

$$y(t) = c_0 H(0) + \sum_{n=1}^{\infty} 2|c_n|\,|H(jn\omega_0)| \cos[n\omega_0 t + \underline{/c_n} + \underline{/H(jn\omega_0)}] \qquad (7.37)$$

as illustrated in Fig. 7.7.

As an example, consider the application of a 1 V square wave to a linear circuit (a low-pass filter) shown in Fig. 7.8(a). The output of this "system" is the voltage across the capacitor, $v(t)$. The phasor transfer function is computed

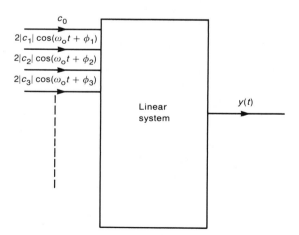

FIGURE 7.7   Illustration of the complete response to a general time-domain signal as the superposition of the responses to the spectral components of the (periodic) input signal.

as in the previous chapter as

$$H(p) = \frac{V(p)}{V_S(c)} \tag{7.38}$$

$$= \frac{1/pC}{R + 1/pC}$$

$$= \frac{1}{1 + pRC}$$

or

$$H(j\omega) = \frac{1}{1 + j\omega RC} \tag{7.39}$$

The transfer function evaluated at the harmonics of $v_S(t)$ is

$$H(jn\omega_o) = \frac{1}{1 + jn\omega_o RC} \tag{7.40}$$

Thus

$$v_n(t) = 2|c_n| |H(jn\omega_o)| \cos[n\omega_o t + \underline{/c_n} + \underline{/H(jn\omega_o)}] \tag{7.41}$$

and the complete steady-state response is

$$v(t) = c_0 H(0) + \sum_{n=1}^{\infty} v_n(t) \tag{7.42}$$

As a numerical example, let us choose $R = 1\ \Omega$ and $C = 1$ F. Thus the time constant of the circuit is $RC = 1$ s. It is expected that choosing the period of $v_S(t)$ to be less than or greater than this time constant will yield quite different responses. Let us choose $V_0 = 1$ V, $T = 2$ s, and $\tau = 1$ s to give a duty cycle of 50%. The transfer function is

$$H(jn\omega_o) = \frac{1}{1 + jn\pi}$$

The complex-exponential Fourier series coefficients are

$$c_n = \frac{V_0 \tau}{T} \frac{\sin(n\pi\tau/T)}{n\pi\tau/T} e^{-jn\pi\tau/T}$$

$$= \frac{1}{2} \frac{\sin(\frac{1}{2}n\pi)}{\frac{1}{2}n\pi} e^{-jn\pi/2}$$

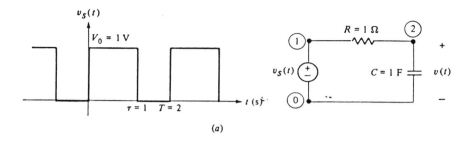

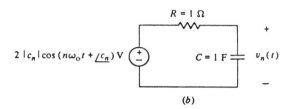

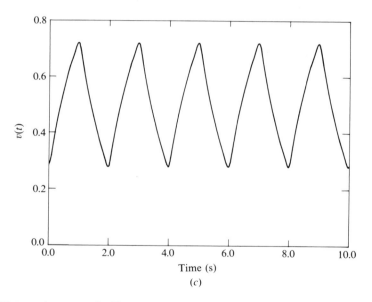

FIGURE 7.8 An example illustrating computation of the steady-state response of a circuit to a periodic signal: (a) the circuit and input signal; (b) computing the response to a spectral component of the input; (c) computed response [1].

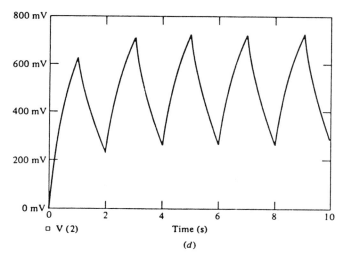

0 V (2)   Time (s)

(d)

FIGURE 7.8 continued   (d) SPICE simulation.

Evaluating these for a few terms gives

$$c_0 = \tfrac{1}{2}, \qquad\qquad\qquad H(0) = 1$$

$$c_1 = \frac{1}{\pi} e^{-j\pi/2}, \qquad H(j\omega_o) = \frac{1}{1 + j\pi} = 0.3033 \; \underline{/-72.3^\circ}$$

$$c_2 = 0,$$

$$c_3 = -\frac{1}{3\pi} e^{-j3\pi/2}, \quad H(j3\omega_o) = \frac{1}{1 + j3\pi} = 0.1055 \; \underline{/-83.9^\circ}$$

$$c_4 = 0,$$

$$c_5 = \frac{1}{5\pi} e^{-j5\pi/2}, \qquad H(j5\omega_o) = \frac{1}{1 + j5\pi} = 0.0635 \; \underline{/-86.4^\circ}$$

$$c_6 = 0,$$

$$c_7 = -\frac{1}{7\pi} e^{-j7\pi/2}, \quad H(j7\omega_o) = \frac{1}{1 + j7\pi} = 0.04543 \; \underline{/-87.4^\circ}$$

Therefore the one-sided spectrum given in (7.27) of the input becomes

$$v_S(t) = \frac{1}{2} + \frac{2}{\pi}\cos(\pi t - 90^\circ) - \frac{2}{3\pi}\cos(3\pi t + 90^\circ) + \frac{2}{5\pi}\cos(5\pi t - 90^\circ)$$

$$- \frac{2}{7\pi}\cos(7\pi t + 90^\circ) + \cdots$$

Computing the sinusoidal steady-state response to each component according to (7.36) and applying superposition gives

$$v(t) = \tfrac{1}{2} + 0.1931 \cos(\pi t - 162.3°) - 0.02239 \cos(3\pi t + 6.1°)$$
$$+ 0.008085 \cos(5\pi t - 176.36°) - 0.0041 \cos(7\pi t + 2.6°) + \cdots$$

A plot of this result using the first seven harmonics is shown in Fig. 7.8(c). Note that no transient time interval appears, since this result is only the steady-state portion of the solution. The complete SPICE simulation is shown in Fig. 7.8(d) using

```
EXAMPLE FIGURE 7.8
V1 1 0 PULSE(0 1 0 0.01 0.01 0.98 2)
R1 1 2 1
C1 2 0 1 IC=0
.TRAN 0.03 6 UIC
.PLOT TRAN V(2)
.END
```

7.1.4 Important Computational Techniques

Although direct computation of the coefficients of either the trigonometric or complex-exponential Fourier series is straightforward using the previous results, it can become tedious for some waveforms. The purpose of this section is to illustrate four important properties than can be used to make the computation of these coefficients virtually trivial for *piecewise linear* waveforms. Piecewise linear waveforms are those that consist of straight-line segments. An example of a periodic, piecewise linear waveform is shown in Fig. 7.9(a).

The first and most important property is that of *linearity*. Any waveform or function can be written as (or decomposed into) a *linear combination* of two or more functions:

$$x(t) = A_1 x_1(t) + A_2 x_2(t) + A_3 x_3(t) + \cdots \tag{7.43}$$

For example, the waveform in Fig. 7.9(a) can be written as the linear combination of two other waveforms: $x(t) = A_1 x_1(t) + A_2 x_2(t)$, where $x_1(t)$ and $x_2(t)$ are shown in Fig. 7.9(b) and $A_1 = A$, $A_2 = A$. Consequently the Fourier series of $x(t)$ can be written as a linear combination of the Fourier series representations of $x_1(t)$, $x_2(t)$, $x_3(t)$, ... according to (7.43). For example, suppose that the complex-exponential forms of $x_1(t)$ and $x_2(t)$ are written as

$$x_1(t) = \sum_{n=-\infty}^{\infty} c_{1n} e^{jn\omega_o t} \tag{7.44a}$$

$$x_2(t) = \sum_{n=-\infty}^{\infty} c_{2n} e^{jn\omega_o t} \tag{7.44b}$$

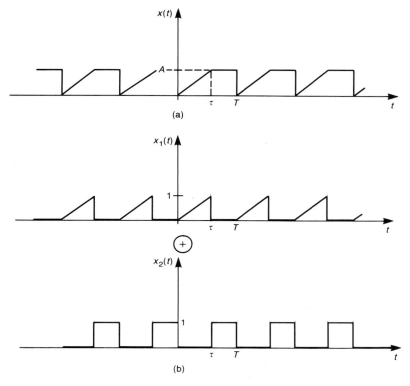

FIGURE 7.9 Illustration of the principle of linear decomposition of a signal: (a) the signal; (b) its decomposition, $x(t) = Ax_1(t) + Ax_2(t)$.

If $x(t) = x_1(t) + x_2(t)$ then

$$x(t) = x_1(t) + x_2(t) \tag{7.45}$$

$$= \sum_{n=-\infty}^{\infty} (c_{1n} + c_{2n})e^{jn\omega_o t}$$

$$= \sum_{n=-\infty}^{\infty} c_n e^{jn\omega_o t}$$

Therefore the expansion coefficient associated with the nth harmonic of $x(t)$ is the sum of the expansion coefficients of that harmonic associated with $x_1(t)$ and $x_2(t)$. Thus we can *decompose* a periodic function into a linear combination of perhaps simpler functions. If it is easier to obtain the expansion coefficients of these simpler functions then the original task of obtaining the expansion coefficients of $x(t)$ will be simplified.

The second important property has to do with *time-shifting a function*. If $x(t)$ is *shifted ahead in t by an amount* α (*delayed in time by* α), it is written as $x(t - \alpha)$.

An example is shown in Fig. 7.10(b). This is easy to remember if we observe that corresponding points on the waveforms of $x(t)$ and $x(t - \alpha)$ occur where the arguments of $x(t)$ and $x(t - \alpha)$ are identical. For example, the point on $x(t)$ at $t = 0$, $x(0)$, corresponds to the point on $x(t - \alpha = 0)$ or $t = \alpha$. Thus $x(t - \alpha)$ is $x(t)$ shifted ahead on the time axis by α (delayed in time). Similarly, $x(t + \alpha)$ is $x(t)$ shifted backward on the time axis by α (advanced in time). An example is shown in Fig. 7.10(c). The Fourier series expansion coefficients of $x(t \pm \alpha)$ can be directly found from the expansion coefficients of $x(t)$ as the following shows. First we recall the complex-exponential Fourier series expansion for $x(t)$ given in (7.19), with the expansion coefficients c_b given in (7.21). Suppose that $x(t)$ is *shifted ahead in t* to give $x(t - \alpha)$. Substituting $t - \alpha$ for t in (7.19) gives

$$x(t - \alpha) = \sum_{n = -\infty}^{\infty} c_n e^{jn\omega_o(t - \alpha)} \qquad (7.46)$$

$$= \sum_{n = -\infty}^{\infty} c_n \underbrace{e^{-jn\omega_o\alpha} e^{jn\omega_o t}}_{c'_n}$$

Therefore *we multiply the expansion coefficients of* $x(t)$ *by* $e^{-jn\omega_o\alpha}$ *to obtain the expansion coefficients of* $x(t - \alpha)$.

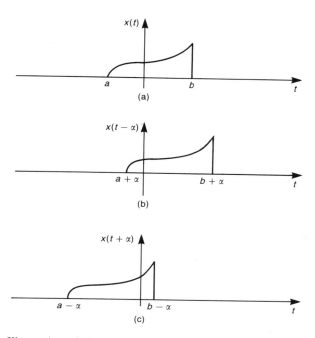

FIGURE 7.10   Illustration of the time-shift principle: (a) $x(t)$; (b) $x(t - \alpha)$; (c) $x(t + \alpha)$.

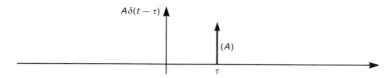

FIGURE 7.11   The impulse function.

The third important property that we will use has to do with the *unit impulse function* $\delta(t)$. This is defined by [1]

$$\delta(t) = \begin{cases} 0 & \text{for } t < 0 \\ 0 & \text{for } t > 0 \\ \int_{0^-}^{0^+} \delta(t)\,dt = 1 \end{cases} \tag{7.47}$$

The notations 0^- and 0^+ denote time immediately prior to and after $t = 0$, respectively. That is, the value of the function is zero except at $t = 0$, where its value is undefined. Intuitively, we say that the unit impulse function has zero width and infinite height such that the *area* under the function is unity [1]. The impulse function is represented by a vertical arrow (whose height is immaterial). The *strength* of the impulse is denoted in parenthenses adjacent to this arrow. Figure 7.11 illustrates $A\delta(t - \tau)$. Consider a periodic train of unit impulse functions

$$x(t) = \delta(t \pm kT), \quad k = 0, \pm 1, \pm 2, \pm 3, \ldots \tag{7.48}$$

shown in Fig. 7.12(a). The expansion coefficients are

$$c_n = \frac{1}{T} \int_0^T \delta(t) e^{-jn\omega_o t}\,dt \tag{7.49}$$

$$= \frac{1}{T} \int_{0^-}^{0^+} \delta(t) e^{-jn\omega_o t}\,dt = \frac{1}{T} \int_{0^-}^{0^+} \delta(t)\,dt$$

$$= \frac{1}{T}$$

If the pulse train is shifted ahead in t by α as shown in Fig. 7.12(b) then, by the time-shift property, the expansion coefficients become

$$c_n = \frac{1}{T} e^{-jn\omega_o \alpha} \tag{7.50}$$

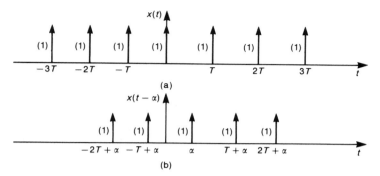

FIGURE 7.12   A periodic train of unit impulses (a) and (b) shifted in time.

The fourth and final property has to do with the relation between the expansion coefficients for a periodic function $x(t)$ and the expansion coefficients of its various *derivatives*. If $x(t)$ is represented with the complex-exponential Fourier series as

$$x(t) = \sum_{n=-\infty}^{\infty} c_n e^{jn\omega_o t} \tag{7.51}$$

and the kth derivative is represented as

$$\frac{d^k x(t)}{dt^k} = \sum_{n=-\infty}^{\infty} c_n^{(k)} e^{jn\omega_o t} \tag{7.52}$$

the expansion coefficients are related by

$$c_n = \frac{1}{(jn\omega_o)^k} c_n^{(k)} \qquad n \neq 0 \tag{7.53}$$

This is simple to show by differentiating (7.51) to yield

$$\frac{d^k x(t)}{dt^k} = \sum_{n=-\infty}^{\infty} \underbrace{(jn\omega_o)^k c_n}_{c_n^{(k)}} e^{jn\omega_o t} \tag{7.54}$$

We are now ready to utilize these four important properties to simplify the computation of the expansion coefficients in the complex-exponential Fourier series for *piecewise linear* functions. The technique is to *repeatedly differentiate the function until the first occurrence of an impulse function*. If the differentiated function does not consist solely of impulse functions, write the result as the sum of the part that contains the impulse functions and a part that contains the remainder. Determine the expansion coefficients for the part containing the

impulse functions using the above results and continue to differentiate the part
that did not contain impulses until impulses occur. Repeat the process until
the expansion is complete. Divide each part by the required power of $jn\omega_o$
according to (7.53) to return to the expansion coefficient of the original function.
For example, consider the function shown in Fig. 7.13(a). Differentiating this
once gives the result in Fig. 7.13(b). This may be represented as the sum of the

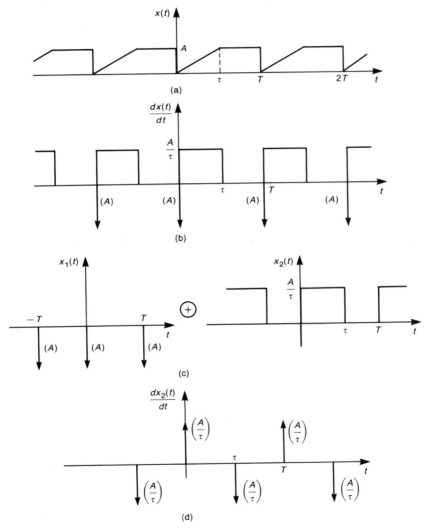

FIGURE 7.13 Illustration of the use of differentiation to compute the expansion
coefficients for a piecewise linear, periodic signal: (a) the signal; (b) the first derivative;
(c) use of linearity; (d) repeated differentiation of a component.

two functions shown in Fig. 7.13(c):

$$\frac{dx(t)}{dt} = x_1(t) + x_2(t) \tag{7.55}$$

The expansion coefficients for $x_1(t)$ are easily determined with the above properties of the impulse train as

$$c_{1n}^{(1)} = -\frac{A}{T} \tag{7.56}$$

The function $x_2(t)$ is differentiated again to yield the result in Fig. 7.13(d). The expansion coefficients for this are

$$c_{2n}^{(2)} = \frac{A}{\tau}\frac{1}{T} - \frac{A}{\tau}\frac{1}{T}e^{-jn\omega_o\tau} \tag{7.57}$$

Notice in this last result we have used the properties of linearity and time shift. The expansion coefficients for the original function are then

$$c_n = \frac{1}{jn\omega_o}c_{1n}^{(1)} + \frac{1}{(jn\omega_o)^2}c_{2n}^{(2)} \qquad n \neq 0 \tag{7.58}$$

$$= -\frac{1}{jn\omega_o}\frac{A}{T} + \frac{1}{(jn\omega_o)^2}\left(\frac{A}{\tau}\frac{1}{T} - \frac{A}{\tau}\frac{1}{T}e^{-jn\omega_o\tau}\right)$$

$$= j\frac{A}{n\omega_o T}\left[1 + j\frac{1}{n\omega_o\tau}(1 - e^{-jn\omega_o\tau})\right]$$

$$= j\frac{A}{n\omega_o T} - j\frac{A}{n\omega_o T}\frac{\sin(\frac{1}{2}n\omega_o\tau)}{\frac{1}{2}n\omega_o\tau}e^{-jn\omega_o\tau/2}$$

And finally we will use these properties to obtain the expansion coefficients for the square wave shown in Fig. 7.4. The derivative of that function gives a periodic train of impulses of strength A occurring at the rising edge of each pulse, $A\delta(t \pm nT)$, and a periodic train of impulses of strength $-A$ occurring at the falling edge of each pulse, $-A\delta(t - \tau \pm nT)$. The expansion coefficients for the derivative of this function are the sums of the expansion coefficients for each pulse train component:

$$c_n^{(1)} = A\frac{1}{T} - A\frac{1}{T}e^{-jn\omega_o\tau} \tag{7.59}$$

$$= \frac{A}{T}(1 - e^{-jn\omega_o\tau})$$

$$= \frac{A}{T}(e^{jn\omega_o\tau/2} - e^{-jn\omega_o\tau/2})e^{-jn\omega_o\tau/2}$$

$$= jn\omega_o\frac{A\tau}{T}\frac{\sin(\frac{1}{2}n\omega_o\tau)}{\frac{1}{2}n\omega_o\tau}e^{-jn\omega_o\tau/2}$$

The expansion coefficients for $x(t)$ are

$$c_n = \frac{1}{jn\omega_o} c_n^{(1)} \qquad (7.60)$$

$$= \frac{A\tau}{T} \frac{\sin(\frac{1}{2}n\omega_o\tau)}{\frac{1}{2}n\omega_o\tau} e^{-jn\omega_o\tau/2}$$

as before.

7.2   SPECTRA OF DIGITAL CIRCUIT WAVEFORMS

As was mentioned earlier, the waveforms of primary importance in digital circuits are those which represent clock and data signals. The clock signals are periodic, deterministic signals and are therefore representable by the methods of the previous sections. Although the clock waveforms resemble square waves as shown in Fig. 7.4, the purpose of this section is to obtain a better representation of them and to simplify that representation to provide insight into the time-domain factors that affect their spectral content. Data waveforms will be considered in Section 7.6.

7.2.1   The Spectrum of Trapezoidal (Clock) Waveforms

Clock waveforms will be represented as periodic trains of trapezoidal-shaped pulses shown in Fig. 7.14. Each pulse is described by an amplitude A, a pulse *risetime* τ_r, a pulse *falltime* τ_f, and a pulse *width* (between 50% points of the waveform amplitude) τ. For our purposes we will represent the pulse rise- and falltimes as being the time required for the signal to transition from 0 to A; that is, from the 0% to the 100% points. Because actual pulses do not transition as sharply as we have shown, it is common in industry to define the rise- and falltimes as being from $0.1A$ to $0.9A$; that is, from the 10% to 90% points. The purpose of this section is to investigate the effect of these pulse parameters on the spectrum of the waveform. We fill find that *the key parameters that contribute*

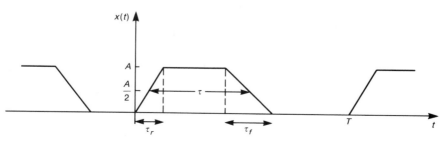

FIGURE 7.14 The periodic, trapezoidal pulse train representing clock and data signals of digital systems.

to the high-frequency spectral content of the waveform are the rise- and falltimes of the pulse. The levels of the emissions in the regulatory frequency range are therefore strongly dependent on the risetimes and falltimes of these pulses.

In order to obtain the complex-exponential Fourier series of this waveform, we will use results of Section 7.1.4. The first derivative of the waveform is shown in Fig. 7.15(a). Differentiating this once again to yield impulses gives the waveform in Fig. 7.15(b). Since this waveform contains impulses, we expand it to give the expansion coefficients

$$
c_n^{(2)} = \frac{1}{T}\frac{A}{\tau_r} - \frac{1}{T}\frac{A}{\tau_r}e^{-jn\omega_o\tau_r} - \frac{1}{T}\frac{A}{\tau_f}e^{-jn\omega_o[\tau + (\tau_r - \tau_f)/2]} \tag{7.61}
$$

$$
+ \frac{1}{T}\frac{A}{\tau_f}e^{-jn\omega_o[\tau + (\tau_r + \tau_f)/2]}
$$

$$
= \frac{A}{T}\left[\frac{1}{\tau_r}e^{-jn\omega_o\tau_r/2}\left(e^{jn\omega_o\tau_r/2} - e^{-jn\omega_o\tau_r/2}\right)\right.
$$

$$
\left. - \frac{1}{\tau_f}e^{-jn\omega_o\tau_r/2}e^{-jn\omega_o\tau}\left(e^{jn\omega_o\tau_f/2} - e^{-jn\omega_o\tau_f/2}\right)\right]
$$

$$
= j\frac{A}{2\pi n}(n\omega_o)^2 e^{-jn\omega_o(\tau + \tau_r)/2}\left[\frac{\sin(\frac{1}{2}n\omega_o\tau_r)}{\frac{1}{2}n\omega_o\tau_r}e^{jn\omega_o\tau/2} - \frac{\sin(\frac{1}{2}n\omega_o\tau_f)}{\frac{1}{2}n\omega_o\tau_f}e^{-jn\omega_o\tau/2}\right]
$$

According to (7.53), the expansion coefficients for the original trapezoidal waveform are

$$
c_n = \frac{1}{(jn\omega_o)^2}c_n^{(2)} \qquad n \neq 0 \tag{7.62}
$$

$$
= -\frac{c_n^{(2)}}{(n\omega_o)^2}
$$

$$
= -j\frac{A}{2\pi n}e^{-jn\omega_o(\tau + \tau_r)/2}\left(\frac{\sin(\frac{1}{2}n\omega_o\tau_r)}{\frac{1}{2}n\omega_o\tau_r}e^{jn\omega_o\tau/2} - \frac{\sin(\frac{1}{2}n\omega_o\tau_f)}{\frac{1}{2}n\omega_o\tau_r}e^{-jn\omega_o\tau/2}\right)
$$

The result in (7.62) cannot be combined any further to yield useful results. However, *if the pulse risetime equals the falltime, $\tau_r = \tau_f$, a very useful can be obtained.* Substituting $\tau_r = \tau_f$ into (7.62) gives the expansion coefficients as

$$
c_n = A\frac{\tau}{T}\frac{\sin(\frac{1}{2}n\omega_o\tau)}{\frac{1}{2}n\omega_o\tau}\frac{\sin(\frac{1}{2}n\omega_o\tau_r)}{\frac{1}{2}n\omega_o\tau_r}e^{-jn\omega_o(\tau + \tau_r)/2} \tag{7.63}
$$

with

$$
\tau_r = \tau_f
$$

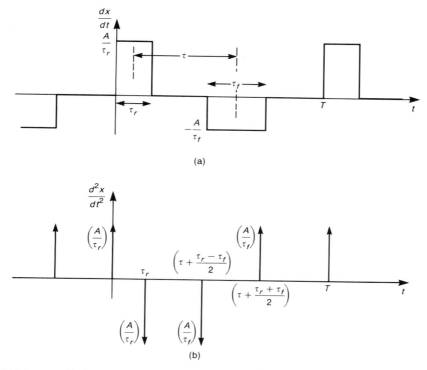

FIGURE 7.15 Various derivatives of the trapezoidal pulse train: (a) first derivative; (b) second derivative.

Notice that the result can be placed in the form of the product of two $(\sin x)/x$ terms. *If the rise- and falltimes are not equal, $\tau_r \neq \tau_f$, the result cannot be placed in the form of the product of two $(\sin x)/x$ terms!* There have been attempts to use the above result for $\tau_r \neq \tau_f$ and to replace the risetime in this expression with an average rise/falltime, $\frac{1}{2}(\tau_r + \tau_f)$. Although this would be desirable and reasonably accurate as an approximation if τ_r and τ_f did not greatly differ, the fact is that *this is not correct!* Nevertheless, assuming the rise- and falltimes to be the same will give important insight for the more general case.

Continuing with the assumption that the pulse rise- and falltimes are the same, we may obtain the expansion coefficients for the *one-sided spectrum* (positive frequencies), c_n^+, where

$$x(t) = c_0 + \sum_{n=1}^{\infty} |c_n^+| \cos(n\omega_0 t + \underline{/c_n}) \tag{7.64a}$$

According to (7.27), to obtain this, we *double all the magnitudes of the expansion coefficients for the two-sided spectrum except the c_0 term.* From the result in

(7.63) we obtain

$$|c_n^+| = 2|c_n| = 2A \frac{\tau}{T} \left| \frac{\sin(n\pi\tau/T)}{n\pi\tau/T} \right| \left| \frac{\sin(n\pi\tau_r/T)}{n\pi\tau_r/T} \right| \quad \text{for } n \neq 0$$

with

$$\tau_r = \tau_f \tag{7.64b}$$

and

$$c_0 = A \frac{\tau}{T}$$

where we have substituted $\omega_o = 2\pi/T$. The angle is

$$\angle c_n = \pm n\pi \frac{\tau + \tau_r}{T} \tag{7.64c}$$

with

$$\tau_r = \tau_f$$

As a check on this result, substitute $\tau_r = 0$, which gives the result in (7.31) for the square wave.

Several important points can be seen from the result in (7.64). First suppose that $\tau = \frac{1}{2}T$; that is, a 50% duty cycle. In this case the first sine term becomes $|\sin(n\pi\tau/T)|/|n\pi\tau/T| = |\sin \frac{1}{2}n\pi|/|\frac{1}{2}n\pi|$, which is zero for even n. Therefore *there are (theoretically) no even harmonics for a 50% duty cycle.* Digital clock signals tend to approach 50% duty cycle. However, in order for the even harmonics to be absent, the duty cycle must be *exactly 50%.* As a practical matter, the even harmonics can never be completely eliminated, since the duty cycle cannot be set to *exactly 50%.* However, the even harmonics will be increasingly smaller than the odd harmonics the closer we approach a 50% duty cycle. The odd-harmonic levels are quite stable for slight variations in duty cycle. This illustrates that *slight variations in duty cycle from an exact 50% duty cycle can cause the even-harmonic levels to vary widely, and these levels may, in a practical situation, be significant!* This points out a potentially serious difficulty in reproducing measured data on the radiated emissions of a product from one measurement to another *for the same product.* If the offending frequency happens to be an *even harmonic* of a clock in the system then it is quite likely that a radiated emission at that frequency measured on one day and repeated measurement of the (identical and unchanged) product on a subsequent day will show significant differences in the level! This variation is likely due to the above phenomenon.

7.2.2   Spectral Bounds for Trapezoidal Waveforms

Although the result in (7.64) gives the one-sided spectrum expansion coefficients, it is desirable to extract more intuitive information than is apparent in these equations. In order to do this, we will generate *bounds on the magnitude spectrum*. Although these are upper bounds (worst case) on the spectral components and are approximate, their utility will be of considerable benefit in understanding the impact of risetime, falltime and pulse width on the spectrum of the waveform.

To begin this discussion, recall the result for a square wave given in (7.31); this is, a trapezoidal pulse train with zero rise/falltime. The magnitudes of the expansion coefficients are in the form of a $(\sin x)/x$ expression. Although the spectral components only exist at frequencies $f = n/T$ for $n = 0, 1, 2, \ldots$, *the envelope of these spectral components follows a form of* $(\sin \pi \tau f)/\pi \tau f$. This is illustrated in Fig. 7.6b. The envelope has zeros at $f = m/\tau$ for $m = 1, 2, 3, \ldots$. This envelope can be bounded as follows [3]. Recall the small-argument expression for $\sin x$, which is that $\sin x \cong x$ for small x. Thus we have

$$\left|\frac{\sin x}{x}\right| \leq \begin{cases} 1 & \text{for small } x \\ \dfrac{1}{|x|} & \text{for large } x \end{cases}$$

This can be drawn as two asymptotes, as shown in Fig. 7.16. The first asymptote is unity and has slope on a logarithmic or *Bode* plot of 0 dB/decade [1]. The second asymptote decreases linearly with x, or -20 dB/decade [1]. The two asymptotes converge at $x = 1$. In the expansion coefficient for the square wave in (7.31), $x = \pi \tau f$, where $f = n/T$. Thus for the square wave the first asymptote has a 0 dB/decade slope out to $f = 1/\pi \tau$ and a -20 dB/decade slope above that frequency, as shown in Fig. 7.16.

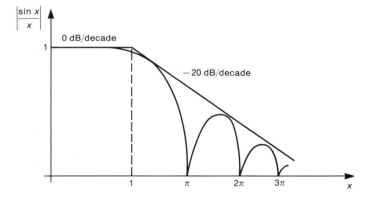

FIGURE 7.16   Bounds on the $(\sin x)/x$ function.

7.2.2.1 Effect of Rise/Falltime on Spectral Content We now are ready to extend these notions to the trapezoidal pulse train. Again, in order to use these results, we must assume that the rise and falltimes are equal, $\tau_r = \tau_f$, so that the expansion coefficient expression can be put in the form of the product of two $(\sin x)/x$ expressions as given in (7.64). Replacing the discrete spectrum with a continuous envelope by substituting $f = n/T$ gives

$$\text{envelope} = 2A\frac{\tau}{T}\left|\frac{\sin(\pi\tau f)}{\pi\tau f}\right|\left|\frac{\sin(\pi\tau_r f)}{\pi\tau_r f}\right| \qquad (7.65)$$

Recall that the dc term or level is $A\tau/T$. In order to generate bounds for this spectrum, we take the logarithm of (7.65) to give

$$20\log_{10}(\text{envelope}) = 20\log_{10}\left(2A\frac{\tau}{T}\right) + 20\log_{10}\left|\frac{\sin(\pi\tau f)}{\pi\tau f}\right| \qquad (7.66)$$
$$+ 20\log_{10}\left|\frac{\sin(\pi\tau_r f)}{\pi\tau_r f}\right|$$

This shows that the composite plot is the sum of three plots:

$$\text{Plot 1} = 20\log_{10}\left(2A\frac{\tau}{T}\right) \qquad (7.67a)$$

$$\text{Plot 2} = 20\log_{10}\left|\frac{\sin(\pi\tau f)}{\pi\tau f}\right| \qquad (7.67b)$$

$$\text{Plot 3} = 20\log_{10}\left|\frac{\sin(\pi\tau_r f)}{\pi\tau_r f}\right| \qquad (7.67c)$$

Plot 1 has 0 dB/decade slope and a level of $2A\tau/T = 2A\tau f_o$. Plot 2 has two asymptotes. The first asymptote has a 0 dB/decade slope as discussed for the square wave and unit level (0 dB). The second asymptote has a -20 dB/decade slope. The two asymptotes join at $f = 1/\pi\tau$. Plot 3 also consists of two asymptotes, one of which has a 0 dB/decade slope and unit level (0 dB) and the other has a -20 dB/decade slope. The asymptotes of the third plot join at $f = 1/\pi\tau_r$. The composite asymptote is the sum of these asymptotes, as shown in Fig. 7.17. The composite asymptote thus consists of three straight-line segments. The first is due to Plot 1 and has a slope of 0 dB/decade and a starting level of $2A\tau/T$. The second segment has a slope of -20 dB/decade and is due to Plot 2. The third segment has a slope of -40 dB/decade and is due to the sum of Plots 2 and 3. It is evident that the pulse width must be greater than or equal to the pulse rise/falltime: $\tau \geq \tau_r$. Thus the first breakpoint in the spectral bound will be related to Plot 2, whose breakpoint is related to the

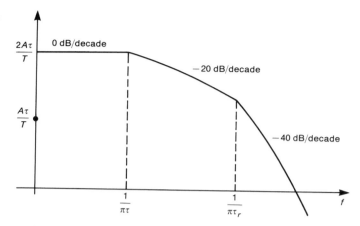

FIGURE 7.17 Bounds on the one-sided magnitude spectrum of a trapezoidal pulse train.

pulse width and is $1/\pi\tau$. The second breakpoint is due to Plot 3, whose breakpoint is related to the pulse rise/falltime and is $1/\pi\tau_r$.

From these spectral bounds it now becomes clear that *the high-frequency content of a trapezoidal pulse train is primarily due to the rise/falltime of the pulse. Pulses having small rise/falltimes will have larger high-frequency spectral content than will pulses having larger rise/falltimes. Thus, in order to reduce the high-frequency spectrum in order to reduce the emissions of a product, increase the rise/falltimes of the clock and/or data pulses.* "Fast" (short) rise/falltimes are the primary contributors to the high-frequency spectral content of the signal and therefore the product's inability to meet the governmental regulatory requirements on radiated and conducted emissions. They are also important in the ability of the product to cause interference. (Recall that the interference potential of a product and whether or not it complies with the regulatory requirements are not necessarily related; that is, a product that complies with the governmental emission requirements may cause interference.) Figure 7.18 shows the exact spectrum along with the spectral bounds for a 1 MHz trapezoidal waveform having a rise/falltime of 20 ns and duty cycles of 50%, 30%, and 10%.

Figure 7.19 shows measured spectra that illustrate these points. Figure 7.19(a) shows the time-domain waveform and associated spectrum (measured with a spectrum analyzer to be discussed) for a 1 V, 10 MHz trapezoidal pulse train having a 50% duty cycle and rise/falltimes of 20 ns. Figure 7.19(b) shows the same waveform, but with the rise/falltimes reduced to 5 ns. Note the dramatic increase in the spectral content at the higher frequencies. The span is from essentially dc up to 500 MHz. The center frequency is 250 MHz. The amplitude setting is the same for both spectral plots. The spectrum of the 20 ns rise/falltime pulse train shows very little spectral content above 250 MHz, whereas the spectrum of the 5 ns pulse train shows a significant spectral content up to

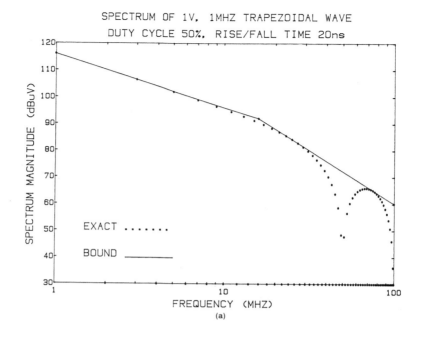

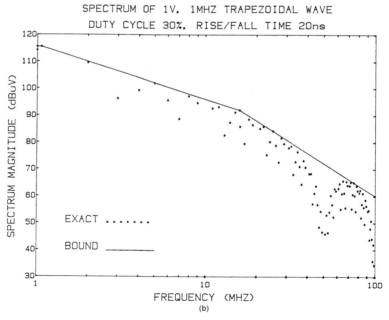

FIGURE 7.18   Examples illustrating the spectral bounds for various duty cycles of a 1 V, 1 MHz, trapezoidal pulse train having rise/falltimes of 20 ns: (a) 50% duty cycle; (b) 30% duty cycle.

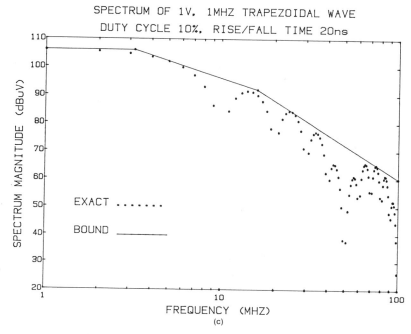

SPECTRUM OF 1V, 1MHZ TRAPEZOIDAL WAVE
DUTY CYCLE 10%, RISE/FALL TIME 20ns

FIGURE 7.18 continued   (c) 10% duty cycle.

500 MHz. Specifically, the 11th harmonic (110 MHz) is increased from 68.0 to 86.1 dBμV, an increase of some 18 dB at 110 MHz! This dramatically illustrates the importance in *using pulses with as large a rise/falltime as possible!*

In order to illustrate this important relationship between pulse rise/falltimes and the spectral content of that waveform, we will consider the following numerical example, which investigates the spectral bounds for the experimental example shown in Fig. 7.19. We will show the spectral bounds for this signal, which is a 10 MHz digital signal having 1 V amplitude, 50% duty cycle, and two rise/falltimes. Figure 7.20(a) shows the spectral bounds for a rise/falltime of 20 ns, and Fig. 7.20(b) shows the spectral bounds for the same signal, but with a rise/falltime of 5 ns. Note that the -40 dB/decade bound has moved out from 15.9 to 63.66 MHz. Thus the high-frequency components above 15.9 MHz will not "roll off" at the -40 dB/decade rate as for the 20 ns pulse, but at only -20 dB/decade up to 63.66 MHz, which will result in a significant increase in the spectral levels above 15.9 MHz.

In order to gauge how much this will incur, we must be able to interpolate on log–log or Bode plots. Consider the log–log plot shown in Fig. 7.21(a). The slope is given as M dB/decade. The equation for this line is

$$\log_{10} Y_2 - \log_{10} Y_1 = M(\log_{10} f_2 - \log_{10} f_1) \qquad (7.68a)$$

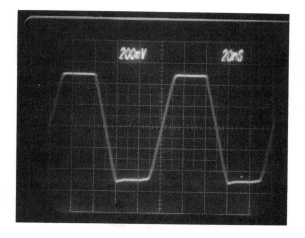

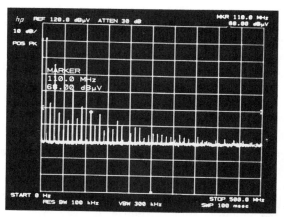

rise/fall time = 20 ns

(a)

FIGURE 7.19 Experimentally measured spectra of 1 V, 10 MHz, 50% duty cycle trapezoidal pulse trains for rise/falltimes of (a) 20 ns.

or

$$\log_{10} Y_2 = \log Y_1 + M \log_{10}\left(\frac{f_2}{f_1}\right) \qquad (7.68b)$$

This simple but important result allows us to estimate the effect of a higher-frequency breakpoint. Consider the trapezoidal spectral bound shown in Fig. 7.21(b). The segment breakpoints occur at f_1 and f_3, and it is desired to determine the reduction in dB from the dc level at the frequencies f_2 and

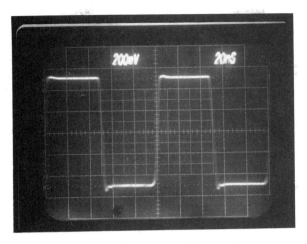

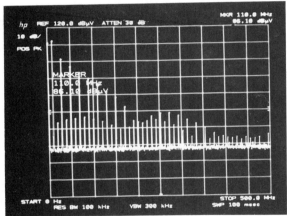

rise/fall time = 5 ns
(b)

FIGURE 7.19 continued   (b) 5 ns.

f_4. The various reductions are

$$\Delta_1 = -20 \log_{10}\left(\frac{f_2}{f_1}\right) \qquad (7.69a)$$

$$\Delta_2 = -20 \log_{10}\left(\frac{f_3}{f_1}\right) \qquad (7.69b)$$

$$\Delta_3 = -40 \log_{10}\left(\frac{f_4}{f_3}\right) \qquad (7.69c)$$

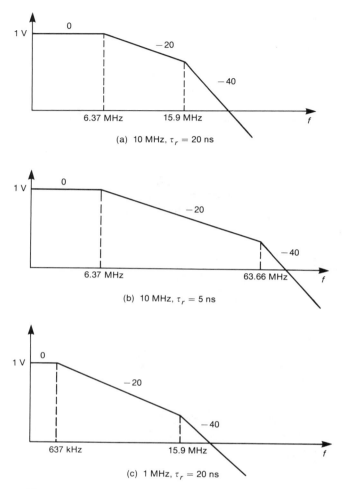

FIGURE 7.20 Illustration of the effect of rise/falltimes and repetition rate on the spectral content of 1 V, 50% duty cycle trapezoidal pulse trains: (a) 10 MHz, $\tau_r = \tau_f = 20$ ns; (b) 10 MHz, $\tau_r = \tau_f = 5$ ns; (c) 1 MHz, $\tau_r = \tau_f = 20$ ns.

Therefore the various levels are

$$K_2 \, \text{dB} = K \, \text{dB} + \Delta_1 \tag{7.70a}$$

$$K_4 \, \text{dB} = K \, \text{dB} + \Delta_2 + \Delta_3 \tag{7.70b}$$

For the 10 MHz, 50% duty cycle trapezoidal waveform whose spectral bounds are shown in Fig. 7.20 and whose experimentally measured spectra are shown in Fig. 7.19, we can compute the level at 110 MHz for the 20 ns

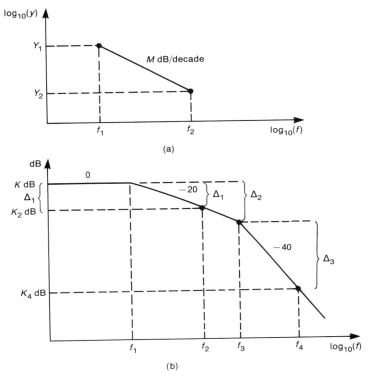

FIGURE 7.21 Illustration of interpolation on log–log (Bode) plots: (a) the general characterization of a linear segment; (b) application to the spectral bound for a trapezoidal pulse train.

rise/falltime to be

$$\text{Level}_{110\text{ MHz}} = 20\log_{10}(1\text{ V}) - 20\log_{10}\left(\frac{15.9\text{ MHz}}{6.37\text{ MHz}}\right) - 40\log_{10}\left(\frac{110\text{ MHz}}{15.9\text{ MHz}}\right)$$

$$= 120\text{ dB}\mu\text{V} - 7.95\text{ dB} - 33.6\text{ dB}$$

$$= 78.45\text{ dB}\mu\text{V}$$

The level at 110 MHz for the 5 ns rise/falltime pulse is

$$\text{Level}_{110\text{ MHz}} = 20\log_{10}(1\text{ V}) - 20\log_{10}\left(\frac{63.7\text{ MHz}}{6.37\text{ MHz}}\right) - 40\log_{10}\left(\frac{110\text{ MHz}}{63.7\text{ MHz}}\right)$$

$$= 120\text{ dB}\mu\text{V} - 20\text{ dB} - 9.49\text{ dB}$$

$$= 90.5\text{ dB}\mu\text{V}$$

which is an increase of 12 dB. The exact values computed using (7.64b) are 73.8 dBμV ($\tau_r = \tau_f = 20$ ns) and 90.4 dBμV ($\tau_r = \tau_f = 5$ ns), which are reasonably close to the bounds. The measured values shown in Fig. 7.19 are reasonably close to these. For example, the measured value for the 11th harmonic (110 MHz) for $\tau_r = \tau_f = 20$ ns is 68.0 dBμV, which is some 6 dB lower than the exact value of 73.8 dBμV. Since the spectrum analyzer measures rms levels, the computed value of 73.8 dBμV is 70.8 dBμV rms, which is within 2 dB of the measured value! Similarly, the measured value for the 11th harmonic (110 MHz) for $\tau_r = \tau_f = 5$ ns is 86.1 dBμV, which is some 4 dB lower than the exact value of 90.4 dBμV. Again, since the spectrum analyzer measures rms levels, the computed value of 90.4 dBμV is 87.4 dBμV rms, which is within 1 dB of the measured value!

7.2.2.2 Effect of Repetition Rate and Duty Cycle One effect of a change in the repetition rate (frequency) of the pulse train is to change the spacing between the discrete harmonics. Also note that the starting or dc level of the spectral bound is $20 \log_{10}(2A\tau/T) = 20 \log(2AD)$, where

$$D = \frac{\tau}{T} \qquad (7.71)$$

is the duty cycle. Writing the one-sided spectrum for the trapezoidal waveform given in (7.64) in terms of the duty cycle D gives

$$|c_n^+| = 2AD \left| \frac{\sin(n\pi D)}{n\pi D} \right| \left| \frac{\sin(n\pi \tau_r f_o)}{n\pi \tau_r f_o} \right| \quad \text{for } n \neq 0 \qquad (7.72)$$

$$c_0 = AD$$

where $f_o = 1/T$ is the fundamental frequency of the waveform. Usually when the frequency of the wave is reduced (the period is increased), the pulse width is similarly increased to maintain the same duty cycle, say, 50%. Therefore reducing the frequency of the pulse train (increasing the period T) does not usually affect the starting level. However, if the frequency of the wave is reduced and the duty cycle remains the same, the pulse width will be increased. This has the effect of moving the first breakpoint in the spectral bound, $1/\pi\tau$, down in frequency, so that part of the spectral content in the region of 0 dB/decade now rolls of at a rate of -20 dB/decade. Figure 7.20(c) shows the effect of reducing the repetition rate of the 10 MHz pulse train to 1 MHz (the rise/falltimes remain at 20 ns and the duty cycle remains at 50%). The level at 110 MHz is

$$\text{Level}_{110\,\text{MHz}} = 20 \log_{10} 1 - 20 \log_{10} \left(\frac{15.9\,\text{MHz}}{637\,\text{kHz}} \right) - 40 \log_{10} \left(\frac{110\,\text{MHz}}{15.9\,\text{MHz}} \right)$$

$$= 120\,\text{dB}\mu\text{V} - 32\,\text{dB} - 33.6\,\text{dB}$$

$$= 54.4\,\text{dB}\mu\text{V}$$

This is a reduction of 78–54 or 24 dB over the 10 MHz, 20 ns pulse train.

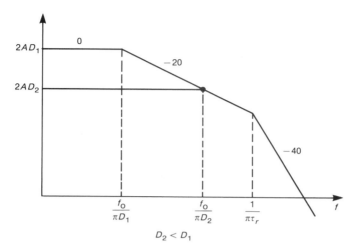

FIGURE 7.22   Illustration of the effect of duty cycle on the spectral bounds of trapezoidal pulse trains.

Figure 7.22 illustrates the effect of the pulse train duty cycle D on the spectral bounds. The first breakpoint in the spectral bound, $1/\pi\tau$, can be written in terms of the duty cycle $D = \tau/T$ and the fundamental frequency of the waveform, $f_o = 1/T$, as $1/\pi\tau = f_o/\pi D$. Recall that the starting level of the 0 dB/decade segment is $2AD$. Therefore, if we reduce the pulse width (reduce the duty cycle), we will lower the starting level and will also move the first breakpoint out in frequency as shown in Fig. 7.22, where $D_1 > D_2$. It is a simple matter to show that the first breakpoint for the smaller duty cycle D_2 will lie on the -20 dB/decade segment for the larger duty cycle D_1, as we have indicated in Fig. 7.22. There *reducing the duty cycle (the pulse width) reduces the low-frequency spectral content of the waveform, but does not affect the high-frequency content.* See the computed results in Fig. 7.18.

7.2.2.3 Effect of Ringing (Undershoot/Overshoot)   Parasitic inductance and capacitance of PCB lands and wires in a digital system can cause a phenomenon referred to as *ringing*. This is illustrated in Fig. 7.23(a) for a square wave. As the signal level transitions from one logic level to another, there is a tendency for the signal level to oscillate about the desired level. Losses tend to damp this *ringing*. This type of waveform can be described mathematically as a function of the form $Ke^{-\alpha t}\sin(\omega_r t + \theta)$, where α is a damping coefficient and $f_r = \omega_r/2\pi$ is the frequency of the ringing. Quite often a discrete resistor is placed in series with the output land of the driving gate to damp this and provide a smooth transition. An example of the utility of this is given in [1, Example 9.11, p. 550].

The purpose of this section is to investigate the effect of such ringing on the spectral content of the waveform. We will find that this ringing tends to

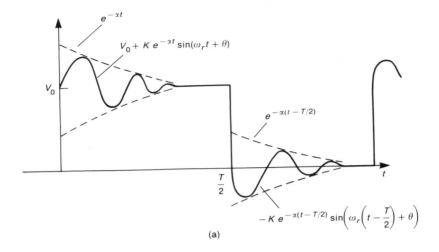

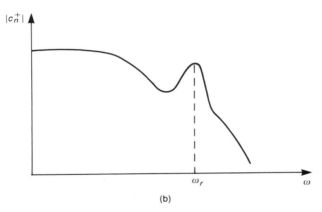

FIGURE 7.23   Illustration of ringing (undershoot/overshoot): (a) time-domain waveform; (b) spectral content.

accentuate or enhance certain regions of the spectrum of the original waveform. This reason combined with the detrimental effect on the functional performance of the system require that we try to prevent this ringing by the use of series resistors or ferrite beads or by matching the transmission line.

Determining the spectrum of the waveform of Fig. 7.23(a) is made simple by recognizing that it is the sum of three, periodic waveforms: a square wave, a damped sinusoid of the form $Ke^{-\alpha t}\sin(\omega_r t + \theta)$, and the same damped sinusoid negated and shifted ahead in t by $\frac{1}{2}T$. Using superposition, the expansion coefficients of the composite waveform are the sums of the expansion coefficients for the three components:

$$c_n = c_{n\ \text{square wave}} + \frac{1}{T}\int_0^{T/2} Ke^{-\alpha t}\sin(\omega_r t + \theta)\, e^{-jn\omega_o t}\, dt \tag{7.73}$$

$$- e^{-jn\omega_o T/2}\left[\frac{1}{T}\int_0^{T/2} Ke^{-\alpha t}\sin(\omega_r t + \theta)\, e^{-jn\omega_o t}\, dt\right]$$

$$= c_{n\ \text{square wave}} + (1 - e^{-jn\omega_o T/2})\frac{1}{T}\int_0^{T/2} Ke^{-\alpha t}\sin(\omega_r t + \theta)\, e^{-jn\omega_o t}\, dt$$

$$= \frac{V_0}{2}\frac{\sin(\tfrac{1}{4}n\omega_o T)}{\tfrac{1}{4}n\omega_o T}e^{-jn\omega_o T/4} + \frac{K}{2}\frac{\sin(\tfrac{1}{4}n\omega_o T)}{\tfrac{1}{4}n\omega_o T}e^{-jn\omega_o T/4}\frac{p\omega_r}{p^2 + 2\alpha p + \alpha^2 + \omega_r^2}$$

$$= \underbrace{\frac{V_0}{2}\frac{\sin(\tfrac{1}{4}n\omega_o T)}{\tfrac{1}{4}n\omega_o T}e^{-jn\omega_o T/4}}_{c_{n\ \text{square wave}}}\frac{p^2 + (2\alpha + (K/V_0)\omega_r)p + \alpha^2 + \omega_r^2}{p^2 + 2\alpha p + \alpha^2 + \omega_r^2}$$

where $p = jn\omega_o$ in the latter expression and we have assumed, to simplify the result, that $\theta = 0$. We have also assumed that $e^{-\alpha T/2} \ll 1$. This last result shows that the expansion coefficients are the sum of those for the square wave with no undershoot/overshoot and a similar waveform that is multiplyed by a *bandpass* transfer function centered about a radian frequency $\omega = \sqrt{\alpha^2 + \omega_r^2} \cong \omega_r$. Consequently, the spectrum of a square wave with undershoot/overshoot has a part of its spectrum *enhanced* or increased about the *ringing frequency* ω_r, as illustrated in Fig. 7.23(b). Consequently, undershoot/overshoot will have the effect of increasing the emissions about the ringing frequency. Quite often we see in the radiated emission profile a seemingly resonant region of enhanced emissions in a narrow frequency band. One possible explanation for this is the undershoot/overshoot present on the digital waveforms. Thus we should try to eliminate this—if not for functional performance reasons then certainly for EMC reasons.

7.2.3   Use of Spectral Bounds in Computing Bounds on the Output Spectrum of a Linear System

It is possible to generate spectral bounds for other types of commonly occuring signals. Nevertheless, even though a periodic signal will only have frequencies that occur at multiples of the basic repetition rate, the use of smooth piecewise linear segments plotted on a log–log format have considerable advantage in estimating the spectral content of the output of a linear system when that signal is the input to the system. This section illustrates this important point and the concepts associated with it.

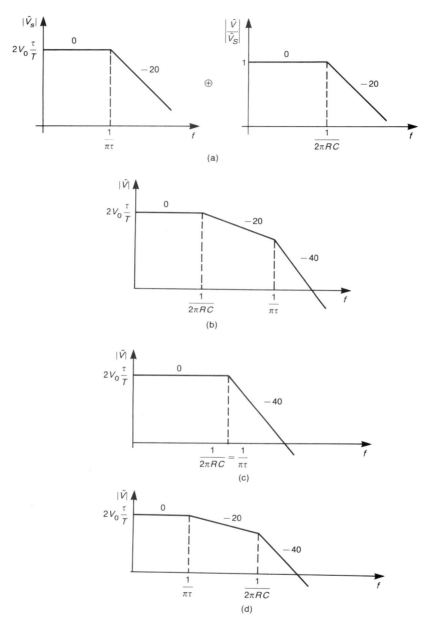

FIGURE 7.24 Illustration of the estimation of processing of a signal by linear systems using spectral bounds and Bode plots for a square-wave input and a low-pass RC circuit: (a) the input and the transfer function; (b) $RC > \frac{1}{2}\tau$; (c) $RC = \frac{1}{2}\tau$; (d) $RC < \frac{1}{2}\tau$.

A linear system having input $x(t)$, output $y(t)$, and impulse response $h(t)$ has an output spectrum

$$Y(jn\omega_o) = H(jn\omega_o)X(jn\omega_o) \tag{7.74}$$

Thus the magnitude and phase spectrum of the output are

$$|Y(jn\omega_o)| = |H(jn\omega_o)| \times |X(jn\omega_o)| \tag{7.75a}$$

$$\underline{/Y(jn\omega_o)} = \underline{/H(jn\omega_o)} + \underline{/X(jn\omega_o)} \tag{7.75b}$$

Equation (7.75a) shows that the magnitude spectrum of the output is the product of the magnitude spectra of the input and impulse response (transfer function). On a log–log or Bode plot the output spectrum is the *sum* of the input and transfer function spectra, since

$$20 \log_{10} |Y(jn\omega_o)| = 20 \log_{10} |H(jn\omega_o)| + 20 \log_{10} |X(jn\omega_o)| \tag{7.76}$$

Therefore we simply add the Bode plots of the input and transfer function magnitude spectra [1].

As an example, consider the circuit shown in Fig. 7.8(a). The bound on the magnitude spectrum of the output is the sum of the input and transfer function magnitude bounds, as shown in Fig. 7.24(a). Depending on the relative breakpoints, $1/\pi\tau$ and $1/2\pi RC$, we have the three possibilities for the output magnitude spectrum shown in Figs. 7.24(b)–(d). From this result it is clear that if we want to significantly reduce the high-frequency of the output, we would choose the time constant much greater than the pulse width: $RC \gg \tau$. This satisfies our intuition, since it will mean that the capacitor has not had sufficient time to charge up to its final value before the pulse switches off. Thus the output will appear as a "sawtooth" waveform rather than a square wave, and will consequently have less high-frequency spectral content.

7.3 SPECTRUM ANALYZERS

Spectrum analyzers are devices that display the *magnitude spectrum for periodic signals*. These devices are basically *radio receivers having a bandpass filter that is swept in time*. (A spectrum analyzer is essentially a superheterodyne receiver wherein the desired signal is mixed with a swept local oscillator and transferred to a lower, fixed intermediate frequency. However, viewing the device as being simply a bandpass filter swept in time gives a simple understanding of the device function.) A photograph of a typical spectrum analyzer is shown in Fig. 7.25(a). Figure 7.25(b) illustrates the point that a bandpass filter whose center frequency is swept in time from the start frequency to the end frequency (chosen by the operator) selects and displays the spectral components of the input signal that

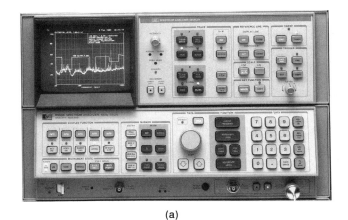

(a)

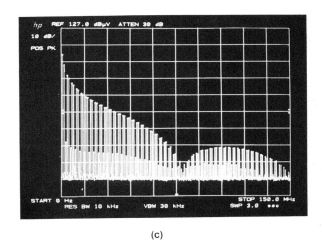

(b)

(c)

FIGURE 7.25 A spectrum analyzer: (a) photograph (courtesy of the Hewlett Packard Company); (b) illustration of the function as a swept bandpass filter; (c) measured spectrum of a 1 V, 1 MHz, 50% duty cycle, $\tau_r = \tau_f = 12.5$ ns, trapezoidal pulse train [1].

are present *within the bandwidth of the instrument at that point in the time of the sweep.* Figure 7.25(c) shows the measured spectrum of a periodic, trapezoidal 1 MHz, 1 V, 50% duty cycle waveform whose rise/falltimes are 12.5 ns (10 ns between the 10% and 90% points). Note the even harmonics of much lower amplitude than the odd harmonics that appear "in the background". Also observe the typical $(\sin x)/x$ behavior of the enveloped. The sweep is from dc to 150 MHz, and a null appears at approximately 80 MHz. This is reasonable to expect, since the magnitude of the one-sided spectrum is given in (7.64). The first $(\sin x)/x$ term is zero where $\pi\tau f = \pi$ or $f = 1/\tau = 2$ MHz. The second $(\sin x)/x$ term is zero where $\pi\tau_r f = \pi$ or $f = 1/\tau_r = 80$ MHz. The deep understanding we now have concerning spectra easily explains seemingly strange results such as these. From the exact result in (7.64) we compute a level at the 15th harmonic (15 MHz) of 92 dBμV, which compares well with the measured value of $87 + 3 = 90$ dBμV (add 3 dB since the spectrum analyzer measures rms values).

7.3.1 Basic Principles

Let us now examine more closely the effect of a swept bandpass filter on the "spectrum" that the spectrum analyzer will display. This is critically important in complying with the regulatory limits on conducted and radiated emissions, since the ultimate test of success is whether any *measured level* exceeds the regulatory limit at that frequency. We will refer to the term "spectrum analyzer" as SA in this section.

A key ingredient in determining the level that is displayed by the SA at that frequency is the *bandwidth* of the SA (chosen by the operator). This is illustrated in Fig. 7.26. The bandwidth is the 6 dB bandwidth, where the response is reduced by 6 dB from its maximum level at the center frequency. Let us "freeze" the sweep of the SA at some point in its cycle. Suppose at this time that three harmonics are within the bandwidth of the filter at this point in the sweep. The *displayed level at the center frequency of the bandwidth will be the sum of the spectral levels that fall within the bandwidth of the filter at that time.* Thus, even though there are no spectral components at the center frequency of the bandwidth, f, the SA will show, *at frequency f*, a level of A + B + C. As the filter moves further to the right in its sweep, level A will "drop out" and it will show a total of B + C. As it moves further to the right, level B will "drop out" and the displayed level will be C. The result is shown in Fig. 7.26(b). This shows the important point that *in order to obtain the lowest possible level on the SA display, we should choose as small a bandwidth as possible!* The regulatory agencies realize this so they set a minimum bandwidth to be used for the measurement. (It would not be sensible to use a SA bandwidth larger than this minimum bandwidth, since the measured levels would then be larger than necessary.) The minimum bandwidths for the FCC and CISPR 22 regulations are given in Tables 7.1 and 7.2. These are 6 dB bandwidths.

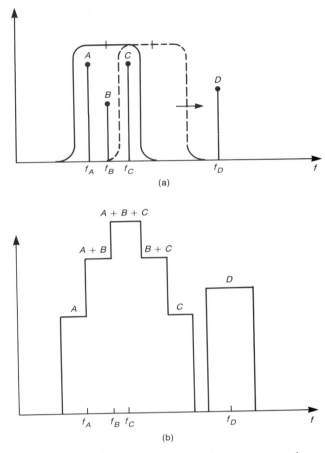

FIGURE 7.26 Illustration of the effect of bandwidth on measured spectrum: (a) a bandwidth that is wide enough to contain several narrowband signals; (b) the resulting display.

TABLE 7.1 FCC Minimum Spectrum Analyzer Bandwidths (6 dB).

Radiated emissions: 30 MHz–40 GHz	100 kHz
Conducted emissions: 450 kHz–30 MHz	9 kHz

TABLE 7.2 CISPR 22 Minimum Spectrum Analyzer Bandwidths (6 dB).

Radiated emissions: 30 MHz–1 GHz	100 kHz
Conducted emissions: 150 kHz–30 MHz	9 kHz

The fact that *the SA adds all spectral levels within the bandwidth of the instrument at that point in its sweep and displays the sum as the level at the center frequency of the filter* illustrates the following important point. *We should attempt to choose clock and data repetition rates such that none of the harmonics of any signal in the system will be closer than the measurement bandwidth of the SA!* As an example, suppose that there are two clock oscillators in the product and we choose both to have a frequency of 10 MHz. Each clock signal may radiate from different parts of the system so that the total received signal at the measurement antenna is the sum of these emissions. Suppose the received levels at the antenna are due to radiations from two different points in the product and are of equal strength. The displayed signal at 10 MHz, 20 MHz, 30 MHz, ... will be 6 dB larger than for one signal. To reduce this, suppose we have an asynchronous communication channel and do not require clocks of the same frequency. If we choose one clock frequency to be 10 MHz and the other to be 15 MHz, we will still have a problem, although it will not be as severe as for identical clock frequencies. The emissions will now occur at 10 MHz, 15 MHz, 20 MHz, 30 MHz, ... Each one of these comes from a different point in the system, and they are separated by more than the required minimum bandwidth of the SA (100 kHz). Upon closer examination, we realize that the ninth harmonic of the 10 MHz oscillator, 90 MHz, and the sixth harmonic of the 15 MHz oscillator, also 90 MHz, will add, causing a 6 dB maximum increase in level over that obtained if they did not add within the bandwidth. Chapter 13 will investigate this important EMC design technique of choice of crystal or oscillator frequencies in a digital system in order to prevent this addition of harmonics.

Even though the amplitudes of the two harmonics that fall within the bandwidth of the SA are not of the exact same level, a potentially significant increase in measured level will be experienced. Table 7.3 illustrates that even though the two levels are widely different, a significant increase may occur. This occurs because the SA adds the absolute levels and converts the sum to dB.

This table is prepared assuming the two signals to be in phase, and thus gives upper bounds on this summation. The results are obtained by first converting the signal amplitudes to absolute levels, adding them, and then converting the result to dB. This represents the way a SA actually adds signals in its bandwidth. Observe that even though two signals differ in level by 10 dB (a ratio of 3.16), they will add to give an increase of 2.39 dB over the level of the larger signal! So it is important to insure that no two harmonics are within the bandwidth of the SA of each other. This will make the job of complying with the regulatory limits easier.

A simple method of determining whether two or more signals are adding within the bandwidth of the SA is to narrow the bandwidth of the receiver. (For the purposes of this test we can lower the bandwidth of the SA below the minimum required by the regulatory agency for this test, since we are only trying to determine whether the viewed signal is due to this addition of two or more harmonics.) If when the bandwidth is reduced, for example from 100 to

TABLE 7.3   The Effect of the Addition of Two Unequal Level Signals.

Difference in Signal Levels (dB)	Increase in dB over the Larger of the Two
0	6.02
1	5.53
2	5.08
3	4.65
4	4.25
5	3.88
6	3.53
7	3.21
8	2.91
9	2.64
10	2.39
18.3	1.0

30 kHz, we see no change in any part of the displayed spectrum then we are assured that no two signal harmonics are adding within the larger bandwidth.

7.3.2   Peak versus Quasi-Peak

Thus far we have been assuming that the *detector* of the SA is set in the *peak mode*. That is, the maximum (actually rms) of the sinusoidal harmonic is displayed. A simple peak detector is shown in Fig. 7.27(a), where the input sinusoid represents a harmonic whose peak level is V_0. (Typical SAs use more sophisticated peak detectors than this.) The regulatory requirements, however, dictate that the level that is to be compared with the limit to determine compliance is to be measured with a *quasi-peak detector*. A simple quasi-peak detector is illustrated in Fig. 7.27(b). (Again, typical quasi-peak detectors are more sophisticated than this, but this circuit will illustrate the essential points and concepts.) Suppose that the received signal consists of "spikes" that are widely separated in time with regard to the time constant RC of the quasi-peak detector. The capacitor will begin to charge until the first spike turns off. It will then discharge through the R. If the next spike occurs after a length of time that allows the capacitor to completely discharge, we will have the first waveform shown at the output of the SA. However, if the spikes occur more closely than the RC time constant, the capacitor will not have fully discharged before the next spike occurs. Thus the output signal will continue to increase to some limit. Although this is a simplistic illustration of the function of a quasi-peak detector, it nevertheless illustrates the important point that *infrequently occurring signals will result in a measured quasi-peak level that is considerably smaller than a peak detector would give*. Thus infrequent events (in relation to the time

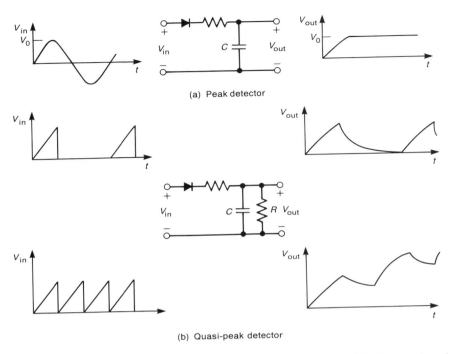

(a) Peak detector

(b) Quasi-peak detector

FIGURE 7.27 Two important detectors: (a) the peak detector; (b) the quasi-peak detector.

constant) may be of sufficient magnitude to give distressingly large received levels on a SA that is set to the peak detector function, yet their quasi-peak levels may not exceed the regulatory limit and are therefore of no consequence. However, if the quasi-peak levels exceed the regulatory limit, the peak levels will surely exceed that limit.

The reason for the use of a quasi-peak detector function relates to the intent of the regulatory limits, which is to prevent interference in radio and wire communication receivers. Infrequent spikes and other events do not substantially prevent the listener from obtaining the desired information. However, a continuous signal modulation results in a continuous detected signal in the radio, and would therefore substantially interfere with the listener's ability to obtain the desired transmitted information.

The CISPR 22 conducted emission limits are given in quasi-peak (QP) and in average (AV) levels. The average levels are obtained with an average detector. The average detector is basically a narrow filter (video filter), placed after the usual envelope detector, that passes only the dc component of the time-varying envelope of the detected waveform. This is useful in uncovering CW (single-frequency or *continuous wave*) signals that are buried in a broadband spectrum, as frequently occurs in switching power supply waveforms.

7.4   REPRESENTATION OF NONPERIODIC WAVEFORMS

A single pulse occurring only once in time is a nonperiodic signal. Although our major interest will be in periodic signals, which represent the major radiation problems in digital systems, it is of interest to determine the spectral content of nonperiodic signals. Even though a spectrum analyzer may record the spectrum of a nonperiodic signal at some point in its sweep, that will occur only once and thus may not be recorded by the quasi-peak detector.

7.4.1   The Fourier Transform

The simplest way to approach this problem of nonperiodic waveforms is to consider a periodic function whose waveshape over one period is the same as the desired nonperiodic signal. Letting the period go to infinity moves the waveforms in adjacent periods further out in time, leaving the desired nonperiodic signal. For example, consider the periodic square-wave pulse train shown in Fig. 7.4, whose spectral components are shown in Fig. 7.6. Suppose we retain the pulse width τ and amplitude A but increase the period T. The complex-exponential Fourier series was obtained earlier. The envelope of spectral components (magnitude) is

$$\text{Envelope} = \frac{A\tau}{T}\frac{\sin(\pi f\tau)}{\pi f\tau} \tag{7.77}$$

Increasing the period lowers the frequency of the fundamental $f_o = 1/T$ and the harmonics. So these spectral components move closer together. The basic shape of the envelope remains the same, since it depends on the pulse width (which is held constant). However, the height of the envelope decreases with increasing T, since it depends on $A\tau/T$. As the period is increased indefinitely, individual spectral components merge into a smooth continuous spectrum where the discrete nature of the spectral content disappears. This is the essential result: *the spectrum of a single pulse is a continuum of frequency components.* Also note that the amplitudes become vanishingly small.

This suggests how to handle the case of a single pulse mathematically. First obtain the complex exponential expansion coefficients, assuming the pulse repeats itself to give a periodic waveform with period T. Then let the period go to infinity, $T \to \infty$, in that result, leaving the single pulse. Carrying out this process gives the *Fourier transform* of the signal as [1]

$$\mathcal{F}\{x(t)\} = X(j\omega) \tag{7.78a}$$

$$= \int_{-\infty}^{\infty} x(t)e^{-j\omega t}\,dt$$

where

$$x(t) = \frac{1}{2\pi} \int_{-\infty}^{\infty} X(j\omega)e^{j\omega t}\, d\omega \qquad (7.78b)$$

As an application, let us determine the Fourier transform of the single pulse shown in Fig. 7.28(a). Directly applying (7.78a) gives

$$X(j\omega) = \int_{0}^{\tau} Ae^{-j\omega t}\, dt \qquad (7.79)$$

$$= -\frac{A}{j\omega}(e^{-j\omega\tau} - 1)$$

$$= -\frac{A}{j\omega}e^{-j\omega\tau/2}(e^{-j\omega\tau/2} - e^{j\omega\tau/2})$$

$$= A\tau \frac{\sin(\frac{1}{2}\omega\tau)}{\frac{1}{2}\omega\tau}e^{-j\omega\tau/2}$$

We have placed the result in the familiar $(\sin x)/x$ form to facilitate plotting. Thus

$$|X(j\omega)| = A\tau \left|\frac{\sin(\frac{1}{2}\omega\tau)}{\frac{1}{2}\omega\tau}\right| \qquad (7.80a)$$

$$\underline{/X(j\omega)} = \pm\tfrac{1}{2}\omega\tau \qquad (7.80b)$$

These are plotted in Figs. 7.28(b) and (c).

By analogy to the Fourier series, we may view the Fourier transform of a nonperiodic signal as resolving a time function $x(t)$ into a continuum of complex sinusoids. But there is one major difference between the two in this analogy: *the amplitudes of the individual sinusoids are infinitesimally small.* So it cannot be said that anything is present at a single frequency. But we can visualize individual amplitudes present over some nonzero frequency range $d\omega$ of amplitude $(1/2\pi)|X(j\omega)|\,d\omega$ according to equation (7.78b).

There are a number of important properties of the Fourier transform that can often be used to simplify its computation. The first of these relates the complex-exponential expansion coefficients and the Fourier transform. *If we know the Fourier transform of a single pulse $X(j\omega)$, we can directly obtain the coefficients of the complex-exponential Fourier series of a periodic train of such pulses by replacing ω in $X(j\omega)$ with $n\omega_o$ and dividing the result by the period T:*

$$c_n = \frac{1}{T}X(j\omega_o) \qquad (7.81)$$

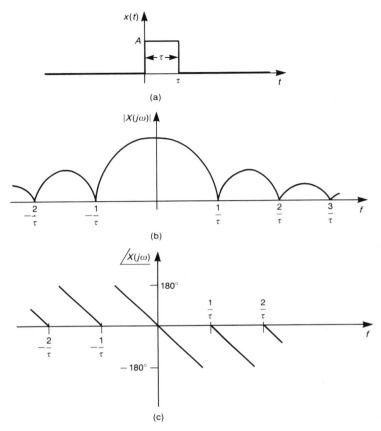

FIGURE 7.28   The Fourier transform of a rectangular pulse: (a) the pulse; (b) the magnitude spectrum; (c) the phase spectrum.

There exist a number of tables of Fourier transforms of various pulse shapes. The result in (7.81) allows us to use those tables to obtain the complex exponential Fourier series expansion coefficients. All of the other properties derived for periodic functions and the Fourier series—linearity, superposition, differentiation, time shifting, impulse functions—apply to the Fourier transform, where we replace the discrete frequency variable $n\omega_0$ with ω and multiply the expansion coefficients c_n by T according to (7.81). Thus the Fourier transform of piecewise linear pulses can be easily determined.

7.4.2   Response of Linear Systems to Nonperiodic Inputs

If we view the Fourier transform of a waveform as resolving that waveform into a continuum of sinusoidal components, it becomes clear that, using superposition, the response of a *linear system* to that waveform is

$$Y(j\omega) = H(j\omega)X(j\omega) \tag{7.82}$$

Therefore *the Fourier transform of the output of a linear system is the product of the Fourier transforms of the input to that system and the impulse response of that system.* It can be shown that this is the *complete response* (*transient plus steady state*) *of the system for zero initial conditions* [1].

7.5   DETERMINING THE TIME-DOMAIN RESPONSE OF LINEAR SYSTEMS FROM THE FREQUENCY-DOMAIN RESPONSE

The frequency-domain impulse response of a linear system is denoted by $H(j\omega)$, and the time-domain impulse response is denoted by $h(t)$. If we can determine $h(t)$ from $H(j\omega)$ then the time-domain output of a linear system can be obtained with the *convolution integral* as [1]

$$y(t) = \int_{\infty}^{\infty} h(t - \tau)x(\tau)\, d\tau \tag{7.83}$$

Determining the frequency-domain impulse response $H(j\omega)$ is simple: apply unit amplitude sinusoids of various frequencies. Determining the time-domain impulse response $h(t)$ is also fairly simple: apply a pulse whose duration is much smaller than the smallest time constant of the system and whose area is unity [1].

Another common way of computing the time-domain response of a *linear system* is the following. First decompose the input signal into its frequency components via the Fourier series for a periodic waveform or the Fourier transform for a nonperiodic waveform. Now, by the principle of superposition, pass each frequency component of the input through the system and combine them at the output to give the spectrum of the output, $Y(j\omega)$. Then apply the inverse Fourier transform to this to give the time-domain output $y(t)$. This is a very common way of determining the time-domain response of a system, but it is only valid for a *linear system*.

In some simple but important cases it is also easy to determine the differential equation relating the input to the output. We will frequently observe this type of (approximate) relationship. For example, consider the frequency-domain impulse responses (transfer functions) of a linear system shown in Fig. 7.29. If the magnitude of the frequency-domain impulse response varies at 20 dB/decade and the angle is $+90°$, as shown in Fig. 7.29(a), $H(j\omega) = j\omega = \omega\,\underline{/90°}$, then the differential equation relating the input and output is $y(t) = (d/dt)x(t)$ [1]. In general we have

$$H(j\omega) = (j\omega)^n \rightarrow y(t) = \frac{d^n}{dt^n}x(t) \tag{7.84}$$

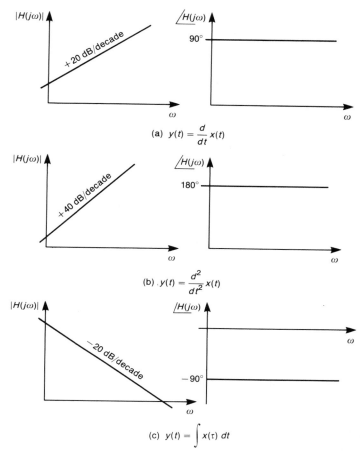

FIGURE 7.29 Illustration of determining the differential equation relating the input and output of a linear system from the frequency-domain transfer function: (a) $y(t) = (d/dt)x(t)$; (b) $y(t) = (d^2/dt^2)x(t)$; (c) $y(t) = \int x(\tau)\, d\tau$.

Similarly,

$$H(j\omega) = \frac{1}{(j\omega)^n} \rightarrow y(t) = \underbrace{\iint \cdots \int x(\tau)\, d\tau}_{n} \tag{7.85}$$

These results are quite simple to see. A slightly more complicated example is where the transfer function forms a low-pass function

$$H(j\omega) = \frac{K}{1 + j\omega\tau} \tag{7.86}$$

Interpreting $j\omega = D$ as a differential operator [1] gives the differential equation as

$$\tau \frac{d}{dt} y(t) + y(t) = Kx(t) \tag{7.87}$$

This can be easily extended, using the differential operator $D^n = d^n/dt^n$, to derive the differential equation relating $y(t)$ to $x(t)$ directly from the transfer function by replacing $j\omega$ with D [1].

7.6   REPRESENTATION OF RANDOM SIGNALS

Thus far we have only considered deterministic signals; that is, signals whose time behavior is known a priori. Random signals are those in which the time behavior is described probabilistically. Digital data waveforms are obviously random signals, otherwise no information would be conveyed. An example of a random signal is the PCM–NRZ waveform shown in Fig. 7.30(a). The term PCM–NRZ refers to pulse code modulation–non return to zero. A PCM–NRZ waveform is one in which two levels are used to represent the two binary states of 0 and 1. The NRZ (non return to zero) designation means that it is not required that the signal return to zero between each state transition. A waveform that transitions between 0 and X_0 can be described as

$$x(t) = \tfrac{1}{2} X_0 [1 + m(t)] \tag{7.88}$$

where $m(t)$ is a random variable that assumes values of ± 1 with equal probability in the bit interval $nT < t < (n + 1)T$. This signal may give a reasonable approximation to certain digital data signals.

The autocorrelation function $R_x(\tau)$ of a random signal $x(t)$ is defined as the expected value of the product of the signal and that signal shifted in time by τ [4]:

$$R_x(\tau) = \overline{x(t)x(t + \tau)} \tag{7.89}$$

$$= \lim_{t \to \infty} \frac{1}{T} \int_{-T/2}^{T/2} x(t)x(t + \tau)\, dt$$

where the overbar indicates the statistical average of all possibilities. Substituting (7.88) into (7.89) gives

$$R_x(\tau) = \tfrac{1}{4} X_0^2 \overline{[1 + m(t)][1 + m(t + \tau)]} \tag{7.90}$$

$$= \tfrac{1}{4} X_0^2 [1 + \overline{m(t)} + \overline{m(t + \tau)} + \overline{m(t)m(t + \tau)}]$$

$$= \tfrac{1}{4} X_0^2 [1 + \overline{m(t)m(t + \tau)}]$$

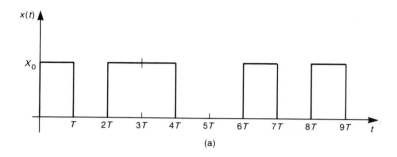

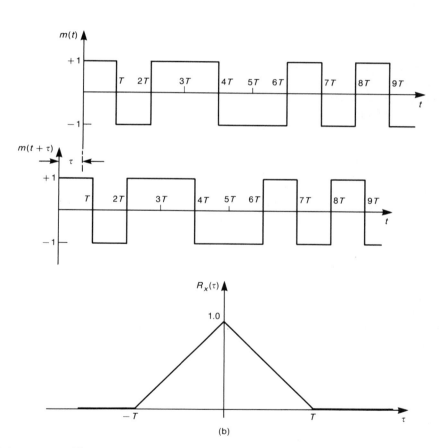

FIGURE 7.30 Illustration of the computation of the power spectral density of a PCM–NRZ signal: (a) a typical waveform; (b) the autocorrelation function.

In the second part of the result we have assumed that the process is stationary, and in the last part we have used the fact that the expected (average) value of $m(t)$ is zero [4]. Computation of the autocorrelation function illustrated in Fig. 7.30 yields

$$R_x(\tau) = 1 - \frac{|\tau|}{T} \quad \text{for } |\tau| < T \tag{7.91}$$

$$= 0 \qquad \text{for } |\tau| > T$$

which is plotted in Fig. 7.30(b).

The characterization of a random signal in the frequency domain is accomplished in terms of the *power spectral density* of the signal with the Wiener–Khinchine theorem, which provides that the power spectral density of the signal is the Fourier transform of the autocorrelation function of the signal [4]:

$$G_x(f) = \int_{-\infty}^{\infty} R_x(\tau)e^{-j\omega\tau}\,d\tau \tag{7.92}$$

The average power associated with the random signal is

$$P_{AV} = \int_{-\infty}^{\infty} G_x(f)\,df \tag{7.93}$$

This is the expected or average power dissipated in a $1\,\Omega$ resistor by this signal if a statistically large number of samples are taken (signals are applied to the

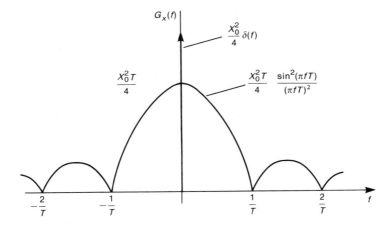

FIGURE 7.31   The power spectral density of the PCM–NRZ signal.

resistor). For the PCM–NRZ waveform the power spectral density is

$$G_x(f) = \frac{X_0^2}{4}\delta(f) + \frac{X_0^2 T}{4}\frac{\sin^2(\pi f T)}{(\pi f T)^2} \tag{7.94}$$

which is plotted in Fig. 7.31. Many other random signals can be characterized in this fashion. The reader is referred to [4] for these other characterizations.

REFERENCES

[1] C.R. Paul, *Analysis of Linear Circuits*, McGraw-Hill, NY (1989).

[2] C.D. McGillem and G.R. Cooper, *Continuous and Discrete Signal and System Analysis*, Holt, Rinehart and Winston, NY (1974).

[3] S.J. Mason and H.J. Zimmerman, *Electronic Circuits, Signals, and Systems*, John Wiley, NY (1960).

[4] A.B. Carlson, *Communication Systems*, McGraw-Hill, NY (1968).

PROBLEMS

7.1 The first eight expansion coefficients of an orthonormal set of basis functions over the unit time interval are shown in Fig. P7.1(a). These are organized according to zero-crossings and are referred to as the Walsh functions. They are easily generated with digital hardware, and the computation of the expansion coefficients is quite simple. Determine the expansion coefficients needed to represent the function shown in Fig. P7.1(b). [$c_0 = 3$, $c_1 = -\frac{3}{2}$, $c_2 = 0$, $c_3 = -\frac{3}{4}$, $c_4 = 0$, $c_5 = 0$, $c_6 = 0$, $c_7 = -\frac{3}{8}$] Determine the ISE for this expansion using the first eight expansion coefficients. [$\frac{1}{256}$]

7.2 The periodic signal shown in Fig. P7.2 is a *half-wave rectified sinusoid* typical of those obtained from linear power supplies (prior to filtering). This is formed by replacing the value of $A \sin(2\pi t/T)$ with zero over alternating half-cycles. Determine the trigonometric Fourier series expansion coefficients for this waveform. [$a_0 = A/\pi$, $a_n = 0$ for odd n, $a_n = 2A/\pi(1 - n^2)$ for even n, $b_1 = \frac{1}{2}A$, $b_n = 0$ for $n \neq 1$] Repeat this for the *full-wave rectified sinusoid* $x(t) = |A \sin(2\pi t/T)|$. [$a_0 = 2A/\pi$, $a_n = 0$ for odd n, $a_n = 4A/\pi(1 - n^2)$ for even n, $b_n = 0$ for all n]

7.3 Determine the trigonometric Fourier series expansion coefficients for the waveform shown in Fig. P7.3. [$a_0 = \frac{1}{2}A$, $a_n = -4A/n^2\pi^2$ for odd n, $a_n = 0$ for even n, $b_n = 0$]

7.4 Determine the trigonometric Fourier series expansion coefficients for the waveform shown in Fig. P7.4. [$a_0 = 0$, $a_n = 0$, $b_n = 0$ for even n, $b_n = 8A/n^2\pi^2$ for $n = 1, 5, 9, \ldots$, $b_n = -8A/n^2\pi^2$ for $n = 3, 7, 11, \ldots$]

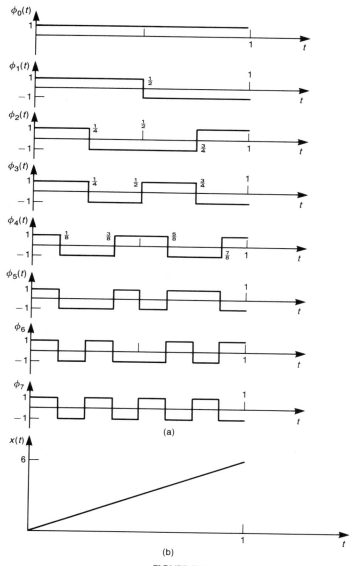

(a)

(b)

FIGURE P7.1

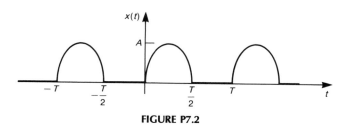

FIGURE P7.2

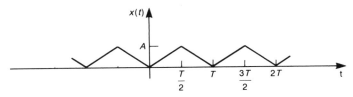

FIGURE P7.3

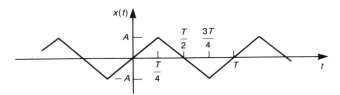

FIGURE P7.4

7.5   Determine the trigonometric Fourier series expansion coefficients for the waveform shown in Fig. P7.5. [$a_0 = 0$, $a_n = 0$, $b_n = 0$ for $n = 4, 8, 12, \ldots$, $b_n = 6/n\pi$ for $n = 1, 3, 5, 7, \ldots$, $b_n = -4/n\pi$ for $n = 2, 6, 10, \ldots$]

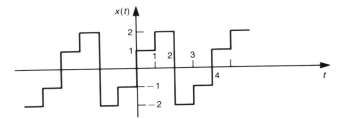

FIGURE P7.5

7.6   Determine the trigonometric Fourier series expansion coefficients for the waveform shown in Fig. P7.6. [$a_0 = 0$, $a_n = -2/n\pi$ for $n = 1, 5, 9, \ldots$, $a_n = 2/n\pi$ for $n = 3, 7, 11, \ldots$, $a_n = 0$ for even n, $b_n = 0$ for even n, $b_n = 6/n\pi$ for odd n]

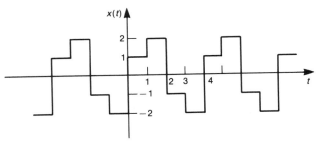

FIGURE P7.6

7.7   Determine the trigonometric Fourier series expansion coefficients for the waveform shown in Fig. P7.7. $[a_0 = \frac{5}{2}, a_n = 0, b_n = -5/n\pi]$

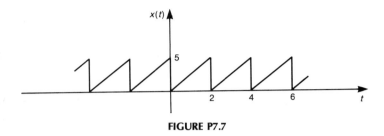

FIGURE P7.7

7.8   Determine the trigonometric Fourier series expansion coefficients for the waveform shown in Fig. P7.8. $[a_0 = \frac{1}{2}A, a_n = 0, b_n = A/n\pi]$

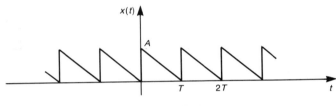

FIGURE P7.8

7.9   Determine the trigonometric Fourier series expansion coefficients a_n and b_n in terms of the complex exponential expansion coefficients c_n. $[a_0 = c_0, a_n = c_n + c_n^*, b_n = j(c_n - c_n^*)]$

7.10 Determine the complex exponential expansion coefficients c_n in terms of the trigonometric Fourier series expansion coefficients a_n and b_n. $[c_0 = a_0, c_n = \frac{1}{2}a_n - j\frac{1}{2}b_n]$

7.11 Determine the complex exponential Fourier series expansion coefficients for the waveform in Fig. P7.2. $[c_0 = A/\pi, c_n = 0$ for odd $n, n \neq 1, c_1 = -\frac{1}{4}jA, c_n = A/\pi(1 - n^2)$ for even $n]$

7.12 Determine the complex exponential Fourier series expansion coefficients for the waveform in Fig. P7.3. $[c_0 = \frac{1}{2}A, c_n = 0$ for even $n, c_n = -2A/n^2\pi^2$ for odd $n]$

7.13 Determine the complex exponential Fourier series expansion coefficients for the waveform in Fig. P7.4. $[c_0 = 0, c_n = 0$ for even $n, c_n = -j4A/n^2\pi^2$ for $n = 1, 5, 9, \ldots, c_n = j4A/n^2\pi^2$ for $n = 3, 7, \ldots]$

7.14 Determine the complex exponential Fourier series expansion coefficients for the waveform in Fig. P7.5. $[c_0 = 0, c_n = -j3/n\pi$ for $n = 1, 3, 5, \ldots, c_n = j2/n\pi$ for $n = 2, 6, 10, \ldots]$

7.15 Determine the complex exponential Fourier series expansion coefficients for the waveform in Fig. P7.6. $[c_0 = 0, c_n = (-1 - j3)/n\pi$ for $n = 1, 5, 9, \ldots, c_n = (1 - j3)/n\pi$ for $n = 3, 7, 1, \ldots, c_n = 0$ for even $n]$

7.16 Determine the complex exponential Fourier series expansion coefficients for the waveform in Fig. P7.7. $[c_0 = 2.5, c_n = j5/2n\pi]$

7.17 Determine the complex exponential Fourier series expansion coefficients for the waveform in Fig. P7.8. $[c_0 = A/T, c_n = -jA/2\pi n]$

7.18 Determine the complex exponential Fourier series expansion coefficients for the waveform in Fig. P7.18.

$$\left[c_0 = \frac{A\tau}{2T}, \quad c_n = -j\frac{A}{2\pi n}\left(1 - \frac{\sin(n\pi\tau/T)}{n\pi\tau/T}e^{-jn\pi\tau/T} \right) \right]$$

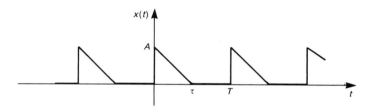

FIGURE P7.18

7.19 Determine the complex exponential Fourier series expansion coefficients for the waveform in Fig. P7.19.

$$\left[c_0 = \frac{A(\tau - \frac{1}{2}\sigma)}{T}, \quad c_n = j\frac{A}{2\pi n}\left(e^{-j2n\pi\tau/T} - \frac{\sin(n\pi\sigma/T)}{n\pi\sigma/T}e^{-jn\pi\sigma/T} \right) \right]$$

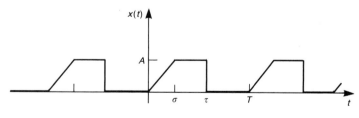

FIGURE P7.19

7.20 Determine the complex exponential Fourier series expansion coefficients for the waveform in Fig. P7.20.

$$\left[c_0 = \frac{A}{\alpha T}(1 - e^{-\alpha T}), \quad c_n = \frac{A}{\alpha T + j2\pi n}(1 - e^{-\alpha T}) \right]$$

This waveform is often observed in dc motor brush arcing.

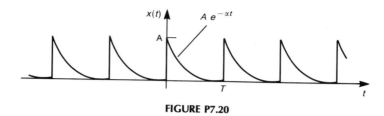

FIGURE P7.20

7.21 A 10 MHz clock oscillator transitioning from 0 to 5 V with $\tau_r = \tau_f = 20$ ns and a 50% duty cycle is connected to a gate as shown in Fig. P7.21. A filter is connected as shown. Determine the level of the 11th harmonic at the gate terminals. Obtain these results by using the exact spectrum and by using the spectral bounds. [73.93 dBμV exact, 78.55 dBμV by interpolation]

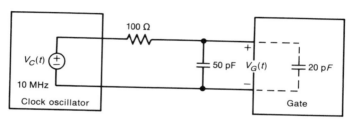

FIGURE P7.21

7.22 Determine the magnitude of the output of the system shown in Fig. P7.22 at $\omega = 50 \times 10^6$ rad/s. [102.04 dBμV]

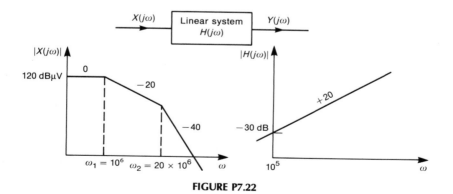

FIGURE P7.22

7.23 A periodic waveform with $\omega_0 = 10^5$ rad/s is applied to a circuit as shown in Fig. P7.23. Determine the level of the 50th harmonic. [83.01 dBμV]

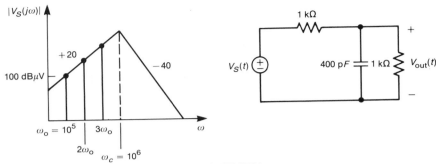

FIGURE P7.23

7.24 A 5 V, 10 MHz oscillator having a rise/falltime of 10 ns and a 50% duty cycle is applied to a gate as shown in Fig. P7.24. A capacitor is applied across the input to the gate. Determine the value of the capacitor such that the fifth-harmonic level is reduced by 20 dB. [63.34 pF]

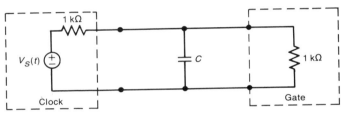

FIGURE P7.24

7.25 A 10 MHz clock oscillator having a 10 Ω internal resistance has an open-circuit waveform as a trapezoidal pulse train with a 50% duty cycle and rise/falltimes of 2 ns. Determine the value of a capacitor (an ideal one) that when placed across the oscillator output will reduce the output voltage of the fifth harmonic by 10 dB. [955 pF] If the open-circuit voltage transitions from 0 to 5 V, estimate the level of this fifth harmonic. [106 dBμV]

7.26 The square-wave current source in Fig. P7.26 is applied to the associated circuit. Determine the current $i(t)$ in the form $i(t) = \sum_{n=1}^{\infty} I_n \sin(n\omega_o t + \theta_n)$. [$I_0 = 2.5$, $I_1 = 0.5$, $\theta_1 = -80.96°$, $I_2 = 0$, $I_3 = 0.0562$, $\theta_3 = -86.96°$, $I_4 = 0$, $I_5 = 0.0203$, $\theta_5 = -88.18°$, $I_6 = 0$, $I_7 = 0.0103$, $\theta_7 = -88.7°$]

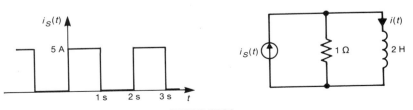

FIGURE P7.26

7.27   A periodic waveform is applied to the circuit of Fig. P7.27. The waveform has the Fourier series representation

$$v_S(t) = 2 + \frac{40}{\pi^2} \sum_{n=1}^{\infty} \frac{1}{(2n-1)^2} \cos[\pi(2n-1)t]$$

Determine the Fourier series of $v(t)$. [$V_0 = 2$, $V_1 = 0.6602$, $\theta_1 = -129.86°$, $V_2 = 0.0099$, $\theta_2 = -161.96°$, $V_3 = 0.0013$, $\theta_3 = -169.1°$]

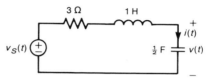

FIGURE P7.27

7.28   A 5 MHz 5 V, 30% duty cycle trapezoidal waveform having $\tau_r = \tau_f = 15$ ns has the spectral bounds given in Fig. P7.28. Determine A, f_1, and f_2. [$A = 129.54$ dBμV, $f_1 = 5.305$ MHz, $f_2 = 21.221$ MHz] Determine the levels of the fundamental, third, fifth, seventh, and ninth harmonics. Give exact and bound values. [128.14 dBμV (129.54), 109.48 dBμV (120.51), 113.97 dBμV (114.65), 98.58 dBμV (108.8), 101.22 dBμV (104.44)]

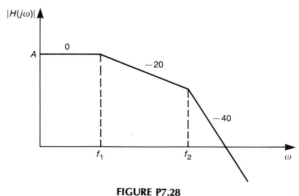

FIGURE P7.28

7.29   Show that when the duty cycle is varied between 0 and 100%, the breakpoint of the 0 dB/decade–20 dB/decade segment, $1/\pi\tau$, is always prior to the breakpoint for the -20 dB/decade–40 dB/decade segment, $1/\pi\tau_r$, as shown in Fig. 7.22.

7.30   For the ringing waveform shown in Fig. 7.23(a) sketch the spectral bounds for $f_o = 1$ MHz, $f_r = 30$ MHz, $V_0 = 5$ V, $K = 1$ V, and $\alpha = 9.5 \times 10^6$.

7.31   A frequently used waveform to represent the time domain of an electromagnetic pulse (EMP) from a nuclear detonation is the double-exponential pulse given by

$$x(t) = \frac{1}{\beta - \alpha}(e^{-\alpha t} - e^{-\beta t}) \quad \text{for } t \geq 0$$

and $x(t) = 0$ for $t < 0$. Determine the Fourier transform of this pulse. $[1/\{\alpha\beta + j(\alpha + \beta)\omega - \omega^2\}]$

7.32 Determine the Fourier transform of a pulse given by the waveform over one period of the waveform of Problem 7.19.

$$\left[X(j\omega) = j\frac{A}{\omega T}\left(e^{-j\omega\tau} - \frac{\sin(\frac{1}{2}\omega\sigma)}{\frac{1}{2}\omega\sigma}e^{-j\omega\sigma/2}\right) \right]$$

7.33 A linear system has the frequency-domain impulse response $H(j\omega) = j\omega M$. Sketch the time-domain output of this system, $y(t)$, if a trapezoidal waveform $x(t)$ is applied. [Differentiate the waveform and multiply the result by M]

7.34 Determine the differential equation relating $v(t)$ to $v_S(t)$ for the circuit of Fig. P7.27.

$$\left[\frac{d^2}{dt^2}v(t) + 3\frac{d}{dt}v(t) + 2v(t) = 2v_S(t) \right]$$

7.35 Determine the differential equation relating $i(t)$ to $v_S(t)$ for the circuit of Fig. P7.27.

$$\left[\frac{d^2}{dt^2}i(t) + 3\frac{d}{dt}i(t) + 2i(t) = \frac{d}{dt}v_S(t) \right]$$

7.36 Determine the differential equation relating $i(t)$ to $i_S(t)$ for the circuit of Fig. P7.26.

$$\left[\frac{d}{dt}i(t) + \frac{1}{2}i(t) = \frac{1}{2}i_S(t) \right]$$

7.37 The "random telegraph" wave is the function that may assume the values of zero or one with equal probability at any instant of time, and it makes random traversals from one value to another. The probability that n traversals occur in a time interval of length T is given by the Poisson distribution $P(n, T) = (aT)^n e^{-aT}/n!$, where a is the average number of traversals per unit of time. Determine and sketch the autocorrelation function. $[R(\tau) = \frac{1}{4}(1 + e^{-2a|\tau|})]$. Determine and sketch the power spectral density of this waveform. $[G(f) = \frac{1}{4}\{\delta(f) + a/(a^2 + \pi^2 f^2)\}]$

7.38 Suppose a random signal $x(t)$ amplitude-modulates a carrier wave as $y(t) = x(t)\cos(\omega_c t + \phi)$. If $x(t)$ is bandlimited, i.e., the spectral extent is less than the carrier frequency f_c, show that the autocorrelation function is $R_y(\tau) = R_x(\tau)\cos\omega_c\tau$, where $R_x(\tau)$ is the autocorrelation function of $x(t)$. Also show that the power spectral density is $G_y(f) = \frac{1}{4}G_x(f - f_c) + \frac{1}{4}G_x(f + f_c)$, where $G_x(f)$ is the power spectral density of $x(t)$.

CHAPTER EIGHT

Radiated Emissions and Susceptibility

In this chapter we will discuss the important mechanisms by which electromagnetic fields are generated in an electronic device and are propagated to a measurement antenna that is used to verify compliance to the governmental regulatory limits. Recall that for domestic emissions the frequency range of measurement is from 30 MHz to perhaps 40 GHz if the product is to be marketed in the US (FCC) and from 30 MHz to 1 GHz if the product is to be marketed outside the US (CISPR 22). The FCC measurement distance is 3 m for Class B products and 10 m for Class A products. For CISPR 22 the measurement distance is 10 m for Class B products and 30 m for Class A products. The lower frequency of 30 MHz is one wavelength at 10 m, whereas the frequency of 1 GHz is one wavelength at 30 cm. The product is therefore in the *near field* of the antenna for certain of the lower-frequency ranges of the regulatory limits and in the *far field* for the higher-frequency ranges. We found in Chapter 5 that the field structure of the emissions in the near field of an emitter is considerably more complicated that in the far field. Certain simplifications valid for the far field are not valid for the near field, although they are frequently used. An example is the inverse-distance rule that is used to translate an emission measured at one distance to another distance. This assumes that the fields increase (decrease) *linearly* with decreasing (increasing) measurement distances, a far-field assumption. We will generate some simple models for first-order predictions of the radiated emissions from wires and PCB lands in this chapter. For simplicity *these models will assume that the measurement antenna is in the far field of the emission (the product)*, although this is not necessarily the case over the entire frequency range of the regulatory limit.

We will also investigate the ability of the product to be susceptible to radiated emissions from other electronic devices by deriving simple models that give the voltages and currents induced in parallel-conductor lines by an incident uniform plane wave. The incident wave is produced by a distant antenna such as a FM

401

radio station. Although only the military places regulatory requirements on the susceptibility of a product at the time of this writing, this aspect of radiated emissions also fits into the company's goal of producing quality products. For example, if a product complies with the relevant regulatory requirements but fails to perform properly in dry climates due to susceptibility to electrostatic discharge (ESD) or will not perform properly when installed in a residence near an airport radar, compliance with the regulatory limits is a moot point.

8.1 SIMPLE EMISSION MODELS FOR WIRES AND PCB LANDS

Our primary interest is to understand the radiation properties of the *unintentional antennas in the system*, which are the wires, PCB lands, and other metallic structures such as cabinets and enclosures. In this section we will formulate some simple models that allow us to understand the factors that cause the radiated emissions from the currents on wires and PCB lands to exceed the regulatory limits. These will be derived for *ideal situations* such as an isolated pair of wires in free space distant from any other obstacles. The sole purpose of these models is to provide insight into the levels and types of currents with regard to their potential for creating radiated emissions.

8.1.1 Differential-Mode versus Common-Mode Currents

Differential-mode and common-mode currents were briefly discussed in Chapter 6 in connection with the use of common-mode chokes for suppression purposes. It is important that we review this concept. Consider the pair of parallel wires or PCB lands of length $\mathcal{L}$ and separation s shown in Fig. 8.1(a). Suppose the currents at the same cross section are directed to the right and denoted as $\hat{I}_1$ and $\hat{I}_2$. We will concentrate on frequency-domain emissions, so that the currents will be the phasor currents. These can be decomposed into differential-mode and common-mode components by writing

$$\hat{I}_1 = \hat{I}_C + \hat{I}_D \tag{8.1a}$$

$$\hat{I}_2 = \hat{I}_C - \hat{I}_D \tag{8.1b}$$

Given the currents $\hat{I}_1$ and $\hat{I}_2$, we can decompose them into their differential-mode component $\hat{I}_D$ and their common-mode component $\hat{I}_C$ by solving (8.1) to give

$$\hat{I}_D = \frac{\hat{I}_1 - \hat{I}_2}{2} \tag{8.2a}$$

$$\hat{I}_C = \frac{\hat{I}_1 + \hat{I}_2}{2} \tag{8.2b}$$

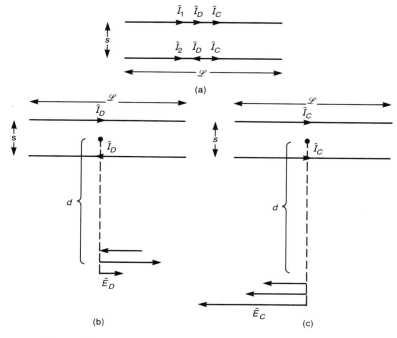

FIGURE 8.1 Illustration of the relative effects of differential-mode currents $\hat{I}_D$ and common-mode currents $\hat{I}_C$ on radiated emissions for parallel conductors: (a) decomposition of the total currents into differential-mode and common-mode components; (b) radiated emissions of differential-mode currents; (c) radiated emissions of common-mode currents.

At a cross section of the line, the differential-mode currents $\hat{I}_D$ are equal in magnitude but opposite in direction. These are the *functional* or desired currents on the line. Ideal models such as the transmission line model in Chapter 4 will predict *only* these differential-mode currents. The common-mode currents $\hat{I}_C$ are undesired currents. They are not necessary for the functional performance of the electronic devices that the line connects. Ideal models such as the transmission line model *will not predict* the common-mode currents [1]. Typically, the common-mode currents will be substantially smaller than the differential-mode currents. However, the reader should not allow this observation to lull him/her into thinking that the radiated emissions of common-mode currents are inconsequential. *Common-mode currents are not inconsequential in typical products, and, moreover, they often produce larger radiated emissions than do the differential-mode currents.*

In order to see why this occurs, let us consider the radiated electric fields in the plane of the wires and at a point midway along the line and a distance d from the line. The configuration for differential-mode currents is illustrated in Fig. 8.1(b). Observe that because the differential-mode currents are equal in magnitude but oppositely directed, the radiated electric fields will also be oppositely directed, and will tend to cancel. They will not exactly cancel, since the wires are not collocated, so the net electric field $\hat{E}_D$ will be the difference between these emission components, as indicated in Fig. 8.1(b). On the other hand, consider the emissions due to the common-mode currents shown in Fig. 8.1(c). Because the common-mode currents are directed in the same direction, their radiated electric field components will add, producing a net radiated electric field $\hat{E}_C$. In the following sections we will show that for a 1 m ribbon cable with wire separation of 50 mils a differential-mode current at 30 MHz of 20 mA will produce a radiated emission just equal to the FCC Class B limit (40 dBμV/m or 100 μV/m from 30 MHz to 88 MHz). On the other hand, a common-mode current of only 8 μA will produce the same emission level! This is a ratio of 2500, or some 68 dB. *Thus seemingly inconsequential common-mode currents are capable of producing significant radiated emission levels.* A number of ideal factors such as proximity to conducting planes and other structural asymmetries can create common-mode currents. The reader is referred to [2] for a more complete discussion of the sources of these common-mode currents.

In this section we will derive simple emission models for a pair of parallel wires or PCB lands due to the currents on those conductors. Although the case of two parallel wires or PCB lands represents only a small subset of the current-carrying conductors of an electronic system, it represents an important and easily analyzed structure. It will therefore provide insight into the radiation mechanism of other structures. Our basic technique to determine the radiated fields of a pair of parallel conductors is to superimpose the radiated fields of each conductor.

This technique of superimposing the fields of each conductor to determine the net radiated field is virtually identical to the determination of the radiated fields of an array of wire antennas considered in Section 5.3. In fact, the two wires of the line shown in Fig. 8.1 can be viewed as an array of current elements (see Fig. 5.7), and we will essentially be computing the array factor. In order to demonstrate this, consider the two wires in the form of an array as shown in Fig. 8.2. The two wires are located on the y axis of a spherical coordinate system and are parallel to the z axis. Wire #1 carrying current $\hat{I}_1(z)$ is located at $y = -\frac{1}{2}s$, and wire #2 carrying current $\hat{I}_2(z)$ is located at $y = \frac{1}{2}s$. Each of these currents is shown as being a function of distance z along the wire, in order to consider various current distributions such as sinusoidal ones. Note that this general problem can be used to characterize the cases of Fig. 8.1. For the case of differential-mode currents we would choose $\hat{I}_1(z) = \hat{I}_D(z)$ and $\hat{I}_2(z) = -\hat{I}_D(z)$. For the case of common-mode currents we would choose $\hat{I}_1(z) = \hat{I}_C(z)$ and $\hat{I}_2(z) = \hat{I}_C(z)$. If we know the *far-field* radiation of each antenna, $\hat{E}_{\theta,\text{ant}}$, then, *assuming that the measurement point is in the far field of the array,* we may determine the *net far-field radiation* by superimposing the fields of each antenna as

$$\hat{E}_\theta = \hat{E}_{\theta,1} + \hat{E}_{\theta,2} \tag{8.3}$$

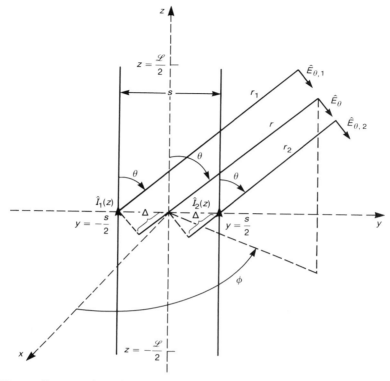

FIGURE 8.2 Computation of the far-field radiated emissions of a two-conductor line as the superposition of the fields due to each conductor.

where the far fields of each antenna are of the form

$$\hat{E}_{\theta,i} = \hat{M}\hat{I}_i \frac{e^{-j\beta_o r_i}}{r_i} F(\theta) \tag{8.4}$$

The term $\hat{I}_i$ is the current at the center of the antenna, and the factor $F(\theta)$ has a maximum value of unity and represents the θ variation of the antenna pattern. Observe that, by symmetry, the pattern of each antenna is independent of ϕ. However, the net pattern of the pair of antennas can be a function of ϕ, as we will see and as discussed in Section 5.3. The term $\hat{M}$ is a function of the antenna type. For example, for a Hertzian dipole (see equation (5.3a)) these are

$$\left.\begin{array}{l} \hat{M} = j\dfrac{\eta_o \beta_o}{4\pi}\mathscr{L} = j2\pi \times 10^{-6} f\mathscr{L} \\[4mm] F(\theta) = \sin\theta \end{array}\right\} \quad \text{(Hertzian dipoles)} \qquad \begin{array}{l} (8.5a) \\[4mm] (8.5b) \end{array}$$

This result is only valid for current segments that are very short electrically, which allows us to assume that the current at all points along the antenna is the same (magnitude and phase). Similarly, for half-wave dipoles that have a sinusoidal current distribution $\hat{M}(\theta)$ and $F(\theta)$ are (see equations (5.18) and (5.21))

$$\left. \begin{aligned} \hat{M} &= j\frac{\eta_o}{2\pi} = j60 \\[2mm] F(\theta) &= \frac{\cos(\frac{1}{2}\pi \cos\theta)}{\sin\theta} \end{aligned} \right\} \quad \text{(half-wave dipoles, } \mathscr{L} = \tfrac{1}{2}\lambda_o\text{)}$$

$$\text{(8.6a)}$$
$$\text{(8.6b)}$$

Assuming that the measurement antenna is physically distant from the array, we may draw the distance vector from the origin of the coordinate system to the measurement point parallel to the distance vectors from the midpoints of the antennas, as shown in Fig. 8.2. Thus we may write the distances r_1 and r_2 from the midpoint of each antenna in terms of the distance r from the midpoint of the array. The distance Δ shown in Fig. 8.2 was determined in Section 5.3 as

$$\Delta = \tfrac{1}{2}s \sin\theta \sin\phi \tag{8.7}$$

Thus

$$r_1 = r + \Delta \tag{8.8a}$$

$$r_2 = r - \Delta \tag{8.8b}$$

Substituting (8.7) and (8.8) into (8.4) and superimposing the results as in (8.3) gives

$$\hat{E}_\theta = \hat{M}\left(\hat{I}_1 \frac{e^{-j\beta_o r_1}}{r_1} + \hat{I}_2 \frac{e^{-j\beta_o r_2}}{r_2} \right) F(\theta) \tag{8.9}$$

$$= \hat{M}\left(\hat{I}_1 \frac{e^{-j\beta_o(r+\Delta)}}{r+\Delta} + \hat{I}_2 \frac{e^{-j\beta_o(r-\Delta)}}{r-\Delta} \right) F(\theta)$$

$$= \hat{M}\frac{e^{-j\beta_o r}}{r}(\hat{I}_1 e^{-j\beta_o\Delta} + \hat{I}_2 e^{j\beta_o\Delta}) F(\theta)$$

where we assume the two antennas are identical (two Hertzian dipoles or two half-wave dipoles, etc.) and have approximated $r \simeq r_1 \simeq r_2$ in the denominator terms. The phase terms depend on electrical distance, $\beta_o r_i = 2\pi r_i/\lambda_o$, so we cannot make a similar approximation in them and must substitute (8.8).

We will specialize this result for the case of differential-mode currents $\hat{I}_1 = \hat{I}_D$ and $\hat{I}_2 = -\hat{I}_D$ and common-mode currents $\hat{I}_1 = \hat{I}_C$ and $\hat{I}_2 = \hat{I}_C$ in the following two sections. It is important to remind the reader once again that the above derivation has the important simplifying *assumption that the measurement point* (*measurement antenna*) *is in the far field of the conductors.* (This is evident in the assumed forms in (8.4), which depend on distance r as $e^{-j\beta_o r}/r$.) If this assumption is violated, the following simple models are not valid. The radiated fields of a model that applies when the measurement antenna is in the near field of the wires are quite complicated, and the insight we will gain from the simple, far-field models is not as readily obtained from those models [3].

8.1.2 Differential-Mode Current Emission Model

In order to simplify the resulting model, we will model each of the wires as a Hertzian dipole. This model makes two important, simplifying assumptions that were used in Chapter 5 to derive the radiated fields of this antenna: (1) the conductor lengths $\mathscr{L}$ are sufficiently short (physically) and the measurement point is sufficiently distant (physically) that the distance vectors from each point on the antenna to the measurement point are approximately parallel; and (2) the current distribution (magnitude and phase) is constant along the line. For a measurement distance of 3 m the first assumption requires that the maximum conductor length be somewhat less than a meter. The assumption of constant distribution of the currents along the conductors is a reasonable approximation so long as the conductors are *electrically short* at the frequency of interest. This simples the results considerably and applies to a large number of problems of practical interest. For example, a two-wire cable of length 1 m is one wavelength at 300 MHz. At 100 MHz the cable is $\frac{1}{3}\lambda_o$ in length, and the current distribution is approximately constant. A PCB land 30 cm in length would be $\frac{1}{10}\lambda_o$ at 100 MHz, and its current distribution would be reasonably constant below 200 MHz. For shorter wires and PCB lands the model's applicability should extend to higher frequencies.

We will also determine the radiated fields at a point that is perpendicular to the line conductors and in the plane containing them, as shown in Fig. 8.3. For a constant current distribution the maximum value of the radiated fields will occur at this location. Intuitively, this makes sense. In addition, the measurement point is at a distance d from the midpoint of the line. Once again, it is important to emphasize that it is assumed that this measurement point is sufficiently distant from the line that the measurement point is in the *far field* of the line.

The radiated fields can be determined, under the assumption of constant current distributions, by treating each wire as a Hertzian dipole and substituting (8.5) into (8.9). Also we substitute $\theta = 90°$ (to give the fields broadside to the line) and $\phi = 90°$ (to give the fields in the plane of the wires) into (8.7), so that $\Delta = \frac{1}{2}s$. And finally, since we are considering differential-mode currents, we substitute

$$\hat{I}_1 = \hat{I}_D \tag{8.10a}$$

$$\hat{I}_2 = -\hat{I}_D \tag{8.10b}$$

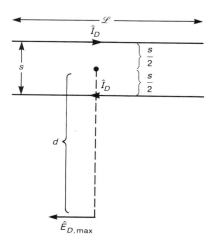

FIGURE 8.3   A simplified estimate of the maximum radiated emissions due to differential-mode currents with constant distribution.

into (8.9). The result becomes

$$\hat{E}_{D,\max} = j2\pi \times 10^{-7} \frac{f \hat{I}_D \mathscr{L}}{d} e^{-j\beta_o d} \{ e^{-j\beta_o s/2} - e^{j\beta_o s/2} \} \qquad (8.11)$$

$$= -4\pi \times 10^{-7} \frac{f \hat{I}_D \mathscr{L}}{d} e^{-j\beta_o d} \sin(\tfrac{1}{2}\beta_o s)$$

where we replace $e^{jA} - e^{-jA} = 2j \sin A$. Substituting $\tfrac{1}{2}\beta_o s = \pi s/\lambda_o = \pi s f/v_o = 1.05 \times 10^{-8} sf$ and assuming that the wire spacing s is electrically small, so that $\sin(\tfrac{1}{2}\beta_o s) \cong \tfrac{1}{2}\beta_o s$, the magnitude of (8.11) reduces to

$$|\hat{E}_{D,\max}| = 1.316 \times 10^{-14} \frac{|\hat{I}_D| f^2 \mathscr{L} s}{d} \qquad (8.12)$$

and is parallel to the wires.

As an example, consider the case of a ribbon cable constructed of #28 gauge wires separated a distance of 50 mils. Suppose the length of the wires is 1 m and that they are carrying a 30 MHz differential-mode current. The level of differential-mode current that will give a radiated emission in the plane of the wires and broadside to the cable (worst case) that just equals the FCC Class B limit (40 dBμV/m or 100 μV/m at 30 MHz) can be obtained by solving (8.12) to give

$$100\,\mu\text{V/m} = 1.316 \times 10^{-14} \frac{|\hat{I}_D|(3 \times 10^7)^2(1)(1.27 \times 10^{-3})}{3}$$

or

$$I_D = 19.95 \text{ mA}$$

Generally, the formula for the maximum emission given in (8.12) is sufficient for estimation purposes [4, 5].

Next we consider the case of a trapezoidal waveform such as a clock or data signal driving a two-wire line as shown in Fig. 8.4. Observe from (8.12) that the *transfer function* relating the maximum received electric field to the current varies with the *loop area* $A = \mathscr{L}s$ and the square of the frequency, so that

$$\left| \frac{\hat{E}_{D,\max}}{\hat{I}_D} \right| = Kf^2 A \tag{8.13}$$

where the constant is $K = 1.316 \times 10^{-14}/d = 4.39 \times 10^{-15}$ for the FCC Class B measurement distance of $d = 3$ m. Thus the frequency response of this transfer function increases at a rate of $+40$ dB/decade. Multiplying this transfer function and the input signal spectrum developed for the trapezoidal waveform in Chapter 7 (adding the Bode plots) gives the resulting spectrum of the received electric field intensity shown in Fig. 8.4. Observe that the resulting electric field spectrum increases at $+40$ dB/decade up to $1/\pi\tau$, then increases at $+20$ dB/decade up to $1/\pi\tau_r$, and is flat above that. As an example, consider a 10 MHz pulse train having a 50% duty cycle and rise/falltimes of 2.5 ns. The various breakpoints are $1/\pi\tau = 6.37$ MHz and $1/\pi\tau_r = 127.3$ MHz. This illustrates that radiated emission problems due to differential-mode currents tend to be confined to the upper frequencies of the radiated emission regulatory limit, typically above 100 MHz.

In summary, the maximum radiation occurs in the plane of the wires broadside to them, as shown in Fig. 8.5(a). At a point equidistant from each wire the radiated fields cancel, as shown in Fig. 8.5(b). Thus the radiated emissions of a pair of parallel wires should be quite sensitive to rotation of the cable. The maximum radiated electric fields vary with (1) the square of the frequency, (2) the loop area $A = \mathscr{L}s$, and (3) the current level $\hat{I}_D$. Therefore, in order to reduce the radiated emissions *at a specific frequency* due to differential-mode currents, we having the following options:

1. *Reduce the current level.*
2. *Reduce the loop area.*

Option 1 can be achieved by reducing the peak levels of the time-domain currents (A in Fig. 8.4). Typically this is not practical since the current levels have been established for functional reasons. Option 1 can also be achieved by slowing (increasing) the pulse rise/falltimes and/or the pulse repetition rate (reducing the pulse train frequency), since these will move the two breakpoints, $1/\pi\tau$ and $1/\pi\tau_r$, of the pulse spectrum shown in Fig. 8.4 lower in frequency,

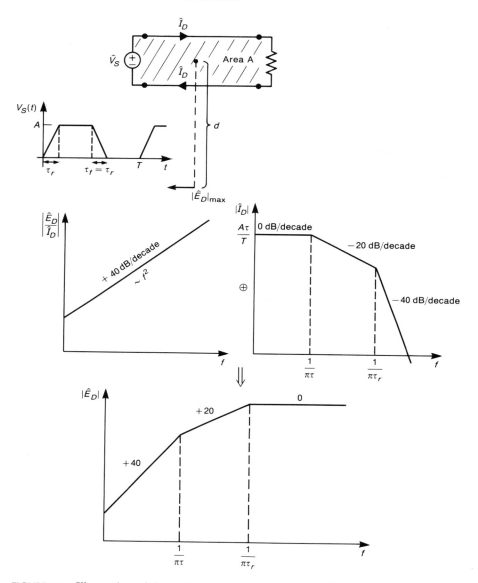

FIGURE 8.4 Illustration of the radiated emissions due to the differential-mode currents for a trapezoidal pulse train applied to a two-conductor line: (a) problem definition; (b) spectral content of the radiated emission.

possibly causing the spectrum to "roll off" at a faster rate at this frequency. Option 2, reduction of the loop area, should be addressed early in the design. This tends to be more of a problem on PCBs than in wiring harnesses. For example, suppose a clock oscillator is to feed an ASIC or a microprocessor.

Routing the clock traces as shown in Fig. 8.6(a) creates a large loop and enhances the differential-mode radiation. Placing the oscillator close to the ASIC tends to reduce the impact of the tendency of the layout personnel to route lands for wiring purposes rather than EMC considerations, thereby inadvertently creating large loop areas. Judiciously choosing pin assignments

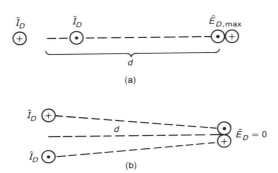

FIGURE 8.5 Illustration of the observation that (a) the fields of differential-mode currents cancel in the plane of the conductors, but (b) is a maximum perpendicular to that plane.

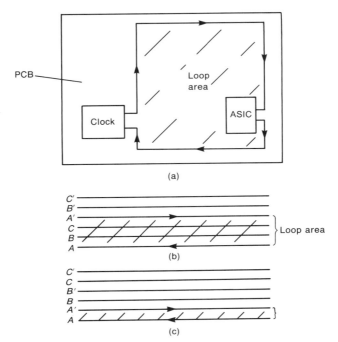

FIGURE 8.6 Common mistakes that lead to unnecessarily large differential-mode emissions: (a) large clock land areas; (b) and (c) choices of connector pin assignments in ribbon cables to minimize loop area.

can also reduce the differential-mode current emissions. For example, consider a ribbon cable carrying signals to a stepper motor, as shown in Fig. 8.6(b). Perhaps for aesthetic reasons, the pin assignments may be chosen as shown creating a loop for phase A consisting of three wire separations. If the pin assignments had been made as shown in Fig. 8.6(c), the radiated emissions due to the differential-mode current of phase A would have been reduced by a factor of 3, or roughly 10 dB! Common-sense considerations such as these provide the EMC designer with "cost-free" methods of reducing radiated emissions.

8.1.3   Common-Mode Current Emission Model

It is quite easy to modify the above results to consider the case of common-mode currents shown in Fig. 8.7. Once again, we assume that the conductors can be approximated as Hertzian dipoles, and the measurement point is (for maximum emissions) in the plane of the conductors and a distance d from a point that is midway between the conductors (and in the far field of the conductors). The common-mode currents at a cross section are equal in magnitude and directed in the same direction:

$$\hat{I}_1 = \hat{I}_C \tag{8.14a}$$

$$\hat{I}_2 = \hat{I}_C \tag{8.14b}$$

Superimposing the fields due to the two Hertzian dipoles as was done above for the differential-mode currents gives (replace the minus sign with a plus sign

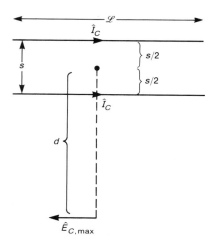

FIGURE 8.7   A simplified estimate of the maximum radiated emissions due to common-mode currents with constant distribution.

in (8.11) since the common-mode currents are directed in the same direction)

$$\hat{E}_{C,\max} = j2\pi \times 10^{-7} \frac{f \hat{I}_C \mathscr{L}}{d} e^{-j\beta_o d} \{ e^{-j\beta_o s/2} + e^{j\beta_o s/2} \} \qquad (8.15)$$

$$= j4\pi \times 10^{-7} \frac{f \hat{I}_C \mathscr{L}}{d} e^{-j\beta_o d} \cos(\tfrac{1}{2}\beta_o s)$$

where we substitute $e^{jA} + e^{-jA} = 2\cos A$. Substituting $\tfrac{1}{2}\beta_o s = \pi s / \lambda_o$ and assuming that the wire spacing s is electrically small, so that $\cos(\tfrac{1}{2}\beta_o s) \cong 1$, the magnitude of (8.15) reduces to

$$|\hat{E}_{C,\max}| = 1.257 \times 10^{-6} \frac{|\hat{I}_C| f \mathscr{L}}{d} \qquad (8.16)$$

and is again parallel to the wires.

As an example, consider the case of a ribbon cable constructed of #28 gauge wires separated a distance of 50 mils that was considered earlier for differential-mode currents. Suppose the length of the wires is 1 m and that they are carrying a 30 MHz common-mode current. The level of common-mode current that will give a radiated emission broadside to the cable (worst case) that just equals the FCC Class B limit (40 dBμV/m or 100 μV/m at 30 MHz) can be obtained by solving (8.16) to give

$$100 \ \mu\text{V/m} = 1.257 \times 10^{-6} \frac{|\hat{I}_C| (3 \times 10^7)(1)}{3}$$

or

$$I_C = 7.96 \ \mu\text{A}$$

Generally, the formula for the maximum emission given in (8.16) is sufficient for estimation purposes [4, 5].

Now we reconsider the case of a trapezoidal waveform such as a clock or data signal driving a two-wire line, as shown in Fig. 8.8. Although it is not necessarily the case, we will assume that the waveshape of the common-mode current is the same as the differential-mode current. (Consider inserting a common-mode choke in the line which presents an inductance to common-mode currents and its transparent to differential-mode currents. This inductance will likely affect the spectrum of the common-mode current.) Observe from (8.16) that the *transfer function* relating the maximum received electric field to the current varies with the line length $\mathscr{L}$ and directly with the frequency, so that

$$\left| \frac{\hat{E}_{C,\max}}{I_C} \right| = Kf\mathscr{L} \qquad (8.17)$$

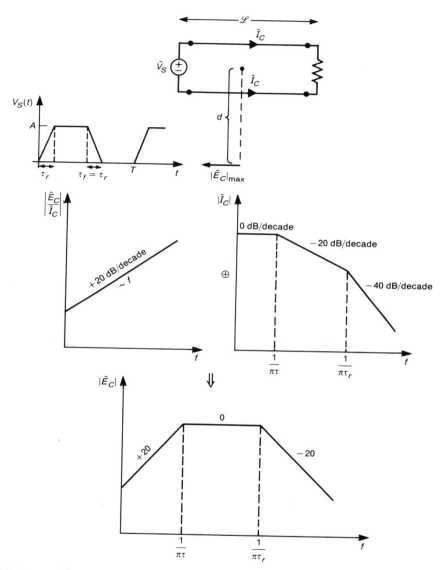

FIGURE 8.8 Illustration of the radiated emissions due to the common-mode currents for a trapezoidal pulse train applied to a two-conductor line: (a) problem definition; (b) spectral content of the radiated emission.

where the constant here is $K = 1.257 \times 10^{-6}/d = 4.19 \times 10^{-7}$ for the FCC Class B measurement distance of $d = 3$ m. Thus the frequency response of this transfer function increases at a rate of $+20$ dB/decade. Multiplying this transfer function and the input signal spectrum developed for the trapezoidal waveform in Chapter 7 (adding the Bode plots) gives the resulting spectrum of the received electric field intensity shown in Fig. 8.8. Observe that the resulting electric field spectrum increases at $+20$ dB/decade up to $1/\pi\tau$, then remains constant up to $1/\pi\tau_r$, and decreases at -20 dB/decade above that. As an example, consider a 10 MHz pulse train having a 50% duty cycle and rise/falltimes of 2.5 ns. The various breakpoints are $1/\pi\tau = 6.37$ MHz and $1/\pi\tau_r = 127.3$ MHz. This illustrates that radiated emission problems due to common-mode currents tend to be confined to the lower frequencies of the radiated emission regulatory limit, typically below 200 MHz.

In summary, the maximum radiation occurs broadside to the wires and is constant around the cable, i.e., is independent of cable rotation. Thus the common-mode current radiated emissions of a pair of parallel wires should not be sensitive to rotation of the cable. The maximum radiated electric fields vary with (1) frequency, (2) the line length $\mathscr{L}$, and (3) the current level $\hat{I}_C$. Therefore, in order to reduce the radiated emissions *at a specific frequency* due to common-mode currents we have the following options:

1. *Reduce the current level.*
2. *Reduce the line length.*

Option 1 can again be achieved by reducing the peak levels of the time-domain currents (A in Fig. 8.8) and/or slowing (increasing) the pulse rise/falltimes and/or the pulse repetition rate (reducing the pulse train frequency), since these will move the two breakpoinds, $1/\pi\tau$ and $1/\pi\tau_r$, of the pulse spectrum shown in Fig. 8.8 lower in frequency, possible causing the spectrum to "roll off" at a faster rate at this frequency. Option 2, reduction of the line length, should again be addressed early in the design. This tends to be more of a problem with wiring harnesses, although long lands on PCBs should also be avoided. Thus, to reduce the clock emissions, place the oscillator (or crystal) close to the module it feeds. Also route the wires to keep their lengths and loop area small. Cable lengths are usually dictated by system considerations such as the necessary lengths to connect to peripheral devices. In this case we have the option of blocking the common-mode currents with a toroid, which was discussed in Chapter 6.

8.1.4   Current Probes

Differential-mode currents are the desired or functional currents in the system and as such can be reliably calculated using transmission-line models or, for electrically short lines, lumped-circuit models [1]. Common-mode currents, on the other hand, are undesired currents and are not necessary for functional performance of the system. They are therefore dependent on nonideal factors such as proximity to nearby ground planes and other metallic objects as well as other asymmetries. Consequently they are difficult to calculate using ideal models [2]. They can, however, be measured using *current probes*. Current probes make use of Ampère's law:

$$\oint_C \vec{H} \cdot d\vec{l} = \int_S \vec{J} \cdot d\vec{s} + \frac{d}{dt} \epsilon \int_S \vec{E} \cdot d\vec{s} \qquad (8.18)$$

where C is the contour bounding the open surface S. Ampère's law shows that a magnetic field can be induced around a contour by either conduction current or displacement current that penetrates the open surface S, as illustrated in Fig. 8.9(a). A time-changing electric field produces a displacement current. If no time-changing electric field penetrates this surface, the induced magnetic field is directly related to the conduction current passing through the loop. Current probes use this principle in order to measure current. A current probe is constructed from a core of ferrite material that is separated into two halves, which are joined by a hinge and closed with a clip. The clip is opened, the core placed around the wire(s) whose current is to be measured, and the probe closed. The current that passes through the loop produces a magnetic field that is concentrated in and circulates around the core. Several turns of wire are wound on the core, so that the time-changing magnetic field that circulates around the core induces, by Faraday's law, an emf that is proportional to this magnetic field. The induced voltage of this loop of wire can therefore be measured and is proportional to the current passing through the probe.

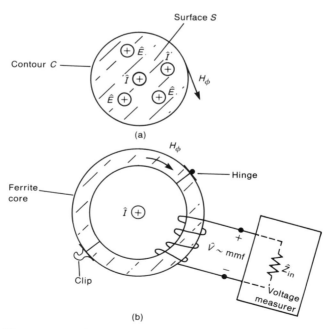

FIGURE 8.9 The current probe: (a) illustration of Ampère's law; (b) use of the current probe to measure currents.

It is not necessary to carry out precise calculations of the resulting fields and induced emf in order to calibrate the probe. Simply pass a current of known magnitude and frequency through the probe and measure the resulting voltage produced at the terminals. The result is a calibration curve that relates the ratio of the voltage $\hat{V}$ to the current $\hat{I}$ as

$$\hat{Z}_T = \frac{\hat{V}}{\hat{I}} \qquad (8.19)$$

The quantity $\hat{Z}_T$ has units of ohms and is referred to as the *transfer impedance* of the current probe. The probe manufacturer provides a calibration chart with the probe that shows the magnitude of the transfer impedance versus frequency. This calibration chart was obtained by passing a current of known amplitude and frequency through the probe and measuring the resulting voltage at the probe terminals. Usually this is given in dB (relative to 1 Ω) as

$$|\hat{Z}_T|_{dB\,\Omega} = |\hat{V}|_{dB\mu V} - |\hat{I}|_{dB\mu A} \qquad (8.20)$$

A typical such plot is shown in Fig. 8.10. For this current probe the transfer impedance is relatively constant at 12 dBΩ from 10 MHz to 100 MHz. In the following we will assume a transfer impedance of 15 dBΩ to simplify the calculations.

There is an important assumption inherent in the transfer impedance calibration curve: the termination impedance of the probe. For example, in the calibration of the probe as illustrated in Fig. 8.9(b) a voltage measurer such as a spectrum analyzer was used to measure the probe voltage in the course of determining the probe transfer impedance. Therefore the load impedance at the terminals of the probe is the input impedance to the measurement device, which is usually 50 Ω. Thus *the calibration curve of the current probe is valid only when the probe is terminated in the same impedance as was used in the course of its calibration (usually 50 Ω).*

For example, consider the problem of determining the level of probe voltage that will correspond to a common-mode current on a cable that will give a radiated emission just meeting a regulatory limit. The probe measures the total or net common-mode current in the cable, and the magnetic fluxes due to the differential-mode currents cancel out in the core. *Thus the current probe will not measure differential-mode current unless it is placed around each individual wire.* Suppose the current probe is clamped around a multiwire peripheral cable of a product and the cable length is $\mathscr{L}$. Lumping the net common-mode current into one wire of length $\mathscr{L}$ and using the results of the previous section gives the net radiated emission a distance d away as (divide (8.16) by 2 since this result is for two wires both carrying a current of $\hat{I}_C$):

$$|\hat{E}_C|_{max} = 6.28 \times 10^{-7} \frac{|\hat{I}_{C,net}| f \mathscr{L}}{d} \qquad (8.21)$$

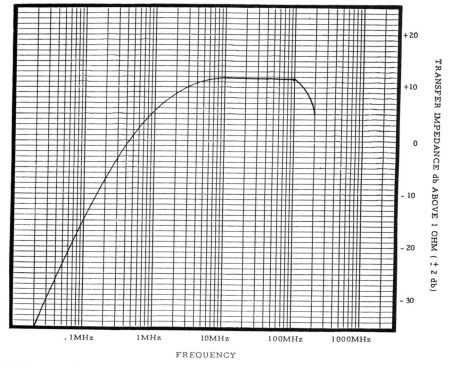

FIGURE 8.10 A typical measured current probe transfer impedance (courtesy of Fischer Custom Communications, Inc.)

This equation can be solved to give the maximum current to produce a radiated emission that equals the FCC Class B limit (or any other regulatory limit). For example, suppose the probe is clamped around a 1 m cable and the probe voltage is measured at 30 MHz. Substituting the FCC Class B limit at 30 MHz of 100 μV/m into equation (8.21) and solving for the common-mode current gives a common-mode current of 15.92 μA or 28 dBμA. For the current probe with transfer impedance given in Fig. 8.10 we would measure a voltage of

$$|\hat{V}_{SA}|_{\text{dB}\mu\text{V}} = |\hat{I}|_{\text{dB}\mu\text{A}} + |\hat{Z}_T|_{\text{dB}\Omega} \qquad (8.22)$$

$$= 24 \text{ dB}\mu\text{A} + 15 \text{ dB}\Omega$$

$$= 39 \text{ dB}\mu\text{V}$$

$$= 89 \text{ }\mu\text{V}$$

where $\hat{V}_{SA}$ denotes the voltage measured by a spectrum analyzer attached to the probe output. Therefore if we measure a probe voltage greater than 39 dBμV at 30 MHz when the probe is clamped around a 1 m cable, the radiated emissions from the net common-mode current on this cable will (ideally) exceed the FCC Class B regulatory limit. It goes without saying that the radiated emissions from this cable must be reduced! Any other emissions from the product are inconsequential, since the emissions from this cable will cause the product to be out of compliance!

In fact, the current probe can be a useful EMC *diagnostic tool* throughout the design of a product. It is a simple matter to measure the net common-mode currents on all peripheral cables of a product or a prototype of the product in the development laboratory using a current probe and an inexpensive spectrum analyzer. This does not require availability of an expensive semianechoic chamber in order to determine whether a particular peripheral cable will create serious radiated emission problems. It is also a simple method for determining whether an anticipated "fix" such as adding a toroid on a peripheral cable to reduce the common-mode currents has, in fact, reduced the common-mode current; simply measure the cable's common-mode current with the probe before and after the installation of the "fix". This is much more efficient than scheduling the semianechoic chamber in the EMC laboratory, taking the product to the lab, setting up the system, and measuring the radiated emissions. It also gives *real-time results* in that one can see rather rapidly if other "fixes" such as attaching grounding straps will affect the cable's common-mode current (they may increase the common-mode current). This author strongly believes in this type of inexpensive diagnostics. A "calibration chart" for determining the level of probe voltage that will correspond to a current that will equal a regulatory limit can be obtained by substituting (8.19) into (8.21) to give

$$|\hat{E}_C|_{\max} = 6.28 \times 10^{-7} \frac{|\hat{V}_{SA}| f \mathscr{L}}{|\hat{Z}_T| d} \tag{8.23}$$

Converting this to dB and solving for the probe voltage gives

$$|\hat{V}_{SA}|_{\mathrm{dB}\mu\mathrm{V}} = |\hat{E}|_{\mathrm{limit,dB}\mu\mathrm{V/m}} + |\hat{Z}_T|_{\mathrm{dB}\Omega} + 20 \log_{10} d \tag{8.24}$$
$$- 20 \log_{10} f_{\mathrm{MHz}} - 20 \log_{10} \mathscr{L} + 4.041$$

For example, consider a 1 m cable, a current probe with calibration chart given in Fig. 8.10. In order to comply with the FCC Class B radiated emission limit ($d = 3$ m) at 30 MHz of 40 dBμV/m, equation (8.24) gives a probe voltage of

$$|\hat{V}_{SA}|_{\mathrm{dB}\mu\mathrm{V}} = 40 \, \mathrm{dB}\mu\mathrm{V/m} + 15 \, \mathrm{dB}\Omega + 20 \log_{10} 3 - 20 \log_{10} 30 - 20 \log_{10} 1$$
$$+ 4.041 = 39 \, \mathrm{dB}\mu\mathrm{V}$$

as before. In fact, a calibration chart for the specific probe and regulatory limit to be met can be prepared from (8.24), so that one only needs to compare the spectrum analyzer reading to this level in order to determine if that cable will produce radiated emission problems. For example, considering meeting the

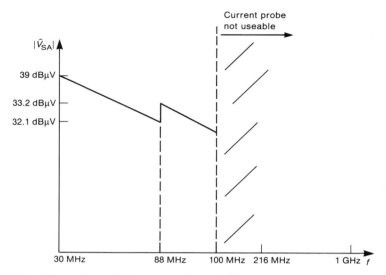

FIGURE 8.11   Illustration of the preparation of diagram to be used with a current probe and a spectrum analyzer to indicate when a measured cable current would cause emissions exceeding the FCC Class B limit.

FCC Class B limit ($d = 3$ m, 40 dBμV/m from 30 MHz to 88 MHz, 43.5 dBμV/m from 88 MHz to 216 MHz, and 46 dBμV/m from 216 MHz to 960 MHz). If we use the current probe whose transfer impedance is given in Fig. 8.10, we can determine that the spectrum analyzer reading must be below the limits shown in Fig. 8.11. Note that this current probe is usable only up to approximately 100 MHz.

8.1.5   Experimental Results

In order to illustrate the relative magnitudes of differential- and common-mode current emissions, as well as to illustrate the prediction accuracy of the above models, we will show experimental results in this section. The reader is again cautioned that the experimental configurations tested are quite simple and do not represent realistic and more complicated electronic products. This is intentional. In order to investigate the accuracy of the prediction models, it is important that the experimental configuration be simple enough so that additional radiation mechanisms will not cloud the interpretation of the data. These data will show that (1) the simple radiated emission prediction models derived in this section are quite accurate *given the currents*, (2) a current probe can be used to provide accurate values of common-mode currents on cables and PCB lands, and (3) a ferrite toroid can be an effective method of reducing the radiated emissions from common-mode currents.

The first experiment is illustrated in Fig. 8.12 and was originally described in [4]. A 10 MHz oscillator packaged in a standard 14-pin dual inline package

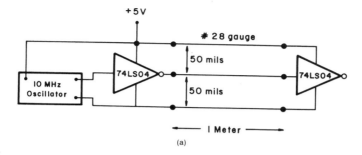

(a)

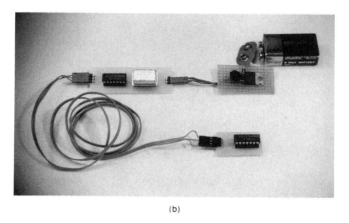

(b)

FIGURE 8.12 An experiment to assess the importance of common-mode currents on cables in the total radiated emissions of the cable: (a) schematic of the device tested; (b) photograph.

(DIP) drives a 74LS04 inverter gate. The output of this gate is attached to the input of another 74LS04 inverter gate via a 1 m, three-wire ribbon cable as shown in Fig. 8.12(a). The ribbon cable wires are #28 gauge (7 × 32) and have center-to-center separations of 50 mils. The middle wire carried the 10 MHz trapezoidal pulse train output of the driven gate to the gate at the other end, which serves as an active load. An outer wire carries the +5 V power for the inverter active load, and the other outer wire serves as the return for both signals. The +5 V power is derived from a 9 V battery that powers a 7805 regulator as shown in Fig. 8.12(b). This provides a compact 5 V source. There is no external connection to the commercial power system. This was intentional, so that radiation from the power cord of a power supply would not contaminate the measurements.

The radiated emissions were measured in a semianechoic chamber that is regularly used for developmental and compliance testing. The measured data to be shown were obtained over the frequency range of 30–200 MHz using a biconical antenna. The antenna and the ribbon cable were positioned parallel to the chamber floor, and both were 1 m above the floor, as shown in Fig. 8.13. The separation between them was 3 m. The antenna was oriented parallel to the ribbon cable in order to obtain the maximum radiated emissions from the cable.

A current probe having the probe transfer impedance given in Fig. 8.10 was used to measure the common-mode current on the cable for the prediction of the common-mode current radiated emissions. The current probe was placed at the midpoint of the cable, and the spectrum of the common-mode current at that point as measured with a spectrum analyzer is shown in Fig. 8.14. Equation (8.22) was used to relate the spectrum analyzer voltage reading to the common-mode current as

$$|\hat{I}_{\text{probe}}|_{\text{dB}\mu\text{A}} = |\hat{V}_{SA}|_{\text{dB}\mu\text{V}} - |\hat{Z}_T|_{\text{dB}\Omega} \qquad (8.25)$$
$$= |\hat{V}_{SA}|_{\text{dB}\mu\text{V}} - 15$$

A 40 foot length of RG-55U coaxial cable connected the probe to the spectrum analyzer. The loss of this cable was measured for each measurement frequency and was included in (8.25). Equation (8.21) was used to provide the predicted electric fields. A correction factor $\hat{F}_{GP}$ to correct for reflections from the ground plane (see Table 5.1) multiplies (8.21), giving

$$|\hat{E}_C|_{\text{max}} = 6.28 \times 10^{-7} \frac{|\hat{I}_{\text{probe}}| f \mathscr{L}}{D} \hat{F}_{GP} \qquad (8.26)$$

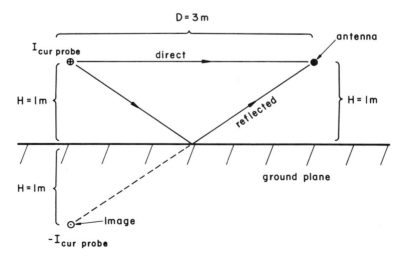

FIGURE 8.13 Physical dimensions of the measurement site, including the effect of the cable image.

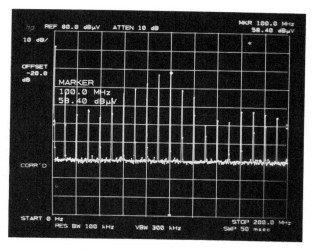

FIGURE 8.14   Measured common-mode current spectral content at the center of the cable.

Combining these gives

$$|\hat{E}_C|_{dB\mu V/m} = |\hat{V}_{SA}|_{dB\mu V} - |\hat{Z}_T|_{dB\Omega} + 20\log_{10} f_{MHz} \qquad (8.27)$$
$$+ |\hat{F}_{GP}|_{dB} + \text{Cable Loss}_{dB} - 13.58$$

The maximum that the ground-plane reflection factor $\hat{F}_{GP}$ can attain is 6 dB (a factor of 2), when the difference in the electrical distances of the path lengths of direct wave and the wave reflected from the ground plane are a multiple of a wavelength so that these fields add at the measurement antenna.

The oscillator has a fundamental frequency of 10 MHz, so only harmonics of 10 MHz will appear in the radiated emissions. A plot of the radiated emissions is shown in Fig. 8.15. The predicted values using (8.27) are shown on the plot and are denoted by X. The predictions are within 3 dB of the measured data, except at 50 MHz, 80 MHz, and 130 MHz.

The common-mode current was then measured at 5 cm intervals along the cable. Table 8.1 shows the levels at these measurement points for the 10th harmonic of 100 MHz. The maximum current of 45.1 dBμA measured at 40 cm corresponds to a current of 180 μA. Note that even though the cable length is $\frac{1}{3}\lambda_o$ at 100 MHz, the common-mode current does not display an extreme variation with position, so that the assumption of a constant current along the cable that is inherent in (8.26) seems to be a reasonable assumption. Results obtained in [2] tend to confirm this behaviour of common-mode currents.

In order to confirm that the dominant radiated emission was due to common-mode currents and that the differential-mode current emissions were smaller than these, the 74LS04 load at the far end of the cable was removed

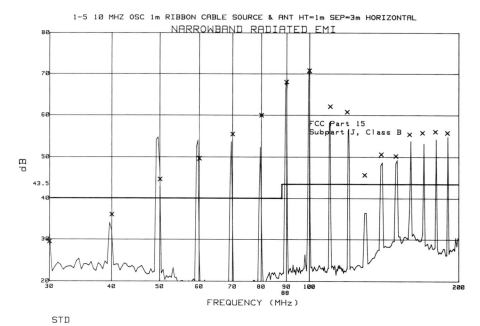

FIGURE 8.15 Measured and predicted emissions of the device of Fig. 8.12.

TABLE 8.1 I_{probe} versus Position on Cable ($f = 100$ MHz).

Position of Current Probe from Oscillator End	$I_{probe,dB\mu A}$
5 cm	38.7
10 cm	40.7
15 cm	41.9
20 cm	42.6
25 cm	43.4
30 cm	44.3
35 cm	44.7
40 cm	45.1
45 cm	44.7
50 cm	44.4
55 cm	43.9
60 cm	43.2
65 cm	41.9
70 cm	41.1
75 cm	40.2
80 cm	39.5
85 cm	38.4
90 cm	35.5
95 cm	34.0

and the radiated emissions remeasured. The results are shown in Fig. 8.16. Comparing Figs. 8.15 and 8.16, we see that the levels are essentially the same with or without the load, which tends to further confirm that *the common-mode currents are the dominant radiation mechanism.* As a further confirmation of this, four turns of the cable were wound through a toroid of NiZn material. The radiated emissions with the toroid inserted and the load attached along with the common-mode current at the midpoint of the cable was remeasured. The measured data along with the predictions are shown in Fig. 8.17. Comparing Figs. 8.17 and 8.15, we observe that the insertion of the toroid has reduced the radiated emissions by over 20 dB at some frequencies. Also, the emissions can be reliably predicted using (8.27) and the measured common-mode current.

The next set of experimental results concern the emissions from a printed circuit board (PCB) and were described in [5]. The reader may be willing to accept the fact as demonstrated above that common-mode currents tend to be the dominant radiation mechanism for long cables. However, the following set of data show that they may also be the dominant radiation mechanism for much shorter lands on PCBs. A parallel pair of 1 ounce copper lands 25 mils in width and 6 inches in length were etched on a glass–epoxy board as shown

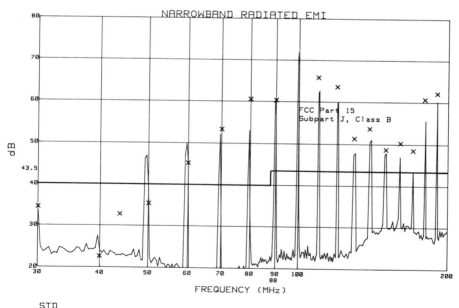

FIGURE 8.16 Measured and predicted emissions of the device of Fig. 8.12, with the load removed. The predicted emissions are virtually unchanged from those with the load attached, illustrating the dominance of common-mode current emissions.

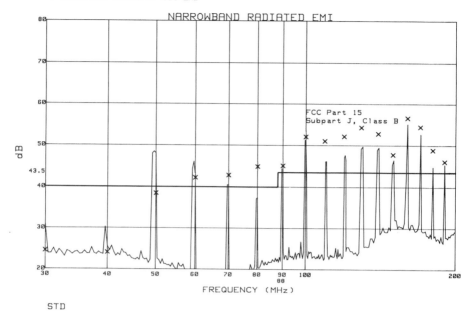

STD

FIGURE 8.17 Measured and predicted emissions of the device of Fig. 8.12 with the cable wrapped four turns through a NiZn ferrite toroid, illustrating the effect of a common-mode choke on the reduction of the radiated emissions and the dominance of those emissions.

in Fig. 8.18. The center-to-center spacing was 380 mils and the board thickness was 62 mils. The characteristic resistance of this configuration was computed to be 342 Ω using a numerical method. A load resistance of 330 Ω was used to provide a matched load. This was confirmed by time-domain measurements of the input and output voltages of the line. The same 10 MHz DIP oscillator and 5 V power supply used in the previous experiment were used to drive this line. This provided a 10 MHz trapezoidal pulse train having a 50% duty cycle, a risetime of approximately 4 ns, and a falltime of approximately 2 ns. It is important to note again that there was no connection to the commercial power system. The entire setup was quite compact and symmetric. One might therefore expect that common-mode currents on the structure would be virtually nonexistent because of this compact size and the symmetry. The measured data will show that it not the case. The radiated emissions and the common-mode current on the lands were measured exactly as for the previous experiment. The measured emissions and the predictions using (8.27) are shown in Fig. 8.19. Note that using (8.27) and the common-mode currents measured with a current probe at the midpoint of the lands provides reasonable predictions of the radiated emissions. The differential-mode currents were measured with this

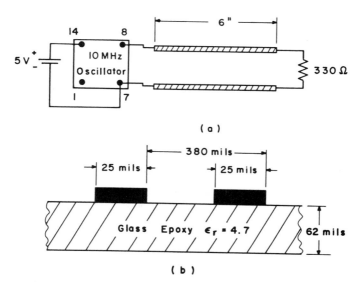

FIGURE 8.18 An experiment illustrating common-mode currents on PCBs: (a) device schematic; (b) the PCB cross-sectional dimensions.

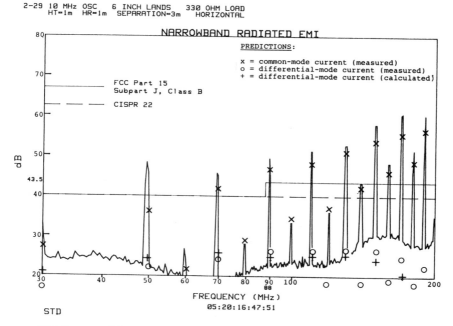

FIGURE 8.19 Measured and predicted emissions of the device of Fig. 8.18.

current probe and also predicted with a transmission-line model. The predictions using the differential-mode current prediction model given in (8.12) are also shown on the plot, and are some 20 dB below the measured radiated emissions. This further confirms that the radiated emission from the common-mode current is the dominant radiation mechanism.

To further confirm that common-mode currents are the dominant contributor to the radiated emissions, the 330 Ω load was removed and the radiated emissions and common-mode current remeasured. In this case the differential-mode currents should be considerably less than with the load attached. (The differential-mode currents will not be zero for an open-circuited load due to the displacement current path between the wires.) The radiated emissions with the load removed are shown in Fig. 8.20, and are virtually the same with the load attached! Again, the predictions using measured common-mode currents are reasonably accurate. And finally, the 330 Ω load was reattached and the board placed on edge with the lands in a vertical plane. The radiated emissions were remeasured, and were virtually the same as in Fig. 8.19. If differential-mode currents provided any significant contributions to the total radiated emissions, this latter test should show significantly lower emissions, since the differential-mode current emissions from the two lands when the board was placed on edge should cancel at the measurement antenna (see Fig. 8.5b). Such was not the case.

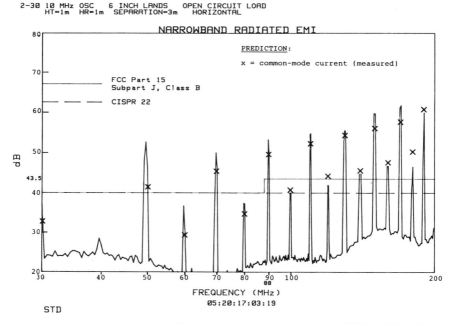

FIGURE 8.20   Measured and predicted emissions of the device of Fig. 8.18, with the load removed. The radiated emissions are virtually unchanged from those with the load attached, illustrating the dominance of common-mode current emissions.

8.2 SIMPLE SUSCEPTIBILITY MODELS FOR WIRES AND PCB LANDS

Complying with the regulatory limits on radiated (and conducted) emissions is an absolute necessity in order to be able to market a digital electronic product. However, as was point out previously, *simply being able to comply with regulatory limits does not represent a complete product design from the standpoint of EMC.* If a product exhibits susceptibility of external disturbances such as radiated fields from radio transmitters and radars or is susceptible to lightning- or electrostatic-discharge (ESD)-induced transients, then unreliable performance will result and customer satisfaction will be impacted.

The model that we will develop is a simplified version of the more exact transmission line model described in [6–13], but it will be suitable for estimation purposes. A digital computer program for the exact transmission line model for general multiconductor lines is described in [9]. We consider a parallel-wire transmission of length $\mathscr{L}$ that has a uniform plane wave incident on it as shown in Fig. 8.21(a). (The model also applies to the case of a pair of parallel lands on a PCB, with the appropriate change in two parameters.) The wires are separated a distance s and have load resistances R_S and R_L. In order to quantify our results, we will place the two wires in the xy plane, with R_S located at $x = 0$ and R_L at $x = \mathscr{L}$. The wires are parallel to the x axis. Our interest is in predicting the terminal voltages $\hat{V}_S$ and $\hat{V}_L$ given the magnitude of a sinusoidal, steady-state incident electric field $\hat{E}_i$ of a uniform plane wave, its polarization, and the direction of propagation of the wave. Two components of the incident wave contribute to the induced voltages. These are the *component of the incident electric field that is transverse to the line axis,* $\hat{E}_t^i = \hat{E}_y^i$ (in the plane of the wires and perpendicular to them), and the *component of the incident magnetic field that is normal to the plane of the wires,* $\hat{H}_n^i = -\hat{H}_z^i$ (perpendicular to the plane of the wires), as shown in Fig. 8.21(b). The line will possess per-unit-length parameters of inductance l and capacitance c, as were discussed in Chapter 4. For the parallel-wire line having wires of radius r_w these per-unit-length parameters were derived in Chapter 4, and are

$$l = \frac{\mu_o}{\pi} \ln\left(\frac{s}{r_w}\right) \quad \text{(in H/m)} \tag{8.28a}$$

$$c = \pi \epsilon_o \epsilon_r \bigg/ \ln\left(\frac{s}{r_w}\right) \quad \text{(in F/m)} \tag{8.28b}$$

where ϵ_r is the relative permeability of the surrounding medium (assumed homogeneous and nonferromagnetic). The essential modification required for the following model to apply to two parallel lands on a PCB are the use of the proper per-unit-length parameters of capacitance and inductance. Those elements are given in Chapter 6. A model of a Δx section of the line is shown in Fig. 8.21(c), where the per-unit-length parameters are multiplied by the length of the section, Δx. The per-unit-length induced sources $\hat{V}_s$ and $\hat{I}_s$ are generated

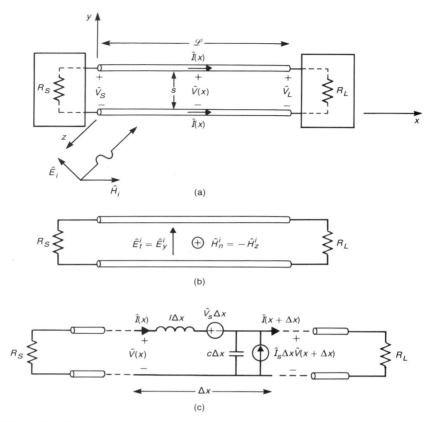

FIGURE 8.21 Modeling a two-conductor line to determine the terminal voltages induced by an incident electromagnetic field: (a) problem definition; (b) effects of the transverse electric field component and the normal magnetic field component; (c) a per-unit-length equivalent circuit.

by the incident wave according to the following considerations. First consider the normal component of the incident magnetic field intensity vector, $\hat{H}_n^i$. Faraday's law shows that this will induce an emf in the loop bounded by the wires as

$$\text{emf} = j\omega \int_S \hat{B}_n^i \, ds \tag{8.29}$$

$$= j\omega\mu_o \int_S \hat{H}_n^i \, ds$$

$$= j\omega\mu_o \, \Delta x \int_{y=0}^{s} \hat{H}_n^i \, dy$$

This induced emf can be viewed as an induced voltage source whose polarity, according to Lenz's law, is such that it tends to produce a current and associated magnetic field that opposes any change in the incident magnetic field. Thus for the incident magnetic field intensity vector in the negative z direction (into the page) the positive terminal of the source will be on the left. For a Δx section the per-unit-length source will be given by dividing the result in (8.29) by Δx to give

$$\hat{V}_s(x) = j\omega\mu_o \int_{y=0}^{s} \hat{H}_n^i \, dy \tag{8.30}$$

$$= -j\omega\mu_o \int_{y=0}^{s} \hat{H}_z^i \, dy$$

The per-unit-length induced current source $\hat{I}_s$ is directed in the $+y$ direction, and is due to the component of the incident electric field intensity vector that is transverse to the line and directed in the $+y$ direction. The result is derived in [6–12] and becomes

$$\hat{I}_s(x) = -j\omega c \int_{y=0}^{s} \hat{E}_t^i \, dy \tag{8.31}$$

$$= -j\omega c \int_{y=0}^{s} \hat{E}_y^i \, dy$$

This source is the dual of $\hat{V}_s$. The incident electric field essentially induces a voltage between the two conductors, which induces a displacement current in the impedance of the per-unit-length capacitance $1/(j\omega c \, \Delta x)$ between the two wires.

The incident fields at the position of the line may be produced by some distant antenna. These incident fields in the vicinity of the line can then be determined using the Friis transmission equation, which was derived in Chapter 5. The antenna producing these incident fields is assumed to be transmitting a radiated power P_T, is located a distance d away, and has a power gain G in the direction of the line. The incident electric field is (see (5.69) of Chapter 5)

$$|\hat{E}^i| = \frac{\sqrt{60 P_T G}}{d} \tag{8.32a}$$

The incident magnetic field, assuming a uniform plane wave, is obtained by dividing the electric field by the intrinsic impedance of free space, $\eta_o = 120\pi = 377$, to give

$$|\hat{H}^i| = \frac{|\hat{E}^i|}{\eta_o} \tag{8.32b}$$

For example, consider a half-wave dipole having a power gain in the main beam of 2.15 dB (1.64 absolute), transmitting 1 kW radiated power at 100 MHz. If the line is located a distance of 3000 m from the antenna, the maximum electric and magnetic fields in the vicinity of the line are

$$|\hat{E}^i|_{max} = \frac{\sqrt{60 \times 1000 \times 1.64}}{3000} = 0.105 \text{ V/m}$$

$$|\hat{H}^i|_{max} = \frac{|\hat{E}^i|_{max}}{120\pi} = 0.277 \text{ mA/m}$$

The model in Fig. 8.21 also applies to incident fields that are not uniform plane waves such as magnetic fields from nearby switching transformers, but the incident fields are more difficult to compute in this case [9].

From the per-unit-length model in Fig. 8.21(c) we may derive the *transmission-line equations* that relate the voltage $\hat{V}(x)$ and current $\hat{I}(x)$ along the line. To do this, we write, from the per-unit-length equivalent circuit of Fig. 8.21(c),

$$\hat{V}(x + \Delta x) - \hat{V}(x) = -j\omega l \, \Delta x \, \hat{I}(x) - \hat{V}_s(x) \, \Delta x \tag{8.33a}$$

$$\hat{I}(x + \Delta x) - \hat{I}(x) = -j\omega c \, \Delta x \, \hat{V}(x + \Delta x) + \hat{I}_s(x) \, \Delta x \tag{8.33b}$$

Dividing both sides by Δx and taking the limit as $\Delta x \to 0$ gives the transmission-line equations:

$$\frac{d\hat{V}(x)}{dx} + j\omega l \hat{I}(x) = -\hat{V}_s(x) = -j\omega\mu_o \int_{y=0}^{s} \hat{H}_n^i \, dy \tag{8.34a}$$

$$\frac{d\hat{I}(x)}{dx} + j\omega c \hat{V}(x) = \hat{I}_s(x) = -j\omega c \int_{y=0}^{s} \hat{E}_t^i \, dy \tag{8.34b}$$

The solution to these equations is described in [6–12]. This exact solution is not necessary for estimation purposes, and we will obtain an approximate solution.

For many cases of practical interest the line length is *electrically short* at the frequency of interest; that is, $\mathscr{L} \ll \lambda_o$. This will be the case of interest here for the purposes of estimating the induced terminal voltages. If the line length is electrically short at the frequency of interest, we may lump the distributed parameters by using one section of the form in Fig. 8.21(c) to represent the entire line and replacing Δx with $\mathscr{L}$. Thus the per-unit-length elements and sources are multiplied by the total line length $\mathscr{L}$. Although the terminal voltages can be calculated from this model for electrically short lines, we will make a final simplification that provides an extremely simple model that is valid for a wide variety of practical situations. In this simple model we ignore the per-unit-length parameters of inductance and capacitance. Neglecting the line

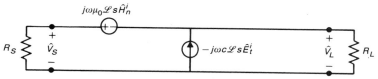

FIGURE 8.22 A simplified, lumped equivalent circuit of the pickup of incident fields for a two-conductor line that is very short, electrically.

inductance and capacitance is typically valid so long as the termination impedances are not extreme values such as short or open circuits. In addition, since the wire separation is much less than the wire length and is therefore also electrically short, the field vectors do not vary appreciably across the wire cross section; that is, with respect to y. Therefore the integrals in the sources with respect to y can be replaced with the wire separation s, giving

$$\hat{V}_s \mathscr{L} \cong j\omega\mu_o \hat{H}_n^i A \tag{8.35a}$$

$$\hat{I}_s \mathscr{L} \cong -j\omega c \hat{E}_t^i A \tag{8.35b}$$

where A is the area of the loop:

$$A = s\mathscr{L} \tag{8.36}$$

The simplified model is shown in Fig. 8.22. From this model it is a simple matter to compute the induced terminal voltages, using superposition, as

$$\hat{V}_s = \frac{R_S}{R_S + R_L}j\omega\mu_o\mathscr{L}s\hat{H}_n^i - \frac{R_S R_L}{R_S + R_L}j\omega c\mathscr{L}s\hat{E}_t^i \tag{8.37a}$$

$$\hat{V}_L = -\frac{R_L}{R_S + R_L}j\omega\mu_o\mathscr{L}s\hat{H}_n^i - \frac{R_S R_L}{R_S + R_L}j\omega c\mathscr{L}s\hat{E}_t^i \tag{8.37b}$$

This is a particularly simple model that will yield useful estimations of the effects of incident fields, as the following examples show.

Consider, as a first example, the 1 m ribbon cable shown in Fig. 8.23(a). The wires are #28 gauge 7×36 ($r_w = 7.5$ mils) and are separated by 50 mils. The termination impedances are $R_S = 50\,\Omega$ and $R_L = 150\,\Omega$. The characteristic impedance of this cable is

$$Z_C = \sqrt{\frac{l}{c}}$$

$$= \frac{1}{\pi}\sqrt{\frac{\mu_o}{\epsilon_o}}\ln\left(\frac{s}{r_w}\right)$$

$$= 120\ln\left(\frac{s}{r_w}\right)$$

$$= 228\,\Omega$$

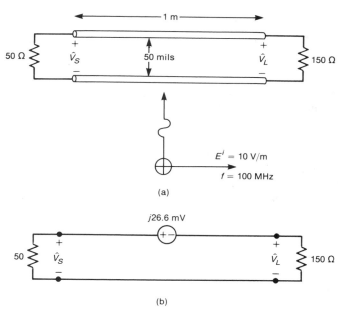

FIGURE 8.23   An example illustrating the computation of induced voltages for a 10 V/m, 100 MHz incident uniform plane wave with broadside incidence: (a) problem definition; (b) the equivalent circuit.

and we have ignored the wire dielectric insulation, $\epsilon_r = 1$. Thus the termination impedances are less than the characteristic resistance of the line. They are therefore classified as "low-impedance loads." The incident uniform plane wave has a frequency of 100 MHz and is traveling in the xy plane in the y direction, and is said to be incident "broadside" to the line. The line is $\frac{1}{3}\lambda_o$ at 100 MHz. This is probably marginal for the line to be considered electrically short, and a transmission line model should perhaps be used. For illustration purposes we will assume that the linear is electrically short and use the simplified model in Fig. 8.22. The electric field intensity vector has a magnitude $E^i = 10$ V/m and is polarized in the x direction. The magnetic field intensity vector is therefore directed in the negative z direction according to the properties of uniform plane waves, and is given by $H^i = E^i/\eta_o = 10/120\pi = 2.65 \times 10^{-2}$ A/m. Thus the component of the electric field transverse to the line is zero, and the component of the magnetic field that is normal to the plane of the wires is the total magnetic field vector. Therefore the induced sources are obtained from (8.35) as

$$\hat{V}_s \mathscr{L} = j\omega\mu_o \hat{H}_n^i A$$
$$= j \times 2\pi \times 10^8 \times 4\pi \times 10^{-7} \times \frac{E^i}{\eta_o} \times 1 \text{ m} \times 1.27 \times 10^{-3}$$
$$= j26.6 \text{ mV}$$

Because there is no component of the electric field that is transverse to the line axis, the current source is absent. The equivalent circuit is shown in Fig. 8.23(b), from which we calculate (by voltage division)

$$\hat{V}_S = \frac{50}{50 + 150} j26.6 \text{ mV}$$

$$= j6.65 \text{ mV}$$

$$\hat{V}_L = -\frac{150}{50 + 150} j26.6 \text{ mV}$$

$$= -j20 \text{ mV}$$

As a second example, consider the same ribbon cable as before but with a different direction of incidence of the wave, as shown in Fig. 8.24(a). The wave is propagating in the x direction, along the line axis, and the electric field intensity vector is polarized in the y direction (transverse to the line). The magnetic field intensity vector is therefore directed in the z direction (out of the page) and is normal to the plane of the wires. Therefore the equivalent sources are given by

$$\hat{V}_s \mathscr{L} = j\omega\mu_o \hat{H}_n^i A$$

$$= j \times 2\pi \times 10^8 \times 4\pi \times 10^{-7} \times \frac{E^i}{\eta_o} \times 1 \text{ m} \times 1.27 \times 10^{-3}$$

$$= j26.6 \text{ mV}$$

$$\hat{I}_s \mathscr{L} = -j\omega c \hat{E}_t^i A$$

$$= -j\omega \frac{\pi\epsilon_o\epsilon_r}{\ln(s/r_w)} E^i \mathscr{L}s$$

$$= -j2\pi \times 10^8 \times 14.64 \text{ pF/m} \times 10 \text{ V/m} \times 1 \text{ m} \times 1.27 \times 10^{-3} \text{ m}$$

$$= -j0.1168 \text{ mA}$$

The equivalent circuit is shown in Fig. 8.24(b). Note the polarity of the equivalent voltage source. The positive terminal of thise source is at the right because the incident magnetic field is directed out of the page. The terminal voltages are given (again by superposition) as

$$\hat{V}_S = -\frac{50}{50 + 150} j26.6 \text{ mV} - \frac{50 \times 150}{50 + 150} j0.1168 \text{ mA}$$

$$= -j6.65 \text{ mV} - j4.38 \text{ mV}$$

$$= -j11.03 \text{ mV}$$

$$\hat{V}_L = \frac{150}{50 + 150} j26.6 \text{ mV} - \frac{50 \times 150}{50 + 150} j0.1168 \text{ mA}$$

$$= j20.0 \text{ mV} - j4.38 \text{ mV}$$

$$= j15.62 \text{ mV}$$

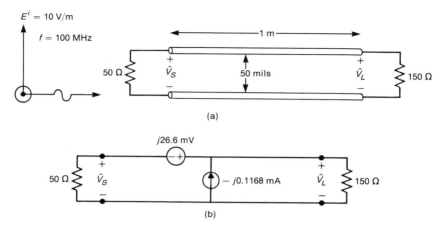

(a)

(b)

FIGURE 8.24 An example illustrating the computation of induced voltages for a 10 V/m, 100 MHz incident uniform plane wave with endfire incidence: (a) problem definition; (b) the equivalent circuit.

Observe that because of the orientation of the incident magnetic field and the resulting equivalent voltage source, the contributions from the two equivalent sources subtract in $\hat{V}_L$ but add in $\hat{V}_S$. We will find this phenomenon to occur in crosstalk between transmission lines, which will be examined in Chapter 10.

8.2.1   Experimental Results

In order to illustrate the prediction accuracy of this simple model, an experiment illustrated in Fig. 8.25 was performed, and is described in [6]. A parallel-plate, transmission-line antenna was used to generate a known, uniform plane wave traveling along the line axis, with the electric field polarized in the plane of the wires, as shown in Fig. 8.25. The parallel-plate antenna consisted of six pieces of 0.125 inch thick PC board material held in place by 1 inch × 4 inch Plexiglas structural members. The top and bottom plates are 29.875 inches wide by 6 feet long, with the excess copper being etched away from the edges, leaving a copper strip 25.625 inches wide. The end pieces are 3 feet long and taper from 25.625 inches wide at one end down to 9 inches at the other. However, the copper is etched away, leaving a triangular shape having a 25.625 inch base and a 3 foot height. All parts of the antenna are held together by large nylon screws ($\frac{1}{4}$ × 20). Wideband baluns at each end of the antenna provided a balanced structure. The antenna was driven with a wideband power amplifier connected to a frequency synthesizer as a source. A small field-strength meter, optically coupled to the power amplifier, completed a feedback loop and kept the electric field within the line constant at 1 V/m.

   The transmission line to be tested consisted of a 1.5 m length of Belden 8285 twin lead, whose characteristic resistance is approximately 300 Ω ($c = 11.8$ pF/m).

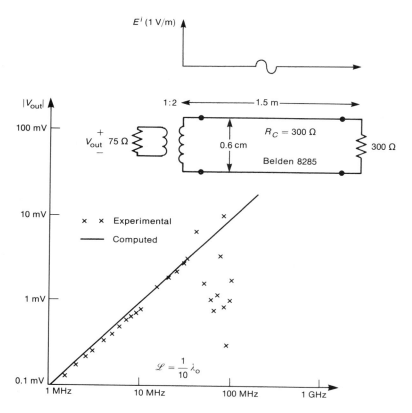

FIGURE 8.25 Measured and predicted results for incident field pickup of a two-conductor line.

One end was terminated with a 300 Ω resistor, whereas the other was connected to a 2:1 wideband balun. The output of the balun was terminated in 75 Ω. The induced voltage across this side of the balun was measured (see Fig. 8.25).

The transmission line was placed midway between the upper and lower plates of the parallel-plate antenna, with the plane of the two wires perpendicular to the walls of the antenna and on the axis of the direction of propagation. Thus the incident wave propagated along the line axis, with the electric field intensity vector in the plane of the wires. A Hewlett Packard 8405 Vector Voltmeter was used to measure the induced voltage across the 75 Ω of the balun. Since this is a relatively high-impedance instrument, the balun at the measured end of the line was terminated with a 75 Ω feedthrough resistor.

These data are shown in Fig. 8.25 along with the predictions of the simple model. The 75 Ω termination reflected through the balun presented a matched 300 Ω load to this end of the transmission line. The received voltage at this end as predicted by the model is reflected through the balun by a factor of $\frac{1}{2}$ to give a received voltage of

$$
\hat{V}_S = \frac{1}{2}\left(-\frac{300}{300+300}j\omega\mu_o\mathscr{L}s\hat{H}_n^i - \frac{300 \times 300}{300+300}j\omega c\mathscr{L}s\hat{E}_t^i \right)
$$

$$
= \frac{1}{2}\left(-\frac{300}{300+300}j\omega\mu_o\mathscr{L}s\frac{\hat{E}_t^i}{\eta_o} - \frac{300 \times 300}{300+300}j\omega c\mathscr{L}s\hat{E}_t^i \right)
$$

$$
= \tfrac{1}{2}j\omega\mathscr{L}s\hat{E}_t^i\left(-\frac{300}{300+300}\frac{\mu_o}{\eta_o} - \frac{300 \times 300}{300+300}c \right)
$$

$$
= \tfrac{1}{2}j2\pi f \times 1.5 \times 0.6 \times 10^{-2} \times 1
$$

$$
\times \left(-\frac{300}{300+300}3.33 \times 10^{-9} - \frac{300 \times 300}{300+300}11.8 \times 10^{-12} \right)
$$

$$
= -j9.71 \times 10^{-11}f
$$

Substituting the first frequency of 1 MHz gives $|\hat{V}_S| = 9.71 \times 10^{-5}$ V, as opposed to a measured value of 9×10^{-5} V. The predictions of the simple model are within 1 dB up to approximately 30 MHz. Above that, standing waves on the line are predominant, causing a deviation from the 20 dB/decade (linear with frequency) variation of the simple model. It is worth noting that the line is $\frac{1}{10}\lambda_o$ at 20 MHz. The measured data clearly follow an increase of 20 dB/decade below 20 MHz, as predicted by the simple model. This confirms that the simple model is valid for electrically short lines and termination impedances that do not differ substantially from the characteristic impedance of the line.

Additional confirmations of the transmission line model for predicting the affects of incident fields were carried out in [8] and [13]. An experimental verification was conducted in [13] and an analytical investigation in [8]. In [8] a numerical analysis code was used to provide the baseline data. Those data show that the transmission-line model is valid for frequencies so long as the wire separation is small, electrically. In this section we have concentrated on a simple, approximate version of the basic transmission-line model. This simple model is valid for lines that are electrically short and termination impedances that do not vary significantly from the line characteristic impedance. Including the line inductance and capacitance relaxes the restriction on the termination impedance values, but retains the restriction that the line be electrically short. Unless extreme termination impedances are encountered, the simple model that we have concentrated on and that is shown in Fig. 8.22 is usually sufficient for estimation purposes.

8.2.2   Shielded Cables and Surface Transfer Impedance

Coaxial cables consist of a concentric shield enclosing an interior wire that is located on the axis of the shield. The intent of the shield is to completely enclose a circuit in order to prevent coupling to the terminations from incident fields

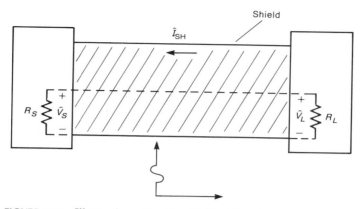

FIGURE 8.26   Illustration of incident field pickup for a shielded cable.

outside the shield, as illustrated in Fig. 8.26(a). If the shield could be constructed of a solid, perfectly conducting material, this would be the case. Even if this was possible, we would have to ensure that there were no breaks or discontinuities in the shield that would allow the incident field to penetrate to the inner wire and induce signals on that wire. This would require connectors at both ends that *peripherally bond* the shield to the enclosure housing the terminations [14–16]. *Pigtails* are breaks in the shield at the terminations that facilitate termination of the shield at the enclosures. These are common, but should be avoided if the full shielding effectiveness of the shield is to be realized. *We will assume that pigtails and other breaks in the shield are not present, so that the only penetration of an external field is through the shield.*

External fields penetrate nonideal shields via *diffusion* of the current that is induced by the external field on the external surface of the shield. A typical way of calculating this interaction is to first calculate the current induced on the shield exterior by the external, incident field, *assuming the shield is a perfect conductor and completely encloses the interior circuitry.* Any interaction between the exterior and the interior of the shield are neglected in this calculation. Once the exterior shield current $\hat{I}_{SH}$ is computed in this fashion, the induced voltages in the terminations $\hat{V}_S$ and $\hat{V}_L$ are computed in the following manner. The shield current *diffuses through the shield wall* to give a voltage drop on the *interior surface of the shield of*

$$d\hat{V} = \hat{Z}_T \hat{I}_{SH}\, dx \qquad (8.38)$$

where the *surface transfer impedance of the shield* is given by [14–16]

$$\hat{Z}_T = \frac{1}{\sigma 2\pi r_{sh} t_{sh}} \frac{\gamma t_{sh}}{\sinh \gamma t_{sh}} (\text{in } \Omega/\text{m}) \qquad (8.39a)$$

and the propagation constant in the shield material is

$$\gamma = \frac{1 + j1}{\delta} \tag{8.39b}$$

and $\delta = 1/\sqrt{\pi f \mu_o \sigma}$ is the skin depth. The shield inner radius is denoted by r_{sh} and the shield thickness is by t_{sh}. A plot of the surface transfer impedance is shown in Fig. 8.27 [16]. This is normalized to the per-unit-length dc resistance of the shield

$$r_{DC} = \frac{1}{\sigma 2\pi r_{sh} t_{sh}} \quad \text{(in } \Omega/\text{m)} \quad \text{for } t_{sh} \ll \delta \tag{8.40}$$

and shows that the shield current on the exterior of the shield completely diffuses through the shield wall for wall thicknesses less than a skin depth, $t_{sh} \ll \delta$, as we would expect. For wall thicknesses greater than a skin depth, the current on the exterior only partially diffuses through the shield wall, and the transfer impedance decreases with decreasing skin depth (increasing frequencies). Equation (8.39) gives the per-unit-length surface transfer impedance for solid shields. Shields are frequently constructed of braids of wire woven in a "herringbone" pattern to give flexibility [14, 15]. For braided shields this

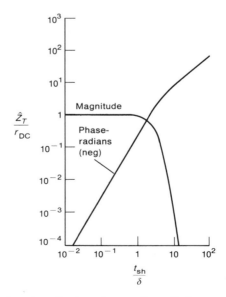

FIGURE 8.27 The surface transfer impedance of a cylinder as a function of the ratio of shield thickness to skin depth [16].

becomes [14, 15]

$$\hat{Z}_T = \frac{1}{\sigma \pi r_{bw}^2 BW \cos \theta_w} \frac{\gamma 2 r_{bw}}{\sinh(\gamma 2 r_{bw})} \quad (\text{in } \Omega/\text{m}) \tag{8.41}$$

This is simply the expression in (8.39) with the dc resistance of a solid shield replaced with the dc resistance of the braid and the thickness of the solid shield wall replaced with the diameter of the braid wires. For these *braided shields* the per-unit-length dc resistance may be approximately computed by considering the braid wires as being simply connected (electrically) in parallel, and becomes

$$r_{DC} = \frac{r_b}{BW \cos \theta_w} \quad (\text{in } \Omega/\text{m}) \tag{8.42}$$

where B is the number of *belts* in the shield braid, W is the number of braid wires per belt, θ_w is the weave angle of these belts, and r_b is the per-unit-length dc resistance of the *braid wires*, which have radii r_{bw}:

$$r_b = \frac{1}{\sigma \pi r_{bw}^2} \quad (\text{in } \Omega/\text{m}) \quad \text{for } r_{bw} \ll \delta \tag{8.43}$$

This voltage drop on the interior surface of the shield acts as a voltage source $\hat{Z}_T \hat{I}_{\text{SH}} \Delta x$ along the longitudinal interior surface of the shield. A per-unit-length equivalent circuit for the circuit enclosed by the shield is shown in Fig. 8.28(a), where r, l, g, and c are the per-unit-length resistance, inductance, conductance, and capacitance of the interior wire-shield circuit. The transmission-line equations can again be derived for this per-unit-length equivalent circuit, giving

$$\frac{d\hat{V}(x)}{dx} + (r + j\omega l)\hat{I}(x) = -\hat{Z}_T \hat{I}_{\text{SH}} \tag{8.44a}$$

$$\frac{d\hat{I}(x)}{dx} + (g + j\omega c)\hat{V}(x) = 0 \tag{8.44b}$$

Observe that the driving source is the distributed source $\hat{Z}_T \hat{I}_{\text{SH}}$. For an electrically short line we can approximate the solution by lumping the source and ignoring the per-unit-length parameters of the inner wire-shield circuit, as shown in Fig. 8.28(b), to give

$$\hat{V}_S = \frac{R_S}{R_S + R_L} \hat{Z}_T \hat{I}_{\text{SH}} \mathscr{L} \tag{8.45a}$$

$$\hat{V}_L = -\frac{R_L}{R_S + R_L} \hat{Z}_T \hat{I}_{\text{SH}} \mathscr{L} \tag{8.45b}$$

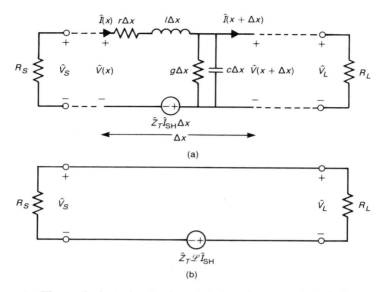

FIGURE 8.28 The equivalent circuit of the interior of a coaxial cable for computing the pickup of external fields: (a) the per-unit-length equivalent circuit; (b) a simplified equivalent circuit for cables that are very short, electrically.

In addition to the diffusion source $\hat{Z}_T \hat{I}_{SH}$, there are two other sources that are induced by the external field penetration through the holes in the braided shield [14–16]. The magnetic field of the incident field penetrates, giving a per-unit-length mutual inductance m_{12}, so that the complete surface transfer impedance becomes

$$\hat{Z}_T = \frac{1}{\sigma \pi r_{bw}^2 BW \cos \theta_w} \frac{\gamma 2 r_{bw}}{\sinh(\gamma 2 r_{bw})} + j\omega m_{12} \quad \text{(in } \Omega/\text{m)} \qquad (8.46)$$

In addition, the electric field of the incident field penetrates through these holes in the braided shield to give a per-unit-length mutual capacitance c_{12}, so that a parallel current source must be added to the per-unit-length equivalent circuit of Fig. 8.28(a), where [14, 15]

$$\hat{Y}_T = \frac{1}{\hat{V}_{SG}} \frac{d\hat{I}}{dx} \qquad (8.47)$$

and $\hat{V}_{SG}$ is the voltage between the shield and the ground plane, and $\hat{I}$ is the current on the interior wire of the shield. The term $\hat{Y}_T$ is referred to as the *surface transfer admittance* of the shield and depends not only on the shield construction but also on the exterior circuit, e.g., the height of the shield above the ground plane. Thus a source $\hat{Y}_T \hat{V}_{SG}$ must be added to the right-hand side of (8.44b). References [14, 15] give the details of this parameter. Typically the surface transfer admittance can be neglected except for very large termination impedances.

REFERENCES

[1] C.R. Paul and S.A. Nasar, *Introduction to Electromagnetic Fields*, second edition, McGraw-Hill, NY (1987).

[2] K.B. Hardin, Decomposition of radiating structures to directly predict asymmetric-mode radiation, PhD Dissertation, University of Kentucky (April 1991).

[3] T.J. Dvorak, Fields at a radiation measuring site, *Proceedings of the 1988 IEEE International Symposium on Electromagnetic Compatibility, Seattle, WA, August 1988.*

[4] C.R. Paul and D.R. Bush, Radiated emissions from common-mode currents, *Proceedings of the 1987 IEEE International Symposium on Electromagnetic Compatibility, Atlanta, GA, August 1987.*

[5] C.R. Paul, A comparison of the contributions of common-mode and differential-mode currents in radiated emissions, *IEEE Trans. on Electromagnetic Compatibility,* **31**, 189–193 (1989).

[6] C.R. Paul and D.R. Bush, Bounds on currents induced in transmission lines by incident fields, *Proceedings of IEEE SouthEasCon, Louisville, KY, April 1984.*

[7] C.R. Paul, Frequency response of multiconductor transmission lines illuminated by an electromagnetic field, *IEEE Trans. on Electromagnetic Compatibility,* **EMC-18**, 183–190 (1976).

[8] C.R. Paul and R.T. Abraham, Coupling of electromagnetic fields to transmission lines, *Proceedings of the IEEE International Symposium on Electromagnetic Compatibility, Boulder, CO, August 1981.*

[9] C.R. Paul, Applications of multiconductor transmission line theory to the prediction of cable coupling, Vol. VI: A digital computer program for determining terminal currents induced in a multiconductor transmission line by an incident electromagnetic field, *Technical Report, Rome Air Development Center, Griffiss AFB, NY,* RADC-TR-76-101, Vol. VI (February 1978).

[10] C.D. Taylor, R.S. Satterwhite, and C.W. Harrison, The response of a terminated two-wire transmission line excited by a nonuniform electromagnetic field, *IEEE Trans. on Antennas and Propagation,* **AP-13**, 987–989 (1965).

[11] A.A. Smith, A more convenient form of the equations for the response of a transmission line excited by nonuniform fields, *IEEE Trans. on Electromagnetic Compatibility,* **EMC-15**, 151–152 (1973).

[12] A.A. Smith, *Coupling of External Electromagnetic Fields to Transmission Lines,* second edition, Interference Control Technologies, Inc., Gainesville, VA (1987).

[13] C.R. Paul and D.F. Herrick, Coupling of electromagnetic fields to transmission lines, *Proceedings of the 1982 IEEE International Symposium on Electromagnetic Compatibility, Santa Clara, CA, August 1982.*

[14] E.F. Vance, *Coupling to Shielded Cables,* John Wiley, NY (1978).

[15] E.F. Vance, Shielding effectiveness of braided-wire shields, *IEEE Trans. on Electromagnetic Compatibility*, **EMC-17**, 71–77 (1975).

[16] L.O. Hoeft and J.S. Hofstra, Measured electromagnetic shielding performance of commonly used cables and connectors, *IEEE Trans. on Electromagnetic Compatibility*, **EMC-30**, 260–275 (1988).

[17] R.F. German, H.W. Ott, and C.R. Paul, Effect of an image plane on radiated emissions, *Proceedings of the 1990 IEEE International Symposium on Electromagnetic Compatibility, Washington, DC, August 1990*.

PROBLEMS

8.1 The radiated emissions of a two-wire cable shown in Fig. P8.1 are to be predicted at 100 MHz. The currents on the wires at 100 MHz are measured as shown. Calculate the radiated electric field at 100 MHz at a distance of 10 m. [89.84 dBμV/m]

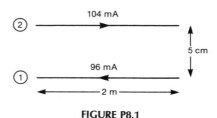

FIGURE P8.1

8.2 The radiated emissions of a cable are being measured as shown in Fig. P8.2 at 100 MHz. Calculate the voltage measured by the spectrum analyzer at 100 MHz if the antenna factor at 100 MHz is 15 dB and the antenna is oriented parallel to and in the plane of the wires. [111.5 dBμV]

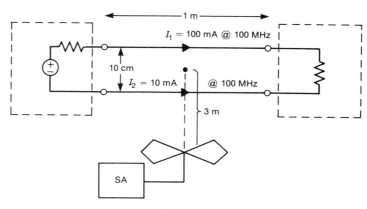

FIGURE P8.2

8.3 A cable carrying a common-mode current I as shown in Fig. P8.3 radiates. If a spectrum analyzer reads 32 dBμV at 100 MHz and the antenna has an antenna factor at 100 MHz of 16 dB, determine the current at 100 MHz. [36.6 μA]

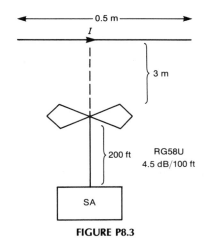

FIGURE P8.3

8.4 A wire carrying a 1 mA current at 100 MHz is parallel to and 1 cm from an infinite, perfectly conducting ground plane as shown in Fig. P8.4. Determine the electric field 3 m above the ground plane. [0.877 mV/m]

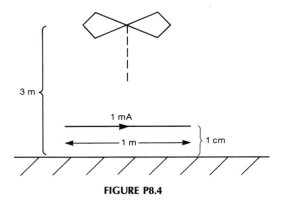

FIGURE P8.4

8.5 A current probe having $Z_T = 15$ dBΩ at 100 MHz measures a current on a 0.5 m wire as shown in Fig. P8.5. The spectrum analyzer is connected to the current probe with a 300 foot length of RG-58U coaxial cable, and reads a level of 20 dBμV. Determine the radiated electric field in a FCC Class B radiated emission test. [38.9 dBμV/m] Will this emission pass Class B? [No]

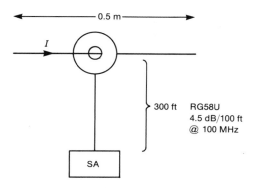

FIGURE P8.5

8.6   Two parallel wires of length 0.5 m and separation of 0.1 mm carry equal and opposite currents (differential mode) of 100 mA at 100 MHz. An antenna with an antenna factor of 20 dB at 100 MHz is used to measure the emissions in an FCC Class B test. If the antenna is parallel to and in the plane of the wires, determine the voltage measured by a spectrum analyzer attached to the antenna. [26.8 dBμV]

8.7   The maximum radiated emissions from differential-mode currents is of the form $E_{D,\max} = KIf^2 \mathscr{L}s/d$. Determine K when d is in m, I in mA, f in MHz, $\mathscr{L}$ in feet and s in mils. [$K = 1.019 \times 10^{-10}$]

8.8   The common-mode current on a 1 m cable is measured, and consists of a 10 MHz trapezoidal pulse train having a 50% duty cycle and rise/ falltimes of 20 ns, as shown in Fig. P8.8. The radiated emissions of this

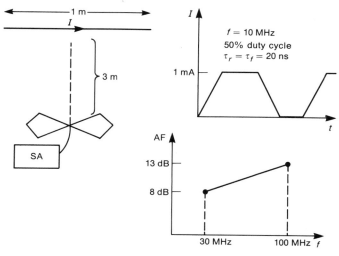

FIGURE P8.8

cable are measured at a distance of 3 m parallel to the wire using an antenna that has an antenna factor of 8 dB at 30 MHz and 13 dB at 100 MHz. Draw the envelope of the emission as measured on the spectrum analyzer between 30 MHz and 100 MHz. [49 dBμV at 30 MHz and 33.5 dBμV at 100 MHz]

8.9 An FM transmitter transmitting at 100 MHz illuminates a two-wire cable in a product at a distance of 300 m away. The transmitter is transmitting an average power of 10 W and the antenna has a power gain in the direction of the product of 20 dB. The cable is 30 cm in length and the wires are separated by 5 cm. The cable per-unit-length capacitance is 20 pF/m. Determine the maximum induced voltage at the cable endpoints if the cable is terminated in 100 Ω resistors at both ends. [21 mV]

8.10 An FM antenna that is a half-wave dipole broadcasts 500 W at 108 MHz. Determine the maximum magnetic field intensity 3 miles away. [121.9 μA/m]

8.11 A 10 V/m, 100 MHz uniform plane wave is incident on a two-wire line as shown in Fig. P8.11. Determine the induced voltage V if the cable has a per-unit-length capacitance of 50 pF/m. [−17.27 mV]

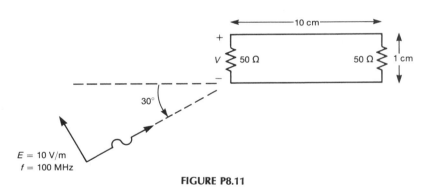

FIGURE P8.11

8.12 An FM transmitter is transmitting 5 kW at 100 MHz. The gain of the transmitting antenna is 14 dB and the pattern is omnidirectional in a plane perpendicular to the antenna. Determine the maximum electric field intensity at a distance of 5000 m from the antenna. [114.8 dBμV/m]

8.13 A 100 MHz, 10 V/m uniform plane wave is propagating parallel to an air-filled two-wire transmission line as shown in Fig. P8.13. The electric field is in the plane of the two wires. Compute the magnitude of the voltage induced across the 50 Ω load. [−9.39 mV]

8.14 A personal computer is to be installed in an office complex that is 1 mile from an FM broadcast antenna. The frequency of the transmitter is 100 MHz, and 100 W of power is being transmitted. For maximum

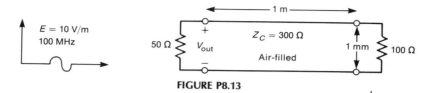

FIGURE P8.13

coverage, the antenna has an omnidirectional pattern. Determine the electric field in the vicinity of the office complex. [93.6 dBμV/m]

8.15 A two-wire ribbon cable is terminated in a 3333 Ω resistor at one end and is open-circuited at the other. The cable is 50 cm in length, and the wires are separated by 1 cm. The per-unit-length capacitance of the cable is 1 pF/m. An airport surveillance radar 1 mile away illuminates this cable. The radar is transmitting an average power of 1 kW, and the gain in the main beam is 20 dB. Assume the transmission to be a sinusoid at the transmission frequency of 1 GHz and determine the maximum voltage at the line terminations. [0.3187 V]

8.16 A 1 m braided shield cable is illuminated by a 1 MHz incident uniform plane wave. The shield is composed of 16 belts with 4 wires per belt of braid wires having radii of 2.5 mils. The weave angle is 30°. The shield interior radius is 35 mils. Determine the net shield resistance (dc). [24.6 mΩ] Determine the surface transfer impedance of the shielf at 1 MHz. [19 mΩ $\angle-64.3°$]. The interior circuit is terminated in 300 and 50 Ω resistors. Determine the voltages induced across the loads if the current induced on the exterior of the shield is 31.5 mA. [513 μV, 85.5 μV]

8.17 German, Ott, and Paul showed [17] that an image plane placed beneath a PCB can reduce the radiated emissions from both differential-mode and common-mode currents on the PCB lands, as shown in Fig. P8.17. Explain how this is possible. [Use image theory]

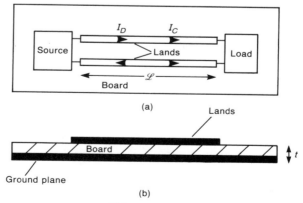

(a)

(b)

FIGURE P8.17

CHAPTER NINE

Conducted Emissions and Susceptibility

In this chapter we will investigate the mechanism by which emissions are generated and are conducted out of the product along the product's ac power cord. Regulatory agencies impose limits on these *conducted emissions* for the reason that they are placed on the commercial power system net of the installation. The commercial power distribution system in an installation is a large array of wires connecting the various power outlets from which the other electronic systems in the installation receive their ac power. It therefore represents a large "antenna" system from which these conducted emissions can radiate quite efficiently, causing interference in the other electronic systems of the installation. Thus the conducted emissions may cause radiated emissions, which may then cause interference. Ordinarily, the reduction of these conducted emissions is somewhat simpler than the reduction of radiated emissions discussed in the previous chapter, since there is only one path for these emissions that needs to be controlled: the unit's power cord. However, it is important to realize that *if a product fails to comply with the limits on conducted emissions, compliance with the limits on radiated emissions is a moot point*! Therefore controlling conduced emissions of a product has equal priority with the control of radiated emissions.

Once again, manufacturers of electronic products realize that simply complying with the regulatory limits on conducted and radiated emissions does not represent a complete design from the standpoing of EMC. A product must be reasonably insensitive to disturbances that are present on the power system net in order to insure reliable operation of the product. For example, lightning strokes may strike the power transmission lines that feed power to the installation. This may cause disturbances that range from a complete loss of commercial power (which no product is expected to withstand) to momentary power loss due to power system circuit breakers attempting to reclose (which a product is expected to withstand without loss of data or function). The regulatory limits on conducted emissions are intended to control the interference potential of the *radiated emissions* due to the noise currents that are placed on the commercial power wiring by their being conducted out the product along its ac power cord. Normally these noise currents are too small to cause direct interference by their conduction into a product along its ac power cord. However, disturbances such as those induced by lightning are of sufficient magnitude to cause interference by their direct conduction into a product via its ac power cord. This type of interference represents a *conducted susceptibility* problem, and is one that manufacturers realize and try to design a product to withstand.

9.1   MEASUREMENT OF CONDUCTED EMISSIONS

It is important to understand the measurement procedure that is used to verify compliance to the conducted emission regulatory limits. This was discussed in Section 2.2.2. The FCC limits on conducted emissions extend from 450 kHz to 30 MHz (see Figs 2.1 and 2.2), whereas the CISPR22 conducted emission limits extend from 150 kHz to 30 MHz (see Fig. 2.7). Emissions measured for verification of compliance with the regulatory limits are to be measured with a *line impedance stabilization network (LISN)* inserted between the commercial power outlet and product's ac power cord. The topologies or structures of these LISNs for FCC and CISPR 22 tests are quite similar; the component values are different. A typical test configuration is illustrated in Fig. 9.1. The ac power cord of the product is plugged into the input of the LISN. The output of the LISN is plugged into the commercial power system outlet. AC power passes through the LISN to power the product. A spectrum analyzer is attached to the LISN and measures the "conducted emissions" of the product. The FCC

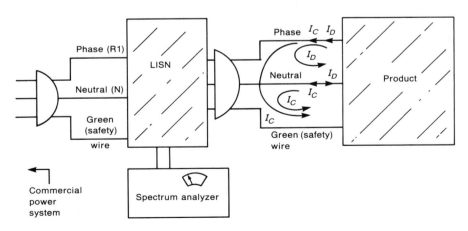

FIGURE 9.1 Illustration of the use of a line impedance stabilization network (LISN) in the measurement of conducted emissions of a product.

measurement procedure is detailed in [1], and the CISPR 22 measurement is covered in [2]. Additional discussions of the measurement procedure are contained in [3, 4]. The military measurement procedure is detailed in MIL-STD-462 [5].

9.1.1 The Line Impedance Stabilization Network (LISN)

The purpose of the conducted emission test is to measure the noise currents that exit the product's ac power cord conductors. At first glance it may appear that these emissions could be simply measured with a current probe. (In fact, the military specifies in MIL-STD-462 that conducted emissions be measured with a current probe [5].) However, the requirement that the measured data be correlatable between measurement sites may render this simple test unrealistic. The impedance seen looking into the ac power system wall outlets varies considerably over the measurement frequency range and from outlet to outlet and building to building [6]. This variability in the loading presented to the product affects the amount of noise that is conducted out the power cord. In order to make this consistent between test sites, the impedance seen by the product looking out the product's ac power cord must be stabilized from measurement site to measurement site. This is the first objective of the LISN—*to present a constant impedance to the product's power cord outlet over the frequency range of the conducted emission test.* Also, the amount of noise that is present on the power system net varies from site to site. This "external" noise enters the product's ac power cord, and, unless it is somehow excluded, will add to the measured conducted emissions. It is desired to measure only those conducted emissions that are due to the product, and this gives the second objective of the LISN—*to block conducted emissions that are not due to the product being tested so that only the conducted emissions of the product are measured.* Therefore *the two objectives of the LISN are (1) to present a constant impedance (50 Ω) between the Phase conductor and the safety wire (the "Green Wire") and between the Neutral conductor and the safety wire, and (2) to prevent external conducted noise on the power system net from contaminating the measurement.* These two objectives are to be satisfied only over the frequency range of the conducted emission test (450 kHz–30 MHz for FCC measurements and 150 kHz–30 MHz for CISPR 22 measurements). Another subtle but unstated requirement for the LISN is that it be able to pass the 60 Hz (50 Hz) power required for operation of the product.

The LISN specified for use in the FCC conducted emission measurement is shown in Fig. 9.2. The purpose of the 1 μF capacitors between Phase and Green Wire and between Neutral and Green Wire on the commercial power side is to divert "external noise" on the commercial power net and prevent that noise from flowing through the measurement device and contaminating the test data. Similarly, the purpose of the 50 μH inductors is to block that noise. The purpose of the other 0.1 μF capacitors is to prevent any dc from overloading

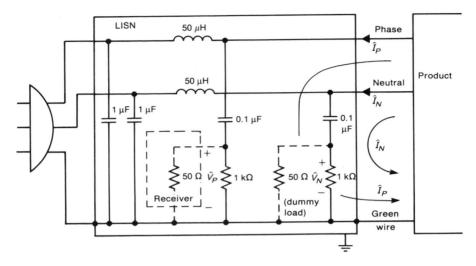

FIGURE 9.2   Illustration of the LISN circuit.

the input of the test receiver. It is instructive to compute the impedances of these elements at the lower frequency limit, 450 kHz, and the upper frequency limit, 30 MHz, of the FCC regulatory limit. These are

Element	$Z_{450 \text{ kHz}}$	$Z_{30 \text{ MHz}}$
50 μH	141.3 Ω	9420 Ω
0.1 μF	3.54 Ω	0.053 Ω
1 μF	0.354 Ω	0.0053 Ω

Thus the capacitors are essentially short circuits over the measurement frequency range, and the inductor presents a large impedance. The 1 kΩ resistors act as static charge paths to discharge the 0.1 μF capacitors in the event that the 50 Ω resistors are removed. Resistances of 50 Ω are placed in parallel with these 1 kΩ resistors become disconnected. One 50 Ω resistor is the input impedance of the test receiver (spectrum analyzer), while the other is a 50 Ω dummy load that insures that the impedance between Phase and the safety wire and between Neutral and the safety wire is approximately 50 Ω at all times. The measured voltages, denoted by $\hat{V}_P$ and $\hat{V}_N$, are measured between the Phase wire and the safety wire and between the Neutral wire and the safety wire. *Both the Phase and the Neutral voltages must be measured over the frequency range of the conducted emission limit, and must be below the specified limit at every frequency in the limit frequency range.* Now we see why the conducted emission limits are specified in terms of *voltages* when, in fact, we are interested in conducted emission *currents*. The Phase current $\hat{I}_P$ and the Neutral current, $\hat{I}_N$ are related to the measured voltages by

$$\hat{V}_P = 50\hat{I}_P \qquad (9.1a)$$

$$\hat{V}_N = 50\hat{I}_N \qquad (9.1b)$$

where we have assumed that the capacitors of the LISN are short circuits and the inductors are open circuits over the frequency range of the measurement. Therefore the measured voltages are directly related to the noise currents that exit the product via the Phase and Neutral wires.

The capacitors (inductors) of the LISN are, as we showed, essentially short (open) circuits throughout the frequency range of the conducted emission test. Therefore the equivalent circuit of the LISN will be 50 Ω resistors between the Phase wire and the safety wire and between the Neutral wire and the safety wire, as shown in Fig. 9.3. At the 60 Hz power frequency the inductors have impedances of 18.8 mΩ, the 0.1 μF capacitors have impedances of 26.5 kΩ, and the 1 μF capacitors have impedances of 2.7 kΩ. Thus at the 60 Hz power frequency the LISN has virtually no affect, and ac power for functional operation is provided to the product.

The frequency range for the CISPR 22 conducted emission test is different than for the FCC conducted emission test (150 kHz–30 MHz). Consequently the LISN capacitors and inductors must have different values so that their impedances will behave as desired. This requires a slightly more complicated LISN topology [2].

And finally it is important to point out that *the object of designing for regulatory compliance is to prevent currents in the frequency range of the regulatory limit from flowing through the 50 Ω resistors of the LISN.* Emissions outside the

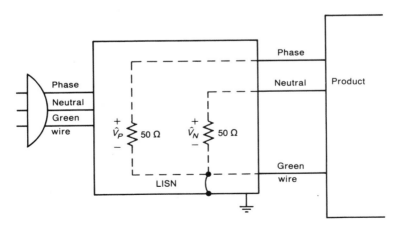

FIGURE 9.3 Equivalent circuit of the LISN as seen by the product over its intended frequency range of use.

frequency range of the regulatory limit are of no concern with regard to meeting the conducted emission limits. They may, however, be important in preventing interference with other products and so cannot be completely disregarded in the course of designing a quality product. *Any current in the frequency range of the regulatory limit that exists on the product's power cord will be measured by the LISN and can contribute to the product's failure to comply with that limit.* A common example is the existence of clock harmonics of the system oscillators on the power cord. For example, suppose that a system clock is 10 MHz. If this signal couples to the ac power cord, it will provide signals to the LISN within the regulatory frequency range (10 MHz, 20 MHz, and 30 MHz). Although the power cord is "not intended" to carry these currents, if they exist on the power cord they will be measured by the LISN and may contribute to the product's failure to comply with the regulatory limits.

9.1.2 Common- and Differential-Mode Currents Again

Representing the LISN as 50 Ω resistors between Phase and Green Wire and between Neutral and Green Wire as in Fig. 9.3 (the ideal behavior of the LISN over the conducted emission regulatory limit) simplifies the analysis of conducted emissions. The voltages that are to be measured for verification of compliance to the regulatory limit are the voltages across these 50 Ω resistors, and are denoted as $\hat{V}_P$ and $\hat{V}_N$. These voltages are related to the emission currents via Ohm's law according to (9.1). As was the case for radiated emissions, we may decompose these currents into *a differential-mode component that flows out through the Phase conductor and returns on the Neutral conductor, and a common-mode component that flows out through the Phase and Neutral conductors and returns on the Green Wire* as shown in Fig. 9.4:

$$\hat{I}_P = \hat{I}_C + \hat{I}_D \tag{9.2a}$$

$$\hat{I}_N = \hat{I}_C - \hat{I}_D \tag{9.2b}$$

Solving these gives

$$\hat{I}_D = \tfrac{1}{2}(\hat{I}_P - \hat{I}_N) \tag{9.3a}$$

$$\hat{I}_C = \tfrac{1}{2}(\hat{I}_P + \hat{I}_N) \tag{9.3b}$$

The measured voltages are

$$\hat{V}_P = 50(\hat{I}_C + \hat{I}_D) \tag{9.4a}$$

$$\hat{V}_N = 50(\hat{I}_C - \hat{I}_D) \tag{9.4b}$$

It is important to understand that, as opposed to radiated emissions, *common-mode currents can be of the order of or exceed differential-mode currents in*

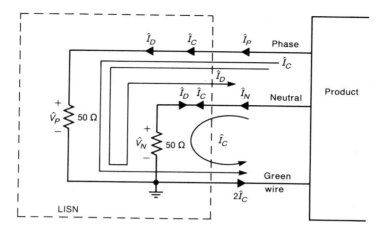

FIGURE 9.4 Illustration of the contributions of differential-mode and common-mode current components on the measured conducted emissions.

conducted emissions! We will show experimental results that confirm this important fact. Therefore *one should not assume that common-mode currents are inconsequential in conducted emissions.* It is also important to remember that differential-mode currents for conducted emission regulatory compliance are not the functional 60 Hz power line currents. Observe that the differential-mode current flows down through one 50 Ω and up through the other, whereas the common-mode currents flow down through both 50 Ω resistors. Therefore *the contributions due to each current add in* $\hat{V}_P$ *and subtract in* $\hat{V}_N$. Therefore if the common- and differential-mode currents are of the same magnitude, the Phase and Neutral voltages will not be the same. Generally, one component dominates the other so that the magnitudes of the Phase and Neutral voltages are the same:

$$\hat{V}_P = 50\hat{I}_C, \quad \hat{I}_C \gg \hat{I}_D \qquad (9.5a)$$
$$\hat{V}_N = 50\hat{I}_C, \quad \hat{I}_C \gg \hat{I}_D \qquad (9.5b)$$

or

$$\hat{V}_P = 50\hat{I}_D, \quad \hat{I}_D \gg \hat{I}_C \qquad (9.6a)$$
$$\hat{V}_N = -50\hat{I}_D, \quad \hat{I}_D \gg \hat{I}_C \qquad (9.6b)$$

Virtually all products contain a power supply filter as the last circuit that noise currents pass through before they exit the product through the power cord and then pass through the LISN. Power supply filters contain components each of which is intended to reduce either differential-mode or common-mode currents. The decomposition of the total currents into common-mode and differential-mode components along with this realization that each element of the power supply filter affects one and only one of these components is the key to designing power supply filters that are effective in the reduction of conducted emissions so that the product will comply with the regulatory limits.

Two common methods of blocking the common-mode current path are shown in Fig. 9.5. In many electronic products an inductor is placed in the Green Wire where it enters the product as shown in Fig. 9.5(a). This tends to present a high impedance to the common-mode currents that are in the frequency range of the conducted emission regulatory limit, yet a path for fault currents to flow still exists that preserves the shock hazard protection of the Green Wire. For safety reasons it is undesirable to physically solder an inductor in the green wire, because the solder joint may become defective, opening the safety wire path and leaving a potential shock hazard. In order to prevent this from happening, the inductor is constructed by winding several turns of the

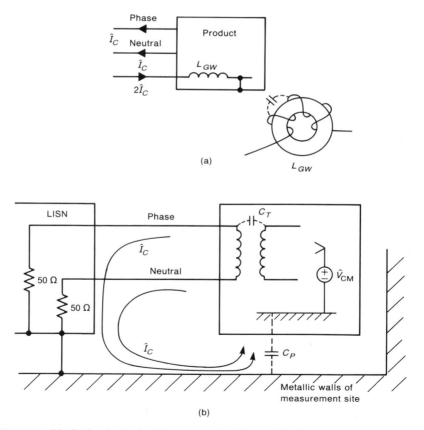

(a)

(b)

FIGURE 9.5   Methods of reducing the common-mode contribution to conducted emissions: (a) a Green Wire inductor; (b) a two-wire product.

Green Wire around a ferrite toroid (that has suitable characteristics over the conducted emission limit frequency range). Typical values of this Green Wire inductance are of order 0.5 mH, which has an impedance of some 1400 Ω at the lower frequency of the FCC regulatory limit (450 kHz). One might expect that this impedance would increase at the upper limit of 30 MHz, but this is not necessarily true. Parasitic capacitances between the windings of the toroid will typically cause its performance to deteriorate at the higher frequencies, as was discussed in Chapter 6.

Another technique for blocking the common-mode path is to construct a so-called "two-wire product." The power cord contains only the Phase and the Neutral wires, and the safety wire is absent. A two-wire product has an inherent shock hazard because the Neutral of the power distribution system is tied directly to earth ground (at the service entrance panel), and the Phase conductor is "hot" with respect to the earth ground. It would not be possible to tie the Neutral to the product chassis, since we would have no assurance that the consumer would plug the product into the correct holes of the power outlet. If the consumer plugged the product into the wrong holes of the power outlet, the chassis would be "hot" with respect to earth ground, setting up a clear shock hazard. Two-wire products invariably combat this problem by placing a 60 Hz transformer at the power entrance of the product, as shown in Fig. 9.5(b). The chassis may be tied to the secondary side of the transformer, and would therefore not be directly connected to either the Phase or the Neutral conductor. The elimination of the Green Wire in this type of product is frequently thought to eliminate common-mode currents. This not necessarily true, for reasons illustrated in Fig. 9.5(b). Stray capacitances between the product chassis and the metallic walls of the test site act to provide the equivalent Green Wire path back to the LISN (which is required to be bonded to the ground plane of the test site). Any common-mode voltage between the electronics of the product and the product frame will tend to drive these common-mode currents through this path. Stray capacitances between primary and secondary of the transformer also exist.

9.2 POWER SUPPLY FILTERS

There are virtually no electronic products today that can comply with the conducted emission regulatory requirements without the use of some form of power supply filter being inserted where the power cord exits the product. Some products may appear not to contain a filter when there is, in fact, one present. An example is the use of a large 60 Hz transformer at the power input of the product in a two-wire product or when using a linear power supply. Properly designed transformers can provide inherent filtering, and so can, in some cases, obviate the need for an "intentional" filter. We will concentrate on the design of intentional power supply filters in this section.

9.2.1   Basic Properties of Filters

We will begin with a discussion of general filter properties. Electric filters occur throughout all branches of electrical engineering, such as communications, signal processing, and automatic controls. There is a wealth of design information available for these type of filters. The reader is cautioned that *power supply filters that are intended to reduce conducted emissions are rarely designed using these traditional filter designs.* Nevertheless, a discussion of these basic principles of the traditional filters serves a useful purpose in the illumination of certain basic principles that are common to all filters.

Filters are typically characterized by their *insertion loss IL*, which is typically stated in dB. Consider the problem of supplying a signal to a load as shown in Fig. 9.6(a). A filter is inserted between the source and the load in order to prevent certain frequency components of the source from reaching the load, as shown in Fig. 9.6(b). The load voltage without the filter inserted is denoted by $\hat{V}_{L,wo}$ and the load voltage with the filter inserted is denoted as $\hat{V}_{L,w}$. The insertion loss of the filter is defined as

$$\text{IL}_{dB} = 10 \log_{10}\left(\frac{P_{L,wo}}{P_{L,w}}\right) \tag{9.7}$$

$$= 10 \log_{10}\left(\frac{V_{L,wo}^2/R_L}{V_{L,w}^2/R_L}\right)$$

$$= 20 \log_{10}\left(\frac{V_{L,wo}}{V_{L,w}}\right)$$

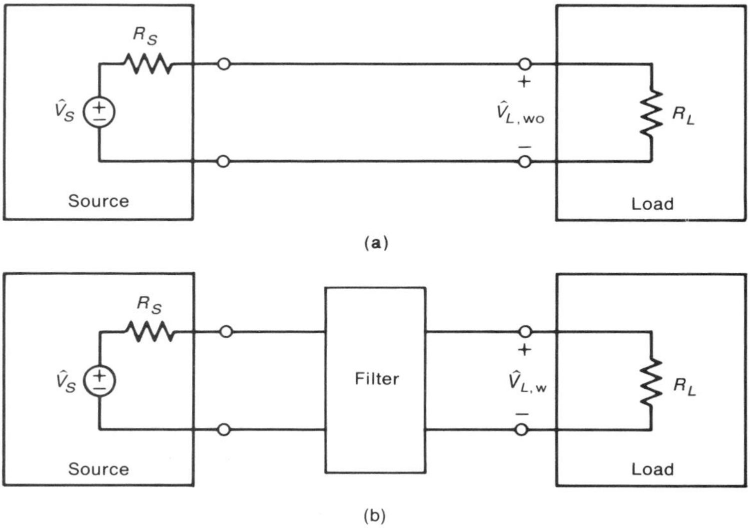

FIGURE 9.6   Definition of the insertion loss of a filter: (a) load voltage without the filter; (b) with the filter inserted.

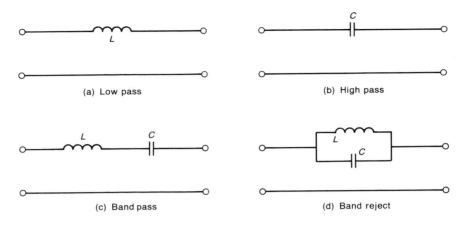

FIGURE 9.7 Four simple filters: (a) low-pass; (b) high-pass; (c) bandpass; (d) band-reject.

Note that the voltages in this expression are not denoted with a caret ($\hat{}$), and are therefore the magnitudes of these voltages. The insertion loss gives the reduction in the load voltage *at the frequency of interest* due to the insertion of the filter. Typically, the insertion loss is displayed as a function of frequency.

Some simple filters are shown in Fig. 9.7. These can be analyzed using the techniques discussed in Chapter 6. For example, let us determine the insertion loss of the simple low-pass filter of Fig. 9.7(a). The load voltage without the filter can be easily determined from Fig. 9.6(a) as

$$\hat{V}_{L,wo} = \frac{R_L}{R_S + R_L} \hat{V}_S \tag{9.8}$$

The load voltage with the filter inserted is

$$\hat{V}_{L,w} = \frac{R_L}{R_S + j\omega L + R_L} \hat{V}_S \tag{9.9}$$

$$= \frac{R_L}{R_L + R_S} \frac{1}{1 + j\omega L/(R_S + R_L)} \hat{V}_S$$

The insertion loss is the ratio of (9.8) and (9.9):

$$IL = 20 \log_{10} \left| 1 + \frac{j\omega L}{R_S + R_L} \right| \tag{9.10}$$

$$= 20 \log_{10} \left[\sqrt{1 + (\omega\tau)^2} \right]$$

$$= 10 \log_{10} [1 + (\omega\tau)^2]$$

where

$$\tau = \frac{L}{R_S + R_L} \tag{9.11}$$

is the time constant of the circuit. A plot of the insertion loss would show 0 dB from dc to the 3 dB point of $\omega_{3\,dB} = 1/\tau$ and an increase at a rate of 20 dB/decade above that. Therefore the low-pass filter passes frequency components of the source from dc to $\omega_{3\,dB}$ and increasingly reduces the components at frequencies above that. For frequencies above the 3 dB point the insertion loss expression simplifies to

$$\mathrm{IL} \cong 10 \log_{10}[(\omega\tau)^2], \quad \omega \gg \tau \tag{9.12}$$

$$= 20 \log_{10} \omega\tau$$

$$= 20 \log_{10}\left(\frac{\omega L}{R_S + R_L}\right)$$

Other filters are analyzed in a similar fashion.

The above example has illustrated an important point: *the insertion loss of a particular filter depends on the source and load impedances, and therefore cannot be stated independently of the termination impedances.* Most filter manufacturers provide frequency response plots of the insertion loss of a particular filter. Since the insertion loss of a filter is dependent on the source and load impedances, what value of source and load impedance is assumed in these specifications? The answer is rather obvious: it is assumed that $R_S = R_L = 50\,\Omega$! This brings up another important point; how does this specification of insertion loss based on $50\,\Omega$ source and load impedances relate to the filter's performance in a conducted emission test? Consider the use of the filter in that test. The "load impedance" corresponds to the $50\,\Omega$ impedances of the LISN between Phase and Green Wire and between Neutral and Green Wire. What is the "source impedance R_S" in this useage? The answer is that we do not know, since this is the source impedance seen looking back into the product's power input terminals. It is doubtful that this will be $50\,\Omega$ and furthermore that it will be constant *over the frequency range of the conducted emission test*! So use of the manufacturer's insertion loss data to assess the performance of the filter in a product may not give realistic results.

Furthermore there are two currents that must be reduced: common-mode and differential-mode. Filter manufacturers typically give separate insertion loss data for these currents. These data are obtained as shown in Fig. 9.8. For the differential-mode insertion loss measurement the Green Wire terminals are left unconnected and the Phase and Neutral wires form the circuit to be tested, as shown in Fig. 9.8(a), since the differential-mode current, by definition, flows out the Phase wire and returns via the Neutral wire, and no differential-mode

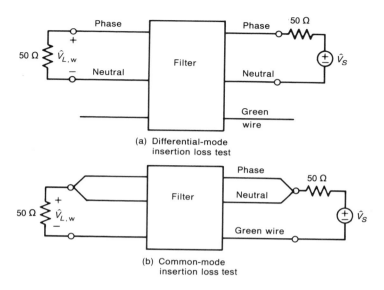

FIGURE 9.8   (a) Differential-mode insertion loss test

(b) Common-mode insertion loss test

FIGURE 9.8   Insertion loss tests: (a) differential mode; (b) common mode.

current returns on the Green Wire. For the common-mode test, the Phase and Neutral wires are tied together and form the test circuit with the Green Wire, as shown in Fig. 9.8(b). Once again, *the source and load impedances for each test are assumed to be 50 Ω.*

9.2.2   A Generic Power Supply Filter Topology

The most common power supply filter topology is some version of the generic filter topology shown in Fig. 9.9. The reader should note that this filter topology resembles a Pi structure. The differential- and common-mode currents at the output of the product (usually the input to the product's power supply) are denoted as $\hat{I}_D$ and $\hat{I}_C$, whereas these currents at the input to the LISN (at the output of the filter) are denoted with primes as $\hat{I}'_D$ and $\hat{I}'_C$. The object of the filter is to reduce the unprimed current levels to the primed levels such that the primed currents $\hat{I}'_D$ and $\hat{I}'_C$ give measured voltages

$$\hat{V}_P = 50(\hat{I}'_C + \hat{I}'_D) \tag{9.13a}$$

$$\hat{V}_N = 50(\hat{I}'_C - \hat{I}'_D) \tag{9.13b}$$

that are below the conducted emission limit *at all frequencies in the frequency range of that limit.*

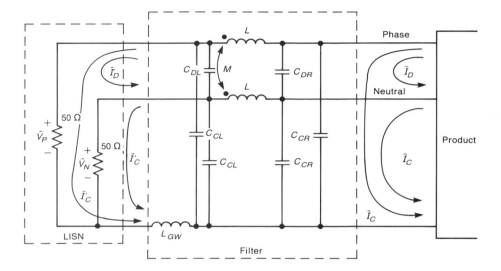

FIGURE 9.9   A typical power supply filter topology.

9.2.3   Effect of the Filter Elements on Common- and Differential-Mode Currents

A Green Wire inductor L_{GW} is included in the Green Wire *between the filter output and the LISN input* to block common-mode currents as discussed above. Capacitors between Phase and Neutral wires, C_{DL} and C_{DR}, are included to *divert differential-mode currents.* These are referred to as *line-to-line capacitors.* Capacitors that have insulation properties approved by safety agencies and are suitable for use as line-to-line capacitors are referred to as "X-caps." The subscripts L and R denote "left" and "right" with regard to the side of the filter on which they are placed. Capacitors C_{CL} and C_{CR} are included between Phase and Green Wire and between Neutral and Green Wire to *divert common-mode currents.* These are referred to as *line-to-ground capacitors.* Capacitors that have insulation properties approved by safety agencies and are suitable for use as line-to-ground capacitors are referred to as "Y-caps." The reason that different capacitors are needed for these tasks is due to safety considerations. For example, suppose one of the line-to-ground Y-caps shorts out. If this capacitor happens to be connected to the Phase wire, 120 V will be tied to the Green Wire, which is usually tied directly to the product frame, presenting an obvious shock potential. Also, the safety agencies such as the Underwriters Laboratory (UL) in the US specify the *maximum leakage current* that may flow through the line-to-ground capacitors at 60 Hz in order to minimize shock hazards due to these leakage currents. The UL limit on leakage current at 60 Hz is of order 1 mA, depending on the UL specification. If we assume a leakage current of 1 mA at 60 Hz then for 120 V, 60 Hz power this limits the maximum line-to-ground capacitance to $C_C < \frac{1}{2} \times 1 \times 10^{-3}/120 \times 2\pi \times 60 = 5526$ pF. The *total leakage current* is specified so each capacitor can carry a maximum of one-half of that total. Some filters include only the capacitors on the left or on the right, and some include both sets. Still other filters may, for example, include only C_{DL} and C_{CR} and omit C_{DR} and C_{CL}. Typical values of these capacitors are $C_D \cong 0.047\ \mu F$ and $C_C \cong 2200$ pF. Observe that the line-to-ground capacitors on the left, C_{CL}, are in parallel with the 50 Ω resistors of the LISN. Therefore if their impedances at the frequency of interest are not significantly lower than 50 Ω then these capacitors will be ineffective in diverting the common-mode current. To judge whether the line-to-ground capacitors on the left will be effective, let us compute their impedance for typical values of $C_{CL} = 2200$ pF. The impedances of these capacitors will equal 50 Ω at 1.45 MHz, and so the capacitors C_{CL} will be effective in diverting common-mode currents from the LISN 50 Ω resistors only above this frequency!

One final element is typically included—the *common-mode choke* represented by the coupled inductors. The self-inductances of each winding are represented by L and the mutual inductance is represented by M. Typically this element consists of two identical windings on a common ferrite core (that has suitable characteristics over the conducted emission frequency range), and so is similar to a transformer, as shown in Fig. 9.10(a). Because the windings are identical and are wound tightly on the same core, the mutual inductance is approximately equal to the self inductance, $L \cong M$, and as such has a coupling coefficient approaching unity:

$$k = \frac{M}{\sqrt{L_1 L_2}} \tag{9.14}$$

$$\cong \frac{M}{L}$$

$$\cong 1$$

The purpose of the common-mode choke is to block common-mode currents. Ideally, *the common-mode choke does not affect differential-mode currents.* This was shown in Chapter 6, but it is worthwhile to repeat that here. Consider only differential-mode currents through the choke, as shown in Fig. 9.10(b). Computing the voltage drop across one side of the choke gives

$$\hat{V} = j\omega L \hat{I}_D - j\omega M \hat{I}_D \tag{9.15}$$

$$= j\omega (L - M)\hat{I}_D$$

Therefore the element inserts an inductance $L - M$ in each lead with regard to differential-mode currents. This is commonly referred to as the *leakage inductance*, and is due to the portion of the magnetic flux that leaks out the core and does not couple between the windings. Ideally this is zero, and the common-mode choke has no effect on differential-mode currents. We will see that this leakage

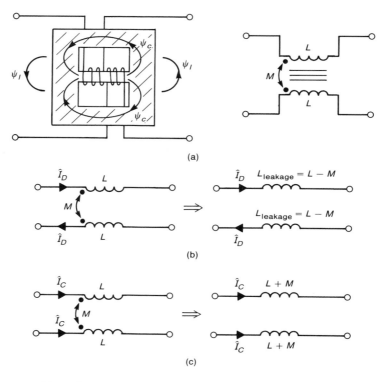

(a)

(b)

(c)

FIGURE 9.10 Use of a common-mode choke to block common-mode conducted emissions: (a) physical construction and equivalent circuit; (b) equivalent circuit for differential-mode currents; (c) equivalent circuit for common-mode currents.

inductance is not zero for actual chokes, and has an important role in the blockage of the differential-mode currents. Now consider the effect of the choke on common-mode currents, shown in Fig. 9.10(c). Computing the voltage drop across one side of the choke gives

$$\hat{V} = j\omega L\hat{I}_C + j\omega M\hat{I}_C \qquad (9.16)$$
$$= j\omega(L + M)\hat{I}_C$$

Therefore the element inserts an inductance $L + M$ in each lead with regard to common-mode currents. Consequently, the common-mode choke tends to block common-mode currents. Typical values for the inductance are of order 10 mH. Thus the common-mode current impedance is $j\omega(L + M) = 56{,}549 \ \Omega$ at 450 kHz and 3.77 MΩ at 30 MHz. It is important to emphasize that these are *ideal* values. Parasitic capacitance between the windings as well as the type of core material strongly influence the frequency behavior of the choke.

It is important to reemphasize another important characteristic of the common-mode choke. In addition to the noise signal in the differential-mode current, there is another component—the high-level, 60 Hz power current. Typically this can be several amps. Currents of this level will easily *saturate* a ferrite core, and will therefore reduce its permeability to values approaching that of air. The ability of the choke to block common-mode currents relies on large values of L and M being obtained, which in turn relies on having a large value of permeability of the core. If the core material were saturated by the high-level, 60 Hz current, we would not obtain sufficiently large inductance to provide blocking of the common-mode currents. On the other hand we do not wish to drop much of the power voltage across the choke. The fact that the differential-mode current flux tends to cancel in the core because of the way the windings are wound on the core means that, ideally, the choke does not provide any impedance to differential-mode currents, and the choke is transparent to these currents (even at 60 Hz). Therefore *the 60 Hz flux cancels in the core and does not saturate it.* This is an additional benefit of a common-mode choke.

Let us now develop some equivalent circuits to represent the effect of the filter on the common- and differential-mode currents. We will assume a filter that is *symmetric* with regard to Phase and Neutral. By this it is meant that the Phase–Green Wire circuit and the Neutral–Green Wire circuit are identical. This is true for the generic filter shown in Fig. 9.9 in that the line-to-ground capacitors between Phase and the Green Wire are identical to those between Neutral and Green Wire, and the self-inductances of both sides of the common-mode choke are identical. This is normally the case, since there seems to be no advantage to constructing an unsymmetrical filter. First consider the effect on common-mode currents, shown in Fig. 9.11. We simulate the

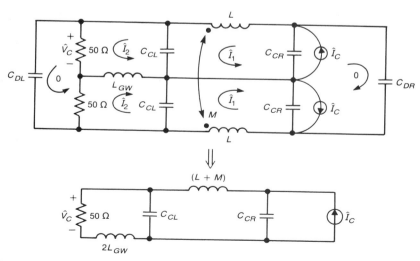

FIGURE 9.11 The equivalent circuit of the filter and LISN for common-mode currents.

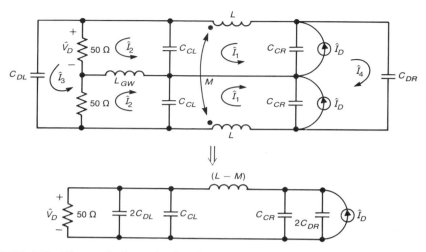

FIGURE 9.12 The equivalent circuit of the filter and LISN for differential-mode currents.

common-mode currents with current sources. By the symmetry of the structure, we can assume certain currents to be identical as shown. Writing mesh equations, we can demonstrate that the equivalent circuit presented to each common-mode current is as shown. This is intuitively obvious in that the choke appears as an inductance $L + M$ and the line-to-line capacitors have no effect. The Green Wire inductor appears twice as large, since $2\hat{I}_C$ flows through it. Next consider the effect on differential-mode currents. We will also simulate these currents as current sources, as shown in Fig. 9.12. Again, the symmetry of the filter dictates that certain mesh currents will be identical. Writing mesh-current equations shows that the equivalent circuit to differential-mode currents appears as shown. Note that the line-to-line capacitors appear twice as large to differential-mode currents. Also note that the line-to-ground capacitors are present, and therefore C_{CL} and C_{CR} also affect differential-mode currents in addition to common-mode currents! This is usually not significant, since the values of the line-to-ground capacitors are typically much smaller than those of the line-to-line capacitors. However, if the line-to-line capacitor with which the line-to-ground capacitor is in parallel is absent then *the line-to-ground capacitor will affect differential-mode current*. Also, for an ideal common-mode choke where $L = M$ the differential-mode currents are completely blocked from appearing in the LISN. This again illustrates the importance of careful design of the common-mode choke.

9.2.4 Separation of the Conducted Emissions into Common- and Differential-Mode Currents for Diagnostic Purposes

The above discussion of the power supply filter concentrated on the *ideal behavior* of the filter. If the line-to-line capacitors behave ideally then little differential-mode current will reach the LISN. If we can obtain sufficiently large values for the common-mode choke inductance, the line-to-ground capacitors, and the Green Wire inductor then very little common-mode current will reach the LISN. In practice this almost never occurs, and the product may still fail to comply with the conducted emission regulatory limit even though a careful design of the filter has been completed. When the product is tested for compliance and found to be out of compliance at certain frequencies in the frequency range of the regulatory limit, the next question is how we shall *effectively and efficiently diagnose and correct the problem and bring the conducted emissions into compliance.* In order to illustrate the necessity for this, it is worthwhile to point out that this author has spent many needless hours changing values of the filter components and observing *no change in the levels of the conducted emissions.* Product development schedules cannot tolerate this inefficiency in the correction of the problem. We must be able to quickly and correctly diagnose the root of the problem. Which element of the power supply filter needs to be changed and what should the element's new value be? There are other ways of reducing the levels of the conducted emissions that will be discussed in Section 9.3, but attacking the power supply filter is a viable first step.

The most important point to realize in the course of changing a power supply element value to effect a reduction in the conducted emissions is illustrated in Fig. 9.13. We have shown a typical plot of the *total* current of the Phase or

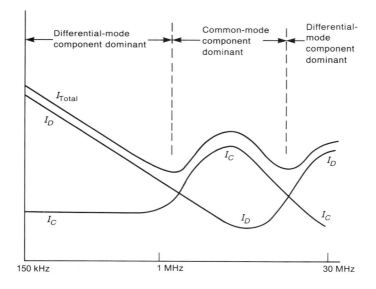

FIGURE 9.13 Illustration of the important observation that one component of current may dominate the other over a particular frequency range of the conducted emission test. In order to reduce the total conducted emission, the dominant component must be reduced.

Neutral wire. This is decomposed into common- and differential-mode components. Although rather obvious, it is important to point out that the total current is the sum or difference of the common-mode and differential-mode components as shown by (9.2):

$$\hat{I}_{\text{Total}} = \hat{I}_C \pm \hat{I}_D \tag{9.17}$$

If one component is larger than the other component, the total current is the dominant component! This seemingly obvious statement illustrates that one component may dominate over certain frequency ranges. If we change the power supply filter so as to reduce the dominant component, we will reduce the total current. On the other hand, if we change the power supply filter so as to reduce the component that is not dominant, we will cause *no reduction in the total!* Therefore, *if we wish to reduce the total conducted emission at a particular frequency, we must reduce the dominant component at that frequency.* It is also important to observe that one component may be dominant over one frequency range yet not be dominant over another portion of the conducted emission frequency range.

Now let us consider how we shall reduce a particular component. We saw previously that each element of the power supply filter affects only one component; either differential- or common-mode. (The line-to-ground capacitors appear in the differential-mode circuit, yet they are usually much smaller than the line-to-line capacitors that are in parallel with them, and as such do not affect the differential-mode current. However, if the line-to-line capacitor with which they are in parallel is absent then they will affect the differential-mode current.) So *if we need to reduce the level of a particular (dominant) component, we must change the value of a power supply filter element that affects that component.* For example, suppose that the common-mode component dominates the differential-mode component at a frequency where the conducted emission exceeds the regulatory limit. Increasing the value of the line-to-line capacitance will only reduce the differential-mode component, and so will not reduce the total conducted emission, since the differential-mode component was not the dominant component at this frequency. Conversely, suppose the differential-mode component dominates the common-mode component at a particular frequency where the conducted emission exceeds the regulatory limit. Increasing the Green Wire inductance by placing more turns on the core will reduce the common-mode component, but will not affect a change in the conducted emissions at this frequency, since the differential-mode component was dominant. These observations easily explain the author's experience that *a radical change in the value of an element of the power supply filter may cause absolutely no change in the total conducted emissions; the component that was not dominant was reduced.*

If we are to be able to rapidly and correctly bring about a reduction in the conducted emissions by changing the values of some elements of the power supply filter, we must know which component is the dominant component in

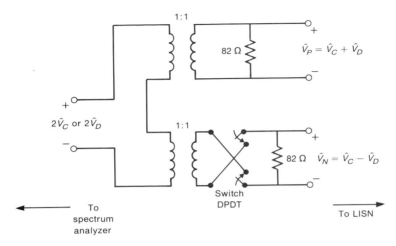

FIGURE 9.14 Schematic of a device to separate the common-mode and differential-mode conducted emission contributions.

FIGURE 9.15   Photograph of the device of Fig. 9.14.

the total conducted emission. We need a *diagnostic tool* that can separate the total conducted emission into its common- and differential-mode components at each frequency of the regulatory limit. Such a device was described in [7]. The schematic of the device is shown in Fig. 9.14 and a photograph in Fig. 9.15. The device basically adds or subtracts the Phase and Neutral voltages of the LISN to give only the differential-mode or only the common-mode component of the total conducted emission. It makes use of two wideband transformers (baluns). The Phase and Neutral output voltages of the LISN are applied to the primaries of the transformers. The secondaries are connected in series, and a switch changes the polarity of the Neutral voltage. Since $\hat{V}_P = \hat{V}_C + \hat{V}_N$ and $\hat{V}_N = \hat{V}_C - \hat{V}_N$, the sum gives $\hat{V}_C$ and the difference gives $\hat{V}_D$:

$$\hat{V}_P + \hat{V}_N = 2\hat{V}_C \qquad (9.18a)$$

$$\hat{V}_P - \hat{V}_N = 2\hat{V}_D \qquad (9.18b)$$

In order to demonstrate the effectiveness of the device and to confirm the above observations about the effect of the individual elements of the power supply filter, we will consider an experiment that was described in [7]. A typical

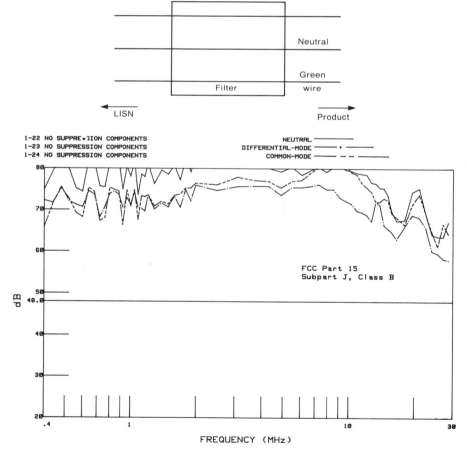

FIGURE 9.16 Measured conducted emissions of a typical product separated into differential- and common-mode components with no power supply filter.

digital device containing a switching power supply and a power supply filter was tested. The elements of the power supply filter were first removed, and the product's conducted emissions were measured. These data are shown in Fig. 9.16. Observe that the conducted emissions exceed the FCC Class B limit by over 30 dB! The device was inserted at the output of the LISN, and the components are also shown in Fig. 9.16. Observe that the common- and differential-mode components of the total emission as measured with the device are the same order of magnitude. The filter elements were next added one by one and their effect observed. First, 3300 pF line-to-ground capacitors were added. The results are shown in Fig. 9.17. Note that the addition of the line-to-ground capacitors has reduced both the common- and differential-mode

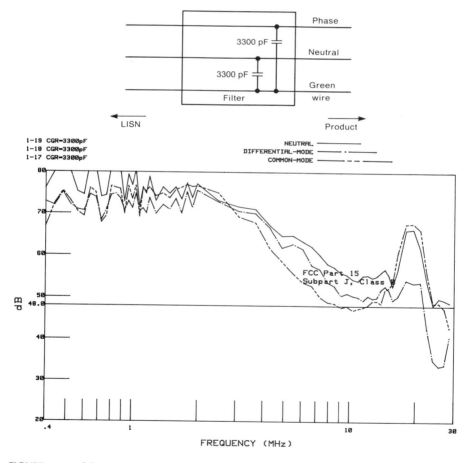

FIGURE 9.17 Measured conducted emissions of a typical product separated into differential- and common-mode components with 3300 pF line-to-ground capacitors added.

components by the same amount, but only above approximately 2 MHz. The 3300 pF line-to-ground capacitors are effectively in parallel with the 50 Ω resistors of the LISN (see Figs. 9.11 and 9.12), and so give a break frequency of order 1 MHz where their impedance equals 50 Ω. Thus they will begin to shunt both common- and differential-mode currents above this frequency. Next we add a 0.1 μF line-to-line capacitor. The results are shown in Fig. 9.18. Note that the common-mode component is unchanged, yet the differential-mode component has been reduced considerably. This is sensible to expect, since this capacitor should affect only the differential-mode component (see Fig. 9.12). According to Fig. 9.12, the break frequency of this effect should be where the impedance of $2C_D + C_C = 0.203$ μF equals 50 Ω. This occurs at 15.7 kHz, which is well below the lower frequency of the plot. Next we add the 1 mH Green

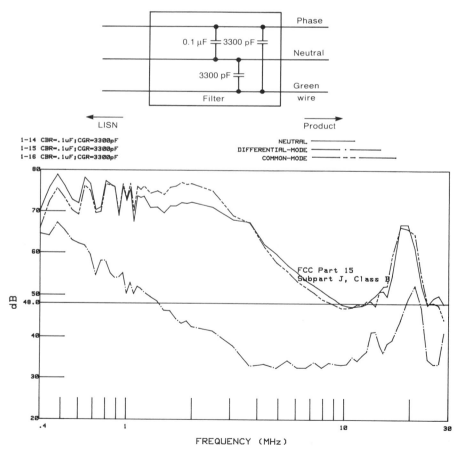

FIGURE 9.18 Measured conducted emissions of a typical product separated into differential- and common-mode components with a 0.1 μF line-to-line capacitor added.

Wire inductor. The results are shown in Fig. 9.19. Observe that the differential-mode components are unchanged, but the common-mode components have been substantially reduced. From Fig. 9.11 this should occur where the impedance of $2L_{GW}$ equals 50 Ω, or approximately 4 kHz. And finally, we add the 28 mH common-mode choke. The results are shown in Fig. 9.20. Here we see that the addition of the common-mode choke does not appear to substantially reduce the common-mode component, yet it drastically reduces the differential-mode component. The reason that it does not seem to reduce the common-mode component, as it was designed to do, is that the Green Wire inductor has already reduced the common-mode component below the noise floor of the spectrum analyzer, so that no further reduction can be seen. The reason that the differential-mode component has been affected is evidently due to the presence

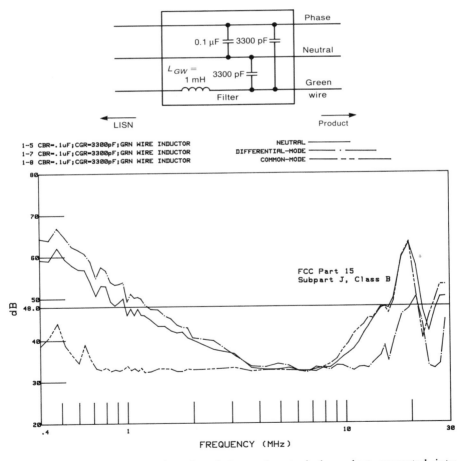

FIGURE 9.19 Measured conducted emissions of a typical product separated into differential- and common-mode components with a Green Wire inductor added.

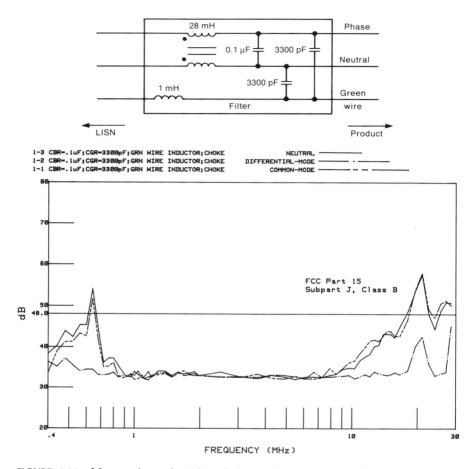

FIGURE 9.20 Measured conducted emissions of a typical product separated into differential- and common-mode components with a common-mode choke added.

of the nonideal leakage inductance of the choke. Therefore we have derived an additional benefit from the nonideal behavior of a component that is not usually desired. Some power supplies include small air-core inductors in Phase and in Neutral to affect the differential-mode current in like fashion. Ferrite core inductors need not be used in this application, since the high-level 60 Hz differential-mode current will saturate the cores, making them no more effective than air core inductors.

9.3 POWER SUPPLIES

The primary source of conducted emissions is generally the power supply of the product. There are some important exceptions to this that we will discuss.

For example, routing wires carrying digital data or clock signals near the output power wires will cause these digital signals to be coupled to the power cord, where they will be measured by the LISN, possibly causing the product to be out of compliance. For the present we will address the noise that is generated in the product's power supply. There are numerous points within a power supply that generate noise measured by the LISN. Each particular type of power supply has unique noise-generating properties. In the previous section we addressed the reduction of conducted emissions by use of a power supply filter. This represents a somewhat "brute force" method of reducing the conducted emissions. However, any power supply filter is only capable of reducing the conducted emissions a certain degree. The most effective method for reducing conducted emissions is to *suppress them at their source*. This should be attempted where possible. But the noise can be reduced at its source only to a certain degree and still retain the functional performance of the supply. Pulses with sharp rise/falltimes have high-frequency spectral content, as was discussed in Chapter 7. Some power supplies such as switched-mode power supplies (switchers) rely on fast-rise/falltime pulses to operate and to reduce power losses in the supply. These types of noise sources can be reduced only to a certain point, so that compromises must be made between retaining the desired functional performance and reduction of the noise source.

It is worthwhile to consider the purpose of the product's power supply. The electronic components of the product (transistors, gates, microprocessors, memory storage, etc.) require dc voltages for proper operation. For example, the digital electronic components require $+5$ V dc for proper operation. This voltage must remain within certain tolerances about the nominal value of 5 V, or the logic function will be impaired. Maintaining this output voltage within certain bounds regardless of the changing load on the power supply as the product performs its required function is an important function of a power supply, and is referred to as *regulation*. Certain linear electronic components such as operational amplifiers, line drivers and receivers, and comparators require dc voltages of ± 12 V. Still other devices within the product, such as dc and stepper motors, require other dc voltages for proper operation. (DC and stepper motors typically require dc voltages of order 30 V.) This process of converting 120 V, 60 Hz commercial ac power to the dc voltage levels required by the product's components and maintaining those levels under varying load conditions are the primary functions of the power supply.

9.3.1 Linear Power Supplies

For many years the *linear power supply* was the predominant method for converting the ac commercial power to the dc voltages needed to power the electronic devices of the product. A typical linear power supply is shown in Fig. 9.21. To begin the discussion, we will disregard the bipolar transistor. A transformer at the input either steps the commercial voltage up or down in magnitude. This is then *rectified* by the two diodes, which form a *full-wave*

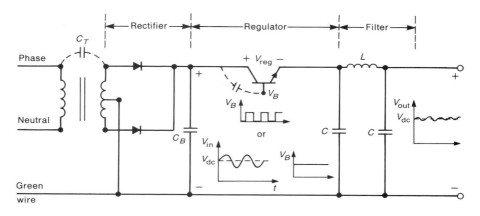

FIGURE 9.21   Illustration of a linear, regulated power supply.

rectifier. The rectifier converts the sinusoidal commercial voltage to pulsating dc [8]. This pulsating dc resembles the ac waveform of the input, except that the negative half-cycles are made positive [8]. This has an average value or, in other words, a dc component V_{DC}. A capacitor C_B (denoting the "bulk capacitor") acts to smooth this pulsating dc, giving an essentially constant waveform V_{in} with some variation and a dc level V_{DC}. A filter shown as a Pi structure consisting of two capacitors C and an inductor L serves to reduce these variations or *ripple* in the output voltage. If we view this waveform in terms of its Fourier components, we see that the filter must be a *low-pass filter* with a cutoff frequency near the fundamental of the pulsating dc waveform (60 Hz) in order to pass the dc level and block the higher harmonics of the waveform. It is not possible to extract only the dc level, so the resulting dc waveform will have a certain amount of *ripple* about the desired dc level. Power supply designs place specifications on the allowed ripple.

If we are satisfied with the level of this dc waveform *and* the load on the supply will remain constant then we do not need the transistor. The transistor allows the output dc level to be changed to some other desired level, and also acts to maintain that output voltage level in the presence of changes in the load on the supply. The process of maintaining a constant output voltage as the output current of the supply (the load) changes is referred to as *regulation*. In order to maintain the desired output voltages under varying load conditions, the transistor acts as a variable resistor to drop a certain voltage across its collector–emitter terminals. A sample of the dc output voltage is feed back to the base terminal of the transistor. If this dc output voltage becomes lower due to increased loading, the transistor is turned on more strongly, resulting in a lower V_{reg} being dropped across its terminals. The output voltage V_{out} of the supply and the output voltage of the rectifier V_{in} are related by

$$V_{out} = V_{in} - V_{reg} \tag{9.19}$$

Thus a reduction in the output voltage is compensated for by less voltage being dropped across the regulating transistor. If the output voltage of the supply increases due to less loading, the transistor drops more voltage across its terminals, reducing the output voltage back to its desired value. Thus regulation is accomplished under varying load conditions. This represents one of the undesirable features of linear power supplies; power is constantly being dissipated in the regulating transistor in order to maintain a constant dc output voltage of the supply.

Linear power supplies tend to be the quietest of all supply types. The switching power supply discussed in the next section is inherently much noisier, but has other advantages over the linear supply from the standpoint of efficiency and lighter weight that make it attractive from a product performance standpoint. From the standpoint of EMC its performance is less desirable than the linear supply, since it is inherently noisier.

9.3.2   Switched-Mode Power Supplies (SMPS)

In this section we will discuss an increasingly popular type of power supply; the *switched-mode power supply (SMPS)*. This is frequently referred to as a "switcher." Linear supplies typically have quite low efficiencies of order 20–40%. Switching power supplies discussed in this section have much higher efficiencies of order 60–90%, which explains their increasing popularity. Switching power supplies also tend to be much lighter in weight than linear power supplies. This is due to the fact that linear supplies require a transformer that will operate efficiently at 60 Hz. Losses due to eddy currents in the transformer are minimized by the use of a large volume of core material. Thus the 60 Hz transformer tends to be heavy. Switching power supplies require transformers to operate at the switching frequency of the supply, which is of order 20–100 kHz. Consequently switching power supplies have transformers that are lighter in weight than those of linear supplies. (There are switchers available that operate at frequencies as high as 1 MHz, which further reduces their required weight.) Therefore switching power supplies are lighter in weight than linear supplies, which is an additional feature that makes them desirable.

There are numerous versions of switchers. However, in order to illustrate the basic principles, we will begin by discussing the basic *buck regulator* shown in Fig. 9.22. A dc voltage V_{DC} is supplied to a switching element. The switching element shown is a p-channel MOSFET. A square-wave pulse train is supplied to the gate of the MOSFET. This waveform has a pulse width τ and a period T that is the inverse of the switching frequency, $f_s = 1/T$. This pulse train turns the MOSFET on and off, which supplies a pulsed voltage of the same duty cycle at V_{in}. This waveform has an average (dc) value given by

$$V_{\text{av}} = \frac{\tau}{T} V_{DC} \qquad (9.20)$$

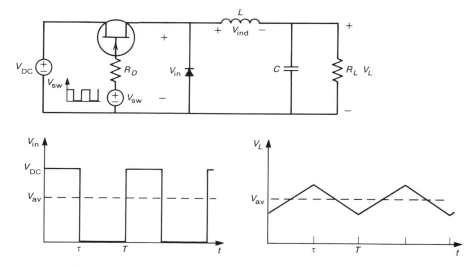

FIGURE 9.22   A simple "buck regulator" switching power supply.

The *duty cycle* of the waveform is the ratio of the pulse width and the period: $D = \tau/T$. Thus this pulsed waveform has an average or dc value that can be changed by changing the duty cycle of the switching waveform that is applied to the gate of the switching element. The L and C form a low-pass filter to remove the dc (desired) component of the waveform, and the diode provides a path to discharge the capacitor when the MOSFET turns off. The cycle of operation is described as follows. When the MOSFET is turned on, the diode is open and the dc voltage is applied to the filter. The inductor begins to store energy in its magnetic field, and the capacitor begins to charge up. When the MOSFET is turned off, the voltage across the inductor, V_{ind}, reverses polarity according to Faraday's law, and the diode closes. The circuit discharges through the diode, giving the waveform across the load resistor shown in Fig. 9.22, which has the desired average or dc value. The advantage of this switcher over a linear regulator is that the switch element, the MOSFET, is either turned full on or full off, and as such dissipates very little power as opposed to the linear regulator, where the transistor is always operated in its linear region, dissipating more power. Regulation is accomplished by simply varying the duty cycle of the switching waveform that is applied to the gate of the MOSFET. Typically a sample of the output voltage V_L is fed to a pulse width modulator (PWM). The output of the PWM is a square wave, which is fed to the gate of the MOSFET. The duty cycle of this waveform is varied by the PWM to give the desired dc output voltage in response to changes in load. Other more practical switchers employ this same basic principle of chopping a dc waveform and varying the duty cycle of the waveform to provide regulation.

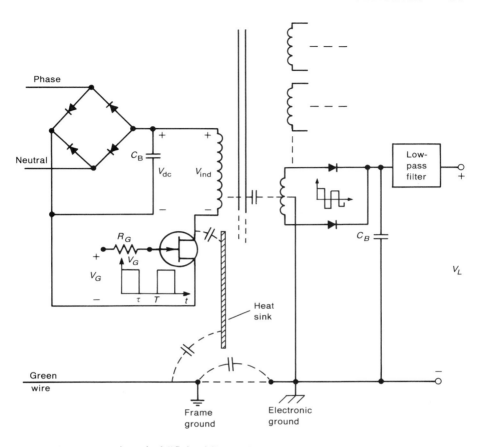

FIGURE 9.23   A typical "flyback" or primary-side switching power supply.

There are two general types of switchers: the primary side switcher and the secondary side switcher. These designations refer to whether the switching occurs on the primary or secondary side of the transformer. For example, the linear regulator shown in Fig. 9.21 can be changed into a *secondary side switcher* by simply supplying a square-wave waveform to the base of the transistor instead of the dc waveform that was used with the linear supply. Regulation is accomplished by varying the duty cycle of this switching waveform in response to changes in the output voltage of the supply. A primary side switcher is shown in Fig. 9.23. The version shown is often referred to as a *flyback converter*. A *fullwave bridge rectifier* rectifies the ac commercial power waveform and produces a pulsating dc waveform, which is smoothed by the bulk capacitor C_B to provide an essentially constant waveform that has the value of the peak commercial voltage waveform. This is applied to a transformer that has multiple "taps" or windings on its secondary. A switching element (usually a power MOSFET) opens or closes the connection to the transformer primary. A variable duty cycle square-wave waveform is applied to the gate of this switching element. Varying the duty cycle of this waveform provides regulation of the output voltages of the supply. Ordinarily a resistor R_G is placed in series with the gate of the MOSFET switch. The effect of this resistor is to change the rise/falltimes of the pulses that are applied to the gate. Increasing the value of this resistor "rounds the sharp edges" of the waveform. This causes the MOSFET to spend more time in the active region, which increases its power dissipation; an undesirable result from the standpoint of thermal considerations. Increasing the rise/falltimes of the pulsed waveform is desirable from the standpoing of EMC, since the spectral content of the noise that is produced by the switching is directly dependent on these rise/falltimes. So there is an apparent tradeoff between reduction of the noise spectral content that will contribute to conducted emissions and the thermal heating of the switching element and the related efficiency of the switcher. A heatsink is usually connected to the body of the MOSFET, which acts to dissipate this heat. Ordinarily this heatsink is not directly connected to the MOSFET, but is insulated from it by a dielectric washer. This produces a parasitic capacitance between the MOSFET and the heatsink. If the heatsink is attached to the Green Wire for safety reasons, a path for common-mode currents is provided.

At the secondary side a pulsating waveform that has alternating polarity pulses is produced. A fullwave rectifier rectifies this waveform, which is smoothed by the bulk capacitor and filtered by the low-pass filter. Because of the ability to have multiple taps on the secondary, numerous dc voltages of different levels can be obtained. For example, we may produce $+5$ V, $+12$ V, -12 V, $+38$ V, etc. Regulation of one of these voltages is accomplished on the primary side of the transformer. Because the transformer is only required to carry the switching frequency (and all its higher harmonics), it can be designed to be smaller and lighter in weight than the 60 Hz transformer. Efficiency is increased over the linear supply due to the switching action (unless the gate resistor is increased to reduce noise emissions). Observe that a major disadvantage of the primary side switcher is that the switching harmonics are fed directly out the product's ac power cord through the bridge rectifier. There is no 60 Hz transformer present to provide any filtering. Any reduction of this noise that cannot be reduced by changing components in the switcher such as R_G must be eliminated by the power supply filter. This places a larger demand on the necessity for and the careful design of the power supply filter.

9.3.3 Effect of Power Supply Components on Conducted Emissions

As was pointed out earlier, *the most efficient method for the reduction of conducted emissions is to reduce them at their source.* For example, increasing the value of the gate resistor R_G in the primary side switcher shown in Fig. 9.23 will increase the rise/falltimes of the switching waveform and hence reduce their spectral content. However, these rise/falltimes can only be increased to a point, because the switching device will spend more time in its active region, which increases its power dissipation.

There are other "noise sources" present in the switcher that should be controlled. One of the major ones comes from the diodes that are used for rectification—particularly the diodes that rectify the switched signal, such as those on the secondary side of the primary side switcher shown in Fig. 9.23. When a diode is forward-biased, charge is stored at its junction in the junction capacitance. Also, charge carriers in one region are injected into the other region. When the diode is turned off, this charge must be removed. This results in the diode current having the waveform shown in Fig. 9.24(a). The diode current passes through zero as the charge is being removed from the junction. Some diodes known as "fast-recovery diodes" snap off sharply. This is referred to as "hard recovery" and is illustrated in Fig. 9.24(a). Other types have soft recoveries where the diode current returns to zero gradually. Obviously, hard-recovery diodes will yield higher-frequency spectra in the current than will soft-recovery diodes due to the sharp edge of the current waveform when the diode current returns to zero. From the standpoint of efficiency, fast-recovery diodes are more desirable than slow-recovery diodes. In order to reduce this undesirable noise generated in the turnoff of the diode, *snubber circuits* illustrated in Fig. 9.24(b) are generally placed in parallel with the diodes (see Section 6.12.3). The snubber circuit consists of a resistor in series with a capacitor that acts as a path to discharge the charge stored at the diode junction when the diode turns off. This tends to smooth the diode current waveform, thereby reducing its high-frequency spectral content. Obviously high-frequency currents will circulate through the snubber circuit, so its leads should be kept short and the elements placed very close to the diode.

Parasitic capacitances provide numerous current paths for noise currents to exit the switcher and contribute to the product's conducted emissions. Capacitance between the switching element and its heatsink will couple switching noise at the element to the heatsink, where it may radiate or be conducted out the line cord. Transformers are used in several places in electronic products. The primary

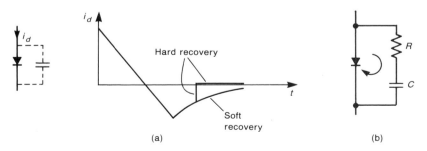

FIGURE 9.24 Illustration of nonideal effects in diodes: (a) various recovery characteristics; (b) an RC "snubber".

use is in the *power supply* for the product. As discussed above, *linear power supplies* use a transformer to *step down* the 120 V, 60 Hz ac (240 V, 50 Hz ac in Europe) to voltage levels that are *rectified* to produce the dc voltages that power the electronic devices in the product. These types of transformers are of necessity heavy and large in order to reduce the *eddy-current* losses in the core caused by the 60 Hz frequency of the primary ac power. *Switching power supplies* chop this primary power at much higher frequencies (25–100 kHz) to produce the required dc voltages, as discussed earlier. Because of the much higher frequency of the flux in the core, transformers in switching power supplies have much lower eddy-current losses than 60 Hz transformers, and can be designed to be much smaller and lighter.

All transformers make use of Faraday's law of induction to change one voltage level to another. Two coils are wound on the same ferromagnetic core as shown in Fig. 9.25(a). The voltage v_p across the *primary* generates a current and associated magnetic flux in that winding. Because of the much lower reluctance of the ferromagnetic core compared with that of the surrounding air, the majority of this flux couples to the *secondary*, inducing, by Faraday's law, a voltage v_s across that winding. The ratio of the two voltages is (ideally) proportional to the ratio of the numbers of the turns in each coil:

$$\frac{v_p}{v_s} = \frac{n_p}{n_s} \qquad (9.21)$$

There are a number of ways to construct these transformers. The 60 Hz transformers in linear supplies are usually constructed from laminations of steel stacked together. The laminations tend to break up the magnetic flux paths and reduce the eddy current losses. High-frequency transformers are usually constructed from ferrite cores. There are several transformer core configurations that are different from the simplified version shown in Fig. 9.25(a) in order to facilitate their assembly. The 60 Hz transformers are constructed from an "I-core" of laminations, with the primary and secondary windings wound on top of each other as shown in Fig. 9.25(b). Outer vertical legs of laminations are connected to complete the flux path. Transformers designed to operate at higher frequencies such as in switching power supplies typically use powdered ferrite cores. These are formed into "bobbins" and are available in various configurations. A common one is the "E-core," in which the two halves are shaped like the letter E. The primary and secondary windings are wound on the center legs of each core, and the two halves are placed together to provide a continuous path for the magnetic flux as shown in Fig. 9.25(c). Ferrite cores used in these types of transformers are also subject to *saturation* due to high levels of flux caused by high current levels. In order to limit this saturation, thin, plastic spacers are placed in the air gaps when the cores are placed together. This acts to increase the reluctance of the magnetic path, which reduces the flux levels and prevents saturation of the core.

Winding the coils on top of each other (lap winding) as with the I-core of

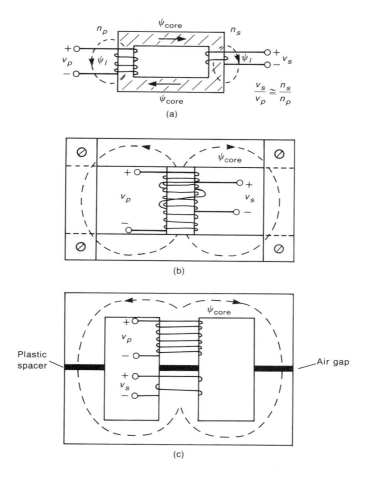

FIGURE 9.25 Construction of transformers: (a) transformer schematic; (b) a 60 Hz transformer; (c) an "E-core" switching power supply transformer.

the 60 Hz transformer introduces a *parasitic capacitance* between the primary and secondary as shown in Fig. 9.26(a). This primary-to-secondary capacitance can introduce an undesired coupling that allows noise on the secondary side (the electronics side) to be more easily coupled to the primary side (where the system power cord is attached). Once the noise is present on the primary side, it passes out through the power cord and is measured as a conducted emission by the LISN unless a power supply filter is inserted between the power cord and the transformer. Evidently the efficiency of this coupling due to the parasitic primary-to-secondary capacitance increases at higher noise frequencies. For example, the system processor clock in a digital system (typically greater than 10 MHz) would couple to the primary more easily than would switcher

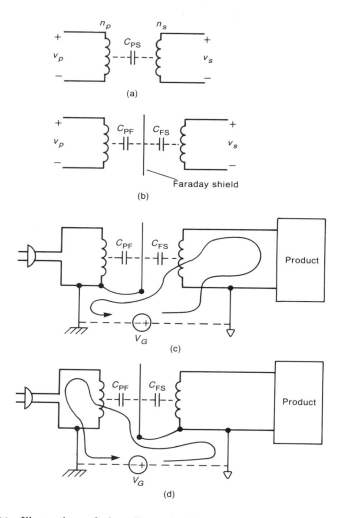

FIGURE 9.26 Illustration of the effect of primary-to-secondary capacitance of a transformer: (a) lumping the parasitic capacitance; (b) use of a Faraday shield; (c) proper connection of the Faraday shield to reduce conducted emissions; (d) improper connection of the Faraday shield.

harmonics. In order to reduce this coupling, the primary and secondary coils that are wound on top of each other may have a cylindrical, metallic shield inserted between them. This is referred to as a *Faraday shield*. This effectively breaks up the primary-to-secondary capacitance, as is illustrated in Fig. 9.26(b). The question always arises as to which side of the transformer, primary or secondary, should we connect the Faraday shield. The answer depends on where we wish to divert any noise current produced by the difference in potentials of the primary and secondary grounds. The difference between the potentials of the secondary and primary grounds are represented by the voltage source V_G. Figure 9.26(c) shows connection of the shield to the primary. For this connection the noise current produced by V_G flows through C_{FS}, the shield, and back through the ground. This is the preferred connection of the shield, since the noise current will not flow through the LISN to produce conducted emission problems. Figure 9.26(d) shows the connection of the shield to the secondary. This connection would cause the noise current to flow through the shield, through C_{PF}, out through the line cord, and back through V_G. Thus the noise currents would flow through the LISN, possibly resulting in conducted emission problems. The same principle applies to a transformer between an amplifier and a receiver. In order to prevent the noise currents from flowing through the receiver input, the shield should be grounded at the receiver input. Evidently, this desired operation depends also on connecting the Faraday shield to the *grounded side* of the transformer. Proper connection of the Faraday shield is important in reducing the effects of the common-mode voltage between primary and secondary. Ideally this is what happens, but realistically we observe less than ideal performance due to the resistance and inductance of the shield.

Ferrite beads tend to be effective in blocking noise currents in power supplies. Recall from Chapter 6 that ferrite beads typically have maximum values of impedance of the order of a few hundred ohms. Therefore, in order for them to be effective, they must be in series with impedances that are no larger than the bead impedance, since otherwise the bead impedance would be overshadowed by this larger impedance. The intent is to use the bead to *block noise currents* by adding a significant impedance to the path. Circuit impedances tend to be small in power supplies as opposed to other electronic circuits. Therefore insertion of a bead tends to provide a significant increase in the circuit impedance in power supply circuits.

9.4   POWER SUPPLY AND FILTER PLACEMENT

Location of components and routing of wires within a product are important considerations in the reduction of conducted and radiated emissions of the product. Being cognizant of some simple notions with regard to the location of components and the routing of wires in a product design ("packaging of the product") can be an essentially cost-free method of reducing conducted and radiated emissions. A common example is the location of the power supply and its associated power supply filter. *The power supply filter should be placed directly at the exit of the power cord from the product.* The power supply should also be placed as close to the filter as possible. The result of not doing this is illustrated in Fig. 9.27(a). There is a wide spectrum of high-frequency noise signals present inside the product. These range from the switching power supply harmonics to harmonics of the clocks used to drive the system's digital elements and processors. Thus the spectral content ranges from 20 kHz to well over

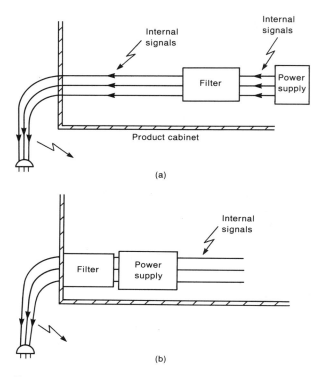

FIGURE 9.27   Illustration of (a) poor filter placement and (b) proper filter placement in the reduction of conducted emissions.

500 MHz. These signals tend to be near-field type, and die out rapidly with distance from the source. Nevertheless, the higher the frequency of the signals, the more efficiently are they coupled to the wires inside the product as we observed in the previous chapter. If the power supply filter is not placed very close to the exit on the power cord, these signals may couple to those wires and exit the product on its power cord. They will therefore be measured as conducted and/or radiated emissions. Note that these signals have bypassed the power supply filter, and as such the filter provides no protection against these signals, causing possible regulatory compliance problems. If the power supply is also placed distant from the power supply filter, these signals may couple to the wires between the filter and the supply. We might expect that the filter would serve to attenuate these. This is not always the case. The power supply filter was designed to attenuate signals in the conducted emission frequency range (450 kHz–30 MHz for the FCC limits). Suppose the system clock has a frequency of 20 MHz. It is doubtful that the power supply filter will provide any filtering to the fundamental frequency of this clock signal, much less to, say, its fifth harmonic of 100 MHz. Once again we must not allow

ourselves to overlook these types of undesired performance of the components of the product. Once the clock signal harmonics are present on the product's power cord they will be measured as conducted emissions (20 MHz) and radiated emissions (40 MHz, 60 MHz, 80 MHz, ...). If the power supply filter and the power supply are placed as close to the power cord exit as possible, as shown in Fig. 9.27(b), this will minimize the coupling of internal signals to their connection wires and will place a considerable number of possibilities for attenuating these signals in their path.

9.5 CONDUCTED SUSCEPTIBILITY

As we have discussed at several points, the regulatory requirements on conducted emissions are intended to control the radiated emissions from noise currents that are placed on the local power distribution system by conduction out through the product's ac power cord. These signals are generally too small to cause interference by direct conduction from the power net into another product via its ac power cord. However, large transients placed on the power distribution net by such phenomena as lightning strokes can cause EMC problems in a product by direct conduction into that product's ac power cord. Equipment manufacturers realize this, and test their products' conducted susceptibility by directly injecting typical such disturbances on the product's ac power cord to insure that the product will operate satisfactorily through such disturbances.

It goes without saying that a well-designed power supply and its proper placement will tend to provide *some protection* to these types of signals on the ac power cord. However, since the spectral content and signal level of these disturbances tends to be larger than the typical conducted emissions, the power supply filter may not be as effective as desired in this case. Testing will generally reveal when additional protection is required.

REFERENCES

[1] FCC, *FCC Methods of Measurement of Radio Noise Emissions from Computing Devices*, FCC/OST MP-4 (July 1987).

[2] *Specifications for Radio Interference Measuring Apparatus and Measurement Methods*, Amendment #1 CISPR Publication 16 (1980).

[3] H.W. Ott, *Noise Reduction Techniques in Electronic Systems*, second edition, John Wiley Interscience, NY (1988).

[4] R.K. Keenan, *Digital Design for Interference Specifications*, The Keenan Corporation, Pinellas Park, FL (1983).

[5] *Electromagnetic Interference Characteristics, Measurement of*, MIL-STD-462 (July 1967).

[6] J.R. Nicholson and J.A. Malack, RF impedance of power lines and line impedance stabilization networks in conducted interference measurements, *IEEE Trans. on Electromagnetic Compatibility*, **EMC-15**, 84–86 (1973).

[7] C.R. Paul and K.B. Hardin, Diagnosis and reduction of conducted noise emissions, *IEEE Trans. on Electromagnetic Compatibility*, **EMC-30**, 553–560 (1988).

[8] C.R. Paul, S.A. Nasar, and L.E. Unnewehr, *Introduction to Electrical Engineering*, McGraw-Hill, NY (1986).

PROBLEMS

9.1   In order to illustrate that the LISN essentially presents 50 Ω impedances between Phase and ground and between Neutral and ground, use SPICE to plot the frequency response of the impedance looking into one side of the LISN shown in Fig. P9.1 over the frequency range of the FCC test for open- and for short-circuit loads (impedance extremes looking into the commercial power outlet). Determine the values at each end of the frequency band. [Short-circuit: 46 Ω and 47 Ω. Open-circuit: 46 Ω and 48 Ω]

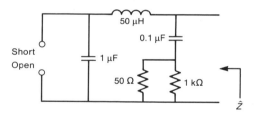

FIGURE P9.1

9.2   In order to investigate the potential for currents on the ac power cord to impact radiated emissions, consider the typical radiated emission test shown in Fig. P9.2. The product has a clock of 10 MHz, and common-mode currents at the harmonics of this clock are inadvertently coupled to the ac power cord. Treat the power cord as a monopole antenna above ground and determine the level of current I_C that will barely exceed the FCC Class B limit at the seventh and ninth harmonics of the clock. Compare the exact expressions derived assuming sinusoidal current distribution and the approximate expression derived assuming a constant current distribution, which were derived in the previous chapter. To simplify this, compute the fields at the surface of the ground plane. [70 MHz: exact, 5.582 μA; approx., 3.409 μA. 90 MHz: exact, 5.732 μA; approx., 3.968 μA]

9.3   The FCC conducted emission limit is 48 dBμV (251 μV). Assume either common-mode or differential-mode currents are dominant and compute the current that corresponds to this limit. [5 μA]

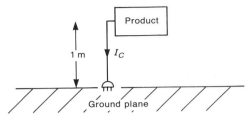

FIGURE P9.2

9.4   Suppose a Green Wire inductor has an inductance of 1 mH and a parasitic capacitance of 10 pF. Determine the resonant frequency of this inductance and its impedance at 30 MHz. [1.6 MHz, 532 Ω]

9.5   Determine an equation for the insertion loss of the high-pass filter shown in Fig. 9.7(b). [$IL_{dB} = 10 \log_{10}\{[1 + (\omega\tau)^2]/(\omega\tau)^2\}$, $\tau = C(R_S + R_L)$]

9.6   Determine an equation for the insertion loss of the bandpass filter shown in Fig. 9.7(c). [$IL_{dB} = 10 \log_{10}\{1 + [(\omega/\omega_0)^2 - 1]^2/(\omega\tau)^2\}$, $\tau = C(R_S + R_L)$, $\omega_0 = 1/\sqrt{LC}$]

9.7   Determine an equation for the insertion loss of the band-reject filter shown in Fig. 9.7(d). [$IL_{dB} = 10 \log_{10}\{1 + (\omega\tau)^2/[(\omega/\omega_0)^2 - 1]^2\}$, $\tau = L/(R_S + R_L)$, $\omega_0 = 1/\sqrt{LC}$]

9.8   Use SPICE to plot the frequency response of the differential-mode insertion loss for a typical power supply filter shown in Fig. P9.8(a). This corresponds to the test shown in Fig. 9.8(a). Repeat this for the common-mode insertion loss test shown in Fig. P9.8(b). Determine the insertion loss at 450 kHz and 30 MHz for both tests. [DM: 79.7 dB and 186.7 dB. CM: 63.4 dB and 167.1 dB]

9.9   Suppose a common-mode choke has self inductances of 28 mH and a coupling coefficient of 0.98. Determine the leakage inductance presented to differential-mode currents. [560 μH] Repeat this for a coupling coefficient of 0.95. [1.4 mH]

9.10 The effect of a power-supply filter on common-mode currents is sometimes characterized by the common-mode impedance as $\hat{Z}_C = \hat{V}_C/\hat{I}_C$ with reference to Fig. 9.11. For $L_{GW} = 1$ mH, $L = 28$ mH, $k = 0.98$, and $C_{CL} = C_{CR} = 3300$ pF use SPICE to plot this from 450 kHz to 30 MHz. Compute this at 450 kHz and 30 MHz. [−64 dBΩ, −211 dBΩ] Repeat this for the differential-mode impedance with respect to Fig. 9.12 with $C_{DR} = C_{DL} = 0.1$ μF. [−54 dBΩ, −165 dBΩ]

9.11 For the specifications of Problem 9.10 suppose that the common-mode and differential-mode currents at the input to the power-supply filter at 450 kHz are equal to 1 mA. Determine the total received voltage across the LISN 50 Ω and which component is dominant. [Diff: 8.4 dBμV]

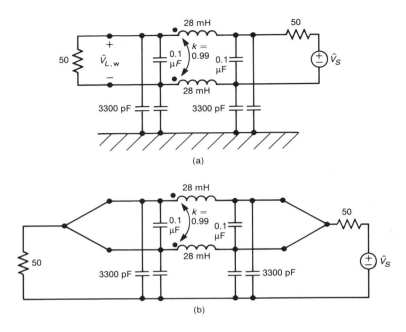

(a)

(b)

FIGURE P9.8

9.12 The buck regulator shown in Fig. 9.22 is to be operated at a switching frequency of 50 kHz, an input voltage of 100 V, and an output voltage of 5 V. Determine the required duty cycle. [0.05] Calculate the level of the 25th harmonic (1.25 MHz) of the ideal chopped voltage V_{in}. (Use the asymptotic approximations developed in Chapter 7 and the exact formula.) [Approx.: 128.1 dBμV. Exact: 125.1 dBμV]

CHAPTER TEN

███████████

Crosstalk

In this chapter we will discuss another important aspect of the design of an electromagnetically compatible product—*crosstalk*. This essentially refers to the *unintended electromagnetic coupling between wires and PCB lands that are in close proximity*. Although the phenomenon is due to the currents and voltages of the wires and is thus similar to the problem of antenna coupling, crosstalk is distinguished from the latter in that is a *near-field coupling problem*. Crosstalk between wires in cables or between lands on PCBs concerns the *intrasystem interference performance* of the product; that is, the *source* of the electromagnetic emission and the *receptor* of this emission are *within the same system*. Thus this reflects the third concern in EMC: *the design of the product such that it does not interfere with itself.*

There are also cases where crosstalk can affect the radiated and/or conducted emissions of the product. Suppose that a ribbon cable internal to a product is placed in close proximity to wires that connect to a peripheral cable that exits the product. Crosstalk between the two cables can induce signals on the peripheral cable that may radiate externally to the product, causing the latter to be out of compliance with the radiated emission regulatory limits. If this internal coupling occurs to the power cord of the product, these coupled signals may cause it to fail the conducted emission regulatory requirements. Crosstalk can also affect the susceptibility of a product to emissions from another product. For example, emissions from some other product that are coupled to a peripheral cable of this product may couple, internal to the latter, to some other cable internal to it where the susceptibility to this signal may be enhanced.

In order to understand how to model crosstalk, it is important to understand the analysis of two-conductor transmission lines. For a two-conductor transmission line there is no crosstalk. *In order to have crosstalk, we must have three or more conductors.* However, the notions involved in two-conductor transmission-line theory carry over to a large degree to the case of multiconductor transmission lines and simplify the understanding of the behavior of those lines. It is for this reason that the reader should review conventional two-conductor transmission line theory given in Chapter 4 and in any undergraduate text on electromagnetic fields, such as [1].

10.1   THREE-CONDUCTOR LINES AND CROSSTALK

Virtually all of the techniques developed in Chapter 4 for the analysis of two-conductor transmission lines can be directly extended to the case of coupled transmission lines that consist of any number of parallel conductors [2]. These types of transmission lines are referred to as multiconductor transmission lines or MTLs. In this section we will consider the first logical extension of the two-conductor line results: a three-conductor transmission line.

10.1.1   Time-Domain Crosstalk versus Frequency-Domain Crosstalk

Adding a third conductor to the two-conductor system provides the possibility of generating interference between the circuits attached to the ends of the line conductors resulting from *crosstalk*. In order to illustrate this important phenomenon, consider the three-conductor line shown in Fig. 10.1. A source consisting of a source resistance R_S and a source voltage $V_S(t)$ is connected to a load R_L via a *generator conductor* and a *reference conductor*. Two other terminations, represented by resistors R_{NE} and R_{FE}, are also connected by a *receptor conductor* and this reference conductor. These terminations represent the input circuitry looking into the terminals of the terminations. Linear, resistive terminations will be shown for illustration, but all results that we will develop will hold for more general terminations, which may include capacitors and/or inductors. The line conductors are assumed to be parallel to the z axis and are of uniform cross section along the line. We will also assume that any surrounding

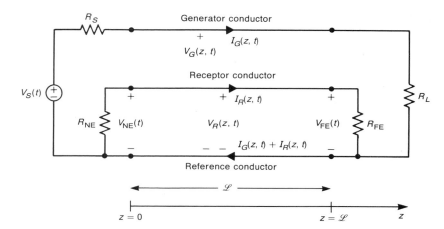

FIGURE 10.1   The general three-conductor transmission line, illustrating crosstalk.

dielectric inhomogeneities also have uniform cross sections along the line axis, so that the lines we will consider will be *uniform lines*. The *generator circuit* consists of the generator conductor and the reference conductor and has current $I_G(z, t)$ along the conductors and voltage $V_G(z, t)$ between them. The current and voltage associated with the generator circuit will generate electromagnetic fields that interact with the *receptor circuit*, which consists of the receptor conductor and the reference conductor. This interaction will induce current $I_R(z, t)$ and voltage $V_R(z, t)$ along the receptor circuit. This induced current and voltage will produce voltages $V_{NE}(t)$ and $V_{FE}(t)$ at the input terminals of the terminations that are attached to the ends of the receptor circuit. The subscripts *NE* and *FE* refer to "near end" and "far end", respectively, with reference to the end of the line adjacent to the end of the generator circuit that contains the driving source $V_S(t)$.

The objective in a crosstalk analysis is to determine (predict) the near-end and far-end voltages $V_{NE}(t)$ and $V_{FE}(t)$ given the line cross-sectional dimensions, and the termination characteristics $V_S(t)$, R_S, R_L, R_{NE}, and R_{FE}. There are two types of analysis that we might be interested in: time-domain analysis and frequency-domain analysis. *Time-domain crosstalk* analysis is the determination of the time form of the receptor terminal voltages $V_{NE}(t)$ and $V_{FE}(t)$ for some general time form of the source voltage $V_S(t)$. *Frequency-domain crosstalk* analysis is the determination of the magnitude and phase of the receptor terminal phasor voltages $\hat{V}_{NE}(j\omega)$ and $\hat{V}_{FE}(j\omega)$ for a sinusoidal source voltage $V_S(t) = V_S \cos(\omega t + \phi)$. Frequency-domain analysis also presumes a steady state, i.e., the source has been attached a sufficient length of time that the transient response has decayed to zero. Of course, these notions are the same as for two-conductor lines, but here we are interested in voltages and currents that are generated in another circuit.

Some typical three-conductor transmission lines to which this analysis will apply are illustrated in Fig. 10.2. Figure 10.2(a) shows a three-conductor line consisting of three wires. A wire serves as the reference conductor for both circuits. Figure 10.2(b) shows two wires above an infinite ground plane that serves as the reference conductor for this configuration. Figure 10.2(c) shows two wires within an overall shield that serves as the reference conductor. For simplicity, we will assume that the medium surrounding the conductors for these configurations in *homogeneous*. Thus dielectric insulations around the wires will be ignored. Again, we will need to obtain the per-unit-length parameters for these configurations in order to characterize them. We will show some simple, closed-form approximate results for the configurations of Fig. 10.2 under the assumption of widely spaced conductors and a homogeneous surrounding medium. It is not possible to obtain exact expressions for the required per-unit-length parameters if either of these restrictions is relaxed. Numerical methods can be employed to obtain these [2, 3].

Some other configurations typically found with PCBs are shown in Fig. 10.3. The configuration shown in Fig. 10.3(a) consists of two parallel lands on a dielectric board that has a ground plane on the opposite side. This is commonly

Cable (Wire) Configurations

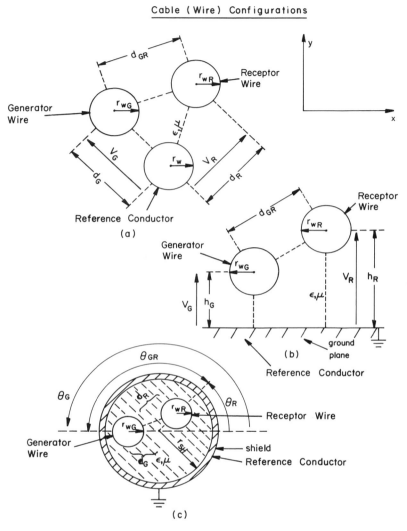

FIGURE 10.2 Wire-type line cross sections whose reference conductors are (a) another wire, (b) a ground plane, or (c) an overall, cylindrical shield.

referred to as the *coupled microstrip* in the microwaves circuit literature [4, 5], and might be used to model an innerplane or multilayer board. The configuration shown in Fig. 10.3(b) consists of three parallel lands on a board. We might refer to this configuration as the *coupled coplanar strips* configuration. In both these configurations the surrounding medium is inhomogeneous, since the fields exist partly in the board dielectric and partly in the surrounding air. The analysis models that we will obtain also apply to the inhomogeneous medium lines in

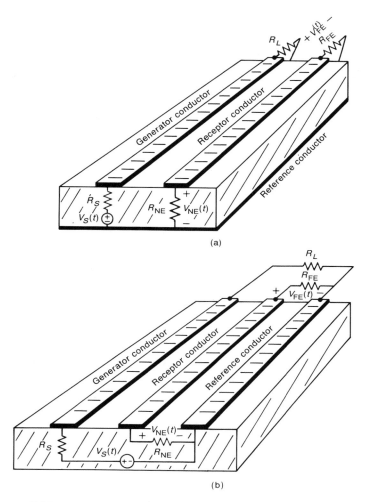

FIGURE 10.3   PCB transmission lines whose reference conductors are (a) a ground plane (coupled microstrip) and (b) another land (coupled coplanar strips).

Fig. 10.3. However, there is an important difference between the analysis of the lines in a homogeneous medium in Fig. 10.2 and the lines in an inhomogeneous medium in Fig. 10.3. Closed-form expressions for the per-unit-length parameters for the lines in a inhomogeneous medium shown in Fig. 10.3 are difficult to determine, whereas we will derive some simple but approximate expressions for the per-unit-length parameters for the lines in a homogeneous medium shown in Fig. 10.2. Generally, numerical methods may be used to obtain the per-unit-length parameters for the configurations in Fig. 10.3 [4, 5]. So the solution of the transmission-line equations for the configurations in Figs. 10.2

and 10.3 is the same; obtaining the per-unit-length parameters makes the application different and possibly difficult. Obtaining the necessary per-unit-length parameters represents the primary problem in the analysis of transmission lines in inhomogeneous media.

10.1.2   The Transmission-Line Equations

The fundamental assumption involved in the analysis of all multiconductor transmission lines is, again, that the *transverse electromagnetic* (*TEM*) mode of propagation is the only field structure present on the line. The TEM field structure assumes that both the electric and the magnetic field vectors lie in the transverse (xy) plane perpendicular to the line (z) axis, i.e., the electric and magnetic fields do not have a component along the line axis. Under the TEM field structure assumption, line voltages $V_G(z, t)$ and $V_R(z, t)$, as well as line currents $I_G(z, t)$ and $I_R(z, t)$, can be uniquely defined for excitation frequencies other than dc. (See Chapter 4 for a proof of this important result.) The total current flowing to the right at any line cross section is zero, so that the currents return through the reference conductor as shown in Fig. 10.1. Furthermore, the TEM field structure is identical to a static (dc) one. This allows the determination of the per-unit-length parameters of inductance and capacitance strictly from dc methods in the cross-section (xy) plane. The pure TEM field structure cannot exist for (1) imperfect line conductors or (2) an inhomogeneous surrounding medium [2]. Nevertheless, for either case the deviation from a TEM field structure is typically small for "good conductors," typical line cross-sectional dimensions, and frequencies below the GHz range. This is referred to as the *quasi-TEM mode assumption*, and will be assumed in our future analyses.

Once again, for the TEM field structure line voltages and currents can be determined uniquely, since the TEM field structure is identical to the field structure for static (dc) excitation. This simplifies the determination of the per-unit-length parameters, since we need only use static field analysis techniques in order to determine those parameters. As in the case of two-conductor lines, we may construct a per-unit-length equivalent circuit for the three-conductor line as shown in Fig. 10.4. The per-unit-length parameters are multiplied by the section length Δz to give the total parameter representing that section. Each conductor is characterized by a per-unit-length resistance r_G, r_R, or r_0, where r_0 is associated with the reference conductor. The generator and receptor circuits have per-unit-length self inductances l_G and l_R associated with them, and a per-unit-length mutual inductance l_m between the two circuits. These inductances characterize the effects of the magnetic fluxes created by the line currents. The electric fields produced by the line voltages generate currents (conduction and displacement) flowing between the line conductors. These are represented by the self elements g_G, c_G, g_R, and c_R, and the mutual elements g_m and c_m. Per-unit-length conductances (due to conduction current in the medium) are denoted as g, and per-unit-length capacitances (due to displacement current in the medium) as c. The transmission-line equations can again be derived by

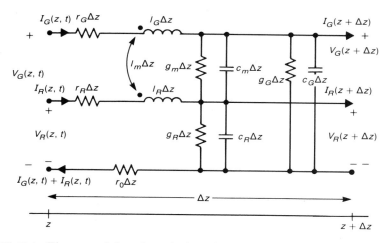

FIGURE 10.4 The per-unit-length equivalent circuit of the TEM mode of propagation on a three-conductor transmission line.

relating the voltages and currents at the two ends of this section and letting the section length Δz go to zero. These become [2]

$$\frac{\partial V_G(z, t)}{\partial z} = -(r_G + r_0)I_G(z, t) - r_0 I_R(z, t) - l_G \frac{\partial I_G(z, t)}{\partial t} - l_m \frac{\partial I_R(z, t)}{\partial t} \qquad (10.1a)$$

$$\frac{\partial V_R(z, t)}{\partial z} = -r_0 I_G(z, t) - (r_R + r_0)I_R(z, t) - l_m \frac{\partial I_G(z, t)}{\partial t} - l_R \frac{\partial I_R(z, t)}{\partial t} \qquad (10.1b)$$

$$\frac{\partial I_G(z, t)}{\partial z} = -(g_G + g_m)V_G(z, t) + g_m V_R(z, t) - (c_G + c_m)\frac{\partial V_G(z, t)}{\partial t} + c_m \frac{\partial V_R(z, t)}{\partial t}$$
$$(10.1c)$$

$$\frac{\partial I_R(z, t)}{\partial z} = g_m V_G(z, t) - (g_R + g_m)V_R(z, t) + c_m \frac{\partial V_G(z, t)}{\partial t} - (c_R + c_m)\frac{\partial V_R(z, t)}{\partial t}$$
$$(10.1d)$$

These partial differential equations can be more compactly displayed in matrix form as

$$\frac{\partial}{\partial z}\mathbf{V}(z, t) = -\mathbf{R}\mathbf{I}(z, t) - \mathbf{L}\frac{\partial}{\partial t}\mathbf{I}(z, t) \qquad (10.2a)$$

$$\frac{\partial}{\partial z}\mathbf{I}(z, t) = -\mathbf{G}\mathbf{V}(z, t) - \mathbf{C}\frac{\partial}{\partial t}\mathbf{V}(z, t) \qquad (10.2b)$$

where

$$\mathbf{V}(z, t) = \begin{bmatrix} V_G(z, t) \\ V_R(z, t) \end{bmatrix} \tag{10.3a}$$

$$\mathbf{I}(z, t) = \begin{bmatrix} I_G(z, t) \\ I_R(z, t) \end{bmatrix} \tag{10.3b}$$

The per-unit-length parameter matrices are

$$\mathbf{R} = \begin{bmatrix} r_G + r_0 & r_0 \\ r_0 & r_R + r_0 \end{bmatrix} \quad (\text{in } \Omega/\text{m}) \tag{10.4a}$$

$$\mathbf{L} = \begin{bmatrix} l_G & l_m \\ l_m & l_R \end{bmatrix} \quad (\text{in } \text{H}/\text{m}) \tag{10.4b}$$

$$\mathbf{G} = \begin{bmatrix} g_G + g_m & -g_m \\ -g_m & g_R + g_m \end{bmatrix} \quad (\text{in } \text{S}/\text{m}) \tag{10.4c}$$

$$\mathbf{C} = \begin{bmatrix} c_G + c_m & -c_m \\ -c_m & c_R + c_m \end{bmatrix} \quad (\text{in } \text{F}/\text{m}) \tag{10.4d}$$

Observe that the form of these equations is identical to those for two-conductor lines considered in Chapter 4, with the addition of matrix notation. In fact, this important observation shows that the form of the transmission line equations for more than three lines (multiconductor transmission lines or MTLs) is the same. This observation provides a general method of solution for MTLs [2].

Solution of the above MTL equations in the time domain is usually a difficult problem. We will develop an exact solution for lossless lines, $\mathbf{R} = \mathbf{G} = \mathbf{0}$, that is easily implementable in SPICE and uses the two-conductor exact solution model that is already present in SPICE. We will also develop a simple, approximate solution that is valid for electrically short lines and is suitable for hand calculations.

The *exact* solution of the above MTL equations is possible in the frequency domain [2]. We will show the *exact* solution for *lossless lines* in the next section, along with a simple solution valid for electrically short lines that is suitable for hand calculations. The above MTL equations for the *sinusoidal steady state* (*frequency domain*) become

$$\frac{d}{dz}\hat{\mathbf{V}}(z) = -\hat{\mathbf{Z}}\hat{\mathbf{I}}(z) \tag{10.5a}$$

$$\frac{d}{dz}\hat{\mathbf{I}}(z) = -\hat{\mathbf{Y}}\hat{\mathbf{V}}(z) \tag{10.5b}$$

where the phasor voltage and current vectors are

$$\hat{\mathbf{V}}(z) = \begin{bmatrix} \hat{V}_G(z) \\ \hat{V}_R(z) \end{bmatrix} \tag{10.6a}$$

$$\hat{\mathbf{I}}(z) = \begin{bmatrix} \hat{I}_G(z) \\ \hat{I}_R(z) \end{bmatrix} \tag{10.6b}$$

The time-domain voltages and currents can be found in the usual fashion from these phasor voltages and currents as

$$\mathbf{V}(z, t) = \mathscr{R}e\{\hat{\mathbf{V}}(z)e^{j\omega t}\} \tag{10.7a}$$

$$\mathbf{I}(z, t) = \mathscr{R}e\{\hat{\mathbf{I}}(z)e^{j\omega t}\} \tag{10.7b}$$

and the per-unit-length *impedance* and *admittance* matrices are given by

$$\hat{\mathbf{Z}} = \mathbf{R} + j\omega\mathbf{L} \tag{10.8a}$$

$$\hat{\mathbf{Y}} = \mathbf{G} + j\omega\mathbf{C} \tag{10.8b}$$

10.1.3   The Per-Unit-Length Parameters

As was pointed out earlier, it is of no use to solve the MTL equations if the per-unit-length parameters cannot be determined for the particular line cross-sectional configuration. The *internal parameters* of per-unit-length resistance r_G, r_R, and r_0, and internal inductance are not dependent on the line configurations if the conductors are separated sufficiently. Therefore the previously derived results given in Chapter 4 for these internal parameters for two-conductor lines can be used for multiconductor lines. Again, we will ignore the per-unit-length internal inductances, since they will be in series with and usually smaller than the external self inductances. Therefore the entires in the per-unit-length inductance matrix $\mathbf{L}$ are the *external inductances*. It therefore remains to determine the *external parameters* of inductance and capacitance, i.e., the entries in $\mathbf{L}$ and $\mathbf{C}$. Approximate equations for the per-unit-length inductances and capacitances for the wire-type lines in a homogeneous medium shown in Fig. 10.2 will be derived for widely separated wires. Corresponding formulae for the PCB-type structures in Fig. 10.3 are difficult to obtain, and generally must be found using approximate, numerical methods [4, 5].

10.1.3.1   Homogeneous versus Inhomogeneous Media   The configurations in Fig. 10.2 are considered to be lines immersed in a *homogeneous medium*. The surrounding medium for the cases of three wires in Fig. 10.2(a) or two wires above a ground plane in Fig. 10.2(b) is logically considered to be free space with parameters ϵ_o and μ_o; that is, the wires are considered to be bare. Dielectric insulations severely complicate the determination of the per-unit-length capacitances for wires, but do not affect the per-unit-length inductances since dielectrics have $\mu = \mu_o$. In order to incorporate these types of inhomogeneous media, numerical methods must be used [2, 3]. For our purposes we will ignore the presence of dielectric insulations and consider the wires to be bare, since the dielectric insulations change the capacitances only slightly for widely separated wires, as we will assume. The case of two wires within an overall shield shown in Fig. 10.2(c) can be analyzed if we assume that the dielectric interior to the shield is homogeneous with ϵ and μ as we will assume. The inhomogeneity caused by a printed circuit board, as with the structures in Fig. 10.3, is more significant, and closed-form solutions for all the entries in **L** and **C** are difficult to obtain for these configurations. One could, as an approximation, obtain the self-inductances and capacitances l_G, l_R, c_G, and c_R for the PCB structures in Fig. 10.3 by treating the two circuits as being isolated and using the results developed in Chapter 6. However, this does not really help in the analysis of crosstalk, since the more important mutual terms l_m and c_m represent the parameters of most interest in the prediction of crosstalk on PCBs.

If the surrounding medium is *homogeneous*, as for the lines in Fig. 10.2, the per-unit-length parameter matrices given in (10.4) have important, special relationships that parallel those found for two-conductor lines. In particular these relationships are [2]

$$\mathbf{LC} = \mathbf{CL} = \mu\epsilon\mathbf{1}_2 \tag{10.9}$$

where the surrounding homogeneous medium is characterized by μ and ϵ, and $\mathbf{1}_2$ is the 2×2 identity matrix:

$$\mathbf{1}_2 = \begin{bmatrix} 1 & 0 \\ 0 & 1 \end{bmatrix} \tag{10.10}$$

Therefore we only need to determine one of the parameter matrices, since the other can be found from (10.9) as

$$\mathbf{C} = \mu\epsilon\mathbf{L}^{-1} \tag{10.11}$$

$$= \frac{1}{v^2}\mathbf{L}^{-1}$$

where $v = 1/\sqrt{\mu\epsilon}$ is the usual phase velocity for uniform plane waves considered in Chapter 3, and is also the velocity of waves on the line. For example, for the three-conductor line we obtain from (10.11)

$$\begin{bmatrix} c_G + c_m & -c_m \\ -c_m & c_R + c_m \end{bmatrix} = \frac{1}{v^2(l_G l_R - l_m^2)}\begin{bmatrix} l_R & -l_m \\ -l_m & l_G \end{bmatrix} \tag{10.12}$$

Comparing the two sides, we obtain the per-unit-length capacitance parameters in terms of the per-unit-length inductance parameters as

$$c_m = \frac{l_m}{v^2(l_G l_R - l_m^2)} \tag{10.13a}$$

$$c_G + c_m = \frac{l_R}{v^2(l_G l_R - l_m^2)} \tag{10.13b}$$

$$c_R + c_m = \frac{l_G}{v^2(l_G l_R - l_m^2)} \tag{10.13c}$$

Thus, as was the case with two-conductor lines, we only need to find one set of parameters. Similarly, it is possible to show that [2]

$$\mathbf{LG} = \mathbf{GL} = \mu\sigma\mathbf{1}_2 \tag{10.14}$$

where the surrounding homogeneous medium has conductivity σ. Therefore the per-unit-length conductance matrix can be found from the per-unit-length inductance matrix as

$$\mathbf{G} = \mu\sigma\mathbf{L}^{-1} \tag{10.15}$$

We will use these properties on numerous occasions to determine the per-unit-length parameters.

In the case of lines in an inhomogeneous medium, such as those shown in Fig. 10.3, the per-unit-length inductance matrix $\mathbf{L}$ is not affected by the dielectric inhomogeneity, since $\mu = \mu_o$. Thus if we designate the per-unit-length capacitance matrix *with the dielectric removed (replaced with free space)* by $\mathbf{C}_o$ then the inductance can be found from (10.9) as $\mathbf{L} = \mu_o \epsilon_o \mathbf{C}_o^{-1}$. Thus for inhomogeneous media we need to determine the per-unit-length capacitance matrices *with and without the dielectric present*, $\mathbf{C}$ and $\mathbf{C}_o$. Numerical methods are frequently used to obtain these [4, 5].

10.1.3.2 *Wide-Separation Approximations for Wires* We will obtain closed-form results for the per-unit-length parameters for three-conductor lines under *the assumption that the wires are separated sufficiently that the charge and current distributions around the peripheries of the wires are essentially uniform.* This is not a serious restriction for practical configurations, as we will see. In order to derive the per-unit-length parameters for wire-type lines in Fig. 10.2, we will rely on the two basic subproblems discussed for two-conductor lines composed of wires in Chapter 4. It is important that the reader understand these fundamental results, since our deviations will rely on their proper application. The first fundamental subproblem concerns the flux due to a current-carrying wire that penetrates a surface of unit length along the wire and whose edges

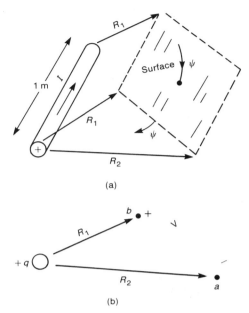

FIGURE 10.5 Illustration of two important subproblems for computing the per-unit-length parameters of wire-type lines: (a) magnetic flux of a current-carrying wire through a surface; (b) voltage between two points for a charge-carrying wire.

are at radial distances R_1 and R_2 from the wire where $R_2 \geq R_1$, as shown in Fig. 10.5(a). The flux with direction shown is given by equation (4.12), and is repeated here:

$$\psi = \frac{\mu_o I}{2\pi} \ln\left(\frac{R_2}{R_1}\right) \tag{10.16a}$$

where the wire carries a current I that is assumed to be uniformly distributed around the wire periphery. *It is very important to determine the correct direction of the resulting flux through the surface.* The second fundamental result concerns the voltage between two points that are at radial distances R_1 and R_2 from a charge-carrying wire, where $R_2 \geq R_1$, as shown in Fig. 10.5(b). The result is given by equation (4.15), and is repeated here:

$$V = \frac{q}{2\pi \epsilon_o} \ln\left(\frac{R_2}{R_1}\right) \tag{10.16b}$$

where the wire carries a per-unit-length charge q that is assumed to be uniformly distributed along the wire and around the wire periphery. Again, as with the determination of the resulting flux, *it is very important to determine the correct direction of the resulting voltage.* The charge distribution on the wire is assumed positive, so that the resulting voltage is positive at the point closest to the wire.

First we will consider the case of three wires shown in Fig. 10.2(a). The per-unit-length external inductance matrix relates the magnetic fluxes penetrating generator and receptor circuits to the currents of those circuits as

$$\psi = LI \tag{10.17a}$$

or

$$\begin{bmatrix} \psi_G \\ \psi_R \end{bmatrix} = \begin{bmatrix} l_G & l_m \\ l_m & l_R \end{bmatrix} \begin{bmatrix} I_G \\ I_R \end{bmatrix} \tag{10.17b}$$

The directions of the required fluxes are shown in Fig. 10.6(a). Each of the

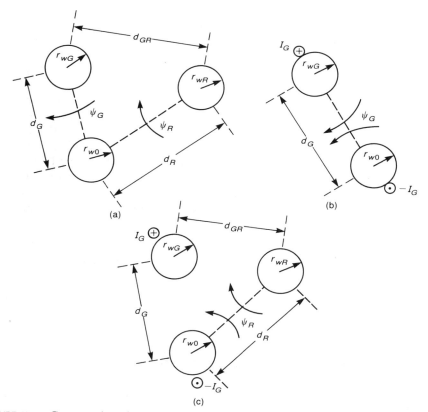

(a)

(b)

(c)

FIGURE 10.6 Computation of the per-unit-length inductances for three wires: (a) definition of circuit magnetic fluxes; (b) computation of the self-inductance; (c) computation of the mutual inductance.

entries in **L** can be obtained in the usual manner for determining two-port parameters by expanding (10.17) to give [6]

$$\psi_G = l_G I_G + l_m I_R \tag{10.18a}$$

$$\psi_R = l_m I_G + l_R I_R \tag{10.18b}$$

First set $I_R = 0$ to give

$$l_G = \frac{\psi_G}{I_G}\bigg|_{I_R = 0} \tag{10.19a}$$

$$l_m = \frac{\psi_R}{I_G}\bigg|_{I_R = 0} \tag{10.19b}$$

Similarly, setting $I_G = 0$ gives

$$l_m = \frac{\psi_G}{I_R}\bigg|_{I_G = 0} \tag{10.19c}$$

$$l_R = \frac{\psi_R}{I_R}\bigg|_{I_G = 0} \tag{10.19d}$$

The results in (10.19) show that the per-unit-length inductances can be found by applying a current on one conductor (and returning on the reference conductor), setting the other current equal to zero, and determining the resulting flux through the appropriate circuit. For example, consider finding the self-inductance of the generator circuit. Equation (10.19a) shows that we place current I_G on the generator circuit, set $I_R = 0$, and determine the resulting flux through the generator circuit as shown in Fig. 10.6(b). Using the fundamental result given in (10.16a), we obtain

$$l_G = \frac{\mu_o}{2\pi} \ln\left(\frac{d_G}{r_{wG}}\right) + \frac{\mu_o}{2\pi} \ln\left(\frac{d_G}{r_{w0}}\right) \tag{10.20}$$

$$= \frac{\mu_o}{2\pi} \ln\left(\frac{d_G^2}{r_{wG} r_{w0}}\right)$$

where we have assumed that the intervening medium is not ferromagnetic, $\mu = \mu_o$. Similarly, the self-inductance of the receptor circuit is

$$l_R = \frac{\mu_o}{2\pi} \ln\left(\frac{d_R^2}{r_{wR} r_{w0}}\right) \tag{10.21}$$

The per-unit-length mutual inductance is found from (10.19b) or (10.19c). We apply a current to one circuit, and determine the flux penetrating the other circuit as shown in Fig. 10.6(c):

$$l_m = \frac{\mu_o}{2\pi} \ln\left(\frac{d_G}{d_{GR}}\right) + \frac{\mu_o}{2\pi} \ln\left(\frac{d_R}{r_{w0}}\right) \tag{10.22}$$

$$= \frac{\mu_o}{2\pi} \ln\left(\frac{d_G d_R}{d_{GR} r_{w0}}\right)$$

The per-unit-length capacitances can be found from the per-unit-length inductances for a homogeneous medium using the result given in (10.9). In order to show the direct derivation, we will use the fundamental result given in (10.16b). First we examine the definition of the per-unit-length capacitance matrix **C**. This relates the per-unit-length charges on the generator and receptor conductors to the voltages of these conductors *with respect to the reference conductor* as

$$\begin{bmatrix} q_G \\ q_R \end{bmatrix} = \begin{bmatrix} c_G + c_m & -c_m \\ -c_m & c_R + c_m \end{bmatrix} \begin{bmatrix} V_G \\ V_R \end{bmatrix} \tag{10.23}$$

This form is not convenient for a direct determination, so we invert the expression to give

$$\begin{bmatrix} V_G \\ V_R \end{bmatrix} = \begin{bmatrix} p_G & p_m \\ p_m & p_R \end{bmatrix} \begin{bmatrix} q_G \\ q_R \end{bmatrix} \tag{10.24a}$$

or

$$\mathbf{V} = \mathbf{Pq} = \mathbf{C}^{-1}\mathbf{q} \tag{10.24b}$$

Expanding this gives

$$V_G = p_G q_G + p_m q_R \tag{10.25a}$$

$$V_R = p_m q_G + p_R q_R \tag{10.25b}$$

The entries can be found by setting each of the charges to zero and determining the ratio of voltage to charge producing it. First set $q_R = 0$ to give

$$p_G = \frac{V_G}{q_G}\bigg|_{q_R = 0} \tag{10.26a}$$

$$p_m = \frac{V_R}{q_G}\bigg|_{q_R = 0} \tag{10.26b}$$

Similarly, setting $q_G = 0$ gives

$$p_m = \frac{V_G}{q_R}\bigg|_{q_G = 0} \tag{10.26c}$$

$$p_R = \frac{V_R}{q_R}\bigg|_{q_G = 0} \tag{10.26d}$$

Inverting (10.24) gives (10.23). In order to obtain the entries in (10.24), we apply the fundamental result given in (10.16b) for the voltage produced by a charge-carrying wire. The self-terms can be found for the case of three wires shown in Fig. 10.2(a) by applying (10.26) as shown in Fig. 10.7. We obtain

$$\begin{aligned} p_G &= \frac{1}{2\pi\epsilon_o} \ln\left(\frac{d_G}{r_{wG}}\right) + \frac{1}{2\pi\epsilon_o} \ln\left(\frac{d_G}{r_{w0}}\right) \\ &= \frac{1}{2\pi\epsilon_o} \ln\left(\frac{d_G^2}{r_{wG}r_{w0}}\right) \end{aligned} \tag{10.27}$$

and

$$\begin{aligned} p_R &= \frac{1}{2\pi\epsilon_o} \ln\left(\frac{d_R}{r_{wR}}\right) + \frac{1}{2\pi\epsilon_o} \ln\left(\frac{d_R}{r_{w0}}\right) \\ &= \frac{1}{2\pi\epsilon_o} \ln\left(\frac{d_R^2}{r_{wR}r_{w0}}\right) \end{aligned} \tag{10.28}$$

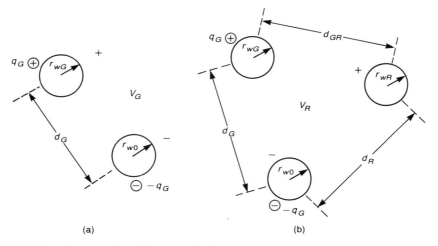

(a)   (b)

FIGURE 10.7 Computation of the per-unit-length coefficients of potential for a three-wire line: (a) self-term; (b) mutual term.

Similarly, from (10.26b) and Fig. 10.7(b) we obtain

$$p_m = \frac{1}{2\pi\epsilon_o} \ln\left(\frac{d_G}{d_{GR}}\right) + \frac{1}{2\pi\epsilon_o} \ln\left(\frac{d_R}{r_{w0}}\right) \qquad (10.29)$$

$$= \frac{1}{2\pi\epsilon_o} \ln\left(\frac{d_G d_R}{d_{GR} r_{w0}}\right)$$

Once the entries in (10.24) are obtained as above, the per-unit-length capacitances in (10.23) can be obtained by inverting (10.24). Comparing the entries in **P** in (10.24) obtained in (10.27), (10.28), and (10.29) with the per-unit-length inductances obtained in (10.20), (10.21), and (10.22), we see that $\mathbf{L} = \mu_o \epsilon_o \mathbf{P}$, which confirms the basic result in (10.9), since $\mathbf{C} = \mathbf{P}^{-1}$.

A common application of these results is for the case of ribbon cables. Consider the three-wire ribbon cable composed of three #28 gauge (7×36) wires whose adjacent separations are 50 mils. It is critical to the correctness of our results that we designate the reference conductor. For our example we will assume in the following derivation that the center wire is to be the reference wire, as shown in Fig. 10.8(a). Determination of the self-inductance l_G is shown

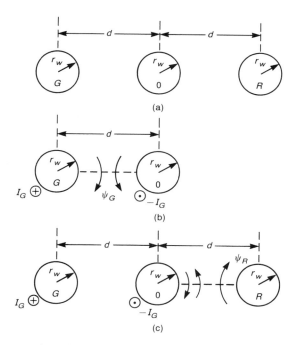

(a)

(b)

(c)

FIGURE 10.8 Computation of the per-unit-length inductances for a ribbon cable where the inner wire is chosen as the reference conductor: (a) problem definition; (b) self-terms; (c) mutual terms.

in Fig. 10.8(b), giving

$$l_G = \frac{\mu_o}{2\pi} \ln\left(\frac{d}{r_w}\right) + \frac{\mu_o}{2\pi} \ln\left(\frac{d}{r_w}\right) \qquad (10.30)$$

$$= \frac{\mu_o}{\pi} \ln\left(\frac{d}{r_w}\right)$$

Similarly, we obtain l_R as

$$l_R = \frac{\mu_o}{\pi} \ln\left(\frac{d}{r_w}\right) \qquad (10.31)$$

The mutual inductance is found from Fig. 10.8(c) by placing a current I_G on the generator wire (and returning on the reference wire) and finding the total flux through the receptor circuit. Note the defined direction of the receptor circuit flux (upward). Applying the fundamental result to this case gives

$$l_m = -\frac{\mu_o}{2\pi} \ln\left(\frac{2d}{d}\right) + \frac{\mu_o}{2\pi} \ln\left(\frac{d}{r_w}\right) \qquad (10.32)$$

$$= \frac{\mu_o}{2\pi} \ln\left(\frac{d}{2r_w}\right)$$

For the ribbon cable with $r_w = 7.5$ mils and $d = 50$ mils we obtain $l_G = l_R = 0.759 \ \mu H/m = 19.3 \ nH/inch$ and $l_m = 0.24 \ \mu H/m = 6.1 \ nH/inch$. We can compute the characteristic impedance of each isolated circuit from $Z_C = v_o l_G = v_o l_R = 227.7 \ \Omega$. The per-unit-length capacitances can be computed from these results, using (10.13), as $c_G = c_R = 11.1 \ pF/m = 0.28 \ pF/inch$ and $c_m = 5.17 \ pF/m = 0.13 \ pF/inch$. We can also compute the characteristic impedance of one circuit *in the presence of the other circuit*, using $Z_C = \sqrt{l_G/c_G} = 261.5 \ \Omega$. Thus the characteristic impedance of one circuit is affected by the presence of the other circuit.

Next consider the case of two wires above an infinite ground plane, as shown in Fig. 10.9. Replacing the ground plane with the wire images as shown in Fig. 10.9 and applying the fundamental result in (10.16a) to (10.19a) gives

$$l_G = \frac{\psi_G}{I_G}\bigg|_{I_R=0} \qquad (10.33)$$

$$= \frac{\mu_o}{2\pi} \ln\left(\frac{h_G}{r_{wG}}\right) + \frac{\mu_o}{2\pi} \ln\left(\frac{2h_G}{h_G}\right)$$

$$= \frac{\mu_o}{2\pi} \ln\left(\frac{2h_G}{r_{wG}}\right)$$

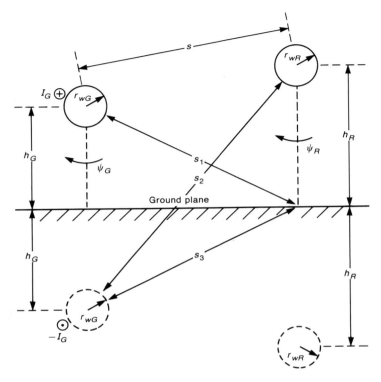

FIGURE 10.9 Computation of the per-unit-length inductances for two wires above a ground plane.

Similarly, (10.19d) yields

$$l_R = \frac{\mu_o}{2\pi} \ln\left(\frac{2h_R}{r_{wR}}\right) \qquad (10.34)$$

The mutual inductance is found from (10.19b) as

$$l_m = \frac{\psi_R}{I_G}\bigg|_{I_R = 0} \qquad (10.35)$$

$$= \frac{\mu_o}{2\pi} \ln\left(\frac{s_1}{s}\right) + \frac{\mu_o}{2\pi} \ln\left(\frac{s_2}{s_3}\right)$$

$$= \frac{\mu_o}{2\pi} \ln\left(\frac{s_2}{s}\right)$$

since $s_1 = s_3$. Substituting the dimensions gives

$$s_2 = \sqrt{s^2 + 4h_G h_R} \tag{10.36}$$

so that

$$l_m = \frac{\mu_o}{4\pi} \ln\left(1 + 4\frac{h_G h_R}{s^2}\right) \tag{10.37}$$

As a numerical example, consider two #20 gauge solid wires ($r_w = 16$ mils) at a height of 2 cm above a ground plane and separated by 2 cm. The per-unit-length inductances are $l_G = l_R = 0.918\ \mu H/m = 23.3$ nH/inch and $l_m = 0.161\ \mu H/m = 4.09$ nH/inch. The characteristic impedance of each isolated circuit is $Z_C = v_o l_G = v_o l_R = 275.4\ \Omega$. The per-unit-length capacitances can be obtained from these values, using (10.13), as $c_G = c_R = 10.3$ pF/m and $c_m = 2.19$ pF/m. We can also compute the characteristic impedance of one circuit *in the presence of the other circuit*, using $Z_C = \sqrt{l_G/c_G} = 298.5\ \Omega$. Thus the characteristic impedance of one circuit is again affected by the presence of the other circuit.

The final configuration to be considered is the case of two wires within an overall shield, as shown in Fig. 10.10. The shield has a radius r_{SH} and the wires are located distances d_G and d_R from the shield center and are separated by an angle of θ_{GR}. We may replace the shield with the wire images. Each image lies

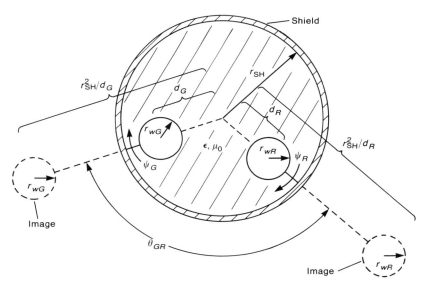

FIGURE 10.10 Computation of the per-unit-length inductances for two wires within an overall, cylindrical shield.

on a radial line from the shield center, and is located a distance r_{SH}^2/d_G and r_{SH}^2/d_R from the shield center [1, 2]. The resulting per-unit-length inductances are derivable in a similar fashion; the derivation is given in [2]. The result is

$$l_G = \frac{\mu_o}{2\pi} \ln\left(\frac{r_{SH}^2 - d_G^2}{r_{SH} r_{wG}}\right) \tag{10.38}$$

$$l_R = \frac{\mu_o}{2\pi} \ln\left(\frac{r_{SH}^2 - d_R^2}{r_{SH} r_{wR}}\right) \tag{10.39}$$

$$l_m = \frac{\mu_o}{2\pi} \ln\left[\frac{d_R}{r_{SH}} \sqrt{\frac{(d_G d_R)^2 + r_{SH}^4 - 2d_G d_R r_{SH}^2 \cos\theta_{GR}}{(d_G d_R)^2 + d_R^4 - 2d_G d_R^3 \cos\theta_{GR}}}\right] \tag{10.40}$$

10.1.4 Frequency-Domain (Sinusoidal, Steady-State) Crosstalk

In this section we will address the solution of the transmission-line equations for sinusoidal, steady-state excitation of the line. This will allow the determination of the *frequency-domain crosstalk*, which is our first objective. We will first investigate the exact solution with no restrictions on the lossless or lossy nature of the line or its electrical length. The solutions we obtain are couched in matrix notation and are, of necessity, not simple. However, the solutions can be programmed on a digital computer and readily solved. FORTRAN programs that accomplish this are described in [7]. Once this exact solution is accomplished, we will obtain some simple results for electrically short lines that provide reasonable approximations and are suitable for hand calculations. Furthermore, these simple results provide considerable insight into the mechanism of crosstalk that is obscured in the numerical solutions. This simple result will divide the crosstalk into *inductive* and *capacitive* coupling contributions that are essentially due to the mutual inductance and mutual capacitance between the two circuits. This will provide the necessary understanding of the effect of shielded wires and twisted pairs on the reduction of crosstalk that will be discussed in Sections 10.2 and 10.3

10.1.4.1 General Solution   The transmission-line equations in the *frequency domain (sinusoidal, steady-state excitation)* are given in (10.5). These are *coupled, first-order differential equations*. We can obtain *uncoupled, second-order* versions by differentiating each equation with respect to z and substituting the other to yield

$$\frac{d^2\hat{V}(z)}{dz^2} = \hat{Z}\hat{Y}\hat{V}(z) \tag{10.41a}$$

$$\frac{d^2\hat{I}(z)}{dz^2} = \hat{Y}\hat{Z}\hat{I}(z) \tag{10.41b}$$

It is important to observe the proper order of multiplication of the per-unit-length matrices $\hat{\mathbf{Z}}$ and $\hat{\mathbf{Y}}$ in these equations, since these matrices do not generally commute. Either of these equations can be solved. We will choose to solve for the form of the solution for (10.41b), and obtain the voltage solution from (10.5b). The general solution can be obtained by defining a transformation to *modal currents* as [2]

$$\hat{\mathbf{I}} = \hat{\mathbf{T}}\hat{\mathbf{I}}_m \qquad (10.42)$$

where the complex transformation matrix is 2×2. Substituting (10.42) into (10.41b) yields

$$\frac{d^2\hat{\mathbf{I}}_m(z)}{dz^2} = \hat{\mathbf{T}}^{-1}\hat{\mathbf{Y}}\hat{\mathbf{Z}}\hat{\mathbf{T}}\hat{\mathbf{I}}_m(z) \qquad (10.43)$$

If a transformation matrix $\hat{\mathbf{T}}$ can be found such that $\hat{\mathbf{Y}}\hat{\mathbf{Z}}$ is *diagonalized*, i.e.,

$$\hat{\mathbf{T}}^{-1}\hat{\mathbf{Y}}\hat{\mathbf{Z}}\hat{\mathbf{T}} = \hat{\gamma}^2 \qquad (10.44)$$

$$= \begin{bmatrix} \hat{\gamma}_G^2 & 0 \\ 0 & \hat{\gamma}_R^2 \end{bmatrix}$$

then the *modal equations* in (10.43) are *uncoupled*:

$$\frac{d^2\hat{I}_{mG}(z)}{dz^2} = \hat{\gamma}_G^2 \hat{I}_{mG}(z) \qquad (10.45a)$$

$$\frac{d^2\hat{I}_{mR}(z)}{dz^2} = \hat{\gamma}_R^2 \hat{I}_{mR}(z) \qquad (10.45b)$$

The terms $\hat{\gamma}_G^2$ and $\hat{\gamma}_R^2$ are the *eigenvalues of* $\hat{\mathbf{Y}}\hat{\mathbf{Z}}$, and $\hat{\gamma}_G$ and $\hat{\gamma}_R$ are referred to as the *propagation constants* [2]. These equations are uncoupled, and the forms of their solutions are the same as for two-conductor lines:

$$\hat{I}_{mG}(z) = e^{-\hat{\gamma}_G z}\hat{I}_{mG}^+ - e^{\hat{\gamma}_G z}\hat{I}_{mG}^- \qquad (10.46a)$$

$$\hat{I}_{mR}(z) = e^{-\hat{\gamma}_R z}\hat{I}_{mR}^+ - e^{\hat{\gamma}_R z}\hat{I}_{mR}^- \qquad (10.46b)$$

where $\hat{I}_{mG}^+$, $\hat{I}_{mG}^-$, $\hat{I}_{mR}^+$, and $\hat{I}_{mR}^-$ are, as yet, undetermined constants. These solutions can be placed in a compact matrix form as

$$\hat{\mathbf{I}}_m(z) = \mathbf{e}^{-\hat{\gamma}z}\hat{\mathbf{I}}_m^+ - \mathbf{e}^{\hat{\gamma}z}\hat{\mathbf{I}}_m^- \qquad (10.47)$$

and

$$\mathbf{e}^{\pm \hat{\gamma} z} = \begin{bmatrix} e^{\pm \hat{\gamma}_G z} & 0 \\ 0 & e^{\pm \hat{\gamma}_R z} \end{bmatrix} \tag{10.48}$$

$$\hat{\mathbf{I}}_m^{\pm} = \begin{bmatrix} \hat{I}_{mG}^{\pm} \\ \hat{I}_{mR}^{\pm} \end{bmatrix} \tag{10.49}$$

The form of the solution for the actual currents can be obtained by multiplying this modal solution by the transformation matrix according to (10.42):

$$\hat{\mathbf{I}}(z) = \hat{\mathbf{T}}\hat{\mathbf{I}}_m \tag{10.50}$$

$$= \hat{\mathbf{T}}(e^{-\hat{\gamma} z}\hat{\mathbf{I}}_m^{+} - e^{\hat{\gamma} z}\hat{\mathbf{I}}_m^{-})$$

The form of the voltage solution can be found from this solution using (10.5b):

$$\hat{\mathbf{V}}(z) = -\hat{\mathbf{Y}}^{-1}\frac{d}{dz}\hat{\mathbf{I}}(z) \tag{10.51}$$

$$= \hat{\mathbf{Y}}^{-1}\hat{\mathbf{T}}\hat{\gamma}(e^{-\hat{\gamma} z}\hat{\mathbf{I}}_m^{+} + e^{\hat{\gamma} z}\hat{\mathbf{I}}_m^{-})$$

From (10.44) we can write

$$\hat{\mathbf{Z}}\hat{\mathbf{T}}\hat{\gamma}^{-1} = \hat{\mathbf{Y}}^{-1}\hat{\mathbf{T}}\hat{\gamma} \tag{10.52}$$

Therefore (10.51) can be written as

$$\hat{\mathbf{V}}(z) = (\hat{\mathbf{Z}}\hat{\mathbf{T}}\hat{\gamma}^{-1}\hat{\mathbf{T}}^{-1})\hat{\mathbf{T}}(e^{-\hat{\gamma} z}\hat{\mathbf{I}}_m^{+} + e^{\hat{\gamma} z}\hat{\mathbf{I}}_m^{-}) \tag{10.53}$$

This suggests that we define the *characteristic impedance matrix* as

$$\hat{\mathbf{Z}}_C = \hat{\mathbf{Z}}\hat{\mathbf{T}}\hat{\gamma}^{-1}\hat{\mathbf{T}}^{-1} \tag{10.54}$$

It is shown in [2, 8] that one can define quantities analogous to the two-conductor case:

$$\hat{\mathbf{Z}}_C = \hat{\mathbf{Z}}\hat{\mathbf{T}}\hat{\gamma}^{-1}\hat{\mathbf{T}}^{-1} \tag{10.55}$$

$$= \hat{\mathbf{Z}}(\sqrt{\hat{\mathbf{Y}}\hat{\mathbf{Z}}})^{-1}$$

$$= \hat{\mathbf{Y}}^{-1}\sqrt{\hat{\mathbf{Y}}\hat{\mathbf{Z}}}$$

which reduce to the familiar scalar case for two-conductor lines: $Z_C = \sqrt{\hat{z}/\hat{y}}$.

Our remaining task is to solve for the undetermined constants in the vectors $\hat{\mathbf{I}}_m^{+}$ and $\hat{\mathbf{I}}_m^{-}$. We essentially need two more vector constraint equations. These

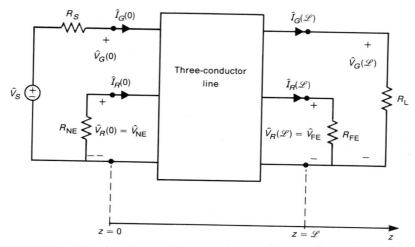

FIGURE 10.11   Viewing the coupled, three-conductor line as a four-port circuit.

are provided by the terminal conditions. Viewing the transmission line as a four-port network as illustrated in Fig. 10.11, we can write these terminal constraints as

$$\hat{\mathbf{V}}(0) = \hat{\mathbf{V}}_S - \hat{\mathbf{Z}}_S\hat{\mathbf{I}}(0) \tag{10.56a}$$

$$\hat{\mathbf{V}}(\mathscr{L}) = \hat{\mathbf{Z}}_L\hat{\mathbf{I}}(\mathscr{L}) \tag{10.56b}$$

where

$$\hat{\mathbf{V}}_S = \begin{bmatrix} \hat{\mathbf{V}}_S \\ 0 \end{bmatrix} \tag{10.57a}$$

$$\hat{\mathbf{Z}}_S = \begin{bmatrix} R_S & 0 \\ 0 & R_{NE} \end{bmatrix} \tag{10.57b}$$

$$\hat{\mathbf{Z}}_L = \begin{bmatrix} R_L & 0 \\ 0 & R_{FE} \end{bmatrix} \tag{10.57c}$$

Evaluating (10.50) and (10.51) at $z = 0$ gives

$$\hat{\mathbf{V}}(0) = \hat{\mathbf{Z}}_C\hat{\mathbf{T}}(\hat{\mathbf{I}}_m^+ + \hat{\mathbf{I}}_m^-) \tag{10.58a}$$

$$\hat{\mathbf{I}}(0) = \hat{\mathbf{T}}(\hat{\mathbf{I}}_m^+ - \hat{\mathbf{I}}_m^-) \tag{10.58b}$$

Substituting these into (10.56) gives

$$\hat{\mathbf{Z}}_C\hat{\mathbf{T}}(\hat{\mathbf{I}}_m^+ + \hat{\mathbf{I}}_m^-) = \hat{\mathbf{V}}_S - \hat{\mathbf{Z}}_S\hat{\mathbf{T}}(\hat{\mathbf{I}}_m^+ - \hat{\mathbf{I}}_m^-) \tag{10.59}$$

Similarly, evaluation of (10.50) and (10.51) at $z = \mathscr{L}$ gives

$$\hat{\mathbf{V}}(\mathscr{L}) = \hat{\mathbf{Z}}_C \hat{\mathbf{T}}(e^{-\hat{\gamma}\mathscr{L}}\hat{\mathbf{I}}_m^+ + e^{\hat{\gamma}\mathscr{L}}\hat{\mathbf{I}}_m^-) \tag{10.60a}$$

$$\hat{\mathbf{I}}(\mathscr{L}) = \hat{\mathbf{T}}(e^{-\hat{\gamma}\mathscr{L}}\hat{\mathbf{I}}_m^+ - e^{\hat{\gamma}\mathscr{L}}\hat{\mathbf{I}}_m^-) \tag{10.60b}$$

Substituting into (10.56b) gives

$$\hat{\mathbf{Z}}_C \hat{\mathbf{T}}(e^{-\hat{\gamma}\mathscr{L}}\hat{\mathbf{I}}_m^+ + e^{\hat{\gamma}\mathscr{L}}\hat{\mathbf{I}}_m^-) = \hat{\mathbf{Z}}_L \hat{\mathbf{T}}(e^{-\hat{\gamma}\mathscr{L}}\hat{\mathbf{I}}_m^+ - e^{\hat{\gamma}\mathscr{L}}\hat{\mathbf{I}}_m^-) \tag{10.61}$$

Equations (10.59) and (10.61) can be combined to give a set of simultaneous equations that can be solved to determine the entries in $\hat{\mathbf{I}}_m^+$ and $\hat{\mathbf{I}}_m^-$ as

$$\begin{bmatrix} (\hat{\mathbf{Z}}_C + \hat{\mathbf{Z}}_S)\hat{\mathbf{T}} & (\hat{\mathbf{Z}}_C - \hat{\mathbf{Z}}_S)\hat{\mathbf{T}} \\ (\hat{\mathbf{Z}}_C - \hat{\mathbf{Z}}_L)\hat{\mathbf{T}}e^{-\hat{\gamma}\mathscr{L}} & (\hat{\mathbf{Z}}_C + \hat{\mathbf{Z}}_L)\hat{\mathbf{T}}e^{\hat{\gamma}\mathscr{L}} \end{bmatrix} \begin{bmatrix} \hat{\mathbf{I}}_m^+ \\ \hat{\mathbf{I}}_m^- \end{bmatrix} = \begin{bmatrix} \hat{\mathbf{V}}_S \\ \mathbf{0} \end{bmatrix} \tag{10.62}$$

Equations (10.62) comprises a set of four equations in the four unknowns $\hat{I}_{mG}^+$, $\hat{I}_{mG}^-$, $\hat{I}_{mR}^+$, and $\hat{I}_{mR}^-$. They can be solved using any convenient Gauss elimination computer subroutine, and the voltages and currents at any point on the line can be obtained by substituting the solutions into (10.50) and (10.51). FORTRAN computer programs that accomplish this are described in [7].

Viewing the line as a four-port network as illustrated in Fig. 10.11 suggests that the port-parameter representation of the line be determined. This is a particularly important representation, since we are usually not interested in the voltages and currents along the line but only in determining their values at the endpoints of the line, in particular $\hat{V}_R(0) = \hat{V}_{NE}$ and $\hat{V}_R(\mathscr{L}) = \hat{V}_{FE}$. In order to do this, we will choose to determine the *chain parameter matrix* of the line, which is

$$\begin{bmatrix} \hat{\mathbf{V}}(\mathscr{L}) \\ \hat{\mathbf{I}}(\mathscr{L}) \end{bmatrix} = \underbrace{\begin{bmatrix} \hat{\boldsymbol{\phi}}_{11} & \hat{\boldsymbol{\phi}}_{12} \\ \hat{\boldsymbol{\phi}}_{21} & \hat{\boldsymbol{\phi}}_{22} \end{bmatrix}}_{\hat{\boldsymbol{\phi}}(\mathscr{L})} \begin{bmatrix} \hat{\mathbf{V}}(0) \\ \hat{\mathbf{I}}(0) \end{bmatrix} \tag{10.63}$$

Eliminating $\hat{\mathbf{I}}_m^+$ and $\hat{\mathbf{I}}_m^-$ from (10.58) and (10.60) gives [2]

$$\hat{\boldsymbol{\phi}}_{11} = \tfrac{1}{2}\hat{\mathbf{Y}}^{-1}\hat{\mathbf{T}}(e^{\hat{\gamma}\mathscr{L}} + e^{-\hat{\gamma}\mathscr{L}})\hat{\mathbf{T}}^{-1}\hat{\mathbf{Y}} \tag{10.64a}$$

$$\hat{\boldsymbol{\phi}}_{12} = -\tfrac{1}{2}\hat{\mathbf{Y}}^{-1}\hat{\mathbf{T}}\hat{\gamma}(e^{\hat{\gamma}\mathscr{L}} - e^{-\hat{\gamma}\mathscr{L}})\hat{\mathbf{T}}^{-1} \tag{10.64b}$$

$$\hat{\boldsymbol{\phi}}_{21} = -\tfrac{1}{2}\hat{\mathbf{T}}(e^{\hat{\gamma}\mathscr{L}} - e^{-\hat{\gamma}\mathscr{L}})\hat{\gamma}^{-1}\hat{\mathbf{T}}^{-1}\hat{\mathbf{Y}} \tag{10.64c}$$

$$\hat{\boldsymbol{\phi}}_{22} = \tfrac{1}{2}\hat{\mathbf{T}}(e^{\hat{\gamma}\mathscr{L}} + e^{-\hat{\gamma}\mathscr{L}})\hat{\mathbf{T}}^{-1} \tag{10.64d}$$

These results can also be placed in forms analogous to the two-conductor results [8]. However, the forms in (10.64) are more suitable for digital computer computation [7].

Substituting the terminal conditions given in (10.56) into (10.63) gives a form of solution for the terminal currents that may be used as an alternative to the solution in (10.62) [2]:

$$(\hat{\boldsymbol{\phi}}_{12} - \hat{\boldsymbol{\phi}}_{11}\hat{\mathbf{Z}}_S - \hat{\mathbf{Z}}_L\hat{\boldsymbol{\phi}}_{22} + \hat{\mathbf{Z}}_L\hat{\boldsymbol{\phi}}_{21}\hat{\mathbf{Z}}_S)\hat{\mathbf{I}}(0) = (\hat{\mathbf{Z}}_L\hat{\boldsymbol{\phi}}_{21} - \hat{\boldsymbol{\phi}}_{11})\hat{\mathbf{V}}_S \quad (10.65a)$$

$$\hat{\mathbf{I}}(\mathscr{L}) = \hat{\boldsymbol{\phi}}_{21}\hat{\mathbf{V}}_S + (\hat{\boldsymbol{\phi}}_{22} - \hat{\boldsymbol{\phi}}_{21}\hat{\mathbf{Z}}_S)\hat{\mathbf{I}}(0) \quad (10.65b)$$

Once (10.65a) is solved for the terminal currents at $z = 0$, $\hat{\mathbf{I}}(0)$, the terminal currents at $z = \mathscr{L}$, $\hat{\mathbf{I}}(\mathscr{L})$, are obtained from (10.65b). This form is particularly useful, and can be programmed in a FORTRAN code for solution with a digital computer [7].

10.1.4.2 Exact Solution for Lossless Lines in Homogeneous Media
The above results, although providing the exact solution, do not provide insight into the mechanism of crosstalk. It is possible to solve these in *literal form*; that is, in terms of symbols rather than numbers, for the special but important case of a *three-conductor, lossless line* that is immersed in a *homogeneous medium*. In other words, we assume that the conductors are perfect conductors and the medium surrounding the conductors is lossless and homogeneous. This approximation gives results that correspond closely to physical results for many practical configurations, and will provide important insight into the crosstalk phenomenon. The following solution was given in [9].

For a *lossless* line $\mathbf{R} = \mathbf{G} = 0$ and $\hat{\mathbf{Z}} = j\omega\mathbf{L}$ and $\hat{\mathbf{Y}} = j\omega\mathbf{C}$. Since the line is immersed in a homogeneous medium, the per-unit-length inductance and capacitance matrices satisfy $\mathbf{LC} = \mu\epsilon\mathbf{1}_2$, which is diagonal. Therefore

$$\hat{\mathbf{Y}}\hat{\mathbf{Z}} = -\beta^2\mathbf{1}_2 \quad (10.66)$$

where β is the phase constant for the medium:

$$\beta = \omega\sqrt{\mu\epsilon} \quad (10.67)$$

Thus the product $\hat{\mathbf{Y}}\hat{\mathbf{Z}}$ is already diagonal, so that

$$\hat{\mathbf{T}} = \mathbf{1}_2 \quad (10.68)$$

and

$$\hat{\boldsymbol{\gamma}} = j\beta\mathbf{1}_2 = j\frac{\omega}{v}\mathbf{1}_2 \quad (10.69)$$

where $v = 1/\sqrt{\mu\epsilon}$ is the velocity of propagation on the line. The entries in the chain parameter matrix in (10.64) reduce to

$$\hat{\Phi}_{11} = \cos \beta \mathscr{L} \, \mathbf{1}_2 \tag{10.70a}$$

$$\hat{\Phi}_{12} = -jv \sin \beta \mathscr{L} \, \mathbf{L} \tag{10.70b}$$

$$= -j\omega \mathscr{L} \frac{\sin \beta \mathscr{L}}{\beta \mathscr{L}} \mathbf{L}$$

$$\hat{\Phi}_{21} = -jv \sin \beta \mathscr{L} \, \mathbf{C} \tag{10.70c}$$

$$= -j\omega \mathscr{L} \frac{\sin \beta \mathscr{L}}{\beta \mathscr{L}} \mathbf{C}$$

$$\hat{\Phi}_{22} = \cos \beta \mathscr{L} \, \mathbf{1}_2 \tag{10.70d}$$

Substituting these into (10.65a) gives

$$\left[\cos \beta \mathscr{L} (\hat{\mathbf{Z}}_S + \hat{\mathbf{Z}}_L) + \frac{\sin \beta \mathscr{L}}{\beta \mathscr{L}} (\hat{\mathbf{Z}}_L j\omega \mathbf{C} \mathscr{L} \hat{\mathbf{Z}}_S + j\omega \mathbf{L}) \right] \hat{\mathbf{I}}(0) \tag{10.71}$$

$$= \left(\cos \beta \mathscr{L} \, \mathbf{1}_2 + \frac{\sin \beta \mathscr{L}}{\beta \mathscr{L}} \hat{\mathbf{Z}}_L j\omega \mathbf{C} \mathscr{L} \right) \hat{\mathbf{V}}_S$$

A similar equation can be used for determining $\hat{\mathbf{I}}(\mathscr{L})$ as an alternative to (10.65b). This can be obtained by observing that the line is *reciprocal*, giving [9]

$$\left[\cos \beta \mathscr{L} (\hat{\mathbf{Z}}_S + \hat{\mathbf{Z}}_L) + \frac{\sin \beta \mathscr{L}}{\beta \mathscr{L}} (\hat{\mathbf{Z}}_S j\omega \mathbf{C} \mathscr{L} \hat{\mathbf{Z}}_L + j\omega \mathbf{L}) \right] \hat{\mathbf{I}}(\mathscr{L}) = \hat{\mathbf{V}}_S \tag{10.72}$$

Equations (10.71) and (10.72) were solved in *literal form* in [9], giving equations for the near-end and far-end crosstalk voltages as

$$\hat{V}_R(0) = \hat{V}_{NE} \tag{10.73a}$$

$$= \frac{S}{\text{Den}} \left[\frac{R_{NE}}{R_{NE} + R_{FE}} j\omega l_m \mathscr{L} \left(C + \frac{j2\pi \mathscr{L}/\lambda}{\sqrt{1 - k^2}} \alpha_{LG} S \right) \hat{I}_{G_{DC}} \right.$$

$$\left. + \frac{R_{NE} R_{FE}}{R_{NE} + R_{FE}} j\omega c_m \mathscr{L} \left(C + \frac{j2\pi \mathscr{L}/\lambda}{\sqrt{1 - k^2}} \frac{1}{\alpha_{LG}} S \right) \hat{V}_{G_{DC}} \right]$$

$$\hat{V}_R(\mathscr{L}) = \hat{V}_{FE} \tag{10.73b}$$

$$= \frac{S}{\text{Den}} \left(-\frac{R_{FE}}{R_{NE} + R_{FE}} j\omega l_m \mathscr{L} \hat{I}_{G_{DC}} + \frac{R_{NE} R_{FE}}{R_{NE} + R_{FE}} j\omega c_m \mathscr{L} \hat{V}_{G_{DC}} \right)$$

and

$$\text{Den} = C^2 - S^2\omega^2\tau_G\tau_R\left[1 - k^2\frac{(1 - \alpha_{SG}\alpha_{LR})(1 - \alpha_{LG}\alpha_{SR})}{(1 + \alpha_{SR}\alpha_{LR})(1 + \alpha_{SG}\alpha_{LG})}\right] + j\omega CS(\tau_G + \tau_R)$$

$$(10.73c)$$

The various parameters in (10.73) are defined in [9] as

$$C = \cos \beta\mathscr{L} \tag{10.74a}$$

$$S = \frac{\sin \beta\mathscr{L}}{\beta\mathscr{L}} \tag{10.74b}$$

$$k = \frac{l_m}{\sqrt{l_G l_R}} = \frac{c_m}{\sqrt{(c_G + c_m)(c_R + c_m)}}, \quad k \leq 1 \tag{10.74c}$$

$$\tau_G = \frac{l_G\mathscr{L}}{R_S + R_L} + (c_G + c_m)\mathscr{L}\frac{R_S R_L}{R_S + R_L} \tag{10.74d}$$

$$\tau_R = \frac{l_R\mathscr{L}}{R_{NE} + R_{FE}} + (c_R + c_m)\mathscr{L}\frac{R_{NE} R_{FE}}{R_{NE} + R_{FE}} \tag{10.74e}$$

$$\hat{V}_{G_{DC}} = \frac{R_L}{R_S + R_L}\hat{V}_S \tag{10.74f}$$

$$\hat{I}_{G_{DC}} = \frac{1}{R_S + R_L}\hat{V}_S \tag{10.74g}$$

$$\hat{Z}_{CG} = \sqrt{\frac{l_G}{c_G + c_m}} = vl_G\sqrt{1 - k^2} \tag{10.74h}$$

$$\hat{Z}_{CR} = \sqrt{\frac{l_R}{c_R + c_m}} = vl_R\sqrt{1 - k^2} \tag{10.74i}$$

The factor k is referred to as the *coupling coefficient* between the generator and receptor circuits, and is always less than unity. This is analogous to the coupling coefficient of two coupled inductors [6]. The two circuits are said to be *weakly coupled* if $k \ll 1$. The quantities $\hat{Z}_{CG}$ and $\hat{Z}_{CR}$ are referred to as the *characteristic impedances of each circuit in the presence of the other circuit*. The quantities τ_G and τ_R are referred to as the *time constants of the circuits*. The voltage and current of the generator circuit *for dc excitation* (no coupling exists to the receptor circuit for dc) are denoted as $\hat{V}_{G_{DC}}$ and $\hat{I}_{G_{DC}}$. And finally, the remaining terms give the ratio of a termination resistance to the characteristic impedance of that line:

$$\alpha_{SG} = \frac{R_S}{Z_{CG}} \tag{10.75a}$$

$$\alpha_{LG} = \frac{R_L}{Z_{CG}} \tag{10.75b}$$

$$\alpha_{SR} = \frac{R_{NE}}{Z_{CR}} \tag{10.75c}$$

$$\alpha_{LR} = \frac{R_{FE}}{Z_{CR}} \tag{10.75d}$$

A particular termination is referred to as being a *low-impedance load* or a *high-impedance load* with reference to whether it is less than or greater than the characteristic of the circuit to which it is attached, i.e., the particular α is less than or greater than unity. We will find this notion of low- and high-impedance terminations to be critical in explaining the crosstalk phenomenon. Although we have used resistive terminations throughout, R_S, R_L, R_{NE}, and R_{FE}, the above results and all other results for the sinusoidal steady state hold for complex-valued terminations, $\hat{Z}_S$, $\hat{Z}_L$, $\hat{Z}_{NE}$, and $\hat{Z}_{FE}$.

10.1.4.3 Inductive and Capacitive Coupling

The above results are an *exact solution*. They may be simplified if we make the following reasonable assumptions. First, let us assume that *the line is electrically short* at the frequency of interest; that is, $\mathscr{L} \ll \lambda$. For an electrically short line we obtain the following approximations

$$C \cong 1 \quad \text{for } \mathscr{L} \ll \lambda \tag{10.76a}$$

$$S \cong 1 \tag{10.76b}$$

Next, we will assume that the generator and receptor circuits are *weakly coupled*; that is, $k \ll 1$. This assumption coupled with the assumption of an electrically short line gives

$$\text{Den} \cong (1 + j\omega\tau_G)(1 + j\omega\tau_R) \quad \text{for small } k, \ \mathscr{L} \ll \lambda \tag{10.77}$$

The exact results in (10.73) simplify to

$$\hat{V}_{NE} = \frac{1}{\text{Den}} \left(\frac{R_{NE}}{R_{NE} + R_{FE}} j\omega l_m \mathscr{L} \hat{I}_{G_{DC}} + \frac{R_{NE}R_{FE}}{R_{NE} + R_{FE}} j\omega c_m \mathscr{L} \hat{V}_{G_{DC}} \right) \tag{10.78a}$$

$$\hat{V}_{FE} = \frac{1}{\text{Den}} \left(-\frac{R_{FE}}{R_{NE} + R_{FE}} j\omega l_m \mathscr{L} \hat{I}_{G_{DC}} + \frac{R_{NE}R_{FE}}{R_{NE} + R_{FE}} j\omega c_m \mathscr{L} \hat{V}_{G_{DC}} \right) \tag{10.78b}$$

This important result suggests that for *weakly coupled, electrically short lines the coupling (crosstalk) is a linear combination of contributions due to the mutual inductance* l_m *between the two circuits (inductive coupling) and the mutual capacitance* c_m *between the two circuits (capacitive coupling)*! This is often referred to as the *principle of the superposition of inductive and capacitive coupling.*

If, in addition to assuming that the line is weakly coupled and electrically short, we also assume that *the frequency of excitation is sufficiently small* that Den $\cong 1$, the results in (10.78) simplify to

$$\hat{V}_{NE} = \frac{R_{NE}}{R_{NE} + R_{FE}} j\omega l_m \mathscr{L} \hat{I}_{G_{DC}} + \frac{R_{NE} R_{FE}}{R_{NE} + R_{FE}} j\omega c_m \mathscr{L} \hat{V}_{G_{DC}} \qquad (10.79a)$$

$$\hat{V}_{FE} = - \frac{R_{FE}}{R_{NE} + R_{FE}} j\omega l_m \mathscr{L} \hat{I}_{G_{DC}} + \frac{R_{NE} R_{FE}}{R_{NE} + R_{FE}} j\omega c_m \mathscr{L} \hat{V}_{G_{DC}} \qquad (10.79b)$$

These formulae can be easily computed from the simple equivalent circuit for the receptor circuit shown in Fig. 10.12. This is also an intuitively reasonable result. The independent voltage source, $j\omega l_m \mathscr{L} \hat{I}_{G_{DC}}$, represents the *induced emf* in the receptor circuit that is due to the flux from the *current of the generator circuit* that penetrates the receptor circuit, according to Faraday's law. The magnetic flux penetrating the receptor circuit is, according to (10.19b), $\psi_R \mathscr{L} = L_m I_G$, where $L_m = l_m \mathscr{L}$ is the *total mutual inductance between the generator and receptor circuits*. The independent current source, $j\omega c_m \mathscr{L} \hat{V}_{G_{DC}}$, represents the *induced* charge in the receptor circuit that is due to the *voltage of the generator circuit*. The charge induced in the receptor circuit is, according to (10.23), $q_R \mathscr{L} = - C_m V_G$, where $C_m = c_m \mathscr{L}$ is the *total mutual capacitance between the generator and receptor circuits*.

This notion of the crosstalk being the superposition of two components, one due to mutual inductance between the two circuits (inductive coupling) and one to mutual capacitance between the two circuits (capacitive coupling), is the key to understanding the phenomenon of crosstalk. Intuitively, we would expect the inductive coupling component to dominate the capacitive coupling component for *low-impedance loads* (high currents, low voltages), and would expect the capacitive coupling component to dominate the inductive coupling component for *high-impedance loads* (low currents, high voltages). From (10.79) we can

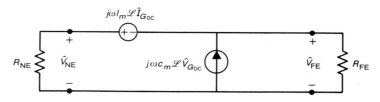

FIGURE 10.12   The simple inductive–capacitive coupling model of the receptor circuit for sinusoidal excitation.

compute that the *inductive coupling* component will *dominate* the *capacitive coupling* component in $\hat{V}_{NE}$ if

$$\frac{R_L}{Z_{CG}} \frac{R_{FE}}{Z_{CR}} < 1 \tag{10.80a}$$

and in $\hat{V}_{FE}$ if

$$\frac{R_L}{Z_{CG}} \frac{R_{NE}}{Z_{CR}} < 1 \tag{10.80b}$$

For this case, (10.79) becomes

$$\hat{V}_{NE}^{IND} = \frac{R_{NE}}{R_{NE} + R_{FE}} j\omega l_m \mathscr{L} \hat{I}_{G_{DC}} \tag{10.81a}$$

$$\hat{V}_{FE}^{IND} = -\frac{R_{FE}}{R_{NE} + R_{FE}} j\omega l_m \mathscr{L} \hat{I}_{G_{DC}} \tag{10.81b}$$

Capacitive coupling dominates inductive coupling if these inequalities are reversed, i.e., in $\hat{V}_{NE}$ if

$$\frac{R_L}{Z_{CG}} \frac{R_{FE}}{Z_{CR}} > 1 \tag{10.82a}$$

and in $\hat{V}_{FE}$ if

$$\frac{R_L}{Z_{CG}} \frac{R_{NE}}{Z_{CR}} > 1 \tag{10.82b}$$

For this case (10.79) becomes

$$\hat{V}_{NE}^{CAP} = \frac{R_{NE} R_{FE}}{R_{NE} + R_{FE}} j\omega c_m \mathscr{L} \hat{V}_{G_{DC}} \tag{10.83a}$$

$$\hat{V}_{FE}^{CAP} = \frac{R_{NE} R_{FE}}{R_{NE} + R_{FE}} j\omega c_m \mathscr{L} \hat{V}_{G_{DC}} \tag{10.83b}$$

The total coupling is the sum of these components:

$$\hat{V}_{NE} = \hat{V}_{NE}^{IND} + \hat{V}_{NE}^{CAP} \tag{10.84a}$$

$$\hat{V}_{FE} = \hat{V}_{FE}^{IND} + \hat{V}_{FE}^{CAP} \tag{10.84b}$$

This supports the notion that *inductive coupling is predominant for low-impedance loads and capacitive coupling is predominant for high-impedance loads.* Observe that the capacitive coupling components are of the same sign and are equal in magnitude, $\hat{V}_{NE}^{CAP} = \hat{V}_{FE}^{CAP}$. The inductive coupling components are not equal, and are of opposite sign. This provides an important observation wherein the termination impedances can be chosen such that the inductive and capacitive coupling components can be made equal in $\hat{V}_{NE}$, and will cancel giving $\hat{V}_{NE} = 0$! This is the basis for the *directional coupler* that is commonly employed in microwave circuits.

The crosstalk can be viewed as a *transfer function* between the input $\hat{V}_S$ and the outputs $\hat{V}_{NE}$ and $\hat{V}_{FE}$. These transfer functions can be obtained by substituting

$$\hat{I}_{G_{DC}} = \frac{1}{R_S + R_L} \hat{V}_S \tag{10.85a}$$

$$\hat{V}_{G_{DC}} = \frac{R_L}{R_S + R_L} \hat{V}_S \tag{10.85b}$$

into (10.79) and factoring out $\hat{V}_S$ to give

$$\frac{\hat{V}_{NE}}{\hat{V}_S} = j\omega \left(\frac{R_{NE}}{R_{NE} + R_{FE}} \frac{L_m}{R_S + R_L} + \frac{R_{NE} R_{FE}}{R_{NE} + R_{FE}} \frac{R_L C_m}{R_S + R_L} \right) \tag{10.86a}$$

$$\frac{\hat{V}_{FE}}{\hat{V}_S} = j\omega \left(-\frac{R_{FE}}{R_{NE} + R_{FE}} \frac{L_m}{R_S + R_L} + \frac{R_{NE} R_{FE}}{R_{NE} + R_{FE}} \frac{R_L C_m}{R_S + R_L} \right) \tag{10.86b}$$

These can be written as

$$\frac{\hat{V}_{NE}}{\hat{V}_S} = j\omega(M_{NE}^{IND} + M_{NE}^{CAP}) \tag{10.87a}$$

$$\frac{\hat{V}_{FE}}{\hat{V}_S} = j\omega(M_{FE}^{IND} + M_{FE}^{CAP}) \tag{10.87b}$$

where

$$M_{NE}^{IND} = \frac{R_{NE}}{R_{NE} + R_{FE}} \frac{L_m}{R_S + R_L} \tag{10.88a}$$

$$M_{NE}^{CAP} = \frac{R_{NE} R_{FE}}{R_{NE} + R_{FE}} \frac{R_L C_m}{R_S + R_L} \tag{10.88b}$$

$$M_{FE}^{IND} = -\frac{R_{FE}}{R_{NE} + R_{FE}} \frac{L_m}{R_S + R_L} \tag{10.88c}$$

$$M_{FE}^{CAP} = M_{NE}^{CAP} = \frac{R_{NE} R_{FE}}{R_{NE} + R_{FE}} \frac{R_L C_m}{R_S + R_L} \tag{10.88d}$$

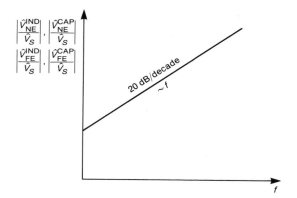

FIGURE 10.13   Frequency response of the crosstalk transfer functions.

Observe that *the inductive and capacitive coupling contributions are direct functions of the excitation frequency*. Therefore they increase linearly with an increase in frequency, so that their frequency responses increase at a rate of +20 dB/decade, as shown in Fig. 10.13. Also *the total crosstalk transfer functions increase at 20 dB/decade. Depending on the load impedances, this frequency response may be due totally to one component or the other, as is illustrated in Fig. 10.14.* This observation will explain the effectiveness (or noneffectiveness) of shielded wires or twisted pairs in the reduction of crosstalk, which will be considered in Sections 10.2 and 10.3.

10.1.4.4 Inclusion of Losses—Common-Impedance Coupling

The foregoing has assumed a lossless line—perfect conductors and a lossless medium. The assumption of a lossless medium is usually a reasonable assumption for frequencies below the GHz range. However, imperfect conductors can produce significant crosstalk *at the lower frequencies*. This is referred to as *common-impedance coupling*, and is easily seen from the following.

Consider the circuit shown in Fig. 10.15(a). For typical loads and low frequencies the resistance of the reference conductor is usually much smaller than these load resistances, so that the majority of the generator wire current returns through the reference conductor. This produces a voltage drop V_0 across that conductor. For an electrically short line we may lump the per-unit-length resistance of the reference conductor as a single resistance $R_0 = r_0 \mathscr{L}$. The voltage drop across the reference conductor is given by

$$\hat{V}_0 = R_0 \hat{I}_G \tag{10.89}$$

$$= \frac{R_0}{R_S + R_L} \hat{V}_S$$

This voltage appears directly in the receptor circuit, producing contributions

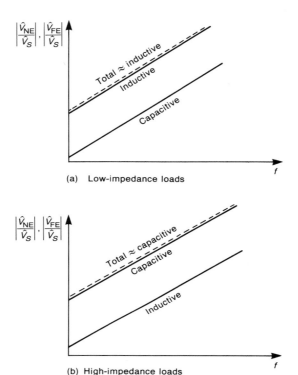

(a)   Low-impedance loads

(b) High-impedance loads

FIGURE 10.14   Effect of load impedance on the dominance of either inductive or capacitive coupling: (a) low-impedance loads; (b) high-impedance loads.

to the crosstalk transfer functions at low frequencies of

$$\frac{\hat{V}_{NE}^{CI}}{\hat{V}_S} = M_{NE}^{CI} \tag{10.90a}$$

$$\frac{\hat{V}_{FE}^{CI}}{\hat{V}_S} = M_{FE}^{CI} \tag{10.90b}$$

where

$$M_{NE}^{CI} = \frac{R_{NE}}{R_{NE} + R_{FE}} \frac{R_0}{R_S + R_L} \tag{10.91a}$$

$$M_{FE}^{CI} = -\frac{R_{FE}}{R_{NE} + R_{FE}} \frac{R_0}{R_S + R_L} \tag{10.91b}$$

This produces an essentially frequency independent "floor" at the lower frequencies, as shown in Fig. 10.15(b).

Common-Impedance Coupling

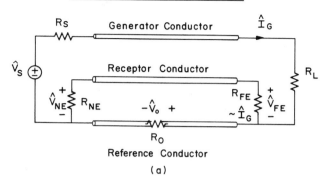

(a)

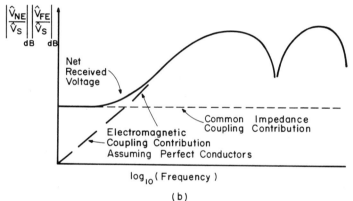

log_{10}(Frequency)

(b)

FIGURE 10.15 Illustration of common-impedance coupling due to nonzero impedance of the reference conductor: (a) model; (b) frequency response of the crosstalk transfer function.

The total coupling is approximately the sum of the inductive, capacitive and common-impedance coupling:

$$\frac{\hat{V}_{NE}}{\hat{V}_S} = j\omega(M_{NE}^{IND} + M_{NE}^{CAP}) + M_{NE}^{CI} \tag{10.92a}$$

$$\frac{\check{V}_{FE}}{\hat{V}_S} = j\omega(M_{FE}^{IND} + M_{FE}^{CAP}) + M_{FE}^{CI} \tag{10.92b}$$

10.1.4.5 *Experimental Results*   As an example, consider the three-wire ribbon cable shown in Fig. 10.16. The cable consists of three #28 gauge (7 × 36) wires of total length $\mathscr{L} = 4.737$ m, which is one wavelength at 63.3 MHz. Thus we might expect the line to be electrically short for frequencies below approximately 6 MHz. The middle wire is the reference conductor. The load resistors will be

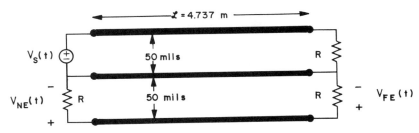

FIGURE 10.16 An experiment illustrating crosstalk using a ribbon cable.

equal, $R_L = R_{NE} = R_{FE} = R$ and $R_S = 0$. We will investigate two values for R: 50 Ω and 1 kΩ. The per-unit-length parameters were computed previously as $l_G = l_R = 0.76$ μH/m, $l_m = 0.24$ μH/m, $c_G = c_R = 11.1$ pF/m, and $c_m = 5.17$ pF/m. The characteristic impedance of each circuit in the presence of the other was computed to be $Z_C = 261.5$ Ω. Therefore we will expect inductive coupling to dominate for $R = 50$ Ω and capacitive coupling for $R = 1$ kΩ. The resistance of the reference wire can be computed by computing the resistance of one of the #36 gauge strands and dividing this by 7 (the number of strands that are electrically in parallel). This gives $r_0 = 0.194$ Ω/m, so that the total resistance is $R_0 = 0.921$ Ω.

For $R = 50$ Ω we compute the contributions of near-end crosstalk transfer function to be

$$M_{NE}^{IND} = \frac{R_{NE}}{R_{NE} + R_{FE}} \frac{L_m}{R_S + R_L}$$

$$= 1.14 \times 10^{-8}$$

$$M_{NE}^{CAP} = \frac{R_{NE} R_{FE}}{R_{NE} + R_{FE}} \frac{R_L C_m}{R_S + R_L}$$

$$= 6.13 \times 10^{-10}$$

$$M_{NE}^{CI} = \frac{R_{NE}}{R_{NE} + R_{FE}} \frac{R_0}{R_S + R_L}$$

$$= 9.21 \times 10^{-3}$$

The inductive coupling component dominates the capacitive coupling component, as expected. The total coupling is

$$\frac{\hat{V}_{NE}}{\hat{V}_S} = j\omega(M_{NE}^{IND} + M_{NE}^{CAP}) + M_{NE}^{CI}$$

$$= j2\pi f(1.14 \times 10^{-8} + 6.13 \times 10^{-10}) + 9.21 \times 10^{-3}$$

$$= j7.55 \times 10^{-8} f + 9.21 \times 10^{-3}$$

The near-end crosstalk transfer function was measured, and the results are shown in Fig. 10.17(a). The exact predictions of the transmission-line model are also shown. Those predictions are within 3 dB of the experimental results for frequencies up to 10 MHz, where $\mathscr{L} \cong \frac{1}{6}\lambda_o$. Observe the common-impedance coupling level of $20 \log_{10}(9.21 \times 10^{-3}) = -40.7$ dB at the lower frequencies. The level of the inductive and capacitive coupling at 1 MHz is computed to be $20 \log_{10}(7.55 \times 10^{-8} \times 10^6) = -22.4$ dB, which is quite close to the experimental value of approximately -23 dB. Also note the $+20$ dB/decade region between 100 kHz and 1 MHz as well as the resonances above 10 MHz, where the line is electrically long.

For $R = 1 \text{ k}\Omega$ we compute the contributions for near-end crosstalk to be

$$M_{NE}^{\text{IND}} = \frac{R_{NE}}{R_{NE} + R_{FE}} \frac{L_m}{R_S + R_L}$$

$$= 5.7 \times 10^{-10}$$

$$M_{NE}^{\text{CAP}} = \frac{R_{NE}R_{FE}}{R_{NE} + R_{FE}} \frac{R_L C_m}{R_S + R_L}$$

$$= 1.23 \times 10^{-8}$$

$$M_{NE}^{\text{CI}} = \frac{R_{NE}}{R_{NE} + R_{FE}} \frac{R_0}{R_S + R_L}$$

$$= 4.61 \times 10^{-4}$$

The capacitive coupling component dominates the inductive coupling component as expected. The total coupling is

$$\frac{\hat{V}_{NE}}{\hat{V}_S} = j\omega(M_{NE}^{\text{IND}} + M_{NE}^{\text{CAP}}) + M_{NE}^{\text{CI}}$$

$$= j2\pi f(5.7 \times 10^{-10} + 1.23 \times 10^{-8}) + 4.61 \times 10^{-4}$$

$$= j8.09 \times 10^{-8}f + 4.61 \times 10^{-4}$$

The near-end crosstalk transfer function was measured, and the results are shown in Fig. 10.17(b). The exact predictions of the transmission-line model are also shown. Those predictions are also within 3 dB up to 10 MHz. Again observe the common-impedance coupling level of $20 \log_{10}(4.61 \times 10^{-4}) = -66.7$ dB at the lower frequencies. The level of the inductive and capacitive coupling at 1 MHz is computed to be $20 \log_{10}(8.09 \times 10^{-8} \times 10^6) = -21.8$ dB, which is quite close to the experimental value of approximately -20 dB. Again note the $+20$ dB/decade region between 10 kHz and 1 MHz as well as the resonances above 10 MHz, where the line is electrically long. Transmission-line modeling of ribbon cables that consist of more than three wires is given in [10].

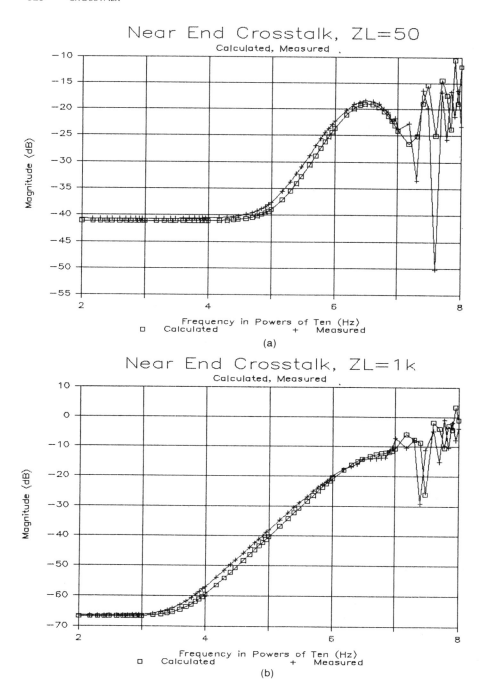

FIGURE 10.17 Frequency response of the near-end crosstalk for the ribbon cable of Fig. 10.16, comparing measured data and the predictions of the transmission-line model for (a) $R = 50\,\Omega$ and (b) $R = 1\,k\Omega$.

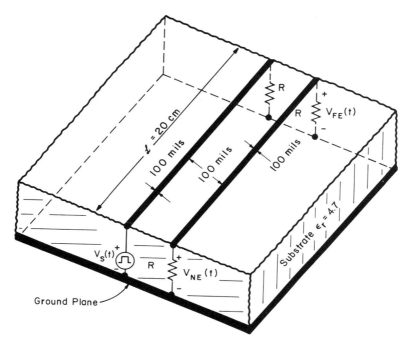

FIGURE 10.18   An experiment illustrating crosstalk using a printed circuit board.

As another example of frequency-domain crosstalk prediction, consider the *coupled microstrip* shown in Fig. 10.18. Two 1 ounce copper lands of width 100 mils and separated (edge to edge) by 100 mils were etched on one side of a glass–epoxy PCB ($\epsilon_r = 4.7$). The board thickness was 62 mils, and a ground plane was placed on the other side. The total length of the lands was 20 cm. The per-unit-length parameters were computed using numerical methods [11, 12]. The per-unit-length capacitance matrix with the dielectric substrate present, **C**, and with the dielectric substrate removed and replaced with air, $\mathbf{C}_0$, were computed. The per-unit-length inductance matrix was then obtained as $\mathbf{L} = \mu_o \epsilon_o \mathbf{C}_0^{-1}$. The effective dielectric constant was found to be $\epsilon_r' \cong 3$. Thus the lands are $\frac{1}{10}\lambda$ at approximately at 86 MHz. Once again, the termination impedances will be equal, $R_L = R_{NE} = R_{FE} = R$ and $R_S = 0$, and two values of R will be investigated: 50 Ω and 1 kΩ. The predictions of the multiconductor transmission-line model (assuming a lossless line) are shown for $R = 50\ \Omega$ in Fig. 10.19(a) and for $R = 1$ kΩ in Fig. 10.19(b) [13]. The prediction accuracies are within 1 dB up to 250 MHz for $R = 50\ \Omega$ and within 3 dB up to 100 MHz for $R = 1$ kΩ. The 1 kΩ loads were constructed by inserting 1 kΩ resistors into BNC cable termination connectors, so that these loads could be easily removed. This places a capacitance of some 10 pF in parallel with the 1 kΩ resistance, which degrades its high-frequency behavior more than for the 50 Ω loads, which were likewise constructed. This explains the seeming deterioration of the

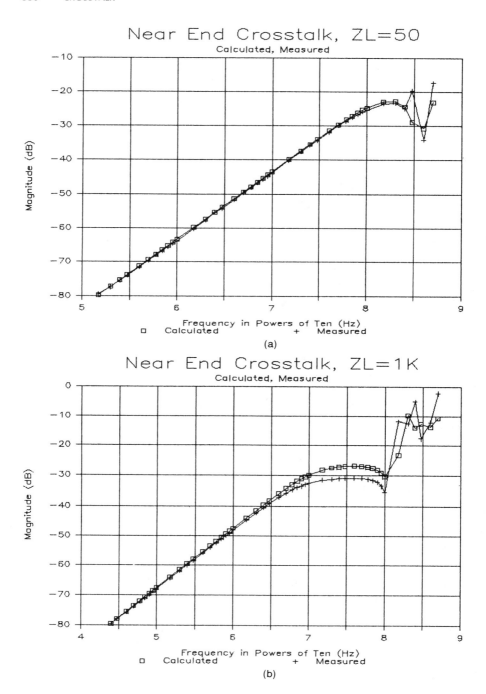

FIGURE 10.19 Frequency response of a near-end crosstalk for the printed circuit board of Fig. 10.18, comparing measured data and the predictions of the transmission-line model for (a) $R = 50 \, \Omega$ and (b) $R = 1 \, k\Omega$.

predictions at a frequency lower than for the 50 Ω loads. Using the measured (or computed) data in the frequency range where the crosstalk varies at 20 dB/decade we can determine that

$$M_{NE} = 1.06 \times 10^{-10} \quad \text{for } R = 50 \, \Omega$$

and

$$M_{NE} = 6.37 \times 10^{-10} \quad \text{for } R = 1 \, \text{k}\Omega$$

These crosstalk coefficients are obtained by observing that in the region where the inductive and capacitive coupling dominates (the frequency response increases at 20 dB/decade) the magnitude of the transfer function is, according to (10.86a) and (10.87a),

$$\left| \frac{\hat{V}_{NE}}{\hat{V}_S} \right| = \omega \left(\frac{L_m}{2R} + \frac{R}{2} C_m \right) \tag{10.93}$$

$$= \omega (M_{NE}^{IND} + M_{NE}^{CAP})$$

$$= 2 \pi f M_{NE}$$

Thus we simply determine the value of the transfer function at a frequency f_o in this range and divide that value by $2\pi f_o$. In fact, using two values of M_{NE} obtained for $R = 50 \, \Omega$ and for $R = 1 \, \text{k}\Omega$ we can determine the total mutual inductance L_m and capacitance C_m from (10.93) and determine the per-unit-length values as

$$l_m = \frac{L_m}{\mathscr{L}}$$

$$= 37.2 \, \text{nH/m}$$

$$= 0.944 \, \text{nH/inch}$$

and

$$c_m = \frac{C_m}{\mathscr{L}}$$

$$= 6.33 \, \text{pF/m}$$

$$= 0.161 \, \text{pF/inch}$$

Predictions of the transmission-line model for PCBs that consist of more than three conductors are given in [14].

These experimental data have shown that the exact multiconductor transmission-line model gives accurate predictions for frequencies where the line is electrically short. Resonances due to standing waves for frequencies where the line is electrically long make the predictions very difficult. The approximate model of consisting of the superposition of inductive, capacitive, and common-impedance coupling contributions gives accurate predictions for almost the same frequency range as the transmission-line model, and the calculations are considerably simpler. There are cases in which the termination impedances differ considerably from the line characteristic impedances where the simple model predictions differ from those of the transmission-line model at a frequency where the line is electrically short [15].

10.1.5 Time-Domain (Transient) Crosstalk

We now address the problem of predicting the near-end and far-end induced voltages and currents for the coupled three-conductor line for general time variation of the source $V_S(t)$. Once again, we assume that the TEM mode of propagation is the only mode of propagation on the line, so that the multiconductor transmission-line equations adequately characterize the line voltages and currents. As for the case of a two-conductor line, line losses, and particularly the skin-effect resistance of the line conductors, complicate the time-domain solution. The losses in the surrounding medium can generally be ignored with no significant loss in accuracy. The skin-effect resistance of the line conductors tends to be important at low and high frequencies. At low frequencies the resistance of the line conductors is typically constant, and contributes a constant level of coupling that is referred to as common-impedance coupling. This will be included in an approximate manner, as was done for frequency-domain crosstalk, where we added the contribution to the inductive and capacitive contributions. The high-frequency resistance of the line conductors tends to influence the sharpness of the pulse transitions, as discussed previously.

Exact solutions for the MTL equations for the frequency-domain crosstalk were given in Section 10.1.4.1. Similar exact solutions of the MTL equations for time-domain crosstalk are much more difficult to formulate. An exact SPICE model is given in Section 10.1.7. In the next section we will examine the time-domain solution using the simple but approximate inductive–capacitive coupling model that was developed for the frequency domain in the preceding sections.

10.1.5.1 Inductive and Capacitive Coupling The *frequency-domain* (*sinusoidal steady state*) transmission-line equations for *lossless, three-conductor lines in homogeneous media* were solved exactly in literal form in Section 10.1.4.2. These were then specialized to the case of an *electrically short line* and *weak coupling* between the generator and receptor circuits and *low frequencies*, which resulted in the simple inductive–capacitive coupling model. For this specialization the equations for the *phasor* near-end and far-end frequency-domain crosstalk voltages become

$$\hat{V}_{NE}(j\omega) = j\omega M_{NE}\,\hat{V}_S(j\omega) \qquad (10.94a)$$

$$\hat{V}_{FE}(j\omega) = j\omega M_{FE}\,\hat{V}_S(j\omega) \qquad (10.94b)$$

where the crosstalk transfer coefficients are

$$M_{NE} = \underbrace{\frac{R_{NE}}{R_{NE} + R_{FE}}\frac{L_m}{R_S + R_L}}_{M_{NE}^{\text{IND}}} + \underbrace{\frac{R_{NE}R_{FE}}{R_{NE} + R_{FE}}\frac{R_L C_m}{R_S + R_L}}_{M_{NE}^{\text{CAP}}} \qquad (10.95a)$$

$$M_{FE} = \underbrace{-\frac{R_{FE}}{R_{NE} + R_{FE}}\frac{L_m}{R_S + R_L}}_{M_{FE}^{\text{IND}}} + \underbrace{\frac{R_{NE}R_{FE}}{R_{NE} + R_{FE}}\frac{R_L C_m}{R_S + R_L}}_{M_{FE}^{\text{CAP}}} \qquad (10.95b)$$

The term $j\omega$ in the frequency domain translates to d/dt in the time domain:

$$j\omega \rightarrow \frac{d}{dt} \qquad (10.96)$$

Therefore the above frequency-domain results translate to the time domain as

$$V_{NE}(t) = M_{NE}\frac{dV_S(t)}{dt} \qquad (10.97a)$$

$$V_{FE}(t) = M_{FE}\frac{dV_S(t)}{dt} \qquad (10.97b)$$

This important result shows that *the frequency components of the input signal for which the line is electrically short are processed by the line to give an output that is the derivative of the source voltage multiplied by the crosstalk coefficients M_{NE} and M_{FE}!* The time-domain crosstalk is the sum of inductive- and capacitive-coupling contributions due to the mutual inductance and capacitance between the generator and receptor circuits. A simple time-domain equivalent circuit for the receptor circuit is shown in Fig. 10.20. Figure 10.21 shows the crosstalk voltages for a periodic, trapezoidal pulse train source voltage that may represent a clock or data signal. Observe that the crosstalk signal appears as pulses occurring during the transitions of the source voltage. During the risetime of the pulse, the crosstalk voltage is a positive pulse if the crosstalk coefficient is positive. During the falltime, where the slope of the source is negative, the crosstalk voltage is a negative pulse if the crosstalk coefficient is positive. The crosstalk coefficient for near-end crosstalk, M_{NE}, is always positive,

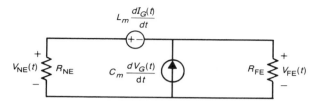

FIGURE 10.20 The simple inductive–capacitive coupling model of the receptor circuit for a general, time-domain excitation.

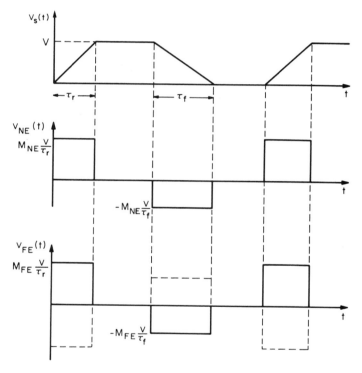

FIGURE 10.21 Time-domain crosstalk prediction of the inductive–capacitive coupling model for a trapezoidal pulse train excitation.

as shown by (10.95a). The far-end crosstalk coefficient M_{FE}, given in (10.95b), is positive if capacitive coupling dominates inductive coupling. If inductive coupling dominates capacitive coupling, the far-end crosstalk coefficient is negative.

The primary restrictions on the use of this simple result are that the line be electrically short and weakly coupled. For a time-domain signal that contains frequency components from dc to (theoretically) infinite frequency only those components of this signal that are below the frequency where the line is electrically short will be correctly processed by this simple model. The higher-frequency components will not be correctly processed. For a given line length and velocity of propagation this places restrictions on the spectral content of the signal. In order to obtain some relation between the frequency-domain and corresponding time-domain restrictions, consider the spectrum of a periodic, trapezoidal pulse train shown in Fig. 10.22 that was obtained in Chapter 7. The actual spectrum consists of discrete frequency components at the fundamental frequency $f_o = 1/T$ and the higher harmonics that follow an envelope that is the product of two $(\sin x)/x$ variations. We developed bounds on that exact

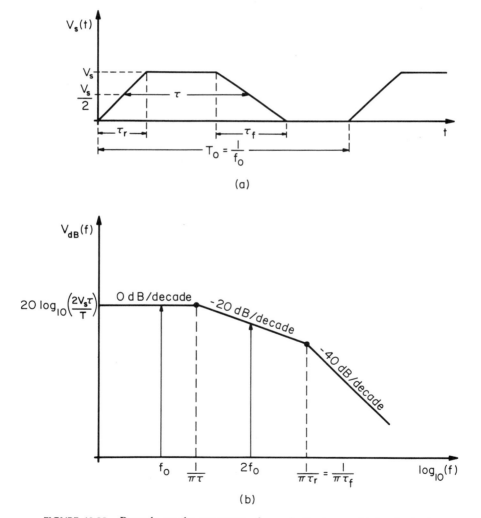

FIGURE 10.22 Bounds on the spectrum of a periodic, trapezoidal pulse train.

spectrum, as shown in Fig. 10.22(b). In order to do that, we assumed that the rise- and falltimes of the pulse are equal, $\tau_r = \tau_f$. The frequency response of the line is shown in Fig. 10.23. Let us assume that the line is "electrically short" for frequencies below f_u. The spectrum of the crosstalk pulse is given by the sum (Bode plots) of the spectrum of the input signal and the spectrum of the transfer function, as shown in Fig. 10.24. It is reasonable to assume that those components of the input spectrum at some point past the second break point in the -40 dB/decade region do not contribute significantly to the overall pulse amplitude. Therefore incorrect processing of these components will not significantly affect the resulting crosstalk prediction. Therefore we will choose to correctly process those frequencies such that

$$f \ll \alpha \frac{1}{\pi \tau_r} \qquad (10.98)$$

where the constant α is chosen greater than unity, $\alpha > 1$. But the line length is

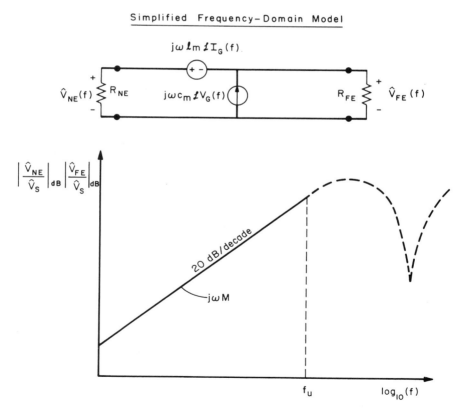

FIGURE 10.23   Frequency response of the simple inductive–capacitive coupling model.

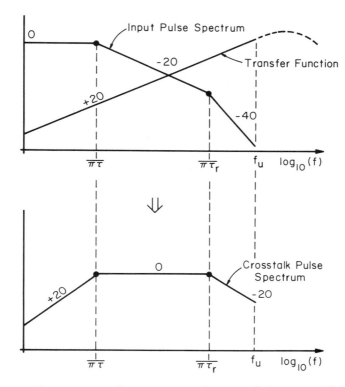

FIGURE 10.24 Output crosstalk spectrum as the sum of the trapezoidal pulse train spectral bounds of Fig. 10.22 and the transfer function of Fig. 10.23.

some fraction of a wavelength at the upper frequency:

$$\mathscr{L} = k\lambda|_{f=f_u} \tag{10.99}$$

Substituting (10.98) into (10.99) along with the fundamental relation $\lambda = v/f$ gives

$$\tau_r > \frac{\alpha}{\pi k}\frac{\mathscr{L}}{v} = \frac{\alpha}{\pi k}T_D \tag{10.100}$$

where $T_D = \mathscr{L}/v$ is the one-way transit time (time delay) of the line. A reasonable choice for these constants would be $\alpha = \pi$ and $k = \frac{1}{10}$. This gives the restriction on the pulse rise/falltime as

$$\tau_r, \tau_f > 10T_D \tag{10.101}$$

This criterion is not meant to be precise, but is only intended to give a notion about those pulse trains that will or will not be processed correctly by the simple model.

10.1.5.2 Inclusion of Losses—Common-Impedance Coupling   Once again, we may include the resistance of the line conductors in an approximate manner by adding the common-impedance coupling coefficients to the above inductive and capacitive coupling contributions:

$$V_{NE}(t) = (M_{NE}^{IND} + M_{NE}^{CAP})\frac{dV_S(t)}{dt} + M_{NE}^{CI} V_S(t) \qquad (10.102a)$$

$$V_{FE}(t) = (M_{FE}^{IND} + M_{FE}^{CAP})\frac{dV_S(t)}{dt} + M_{FE}^{CI} V_S(t) \qquad (10.102b)$$

where the common-impedance coupling M_{NE}^{CI} and M_{FE}^{CI} are given by (10.91):

$$M_{NE}^{CI} = \frac{R_{NE}}{R_{NE} + R_{FE}}\frac{R_0}{R_S + R_L} \qquad (10.103a)$$

$$M_{FE}^{CI} = -\frac{R_{FE}}{R_{NE} + R_{FE}}\frac{R_0}{R_S + R_L} \qquad (10.103b)$$

The effect is to add a reduced replica of $V_S(t)$, $M_{NE}^{CI} V_S(t)$ and $M_{FE}^{CI} V_S(t)$, to the weighted derivative of that waveform, $(M_{NE}^{IND} + M_{NE}^{CAP})\,dV_S(t)/dt$ and $(M_{FE}^{IND} + M_{FE}^{CAP})\,dV_S(t)/dt$.

10.1.5.3 Experimental Results   We will now show some experimental results that illustrate the prediction accuracy of the model. These are the configurations that were considered for the frequency-domain results in Section 10.1.4.5. First consider the ribbon cable shown in Fig. 10.16. The source voltage $V_S(t)$ will be a 2.5 V, 20 kHz trapezoidal pulse train having a 50% duty cycle and rise/falltimes of 400 ns. This is typical of a RS-232 serial data stream. The one-way time delay is

$$T_D = \frac{\mathscr{L}}{v_o} = \frac{4.737}{3 \times 10^8} = 15.8 \text{ ns}$$

The maximum rise/falltimes that may be considered are

$$\tau_r, \tau_f \gtrsim 150 \text{ ns}$$

which this waveform satisfies. The near-end crosstalk coefficients were computed for this case previously for $R_S = 0$ and $R_L = R_{NE} = R_{FE} = 50\ \Omega$ as

$$M_{NE}^{IND} = 1.14 \times 10^{-8}$$

$$M_{NE}^{CAP} = 6.13 \times 10^{-10}$$

$$M_{NE}^{CI} = 9.21 \times 10^{-3}$$

Therefore the near-end crosstalk waveform is given by

$$V_{NE}(t) = 1.20 \times 10^{-8} \frac{dV_S(t)}{dt} + 9.21 \times 10^{-3}\ V_S(t)$$

The "slew rate" of the pulse train is

$$\left| \frac{dV_S(t)}{dt} \right| = \frac{2.5\ \text{V}}{400\ \text{ns}} = 6.25 \times 10^6\ \text{V/s}$$

which gives peak pulse amplitudes (ignoring losses) of 75 mV. The total coupling is sketched in Fig. 10.25(a), and the predicted values are shown circled. The experimentally observed waveform is shown in Fig. 10.25(b). Comparing the two shows good prediction, which is a direct result of the choice of rise/falltimes. Observe the crosstalk pulses occurring during the transition of the source waveform. Also observe the 23 mV "offset" created by the common-impedance coupling during the time where the input pulse is at its maximum level of 2.5 V.

We next consider the ribbon cable with 1 kΩ loads: $R_S = 0$ and $R_L = R_{NE} = R_{FE} = 1\ \text{k}\Omega$. The coupling coefficients were computed previously as

$$M_{NE}^{IND} = 5.7 \times 10^{-10}$$

$$M_{NE}^{CAP} = 1.23 \times 10^{-8}$$

$$M_{NE}^{CI} = 4.61 \times 10^{-4}$$

Therefore the near-end crosstalk waveform is given by

$$V_{NE}(t) = 1.29 \times 10^{-8} \frac{dV_S(t)}{dt} + 4.61 \times 10^{-4}\ V_S(t)$$

The "slew rate" of the pulse train remains

$$\left| \frac{dV_S(t)}{dt} \right| = \frac{2.5\ \text{V}}{400\ \text{ns}} = 6.25 \times 10^6\ \text{V/s}$$

which gives peak pulse amplitudes (ignoring losses) of 80.4 mV. The total coupling is sketched in Fig. 10.26(a), and the predicted values are shown circled.

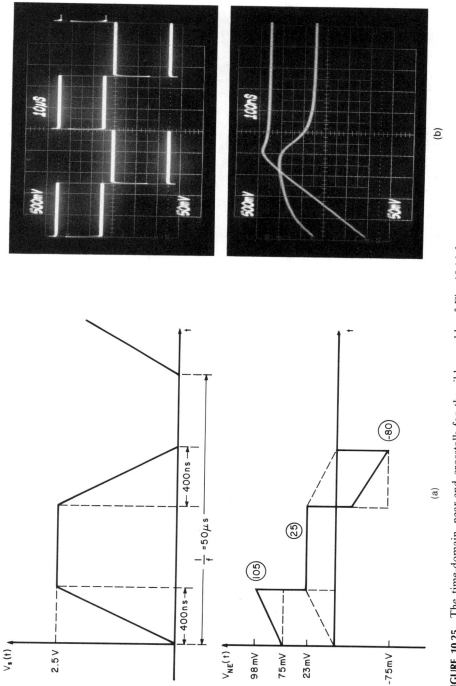

FIGURE 10.25 The time-domain, near-end crosstalk for the ribbon cable of Fig. 10.16 for a trapezoidal pulse train input for $R = 50\ \Omega$: (a) predicted waveform; (b) measured waveform.

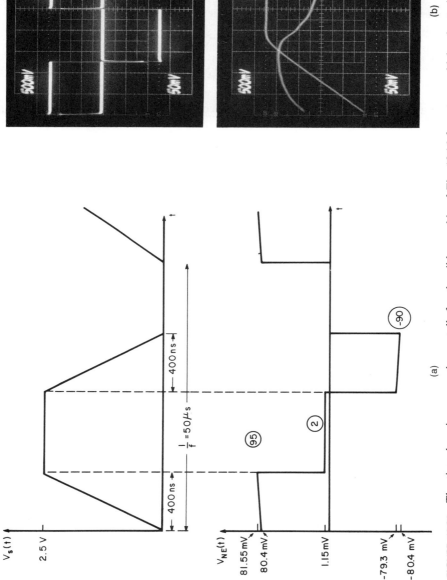

FIGURE 10.26 The time-domain, near-end crosstalk for the ribbon cable of Fig. 10.16 for a trapezoidal pulse train input for $R = 1 \text{ k}\Omega$: (a) predicted waveform; (b) measured waveform.

The experimentally observed waveform is shown in Fig. 10.26(b). Comparing the two shows good prediction, which again is a direct result of the choice of rise/falltimes. Observe the crosstalk pulses occurring during the transition of the source waveform. Also observe that the small 1.15 mV "offset" created by the common-impedance coupling during the time where the input pulse is at its maximum level of 2.5 V is too small to be observed.

Next we will consider the printed circuit board shown in Fig. 10.18 that was considered previously. The source voltage $V_S(t)$ will be a 2.5 V, 1 MHz trapezoidal pulse train having a 50% duty cycle and rise/falltimes of 50 ns. The one-way delay is

$$T_D = \frac{\mathcal{L}}{v_o} = \frac{20 \text{ cm}}{3 \times 10^8} = 1.16 \text{ ns}$$

The maximum rise/falltimes that may be considered are

$$\tau_r, \tau_f \gtrsim 10 \text{ ns}$$

which this waveform satisfies. The near-end crosstalk coefficient was obtained from the frequency-domain experimental results for this case previously for $R_S = 0$ and $R_L = R_{NE} = R_{FE} = 50 \ \Omega$ as

$$M_{NE} = 1.06 \times 10^{-10}$$

The common-impedance coupling was too small to observe, so it will be assumed to be zero. Therefore the near-end crosstalk waveform is given by

$$V_{NE}(t) = 1.06 \times 10^{-10} \frac{dV_S(t)}{dt}$$

The "slew rate" of the pulse train is

$$\left| \frac{dV_S(t)}{dt} \right| = \frac{2.5 \text{ V}}{50 \text{ ns}} = 5.0 \times 10^7 \text{ V/s}$$

which gives peak pulse amplitudes (ignoring losses) of 5.3 mV. The experimentally observed waveform is shown in Fig. 10.27. Observe the crosstalk pulses occurring during the transition of the source waveform. The peak measured voltage is 5.5 mV.

We next consider the PCB with 1 kΩ loads: $R_S = 0$ and $R_L = R_{NE} = R_{FE} = 1 \text{ k}\Omega$. The coupling coefficient was obtained previously from the measured frequency-domain data as

$$M_{NE} = 6.37 \times 10^{-10}$$

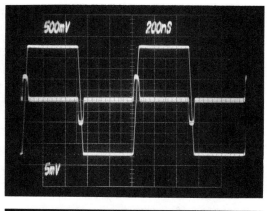

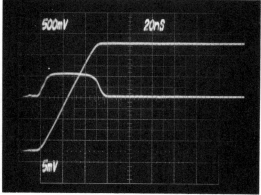

FIGURE 10.27 Measured near-end crosstalk waveforms for the printed circuit board of Fig. 10.18 for $R = 50\,\Omega$.

Therefore the near-end crosstalk waveform is given by

$$V_{NE}(t) = 6.37 \times 10^{-10} \frac{dV_S(t)}{dt}$$

The "slew rate" of the pulse train remains

$$\left| \frac{dV_S(t)}{dt} \right| = \frac{2.5\ \text{V}}{50\ \text{ns}} = 5.0 \times 10^7\ \text{V/s}$$

which gives peak pulse amplitudes (ignoring losses) of 31.9 mV. The experimentally observed waveform is shown in Fig. 10.28. Observe the crosstalk pulses occurring during the transition of the source waveform. The measured peak voltage is 24 mV, which is somewhat lower than the predicted value of 31.9 mV. This

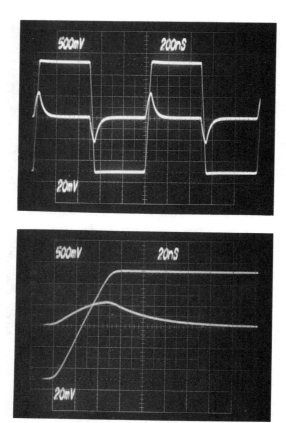

FIGURE 10.28 Measured near-end crosstalk waveforms for the printed circuit board of Fig. 10.18 for $R = 1 \text{ k}\Omega$.

error is primarily due to the difficulty in constructing a high-impedance 1 kΩ load.

10.1.6 Lumped-Circuit Approximate Models

The SPICE equivalent circuit described in the next section is an *exact* implementation of the solution of the transmission-line equations for a *lossless* line. A frequently used alternative is the lumped-circuit approximate model. These are similar to those developed for two-conductor lines, and the lumped-Pi and lumped-Tee models for three-conductor lines are shown in Fig. 10.29. These can be used in lumped-circuit programs such as SPICE for either time-domain or frequency-domain analyses.

As an example, consider the ribbon cable shown in Fig. 10.16 that was considered earlier. The frequency-domain predictions of the near-end crosstalk transfer function are compared with the predictions of the lumped-Pi model

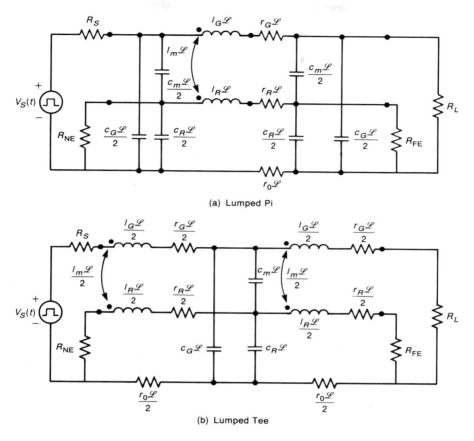

(a) Lumped Pi

(b) Lumped Tee

FIGURE 10.29   Lumped equivalent circuits for computing crosstalk in a three-conductor line: (a) lumped Pi; (b) lumped Tee.

using one and five Pi sections to model the entire line in Fig. 10.30. Similar predictions for the coupled microstrip line of Fig. 10.18 are shown in Fig. 10.31.

10.1.7   An Exact SPICE Model for Lossless, Coupled Lines

In this section we will give a simple but *exact* solution of the transmission-line equations for *lossless, coupled lines* that is implementable in the SPICE circuit analysis program [16]. The method will handle lines in homogeneous or inhomogeneous surrounding media. The multiconductor transmission-line equations derived previously can be written, for *lossless lines*, in matrix form as

$$\frac{\partial}{\partial z}\mathbf{V}(z,t) = -\mathbf{L}\frac{\partial}{\partial t}\mathbf{I}(z,t) \qquad (10.104a)$$

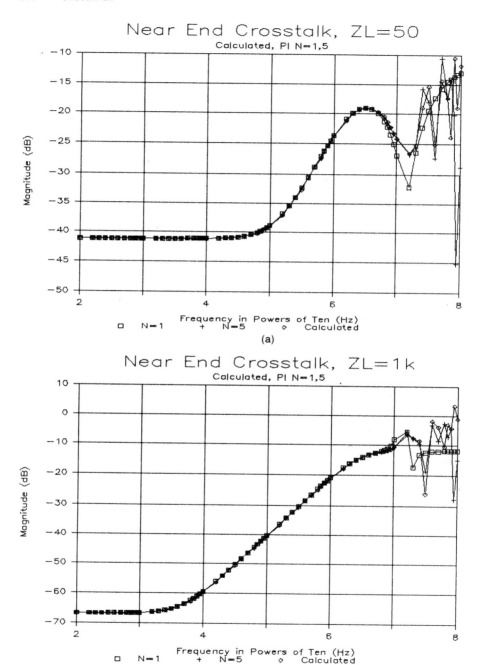

FIGURE 10.30 Predicted crosstalk for the ribbon cable of Fig. 10.16 using the transmission-line model and using the lumped Pi model with one and five Pi sections for (a) $R = 50\ \Omega$ and (b) $R = 1\ k\Omega$.

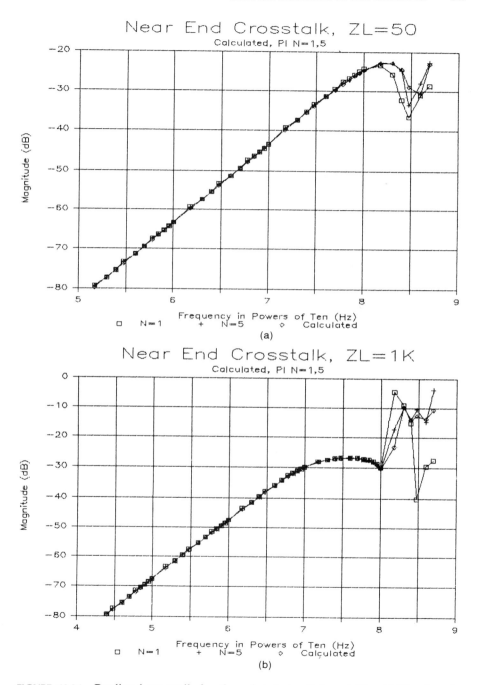

FIGURE 10.31 Predicted crosstalk for the printed circuit board Fig. 10.18 using the transmission-line and using the lumped Pi model with one and five Pi sections for (a) $R = 50\ \Omega$ and (b) $R = 1\ k\Omega$.

$$\frac{\partial}{\partial z}\mathbf{I}(z, t) = -\mathbf{C}\frac{\partial}{\partial t}\mathbf{V}(z, t) \qquad (10.104\text{b})$$

where the entries in the voltage and current vectors $\mathbf{V}(z, t)$ and $\mathbf{I}(z, t)$ are again given as

$$\mathbf{V}(z, t) = \begin{bmatrix} V_G(z, t) \\ V_R(z, t) \end{bmatrix} \qquad (10.105\text{a})$$

and

$$\mathbf{I}(z, t) = \begin{bmatrix} \mathbf{I}_G(z, t) \\ \mathbf{I}_R(z, t) \end{bmatrix} \qquad (10.105\text{b})$$

and

$$\mathbf{L} = \begin{bmatrix} l_G & l_m \\ l_m & l_R \end{bmatrix} \qquad (10.106\text{a})$$

and

$$\mathbf{C} = \begin{bmatrix} c_G + c_m & -c_m \\ -c_m & c_R + c_m \end{bmatrix} \qquad (10.106\text{b})$$

The key to solving these equations in the time domain is to decouple them; that is, reduce the coupled pairs of line to a set of two-conductor lines that do not interact. In order to do this, we define *transformations* that convert these desired line voltages and currents to *mode voltages and currents* $\mathbf{V}_m(z, t)$ and $\mathbf{I}_m(z, t)$ by determining the 2×2 transformation matrices $\mathbf{T}_V$ and $\mathbf{T}_I$ as

$$\mathbf{V}(z, t) = \mathbf{T}_V \mathbf{V}_m(z, t) \qquad (10.107\text{a})$$

$$\mathbf{I}(z, t) = \mathbf{T}_I \mathbf{I}_m(z, t) \qquad (10.107\text{b})$$

where $\mathbf{V}_m$ and $\mathbf{I}_m$ are vectors of *mode voltages and currents*, respectively:

$$\mathbf{V}_m(z, t) = \begin{bmatrix} V_{mG}(z, t) \\ V_{mR}(z, t) \end{bmatrix} \qquad (10.108\text{a})$$

$$\mathbf{I}_m(z, t) = \begin{bmatrix} I_{mG}(z, t) \\ I_{mR}(z, t) \end{bmatrix} \qquad (10.108\text{b})$$

and

$$\mathbf{T}_V = \begin{bmatrix} T_{VGG} & T_{VGR} \\ T_{VRG} & T_{VRR} \end{bmatrix} \tag{10.109a}$$

$$\mathbf{T}_I = \begin{bmatrix} T_{IGG} & T_{IGR} \\ T_{IRG} & T_{IRR} \end{bmatrix} \tag{10.109b}$$

These transformations matrices may not be symmetric, as we will see. Substituting (10.107) into (10.104) gives the transmission-line equations for these mode quantities as

$$\frac{\partial}{\partial z}\mathbf{V}_m(z, t) = -\mathbf{T}_V^{-1}\mathbf{L}\mathbf{T}_I\frac{\partial}{\partial t}\mathbf{I}_m(z, t) \tag{10.110a}$$

$$\frac{\partial}{\partial z}\mathbf{I}_m(z, t) = -\mathbf{T}_I^{-1}\mathbf{C}\mathbf{T}_V\frac{\partial}{\partial t}\mathbf{V}_m(z, t) \tag{10.110b}$$

Suppose we can find these transformation matrices such that they *simultaneously diagonalize the per-unit-length inductance and capacitance matrices* as

$$\mathbf{T}_V^{-1}\mathbf{L}\mathbf{T}_I = \mathbf{l}_m \tag{10.111a}$$

$$= \begin{bmatrix} l_{mG} & 0 \\ 0 & l_{mR} \end{bmatrix}$$

$$\mathbf{T}_I^{-1}\mathbf{C}\mathbf{T}_V = \mathbf{c}_m \tag{10.111b}$$

$$= \begin{bmatrix} c_{mG} & 0 \\ 0 & c_{mR} \end{bmatrix}$$

If this can be done, the transmission-line equations for the mode voltages and currents are *uncoupled* as

$$\frac{\partial}{\partial z}V_{mG}(z, t) = -l_{mG}\frac{\partial}{\partial t}I_{mG}(z, t) \tag{10.112a}$$

$$\frac{\partial}{\partial z}I_{mG}(z, t) = -c_{mG}\frac{\partial}{\partial t}V_{mG}(z, t)$$

and

$$\frac{\partial}{\partial z}V_{mR}(z, t) = -l_{mR}\frac{\partial}{\partial t}I_{mR}(z, t) \tag{10.112b}$$

$$\frac{\partial}{\partial z}I_{mR}(z, t) = -c_{mR}\frac{\partial}{\partial t}V_{mR}(z, t)$$

These transmission-line equations for the mode quantities represent *two uncoupled two-conductor transmission lines* that have characteristic impedances

$$Z_{CmG} = \sqrt{l_{mG}/c_{mG}} \tag{10.113a}$$

$$Z_{CmR} = \sqrt{l_{mR}/c_{mR}} \tag{10.113b}$$

and velocities of propagation

$$v_{mG} = \frac{1}{\sqrt{l_{mG}c_{mG}}} \tag{10.114a}$$

$$v_{mR} = \frac{1}{\sqrt{l_{mR}c_{mR}}} \tag{10.114b}$$

These uncoupled *mode lines* can be modeled in the SPICE program using the exact, two-conductor line model that was discussed in Chapter 4. The SPICE model for the mode voltages and currents along the line is solved, and the mode currents and voltages at the endpoints of the line, $z = 0$ and $z = \mathscr{L}$, can be converted to the actual line currents and voltages by implementing the transformations given in (10.107) using controlled source models. For example, writing out (10.107a) gives

$$V_G = T_{VGG}V_{mG} + T_{VGR}V_{mR} \tag{10.115a}$$

$$V_R = T_{VRG}V_{mG} + T_{VRR}V_{mR} \tag{10.115b}$$

This can be implemented with voltage-controlled voltage sources as shown in Fig. 10.32. Equation (10.107b) can be implemented by inverting this relationship to yield

$$\mathbf{I}_m(z, t) = \mathbf{T}_I^{-1}\mathbf{I}(z, t) \tag{10.116}$$

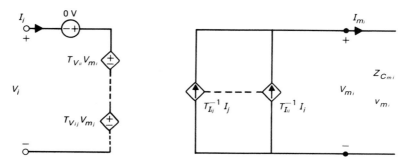

FIGURE 10.32   An equivalent circuit for coupled, lossless transmissions suitable for SPICE implementation.

Writing this out gives

$$I_{mG} = T_{IGG}^{-1}I_G + T_{IGR}^{-1}I_R \qquad (10.117a)$$

$$I_{mR} = T_{IRG}^{-1}I_G + T_{IRR}^{-1}I_R \qquad (10.117b)$$

where we denote

$$\mathbf{T}_I^{-1} = \begin{bmatrix} T_{IGG}^{-1} & T_{IGR}^{-1} \\ T_{IRG}^{-1} & T_{IRR}^{-1} \end{bmatrix} \qquad (10.118)$$

This can be implemented using current-controlled current sources as shown in Fig. 10.32. Zero-volt voltage sources are necessary in SPICE to sample the controlling current for current-controlled sources. The complete model of the coupled lines is shown in Fig. 10.33. In this model

$$V_{C1} = T_{VGG}V_{mG}(0, t) + T_{VGR}V_{mR}(0, t) \qquad (10.119a)$$

$$V_{C2} = T_{VRG}V_{mG}(0, t) + T_{VRR}V_{mR}(0, t) \qquad (10.119b)$$

$$V_{C3} = T_{VGG}V_{mG}(\mathscr{L}, t) + T_{VGR}V_{mR}(\mathscr{L}, t) \qquad (10.119c)$$

$$V_{C4} = T_{VRG}V_{mG}(\mathscr{L}, t) + T_{VRR}V_{mR}(\ell, t) \qquad (10.119d)$$

according to (10.115) and, according to (10.117),

$$I_{mC1} = T_{IGG}^{-1}I_G(0, t) + T_{IGR}^{-1}I_R(0, t) \qquad (10.120a)$$

$$I_{mC2} = T_{IRG}^{-1}I_G(0, t) + T_{IRR}^{-1}I_R(0, t) \qquad (10.120b)$$

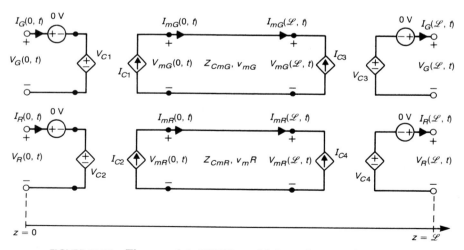

FIGURE 10.33   The complete SPICE model for a three-conductor line.

$$I_{mC3} = T_{IGG}^{-1} I_G(\mathcal{L}, t) + T_{IGR}^{-1} I_R(\mathcal{L}, t) \qquad (10.120c)$$

$$I_{mC4} = T_{IRG}^{-1} I_G(\mathcal{L}, t) + T_{IRR}^{-1} I_R(\mathcal{L}, t) \qquad (10.120d)$$

10.1.7.1 Lossless Lines in Homogeneous Media The per-unit-length inductance and capacitance matrices can be shown to be symmetric using energy considerations [2]. Any real, symmetric matrix can always be diagonalized by a matrix **T** that is real and symmetric [17]. Furthermore, the inverse of the transformation matrix is its transpose. Therefore it is always possible to find a **T** such that

$$\mathbf{T}^t \mathbf{L} \mathbf{T} = \mathbf{1}_m \qquad (10.121)$$

and

$$\mathbf{T}^{-1} = \mathbf{T}^t \qquad (10.122)$$

where we denote the *transpose* of a matrix **M** by **M**^{t}. For a homogeneous medium we also have the important identity

$$\mathbf{L}\mathbf{C} = \mu\epsilon \mathbf{1}_2 \qquad (10.123a)$$

or

$$\mathbf{L} = \mu\epsilon \mathbf{C}^{-1} \qquad (10.123b)$$

Substituting (10.123) into (10.121) gives

$$\mathbf{T}^t \mathbf{C}^{-1} \mathbf{T} = \frac{1}{\mu\epsilon} \mathbf{1}_m \qquad (10.124a)$$

Taking the inverse of both sides and using (10.122) gives

$$\mathbf{T}^t \mathbf{C} \mathbf{T} = \mu\epsilon \mathbf{1}_m^{-1} \qquad (10.124b)$$

Comparing (10.121) and (10.122) with (10.111) shows that

$$\mathbf{T}_I = \mathbf{T} \qquad (10.125a)$$

$$\mathbf{T}_V = \mathbf{T} \qquad (10.125b)$$

$$\mathbf{T}_I^{-1} = \mathbf{T}^t \qquad (10.125c)$$

As an example of a line in a homogeneous medium, consider the case of two bare #20 gauge solid wires separated by 2 cm and suspended 2 cm above a ground plane which is the reference conductor as shown in Fig. 10.34. The line

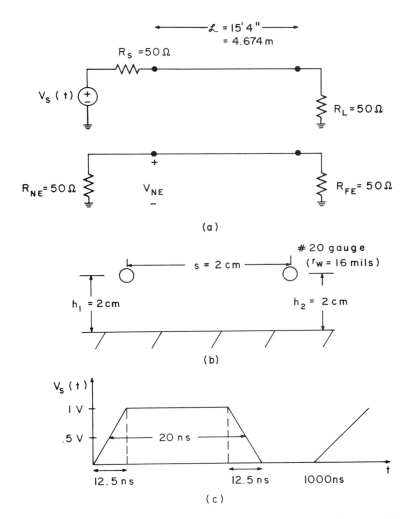

FIGURE 10.34 An experiment consisting of two wires above a ground plane to illustrate the prediction accuracy of the SPICE model of Fig. 10.33 for a homogeneous medium: (a) physical configuration; (b) cross-sectional configuration; (c) input voltage specification.

length is 4.674 m. The dielectric insulation of the wires is ignored, so that this becomes a problem of a homogeneous medium. The per-unit-length inductances and capacitances were computed in Section 10.1.3.2 as

$$l_G = l_R = 0.918 \ \mu\text{H}/\text{m}$$
$$l_m = 0.161 \ \mu\text{H}/\text{m}$$
$$c_G = c_R = 10.3 \ \text{pF}/\text{m}$$
$$c_m = 2.19 \ \text{pF}/\text{m}$$

and

$$Z_C = 298.5 \ \Omega$$

Therefore

$$\mathbf{L} = \begin{bmatrix} 0.918 \times 10^{-6} & 0.161 \times 10^{-6} \\ 0.161 \times 10^{-6} & 0.918 \times 10^{-6} \end{bmatrix}$$

The computation of the diagonalization matrix in (10.121) is simple for the case of a 2 × 2 matrix [17]:

$$\mathbf{T} = \begin{bmatrix} \cos \theta & -\sin \theta \\ \sin \theta & \cos \theta \end{bmatrix} \tag{10.126a}$$

where

$$\tan 2\theta = \frac{2l_m}{l_G - l_R} \tag{10.126b}$$

For this case we obtain $\theta = 45°$, since $l_G = l_R$, so that

$$\mathbf{T} = \begin{bmatrix} \sqrt{\frac{1}{2}} & -\sqrt{\frac{1}{2}} \\ \sqrt{\frac{1}{2}} & \sqrt{\frac{1}{2}} \end{bmatrix}$$

$$\mathbf{T}^t = \mathbf{T}^{-1} = \begin{bmatrix} \sqrt{\frac{1}{2}} & \sqrt{\frac{1}{2}} \\ -\sqrt{\frac{1}{2}} & \sqrt{\frac{1}{2}} \end{bmatrix}$$

The mode-per-unit length inductances in (10.121) become [17]

$$l_{mG} = l_G \cos^2 \theta + 2l_m \cos \theta \sin \theta + l_R \sin^2 \theta$$
$$= l_G \ (= l_R) + l_m$$
$$= 1.079 \times 10^{-6}$$

$$l_{mR} = l_G \sin^2 \theta - 2l_m \sin \theta \cos \theta + l_R \cos^2 \theta$$
$$= l_G \ (= l_R) - l_m$$
$$= 7.569 \times 10^{-7}$$

The mode velocities are that of free space:

$$v_{mG} = v_{mR} = v_0 = 3 \times 10^8$$

The one-way transit times of each mode line are

$$T_G = T_R = \mathscr{L}/v_0 = 15.58 \text{ ns}$$

The mode characteristic impedances are

$$Z_{CG} = v_0 l_{mG}$$
$$= 323.70 \text{ } \Omega$$

$$Z_{CR} = v_0 l_{mR}$$
$$= 227.07 \text{ } \Omega$$

The open-circuit voltage of the source, $V_S(t)$, is a 1 MHz trapezoidal pulse train having a 12.5 ns rise/falltime, a pulse width of 20 ns, and an amplitude of 1 V, as shown in Fig. 10.34(c). SPICE allows the use of polynomial controlled sources, which simplifies the implementation of the mode transformation controlled sources given by (10.119) and (10.120). The SPICE equivalent circuit is given in Fig. 10.35, and the SPICE program is

```
SPICE MTL MODEL; HOMOGENEOUS MEDIUM; FIGURE 10.34
VS 1 0 PULSE(0 1 0 12.5N 12.5N 7.5N 1000N)
RS 1 2 50
V1 2 3
RL 7 0 50
V3 7 6
RNE 13 0 50
V2 13 12
RFE 8 0 50
V4 8 9
EC1 3 0 POLY(2) (4,0) (11,0) 0 .707 -.707
EC2 12 0 POLY(2) (4,0) (11,0) 0 .707 .707
EC3 6 0 POLY(2) (5,0) (10,0) 0 .707 -.707
EC4 9 0 POLY(2) (5,0) (10,0) 0 .707 .707
FC1 0 4 POLY(2) V1 V2 0 .707 .707
FC2 0 11 POLY(2) V1 V2 0 -.707 .707
FC3 0 5 POLY(2) V3 V4 0 .707 .707
FC4 0 10 POLY(2) V3 V4 0 -.707 .707
T1 4 0 5 0 Z0=323.70 TD=15.58N
T2 11 0 10 0 Z0=227.07 TD=15.58N
.TRAN .5N 400N
.OPTIONS LIMPTS=801
.PRINT TRAN V(2) V(7) V(13) V(8)
.PLOT TRAN V(2) V(7) V(13) V(8)
.END
```

The predictions of the near-end crosstalk $V_{NE}(t)$ are shown in Fig. 10.36(a), and the experimentally observed data are shown in Fig. 10.36(b). Comparing these data, we observe excellent prediction of the time-domain, near-end crosstalk.

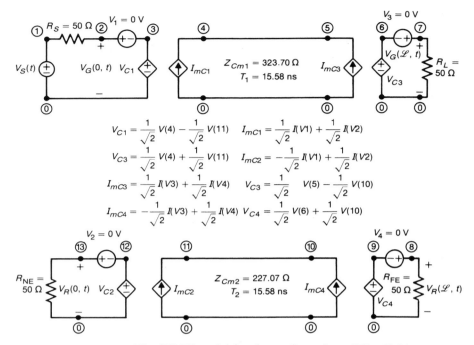

FIGURE 10.35   The SPICE model for the configuration of Fig. 10.34.

The above SPICE exact model for lossless lines can also be used to compute the frequency-domain transfer function. For the circuit of Fig. 10.34, replace $V_S(t)$ with a 1 V phasor and iterate the frequency using the .AC option of SPICE. The above SPICE program requires two changes:

```
VS 1 0 AC 1
```

and

```
.AC DEC 50 1K 100MEG
```

The frequency response is obtained as the ratio of node voltage 13 and node voltage 2:

```
VDB(13)-VDB(2)
```

and is plotted in Fig. 10.37(a). Observe the region from 1 kHz to 1 MHz, where the response increases at 20 dB/decade. The experimental results are shown in Fig. 10.37(b). The low-frequency model developed in the previous section gives

$$\left|\frac{\hat{V}_{NE}}{\hat{V}_G(0)}\right| = 2\pi f\left(\frac{L_m}{2R} + \frac{R}{2}C_m\right)$$
$$= 2\pi f(7.53 \times 10^{-9} + 2.56 \times 10^{-10})$$
$$= 4.89 \times 10^{-8} f$$

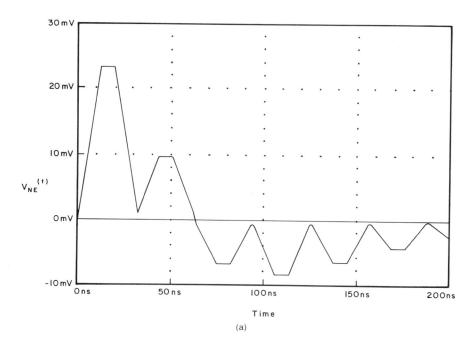

(a)

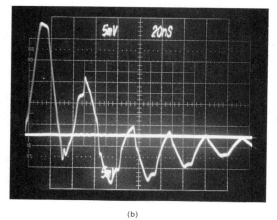

(b)

FIGURE 10.36 Time-domain near-end crosstalk for the configuration of Fig. 10.34: (a) prediction of the SPICE circuit of Fig. 10.35; (b) measured.

where $l_m = 0.161\ \mu H/m$ and $c_m = 2.19\ pF/m$. This gives a value of $4.89 \times 10^{-4} = -66\ dB$ at 10 kHz, which is exactly the predicted value.

The predictions of the exact transmission line model whose literal solution was obtained in Section 10.1.4.2 are also shown in Fig. 10.37(b). These predictions are, as expected, also quite accurate.

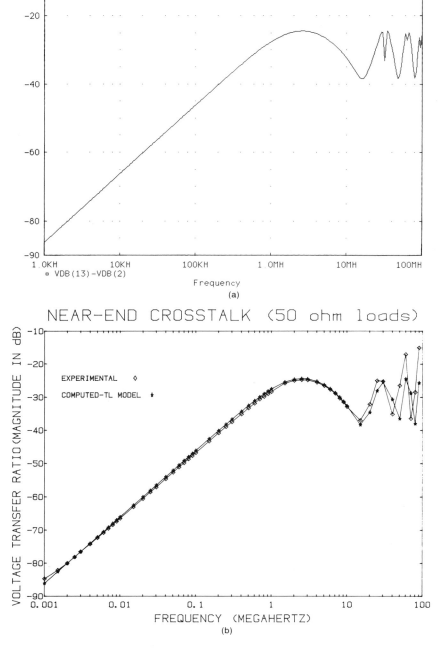

FIGURE 10.37   Frequency-domain predictions of the near-end crosstalk transfer function for the configuration of Fig. 10.34 using (a) the SPICE model of Fig. 10.35 and (b) predictions of the transmission-line model versus measured results.

10.1.7.2 Lossless Lines in Inhomogeneous Media Since the per-unit-length capacitance matrix $\mathbf{C}$ is real and symmetric, it can be diagonalized by a 2×2 matrix $\mathbf{U}$ that is real and for which $\mathbf{U}^{-1} = \mathbf{U}^{\mathrm{t}}$ [2]:

$$\mathbf{U}^{\mathrm{t}}\mathbf{C}\mathbf{U} = \mathbf{D} \tag{10.127}$$

where $\mathbf{D}$ is a diagonal matrix. The per-unit-length capacitance matrix $\mathbf{C}$ is also *positive definite* [2], so that the eigenvalues of $\mathbf{C}$ (diagonal elements of $\mathbf{D}$) are real, positive, and nonzero. Therefore it is sensible to form its square root, $\mathbf{D}^{1/2}$, which has the square root of the eigenvalues of $\mathbf{D}$ on its main diagonal [2]. Next form the matrix

$$\mathbf{S}^{\mathrm{t}}(\mathbf{D}^{1/2}\mathbf{U}^{\mathrm{t}}\mathbf{L}\mathbf{U}\mathbf{D}^{1/2})\mathbf{S} = \mathbf{K} \quad \text{(diagonal)} \tag{10.128}$$

The matrix in parentheses is also diagonalizable as in (10.127) by the matrix $\mathbf{S}$, since it is real and symmetric. Define the 2×2 matrix

$$\mathbf{T} = \mathbf{U}\mathbf{D}^{1/2}\mathbf{S} \tag{10.129}$$

The columns of $\mathbf{T}$ can be normalized to a Euclidean length of unity as

$$\mathbf{T}_{\mathrm{norm}} = \mathbf{T}\boldsymbol{\alpha} \tag{10.130}$$

where $\boldsymbol{\alpha}$ is the 2×2 diagonal matrix with entries

$$\alpha_{ii} = \frac{1}{\sqrt{\displaystyle\sum_{k=1}^{2} T_{ki}^2}} \tag{10.131a}$$

$$\alpha_{ij} = 0 \tag{10.131b}$$

The mode transformations in (10.107) can then be defined as

$$\mathbf{T}_I = \mathbf{U}\mathbf{D}^{1/2}\mathbf{S}\boldsymbol{\alpha} \tag{10.132a}$$

$$= \mathbf{T}_{\mathrm{norm}}$$

$$\mathbf{T}_V = \mathbf{U}\mathbf{D}^{-1/2}\mathbf{S}\boldsymbol{\alpha}^{-1} \tag{10.132b}$$

$$= (\mathbf{T}_{\mathrm{norm}}^{-1})^{\mathrm{t}}$$

Also

$$\mathbf{T}_I^{-1} = \boldsymbol{\alpha}^{-1}\mathbf{S}^{\mathrm{t}}\mathbf{D}^{-1/2}\mathbf{U}^{\mathrm{t}} \tag{10.133a}$$

$$= \mathbf{T}_V^{\mathrm{t}}$$

$$\mathbf{T}_V^{-1} = \boldsymbol{\alpha} \mathbf{S}^t \mathbf{D}^{1/2} \mathbf{U}^t \tag{10.133b}$$

$$= \mathbf{T}_I^t$$

Substituting (10.132) and (10.133) into (10.111) gives

$$\mathbf{T}_V^{-1} \mathbf{L} \mathbf{T}_I = \boldsymbol{\alpha} \mathbf{S}^t \mathbf{D}^{1/2} \mathbf{U}^t \mathbf{L} \mathbf{U} \mathbf{D}^{1/2} \mathbf{S} \boldsymbol{\alpha} \tag{10.134a}$$

$$= \boldsymbol{\alpha} \mathbf{K} \boldsymbol{\alpha}$$

$$\mathbf{T}_I^{-1} \mathbf{C} \mathbf{T}_V = \boldsymbol{\alpha}^{-1} \mathbf{S}^t \mathbf{D}^{-1/2} \mathbf{U}^t \mathbf{C} \mathbf{U} \mathbf{D}^{-1/2} \mathbf{S} \boldsymbol{\alpha}^{-1} \tag{10.134b}$$

$$= \boldsymbol{\alpha}^{-2}$$

Comparing (10.134) and (10.111) shows that

$$\mathbf{l}_m = \boldsymbol{\alpha} \mathbf{K} \boldsymbol{\alpha} \tag{10.135a}$$

$$\mathbf{c}_m = \boldsymbol{\alpha}^{-2} \tag{10.135b}$$

Since $\boldsymbol{\alpha}$ and $\mathbf{K}$ are diagonal, the mode characteristic impedances and velocities of propagation are given by

$$Z_{CmG} = \sqrt{\frac{l_{mG}}{c_{mG}}} \tag{10.136a}$$

$$= \alpha_{11}^2 \sqrt{K_{11}}$$

$$Z_{CmR} = \sqrt{\frac{l_{mR}}{c_{mR}}} \tag{10.136b}$$

$$= \alpha_{22}^2 \sqrt{K_{22}}$$

and

$$v_{mG} = \frac{1}{\sqrt{l_{mG} c_{mG}}} \tag{10.137a}$$

$$= \frac{1}{\sqrt{K_{11}}}$$

$$v_{mR} = \frac{1}{\sqrt{l_{mR} c_{mR}}} \tag{10.137b}$$

$$= \frac{1}{\sqrt{K_{22}}}$$

The desired mode transformation matrices are

$$\mathbf{T}_V = \mathbf{U}\mathbf{D}^{-1/2}\mathbf{S}\boldsymbol{\alpha}^{-1} \tag{10.138a}$$

$$= (\mathbf{T}_{\text{norm}}^{-1})^{\text{t}}$$

$$\mathbf{T}_I^{-1} = \boldsymbol{\alpha}^{-1}\mathbf{S}^{\text{t}}\mathbf{D}^{-1/2}\mathbf{U}^{\text{t}} \tag{10.138b}$$

$$= \mathbf{T}_{\text{norm}}^{-1}$$

As an example of the use of this model for an inhomogeneous medium, consider the PCB (coupled coplanar strips) shown in Fig. 10.38. This experimental example was studied in [14]. The board is glass–epoxy, and the thickness is 47 mils. The 1 ounce copper land widths are 15 mils, and the lands have a center-to-center separation of 30 mils and are 10 inches in length. The length of the lands is 10 inches or 25.4 cm. The source voltage is a 10 MHz, 1 V trapezoidal pulse train with a 50% duty cycle and rise/falltimes of 6.25 ns (5 ns 10–90%). The per-unit-length capacitance matrix with the board present, $\mathbf{C}$, and with the board removed, $\mathbf{C}_0$, were computed using the numerical methods of [11, 12]. The per-unit-length inductance matrix was obtained as

$$\mathbf{L} = \mu_0 \epsilon_0 \mathbf{C}_0^{-1} \tag{10.139}$$

as

$$\mathbf{C} = \begin{bmatrix} 30.5022 & -20.8827 \\ -20.8827 & 41.7655 \end{bmatrix} \text{pF/m}$$

$$\mathbf{L} = \begin{bmatrix} 1.3067 & 0.65335 \\ 0.65335 & 1.0340 \end{bmatrix} \mu\text{H/m}$$

Using the algorithm of this section gives

$$\mathbf{T}_V = \begin{bmatrix} 1.0 & 2.056 \times 10^{-5} \\ 0.5 & 1.118 \end{bmatrix}$$

$$\mathbf{T}_I^{-1} = \begin{bmatrix} 1.0 & 0.5 \\ 2.056 \times 10^{-5} & 1.118 \end{bmatrix}$$

$$Z_{CmG} = 255.215 \,\Omega$$

$$Z_{CmR} = 104.110 \,\Omega$$

$$v_{mG} = 1.95316 \times 10^8 \text{ m/s}$$

$$v_{mR} = 1.83985 \times 10^8 \text{ m/s}$$

From these velocities of propagation of the modes we compute the one-way

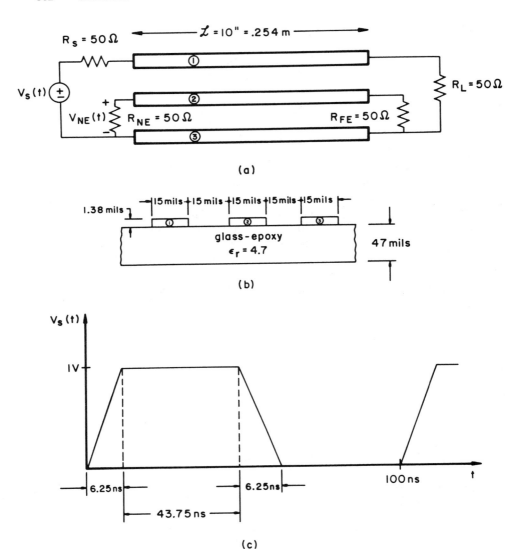

FIGURE 10.38 An experiment consisting of three lands on a PCB to illustrate the prediction accuracy of the SPICE model of Fig. 10.33 for an inhomogeneous medium: (a) physical configuration; (b) cross-sectional configuration; (c) input voltage specification.

transit times of the modes as

$$T_G = 1.30046 \text{ ns}$$

$$T_R = 1.38056 \text{ ns}$$

The SPICE equivalent circuit of Fig. 10.33 and given for the homogeneous medium example in Fig. 10.35 was again used, and (10.119) and (10.120) become

$$V_{C1} = 1.0V_{mG}(0, t) + 2.056 \times 10^{-5}V_{mR}(0, t)$$

$$V_{C2} = 0.5V_{mG}(0, t) + 1.118V_{mR}(0, t)$$

$$V_{C3} = 1.0V_{mG}(\mathscr{L}, t) + 2.056 \times 10^{-5}V_{mR}(\mathscr{L}, t)$$

$$V_{C4} = 0.5V_{mG}(\mathscr{L}, t) + 1.18V_{mR}(\mathscr{L}, t)$$

and

$$I_{C1} = 1.0I_G(0, t) + 0.5I_R(0, t)$$

$$I_{C2} = 2.056 \times 10^{-5}I_G(0, t) + 1.118I_R(0, t)$$

$$I_{C3} = 1.0I_G(\mathscr{L}, t) + 0.5I_R(\mathscr{L}, t)$$

$$I_{C4} = 2.056 \times 10^{-5}I_G(\mathscr{L}, t) + 1.18I_R(\mathscr{L}, t)$$

The SPICE equivalent circuit is the same as for the homogeneous medium example given in Fig. 10.35 with the above controlled-source parameters. The SPICE program is

```
SPICE MTL MODEL; INHOMOGENEOUS MEDIUM; FIGURE 10.38
VS 1 0 PULSE(0 1 0 6.25N 6.25N 43.75N 100N)
RS 1 2 50
V1 2 3
RL 7 0 50
V3 7 6
RNE 13 0 50
V2 13 12
RFE 8 0 50
V4 8 9
EC1 3 0 POLY(2) (4,0) (11,0) 0 1.0 2.056E-5
EC2 12 0 POLY(2) (4,0) (11,0) 0 .5 1.118
EC3 6 0 POLY(2) (5,0) (10,0) 0 1.0 2.056E-5
EC4 9 0 POLY(2) (5,0) (10,0) 0 .5 1.118
FC1 0 4 POLY(2) V1 V2 0 1.0 0.5
FC2 0 11 POLY(2) V1 V2 0 2.056E-5 1.118
FC3 0 5 POLY(2) V3 V4 0 1.0 0.5
FC4 0 10 POLY(2) V3 V4 0 2.056E-5 1.118
T1 4 0 5 0 Z0=255.215 TD=1.30046N
T2 11 0 10 0 Z0=104.110 TD=1.38055N
.TRAN .1N 20N 0 .05N
.PRINT TRAN V(2) V(7) V(13) V(8)
.PLOT TRAN V(2) V(7) V(13) V(8)
.END
```

The SPICE predictions are shown in Fig. 10.39(a), and the experimentally observed data are shown in Fig. 10.39(b). Comparing these, we again observe excellent predictions.

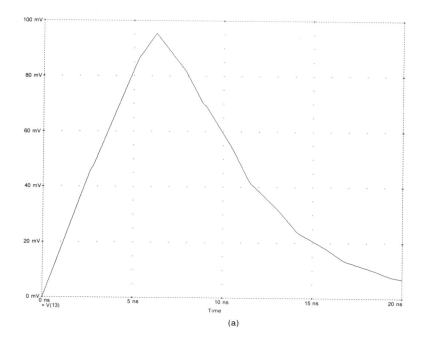

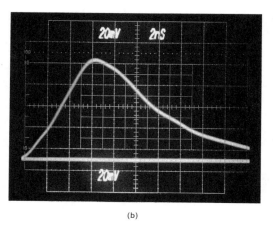

FIGURE 10.39   Time-domain near-end crosstalk for the configuration of Fig. 10.38: (a) SPICE prediction; (b) measured.

The above SPICE exact model for lossless lines can also be used to compute the frequency-domain transfer function. In the previous SPICE time-domain circuit replace $V_S(t)$ with a 1 V phasor and iterate the frequency using the .AC option of SPICE. The above SPICE program requires two changes:

```
VS 1 0 AC 1
```

and

```
.AC DEC 50 10K 1000MEG
```

The frequency response is obtained as the ratio of node voltage 13 and node voltage 2:

```
VDB(13)-VDB(2)
```

and is plotted in Fig. 10.40(a). Observe the region from 10 kHz to 10 MHz, where the response increase at 20 dB/decade. The experimental results are shown in Fig. 10.40(b). The low-frequency model developed in the previous section gives

$$\left|\frac{\hat{V}_{NE}}{\hat{V}_G(0)}\right| = 2\pi f\left(\frac{L_m}{2R} + \frac{R}{2}C_m\right)$$
$$= 2\pi f(1.66 \times 10^{-9} + 1.326 \times 10^{-10})$$
$$= 1.126 \times 10^{-8} f$$

where $l_m = 0.65335\ \mu H/m$ and $c_m = 2.8827\ pF/m$. This gives a value of $1.126 \times 10^{-2} = -39$ dB at 1 MHz, which is exactly the predicted value.

Observe that the experimental results show a low-frequency region below approximately 100 kHz where the resistance of the reference land becomes important and common-impedance coupling becomes dominant. The above SPICE model cannot be readily modified to accommodate losses so that it cannot predict this important aspect of the crosstalk. However, the transmission line model solution for the frequency domain can be modified to include losses [7]. The predictions of this transmission line model are plotted against the experimental data in Fig. 10.40(b) and show reasonable correlation. Also shown in Fig. 10.40(b) are the predictions of the lumped-circuit approximate model. The predictions using one and two Pi sections to represent the entire line show the same prediction accuracy as the transmission line model for frequencies where the line is electrically short (below 100 MHz).

10.2 SHIELDED WIRES

We now consider methods for reducing the crosstalk in a three-conductor line. Suppose the near- or far-end crosstalk in the previous three-conductor line exceeds desired levels, causing interference with the terminations at the ends of the receptor circuit. There are two common methods for reducing the crosstalk; replace the generator and/or receptor wire with a shielded wire or a twisted pair. Consider replacing the receptor wire with a shielded wire, as shown in Fig. 10.41(a). There are two common reference conductors: a ground plane

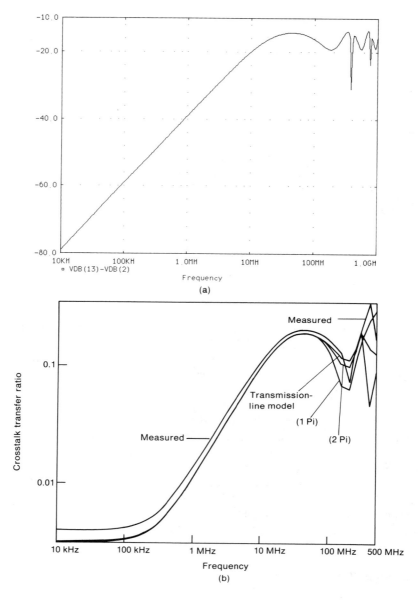

FIGURE 10.40   Frequency-domain predictions of the near-end crosstalk transfer function for the configuration of Fig. 10.38 using (a) the SPICE model and (b) predictions of the transmission-line model and the lumped Pi model using one and two sections versus measured results.

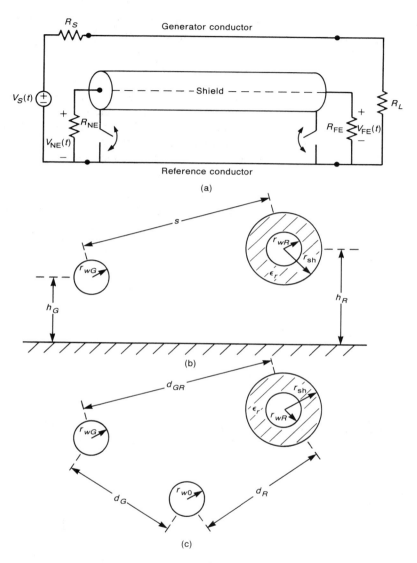

FIGURE 10.41 Addition of a cylindrical shield around the receptor wire: (a) longitudinal configuration; (b) cross-sectional configuration for a ground plane as the reference conductor; (c) cross-sectional configuration for another wire as the reference conductor.

shown in Fig. 10.41(b) and another wire shown in Fig. 10.41(c). Observe that *the configurations in Fig. 10.41 are four-conductor lines.* Thus the results developed previously for three-conductor lines do not apply here. Nevertheless, the general formulations for multiconductor lines developed in [2] can be applied here to develop computer programs to analyze this configuration in an exact fashion [18–23]. We will develop approximate models using the concept of the superposition of inductive and capacitive coupling that has been verified by obtaining the exact literal (symbolic solution) of the transmission-line equations for the shielded line of Fig. 10.41 [24].

We will show the MTL equations for the line in order to illustrate the calculation of the per-unit-length parameters. We will not solve these equations, but will develop approximate solutions. The reader is referred to [18–23] for computer programs that implement the exact solution. The per-unit-length equivalent circuit for a Δz section of the line is shown in Fig. 10.42. The multiconductor transmission-line equations in the *frequency domain* are again of the form [18, 21, 22]

$$\frac{d}{dz}\hat{\mathbf{V}}(z) = -(\mathbf{R} + j\omega\mathbf{L})\hat{\mathbf{I}}(z) \qquad (10.140a)$$

$$\frac{d}{dz}\hat{\mathbf{I}}(z) = -j\omega\mathbf{C}\hat{\mathbf{V}}(z) \qquad (10.140b)$$

where

$$\hat{\mathbf{V}}(z) = \begin{bmatrix} \hat{V}_G(z) \\ \hat{V}_S(z) \\ \hat{V}_R(z) \end{bmatrix} \qquad (10.141a)$$

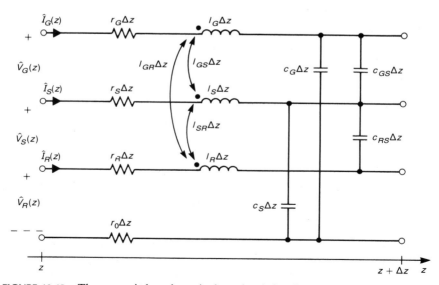

FIGURE 10.42   The per-unit-length equivalent circuit for the TEM mode of propagation for the shielded receptor wire of Fig. 10.41.

$$\hat{\mathbf{I}}(z) = \begin{bmatrix} \hat{I}_G(z) \\ \hat{I}_S(z) \\ \hat{I}_R(z) \end{bmatrix}$$

(10.141b)

$$\mathbf{R} = \begin{bmatrix} r_G + r_0 & r_0 & r_0 \\ r_0 & r_S + r_0 & r_0 \\ r_0 & r_0 & r_R + r_0 \end{bmatrix}$$

(10.141c)

$$\mathbf{L} = \begin{bmatrix} l_G & l_{GS} & l_{GR} \\ l_{GS} & l_S & l_{RS} \\ l_{GR} & l_{RS} & l_R \end{bmatrix}$$

(10.141d)

$$\mathbf{C} = \begin{bmatrix} c_G + c_{GS} & -c_{GS} & 0 \\ -c_{GS} & c_S + c_{GS} + c_{RS} & -c_{RS} \\ 0 & -c_{RS} & c_{RS} \end{bmatrix}$$

(10.141e)

and we neglect any losses in the medium; i.e., $\mathbf{G} = \mathbf{0}$. The voltages are defined with respect to the reference conductor, and the currents return through the reference conductor. Observe that there are two per-unit-length parameters that are missing from this circuit: the mutual capacitance between the generator wire and the shielded wire, c_{GR}, and the self-capacitance between the receptor wire and the reference conductor, c_R [18, 21]. The shield effectively eliminates these per-unit-length parameters, as illustrated in Fig. 10.43. This allows the shield to eliminate capacitive coupling to the receptor circuit as we will see.

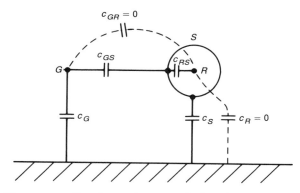

FIGURE 10.43   The cross-sectional capacitance equivalent for the shielded receptor wire of Fig. 10.41.

10.2.1 Per-Unit-Length Parameters

The per-unit-length resistances of the generator and receptor wires, r_G and r_R, can be computed in the usual manner. The per-unit-length resistance of the shield, r_S, depends on the construction technique. For braided-wire shields we can compute the resistance by computing the resistance of one of the braid wires and placing all braid wires in parallel. The result is [21, 22]

$$r_S = \frac{r_b}{BW \cos \theta_w} \tag{10.142}$$

where r_b is the resistance of a braid wire, θ_w is the weave angle, B is the number of belts, and W is the number of braid wires per belt. The per-unit-length resistance of a solid shield such as are used in flatpack, coaxial cables is computed as a solid wire with well-developed skin effect [20, 23]. In other words, we assume the shield current to be uniformly distributed over the shield cross section, so that the shield resistance is its dc value:

$$r_S = \frac{1}{\sigma 2\pi r_{sh} t_{sh}} \tag{10.143}$$

where r_{sh} is the shield interior radius and t_{sh} the shield thickness.

The per-unit-length inductance parameters are simply computed from the flux-current methods used earlier. We will compute these parameters for the ground plane reference conductor shown in Fig. 10.41(b). Computation of these parameters for a wire as the reference conductor as in Fig. 10.41(c) follows a similar pattern. The self-inductances are computed by placing a current on that wire (or shield) and returning through its image, and computing the magnetic flux passing through the circuit (between that wire and the ground plane). The self-inductance of the generator circuit is simply that of a wire above a ground plane computed earlier (see (10.33)):

$$l_G = \frac{\mu_0}{2\pi} \ln \left(\frac{2h_G}{r_{wG}} \right) \tag{10.144}$$

Similarly, the self-inductance of the shield–ground plane circuit is computed as that of a "fat wire" above ground:

$$l_S = \frac{\mu_0}{2\pi} \ln \left(\frac{2h_R}{r_{sh} + t_{sh}} \right) \tag{10.145}$$

where r_{sh} is the shield interior radius and t_{sh} the shield thickness. The self-inductance of the receptor circuit is computed as the self-inductance of a wire above ground (the receptor voltage is defined between the receptor wire and the ground plane and the receptor current returns through the ground plane):

$$l_R = \frac{\mu_0}{2\pi} \ln\left(\frac{2h_R}{r_{wR}}\right) \tag{10.146}$$

The mutual inductance between the generator wire and the shield, l_{GS}, and that between the generator wire and the receptor wire, l_{GR}, are computed in the previous fashion as being between two wires above ground (see (10.37)):

$$l_{GS} = \frac{\mu_0}{4\pi} \ln\left(1 + 4\frac{h_G h_R}{s^2}\right) \tag{10.147}$$

$$= l_{GR}$$

In order for these to be equal, we assume that the shield and generator wire are widely separated which is implicit in all these results. The mutual inductance between the shield and receptor wire circuits is

$$l_{RS} = \frac{\mu_0}{2\pi} \ln\left(\frac{2h_R}{r_{sh} + t_{sh}}\right) \tag{10.148}$$

$$= l_S$$

This result is very important, and deserves further explanation. Consider the shield–receptor wire above ground shown in Fig. 10.44. The shield circuit is

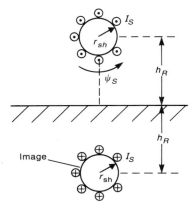

FIGURE 10.44 Calculation of the per-unit-length self inductance of the shield–ground plane circuit for a shielded wire above a ground plane.

between the shield and the ground plane. The mutual inductance between the shield and receptor circuits can be obtained by placing a current on the shield and determining the magnetic flux through the receptor circuit or by placing a current on the receptor wire and determining the magnetic flux through the shield circuit:

$$l_{RS} = \frac{\psi_S}{I_R}\bigg|_{I_S = 0} \qquad (10.149)$$

$$= \frac{\psi_R}{I_S}\bigg|_{I_R = 0}$$

Consider placing a current on the receptor wire and computing the magnetic flux through the shield–ground plane circuit. This can be obtained by placing all of the receptor wire current on the shield, and thus $l_{RS} = l_S$. This important observation allows a shield to eliminate inductive coupling as we will see.

The capacitances are obtainable through the use of the important reciprocal relationship between the inductance and capacitance matrices given in (10.9). The medium interior to the shield can have $\epsilon_r \neq 1$, whereas the medium outside is logically assumed to be free space where we neglect any dielectric insulation around the generator wire and the shield. Thus the capacitance between the shield and receptor wire is the same as for a coaxial cable given in (4.35) of Chapter 4:

$$c_{RS} = \frac{2\pi \epsilon_0 \epsilon_r}{\ln(r_{sh}/r_{wR})} \qquad (10.150)$$

The other capacitances can be found using the reciprocal relationship for conductors in a homogeneous medium by considering only the generator wire and the shield as

$$\begin{bmatrix} c_G + c_{GS} & -c_{GS} \\ -c_{GS} & c_S + c_{GS} \end{bmatrix} = \mu_0 \epsilon_0 \begin{bmatrix} l_G & l_{GS} \\ l_{GS} & l_S \end{bmatrix}^{-1} \qquad (10.151)$$

Using these relationships it can be shown that

$$\mathbf{LC} = \mu_0 \epsilon_0 \begin{bmatrix} 1 & 0 & 0 \\ 0 & \epsilon_r & \epsilon_r - 1 \\ 0 & 0 & 1 \end{bmatrix} \qquad (10.152)$$

10.2.2 Inductive and Capacitive Coupling

The MTL equations above were solved in literal (symbolic) form in [24], showing that the notion of the near-end and far-end crosstalk voltages being composed of inductive and capacitive coupling contributions is valid for *weakly coupled, electrically short* lines. We will give the results for a sufficiently small frequency.

First consider the capacitive coupling with the circuit shown in Fig. 10.45. The element values are the per-unit-length values multiplied by the line length: $C_{RS} = c_{RS}\mathcal{L}$ and $C_{GS} = c_{GS}\mathcal{L}$. The capacitive coupling is

$$\hat{V}_{NE}^{CAP} = \hat{V}_{FE}^{CAP} \qquad (10.153)$$

$$= \frac{j\omega R C_{RS} \| C_{GS}}{1 + j\omega R C_{RS} \| C_{GS}} \hat{V}_G$$

where

$$R = \frac{R_{NE} R_{FE}}{R_{NE} + R_{FE}} \qquad (10.154a)$$

$$C_{RS} \| C_{GS} = \frac{C_{RS} C_{GS}}{C_{RS} + C_{GS}} \qquad (10.154b)$$

Actually, this result could be easily seen by reducing the two capacitors to an equivalent, and capacitors in series add like resistors in parallel [6]. For a

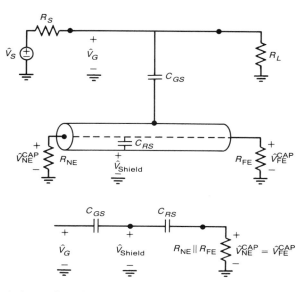

FIGURE 10.45 A lumped equivalent circuit for capacitive coupling for the shielded receptor wire of Fig. 10.41.

sufficiently small frequency this reduces to

$$\hat{V}_{NE}^{CAP} = \hat{V}_{FE}^{CAP}$$ (10.155a)

$$\cong j\omega \frac{R_{NE}R_{FE}}{R_{NE} + R_{FE}} \frac{C_{RS}C_{GS}}{C_{RS} + C_{GS}} V_{G_{DC}}$$

and

$$V_{G_{DC}} = \frac{R_L}{R_S + R_L} V_S$$ (10.155b)

is the dc or "low-frequency" value of the generator line voltage. In effect, this is the same as the capacitive coupling between two wires where the mutual capacitance between them is $C_{RS} \| C_{GS}$. Observe that the capacitive coupling contribution in (10.155a) increases at 20 dB/decade as for unshielded wires. Also, for typical shielded wires $C_{RS} \gg C_{GS}$, so that $C_{RS} \| C_{GS} \cong C_{GS} \cong C_{GR}$. Thus the capacitive coupling is basically unchanged from the unshielded case.

Typically, the shield is connected to the reference conductor ("grounded") at one end or at both ends. Observe that *if the shield is connected to the reference conductor at either end, the shield voltage is reduced to zero and the capacitive coupling contribution is removed*:

$$\hat{V}_{NE}^{CAP} = \hat{V}_{FE}^{CAP}$$ (10.156)

$$= 0 \quad \text{(shield connected to the}$$
reference conductor
at either end)

This is the origin of the notion that a shielded wire eliminates electric field or capacitive coupling wherein the electric field lines from the generator circuit terminate on the shield and not on the receptor line. In order for the shield to eliminate capacitive coupling, the shield voltage $\hat{V}_{shield}$ must be zero. For an electrically short line, grounding the shield at either end will cause the voltage along the shield to be approximately zero. As the line length increases, electrically, the shield must be grounded at multiple points spaced some $\frac{1}{10}\lambda$ along it in order to approximate this.

Next we consider inductive coupling. We have seen that the shield inherently eliminates capacitive coupling so long as it is "grounded" at either end. *A shielded wire must be grounded at both ends in order to eliminate inductive coupling.* In order to show this, consider the magnetic fields generated by the generator wire current as shown in Fig. 10.46. The generator wire current $\hat{I}_G$ produces a magnetic flux ψ_G in the shield–ground plane circuit. This induces, by Faraday's law, an emf in the shield circuit that produces a secondary current $\hat{I}_S$ flowing back along the shield. The flux of this induced shield current tends to cancel

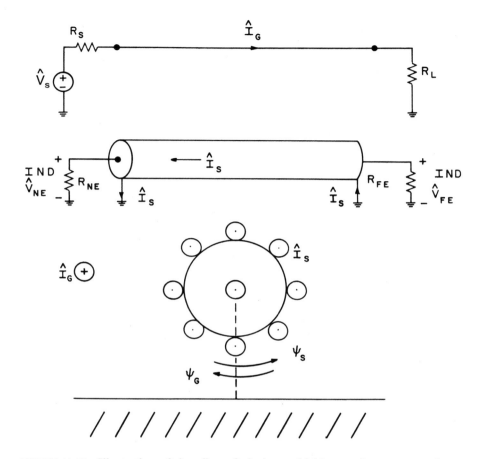

FIGURE 10.46 Illustration of the effect of placing a shield around a receptor wire on inductive coupling.

that of the generator wire current. It is this process that allows a shielded wire to eliminate the inductive or magnetic field coupling.

In order to quantify this effect, consider the inductive equivalent circuit shown in Fig. 10.47. The per-unit-length elements are multiplied by the line length to give $L_{GS} = l_{GS}\mathscr{L} = L_{GR} = l_{GR}\mathscr{L}$, $L_{RS} = l_{RS}\mathscr{L} = L_{SH} = l_{S}\mathscr{L}$, and $R_{SH} = r_{S}\mathscr{L}$. The near-end and far-end crosstalk voltages can be easily obtained from this circuit as

$$\hat{V}_{NE}^{\text{IND}} = \frac{R_{NE}}{R_{FE} + R_{NE}} j\omega(L_{GR}\hat{I}_{G} - L_{RS}\hat{I}_{S}) \qquad (10.157)$$

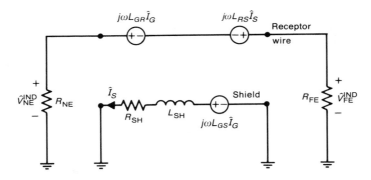

FIGURE 10.47 A lumped equivalent circuit for inductive coupling for the shielded receptor wire of Fig. 10.41.

and the shield current is

$$\hat{I}_S = \frac{j\omega L_{GS}}{R_{SH} + j\omega L_{SH}} \hat{I}_G \qquad (10.158)$$

Substituting (10.158) into (10.157) gives

$$\hat{V}_{NE}^{IND} = \frac{R_{NE}}{R_{FE} + R_{NE}} \frac{j\omega R_{SH} L_{GR} + \omega^2 (L_{GS} L_{RS} - L_{GR} L_{SH})}{R_{SH} + j\omega L_{SH}} \hat{I}_G \qquad (10.159)$$

Substituting the important relationships previously determined:

$$L_{GR} = L_{GS} \qquad (10.160a)$$

$$L_{RS} = L_{SH} \qquad (10.160b)$$

gives

$$\hat{V}_{NE}^{IND} = \underbrace{\frac{R_{NE}}{R_{FE} + R_{NE}} j\omega L_{GR} \hat{I}_G}_{\substack{\text{crosstalk} \\ \text{with shield} \\ \text{removed}}} \underbrace{\frac{R_{SH}}{R_{SH} + j\omega L_{SH}}}_{\substack{\text{effect of} \\ \text{shield}}} \qquad (10.161a)$$

$$\hat{V}_{FE}^{IND} = -\underbrace{\frac{R_{FE}}{R_{FE} + R_{NE}} j\omega L_{GR} \hat{I}_G}_{\substack{\text{crosstalk} \\ \text{with shield} \\ \text{removed}}} \underbrace{\frac{R_{SH}}{R_{SH} + j\omega L_{SH}}}_{\substack{\text{effect of} \\ \text{shield}}} \qquad (10.161b)$$

which is the result for the same case with the shield removed but multiplied by the factor

$$SF = \frac{R_{\text{SH}}}{R_{\text{SH}} + j\omega L_{\text{SH}}} \qquad (10.162)$$

This factor can be written as

$$SF = \frac{1}{1 + j\omega\tau_S} \qquad (10.163\text{a})$$

where the shield *time constant* is given by

$$\tau_S = \frac{L_{\text{SH}}}{R_{\text{SH}}} \qquad (10.163\text{b})$$

Thus the factor is approximated by

$$SF = \begin{cases} 1 & \text{for } \omega < 1/\tau_S \qquad (10.164\text{a}) \\[2mm] \dfrac{R_{\text{SH}}}{j\omega L_{\text{SH}}} & \text{for } \omega > 1/\tau_S \qquad (10.164\text{b}) \end{cases}$$

Observe that for $\omega < 1/\tau_S$ the crosstalk increases at 20 dB/decade and is unaffected by the shield. For $\omega > 1/\tau_S$ the crosstalk is constant with frequency. This is illustrated in Fig. 10.48. The *break frequency* where the shield becomes effective in reducing inductive coupling is denoted as $f_0 = 1/2\pi\tau_S = R_{\text{SH}}/2\pi L_{\text{SH}}$.

A qualitative description of this effect is as follows. For $\omega < 1/\tau_S$ the generator current finds the lowest-impedance path to be through the ground plane, and so the flux from this current threads the entire receptor circuit. For $\omega > 1/\tau_S$ the generator current finds the lowest-impedance path to be through the shield, $\hat{I}_S = \hat{I}_G$ [18–23]. Thus the generator current returns along the shield instead of through the ground plane, resulting in no net magnetic flux threading the receptor circuit. Thus *if the shield is grounded at both ends*, the inductive coupling contribution to the crosstalk transfer function is

$$\frac{\hat{V}_{NE}^{\text{IND}}}{\hat{V}_S} = \frac{R_{NE}}{R_{FE} + R_{NE}} j\omega L_{GR} \frac{R_{\text{SH}}}{R_{\text{SH}} + j\omega L_{\text{SH}}} \frac{1}{R_S + R_L} \qquad (10.165\text{a})$$

$$\frac{\hat{V}_{FE}^{\text{IND}}}{\hat{V}_S} = -\frac{R_{FE}}{R_{FE} + R_{NE}} j\omega L_{GR} \frac{R_{\text{SH}}}{R_{\text{SH}} + j\omega L_{\text{SH}}} \frac{1}{R_S + R_L} \qquad (10.165\text{b})$$

and we have substituted the relation between the low-frequency generator line

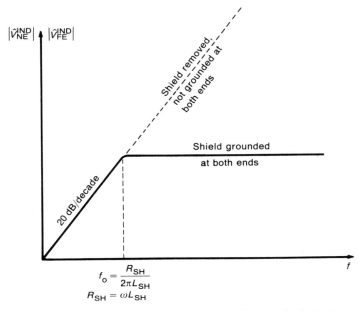

FIGURE 10.48   Illustration of the effect of shield grounding on the inductive coupling to a shielded wire.

current $\hat{I}_G$ and the source voltage $\hat{V}_S$:

$$\hat{I}_G = \frac{1}{R_S + R_L} \hat{V}_S \tag{10.166}$$

The total crosstalk transfer function is the sum of the inductive and capacitive coupling contributions:

$$\frac{\hat{V}_{NE}}{\hat{V}_S} = \frac{\hat{V}_{NE}^{IND}}{\hat{V}_S} + \frac{\hat{V}_{NE}^{CAP}}{\hat{V}_S} \tag{10.167a}$$

$$\frac{\hat{V}_{FE}}{\hat{V}_S} = \frac{\hat{V}_{FE}^{IND}}{\hat{V}_S} + \frac{\hat{V}_{FE}^{CAP}}{\hat{V}_S} \tag{10.167b}$$

In summary, *if the shield is grounded at at least one end, the capacitive coupling contribution is zero, and the inductive coupling is affected by the shield only if the shield is grounded at both ends and the frequency is greater than the reciprocal of the shield time constant.*

In order to illustrate these results, we will consider experimental results. Consider the shielded receptor circuit shown in Fig. 10.49. The line length is 3.6576 m and the conductors are suspended a height of 1.5 cm above a ground

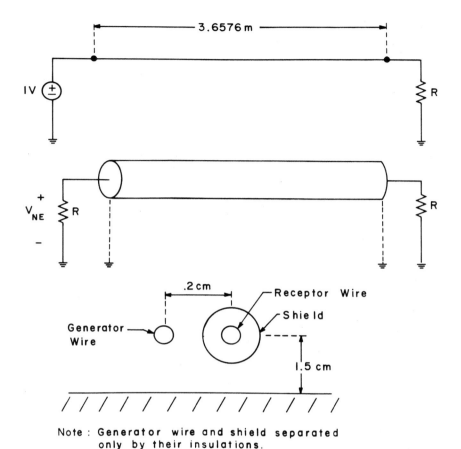

FIGURE 10.49 An experiment to illustrate the effect of shield grounding on crosstalk to a shielded wire.

plane. Two values of termination resistance will be used to accentuate the inductive and capacitive coupling: $R = 50 \ \Omega$ and $1 \ k\Omega$. The generator wire and shield insulations are in contact, giving a separation of approximately 0.2 cm. This will be referred to on the plots as SEPARATION: TOUCHING. The generator wire is a #20 gauge solid wire ($r_{wG} = 16$ mils). The characteristics of the shielded wire are $r_{wR} = 15$ mils (#22 gauge stranded, 7 × 30), $\epsilon_r = 2.1$ (Teflon), $r_{sh} = 35$ mils, $r_{braid} = 2.5$ mils (#36 gauge wires), $\theta_w = 30°$, $B = 16$, $W = 4$, and $t_{sh} \cong 5$ mils $= 2r_{braid}$. The resistance of the shield is

$$R_{SH} = \frac{\mathscr{L}}{\sigma \pi r_{braid}^2 BW \cos \theta_w}$$

$$= 89.8 \ m\Omega$$

The inductances are obtained by multiplying the per-unit-length inductances given by (10.144)–(10.148) by the line length to give

$$L_G = 3.15\ \mu\text{H}$$

$$L_R = 3.19\ \mu\text{H}$$

$$L_{\text{SH}} = 2.48\ \mu\text{H}$$

$$L_{GR} = 1.98\ \mu\text{H}$$

$$= L_{GS}$$

The capacitance C_{RS} (although not needed) is obtained by multiplying (10.150) by the line length to yield

$$C_{RS} = 503.6\ \text{pF}$$

The mutual capacitance C_{GS} is found from (10.151) as

$$C_{GS} = \frac{l_{GS}\mathscr{L}}{v_0^2(l_G l_S - l_{GS}^2)}$$

$$= 76.3\ \text{pF}$$

The shield time constant gives a break frequency of

$$f = \frac{R_{\text{SH}}}{2\pi L_{\text{SH}}}$$

$$= 5.8\ \text{kHz}$$

The near-end crosstalk transfer ratio *with the shield removed* is

$$\left|\frac{\hat{V}_{NE}}{\hat{V}_S}\right| = 2\pi f\left(\frac{1}{2R}L_{GR} + \frac{R}{2}C_{GR}\right)$$

With the shield removed, the mutual inductance between the generator and receptor wires is unchanged, $L_{GR} = 1.98\ \mu\text{H}$, but the mutual capacitance must be recomputed as

$$C_{GR} = \frac{l_{GR}\mathscr{L}}{v_0^2(l_G l_R - l_{GR}^2)}$$

$$= 48.2\ \text{pF}$$

Substitution of these values gives

$$\left|\frac{\hat{V}_{NE}}{\hat{V}_S}\right| = (1.24 \times 10^{-7} + 7.58 \times 10^{-9})f \quad \text{for } R = 50\,\Omega$$

$$= 1.32 \times 10^{-7}f$$

$$\left|\frac{\hat{V}_{NE}}{\hat{V}_S}\right| = (6.22 \times 10^{-9} + 1.51 \times 10^{-7})f \quad \text{for } R = 1\,\text{k}\Omega$$

$$= 1.58 \times 10^{-7}f$$

The experimental results are shown in Fig. 10.50. The results show the transfer function with the shield present but ungrounded at either end (OO), the shield grounded at the near end (SO), the shield grounded at the far end (OS), and with the shield grounded at the near and far ends (SS). The near-end crosstalk transfer function is measured from 100 Hz to 100 MHz. For $R = 50\,\Omega$ the measured crosstalk transfer function is 1.7×10^{-5} at 100 Hz, which is quite close to the calculated value of 1.32×10^{-5}. For $R = 1\,\text{k}\Omega$ the measured crosstalk transfer function is 4×10^{-5} at 100 Hz, which is slightly larger than the calculated value of 1.58×10^{-5}. The fact that the experimental and computed values agree well for low-impedance loads but the calculated value is below the experimental value for high-impedance loads can be explained by the fact that the wire dielectric insulations were neglected in the computation of the mutual capacitance. For high-impedance terminations where the capacitive coupling dominates the inductive coupling omitting the dielectric insulations for this very close wire spacing should cause the predictions to be lower than the experimental results.

The experimental results given in Fig. 10.50 can be explained with reference to Fig. 10.51. First consider the case for $R = 50\,\Omega$. The characteristic impedance of the isolated generator wire above ground can be computed to be $Z_{CG} = v_0 l_G = 258\,\Omega$. Thus inductive coupling is dominant for $R = 50\,\Omega$. Capacitive coupling is well below inductive coupling for the shield untrounded (OO). Thus when the shield is grounded at only one end (near end or far end) (SO or OS), the capacitive coupling is reduced substantially (theoretically eliminated), but no reduction in the total crosstalk is observed, since inductive coupling was dominant. When the shield is grounded at both ends (SS), the coupling is inductive, but becomes constant above the break frequency, as illustrated in Fig. 10.48. On the other hand, for $R = 1\,\text{k}\Omega$ and the shield ungrounded at either end (OO) capacitive coupling dominates inductive coupling. When the shield is grounded at one end (SO or OS), the capacitive coupling is virtually eliminated, but inductive coupling is unaffected and the total coupling drops to the inductive coupling "floor." When the shield is grounded at both ends (SS), the remaining inductive coupling is unaffected below the break frequency, but flattens out above it. There is a slight increase above this break frequency that is due to the "small" exposure of the receptor wire via a 0.5 cm break in the shield at

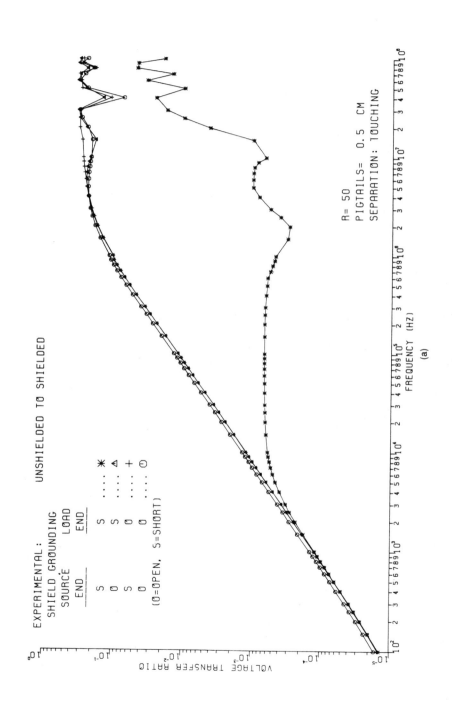

(a)

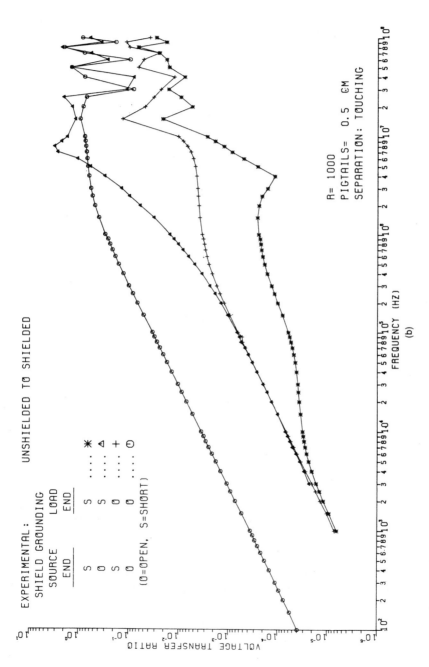

FIGURE 10.50 Measured near-end crosstalk for the configuration of Fig. 10.49 for the shield ungrounded, grounded at the left end, grounded at the right end, and grounded at both ends for (a) $R = 50\ \Omega$ and (b) $R = 1\ \text{k}\Omega$.

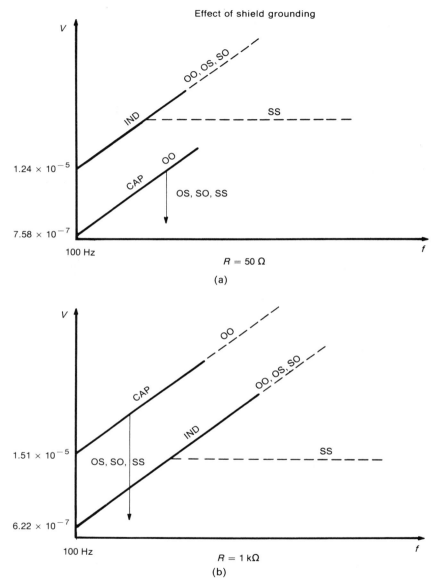

FIGURE 10.51 Explanation of the effect of shield grounding in the experimental results of Fig. 10.50 in terms of inductive and capacitive coupling for (a) $R = 50\ \Omega$ and (b) $R = 1\ k\Omega$.

either end (*pigtail*), necessitated by the need to connect the shield to the ground plane, as we will see in the next section.

The above results can be shown to remain true whether the shield is placed around the receptor wire or around the generator wire [21, 22, 24]. If a shield is placed around the generator wire and around the receptor wire, the results are essentially unchanged in that capacitive coupling can be eliminated by grounding *either shield* at *either end.* The inductive coupling is the product of the inductive coupling with both shields removed multiplied by *two shield factors* of the form of (10.162) [21, 22]. In summary, *connecting either end of a shield to the reference conductor eliminates capacitive coupling. In order to affect inductive coupling, the shield must be connected to the reference conductor at both ends. The inductive coupling will become constant with frequency above the shield break frequency given by* $f_0 = R_{SH}/2\pi L_{SH}$.

10.2.3 Effect of Pigtails

The term *pigtail* is commonly used to refer to the break in a shield required to terminate it to a "grounding point." Shield connections are commonly passed through a connector by connecting the shield to a connector pin with another wire (the pigtail wire), as shown in Fig. 10.52. This exposes the interior, shielded wire over the length of the pigtail wire. Pigtail lengths of over 5 inches are not uncommon.

The effect of the exposed, pigtail section is to allow the direct coupling to the interior shielded wire over the length of the pigtail section. For electrically short lines we can superimpose the coupling over the two pigtail sections at the ends of the cable and the coupling over the shielded section (which we considered in the previous section), as illustrated in Fig. 10.53. If the line is electrically short, we can reflect the termination impedances to the ends of each

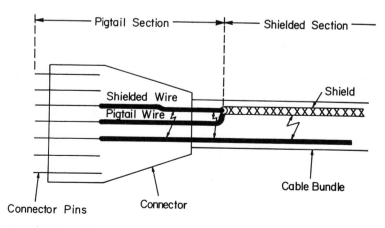

FIGURE 10.52 Illustration of "pigtails" used to terminate shields in a cable connector.

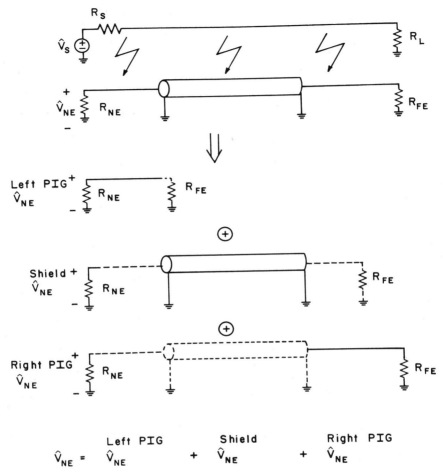

FIGURE 10.53 An approximate method of computing the effect of pigtails on crosstalk by superimposing the inductive and capacitive coupling over each section.

section and treat the individual couplings as separate problems, e.g. [18, 25],

$$\hat{V}_{NE} = \hat{V}_{NE}^{\text{Left Pigtail}} + \hat{V}_{NE}^{\text{Shielded Section}} + \hat{V}_{NE}^{\text{Right Pigtail}} \qquad (10.168)$$

We can compute each of these contributions using the previous methods. Each contribution is the sum of a capacitive and inductive coupling component. If the shield is grounded at either end, the capacitive coupling is eliminated from $\hat{V}_{NE}^{\text{Shielded Section}}$.

As an experimental example, consider the experiment considered previously for the illustration of shielding. We will investigate the near-end crosstalk transfer function for the shield grounded at both ends and for two values of terminations impedances, $R = 50\,\Omega$ and $1\,k\Omega$. Three lengths of pigtail sections on both ends will be used, as illustrated in Fig. 10.54. The first length is 0.5 cm, and is as small as reasonably possible without using a peripheral bonding of the shield. The second length is 3 cm, which is only 1.6% of the total line length. The third length is 8 cm, which is only 4.4% of the total line length. The experimental data are shown in Fig. 10.55. Observe that for $R = 50\,\Omega$ and frequencies above 1 MHz the longest length of pigtail causes the crosstalk to be as much as some 30 dB larger than for the shortest pigtail length. The reason for this is illustrated in Fig. 10.56, where we have shown the components given in (10.168). Observe that the contributions over the pigtail sections are inductive and much smaller than the contribution over the shielded section below 100 kHz. Above the shield break frequency (6 kHz) the generator wire current flows back along the shield,

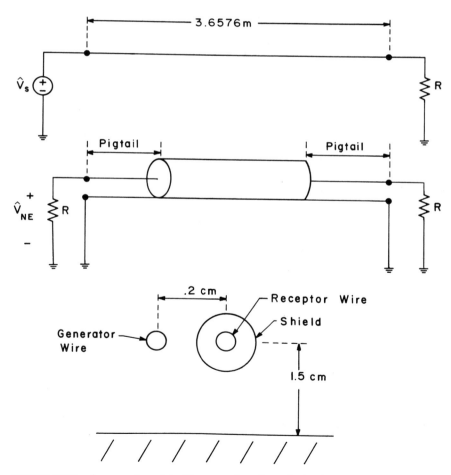

FIGURE 10.54   An experiment to illustrate the effect of pigtail lengths on crosstalk.

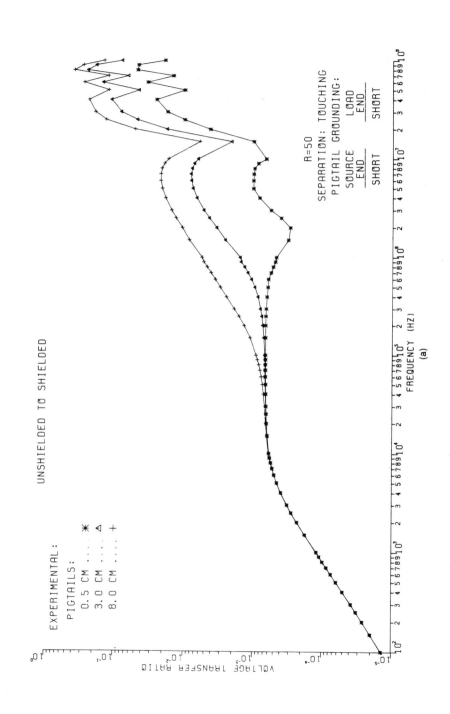

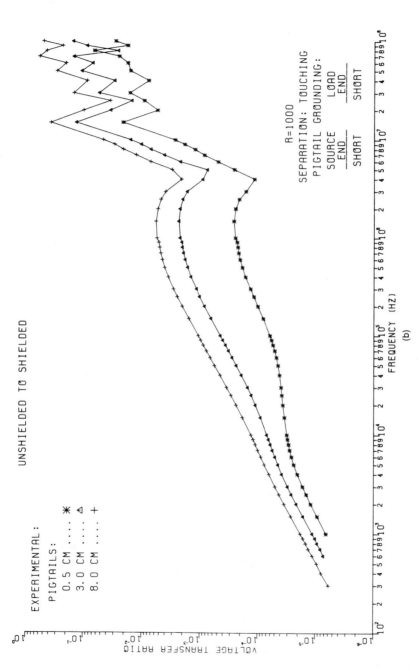

FIGURE 10.55 Experimental results for the configuration of Fig. 10.54 for pigtail lengths of 0.5 cm, 3 cm, and 8 cm for (a) $R = 50 \, \Omega$ and (b) $R = 1 \, k\Omega$.

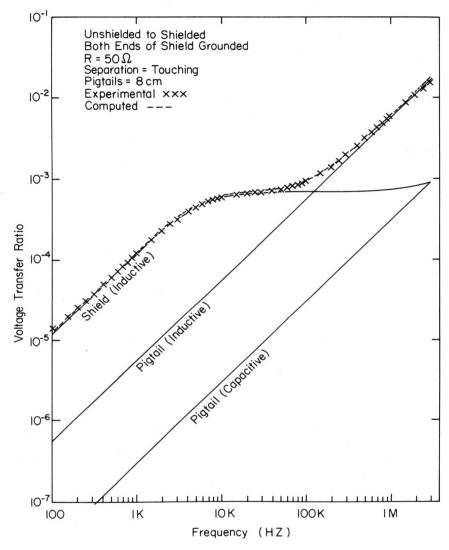

FIGURE 10.56   The near-end crosstalk for the configuration of Fig. 10.54 for 8 cm pigtail lengths and $R = 50\ \Omega$ in terms of contributions according to Fig. 10.53.

and the coupling over the shielded section flattens out. The pigtail coupling continues to increase at 20 dB/decade, and becomes greater than the coupling over the shielded section above 100 kHz. For the case of $R = 1\ k\Omega$ shown in Fig. 10.55 we see a similar effect, but it extends to a much lower frequency than for $R = 50\ \Omega$. This is because the inductive coupling over the shielded section is reduced to a much lower level for $R = 1\,\text{k}\Omega$, and the pigtail section contributions begin to dominate at a much lower frequency.

The result is that *pigtails reduce the shielding effectiveness of a shielded wire over that which could be obtained if the shield were peripherily bonded at each end, thereby causing no exposure of the interior wire.* Pigtails do not eliminate the effect of a shield; they just reduce it from its ideal effect.

10.2.4   Effects of Multiple Shields

We observed that the effect of placing a shield around a wire is to eliminate capacitive coupling if the shield is terminated to the reference conductor at one or both ends. The shield acts to reduce inductive coupling above the break frequency $f_o = R_{\text{SH}}/2\pi L_{\text{SH}}$ only if the shield is terminated to the reference conductor at *both ends.*

Suppose we place shields around *both* the generator *and* receptor wires. Also suppose we terminate both ends of each shield to the reference conductor. The capacitive coupling will be eliminated, and we are left with the inductive coupling to consider. The inductive coupling *with the shields removed* will be multiplied by two terms, each of the form of (10.162). Thus the total coupling is inductive and is given by [18, 21, 22, 25]

$$\frac{\hat{V}_{NE,FE}}{\hat{V}_S} = \underbrace{\frac{\hat{V}_{NE,FE}^{\text{IND}}}{\hat{V}_S}}_{\substack{\text{with both} \\ \text{shields} \\ \text{removed}}} \frac{R_{\text{SHG}}}{R_{\text{SHG}} + j\omega L_{\text{SHG}}} \frac{R_{\text{SHR}}}{R_{\text{SHR}} + j\omega L_{\text{SHR}}} \qquad (10.169)$$

where R_{SHG} and L_{SHG} (R_{SHR} and L_{SHR}) are the shield total resistance and self-inductance of the generator circuit (receptor circuit) shield. Thus the crosstalk transfer ratio will be as shown in Fig. 10.57, where the break frequencies are given by

$$f_{oG} = \frac{R_{\text{SHG}}}{2\pi L_{\text{SHG}}} \qquad (10.170\text{a})$$

$$f_{oR} = \frac{R_{\text{SHR}}}{2\pi L_{\text{SHR}}} \qquad (10.170\text{b})$$

Therefore the presence of shields on both wires (and grounded at both ends) causes the crosstalk to roll off at $-20\,\text{dB/decade}$ above the second break frequency.

In order to illustrate this effect, consider the previous experiment illustrated in Fig. 10.54. We will show the experimentally observed results for (a) no shields, (b) one shield around the receptor wire, and (c) shields on both the generator

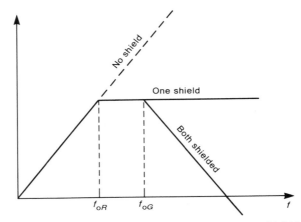

FIGURE 10.57   Frequency response of crosstalk for no shields, one shield (on the generator wire or on the receptor wire), and shields on both wires.

and the receptor wires. The experimental cross-sectional configurations are shown in Fig. 10.58. The separation between generator and receptor wires is $s = 1.5$ cm, which is denoted on the graphs as SEPARATION: WIDE. The pigtail lengths for both shields are 8 cm. Data for other separations and pigtail lengths are given in [21, 22]. The experimental data are shown in Fig. 10.59. Since both shields are identical and are at the same height above the ground plane ($h = 1.5$ cm), $R_{SHG} = R_{SHR}$ and $L_{SHG} = L_{SHR}$, so that the break frequencies are the same, $f_{oG} = f_{oR} \cong 6$ kHz. The behavior of the inductive coupling illustrated in Fig. 10.48 for one shield and in Fig. 10.57 for shields on both wires is evident in the experimental data. Observe that for shields on both wires the -20 dB/decade behavior extends to 100 kHz. Above this, the response increases at $+20$ dB/decade. This is clearly due to the coupling over the 8 cm pigtail sections becoming dominant.

10.2.5   MTL Model Predictions

The exact solution of the multiconductor transmission-line (MTL) equations is described in [18, 21, 22, 25]. The solution technique is to treat the two pigtail sections and the shielded section as cascaded transmission lines and determine the exact chain parameter (CPM) matrix of each section. Multiplying these chain parameter matrices together (in the proper order) gives the CPM of the overall line. The terminal conditions are incorporated [2, 22, 25] in order to solve for the terminal voltages. This has been implemented in a FORTRAN computer program described in [22] that considers a number of nonideal effects such as coupling through holes in the shield. In order to illustrate the prediction accuracy of this exact solution method, we will show the predictions for the data that were given in Fig. 10.59. These results are given in Fig. 10.60(a) for one shield around the receptor wire and in Fig. 10.60(b) for shields around

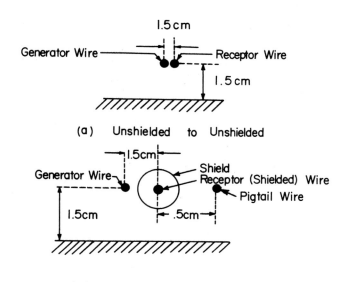

(a)   Unshielded   to   Unshielded

(b)   Unshielded   to   Shielded

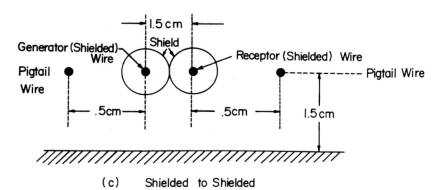

(c)   Shielded   to   Shielded

FIGURE 10.58 Cross section of an experiment to illustrate the effect of placing shields on neither wire, one wire, or both wires.

generator and receptor wires. Both shields are grounded at both ends, the pigtail lengths at each end are 8 cm, and the generator and receptor circuits are separated by 1.5 cm (SEPARATION: WIDE). The MTL model shows excellent predictions of the experimental results.

10.3 TWISTED WIRES

We now turn our attention to the use of a *twisted pair of wires* to reduce crosstalk. A twisted pair is the *dual* to a shielded wire in the following sense. We observed that placing a shield around a receptor wire inherently tends to

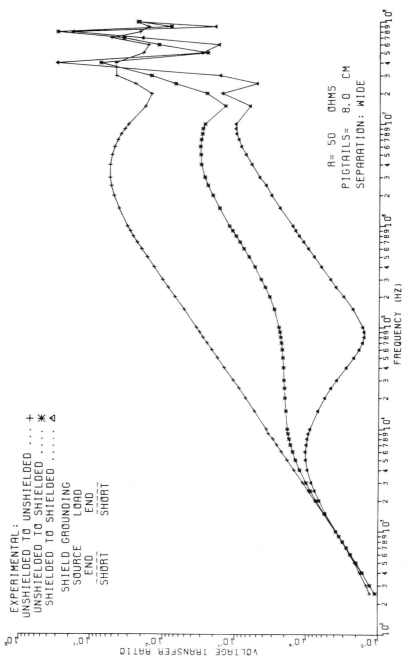

FIGURE 10.59 Experimental results for $R = 50\,\Omega$ and 8 cm pigtails for the configurations of Fig. 10.58.

reduce *capacitive or electric field coupling* so long as either end of the shield is attached to the reference conductor. In order to reduce inductive coupling, the shield must be attached to the reference conductor at *both ends* in order to allow a current to flow back along the shield to produce a counteracting magnetic flux to cancel the flux from the generator current. Replacing a receptor wire with a twisted pair and using one wire of the pair as the return for the receptor circuit inherently reduces *inductive or magnetic field coupling* because of the twist. The twisted pair reduces capacitive coupling only if the terminations at both ends are *balanced with respect to the reference conductor*.

In reality, a twisted pair of wires is a *bifilar helix*. In order to model the twisted pair, we will approximate it as a cascade of alternating loops as shown in Fig. 10.61. The essential way in which the twisted pair reduces *inductive or magnetic field coupling* is as follows. Consider the magnetic flux from the generator wire current. This flux threads the loops of the twisted pair, incuding emfs in each loop. But because the loops alternate in polarity, the induced emfs tend to cancel in the adjacent loops. Thus the net induced emf induced in the receptor circuit (the twisted pair) is that of one half-twist. (A loop is referred to here as a *half-twist*.) The twist does not, inherently, affect capacitive coupling.

A simple *topological model* for the twisted pair is shown in Fig. 10.62. This model assumes that the twisted pair can be approximated as a cascade of alternating rectangular loops in the vertical plane (or in the horizontal plane) The *exact transmission-line model* for this *approximation of the bifilar helix* is solved in [26, 27, 28] by modeling each half-twist whose length is denoted by $\mathscr{L}_{HT}$ along with the adjacent generator wire with a chain parameter matrix (CPM) for this section. The overall chain parameter matrix of the cascade is the product of all these chain parameter matrices of the half-twist sections with an appropriate reversal of the twisted wires at the junctions between the half-twists. This is similar to the process used in the previous section to incorporate pigtail sections of a shielded-wire line. The reader is referred to the implementation of this transmission-line model to [26, 28]. Predictions of the exact MTL model will be shown in Section 10.3.3.

The following approximate model makes use of the notions of inductive and capacitive coupling described previously, and was originally described in [27, 28]. We assume once again that the line is *electrically short* in using the following lumped model. Also we assume that the generator and receptor circuits are weakly coupled. Now consider each loop (half-twist) as a parallel-wire line and compute the mutual inductances from the generator circuit to *each circuit formed between each wire of the twisted pair and the reference conductor*. This gives per-unit-length mutual inductances l_{m1} and l_{m2} and per-unit-length mutual capacitances c_{m1} and c_{m2}. The effects of these mutual elements are represented in the usual fashion as voltage and current sources whose values depend on the mutual element *for that length of the half-twist* and the generator circuit voltage or current, as shown in Fig. 10.63. The total inductance or capacitance parameter is the per-unit-length value multiplied by the length of the half-twist (the loop), denoted by $\mathscr{L}_{HT}$.

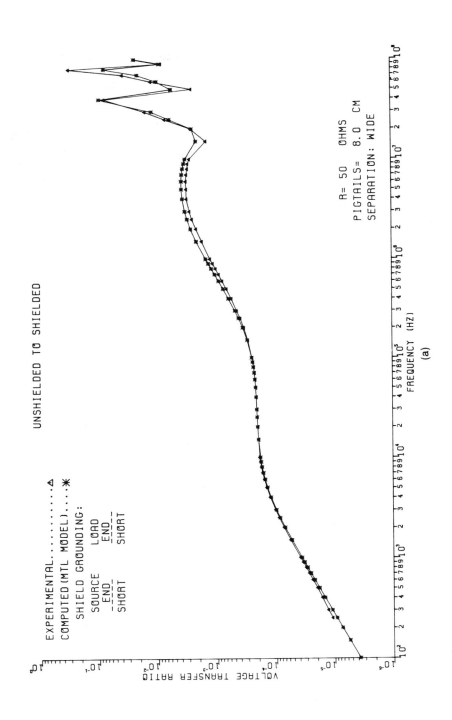

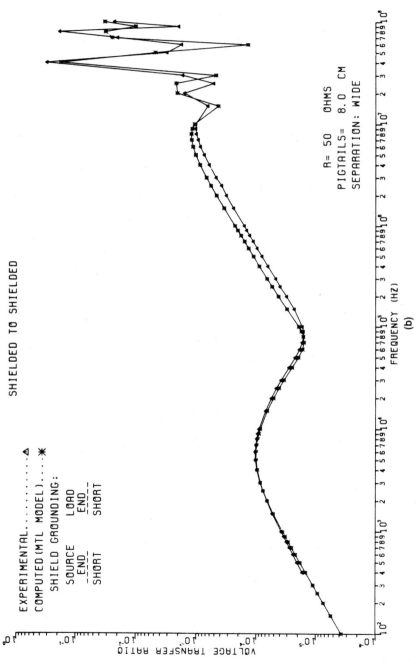

FIGURE 10.60 Predictions of the transmission-line model versus experimental results for the configuration of Fig. 10.58, for (a) a shield on the receptor wire and (b) shields on both the generator and the receptor wires.

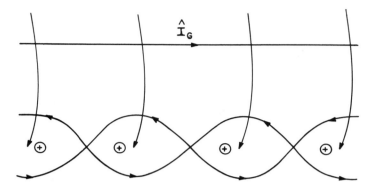

FIGURE 10.61 Illustration of the effect of a twisted pair of receptor circuit wires on magnetic field (inductive) coupling.

Simplified model of twisted pair

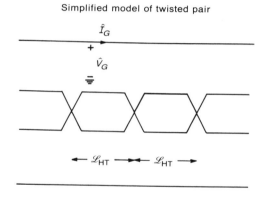

FIGURE 10.62 A simple "abrupt loop" model of a twisted pair of receptor circuit wires.

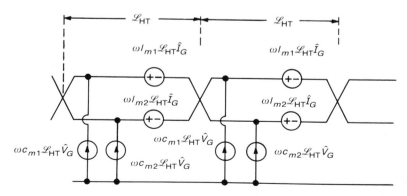

FIGURE 10.63 The simple inductive-capacitive coupling model for the twisted pair of receptor circuit wires of Fig. 10.62.

10.3.1 Per-Unit-Length Parameters

These mutual inductances are computed as shown in Fig. 10.64. For example, consider a generator wire and a twisted pair above a ground plane (the reference conductor), as shown in Fig. 10.65. The generator wire and the twisted pair are at a height h above the ground plane, and the horizontal separation between the generator wire and the twisted pair is d. The separation between the two wires of the twisted pair is $2\,\Delta h$. Assume for illustration that the wires are identical, with radii r_w. The mutual inductances are obtained by treating the each wire of the twisted pair with the ground plane as a circuit and using our previously derived results (see (10.37)) to obtain

$$l_{m1} = \frac{\mu_0}{4\pi} \ln\left[1 + \frac{4h(h + \Delta h)}{d^2 + \Delta h^2} \right] \tag{10.171a}$$

$$l_{m2} = \frac{\mu_0}{4\pi} \ln\left[1 + \frac{4h(h - \Delta h)}{d^2 + \Delta h^2} \right] \tag{10.171b}$$

In order to determine the mutual capacitances, we invert the per-unit-length inductance of the line. The self-inductances are (see (10.33) and (10.34))

$$l_G = \frac{\mu_0}{2\pi} \ln\left(\frac{2h}{r_w} \right) \tag{10.171c}$$

$$l_{R1} = \frac{\mu_0}{2\pi} \ln\left[\frac{2(h + \Delta h)}{r_w} \right] \tag{10.171d}$$

$$l_{R2} = \frac{\mu_0}{2\pi} \ln\left[\frac{2(h - \Delta h)}{r_w} \right] \tag{10.171e}$$

The remaining mutual inductance is computed by placing a current on one

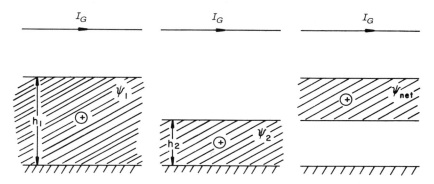

FIGURE 10.64 Illustration of the per-unit-length mutual inductance to a twisted pair of receptor wires.

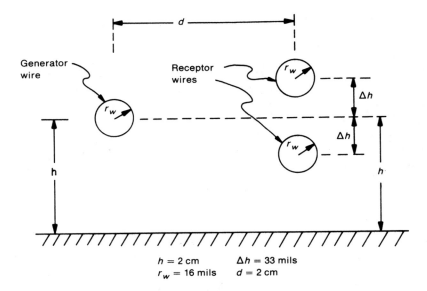

$h = 2$ cm $\Delta h = 33$ mils
$r_w = 16$ mils $d = 2$ cm

FIGURE 10.65 Cross-sectional dimensions of a twisted receptor wire pair for computing the per-unit-length inductances.

wire of the twisted pair (and returning in the image), and computing the flux through the circuit formed by the other wire and the ground plane using the fundamental result in (10.16a):

$$l_{R1R2} = \frac{\mu_0}{2\pi} \ln\left(\frac{h + \Delta h}{h - \Delta h}\right) + \frac{\mu_0}{2\pi} \ln\left[\frac{(h + \Delta h) + (h - \Delta h)}{h + \Delta h}\right] \quad (10.171f)$$

$$= \frac{\mu_0}{2\pi} \ln\left(1 + \frac{h + \Delta h}{h - \Delta h}\right)$$

For the dimensions of $d = 2$ cm, $h = 2$ cm, $r_w = 16$ mils (#20 gauge, solid), and $\Delta h = 33$ mils, we compute

$$l_{m1} = 1.641 \times 10^{-7} \text{ H/m}$$

$$l_{m2} = 1.574 \times 10^{-7} \text{ H/m}$$

$$l_G = 9.179 \times 10^{-7} \text{ H/m}$$

$$l_{R1} = 9.261 \times 10^{-7} \text{ H/m}$$

$$l_{R2} = 9.093 \times 10^{-7} \text{ H/m}$$

$$l_{R1R2} = 1.472 \times 10^{-7} \text{ H/m}$$

The difference between l_{m1} and l_{m2} will be needed:

$$l_{m1} - l_{m2} = 6.706 \times 10^{-9}$$

The mutual capacitances are computed by ignoring the dielectric insulations as (see Fig. 10.66)

$$
\begin{bmatrix}
c_G + c_{m1} + c_{m2} & -c_{m1} & -c_{m2} \\
-c_{m1} & c_{R1R2} + c_{m1} + c_{R1} & -c_{R1R2} \\
-c_{m2} & -c_{R1R2} & c_{R1R2} + c_{m2} + c_{R2}
\end{bmatrix}
$$

$$
= \mu_0 \epsilon_0 \begin{bmatrix}
l_G & l_{m1} & l_{m2} \\
l_{m1} & l_{R1} & l_{R1R2} \\
l_{m2} & l_{R1R2} & l_{R2}
\end{bmatrix}^{-1}
\tag{10.172}
$$

This gives

$$
c_{m1} = 1.964 \text{ pF/m}
$$

$$
c_{m2} = 1.895 \text{ pF/m}
$$

We will also need the sum and difference of c_{m1} and c_{m2}:

$$
c_{m1} + c_{m2} = 3.858 \text{ pF/m}
$$

$$
c_{m1} - c_{m2} = 0.06924 \text{ pF/m}
$$

10.3.2 Inductive and Capacitive Coupling

The resulting model thus consists of a sequence of sources shown in Fig. 10.67 [26–33]. The induced voltage sources $\hat{E}_1$ and $\hat{E}_2$ are due to the mutual inductance between the generator and the twisted pair receptor and are

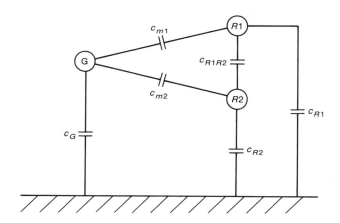

FIGURE 10.66 The per-unit-length capacitances for a twisted receptor pair.

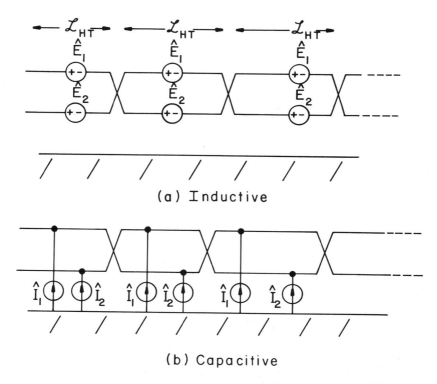

FIGURE 10.67 The simple inductive-capacitive coupling model of a twisted receptor pair: (a) inductive coupling and (b) capacitive coupling.

essentially induced emfs resulting from Faraday's law. The contributions of these voltages sources to the crosstalk voltages at the ends of the twisted pair are referred to as *inductive coupling*. The induced current sources $\hat{I}_1$ and $\hat{I}_2$ are due to mutual capacitance between the generator and twisted-pair receptor, and their contributions to the crosstalk voltages are referred to as *capacitive coupling*. From Fig. 10.63 these sources are given by

$$\hat{E}_2 = j\omega l_{m1} \mathscr{L}_{HT} \hat{I}_G \qquad (10.173a)$$

$$\hat{E}_2 = j\omega l_{m2} \mathscr{L}_{HT} \hat{I}_G \qquad (10.173b)$$

$$\hat{I}_1 = j\omega c_{m1} \mathscr{L}_{HT} \hat{V}_G \qquad (10.173c)$$

$$\hat{I}_2 = j\omega c_{m2} \mathscr{L}_{HT} \hat{V}_G \qquad (10.173d)$$

where $\hat{V}_G$ and $\hat{I}_G$ are the *low-frequency* values of the generator circuit voltage and current. These may be computed as dc values. Now let us "untwist" the wires, giving the circuit for two adjacent half-twists (one full twist) shown in Fig. 10.68.

Now untwist

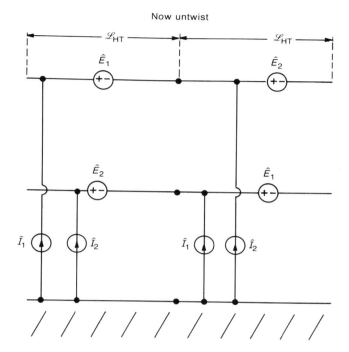

FIGURE 10.68   "Untwisting" the model of Fig. 10.67.

We next must consider the termination configurations. There are generally two methods for terminating twisted pairs, as shown in Fig. 10.69. The circuit in Fig. 10.69(a) is said to be *unbalanced* in the sense that the impedances seen between each wire and ground are not the same. One wire of the twisted pair is connected to the reference conductor at the near end, while the other end is not connected to the reference conductor in order to avoid *ground loops* between that wire and the reference conductor, which will allow circulating currents to flow in that loop. The circuit shown in Fig. 10.69(b) is said to be *balanced* in the sense that the impedances seen between each wire and the reference conductor are equal, which is due to the use of a center-tapped transformed. Balanced line drivers and receivers will also produce this effect.

We will consider the unbalanced case shown in Fig. 10.69(a) in this section, while the balanced case will be considered in Section 10.3.4. The equivalent circuit for the entire line, substituting the simple model for the twisted pair, is given in Fig. 10.70 and shows that the near-end crosstalk is the superposition of the inductive and capacitive coupling contributions. The values of the sources are given in (10.173). Untwisting the twisted pair gives the circuit in Fig. 10.71. This equivalent circuit shows that *the net induced emf in the twisted-pair loop is that of one half-twist for an odd number of half-twists and is zero for an even number of half-twists.* Observe that the current sources that are attached to the

(a)

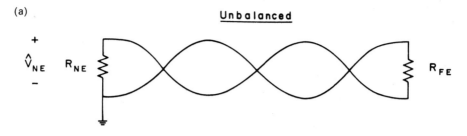

Note: Grounded at only one end to avoid
ground loops.

(b)

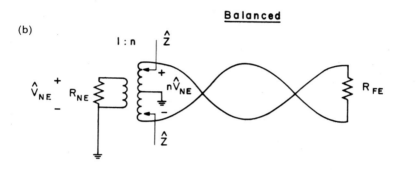

Each wire of twisted pair sees same
impedance to ground at each end.

FIGURE 10.69 Terminating a twisted pair: (a) unbalanced; (b) balanced.

wire of the twisted pair that is grounded at the near end are "shorted out," and thus do not contribute to the solution. The net current source attached to the wire that is not grounded over one complete twist is $\hat{I}_1 + \hat{I}_2 = j\omega c_{m1} \mathscr{L}_{HT} \hat{V}_G + j\omega c_{m2} \mathscr{L}_{HT} \hat{V}_G$, according to Fig. 10.63, where N is the total number of half-twists. Since the two twisted wires are very close together, $c_{m1} \cong c_{m2}$. Also, $\mathscr{L} = N\mathscr{L}_{HT}$, where N is the total number of twists. Thus the net current source attached to the ungrounded wire is approximately $N(\hat{I}_1 + \hat{I}_2) = j\omega c_m \mathscr{L}$, where we may use c_{m1} or c_{m2} for c_m. Essentially this gives the same capacitive coupling as that of the *untwisted pair*, which we will refer to as the *straight-wire pair* or *SWP*. The *twisted-wire pair* will be referred to as *TWP*. From this circuit we can determine the near-end and far-end crosstalk voltage transfer ratios in the usual fashion as

$$\frac{\hat{V}_{NE}}{\hat{V}_S} = \frac{R_{NE}}{R_{NE} + R_{FE}} j\omega(l_{m1} - l_{m2})\mathscr{L}_{HT} \frac{1}{R_S + R_L} \qquad (10.174a)$$

$$+ \frac{R_{NE}R_{FE}}{R_{NE} + R_{FE}} j\omega(c_m)\mathscr{L} \frac{R_L}{R_S + R_L}$$

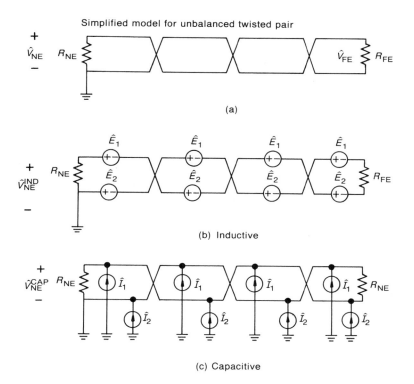

FIGURE 10.70 A simplified model for the unbalanced twisted receptor wire pair: (a) physical configuration; (b) inductive coupling model; (c) capacitive coupling model.

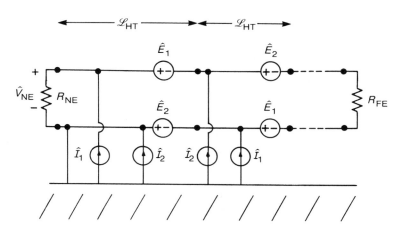

FIGURE 10.71 The inductive–capacitive coupling model for the unbalanced twisted receptor wire pair of Fig. 10.70 obtained by "untwisting" the wires.

$$\frac{\hat{V}_{FE}}{\hat{V}_S} = -\frac{R_{FE}}{R_{NE} + R_{FE}} j\omega(l_{m1} - l_{m2})\mathscr{L}_{HT}\frac{1}{R_S + R_L} \quad (10.174b)$$

$$+ \frac{R_{NE}R_{FE}}{R_{NE} + R_{FE}} j\omega(c_m)\mathscr{L}\frac{R_L}{R_S + R_L}$$

where we have substituted

$$\hat{I}_G = \frac{R_S}{R_S + R_L}\hat{V}_S \quad (10.175a)$$

$$\hat{V}_G = \frac{1}{R_S + R_L}\hat{V}_S \quad (10.175b)$$

Observe in (10.174) that the net mutual inductance is the *difference* of the per-unit-length mutual inductances between the generator circuit and the circuits formed from each wire of the twisted pair with the ground plane *multiplied by the length of a half-twist*. Thus the *inductive coupling is the same as to a straight-wire-pair of total length equal to one half-twist* (a "short line"). However, the net mutual capacitance is the per-unit-length mutual capacitance to the ungrounded wire of a straight-wire-pair *multiplied by the total line length* $\mathscr{L}$. Therefore the near-end or far-end crosstalk becomes

$$\frac{\hat{V}_{NE,FE}}{\hat{V}_S} = \underbrace{\frac{\hat{V}_{NE,FE}^{IND}}{\hat{V}_S}\bigg|_{\mathscr{L}_{SWP} = \mathscr{L}_{HT}}}_{\substack{\text{inductive coupling} \\ \text{for SWP whose} \\ \text{length is one} \\ \text{half-twist}}} + \underbrace{\frac{\hat{V}_{NE,FE}^{CAP}}{\hat{V}_S}\bigg|_{\mathscr{L}_{SWP} = \mathscr{L}}}_{\substack{\text{capacitive coupling} \\ \text{for SWP whose length} \\ \text{is the total line length}}} \quad (10.176)$$

This assumes that there is an *odd number of half-twists*. If the number of half-twists is even, all adjacent induced voltages will cancel, and the total inductive coupling is reduced to zero. In either event, *for this unbalanced termination the capacitive coupling is unaffected by the twist, i.e., it is the same as to a straight-wire pair (SWP) whose length is the total length of the twisted pair.*

With this understanding, let us consider the significance of the twist illustrated in Fig. 10.72. Consider an untwisted receptor pair (SWP). Suppose that inductive coupling dominates capacitive coupling, as shown in Fig. 10.72(a). If we replace the straight-wire pair with a twisted pair, the inductive coupling is reduced to that of one half-twist (if there are an odd number of half-twists), but the capacitive coupling is essentially unaffected. Since inductive coupling dominated capacitive coupling prior to the twisting of the wires, the total crosstalk will drop to the level of the capacitive coupling of the untwisted pair. On the other hand, consider the case for high-impedance loads where capacitive coupling dominates inductive

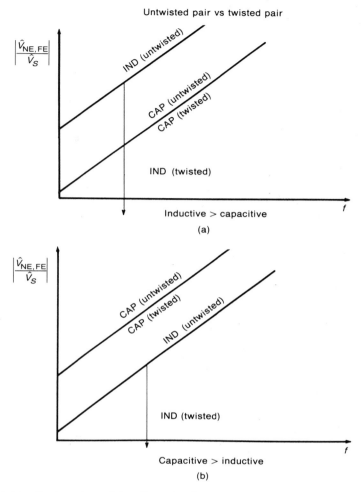

FIGURE 10.72 Explanation of the effect of twist on crosstalk to an unbalanced twisted pair for (a) inductive coupling dominant before twisting and (b) capacitive coupling dominant before twisting.

coupling for the untwisted pair. In this case twisting the wires will drop the inductive coupling contribution, and the capacitive coupling is again unaffected. But since capacitive coupling dominated inductive coupling prior to twisting the wires, *no reduction in the total coupling will be observed.* This illustrates that *for the unbalanced termination twisting the pair of receptor wires will reduce the total coupling for low-impedance terminations, but will not change it for high-impedance terminations!* This shows that twisted wires may or may not be effective in reducing crosstalk. They will usually be effective only in *low-impedance circuits.* This is why twisted pairs are typically effective in power distribution circuits, which typically have low-impedance terminations.

10.3.3 Effects of Twist

In order to illustrate the effectiveness or ineffectiveness of a twisted pair in the reduction of crosstalk, we will show some experimental results. We will compare the near-end crosstalk voltage transfer ratio for three configurations shown in Fig. 10.73. The *single-receptor-wire* configuration shown in Fig. 10.73(a) was considered earlier, and should give the largest crosstalk. The *straight-wire pair* (*SWP*) shown in Fig. 10.73(b) should reduce inductive coupling, since the receptor circuit loop area is reduced. The capacitive coupling should not be reduced, since it results from the current source attached to the ungrounded wire of the pair, which is essentially that of the single receptor wire (see

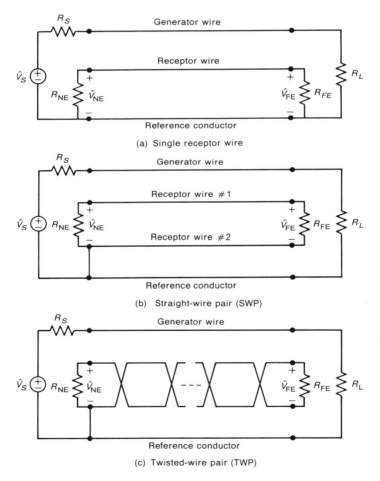

FIGURE 10.73 The three levels of reducing inductive crosstalk; (a) single receptor wire; (b) straight-wire receptor pair (SWP); (c) twisted-wire receptor pair (TWP).

Fig. 10.71). The *twisted-wire pair* (*TWP*) shown in Fig. 10.73(c) should only reduce inductive coupling.

The experimental configuration is shown in Fig. 10.74, and consists of a line of total length 4.705 m above a ground plane. The wires are #20 gauge and

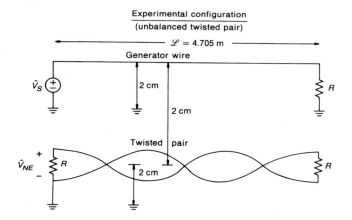

FIGURE 10.74 An experiment to illustrate the effect of a twisted pair on crosstalk for unbalanced terminations.

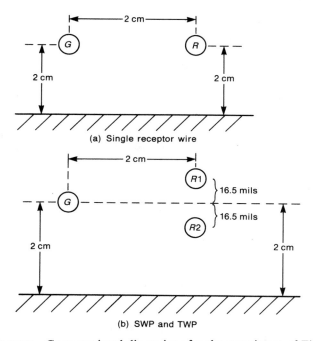

FIGURE 10.75 Cross-sectional dimensions for the experiment of Fig. 10.74.

are separated by 2 cm and suspended above the ground plane at a height of 2 cm. The cross-sectional configurations are shown in Fig. 10.75, and the per-unit-length parameters for the SWP and TWP configurations were computed in Section 10.3.1. The number of half-twists was $N = 225$, so that the length of a half-twist was $\mathscr{L}_{HT} = 2.09$ cm. This gives a twist rate of slightly over 7 twists/foot, which is typical. We will measure the near-end crosstalk voltage transfer ratio for $R_S = 0$ and $R_L = R_{NE} = R_{FE} = R$, where three values of R will be used: $R = 1$ kΩ, 50 Ω, and 1 Ω. The experimental results will also be compared with the predictions of the multiconductor transmission-line model, which forms the overall chain-parameter matrix of the line from the products of the chain-parameter matrices of the half-twists with an appropriate interchange of the wires of the twisted pair at the junctions [26]. The inductive/capacitive coupling model in (10.174) gives

$$\left|\frac{\hat{V}_{NE}}{\hat{V}_S}\right| = 2\pi f\left[\frac{1}{2R}(l_{m1} - l_{m2})\mathscr{L}_{HT} + \frac{R}{2}(c_m)\mathscr{L}\right] \tag{10.177}$$

where $\mathscr{L}_{HT} = 2.09$ cm and $\mathscr{L} = 4.705$ m. The data for $R = 1$ kΩ are shown in Fig. 10.76(a). Observe that there is little difference in the crosstalk for all three configurations. This confirms our notion that capacitive coupling is dominant for the single-receptor-wire configuration and is unchanged for the SWP and TWP configurations. Equation (10.177) gives

$$\left|\frac{\hat{V}_{NE}}{\hat{V}_S}\right|_{R=1\text{ k}\Omega} = 2\pi f\left[\frac{1}{2R}(l_{m1} - l_{m2})\mathscr{L}_{HT} + \frac{R}{2}c_m\mathscr{L}\right]$$

$$= f(4.403 \times 10^{-13} + 2.852 \times 10^{-8})$$

where we have used $c_m = \frac{1}{2}(c_{m1} + c_{m2})$. At the lowest measured frequency of 1 kHz this gives a prediction of 2.9×10^{-5}, which compares well with the measured value of 2×10^{-5}. The data for $R = 50$ Ω are shown in Fig. 10.76(b). Here we see that the SWP configuration reduces the crosstalk by some 20 dB, and the TWP configuration further reduces that by some 10 dB. Equation (10.177) gives

$$\left|\frac{\hat{V}_{NE}}{\hat{V}_S}\right|_{R=50\text{ }\Omega} = 2\pi f\left[\frac{1}{2R}(l_{m1} - l_{m2})\mathscr{L}_{HT} + \frac{R}{2}c_m\mathscr{L}\right]$$

$$= f(8.81 \times 10^{-12} + 1.43 \times 10^{-9})$$

This gives a prediction of 1.44×10^{-6} at 1 kHz, which matches the measured value of 1.8×10^{-6} rather well. Figure 10.76(c) shows the results for $R = 1$ Ω. Observe that the SWP configuration reduces the crosstalk by some 26 dB, whereas the TWP configuration further reduces the crosstalk by some 80 dB. The crosstalk levels for the TWP configuration and $R = 1$ Ω shown in

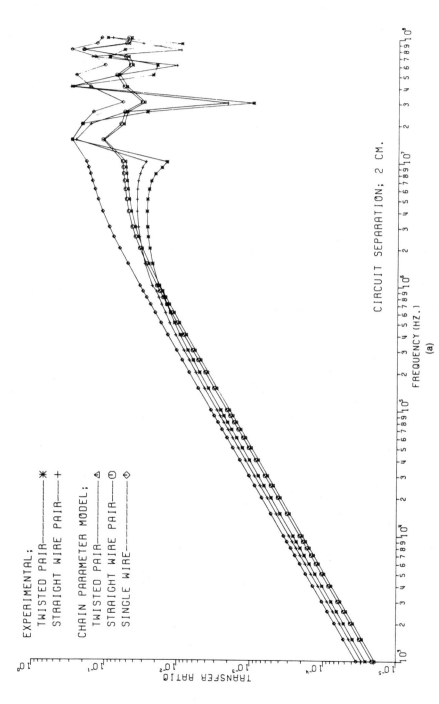

FIGURE 10.76 Experimental results for the experiment of Fig. 10.74, comparing the predictions of the transmission-line model (chain parameter model) to measured data for a single receptor wire, a straight-wire receptor pair, and a twisted-wire receptor pair, for (a) $R = 1$ kΩ.

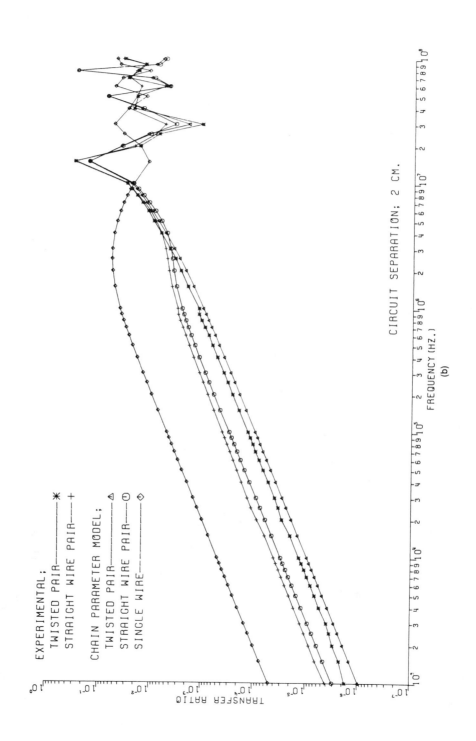

EXPERIMENTAL;
TWISTED PAIR———✳
STRAIGHT WIRE PAIR———+

CHAIN PARAMETER MODEL;
TWISTED PAIR———△
STRAIGHT WIRE PAIR———⊙
SINGLE WIRE———◇

TRANSFER RATIO

FREQUENCY (HZ.)

(b)

CIRCUIT SEPARATION; 2 CM.

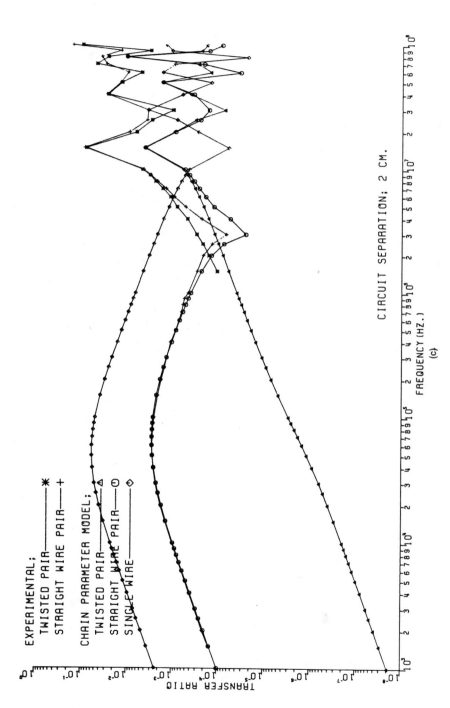

FIGURE 10.76 continued   (b) $R = 50\,\Omega$, and (c) $R = 1\,\Omega$.

Fig. 10.76(c) were too small to be measured below 1.5 MHz. The trend of the data at this lowest measurement frequency indicates that the predictions of the MTL model that are shown are reasonable and that there is a large reduction for $R = 1\,\Omega$. Measurements for $R = 3\,\Omega$ and given in [28] confirm this behavior for very low-impedance loads. Equation (10.177) gives

$$\left|\frac{\hat{V}_{NE}}{\hat{V}_S}\right|_{R=1\,\Omega} = 2\pi f\left[\frac{1}{2R}(l_{m1} - l_{m2})\mathscr{L}_{HT} + \frac{R}{2}c_m\mathscr{L}\right]$$
$$= f(4.40 \times 10^{-10} + 2.90 \times 10^{-11})$$

This gives a predicted value of 4.69×10^{-7} at 1 kHz. The calculated value from the MTL model of 2×10^{-8} is for an even number of half-twists, so that the inductive coupling is not included. Thus the above capacitive coupling of 2.9×10^{-8} at 1 kHz matches the prediction rather well.

The explanation for this effectiveness or ineffectiveness of the TWP in the reduction of crosstalk is illustrated in Fig. 10.77, and results from our notion

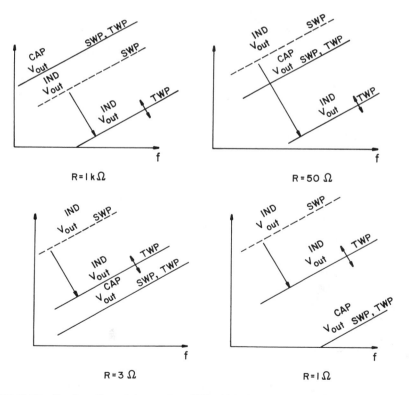

FIGURE 10.77 Explanation of the results of Fig. 10.76 in terms of inductive and capacitive coupling.

that the capacitive coupling is the same for all three configurations. The inductive coupling for the SWP configuration is smaller than for the single receptor wire, and is still smaller for the TWP configuration (equivalent to that of one half-twist at most). For $R = 1$ kΩ capacitive coupling dominates inductive coupling for the single-receptor-wire configuration. So replacing the receptor wire with a straight-wire pair or a twisted-wire pair should not significantly reduce the *total crosstalk*, as is evident from the experimental data. For $R = 50$ Ω inductive coupling dominates capacitive coupling for the single-receptor-wire case. Replacing the single receptor wire with a straight-wire pair reduces the inductive coupling, so the total crosstalk reduces to the capacitive coupling level. For $R = 3$ Ω the same result is observed, except that the capacitive coupling "floor" is much smaller than for $R = 50$ Ω, so that the total crosstalk drops more but is still restricted to being no less than the capacitive coupling. For $R = 1$ Ω the capacitive coupling floor is reduced considerably, so that the inductive coupling reduction caused by the twist can be fully realized. These observations are confirmed with experimental results in [28].

This brings up an important notion: *sensitivity of crosstalk in twisted-pairs to line twist* [29–31]. Consider the previous experiment. For the extremely low impedance of $R = 1$ Ω we pointed out that the capacitive coupling "floor" is reduced substantially, and the inductive coupling of one half-twist, although small, is larger than this floor. We also pointed out that, according to our model of a twisted pair as being a sequence of alternating loops, if the twisted pair consists of an *even number of half-twists* then the inductive coupling should completely cancel out, leaving *an inductive coupling of zero*! This suggests that for the case of $R = 1$ Ω, where the inductive coupling of one half-twist dominates the capacitive coupling floor, we should see an extreme sensitivity of the crosstalk to whether the line consists of an *odd or even number of half-twists*. In order to assess this, we rotated the far end of the twisted pair to give the minimum and maximum crosstalk at each frequency. The data for $R = 3$ Ω are plotted in Fig. 10.78(a) and those for $R = 1$ Ω in Fig. 10.78(b). These data show this extreme sensitivity to line twist (as much as 40 dB)! This suggests that *precise prediction of crosstalk is not feasible for the very small values of termination impedances.* The low-frequency predictions for the SWP configuration are shown on these data, and indicate that the minimum coupling is not zero but is restricted to the capacitive coupling floor, which is quite small. No such sensitivity to twist was observed for the high-impedance terminations of $R = 1$ kΩ or the case of $R = 50$ Ω. This is sensible to expect based on Fig. 10.77, since the capacitive coupling floor is dominant for the twisted pair configuration for $R = 1$ kΩ and 50 Ω.

10.3.4 Effects of Balancing

We will now consider the effect of *balanced loads* on the twisted pair as illustrated in Fig. 10.69(b). The model for the twisted-pair of receptor wires is unchanged, but the terminations affect the resulting crosstalk. The inductive and capacitive

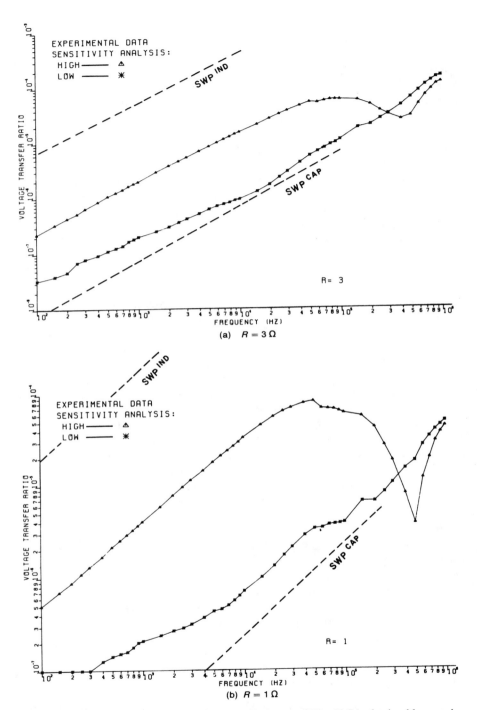

FIGURE 10.78 Experimental results for the experiment of Fig. 10.74, obtained by varying the number of twists, showing that for low-impedance loads the inductive coupling can be dominant for an odd number of half-twists and eliminated for an even number: (a) $R = 3\,\Omega$; (b) $R = 1\,\Omega$.

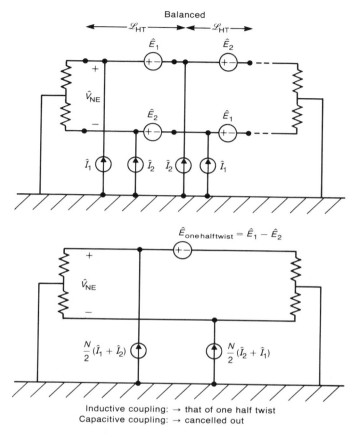

Inductive coupling: → that of one half twist
Capacitive coupling: → cancelled out

FIGURE 10.79   The inductive–capacitive coupling model for a twisted receptor wire pair and balanced terminations.

coupling models are shown in Fig. 10.79. The inductive coupling is unchanged from the unbalanced case, since the total induced emf around the receptor wire loop is the same. However, *due to the balanced loads, the capacitive coupling contributions cancel out*. Thus *the effect of balancing is to eliminate the capacitive coupling contribution!* The resulting crosstalk voltage transfer ratio is given by the unbalanced case with the capacitive coupling contribution removed:

$$\frac{\hat{V}_{NE,FE}}{\hat{V}_S} = \underbrace{\left.\frac{\hat{V}_{NE,FE}^{IND}}{\hat{V}_S}\right|_{\mathscr{L}_{SWP} = \mathscr{L}_{HT}}^{SWP}}_{\substack{\text{inductive coupling} \\ \text{for SWP whose} \\ \text{length is one} \\ \text{half-twist}}} \tag{10.178}$$

where the inductive coupling contribution is given by the inductive coupling portion of (10.174).

The previous experiment was repeated for a balanced termination at the near end of the twisted pair. A center-tapped 1:2 transformer was used to provide the balance, as shown in Fig. 10.69(b). Figure 10.80 illustrates the effect of balance. For $R = 1\ \Omega$ and 3 Ω, where inductive coupling dominated capacitive coupling for unbalanced terminations, eliminating the capacitive coupling contribution by balancing the loads should have no effect on either the crosstalk levels or the sensitivity to line twist. For $R = 1\ \text{k}\Omega$ and 50 Ω, where capacitive coupling dominated inductive coupling for the unbalanced case, balancing the twisted pair terminations and thus eliminating capacitive coupling should affect the crosstalk level and the sensitivity to line twist. Experimental data shown in

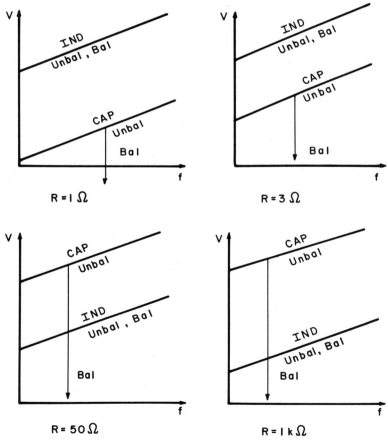

FIGURE 10.80 Explanation of the effect of balanced versus unbalanced terminations on the crosstalk to a twisted receptor wire pair for $R = 1\ \Omega$, 3 Ω, 50 Ω, and 1 kΩ.

[32, 33] confirm these observations. This suggests that *sensitivity to line twist should be a significant problem for balanced, twisted pairs for a wide range of termination impedance levels.*

10.4 MULTICONDUCTOR LINES AND EFFECTS OF INCIDENT FIELDS

Virtually all of the above transmission-line models can be extended to lines that consist of a large number of conductors [2, 7, 10, 22, 34–36]. The multiconductor transmission-line (MTL) equations are the same as before; the size of the vectors and matrices are increased to $n \times 1$ and $n \times n$ where the line consists of n conductors and a reference conductor. The only difficulty in solving these equations is their implementation in a digital computer program and the computation of the per-unit-length parameters. References [7, 22, 23, 37] contain frequency-domain codes for performing these calculations. Computation of the per-unit-length parameters for dielectric-insulated wires such as are found in ribbon cables is addressed in [3, 10, 38–41].

The effects of incident electromagnetic fields on MTLs can be addressed by including forcing functions in the per-unit-length model as discussed for two-conductor lines in Chapter 8. The MTL equations become [2, 37, 42–45]:

$$\frac{d}{dz}\hat{\mathbf{V}}(z) = -\hat{\mathbf{Z}}\hat{\mathbf{I}}(z) + \hat{\mathbf{V}}_s(z) \qquad (10.179a)$$

$$\frac{d}{dz}\hat{\mathbf{I}}(z) = -\hat{\mathbf{Y}}\hat{\mathbf{V}}(z) + \hat{\mathbf{I}}_s(z) \qquad (10.179b)$$

The sources $\hat{\mathbf{V}}_s(z)$ and $\hat{\mathbf{I}}_s(z)$ represent per-unit-length induced voltage and current sources in each line. The induced voltage sources are the induced emfs in that circuit due to the incident magnetic field that is normal to the loop. The induced current sources are due to the incident electric field that is transverse to the line and between the reference conductor and the conductor of that circuit, as discussed in Chapter 8. The calculation of these sources and the solution of the MTL equations are straightforward, and are documented for wire lines in [37]. A FORTRAN computer code is given in [37].

REFERENCES

[1]   C.R. Paul and S.A. Nasar, *Introduction to Electromagnetic Fields*, second edition, McGraw-Hill, NY (1987).

[2]   C.R. Paul, Application of multiconductor transmission line theory to the prediction of cable coupling, Vol. I: Multiconductor transmission line theory, *Technical Report, Rome Air Development Center, Griffiss AFB, NY*, RADC-TR-76-101 (April 1976). (A025028)

[3] J.C. Clements, C.R. Paul and A.T. Adams, Two-dimensional systems of dielectric-coated, cylindrical conductors, *IEEE Trans. on Electromagnetic Compatibility*, EMC-17, 238–248 (1975).

[4] K.C. Gupta, R. Garg, and I.J. Bahl, *Microstrip Lines and Slotlines*, Artech House, Dedham, MA (1979).

[5] K.C. Gupta, R. Garg, and R. Chadha, *Computer-Aided Design of Microwave Circuits*, Artech House, Dedham, MA (1981).

[6] C.R. Paul, *Analysis of Linear Circuits*, McGraw-Hill, NY (1989).

[7] C.R. Paul, Applications of multiconductor transmission line theory to the prediction of cable coupling, Vol. VII: Digital computer programs for the analysis of multiconductor transmission lines, *Technical Report, Rome Air Development Center, Griffiss AFB, NY*, RADC-TR-76-101 (July 1977). (A046662)

[8] C.R. Paul, Useful matrix chain parameter identities for the analysis of multiconductor transmission lines, *IEEE Trans. on Microwave Theory and Techniques*, MTT-23, 756–760 (1975).

[9] C.R. Paul, Solution of the transmission line equations for three-conductor lines in homogeneous media, *IEEE Trans. on Electromagnetic Compatibility*, EMC-20, 216–222 (1978).

[10] C.R. Paul, Prediction of crosstalk in ribbon cables: comparison of model predictions and experimental results, *IEEE Trans. on Electromagnetic Compatibility*, EMC-20, 394–406 (1978).

[11] A.E. Ruehli and P.A. Brennan, Efficient capacitance calculations for three-dimensional multiconductor systems, *IEEE Trans. on Microwave Theory and Techniques*, MTT-21, 76–82 (1973).

[12] A.E. Ruehli and P.A. Brennan, Capacitance models for integrated circuit metallization wires, *IEEE J. Solid-State Circuits*, SC-10, 530–536 (1975).

[13] C.R. Paul and W.W. Everett, III, Modeling crosstalk on printed circuit boards, *Technical Report, Rome Air Development Center, Griffiss, AFB, NY*, RADC-TR-85-107 (July 1985).

[14] C.R. Paul, Modeling of printed circuit boards for the prediction of crosstalk and ground drop, *IBM J. Research and Development*, 33, pp. 33–50 (1989).

[15] C.R. Paul, On the superposition of inductive and capacitive coupling in crosstalk prediction models, *IEEE Trans. on Electromagnetic Compatibility*, EMC-24, 335–343 (1982).

[16] C.R. Paul, A simple SPICE model for coupled transmission lines, *1988 IEEE International Symposium on Electromagnetic Compatibility*, Seattle, WA, September 1988.

[17] A. Ralston, *A First Course in Numerical Analysis*, McGraw-Hill, NY (1965).

[18] C.R. Paul, Transmission-line modeling of shielded wires for crosstalk prediction, *IEEE Trans. on Electromagnetic Compatibility*, EMC-23, 345–351 (1981).

[19] C.R. Paul, SHIELD—a digital computer for the prediction of crosstalk to shielded cables, *1983 International Symposium and Technical Exhibition on Electromagnetic Compatibility*, Zurich, Switzerland, March 1983.

[20] C.R. Paul and W.E. Beech, Prediction of crosstalk in flatpack, coaxial cables, *1984 IEEE International Symposium on Electromagnetic Compatibility*, San Antonio, TX, April 1984.

[21] C.R. Paul, Applications of multiconductor transmission line theory to the prediction of cable coupling, Vol. VIII: Prediction of crosstalk involving braided-shield cables, *Technical Report, Rome Air Development Center, Griffiss AFB, NY*, RADC-TR-76-101 (August 1980).

[22] C.R. Paul, SHIELD, a digital computer program for computing crosstalk between shielded cables, *Technical Report, Rome Air Development Center, Griffiss AFB, NY*, RADC-TR-82-286, Vol. IV B (November 1982).

[23] C.R. Paul and W.E. Beech, Prediction of crosstalk in flatpack coaxial cables, *Technical Report, Rome Air Development Center, Griffiss AFB, NY*, RADC-TR-82-286, Vol. IV F (December 1984).

[24] C.R. Paul and B.A. Bowles, Symbolic solution of the multiconductor transmission-line equations for lines containing shielded wires, *IEEE Trans. on Electromagnetic Compatibility*, **EMC-33**, 149–162 (1991). See also B.A. Bowles, Literal solution to the transmission line equations for shielded wires, MSEE Thesis, University of Kentucky (December 1990).

[25] C.R. Paul, Effect of pigtails on crosstalk to braided-shield cables, *IEEE Trans. on Electromagnetic Compatibility*, **EMC-22**, 161–172 (1980).

[26] C.R. Paul and J.A. McKnight, Prediction of crosstalk involving twisted pairs of wires, Part I: A transmission line model for twisted wire pairs, *IEEE Trans. on Electromagnetic Compatibility*, **EMC-21**, 92–105 (1979).

[27] C.R. Paul and J.A. McKnight, Prediction of crosstalk involving twisted pairs of wires, Part II: A simplified, low-frequency prediction model, *IEEE Trans. on Electromagnetic Compatibility*, **EMC-21**, 105–114 (1979).

[28] C.R. Paul and J.A. McKnight, Applications of multiconductor transmission line theory to the predictions of cable coupling, Vol. V: Prediction of crosstalk involving twisted wire pairs, *Technical Report, Rome Air Development Center, Griffiss AFB, NY*, RADC-TR-76-101 (February 1978). (A053559)

[29] C.R. Paul and M.B. Jolly, Sensitivity of crosstalk in twisted-pair circuits to line twist, *IEEE Trans. on Electromagnetic Compatibility*, **EMC-24**, 359–364 (1982).

[30] C.R. Paul and M.B. Jolly, Crosstalk in balanced, twisted-pair circuits, *1981 IEEE International Symposium on Electromagnetic Compatibility*, Boulder, CO, August 1981.

[31] C.R. Paul and M.B. Jolly, Crosstalk in twisted-wire circuits, *Technical Report, Rome Air Development Center, Griffiss AFB, NY*, RADC-TR-82-286, Vol. IV C (November 1982).

[32] C.R. Paul and D. Koopman, Sensitivity of coupling to balanced, twisted pair lines to line twist, *1983 International Symposium and Technical Exhibition on Electromagnetic Compatibility*, Zurich, Switzerland, March 1983.

[33] C.R. Paul and D. Koopman, Prediction of crosstalk in balanced, twisted pair circuits, *Technical Report, Rome Air Development Center, Griffiss AFB, NY*, RADC-TR-82-286, Vol. IV D (August 1984).

[34] C.R. Paul, Computation of crosstalk in a multiconductor transmission line, *IEEE Trans. on Electromagnetic Compatibility*, **EMC-23**, 352–358 (1981).

[35] C.R. Paul, Analysis of electromagnetic coupling in branched cables, *IEEE International Symposium on Electromagnetic Compatibility*, San Diego, CA, October 1979.

[36] C.R. Paul, Applications of multiconductor transmission line theory to the prediction of cable coupling, Vol. IV: Prediction of crosstalk in ribbon cables, *Technical Report, Rome Air Development Center, Griffiss AFB, NY*, RADC-TR-76-101 (February 1978). (A053548)

[37] C.R. Paul, Applications of multiconductor transmission line theory to the prediction of cable coupling, Vol. VI: A digital computer program for determining terminal currents induced in a multiconductor transmission line by an incident electromagnetic field, *Technical Report, Rome Air Development Center, Griffiss AFB, NY*, RADC-TR-76-101 (February 1978). (A053560)

[38] C.R. Paul and A.E. Feather, Application of moment methods to the characterization of ribbon cables, *Computers and Electrical Engineering*, **4**, 173–184 (1977).

[39] C.R. Paul and A.E. Feather, Computation of the transmission line inductance and capacitance matrices from the generalized capacitance matrix, *IEEE Trans. on Electromagnetic Compatibility*, **EMC-18**, 175–183 (1976).

[40] C.R. Paul, Applications of multiconductor transmission line theory to the prediction of cable coupling, Vol. II: Computation of the capacitance matrices for ribbon cables, *Technical Report, Rome Air Development Center, Griffiss AFB, NY*, RADC-TR-76-101 (April 1976). (A025029)

[41] C.R. Paul, Applications of multiconductor transmission line theory to the prediction of cable coupling, Vol. III: Prediction of crosstalk in random cable bundles, *Technical Report, Rome Air Development Center, Griffiss AFB, NY*, RADC-TR-76-101 (February 1977). (A038316)

[42] C.R. Paul, Frequency response of multiconductor transmission lines illuminated by an incident electromagnetic field, *IEEE Trans. on Electromagnetic Compatibility*, **EMC-18**, 183–190 (1976).

[43] C.R. Paul and D.F. Herrick, Coupling of electromagnetic fields to transmission lines, *1982 IEEE International Symposium on Electromagnetic Compatibility*, Santa Clara, CA, September 1982.

[44] C.R. Paul and R.T. Abraham, Coupling of electromagnetic fields onto transmission lines: a comparison of the transmission line model and the method of moments, *Technical Report, Rome Air Development Center, Griffiss AFB, NY*, RADC-TR-82-286, Vol. IV A (November 1982).

[45] C.R. Paul and R.T. Abraham, Coupling of electromagnetic fields to transmission lines, *1981 IEEE International Symposium on Electromagnetic Compatibility*, Boulder, CO, August 1981.

PROBLEMS

10.1   Three bare, #20 gauge solid wires have equal adjacent spacings, so that their cross-sectional locations are at the vertices of a triangle having equal-length sides. Determine the per-unit-length inductances and capacitances if the separation is 50 mils. $[l_G = l_R = 0.456 \ \mu\text{H/m}, l_m = 0.228 \ \mu\text{H/m}, c_G = c_R = c_m = 16.25 \ \text{pF/m}]$

10.2   A #20 gauge, bare generator wire is suspended $\frac{1}{4}$ inch above a ground plane. Another identical wire is suspended $\frac{1}{2}$ inch above the ground plane and separated horizontally by $\frac{3}{8}$ inch from the generator wire. Determine the per-unit-length inductances and capacitances. [$l_G = 0.688 \mu H/m$, $l_R = 0.827 \mu H/m$, $l_m = 0.124 \mu H/m$, $c_m = 2.49 pF/m$, $c_G = 14.1 pF/m$, $c_R = 11.32 pF/m$]

10.3   Figure P10.3 shows a ribbon cable transmission line where an outside wire is used as the reference wire. Assume that all wires are #28 gauge stranded (7×32) and the adjacent wire spacings are 50 mils. Compute the per-unit-length inductances and capacitances, neglecting the wire insulations. [$l_G = 1.036 \mu H/m$, $l_R = 0.76 \mu H/m$, $l_m = 0.518 \mu H/m$, $c_m = 11.12 pF/m$, $c_G = 5.17 pF/m$, $c_R = 11.12 pF/m$]

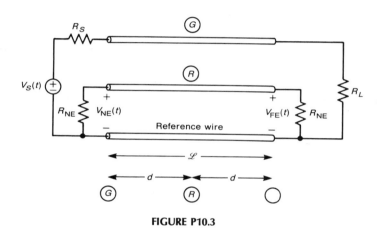

FIGURE P10.3

10.4   Two #20 gauge solid wires are placed inside a shield that is filled with polyvinylchloride (pvc) insulation ($\epsilon_r = 4$). The shield interior radius is 250 mils, and each wire is placed at a distance of 100 mils from the shield center and on a line through the shield center ($\theta_{GR} = 180°$). Determine the per-unit-length inductances and capacitances. [$l_G = 0.515 \mu H/m$, $l_R = 0.515 \mu H/m$, $l_m = 74.3 nH/m$, $c_m = 6.36 pF/m$, $c_G = 37.7 pF/m$, $c_R = 37.7 pF/m$]

10.5   The self and mutual inductances of wire lines are relatively insensitive to changes in separation because the dimensions are involved in a natural logarithm term. To illustrate this, consider two #20 gauge solid wires both located 1 inch above an infinite ground plane. Compute and plot the per-unit-length mutual inductance for various spacings from $\frac{1}{16}$ inch to 2 inches. [$\frac{1}{16}$ inch, 0.69 $\mu H/m$; $\frac{1}{8}$ inch, 0.555 $\mu H/m$; $\frac{1}{4}$ inch, 0.4174 $\mu H/m$; 1 inch, 0.161 $\mu H/m$]

10.6   For a lossless transmission line immersed in a homogeneous medium the characteristic impedance matrix given in (10.55) simplifies, using the fundamental identity for lines in a homogeneous medium given in (10.9). For a three-conductor lossless line immersed in a homogeneous medium determine the diagonalization matrix given in (10.44) and the characteristic impedance matrix given in (10.55). $[\hat{\mathbf{T}} = \mathbf{1}_2, \gamma = j\beta\mathbf{1}_2, \hat{\mathbf{Z}}_C = v\mathbf{L}]$ Determine the characteristic impedance matrix for the ribbon cable of Problem 10.3 by neglecting the wire insulations and assuming the surrounding medium is free space. $[Z_{C11} = 310.8 \,\Omega, \; Z_{C22} = 228.0 \,\Omega, \; Z_{C12} = 155.4 \,\Omega]$ This shows that a line consisting of more than two conductors cannot be *matched* by terminating the each conductor to the reference conductor. For this problem determine a matching network consisting of resistor R_G between the generator wire and the reference wire, resistor R_R between the receptor wire and the reference wire, and resistor R_m between the generator and receptor wires. $[\hat{\mathbf{Y}}_C = \hat{\mathbf{Z}}_C^{-1}, \; R_m = 300.6 \,\Omega, \; R_G = 643.4 \,\Omega, \; R_R = 300.6 \,\Omega]$

10.7   Termination networks can be characterized with a generalized Norton equivalent circuit where $\hat{\mathbf{I}}(0) = \hat{\mathbf{I}}_S - \hat{\mathbf{Y}}_S\hat{\mathbf{V}}(0)$ and $\hat{\mathbf{I}}(\mathscr{L}) = \hat{\mathbf{Y}}_{\mathscr{L}}\hat{\mathbf{V}}(\mathscr{L})$ in addition to the dual characterization of a generalized Thévenin equivalent shown in equation (10.56). Sometimes this is more easily derived for a given termination network. For example, determine the generalized Norton equivalent for the network shown in Fig. P10.7.

$$\left[\hat{\mathbf{I}}_S = \begin{bmatrix} V_1/R_1 \\ V_2/R_2 \end{bmatrix}, \; \hat{\mathbf{Y}}_S = -\begin{bmatrix} 1/R_1 + 1/R_3 & -1/R_3 \\ -1/R_3 & 1/R_2 + 1/R_3 \end{bmatrix}, \right.$$
$$\left. \hat{\mathbf{Y}}_{\mathscr{L}} = -\begin{bmatrix} 1/R_4 + 1/R_6 & -1/R_6 \\ -1/R_6 & 1/R_5 + 1/R_6 \end{bmatrix}\right]$$

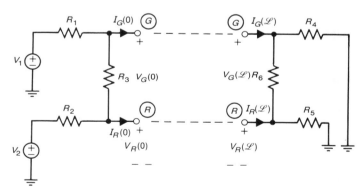

FIGURE P10.7

10.8   Determine the generalized Thévenin equivalent characterization given in equations (10.56) of the termination networks shown in Fig. P10.8.

$$\left[\hat{\mathbf{V}}_S = \begin{bmatrix} V_1 \\ V_2 \end{bmatrix}, \; \hat{\mathbf{Z}}_S = \begin{bmatrix} R_1 + R_3 & R_3 \\ R_3 & R_2 + R_3 \end{bmatrix}, \; \hat{\mathbf{Z}}_{\mathscr{L}} = \begin{bmatrix} R_4 + R_6 & R_6 \\ R_6 & R_5 + R_6 \end{bmatrix}\right.$$

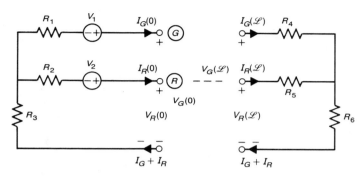

FIGURE P10.8

10.9 In order to demonstrate the correctness of the exact solution for a three-conductor lossless line in a homogeneous medium given in Section 10.1.4.2 in equations (10.73), (10.74), and (10.75), prepare a FORTRAN computer program to plot the frequency response of the near-end crosstalk transfer ratio for the problem of two wires above a ground plane shown in Fig. 10.34. The frequency response is plotted using the exact SPICE program in Fig. 10.37(a).

10.10 Plot the frequency response of the near-end crosstalk transfer ratio for the problem shown in Fig. 10.34 using the inductive–capacitive coupling approximate model described in Section 10.1.4.3. Determine the line lengths in wavelengths where these two prediction models diverge. $[\mathscr{L} \cong \frac{1}{64}\lambda]$

10.11 For the ribbon cable shown in Fig. P10.3 assume the total mutual inductance and total mutual capacitance to be $L_m = 1\ \mu\text{H}$ and $C_m = 250\ \text{pF}$. If $V_S(t)$ is a 1 MHz sinusoid of magnitude 1 V, calculate the magnitude of the far-end crosstalk if the termination impedances are $R_S = 50\ \Omega$, $R_L = 50\ \Omega$, $R_{NE} = 100\ \Omega$, and $R_{FE} = 100\ \Omega$. [7.854 mV] Determine the near-end and far-end inductive and capacitive coupling coefficients. $[M_{NE}^{\text{IND}} = 5 \times 10^{-9}$, $M_{FE}^{\text{IND}} = -5 \times 10^{-9}$, $M_{NE}^{\text{CAP}} = 6.25 \times 10^{-9} = M_{FE}^{\text{CAP}}]$

10.12 For the ribbon cable shown in Fig. P10.3 assume the total mutual inductance and total mutual capacitance to be $L_m = 0.4\ \mu\text{H}$ and $C_m = 400\ \text{pF}$. If $V_S(t)$ is a 1 MHz sinusoid of magnitude 1 V, calculate the magnitude of the near-end crosstalk if the termination impedances are $R_S = 50\ \Omega$, $R_L = 50\ \Omega$, $R_{NE} = 50\ \Omega$, and $R_{FE} = 50\ \Omega$. [44 mV] Determine the near-end and far-end inductive and capacitive coupling coefficients. $[M_{NE}^{\text{IND}} = 2 \times 10^{-9}$, $M_{FE}^{\text{IND}} = -2 \times 10^{-9}$, $M_{NE}^{\text{CAP}} = 5 \times 10^{-9} = M_{FE}^{\text{CAP}}]$

10.13 For the ribbon cable shown in Fig. P10.3, assume the total mutual inductance and total mutual capacitance to be $L_m = 1\ \mu\text{H}$ and $C_m = 250\ \text{pF}$. Suppose the termination impedances are equal: $R_S =$

$R_L = R_{NE} = R_{FE} = R$. Determine the value of R for which the inductive and capacitive coupling contributions are exactly equal. [$R = \sqrt{L_m/C_m} = 63.25\ \Omega$]

10.14 Consider the case of two wires above a ground plane shown in Fig. P10.14. Suppose $r_{wG} = 20$ mils, $r_{wR} = 20$ mils, $h_G = 2$ inches, $h_R = 1$ inch, $s = 0.5$ inch, and $\mathscr{L} = 10$ feet. Determine the total inductances and capacitances of the line. [$L_G = 3.23\ \mu H$, $L_R = 2.81\ \mu H$, $L_m = 1.07\ \mu H$, $C_G = 22.7$ pF, $C_R = 28.2$ pF, $C_m = 13.9$ pF]

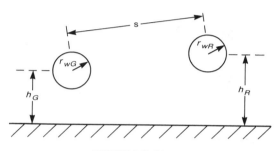

FIGURE P10.14

10.15 Consider the case of two wires above a ground plane shown in Fig. P10.14. Suppose $l_m = 2$ nH/m, $c_m = 0.6$ pF/m, $\mathscr{L} = 2$ m, $R_S = 0$, $R_L = 50\ \Omega$, $R_{NE} = 100\ \Omega$, $R_{FE} = 200\ \Omega$, and $V_S(t) = 1 \cos \omega t\ V$, where $f = 1$ MHz. Determine the time-domain far-end and near-end crosstalk voltages. [$V_{FE}(t) = 0.167 \cos(\omega t + 90°)\ mV$, $V_{NE}(t) = 0.67 \cos(\omega t + 90°)\ mV$] Determine the near-end and far-end inductive and capacitive coupling coefficients. [$M_{NE}^{IND} = 2.67 \times 10^{-11}$, $M_{FE}^{IND} = -5.33 \times 10^{-11}$, $M_{NE}^{CAP} = 8 \times 10^{-11} = M_{FE}^{CAP}$]

10.16 For the ribbon cable of Problem 10.11, suppose the wires are #28 (7×36) stranded. Determine the common-impedance coupling level for the near-end crosstalk voltage if the total line length is 3 m and the frequency where this level equals the inductive–capacitive coupling level. [$M_{NE}^{CI} = 2.92 \times 10^{-3}$, $M_{FE}^{CI} = -2.92 \times 10^{-3}$, $f_{NE} = 41.26$ kHz, $f_{FE} = 371.3$ kHz]

10.17 For the ribbon cable of Problem 10.12 suppose the wires are #24 (7×32) stranded. Determine the common-impedance coupling level for the near-end crosstalk voltage if the total line length is 2 m and the frequency where this level equals the inductive–capacitive coupling level. [$M_{NE}^{CI} = 7.6 \times 10^{-4}$, $M_{FE}^{CI} = -7.6 \times 10^{-4}$, $f_{NE} = 17.27$ kHz, $f_{FE} = 40.3$ kHz]

10.18 For the case of two wires above a ground plane in Problem 10.15 suppose the per-unit-length resistance of the ground plane is 0.001 Ω/m. Determine the near-end and far-end common impedance coupling coefficients and the frequency where this level equals the inductive–capacitive coupling level. $[M_{NE}^{CI} = 1.33 \times 10^{-5}, M_{FE}^{CI} = -2.67 \times 10^{-5}, f_{NE} = 19.89 \text{ kHz}, f_{FE} = 159.15 \text{ kHz}]$

10.19 For the PCB experiment shown in Fig. 10.38 determine from the measured data shown in Fig. 10.40(b) and computed data shown in Fig. 10.40(a) the near-end inductive–capacitive coupling coefficient. (Note that Fig. 10.40 gives the ratio of the near-end voltage to the voltage at the input to the generator line, not the source voltage. To convert this to the usual crosstalk ratio, assume the input impedance to the generator line equals the load resistance. Since $R_S = R_L$, multiply the crosstalk ratio obtained from Fig. 10.40 by a factor of $\frac{1}{2}$.) $[M_{NE} = 8.93 \times 10^{-10}]$ Check this by direct computation from the per-unit-length parameters given after equation (10.139). $[M_{NE} = 8.96 \times 10^{-10}, M_{NE}^{IND} = 8.3 \times 10^{-10}, M_{NE}^{CAP} = 6.63 \times 10^{-11}]$

10.20 For the case of two wires above a ground plane in Problem 10.14 determine the transformation matrix for the exact SPICE model given in (10.126).
$$\left[\mathbf{T} = \begin{bmatrix} 0.77 & -0.63 \\ 0.63 & 0.77 \end{bmatrix} \right]$$
Determine the mode velocities, transit times, and characteristic impedances. $[v_{mG} = 3 \times 10^8 = v_{mR}, \ T_G = T_R = 10.2 \text{ ns}, \ Z_{CG} = 404.04 \ \Omega, \ Z_{CR} = 190.17 \ \Omega]$

10.21 Devise a SPICE model for the ribbon cable experiment shown in Fig. 10.16, and verify the frequency-domain crosstalk shown in Fig. 10.17(a) and (b).

10.22 Apply a 5 V, 1 MHz, 50% duty cycle trapezoidal pulse train having rise/falltimes of 50 ns to the ribbon cable SPICE model of Problem 10.21 and determine the near-end and far-end time-domain crosstalk waveforms.

10.23 For the experimental results for two wires above a ground plane shown in Fig. 10.34 estimate the near-end time-domain crosstalk voltage using the simple inductive–capacitive coupling model. For this combination of rise/falltimes and line length is the simple inductive–capacitive coupling model adequate? [No] Why? $[\tau_r = 1.246 T_D]$

10.24 Repeat Problem 10.23 for the PCB experiment shown in Fig. 10.38.

10.25 For the ribbon cable of Problem 10.11 suppose $V_S(t)$ is replaced with a 1 MHz, 5 V, 50% duty cycle trapezoidal pulse train having rise/falltimes of 50 ns. Compute the maximum near- and far-end crosstalk voltages. $[V_{NE,\max} = 1.13 \text{ V}, V_{FE,\max} = 125 \text{ mV}]$

10.26 Repeat Problem 10.25 for the ribbon cable of Problem 10.12. $[V_{NE,\max} = 0.7 \text{ V}, V_{FE,\max} = 0.3 \text{ V}]$

10.27   Repeat Problem 10.25 for the problem of two wires above a ground plane in Problem 10.15. [$V_{NE,max} = 10.67$ mV, $V_{FE,max} = 2.67$ mV]

10.28   Consider the case of three identical wires having radii of 10 mils and arranged on the vertices of an equilateral triangle with 100 mil sides as shown in Fig. P10.28(a). The wires are immersed in a homogeneous medium having $\epsilon_r = 2.1$. It is desired to model a 2 m length of this line using a lumped-Pi circuit as shown in Fig. P10.28(b). Determine element values in this equivalent circuit. [$L_G = 1.84$ μH, $L_R = 1.84$ μH, $L_m = 0.921$ μH, $C_G = 16.9$ pF, $C_R = 16.9$ pF, $C_m = 16.9$ pF] Above what frequency would you expect this model to be invalid for sinusoidal inputs to the line? [10.35 MHz] For trapezoidal pulse train inputs what would be the rise/falltime of the fastest allowable pulse train for which this model would be expected to yield reasonable predictions? [96.6 ns]

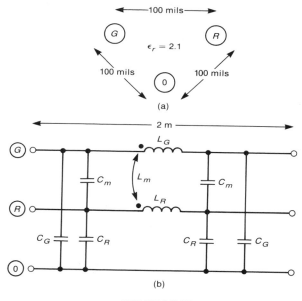

FIGURE P10.28

10.29   Generate a lumped-Pi equivalent circuit to predict the near-end crosstalk for the ribbon cable experiment shown in Fig. 10.16, and compare with the experimental results shown in Figs. 10.17(a), 10.17(b), 10.25(b), and 10.26(b).

10.30   Generate a lumped-Pi equivalent circuit to predict the near-end crosstalk for the experiment shown in Fig. 10.34, and compare with the experimental results shown in Figs. 10.36 and 10.37.

10.31   Generate a lumped-Pi equivalent circuit to predict the near-end crosstalk for the experiment shown in Fig. 10.38, and compare with the experimental results shown in Figs. 10.39 and 10.40.

10.32   Consider the ribbon cable shown in Fig. P10.3. The total mutual inductance is $L_m = 1\ \mu\text{H}$, and the total mutual capacitance is $C_m = 25\ \text{pF}$. If $R_S = 0$, $R_L = R_{NE} = R_{FE} = 100\ \Omega$, and the pulse waveform shown in Fig. P10.32 is applied, sketch the time-domain near-end crosstalk and determine the maximum and minimum crosstalk voltage levels. [62.5 mV, −62.5 mV]

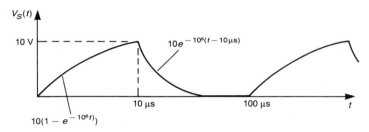

FIGURE P10.32

10.33   Consider the case of two wires above a ground plane, with the receptor wire shielded, as shown in Fig. 10.41(b). Assume that the generator wire is a #22 (7 × 30) stranded wire. The receptor conductor is an RG-58U coaxial cable having an inner, solid #20 gauge wire, a shield interior radius of 64 mils, an interior dielectric of polyethylene giving the cable a nominal velocity of propagation of 66% of free space, and a shield thickness of approximately 15 mils (assume solid copper). The generator wire is suspended a height of 1 inch and the coaxial cable is suspended a height of 1.5 inches. The separation between the generator wire and the cable is $\frac{1}{4}$ inch. Determine the per-unit-length parameters of the line. [$r_S = 4.43\ \text{m}\Omega/\text{m}$, $l_G = 0.979\ \mu\text{H}/\text{m}$, $l_S = 0.727\ \mu\text{H}/\text{m}$, $l_R = 1.05\ \mu\text{H}/\text{m}$, $l_{GS} = 0.4575\ \mu\text{H}/\text{m}$, $c_{RS} = 92.17\ \text{pF}/\text{m}$, $c_{GS} = 10.12\ \text{pF}/\text{m}$, $c_G = 5.97\ \text{pF}/\text{m}$, $c_S = 11.52\ \text{pF}/\text{m}$]

10.34   For the line in Problem 10.33 determine the frequency where the shield will affect the inductive coupling when it is grounded at both ends. [970 Hz]

10.35   If the shield in Problem 10.33 is grounded at both ends, the line length is 2 m and $R_S = 0$, $R_L = 1\ \text{k}\Omega$, $R_{NE} = 100\ \Omega$, and $R_{FE} = 50\ \Omega$, determine the near-end crosstalk transfer ratio at 100 Hz, 1 kHz, 100 kHz, and 10 MHz. [0.38×10^{-6}, 2.67×10^{-6}, 3.72×10^{-6}, 3.72×10^{-6}]

10.36   For the ribbon cable shown in Fig. P10.3 assume the total mutual inductance and mutual capacitance to be $L_m = 0.4\ \mu\text{H}$ and $C_m = 400\ \text{pF}$.

If $V_S(t)$ is a 1 MHz sinusoid of magnitude 1 V, and the termination impedances are $R_S = R_L = R_{NE} = R_{FE} = 50\,\Omega$, determine the near-end crosstalk if a shield is placed around wires #2 and #3 and the shield is only connected to the near end of wire #3. [12.57 mV] How much does the shield reduce the crosstalk? [10.88 dB]

10.37   Consider the case of two wires above a ground plane shown in Fig. P10.14. The line has parameters of $l_m = 2$ nH/m, $c_m = 0.6$ pF/m, $V_S(t) = 1 \cos \omega t$ V, $f = 1$ MHz, $\mathscr{L} = 2$ m, $R_S = 0$, $R_L = 50\,\Omega$, $R_{NE} = 100\,\Omega$, and $R_{FE} = 200\,\Omega$. A shield is placed around the receptor wire, and is connected to the ground plane only at the near end. Determine the far-end crosstalk voltage. [$0.335 \cos(\omega t - 90°)$ mV] By how much does the shield reduce the crosstalk? [6 dB]

10.38   The shield of Problem 10.37 is connected to the ground plane at both ends, and has a per-unit-length resistance of 1 Ω/m and per-unit-length self inductance of $l_S = 16\,\mu$H/m. Determine the near-end crosstalk voltage. [$1.667 \cos(\omega t + 0°)\,\mu$V]

10.39   Consider a shielded receptor wire where the shield is connected to the reference conductor at both ends. The capacitive coupling is essentially eliminated, and the near-end and far-end crosstalk transfer ratios are due solely to inductive coupling such that they are of the form $M_{NE,FE} = j\omega A/(1 + j\omega B)$. (See equations (10.165).) This is essentially the form of a high-pass filter with gain. Construct a lumped circuit that is suitable for use with SPICE and can model both frequency-domain and time-domain sources on the generator line. [Place a voltage-controlled voltage source in series with a resistor R and a capacitor C and take the near-end or far-end crosstalk voltage across the resistor. Choose $RC = B$ and the controlled source parameter to be $(A/B)V_S(t)$] Apply this model to the experimental circuit of Fig. 10.49 to predict the frequency-domain near-end crosstalk shown in Fig. 10.50. Apply a 1 MHz, 5 V, 50% duty cycle trapezoidal waveform having rise/falltimes of 50 ns and compute the near-end time-domain crosstalk waveform.

10.40   Consider the case of an unbalanced twisted generator pair shown in Fig. P10.40. Assume that the generator wires are #28 gauge stranded (7×36), as is the receptor wire, and the heights above the ground plane are 1 inch. The separation between the generator and receptor circuits is $s = \frac{1}{4}$ inch and the generator wires are separated only by their insulations, giving $d = 35$ mils. Ignore the effect of the wire insulations and compute the per-unit-length parameters. [$l_{G1R} = (\mu_0/4\pi) \ln[1 + 4h_G h_R/(s + \Delta d)^2] = 0.404\,\mu$H/m, $l_{G2R} = (\mu_0/4\pi) \ln[1 + 4h_G h_R/(s - \Delta d)^2] = 0.432\,\mu$H/m, $l_{G1G2} = (\mu_0/4\pi) \ln(1 + 4h_G^2/d^2) = 0.809\,\mu$H/m, $l_{G1} = (\mu_0/2\pi) \ln(2h_{G1}/r_{wG1}) = 1.12\,\mu$H/m, $l_{G2} = (\mu_0/2\pi) \ln(2h_{G2}/r_{wG2}) = 1.12\,\mu$H/m, $l_R = (\mu_0/2\pi) \ln(2h_R/r_{wR}) = 1.12\,\mu$H/m,

$c_{G1R} = 2.046$ pF/m, $c_{G2R} = 3.112$ pF/m, $c_{G1G2} = 14.61$ pF/m, $c_{G1} =$
4.61 pF/m, $c_{G2} = 4.006$ pF/m, $c_R = 6.73$ pF/m]

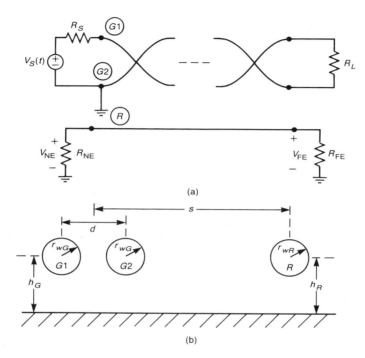

(a)

(b)

FIGURE P10.40

10.41 For the unbalanced twisted pair of Problem 10.40 determine the near-end
and far-end crosstalk transfer ratios if $R_S = R_L = R_{NE} = R_{FE} = 10\ \Omega$
and the line length is 10 m and the twist rate 10 twists/m. [$M_{NE}^{IND} =$
-3.45×10^{-11}, $M_{FE}^{IND} = 3.45 \times 10^{-11}$, $M_{NE}^{CAP} = M_{FE}^{CAP} = 5.12 \times 10^{-11}$]
If $V_S(t)$ is a 5 V, 1 MHz sinusoid, determine the near-end and far-end
crosstalk voltages. [$V_{NE} = 0.52$ mV, $V_{FE} = 2.69$ mV]

10.42 Suppose the generator twisted pair in Problem 10.41 is balanced.
Determine the crosstalk transfer ratios. [$M_{FE}^{IND} = -3.45 \times 10^{-11}, M_{FE}^{IND} =$
3.45×10^{-11}, $M_{NE}^{CAP} = M_{FE}^{CAP} = 0$] If $V_S(t)$ is a 5 V, 1 MHz sinusoid,
determine the near-end and far-end crosstalk voltages. [$V_{NE} =$
1.084 mV $= V_{FE}$]

CHAPTER ELEVEN

Shielding

This chapter addresses the concept of *shielding* of electronic circuits. The term *shield* usually refers to a metallic enclosure that *completely encloses* an electronic product or a portion of that product. There are two purposes of a shield, as illustrated in Fig. 11.1. The first, as shown in Fig. 11.1(a), is to prevent the emissions of the electronics of the product or a portion of those electronics from radiating outside the boundaries of the product. The motivation here is to either prevent those emissions from causing the product to fail to comply with the radiated emissions limits or to prevent the product from causing interference with other electronic products. The second purpose of a shield, as shown in Fig. 11.1(b), is to prevent radiated emissions external to the product from coupling to the product's electronics, which may cause interference in the product. As an example, shielding may be used to reduce the *susceptibility* to external signals such as high-power radars or radio and TV transmitters.

Therefore a shield is, conceptually, a barrier to the transmission of electromagnetic fields. In a general sense the general definition of a shield may be extended to include conducted emissions. In this sense a power-line filter may be considered to be a shield. However, we will restrict our notion of a shield to being a barrier to *radiated emissions*. We may view the *effectiveness*

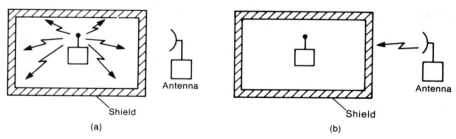

FIGURE 11.1 Illustration of the use of a shielded enclosure: (a) to contain radiated emissions and (b) to exclude radiated emissions.

of a shield as being the ratio of the magnitude of the electric (magnetic) field that is incident on the barrier to the magnitude of the electric (magnetic) field that is transmitted through the barrier. Alternatively, we may view this as the ratio of the electric (magnetic) field incident on the product's electronics with the shield removed to that with the shield in place. In this latter sense we may view the quantification of *shielding effectiveness* as being equivalent to an *insertion loss*, as was discussed for filters in Chapter 9. These notions give a qualitative idea about the meaning of the term shield and will be made more precise, quantitatively, in the following sections.

We will compute *ideal values of shielding effectiveness* in the following sections and will obtain some rather large values of shielding effectiveness of the order of hundreds of dB. A shielding effectiveness of 100 dB means that the incident field has been reduced by a factor of 100,000 as it exits the shield! In order to realize these ideal and extremely large values of shielding effectiveness, *the shield must completely enclose the electronics and must have no penetrations such as holes, seams, slots or cables. Any penetrations in a shield unless properly treated, may drastically reduce the effectiveness of the shield.* For example, consider a closed metallic box (shield) that has a wire penetrating it as shown in Fig. 11.2(a). Suppose a nearby radiating source such as an antenna radiates an electromagnetic field. This field will be coupled to the wire, generating a current in the latter. This current will flow unimpeded into the enclosure and couple to the internal electronics. The converse is also true; noise internal to the shield will couple to the wire, flow out the enclosure on the wire, and radiate. This type of penetration of a shield will virtually eliminate any effectiveness of the shield. Cable penetrations such as these must be properly treated in order to preserve the effectiveness of the shield. Some common methods are to provide filtering of the cable at its entry/exit point or to use shielded cables whose shields are peripherally bonded to the product shield as shown in Fig. 11.2(b) [1]. Observe that the cable will have currents induced on it by the external field. Simply connecting the cable shield to the product shield with another wire as shown in Fig. 11.2(b) may cause the currents on the cable shield to be conducted into the interior of the product's shield surface, where they may, once again, radiate to the internal electronics of the product, thus reducing the effectiveness of the product shield.

Again, the converse is also true; unless the cable shield is peripherally bonded to the enclosure, the noise currents internal to the enclosure may flow out the enclosure along the exterior of the shield, where they may radiate. *Removing an overall shield from a peripheral cable may actually decrease the radiated emissions of a cable!* As a general rule, *a shield placed around cable wires will not necessarily reduce the radiated emissions of the cable.* The explanation of this phenomenon is illustrated in Fig. 11.2(c). *In order to realize the shielding effectiveness of a cable shield in reducing the radiated emissions of the cable, the cable shield must be attached to a zero-potential point (an ideal ground).* If the voltage of the point at which the shield "pigtail" is attached, say, the logic ground of an electronics PCB, is varying, as illustrated in Fig. 11.2(c), then we

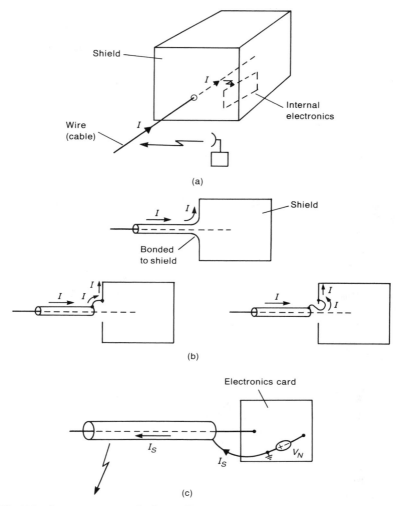

FIGURE 11.2 Important practical considerations that seriously degrade shielding effectiveness: (a) penetration of an enclosure by a cable allowing direct entry of external fields; (b) pigtail termination of a cable shield at the entry point to a shielded enclosure; (c) termination of a cable shield to a noisy point causing the shield to radiate.

have produced a monopole antenna (the cable shield). If the length of the cable shield of order $\frac{1}{4}\lambda_o$, the shield becomes an effective radiator. Peripheral cables such as printer cables for personal computers tend to have lengths of order 1.5 m, which is a quarter-wavelength at 50 MHz. Resonances in the radiated emissions of a product due to common-mode currents on these types of peripheral cables are frequently observed in the frequency range of 50–100 MHz. Disconnecting the peripheral cable from the product usually removes these resonances and their accentuated radiated emissions.

Another common penetration is that of a hole or *aperture* in the shield wall. Fields interior to the shield or exterior to it will radiate through this aperture, thus reducing the effectiveness of the shield. This is due, in part, to an important theorem known as *Babinet's principle* [2]. In order to illustrate this principle, consider a slot that is cut in a perfectly conducting screen as shown in Fig. 11.3. A transmission line is connected across two points of the slot, and excites it. The far fields that are radiated by this "slot antenna" are denoted by $E_{\theta s}$, $E_{\phi s}$, $H_{\theta s}$, and $H_{\phi s}$. Now consider the *complementary structure* shown in Fig. 11.3, which consists of the replacement of the screen with free space and the replacement of the slot with a perfect conductor of the same shape as the slot. The antenna is fed, again, with the transmission line attached to the two halves. The far fields radiated by this complementary structure are denoted by $E_{\theta c}$, $E_{\phi c}$, $H_{\theta c}$, and $H_{\phi c}$. *Babinet's principle states that the far fields radiated by the original screen with the slot and those radiated by the complementary structure are related by* [2]

$$E_{\theta s} = H_{\theta c} \tag{11.1a}$$

$$E_{\phi s} = H_{\phi c} \tag{11.1b}$$

$$H_{\theta s} = -\frac{E_{\theta c}}{\eta_o^2} \tag{11.1c}$$

$$H_{\phi s} = -\frac{E_{\phi c}}{\eta_o^2} \tag{11.1d}$$

This illustrates that *apertures can be as effective radiators as antennas whose conductor dimensions are those of the aperture.*

All these considerations should alert the reader to the fact that shielding should not be relied on to completely eliminate radiated emissions of the product.

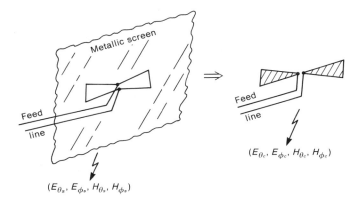

FIGURE 11.3   Illustration of the effect of apertures in a shield (illustration of Babinet's principle).

There are numerous examples of electronic products that do not employ shielded enclosures, yet they are able to comply with the regulatory limits on radiated and conducted emissions. Electronic typewriters are contained in plastic enclosures and have clock frequencies of order 10 MHz and higher. The expense and impracticality of a contiguous metallic enclosure rules out their use for these types of products. On the other hand, personal computers and large mainframe computers employ metallic enclosures. Effectively treating all penetrations provides effective use of shields for these types of products. As a cardinal rule, *the EMC designer should incorporate the same EMC design into a product whether that product is to be shielded or not.* The following discussion of the principles of shielding will nevertheless serve to illustrate the quantitative aspects of shielding.

11.1 SHIELDING EFFECTIVENESS

In this section we will begin the quantitative discussion of the shielding effectiveness of a metallic shield. The general notion of shielding effectiveness was discussed above. In order to quantify these notions, we consider the general problem of a metallic barrier of thickness t, conductivity σ, relative permittivity ϵ_r, and relative permeability μ_r, shown in Fig. 11.4. An electromagnetic wave is incident on this barrier. A reflected wave is produced, and a portion of this

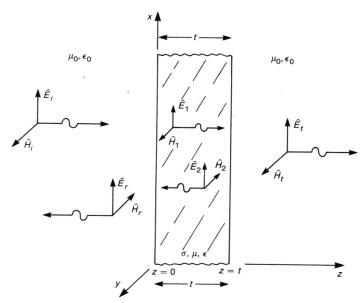

FIGURE 11.4 Illustration of the shielding effectiveness of a conducting barrier to a normal-incidence uniform plane wave.

incident wave is transmitted through the barrier. The *shielding effectiveness* of the barrier is defined for the electric field, in decibels, as

$$SE = 20 \log_{10} \left| \frac{\hat{E}_i}{\hat{E}_t} \right| \tag{11.2}$$

Note that this will be a positive result, since the incident field is expected to be greater than the field that exits the barrier. For example, a shielding effectiveness of 120 dB means that the magnitude of the transmitted field is reduced from the magnitude of the incident field by a factor of 10^6. Some definitions of shielding effectiveness are in terms of the ratio of the transmitted field to the incident field. This definition would give a negative result in dB, which is the negative of (11.2). In terms of the magnetic field, the shielding effectiveness could be defined as

$$SE = 20 \log_{10} \left| \frac{\hat{H}_i}{\hat{H}_t} \right| \tag{11.3}$$

If the incident field is a uniform plane wave and the media on each side of the barrier are identical then the two definitions are identical, since the electric and magnetic fields are related by the intrinsic impedance of the medium. For near fields and/or different media on the two sides of the boundary the two are not equivalent. However, the definition of the shielding effectiveness in terms of the electric field in (11.2) is usually taken to be the standard for either situation.

There are several pheomena that contribute to the reduction of the incident field as it passes through the barrier. Consider the diagram shown in Fig. 11.5, which shows these effects. The first effect is *reflection* at the left surface of the barrier. The portion of the incident electric field that is reflected is given by the

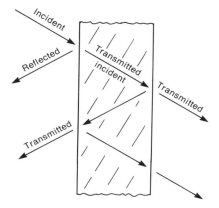

FIGURE 11.5   Illustration of multiple reflections within a shield.

reflection coefficient for that surface. The portion of the wave that crosses this surface proceeds through the shield wall. As it passes through this conductive medium, its amplitude is attenuated according to the factor $e^{-\alpha z}$, where α is the attentuation constant of the material discussed in Chapter 3. This is referred to as *absorption loss*. For barrier materials that constitute good conductors (as is usually the case) the attenuation constant α is related to the skin depth of the material, δ, as $\alpha = 1/\delta$. Therefore the amplitudes of the fields are attenuated according to the factor $e^{-z/\delta}$. If the barrier thickness t is much greater than the skin depth of the barrier material at the frequency of the incident wave then this wave that is transmitted through the first interface is greatly attenuated when it strikes the right interface. This becomes the incident wave for the right interface, and is incident on this interface from the metal. A portion of this incident wave is reflected, and a portion transmitted across the barrier into the medium on the right of the barrier. The reflected portion of this wave is transmitted back through the barrier and strikes the first interface, being incident from the metal. Once again, a portion of this wave is transmitted through the left interface and adds to the total reflected field in the left medium, and a portion is reflected and proceeds to the right. This portion is again attenuated as it passes through the barrier. Once it has passed through the barrier and strikes the right interface, a portion is reflected and a portion is transmitted through the right interface. The portion transmitted through the right interface adds to the total field that is transmitted through the shield. The process continues in like fashion, but the additional reflected and transmitted fields are progressively attenuated by their travel through the conductive barrier. If a shield is designed to have a thickness that is much greater than the skin depth of the material at the frequency of the anticipated incident field, there is little consequence to the continued re-reflection at the interior surfaces of the barrier. These *multiple reflections and transmissions* can therefore generally be disregarded for shield thicknesses that are much greater than a skin depth, and only the initial reflection and transmission at the left and right interfaces need be considered. Nevertheless, the shielding effectiveness given in (11.2) can be broken into the product of three terms each representing one of the phenomena of *reflection loss, absorption loss, and multiple reflections.* In decibels these factors add to give

$$\text{SE}_{\text{dB}} = R_{\text{dB}} + A_{\text{dB}} + M_{\text{dB}} \tag{11.4}$$

where R represents the reflection loss caused by reflection at the left and right interfaces, A represents the absorption loss of the wave as it proceeds through the barrier, and M represents the additional effects of multiple re-reflections and transmissions. Observe that the re-reflections will create fields that will add to the initial field transmitted across the right interface. Thus the multiple-reflection factor M will be a negative number and will, in general, reduce the shielding effectiveness (since R and A will be positive). We now embark on a quantitative determination of these factors that contribute to the shielding effectiveness of a barrier. In addition to the following derivations, the reader is referred to [3–7] for similar developments.

11.2   SHIELDING EFFECTIVENESS—FAR-FIELD SOURCES

In this section we will assume that the source for the field that is incident on the barrier is sufficiently distant from the barrier that the incident field resembles a uniform plane wave, whose properties were discussed in Chapter 3. We first determine the exact solution for the shielding effectiveness, and will then determine this in an approximate fashion to show that the two methods yield the same results for shields that are constructed from "good conductors" whose thickness t is much greater than a skin depth at the frequency of the incident wave.

11.2.1   Exact Solution

In order to obtain the exact solution for the shielding effectiveness of a metallic barrier, we solve the problem illustrated in Fig. 11.4. A shield of thickness t, conductivity σ, permittivity ϵ, and permeability μ has an incident uniform plane wave incident on its leftmost surface. The medium on either side of the shield is assumed, for practical reasons, to be free space. A rectangular coordinate system is used to define the problem, with the left surface lying in the xy plane at $z = 0$ and the right surface located at $z = t$. Forward- and backward-traveling waves are present in the left medium and in the shield according to the general properties of the solution of Maxwell's equations. Only a forward-traveling wave is postulated in the medium to the right of the shield, since we reason that there is no additional barrier to create a reflected field. The general forms of these fields are (see Section 5.6.2)

$$\vec{E}_i = \hat{E}_i e^{-j\beta_o z} \vec{a}_x \tag{11.5a}$$

$$\vec{H}_i = \frac{\hat{E}_i}{\eta_o} e^{-j\beta_o z} \vec{a}_y \tag{11.5b}$$

$$\vec{E}_r = \hat{E}_r e^{j\beta_o z} \vec{a}_x \tag{11.5c}$$

$$\vec{H}_r = -\frac{\hat{E}_r}{\eta_o} e^{j\beta_o z} \vec{a}_y \tag{11.5d}$$

$$\vec{E}_1 = \hat{E}_1 e^{-\hat{\gamma} z} \vec{a}_x \tag{11.5e}$$

$$\vec{H}_1 = \frac{\hat{E}_1}{\hat{\eta}} e^{-\hat{\gamma} z} \vec{a}_y \tag{11.5f}$$

$$\vec{E}_2 = \hat{E}_2 e^{\hat{\gamma} z} \vec{a}_x \tag{11.5g}$$

$$\vec{H}_2 = -\frac{\hat{E}_2}{\hat{\eta}} e^{\hat{\gamma}z}\vec{a}_y \tag{11.5h}$$

$$\vec{E}_t = \hat{E}_t e^{-j\beta_o z}\vec{a}_x \tag{11.5i}$$

$$\vec{H}_t = \frac{\hat{E}_t}{\eta_o} e^{-j\beta_o z}\vec{a}_y \tag{11.5k}$$

where the various quantities are given by

$$\beta_o = \omega\sqrt{\mu_o \epsilon_o} \tag{11.6a}$$

$$\eta_o = \sqrt{\frac{\mu_o}{\epsilon_o}} \tag{11.6b}$$

and

$$\hat{\gamma} = \sqrt{j\omega\mu(\sigma + j\omega\epsilon)} \tag{11.7a}$$

$$= \alpha + j\beta$$

$$\hat{\eta} = \sqrt{\frac{j\omega\mu}{\sigma + j\omega\epsilon}} \tag{11.7b}$$

$$= \eta \angle\theta_\eta$$

The magnitude of the incident field $\hat{E}_i$ is assumed known. In order to determine the remaining amplitudes $\hat{E}_r$, $\hat{E}_1$, $\hat{E}_2$, and $\hat{E}_t$, we need four equations. These are generated by enforcing the boundary conditions on the field vectors at the two boundaries, $z = 0$ and $z = t$. Continuity of the tangential components of the electric field at the two interfaces gives

$$\vec{E}_i|_{z=0} + \vec{E}_r|_{z=0} = \vec{E}_1|_{z=0} + \vec{E}_2|_{z=0} \tag{11.8a}$$

$$\vec{E}_1|_{z=t} + \vec{E}_2|_{z=t} = \vec{E}_t|_{z=t} \tag{11.8b}$$

Continuity of the tangential components of the magnetic field at the two interfaces gives

$$\vec{H}_i|_{z=0} + \vec{H}_r|_{z=0} = \vec{H}_1|_{z=0} + \vec{H}_2|_{z=0} \tag{11.9a}$$

$$\vec{H}_1|_{z=t} + \vec{H}_2|_{z=t} = \vec{H}_t|_{z=t} \tag{11.9b}$$

Substituting the forms given in (11.5) gives the required four equations as

$$\hat{E}_i + \hat{E}_r = \hat{E}_1 + \hat{E}_2 \tag{11.10a}$$

$$\hat{E}_1 e^{-\hat{\gamma}t} + \hat{E}_2 e^{\hat{\gamma}t} = \hat{E}_t e^{-j\beta_o t} \tag{11.10b}$$

$$\frac{\hat{E}_i}{\eta_o} - \frac{\hat{E}_r}{\eta_o} = \frac{\hat{E}_1}{\hat{\eta}} - \frac{\hat{E}_2}{\hat{\eta}} \tag{11.10c}$$

$$\frac{\hat{E}_1}{\hat{\eta}} e^{-\hat{\gamma}t} - \frac{\hat{E}_2}{\hat{\eta}} e^{\hat{\gamma}t} = \frac{\hat{E}_t}{\eta_o} e^{-j\beta_o t} \tag{11.10d}$$

Solving these equations gives the ratio of the incident and transmitted waves as

$$\frac{\hat{E}_i}{\hat{E}_t} = \frac{(\eta_o + \hat{\eta})^2}{4\eta_o\hat{\eta}} \left[1 - \left(\frac{\eta_o - \hat{\eta}}{\eta_o + \hat{\eta}}\right)^2 e^{-2t/\delta} e^{-j2\beta t} \right] e^{t/\delta} e^{j\beta t} e^{-j\beta_o t} \tag{11.11}$$

Equation (11.11) is the *exact expression* for the ratio of the total electric field that is incident on the boundary and the electric field that is transmitted through the boundary. We have substituted the relation $\hat{\gamma} = \alpha + j\beta$ from (11.7a) and also $\alpha = 1/\delta$ (assuming the barrier material is a good conductor), where δ is the skin depth for the barrier material at the frequency of the incident wave. We can, however, make some reasonable approximations to reduce this to a result derived by approximate means in the following sections. This will not only simplify the result, but will also demonstrate that the same result can be derived by approximate methods without any significant loss in accuracy, as we will do in the next section.

In order to simplify (11.11), we will assume that the barrier is constructed from a "good conductor," so that the intrinsic impedance of the conductor is much less than that of free space:

$$\hat{\eta} \ll \eta_o \tag{11.12}$$

Therefore we may approximate

$$\frac{\eta_o - \hat{\eta}}{\eta_o + \hat{\eta}} \cong 1 \tag{11.13}$$

Also we assume that the skin depth δ is much less than the barrier thickness t. Thus

$$e^{-\hat{\gamma}t} = e^{-\alpha t} e^{-j\beta t} \tag{11.14}$$

$$= e^{-t/\delta} e^{-j\beta t}$$

$$\ll 1 \quad \text{for } t \gg \delta$$

Substituting these into the exact result given in (11.11) and taking the absolute value of the result gives

$$\left|\frac{\hat{E}_i}{\hat{E}_t}\right| = \left|\frac{(\eta_o + \hat{\eta})^2}{4\eta_o\hat{\eta}}\right| e^{t/\delta} \tag{11.15}$$

$$\cong \left|\frac{\eta_o}{4\hat{\eta}}\right| e^{t/\delta}$$

Taking the logarithm of this result in order to express the shielding effectiveness in dB in accordance with (11.2) gives

$$\underbrace{SE_{dB} \cong 20 \log_{10} \left| \frac{\eta_o}{4\hat{\eta}} \right|}_{R_{dB}} + \underbrace{20 \log_{10} e^{t/\delta} + M_{dB}}_{A_{dB}} \tag{11.16a}$$

The multiple-reflection loss in (11.4) is evidently the middle term of (11.11):

$$M_{dB} = 20 \log_{10} \left| 1 - \left(\frac{\eta_o - \hat{\eta}}{\eta_o + \hat{\eta}} \right)^2 e^{-2t/\delta} e^{-j2\beta t} \right| \tag{11.16b}$$

$$\cong 20 \log_{10} |1 - e^{-2t/\delta} e^{-j2t/\delta}|$$

which can be neglected for shields that are constructed of good conductors, $\hat{\eta} \ll \eta_o$, and whose thicknesses are much greater than a skin depth, $t \gg \delta$. We have also substituted $\beta = \alpha = 1/\delta$, assuming the barrier is constructed from a good conductor. (See Section 3.6.4.) Observe that this term is of the form $1 - \hat{\Gamma}_{in}^2$, where $\hat{\Gamma}_{in} = [(\eta_o - \hat{\eta})/(\eta_o + \hat{\eta})]e^{-2\hat{\gamma}t}$ is the reflection coefficient at the right boundary referred to the left boundary. The multiple-reflection term is approximately unity ($M_{dB} \simeq 0$) for barrier thicknesses that are thick compared with a skin depth, $t \gg \delta$, and is of no consequence. However, for barrier thicknesses that are thin compared with a skin depth, $t \ll \delta$, the multiple-reflection factor is negative (in dB). In this case, multiple reflections reduce the shielding effectiveness of the barrier. For example, for $t/\delta = 0.1$, equation (11.16b) gives $M_{dB} = -11.8$ dB.

The separation of the exact result into a component due to reflection, a component due to absorption, and a component due to multiple reflections as in equation (11.4) is evident in (11.16a). This result will be derived by approximate methods in the following section.

11.2.2 Approximate Solution

We now consider deriving the previous result *under the assumption that the barrier is constructed of a good conductor, $\hat{\eta} \ll \eta_o$, and the barrier thickness is much greater than a skin depth at the frequency of the incident wave, i.e., $t \gg \delta$.* These assumptions are usually inherent in a well-designed shield and thus are not restrictive from a practical standpoint. The basic idea is illustrated in Fig. 11.6. First of all, it is worth noting that this approximate solution is analogous to the problem of analyzing the overall gain of cascaded amplifiers. In that problem we compute the input impedance of the first stage, using the input impedance of the second stage as the load for the first. Then we can compute the ratio of the output voltage of the first stage to its input voltage. Next we compute the ratio of the output voltage of the second stage to its input

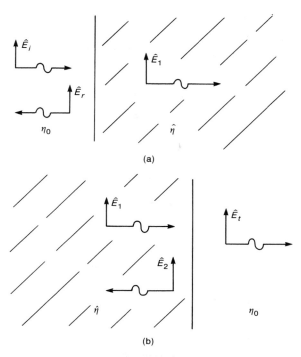

FIGURE 11.6   Approximate calculation of shielding effectiveness for uniform plane waves.

voltage, using the input impedance of the third stage as the load for the second stage. This process continues until we finally compute the gain of the last stage. The overall gain of the cascade is then the product of the gains of the individual stages. This technique takes into account the loading of each stage on the preceeding stage, and this loading generally cannot be neglected. However, if the input impedances of the individual stages are quite large, as is generally the case for FET and vacuum-tube amplifiers, then this loading can be ignored and the overall gain of the cascade can be computed as the gains of the individual, *isolated* stages.

11.2.2.1 Reflection Loss   The approximate analysis technique we will use is the direct analogy of the method for analyzing cascaded amplifiers described above. Assuming that the barrier thickness is much greater than a skin depth at the frequency of the incident wave, the portion of the incident wave that is transmitted across the left interface in Fig. 11.4, $\hat{E}_1$, is greatly attenuated by the time it reaches the right interface. Thus the reflected wave $\hat{E}_2$, when it arrives at the left interface, is not of much consequence and so contributes little to the total reflected wave $\hat{E}_r$. ($\hat{E}_2$ is also greatly attenuated as it travels from the second interface back to the left interface.) Therefore we can approximately compute the portion of the incident wave that is transmitted across the left interface, $\hat{E}_1$, by assuming that the barrier is infinitely thick and therefore assuming that $\hat{E}_2 = 0$. This then becomes the basic problem considered in Section 5.6.2, and is illustrated in Fig. 11.6(a). The transmission coefficient becomes

$$\frac{\hat{E}_1}{\hat{E}_i} \cong \frac{2\hat{\eta}}{\eta_o + \hat{\eta}} \tag{11.17}$$

The next basic problem occurs at the right interface, as illustrated in Fig. 11.6(b), and is again related to the basic problem considered in Section 5.6.2. The transmission coefficient for this case gives

$$\frac{\hat{E}_t}{\hat{E}_1} \cong \frac{2\eta_o}{\eta_o + \hat{\eta}} \tag{11.18}$$

Note that for this case the intrinsic impedance of the medium for the transmitted wave is η_o and the intrinsic impedance for the incident wave is $\hat{\eta}$. For the first half of this problem the intrinsic impedance of the medium for the transmitted wave is $\hat{\eta}$ and the intrinsic impedance for the incident wave is η_o. Taking the product of (11.17) and (11.18) gives the ratio of the transmitted field and the incident field *in the absence of attenuation* as

$$\frac{\hat{E}_t}{\hat{E}_i} = \frac{\hat{E}_t}{\hat{E}_1} \frac{\hat{E}_1}{\hat{E}_i} \tag{11.19}$$

$$= \frac{2\eta_o}{\eta_o + \hat{\eta}} \frac{2\hat{\eta}}{\eta_o + \hat{\eta}}$$

$$= \frac{4\eta_o \hat{\eta}}{(\eta_o + \hat{\eta})^2}$$

Note that because $\hat{\eta} \ll \eta_o$, (11.17) is much smaller than (11.18). Thus *the transmission coefficient is very small at the first boundary, and is approximately two at the second boundary.* Thus *very little of the electric field is transmitted through the first (left) boundary.* The reflection coefficient at the first (left) interface is $\Gamma_1 = (\hat{\eta} - \eta_o)/(\hat{\eta} + \eta_o) \cong -1$, and the electric field is effectively "shorted out" by the good conductor. The reflection coefficient at the second (right) boundary is $\Gamma_2 = (\eta_o - \hat{\eta})/(\eta_o + \hat{\eta}) \cong +1$. These are analogous to the voltage reflections at the end of a short-circuited (left boundary) or open-circuited (right boundary) transmission line. Thus the majority of the electric field that is incident on each interface is reflected. However, because very little of the electric field is transmitted through the first boundary, it is of little consequence that the reflection coefficient at the second boundary is approximately unity! The reflection loss term in (11.4) is therefore

$$R_{dB} = 20 \log_{10} \left| \frac{\hat{E}_i}{\hat{E}_t} \right| \tag{11.20}$$

$$= 20 \log_{10} \left| \frac{(\eta_o + \hat{\eta})^2}{4\eta_o \hat{\eta}} \right|$$

$$\cong 20 \log_{10} \left| \frac{\eta_o}{4\hat{\eta}} \right|$$

where we have substituted the approximation $\hat{\eta} \ll \eta_o$.

It is instructive to consider the magnetic field transmissions. Recall from Chapter 5 that the reflection and transmission coefficients were derived for the electric field only, and could not be used for the magnetic field. If we wish to determine the reflected and transmitted magnetic fields, we need to divide the electric fields by the appropriate intrinsic impedances to give

$$\frac{\hat{H}_1}{\hat{H}_i} = \frac{\hat{E}_1/\hat{\eta}}{\hat{E}_i/\eta_o} \tag{11.21}$$

$$= \frac{\hat{E}_1}{\hat{E}_i} \frac{\eta_o}{\hat{\eta}}$$

$$= \frac{2\eta_o}{\eta_o + \hat{\eta}}$$

Similarly, we obtain

$$\frac{\hat{H}_t}{\hat{H}_1} = \frac{\hat{E}_t/\eta_o}{\hat{E}_1/\hat{\eta}} \tag{11.22}$$

$$= \frac{\hat{E}_t}{\hat{E}_1} \frac{\hat{\eta}}{\eta_o}$$

$$= \frac{2\hat{\eta}}{\eta_o + \hat{\eta}}$$

Taking the product of (11.21) and (11.22) gives the ratio of the transmitted and incident magnetic field intensities:

$$\frac{\hat{H}_t}{\hat{H}_i} = \frac{\hat{H}_t}{\hat{H}_1} \frac{\hat{H}_1}{\hat{H}_i} \tag{11.23}$$

$$= \frac{2\hat{\eta}}{\eta_o + \hat{\eta}} \frac{2\eta_o}{\eta_o + \hat{\eta}}$$

$$= \frac{4\eta_o \hat{\eta}}{(\eta_o + \hat{\eta})^2}$$

Comparing (11.23) and (11.19) shows that *the ratio of the transmitted and incident electric fields are identical to the ratio of the transmitted and incident magnetic fields.* However, there is one difference: *the primary transmission of the magnetic field occurs at the left interface, whereas the primary transmission of the electric field occurs at the right interface.* (See (11.17), (11.18), (11.21), and (11.22).) Therefore the attenuation of the magnetic field as it passes through the boundary is more important than is the attenuation of the electric field. This points out that "thick" boundaries have more effect on shielding against magnetic fields than to electric fields (because of this attenuation of the magnetic field as it travels through the boundary).

Since the primary transmission of the electric field occurs at the second boundary, shield thickness is not of as much importance as it is for magnetic field shielding, wherein the primary transmission occurs at the first boundary. Attenuation of the barrier is of more consequence in magnetic field shielding, since there is considerable transmission of the magnetic field at the first boundary. Therefore effective shields for electric fields can be constructed from thin shields, which effectively "short out" the electric field at the first boundary.

11.2.2.2 Absorption Loss This previous result assumed that the barrier thickness was much greater than a skin depth, so that we could "uncouple" the calculation of the reflections and transmissions at the two interfaces. However, in taking the product of the two transmission coefficients as in (11.19) we are assuming that $\hat{E}_1$ is the same amplitude at the left and right interfaces. But the magnitude of $\hat{E}_1$ at the right interface will be reduced substantially from its value at the left interface by the factor $e^{-t/\delta}$. This attenuation can be easily accounted for: simply multiply (11.19) by $e^{-t/\delta}$. Thus the *absorption factor* accounting for attenuation becomes

$$A = e^{t/\delta} \tag{11.24}$$

In decibels this becomes

$$A_{\mathrm{dB}} = 20 \log_{10} e^{t/\delta} \tag{11.25}$$

11.2.2.3 Multiple Reflection Loss In the previous approximate calculations we have assumed that any "secondary reflections" are of no consequence, since they will have suffered substantial attenuation as they travel back and forth through the barrier. If the barrier thickness is not much greater than a skin depth, as was assumed, then the re-reflections and transmissions may be important. This is particularly true for magnetic fields, since the primary transmission occurs at the first boundary, and thus these multiple reflections can be more significant for magnetic field shielding. In the case of multiple reflections that are significant they are accounted for with a *multiple-reflection factor* given in (11.16b) and illustrated in Fig. 11.7(a). The total transmitted

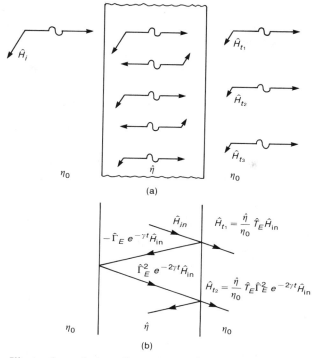

FIGURE 11.7   Illustration of the effect of multiple reflections within the barrier: (a) combining multiple transmissions; (b) calculation in terms of reflection and transmission coefficients.

magnetic field is the sum of the secondary transmissions as

$$\hat{H}_t = \hat{H}_{t_1} + \hat{H}_{t_2} + \hat{H}_{t_3} + \cdots \tag{11.26}$$
$$= \hat{H}_{t_1}(1 + \Delta_2 + \Delta_3 + \cdots)$$

where $\hat{H}_{t_1}$ is the magnetic field transmitted across the second boundary, which was considered to be the total transmitted field in the previous approximate solution. The shielding effectiveness for the magnetic field is

$$\mathrm{SE_{dB}} = 20 \log_{10} \left| \frac{\hat{H}_i}{\hat{H}_t} \right| \tag{11.27}$$

$$= -20 \log_{10} \left| \frac{\hat{H}_t}{\hat{H}_i} \right|$$

$$= -20 \log_{10} \left| \frac{\hat{H}_{t_1}}{\hat{H}_i} \right| \underbrace{-20 \log_{10} |1 + \Delta_2 + \Delta_3 + \cdots|}_{\mathrm{M_{dB}}}$$

The first term is positive, since the transmitted field is expected to be smaller than the incident field. In the following we will demonstrate that the second term is the *multiple-reflection factor* given in (11.16b).

It is a simple matter to show that the multiple-reflection factor in (11.16b) is equivalent to (11.27). (Although we will show this for the magnetic field, recall that under our assumption of uniform plane waves this also directly applies to the electric fields.) Consider Fig. 11.7(b). The portion of the magnetic field incident at the right boundary, $\hat{H}_{\text{in}}$, that is transmitted across the boundary is $\hat{H}_{t_1} = \hat{T}_H \hat{H}_{\text{in}}$, where $\hat{T}_H = 2\hat{\eta}/(\hat{\eta} + \eta_0)$. The portion of $\hat{H}_{\text{in}}$ that is reflected is given by $-\hat{\Gamma}_E \hat{H}_{\text{in}}$, where $\hat{\Gamma}_E = (\eta_0 - \hat{\eta})/(\eta_0 + \hat{\eta})$. This traverses the boundary moving to the left, and suffers attenuation represented by a multiplication by $e^{-\hat{\gamma}t}$. A portion $-(\eta_0 - \hat{\eta})/(\eta_0 + \hat{\eta}) = -\hat{\Gamma}_E$ is reflected at the left boundary. It too suffers attenuation represented by the factor $e^{-\hat{\gamma}t}$ as it moves through the boundary to the right. Therefore the portion incident at the right boundary is $(-\hat{\Gamma}_E e^{-\hat{\gamma}t})^2 = \hat{\Gamma}_E^2 e^{-2\hat{\gamma}t}$. A portion of this, $\hat{T}_H = 2\hat{\eta}/(\hat{\eta} + \eta_0)$, is transmitted across the boundary to give $\hat{H}_{t_2} = \hat{T}_H \hat{\Gamma}_E^2 e^{-2\hat{\gamma}t} \hat{H}_{\text{in}}$. Thus Δ_2 in (11.26) is Δ^2, where $\Delta = \hat{\Gamma}_E e^{-\hat{\gamma}t} = [(\eta_0 - \hat{\eta})/(\eta_0 + \hat{\eta})]e^{-\hat{\gamma}t}$, since $\hat{H}_{t_1} = \hat{T}_H \hat{H}_{\text{in}}$. Continuing in like fashion gives the total field as $\hat{H}_t = \hat{H}_{t_1}(1 + \Delta^2 + \Delta^4 + \Delta^6 + \cdots) = \hat{H}_{t_1}/(1 - \Delta^2)$: a summation that is valid for $|\Delta| < 1$, as is the case here. This process shows that the middle term of (11.11) given in (11.16b) is indeed this multiple-reflection factor.

11.2.2.4   Total Loss   Combining the above results gives the three components of the shielding effectiveness given in (11.4). The reflection loss is given in (11.20). Substituting the approximation for the intrinsic impedance of a good conductor as

$$\hat{\eta} = \sqrt{\frac{j\omega\mu}{\sigma + j\omega\epsilon}} \tag{11.28}$$

$$= \sqrt{\frac{j\omega\mu}{\sigma}} \sqrt{\frac{1}{1 + j\omega\epsilon/\sigma}}$$

$$\cong \sqrt{\frac{j\omega\mu}{\sigma}}$$

$$= \sqrt{\frac{\omega\mu}{\sigma}} \; \underline{/45°}$$

and

$$\eta_0 = \sqrt{\frac{\mu_0}{\epsilon_0}} \tag{11.29}$$

into (11.20) gives

$$R_{dB} = 20 \log_{10}\left(\frac{1}{4}\sqrt{\frac{\sigma}{\omega\mu_r\epsilon_o}}\right) \qquad (11.30)$$

where we have assumed $\mu = \mu_o\mu_r$ and $\epsilon = \epsilon_o$. It is customary to refer the conductivity of metals to that of copper, which has a conductivity $\sigma_{Cu} = 5.8 \times 10^7$ S/m. Thus the conductivity of other metals is written as $\sigma = \sigma_{Cu}\sigma_r$, where σ_r is the *conductivity relative to copper*. Substituting this into (11.30) gives

$$R_{dB} = 168 + 10 \log_{10}\left(\frac{\sigma_r}{\mu_r f}\right) \qquad (11.31)$$

Observe that *the reflection loss is greatest at low frequencies and for high-conductivity metals*. It decreases at a rate of -10 dB/decade with frequency. As an example, consider a shield constructed of copper ($\mu_r = 1$). The reflection loss at 1 kHz is 138 dB. At 10 MHz the reflection loss is 98 dB. On the other hand, sheet steel has $\mu_r = 1000$ and $\sigma_r = 0.1$. At 1 kHz the reflection loss is 98 dB, and at 10 MHz it is reduced to 58 dB.

The absorption loss is given by (11.25). This can also be simplified. The skin depth is

$$\delta = \frac{1}{\sqrt{\pi f \mu \sigma}} \qquad (11.32)$$

$$= \frac{0.06609}{\sqrt{f\mu_r\sigma_r}} \text{ m}$$

$$= \frac{2.6}{\sqrt{f\mu_r\sigma_r}} \text{ inches}$$

$$= \frac{2602}{\sqrt{f\mu_r\sigma_r}} \text{ mils}$$

where we have written the result in various units. Substituting (11.32) into (11.25) gives

$$A_{dB} = 20 \log_{10} e^{t/\delta} \qquad (11.33)$$

$$= 20t/\delta \log_{10} e$$

$$= 8.6859 t/\delta$$

$$= 131.4 t \sqrt{f\mu_r\sigma_r} \quad (t \text{ in m})$$

$$= 3.338 t \sqrt{f\mu_r\sigma_r} \quad (t \text{ in inches})$$

Equation (11.33) shows that the absorption loss increases with increasing frequency as $\sqrt{f}$ *on a decibel scale*. This is quite different from the absorption loss being proportional to the square root of frequency so that it increases at a rate of 10 dB/decade on a decibel scale. Therefore the absorption loss increases quite rapidly with increasing frequency. Ferromagnetic materials where $\mu_r \gg 1$ increase this loss over copper (assuming that $\mu_r \sigma_r \gg 1$). The absorption loss can also be understood in terms of the thickness of the shield relative to a skin depth, as is evident in (11.33):

$$A_{dB} = 8.6859t/\delta \tag{11.34}$$

$$= 8.7 \text{ dB} \quad \text{for } t/\delta = 1$$

$$= 17.4 \text{ dB} \quad \text{for } t/\delta = 2$$

This illustrates the importance of skin depth in absorption loss.

Observe that the reflection loss is a function of the ratio σ_r/μ_r, whereas the absorption loss is a function of the product $\sigma_r\mu_r$. Table 11.1 shows these factors for various materials.

Figure 11.8 shows the components of the shielding effectiveness for a 20 mil thickness of copper as a function of frequency from 10 Hz to 10 MHz. Observe that the absorption loss is dominant above 2 MHz. Figure 11.9 shows the same data for Steel (SAE 1045) for a 20 mil thickness. These data are plotted from 10 Hz to only 1 MHz. Note that for this material reflection loss dominates only below 20 kHz. These data indicate that reflection loss is the primary contributor to the shielding effectiveness at low frequencies for either ferrous or nonferrous shielding materials. At the higher frequencies ferrous materials increase the absorption loss and the total shielding effectiveness. It is worthwhile reiterating that for electric fields the primary transmission occurs at the second boundary,

TABLE 11.1

Material	σ_r	μ_r	$\mu_r\sigma_r$	σ_r/μ_r
Silver	1.05	1	1.05	1.05
Copper	1	1	1	1
Gold	0.7	1	0.7	0.7
Aluminum	0.61	1	0.61	0.61
Brass	0.26	1	0.26	0.26
Bronze	0.18	1	0.18	0.18
Tin	0.15	1	0.15	0.15
Lead	0.08	1	0.08	0.08
Nickel	0.2	100	20	2×10^{-3}
Stainless steel (430)	0.02	500	10	4×10^{-5}
Steel (SAE 1045)	0.1	1000	100	1×10^{-4}
Mumetal (at 1 kHz)	0.03	20,000	600	1.5×10^{-6}
Superpermalloy (at 1 kHz)	0.03	100,000	3000	3×10^{-7}

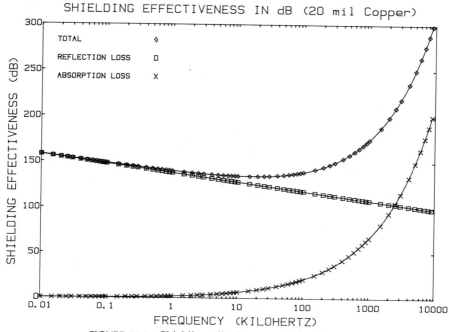

FIGURE 11.8   Shielding effectiveness of 20 mil copper.

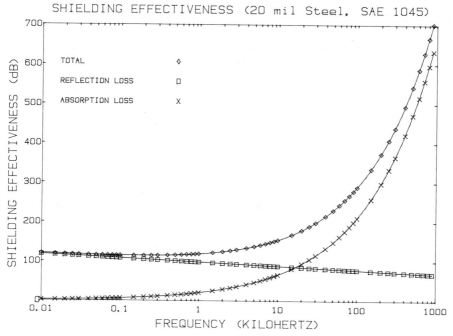

FIGURE 11.9   Shielding effectiveness of 20 mil steel (SAE 1045).

whereas for magnetic fields it occurs at the first boundary, so that absorption is more important for the reduction of magnetic fields.

11.3 SHIELDING EFFECTIVENESS—NEAR-FIELD SOURCES

The previous analysis of shielding effectiveness assumed a uniform plane wave incident normal to the surface of the shield. However, the analysis relied on a uniform plane wave being incident on the shield which assumes that the source of the incident field is sufficiently distant from the shield. In this section we will consider *near-field sources*. We will find that the techniques for shielding depend on the *type of source*; whether the source is a *magnetic field source* or an *electric field source*.

11.3.1 Near Field versus Far Field

In order to demonstrate why it is important to differentiate between types of sources when they are close to the shield, consider the elemental electric (Hertzian) dipole considered in Section of 5.1.1. We found that at sufficiently distant points from this source, the fields resembled plane waves in that (1) the far-field components E_θ and H_ϕ are orthogonal and (2) the ratio of these two field vectors is the intrinsic impedance of the medium $E_\theta/H_\phi = \eta_o$. In the "near field" of this source these conditions are not satisfied. In particular, one must be of order $3\lambda_o$ from this source in order for these two characteristics to hold, i.e., the far field. In general, the near fields have more field components than these. Furthermore, the field components do not vary simply as inverse distance $1/r$, but depend on $1/r^2$ and $1/r^3$. The $1/r$ terms equal the $1/r^2$ and $1/r^3$ terms at $r = \lambda_o/2\pi$ or about $\frac{1}{6}\lambda_o$. A reasonable criterion for the near-field/far-field boundary would be where the ratio of E_θ to H_ϕ is approximately η_o. This ratio is referred to as the *wave impedance*:

$$\hat{Z}_w = \frac{\hat{E}_\theta}{\hat{H}_\phi} \tag{11.35}$$

It is only in the far field where it is appropriate to use the term "intrinsic impedance" to characterize the wave impedance. The wave impedance is obtained from equations (5.1) and (5.2) as the ratio of the total fields:

$$\hat{Z}_w = \eta_o \frac{j/\beta_o r + 1/(\beta_o r)^2 - j/(\beta_o r)^3}{j/\beta_o r + 1/(\beta_o r)^2} \tag{11.36}$$

The magnitude of the wave impedance is plotted versus distance from the source in Fig. 11.10(a). In the far field the $1/r$ terms dominate, giving $\hat{Z}_w \cong \eta_o$. Equation

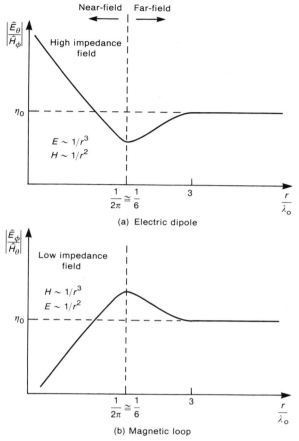

FIGURE 11.10   Wave impedance of (a) the electric (Hertzian) dipole and (b) the magnetic (loop) dipole.

(11.36) reduces in the near field to

$$\hat{Z}_w \cong \eta_0 \left(-j \frac{1}{\beta_o r} \right) \quad \text{(near field, } \beta_o r \ll 1) \tag{11.37}$$

$$\cong \frac{\eta_o}{\beta_o r} \underline{/-90^\circ}$$

In the near field the electric field is proportional to $1/r^3$ while the magnetic field is proportional to $1/r^2$:

$$\left. \begin{array}{l} \hat{E}_\theta \sim \dfrac{1}{r^3} \\[2ex] \hat{H}_\phi \sim \dfrac{1}{r^2} \end{array} \right\} \quad \text{(electric source, near field)}$$

$$\tag{11.38a}$$

$$\tag{11.38b}$$

Also, in the near field of the electric dipole the wave impedance is greater than the intrinsic impedance of the medium. Therefore the electric dipole is referred to as a *high-impedance source*. The magnitude of the wave impedance for an *electric field source* is therefore

$$|\hat{Z}_w|_e = \frac{1}{2\pi f \epsilon_o r} \tag{11.39}$$

$$= 60 \frac{\lambda_o}{r}$$

where the subscript e denotes an electric field source (the Hertzian dipole).

The elemental magnetic dipole (loop) is the dual of the elemental electric dipole in that we can interchange the electric and magnetic field quantities and obtain corresponding results. The fields of the elemental magnetic dipole (loop) are given in equations (5.9) and (5.10). The far-field components for the elemental magnetic dipole (loop) are E_ϕ and H_θ. The wave impedance for this source is therefore defined as

$$\hat{Z}_w = \frac{\hat{E}_\phi}{\hat{H}_\theta} \tag{11.40}$$

and is obtained from (5.9) and (5.10) as

$$\hat{Z}_w = -\eta_o \frac{j/\beta_o r + 1/(\beta_o r)^2}{j/\beta_o r + 1/(\beta_o r)^2 - j/(\beta_o r)^3} \tag{11.41}$$

The magnitude of the wave impedance is plotted versus distance from the source in Fig. 11.10(b). In the far field the $1/r$ terms dominate, giving $\hat{Z}_w \cong \eta_o$. Equation (11.41) reduces in the near field to

$$\hat{Z}_w \cong -j\eta_o \beta_o r \quad \text{(near field, } \beta_o r \ll 1) \tag{11.42}$$

$$\cong \eta_o \beta_o r \;\underline{/-90°}$$

In the near field the magnetic field is proportional to $1/r^3$ while the electric field is proportional to $1/r^2$:

$$\left. \begin{aligned} \hat{H}_\theta &\sim \frac{1}{r^3} \\[2em] \hat{E}_\phi &\sim \frac{1}{r^2} \end{aligned} \right\} \quad \text{(magnetic source, near field)} \quad \begin{aligned} &(11.43a) \\[2em] &(11.43b) \end{aligned}$$

Also, in the near field of the magnetic dipole the wave impedance is less than the intrinsic impedance of the medium. Therefore the magnetic dipole is referred to as a *low-impedance source*. The magnitude of the wave impedance for an *magnetic field source* is therefore

$$|\hat{Z}_w|_m = 2\pi f \mu_0 r \tag{11.44}$$

$$= 2369 \, \frac{r}{\lambda_0}$$

where the subscript m denotes a magnetic field source (the magnetic loop).

The distinction between electric and magnetic sources will allow us to translate much of our results obtained for far-field sources to the case of near-field sources. There are numerous examples of such sources. For example, a transformer is constructed of turns of wire wound on a magnetic core. The electromagnetic field in the vicinity of this source tends to be predominantly magnetic. In fact, the transformer resembles the magnetic loop. For this source the near fields have the properties that the wave impedance is much less than η_0, and the electric field varies with distance as $1/r^2$ while the magnetic field varies as $1/r^3$. Examples of electric field sources are spark gaps and other points where arcing takes place, such as at the brushes of a dc motor. For this source the near fields have the properties that the wave impedance is much greater than η_0, and the magnetic field varies with distance as $1/r^2$ while the electric field varies as $1/r^3$.

11.3.2 Electric Sources

The basic mechanisms of shielding observed for far-field sources are prevalent for near-field sources, but the type of source is critical to determining effective shielding methodologies. An exact solution for this problem is considerably more difficult than for the uniform plane-wave source case [7]. As an approximation we write the shielding effectiveness as the product of a reflection term, an absorption term, and a multiple-reflection term, and obtain each factor using the previous results but substituting the wave impedance $\hat{Z}_w$ for η_0 in those equations. The absorption loss term is unaffected by the type of source.

The reflection loss is obtained by substituting the wave impedance for intrinsic impedance of the free space in (11.20):

$$R_{dB} = 20 \log_{10} \left| \frac{(\hat{Z}_w + \hat{\eta})^2}{4\hat{Z}_w \hat{\eta}} \right| \tag{11.45}$$

$$\cong 20 \log_{10} \left| \frac{\hat{Z}_w}{4\hat{\eta}} \right|$$

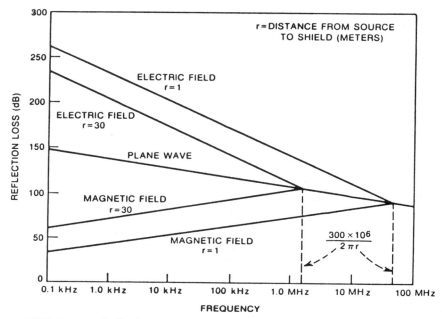

FIGURE 11.11 Reflection loss of near-field electric and magnetic sources [3].

Substituting the wave impedance for electric field sources from (11.39) and the approximation for a good conductor from (11.28) gives

$$R_{e,\text{dB}} = 322 + 10 \log_{10}\left(\frac{\sigma_r}{\mu_r f^3 r^2}\right) \tag{11.46}$$

The reflection loss for an electric field source is plotted in Fig. 11.11 for a copper shield and various distances from the source to the shield [3]. The reflection loss for a uniform plane-wave source is shown to illustrate the asymptotic convergence as the source-to-shield distance is increased. Observe in this figure that the reflection loss for near-field electric sources is considerably higher than for a uniform plane-wave source. It also increases with decreasing distance between the source and the shield.

11.3.3 Magnetic Sources

Again, the absorption loss is the same as for uniform plane-wave sources. The reflection loss for near-field magnetic sources is obtained by substituting the wave impedance from (11.44) into (11.45) to give

$$R_{m,\text{dB}} = 14.57 + 10 \log_{10}\left(\frac{f r^2 \sigma_r}{\mu_r}\right) \tag{11.47}$$

Figure 11.11 also shows the reflection loss for near-field, magnetic sources for various source–shield distances and frequency. Observe that the reflection loss decreases for decreasing frequencies, and is less than for the plane-wave reflection loss. Consequently, reflection loss is usually negligible for lower frequencies. Also, the absorption loss is small for low frequencies, so that other techniques must be used for shielding against low-frequency, near-field magnetic sources.

11.4 LOW-FREQUENCY, MAGNETIC FIELD SHIELDING

In the previous sections we have seen that *for far-field (uniform plane-wave) sources reflection loss is the predominant shielding mechanism at the lower frequencies, while absorption loss is the predominant shielding mechanism at the higher frequencies.* For near-field, electric sources, the situation is to a large degree unchanged from the uniform plane-wave case: *reflection loss is predominant at the lower frequencies, while absorption loss is predominant at the higher frequencies.* For near-field, magnetic sources the situation is quite different at low frequencies. *Absorption loss is the dominant shielding mechanism for near-field, magnetic sources at all frequencies.* However, both reflection and absorption loss are quite small for near-field, magnetic sources at low frequencies, so that other, more effective, methods of shielding against low-frequency magnetic sources must be used.

There are two basic methods for shielding against low-frequency, magnetic sources: diversion of the magnetic flux with high-permeability materials and the generation of opposing flux via Faraday's law, commonly known as the "shorted turn method". The diversion of magnetic flux with a low-reluctance (high-permeability) path is illustrated in Fig. 11.12(a). Assuming the external medium is free space with $\mu = \mu_0$ and the shield is constructed of a ferromagnetic material having $\mu = \mu_r \mu_0$ with $\mu_r \gg 1$, the magnetic field will tend to concentrate in the low-reluctance ferromagnetic path, and as such will be diverted from affecting the region interior to the shield. In the "shorted-turn" method illustrated in Fig. 11.12(b) a conductor loop such as a wire is placed such that the incident

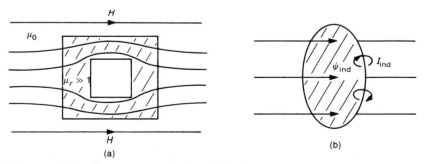

FIGURE 11.12 Two important methods of shielding against low-frequency magnetic fields: (a) using a highly permeable ferromagnetic material to divert the magnetic field; (b) using "bands" to generate an opposing magnetic field.

magnetic field penetrates the surface bounded by the loop, thereby inducing, according to Faraday's law, a current I_{ind} in the loop, and associated magnetic flux ψ_{ind}. This induced magnetic flux is of a polarity or direction as to counteract the original incident magnetic field, and so the net magnetic field in the vicinity of the loop is reduced. There are numerous applications of these two notions. These represent the majority of situations where shielding can be effective in the reduction of the effect of low-frequency magnetic fields.

There are two factors, however, that may degrade the effectiveness of the flux diversion technique and must be kept in mind. These are:

1. *The permeability of ferromagnetic materials decreases with increasing frequency.*

2. *The permeability of ferromagnetic materials decreases with increasing magnetic field strength.*

Manufacturers of ferromagnetic materials tend to specify the relative permeability of the material at a low frequency such as 1 kHz so that the stated value of μ_r may be the largest that will be obtained. For example, Mumetal has a relative permeability of over 10,000 from dc up to around 1 kHz, as shown in Fig. 11.13 [3]. Above 1 kHz the relative permeability of Mumetal decreases dramatically, and above around 20 kHz it is no greater than that of cold-rolled steel. Therefore

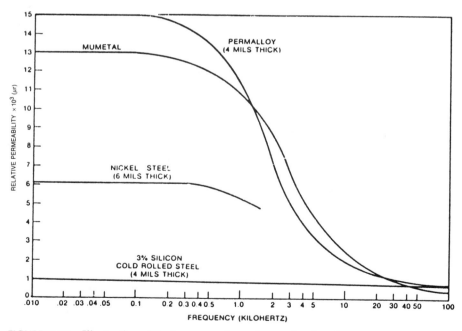

FIGURE 11.13 Illustration of the frequency dependence of various ferromagnetic materials [3].

high-permeability materials such as Mumetal are only effective for magnetic fields below 1 kHz. In order to shield against magnetic fields above 20 kHz, steel is as effective as these high-permeability materials. This is why *shielding enclosures for switching power supplies are constructed from steel rather than Mumetal.* Steel is less expensive than Mumetal and is as effective at the switcher fundamental frequency (20–100 kHz) and the harmonics of the switcher. Shielding the power supply prevents the low-frequency, high-level magnetic fields of the switching transformer from radiating to other PCBs or wires in the product, where they may cause functional problems or be conducted out the power cord, causing conducted emission problems.

On the other hand, shielding against 60 Hz interference is more effective with Mumetal if the field strengths are not too large that they *saturate* the material. This phenomenon of saturation of ferromagnetic materials by high-level magnetic fields was discussed in Chapter 6, and is illustrated in Fig. 11.14(a). The slope of the $B–H$ curve is proportional to the relative permeability of the material. The magnetic field intensity is proportional to the ampere turns if a series of turns of wire carrying current I is wound around the material. Thus high currents tend to give high levels of H where the slope of the curve is flatter, resulting in a lowering of μ_r for high levels of magnetic fields. Thus, even though Mumetal may appear to be an effective shielding material for 60 Hz magnetic fields due to its high relative permeability at this low frequency, this may not be realized, since high currents are usually associated with the 60 Hz power

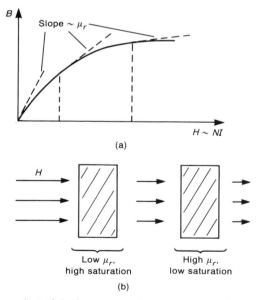

FIGURE 11.14 Illustration of the important phenomenon of saturation of ferromagnetic materials: (a) the $B–H$ curve; (b) use of a multilayer magnetic shield to reduce the effects of saturation.

frequencies. Manufacturers of ferromagnetic materials typically state the *initial* relative permeability of their materials at (1) a low frequency, typically 1 kHz, and (2) low field levels. A typical way of minimizing the saturation effect is to use two shields, as shown in Fig. 11.14(b). The first shield has a low μ_r and low susceptibility to saturation. The purpose of this shield is to reduce the incident magnetic field so that it does not saturate the second shield, which has a high μ_r and a high potential for saturation. Typically, the first layer provides some reflection loss for the electric field.

Although not discussed in Chapter 2, there are some regulatory agencies that impose limits on the low-frequency magnetic field emissions of a product. An example is the VDE regulation for products marketed in Germany. This regulation contains a limit on the magnetic fields radiated from the product. The radiated magnetic fields are to be measured at a distance of 3 m with a loop antenna from 10 kHz to 30 MHz. Low-frequency radiated magnetic fields from transformers of switching power supplies tend to be among the major problems in complying with this legal requirement.

A common application of the "shorted turn" effect in the reduction of magnetic fields is with transformers. A conductive "turn" consisting of a contiguous strip of copper tape is wrapped around the transformer as shown in Fig. 11.15(a). The objective of this "shorted turn" or "band" is to reduce the radiated magnetic field of the leakage flux of the transformer. It is important to place the loop such that the surface bounded by the loop is as perpendicular as possible to the flux that it is intended to cancel, so that the maximum emf will be induced in the band. There are cases where two, orthogonal bands must

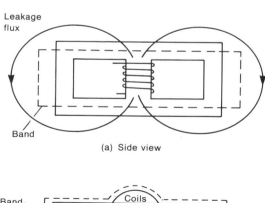

(a) Side view

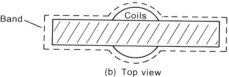

(b) Top view

FIGURE 11.15 Use of the shorted turn effect (bands) to reduce the radiated magnetic fields of transformer leakage flux: (a) side view; (b) top view.

be used. The transformer core is frequently "gapped" in order to reduce the magnetic flux levels and prevent saturation of the core, as discussed in Section 9.3.3. The leakage flux at this gap (which can be quite intense) may lie orthogonal to that from the windings, depending on the location of the gap. If this leakage flux does not penetrate the area bounded by the first band orthogonal to that area, a second band may be needed orthogonal to the first band. This use of bands on transformers, in particular switching transformers of switching power supplies, is quite effective in allowing the product to meet the VDE limit on low-frequency magnetic fields. It is also effective in preventing interference from the transformer, such as with a video monitor that may be placed by the consumer on top of the product and close to the transformer.

Shielding can be an effective suppression method if it is used properly and fits the problem at hand, such as in low-frequency magnetic shielding for power supply enclosures and bands on transformers. Shielding should not be relied on in all instances, since it is all too often misapplied and too much is expected of it.

11.5 EFFECTS OF APERTURES

As was pointed out earlier, there are numerous cases where openings in an otherwise contiguous shield cannot be avoided for practical reasons. One of the more common ones is the need to ventilate the internal electronics for thermal reasons. Fans are frequently employed to move hot air inside the shield to the outside. The reader will observe that these types of openings are frequently in the form of a large number of small holes rather than one large hole. There is an important reason for this, which is illustrated in Fig. 11.16. Consider the

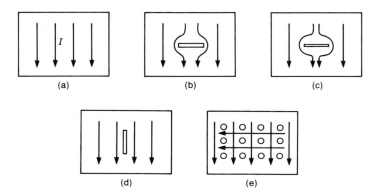

FIGURE 11.16 Illustration of the effects of slots on induced currents on shields. Many small holes provide as much ventilation as one long slit, but perturb the induced current much less, thereby reducing the degradation caused by the aperture.

solid shield shown in Fig. 11.16(a). Currents are induced to flow in this shield, and it is these currents and their associated fields that generate "scattered" fields, which counteract or reduce the effects of incident fields. This was illustrated in Section 5.6.2, where we considered a uniform plane wave incident normally to the surface of a perfect conductor. The incident field induces a surface current, which may be thought of as producing the reflected field. The reflected field is of a polarity such that it tends to cancel the incident field in order to satisfy the boundary condition that the *total electric field* tangent to a perfect conductor must be zero. In order for the shield to perform this cancellation, the induced currents must be allowed to flow unimpeded. Suppose we place a slot in the shield perpendicular to the direction of these induced currents. The slot will interrupt the current flow, and will tend to reduce the shielding effectiveness. The width of the slot does not significantly affect this, as illustrated in Figs. 11.16(b) and (c). On the other hand, if we orient the slot parallel to the diretion of the induced current, the slot will have much less effect on the shielding, as illustrated in Fig. 11.16(d). Since it is not feasible to determine the direction of the induced current and place the slot direction appropriately, a large number of small holes are used instead, as illustrated in Fig. 11.16(e).

It is also necessary to have access doors to allow entry into the shielded enclosure. Where these doors ae closed, a gap is produced around the door that can act as a slot antenna, as discussed previously. Even though the gap opening may be quite small, the radiation potential of the gap can be quite large, as is illustrated by Babinet's principle discussed earlier. For example, suppose a lid is required on top of a shielded enclosure as illustrated in Fig. 11.17(a). Babinet's principle allows the replacement of the gap with a solid conductor whose dimensions are those of the gap. This illustrates rather dramatically that the *length of the gap is more important than its thickness in determining the radiated emissions of the gap.* If the gap length happens to be of the order of a half-wavelength, it is clear from Babinet's principle that the gap has similar radiation potential to that of a half-wave dipole antenna. This is why it is necessary to place many screws at frequent intervals around a lid in order to break up these potential slot antennas, as illustrated in Fig. 11.17(b), since shorter linear antennas tend to radiate less efficiently than long ones. Metallic *gaskets* are frequently used to close gaps as illustrated in Fig. 11.17(c). These are of the form of wire knit mesh or beryllium copper "finger stock." Gaskets should be placed on the inside of any securing screws, since if they are placed outside these screws, radiation from the screw holes will not be protected against.

Openings in shielded enclosures can also be protected by using the "waveguide above cutoff" principle [8]. A square waveguide with side dimensions d has a cutoff frequency for the propagation of higher-order modes TE_{mn} and TM_{mn} given by [8]

$$f_{c,mn} = \frac{v_o}{2d} \sqrt{m^2 + n^2}$$

(11.48)

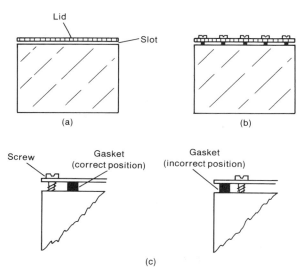

FIGURE 11.17 Illustration of the effects of slots at doors and covers: (a) the slot radiates, according to Babinet's principle, as efficiently as a conductor of the same dimensions as the slot; (b) many closely spaced screws break up the slot length; (c) proper and improper gasket placement to reduce the radiated emissions from screws.

The lowest-order propagating mode is the TE_{10} mode, with cutoff frequency

$$f_{c,10} = \frac{v_o}{2d} \tag{11.49}$$

$$= \frac{1.5 \times 10^8}{d} \quad (d \text{ in m})$$

$$= \frac{5.9 \times 10^9}{d} \quad (d \text{ in inches})$$

The attenuation of a rectangular waveguide to these higher-order modes can be computed by determining the effective attenuation constant for this guide from [8]

$$\alpha_{mn} = \omega \sqrt{\mu_o \epsilon_o} \sqrt{\left(\frac{f_{c,mn}}{f}\right)^2 - 1} \tag{11.50}$$

where $f_{c,mn}$ is the cutoff frequency of the particular mode. Assuming that the frequency of the incident wave is much less than the cutoff frequency for the mode, (11.50) simplifies to

$$\alpha_{mn} \cong \frac{2\pi f f_{c,\,mn}}{v_0} \frac{}{f}$$

(11.51)

$$= \frac{2\pi f_{c,\,mn}}{v_0}$$

Substituting the relation for the cutoff frequency for the lowest-order TE_{10} mode given in (11.49) gives

$$\alpha_{10} = \frac{\pi}{d}$$

(11.52)

The attenuation of a guide of length l is proportional to $e^{-\alpha l}$. Thus the attenuation or shielding effectiveness afforded by one guide is

$$SE_{dB} = 20 \log_{10} e^{\alpha_{10}l}$$

(11.53)

$$= \alpha_{10} l 20 \log_{10} e$$

$$= 27.3 \frac{l}{d}$$

This rather simple result shows that TE and TM modes are strongly attenuated in direct proportion to the guide length as they travel along the guide. This is the basis for the use of *waveguides below cutoff* to allow air flow into a shielded enclosure and at the same time prevent the propagation of high frequencies into the enclosure. Many small waveguides are welded together in a "honeycomb" fashion as illustrated in Fig. 11.18 in order to provide sufficient volume of air flow and to give the appropriate waveguide dimensions.

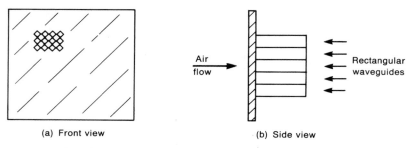

(a) Front view   (b) Side view

FIGURE 11.18 Use of the waveguide-below-cutoff principle to provide ventilation of an enclosure and minimize the electromagnetic field penetrations: (a) front view; (b) side view.

REFERENCES

[1]   E.F. Vance, Electromagnetic-interference control, *IEEE Trans. on Electromagnetic Compatibility*, **EMC-22**, 319–328 (1980).

[2]   C.A. Balanis, *Antenna Theory Analysis and Design*, Harper & Row, NY (1982).

[3]   H.W. Ott, *Noise Reduction Techniques in Electronic Systems*, second edition, John Wiley Interscience, NY (1988).

[4]   S.A. Schelkunoff, *Electromagnetic Waves*, Van Nostrand, NJ (1943).

[5]   R.B. Schulz, V.C. Plantz, and D.R. Brush, Shielding theory and practice, *IEEE Trans. on Electromagnetic Compatibility*, **EMC-30**, 187–201 (1988).

[6]   R.K. Kennan, *Digital Design for Interference Specifications*, The Keenan Corporation, Pinellas Park, FL (1983).

[7]   P.R. Bannister, New theoretical expressions for predicting shielding effectiveness for the plane shield case, *IEEE Trans. on Electromagnetic Compatibility*, **EMC-10**, 1–7 (1968).

[8]   C.R. Paul and S.A. Nasar, *Introduction to Electromagnetic Fields*, second edition, McGraw-Hill, NY (1987).

[9]   E.C. Jordan and K.G. Balmain, *Electromagnetic Waves and Radiating Systems*, second edition, Prentice-Hall, Englewood Cliffs, NJ (1968).

PROBLEMS

11.1   Examine a typical personal computer having a metallic enclosure and list the "penetrations" that allow signals from the inside to pass to the outside, avoiding the shielding of the metallic enclosure.

11.2   Model the shield on a 2 m personal computer printer cable as a monopole antenna (with the metallic structure of he computer as the ground plane) and estimate the maximum radiated emissions at a measurement distance of 3 m if the voltage of the shield attachment point with respect to the metallic structure is a 1 mV, 37.5 MHz signal.[53.5 dBμV/m broadside and parallel to the cable]

11.3   Consult [9] to show that for the complementary slot problem shown in Fig. P11.3 that the impedance of the complementary dipole, Z_d, and that of the slot, Z_s, are related by $Z_s Z_d = \frac{1}{4}\eta^2$ where the surrounding medium has intrinsic impedance η. Use this result to determine the impedance of the slot if $L = \frac{1}{2}\lambda$ and the slot width is infinitesimally small. [$Z_s = (363 - j211)\,\Omega$]

11.4   Determine the skin depth at 30 MHz, 100 MHz, and 1 GHz for steel (SAE 1045). [0.048 mils, 0.026 mils, 0.0082 mils] Repeat for nickel and brass. [0.106 mils, 0.058 mils, 0.0184 mils, 0.93 mils, 0.51 mils, 0.16 mils]

11.5   Compute the intrinsic impedance of steel (SAE 1045) at 30 MHz, 100 MHz, and 1 GHz. [0.202 $\underline{/45°}\,\Omega$, 0.369 $\underline{/45°}\,\Omega$, 1.17 $\underline{/45°}\,\Omega$] Repeat for brass. [3.96 × 10^{-3} $\underline{/45°}\,\Omega$, 7.24 × 10^{-3} $\underline{/45°}\,\Omega$, 2.29 × 10^{-2} $\underline{/45°}\,\Omega$]

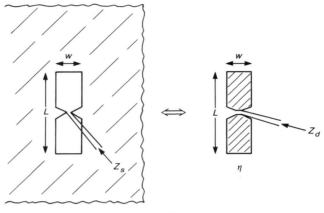

FIGURE P11.3

11.6   Compute the electric field reflection coefficient for an air–steel interface at 30 MHz, 100 MHz, and 1 GHz. [0.999 $\underline{/180°}$ = −0.999, 0.999 $\underline{/180°}$ = −0.999, 0.996 $\underline{/180°}$] Compute the electric field transmission coefficient for an air–steel interface at 30 MHz, 100 MHz, and 1 GHz. [1.07 × 10^{-3} $\underline{/45°}$, 1.96 × 10^{-3} $\underline{/45°}$, 6.2 × 10^{-3} $\underline{/44.9°}$]

11.7   Compute the reflection loss and absorption loss for a 20 mil steel (SAE 1045) barrier at 30 MHz, 100 MHz, and 1 GHz, assuming a far-field source. [53.23 dB, 3656.6 dB, 48 dB, 6676.0 dB, 38 dB, 21,111 dB]

11.8   Compute the reflection loss and absorption loss for a 20 mil steel (SAE 1045) barrier at 10 kHz, 100 kHz, and 1 MHz for a near-field electric source that is a distance of 5 cm from the shield. [188.02 dB, 66.76 dB, 158.02 dB, 211.11 dB, 128.02 dB, 667.6 dB]

11.9   Compute the reflection loss and absorption loss for a 20 mil steel (SAE 1045) barrier at 10 kHz, 100 kHz, and 1 MHz for a near-field magnetic source that is a distance of 5 cm from the shield. [−11.45 dB use $R = 0$, 66.76 dB, −1.45 dB use $R = 0$, 211.11 dB, 8.55 dB, 667.6 dB]

11.10   Discuss the advantageous shielding aspects of a typical paint can that has a press-on lid, a zinc coating, and a tin metal can.

11.11   A transformer consists of 2000 turns of wire with a radius of 5 cm and carries a 50 kHz current at a level of 1 A. Model this as a small loop and compute the magnetic flux density at a distance of 3 m from the loop center in the plane of the loop. [5.8 × 10^{-8} Wb/m^2]

11.12   Determine the length of a waveguide to provide 100 dB attenuation if the guide dimensions are 100 mils × 100 mils. [9.3 mm] What range of frequencies are attenuated by this guide (at least in the waveguide propagation mode)? [dc to 59 GHz]

CHAPTER TWELVE

Electrostatic Discharge (ESD)

Electrostatic discharge (ESD) results from the separation of static charge. There are several ways of producing a separation of static charge. Rubbing two types of insulating materials together can cause charge to be transferred from one material to the other. As the two materials are separated, the charge separation creates intense electric fields and consequently a voltage difference between the two materials. This may result in a breakdown of the intervening air, resulting in an intense arc such as occurs in a "showering arc" when opening a mechanical switch, as discussed in Chapter 6. A direct conduction path may also result if one material comes in contact with a metallic conductor. This phenomenon is a familiar one when we walk across a carpet on a dry day and touch a metallic doorknob. If the resulting charge discharge or arc current enter sensitive electronic circuits, effects ranging from corruption of data to permanent damage may result. Earlier electronic circuitry consisting of vacuum tubes had an inherent high level of immunity to these *ESD events*. The semiconductor integrated circuits in today's electronic circuitry are considerably more susceptible to the effects of ESD. Permanent destruction of the electronic components as well as less destructive but equally troubling loss of data, improper function, etc. are becoming an important problem.

Susceptibility of a product to ESD is as important to the overall EMC of the product as its ability to comply with regulatory requirements on radiated and conducted emissions. A product's successful compliance with the regulatory requirements is of little importance if it fails to operate reliably due to a susceptibility to ESD. Designing ESD immunity into a product is evidently another important aspect of a quality design. In this chapter we will investigate the mechanisms by which the ESD event is produced, as well as it's effects on a product's electronics. We also investigate the various methods for minimizing the effects of the ESD event on the product's electronics.

12.1 ORIGIN OF THE ESD EVENT

If two initially neutral insulators are placed in contact, charge may be transferred from one material to the other. When the materials are separated, they become charged: one negatively and the other positively. The degree to which charge is transferred depends on numerous factors. A general idea of the degree of charge transfer is determined by the *triboelectric series* given in Table 12.1 [1].

TABLE 12.1   The Triboelectric Series.

POSITIVE

1	Air
2	Human skin
3	Asbestos
4	Glass
5	Mica
6	Human hair
7	Nylon
8	Wool
9	Fur
10	Lead
11	Silk
12	Aluminum
13	Paper
14	Cotton
15	Wood
16	Steel
17	Sealing wax
18	Hard rubber
19	Mylar
20	Epoxy–glass
21	Nickel, copper
22	Brass, silver
23	Gold, platinum
24	Polystyrene foam
25	Acrylic
26	Polyester
27	Celluloid
28	Orlon
29	Polyurethane foam
30	Polyethylene
31	Polypropylene
32	Polyvinylchloride (PVC)
33	Silicon
34	Teflon

NEGATIVE

The triboelectric series shows which materials tend to give up electrons easily and become positively charged (those at the top or positive end of the chart) and which tend to accept electrons and become negatively charged (those at the bottom or negative end of the chart). For example, rubbing nylon against Teflon can cause electrons to be transferred from the surface of the nylon to the surface of the Teflon. Therefore the nylon may acquire a positive charge, whereas the Teflon may acquire a negative charge. The degree to which this charge transfer takes place depends on a number of factors, and the triboelectric series is only a rough indicator of this. The order of the two materials in the triboelectric series is an important factor, but it does not completely determine the degree of charge separation. Other factors such as the smoothness of the surface, surface cleanliness, contact surface area, contact pressure, degree of rubbing, and the speed of separation are more important. Charge can also be separated when two like materials are in contact, as with the opening of a plastic bag used to carry produce in a grocery store.

Touching an insulator to a conductor can also create charge separation. However, the degree of resulting charge separation is less than for two insulators because as the conductor is separated from the insulator, the charge tends to redistribute itself over the conductor in relation to the relaxation time of the conductor material. Static charge is stored on the surface of materials. With an insulator the charge tends to remain in the vicinity of the attachment point; whereas for a conductor the charge tends to distribute evenly over the surface. Grounding a conductor will "bleed off" the charge, but grounding an insulator will not. *ESD wrist straps* are often worn to prevent the buildup of charge and hence the possibility of ESD damage to electronic components during their manufacture or installation. A large (1 MΩ) resistance connects the installer's wrist to earth ground. Any static charge stored on the body's skin (a conductor) is discharged to ground through the connection. The large resistance will effectively bleed off the charge, but will prevent large transient discharge currents.

The voltage developed between two objects is a function of the charge on the objects and the capacitance between the two objects according to the usual capacitance relation [2]

$$V = \frac{Q}{C} \tag{12.1}$$

Suppose two materials are in contact, and charge is transferred from one to the other. If the materials are drawn apart, the charge remains on the materials, but the voltage between the two materials changes in accordance with the change in capacitance between the bodies according to (12.1). For close separations the capacitance is large, and hence the voltage between the two objects is small. Increasing the separation decreases the capacitance between them, and hence increases the voltage between them. A charge separation of 1 μC associated with a capacitance of 100 pF will produce a voltage difference between the two materials of 10 kV! At certain separations the intervening air may break down, since the breakdown electric field strength for air is of order 30 kV/cm.

Capacitance is normally thought of as being between two bodies. For example, the capacitance between two parallel plates of area A that are separated a distance d in air is approximately (neglecting fringing of the field at the plate edges) [2]

$$C = \epsilon_o \frac{A}{d} \qquad (12.2)$$

$$= 8.85 \frac{A}{d} \quad \text{(in pF)}$$

For example, two 1 square inch plates separated by 20 mils gives a capacitance of approximately 11 pF. Two concentric conducting spheres of radii R_1 and R_2 with $R_2 > R_1$ have a capacitance between them given by [2]

$$C = \frac{4\pi\epsilon_o}{1/R_1 - 1/R_2} \qquad (12.3)$$

The capacitance of an isolated body can also be defined and calculated. Essentially, when we speak of the capacitance of a single conductor, we are discussing the capacitance between this conductor and infinity. Nevertheless, an estimate of the capacitance of a single conductor can be obtained from the result in (12.3) by allowing the outer sphere to recede to infinity, i.e., allowing $R_2 \to \infty$, giving

$$C = 4\pi\epsilon_o R \qquad (12.4)$$

$$= 111 \, R \quad \text{(in pF)}$$

An object the size of a marble has a capacitance of approximately 1 pF (with respect to infinity). If we approximate the human body as a sphere of diameter 1 m, it has a capacitance of some 50 pF. Typical values of the capacitance of the human body range from 50 pF to 100 pF. If the isolated body is placed close to other objects, there will be additional capacitances between points on the body and those objects (mutual capacitances).

Charged insulators are not in themselves the problem, because the charge on an insulator is not free to move. The problem in ESD is the inducement of charge on a conductor and having that conductor approach another conductor, resulting in an intense arc discharge between the two conductors. As indicated earlier, charge is not very mobile on the surface of an insulator. However, charge on the surface of a conductor is very mobile, and tends to move about to

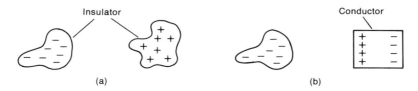

FIGURE 12.1   Illustration of the charging of a conductor by induction: (a) charge separation by contact between two insulators; (b) charge separation on a conductor by induction.

distribute the charge equally over that surface. Charge separation on the surface of a conductor can be caused by *induction* by bringing a charged body near the conductor. Consider charging two insulators by contact and then separating them as shown in Fig. 12.1(a). Now suppose the negatively charged body is placed next to a neutral conductor as in Fig. 12.1(b). The electric field produced by the charge on the charged body will induce a separation of charge on the surface of the neutral conductor. Charge of polarity opposite to that on the insulator will move to the side of the conductor facing the charged insulator. Opposite charge will move to the opposite side of the conductor. This creates separation of charge on the surface of the conductor.

The ESD event is generally a result of the following sequence [3, 4]. First, charge is placed on an insulator by contact. The charged insulator induces a charge separation on a conductor by induction. This charged conductor is moved near another conductor (grounded or not), and a discharge or arc occurs from one conductor to the other. This is illustrated by walking across a nylon carpet with rubber-soled shoes and then touching a computer keyboard. Electrons are transferred from the carpet to the rubber shoes, leaving a positively charged footprint on the carpet and a negative charge on the soles of the shoes. This induces a charge separation (by induction) on the body (a conductor). Positive charge is induced on the soles of the feet in response to the nearby negative charge on the soles of the shoes, and negative charge moves to the upper parts of the body such as the hand. When the negatively charged finger approaches the keyboard, electrons will be drawn through the ground of the keyboard cable and the Green Wire of the power cord, leaving a net positive charge on the keyboard. As the finger approaches the keyboard, the charge separation between the finger and the keyboard will produce an intense electrostatic field, which can induce voltage differences between points in the product, causing destruction of electronic components. An arc may also be initiated from the finger to the product similar to a miniature lightning stroke as the intense electric field causes the air to break down. This discharge current may pass through the computer and its internal circuitry, resulting in possible damage to its components or degradation of its function. The speed of approach is an important factor in determining the intensity of the discharge. Formation of the arc requires more time than the discharge. During the formation of the arc, a fast approach will give a narrower arc gap length, resulting in a more intense discharge with faster rise times and peak currents. It is important to note that ESD voltage discharges of less than 3500 V cannot be felt or seen by humans yet are capable of creating functional problems or even component destruction in electronic circuits.

Figure 12.2 shows typical waveforms of the discharge current in an ESD event. The discharge circuit is represented by a charged capacitance C with an initial voltage V_0 and resistance R of the discharging body. Inductances of the charged body as well as the Green Wire of the product are represented by inductor L. These inductances as well as the resistance cause the arc current waveform to be either overdamped or underdamped. The circuit shown in Fig. 12.2(a) is a somewhat simplistic model, but is sufficient for illustrating the basic phenomenon. More elaborate models are given in the literature [3]. We have been discussing "personnel discharge," i.e., from the human hand to the product. There are other types of discharge that occur when a metallic chair is moved across a carpet, creating a separation of charge on the chair. When the chair approaches either the product or a metallic table on which the product is placed, a "furniture discharge" may occur. Typical immunity tests try to create personnel as well as furniture discharges [5]. A furniture discharge typically has smaller resistance in Fig. 12.2(a), resulting in underdamped waveforms, whereas personnel discharge typically results in overdamped

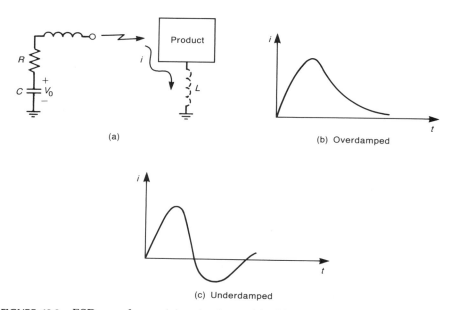

(a)

(b) Overdamped

(c) Underdamped

FIGURE 12.2 ESD waveforms: (a) a simple model of human body discharge; (b) the overdamped waveform (typical of human discharges); (c) the underdamped waveform (typical of furniture discharges).

waveforms because of the higher resistance. Typical rise times are of order 200 ps–70 ns, with a total duration of around 100 ns–2 μs [3]. The peak current levels may approach tens of amps for a voltage difference of 10 kV. This indicates that *the spectral content of the arc may have large amplitudes, and can extend well into the GHz frequency range!* Problems 12.4 and 12.8 investigate the spectral content of typical ESD waveforms.

The simple R–L–C circuit of Fig. 12.2(a) is said to be a single-discharge model. It has been observed that typical ESD discharges can have multiple discharges within an event. It is interesting to conjecture that this multiple-discharge ESD event is similar to the showering arc discharge associated with the opening or closing of a mechanical switch discussed in Chapter 6.

This simple scenario represents the typical ESD event. Several mitigating effects will cause the precise discharge generated by one person to be different from that produced by another. For example, as the person walks across the carpet, contact resistance between the soles of the shoes and the carpet will cause a reverse charge flow to occur, which will reduce the charge separated. High humidity will decrease the contact resistance of this path, and hence increase the reverse charge flow. Thus less charge is built up in a humid environment, and hence the ESD event will be of a smaller degree. Antistatic sprays are available that when sprayed on a carpet will tend to increase the conductivity of the shoe—carpet path, facilitating the reduction of stored charge on the body in the same way as high humidity. Nevertheless, the engineer cannot rely on the mitigating effects of such factors as humidity in all potential product installations, so that ESD protection must be included in the product design.

12.2 EFFECTS OF THE ESD EVENT

There are basically two primary effects associated with an ESD event:

1. *The intense electrostatic field created by the charge separation prior to the ESD arc.*
2. *The intense arc discharge current.*

The extreme differences of potential caused by the intense electrostatic field created by the charge separation prior to the ESD arc can overstress dielectric insulations of electronic components causing their destruction. We will focus on the effects of the arc discharge.

The intense arc discharge current can cause problems ranging from functional upset to component destruction via four secondary processes:

1. *Direct conduction through the electronics.*
2. *Secondary arcs or discharges.*

3. *Capacitive coupling to the product's electronic circuits.*
4. *Inductive coupling to the product's electronic circuits.*

Discharging a large ESD current through the electronics can evidently cause direct damage through thermal heating or can produce large potential differences that can also cause dielectric breakdown, resulting in component destruction. Arc discharges to exposed metallic parts of the product enclosure can result in *secondary discharges* to the interior electronics. The arc current also creates electric and magnetic fields that couple to and induce voltages and currents in the circuits on PCBs and cabling within the product. These radiated emissions are predominantly a near-field phenomenon due to the proximity of the components to the arc. In high-impedance circuits the large voltages create capacitive coupling to the electronics. In low-impedance circuits the large discharge currents create inductive coupling to the electronics. This represents the two mechanisms by which an ESD arc discharge can create functional problems:

1. *Conduction.*
2. *Radiation.*

Conduction tends to produce both malfunction and damage in the electronics. Radiation (near-field radiation here) tends to only cause upset, but may also cause damage. The electromagnetic wave created by the arc discharge current may also couple to any peripheral cables, and be subsequently conducted into the internal electronics. Thus there may be a combination of these two mechanisms: radiation followed by conduction. Ordinarily, the term conduction refers to the direct conduction of the arc discharge current through the electronics.

12.3 MITIGATION DESIGN TECHNIQUES

If it were possible to place the electronics in a contiguous metal box having no points of entry such as cables or apertures, the arc discharge would pass along the box exterior through the Green Wire connection to ground, thus causing no interference or damage to the components inside the box. In reality, the green wire connection will have a substantial amount of inductance associated with it due to its length. Thus the potential of the box will rise with respect to ground due to the voltage drop across the Green Wire inductance as the discharge current flows through it. If the enclosure is connected to ground via the Green Wire, the voltage of the enclosure can rise to several thousand volts. If it is not grounded, the voltage of the enclosure can rise to that of the source, which is at most about 25 kV. However, the interior circuitry will also rise to this potential if it is surrounded by a contiguous metallic enclosure having no penetrations. Because there is no difference in potential between different parts of the circuitry, no functional upset would occur. Also, the arc current does not pass through the circuitry, so no damage would occur. Contiguous plastic enclosures having no penetrations also do not permit entry of an arc discharge, and therefore protect against it.

In practice the enclosure will have numerous penetrations or points of entry such as cables, the power cord, and air vents, as illustrated in Fig. 12.3(a). All of these points of entry allow the effects of the ESD event to pass into the interior of the enclosure. Intense electric and magnetic fields around these penetrations may result in secondary discharges or field coupling to the internal circuitry. The ESD discharge current may find a lower-impedance path to ground through the electronic circuitry, resulting in damage. Even if the discharge current does not pass directly through the electronic circuitry, the fields may couple to the circuitry via capacitive and/or inductive coupling, causing operational problems.

Today's electronic products are often housed in plastic rather than metallic enclosures. These plastic enclosures can be coated along their interior with a conductive paint or have metallic fibers imbedded to give some degree of shielding. For those products where this is not feasible or effective the internal circuitry is exposed to the intense electromagnetic fields generated by the event, as shown in Fig. 12.3(b). Arc discharge to keyboards and other exposed parts

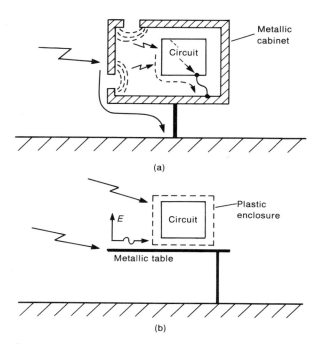

(a)

(b)

FIGURE 12.3 Illustration of the coupling of an ESD discharge to an electronic circuit for (a) a metal enclosure with apertures and (b) a plastic enclosure.

of the system can conduct the arc current into and through the electronic circuitry, resulting in malfunction or the destruction of components. Even if the discharge does not pass directly through the circuitry, the event essentially bathes it in a travelling wave that couples into that circuitry, possibly causing functional upset.

There are essentially three techniques for preventing problems caused by an ESD event:

1. *Prevent occurrence of the ESD event.*
2. *Prevent or reduce the coupling (conduction or radiation) to the electronic circuitry of the product (hardware immunity).*
3. *Create an inherent immunity to the ESD event in the electronic circuitry through software (software immunity).*

Some or all of these techniques may be appropriate for a system, whereas some may be impractical from the standpoint of cost to implement or the anticipated environment of the installation.

12.3.1   Preventing the ESD Event

Electronic components such as ICs are placed in pink polyethylene bags or have their pins inserted in antistatic foam for transport. The pink polyethylene bags have a lower surface resistivity (of order $10^{12}\ \Omega/\text{square}$) than ordinary insulating materials, which allows them to redistribute charge rapidly. The static charge distributes itself over the bag surface quickly, preventing concentration of charge. Insulators typically have surface resistivities greater than $10^{14}\ \Omega/\text{square}$ and cannot redistribute the charge rapidly, resulting in charge separation, which may cause an ESD event by, for example induction.

Some products can utilize charge generation prevention techniques. For example, printers constantly roll paper around a rubber platen. This causes charge to be stripped off the paper, resulting in a buildup of static charge on the rubber platen. Wire brushes contacting the paper or passive ionizers prevent this charge buildup. In many other applications, preventing charge buildup is usually not a practical measure, since the engineer has little control over the installation, and so the design must mitigate the effects of the ESD event.

12.3.2   Hardware Immunity

The various hardware immunity techniques attempt to prevent or reduce the effects of the four basic ESD coupling mechanisms:

1. *Secondary arc discharges.*
2. *Direct conduction.*
3. *Electric field (capacitive) coupling.*
4. *Magnetic field (inductive) coupling.*

ESD hardware immunity design generally incorporates all of these considerations.

Secondary arc dischargees are prevented by (a) grounding exposed metallic parts of the enclosure to chassis ground and/or (b) *insulating* the exposed part from the nearby electronics. *All metallic parts of the enclosure that are exposed should be connected to chassis (Green Wire) ground to prevent secondary arc discharges. Interior electronics should be separated from ungrounded parts of the enclosure by 1 cm and from grounded parts by 1 mm to further prevent secondary arc discharges to the electronics.* These recommendations result from the following considerations. Suppose a metallic item of the enclosure such as a decal or nameplate is isolated. If the operator touches this part, the resulting charge transfer raises its potential, creating a potentially large electric field between the part and nearby electronics. If the electronics is sufficiently close to the part, the intervening air may breakdown, resulting in a *secondary arc discharge* to the electronics. Secondary discharges of this type can result in more severe arc currents than can arc discharges from the human body, since the resistance of the metallic part is much smaller than that of the human body. The breakdown electric field strength in air is of order 30 kV/cm, and the human body can be charged to at most around 25 kV. An ungrounded part can rise in potential to the potential of the charged body. Therefore the maximum voltage between the exposed metal part and nearby electronics is some 25 kV. In order to prevent air breakdown between these parts and adjacent electronics and a secondary discharge, the electronics should be separated by about $d_{min} = 25\text{ kV}/(30\text{ kV/cm}) \cong 1$ cm, as illustrated in Fig. 12.4(a). If the metal part is grounded, the voltage across the inductance of the Green Wire ground due to the passage of the ESD discharge current through it and consequently the voltage of the part may rise to about 1500 V. Therefore the minimum separation should be of order $d_{min} = 1500\text{ V}/(30\text{ kV/cm}) \cong 1$ mm. These separation distances were predicated on air being the insulation medium. Other insulation media (such as Mylar) have much higher breakdown voltages, and so the minimum separation can be reduced. Another way of preventing this secondary arc discharge is to lengthen the discharge path, as with overlapping joints as illustrated in Fig. 12.4(b). Still another technique is to use a secondary shield connected to circuit ground to break up the capacitance between the part and the adjacent electronics, as shown in Figure 12.4(c). Plastic parts such as knobs can accumulate charge. The effect of secondary arc discharges from these parts is reduced by placing a grounded metallic shield behind them to safely conduct any discharge away from the sensitive electronics, as Fig. 12.4(d) illustrates. These are frequently referred to as *spark arrestors* [1].

The first priority in designing for ESD hardware immunity is to *prevent the ESD discharge current from flowing through sensitive circuitry by direct conduction.* There are essentially two ways to accomplish this goal. The first method is to *block the discharge path through the circuitry.* Insulating the circuitry as discussed above is one way of blocking this path. The second method is to *divert the arc discharge current around the electronics to prevent it from flowimng through the circuitry.* If the product enclosure is metallic, the enclosure can be used to *divert the discharge current to ground* as depicted in Fig. 12.3(a). Any penetrations

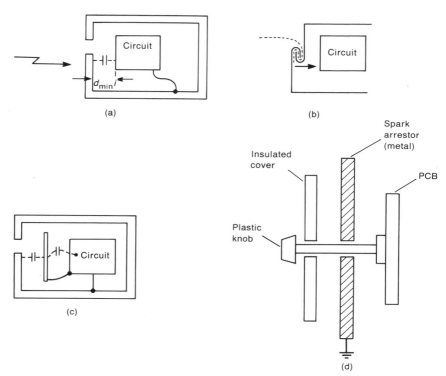

FIGURE 12.4   Methods of reducing secondary arcing: (a) circuit insulation by separation; (b) lengthening the discharge path; (c) use of a secondary shield; (d) a spark arrestor.

(e.g., apertures and cables) create possible paths through the interior circuitry for the arc current. Properly treating penetrations in metallic enclosures is critical to making the best use of the metallic enclosure. This is virtually no different than preventing electromagnetic fields internal to a shielded region from exiting the shield, as discussed in Chapter 11. Holes in the shield for air vents, etc. should be replaced with many small holes. The degree of coupling through a hole is again more dependent on the maximum dimension than on the hole area. Thus long, narrow slots such as at panels and doors should be broken into many shorter slots with closely spaced screws or with metallic gaskets. The electromagnetic fields generated by the ESD event as the current flows on the surface of the enclosure tend to be the most intense around apertures. Therefore sensitive circuitry should not be placed near any aperture regardless of whether the aperture has been properly treated or not. Virtually all of the techniques that were discussed in Chapter 11 to prevent the degradation of shielding effectiveness should be used and are effective in preventing ESD coupling to the interior of the enclosure. Once these penetrations are properly treated, the only other access points are via the cables.

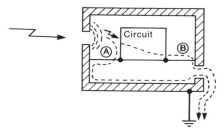

FIGURE 12.5   Effect of geometry on the discharge path.

Discharges that penetrate a metallic enclosure due to improperly treated apertures have the potential for passing through the electronic circuitry causing damage. The paths for these currents are quite complicated and not immediately obvious. The problem is more one of *geometry* [6]. For example, consider Fig. 12.5, where the circuit internal to a metallic enclosure has two possible connections of circuit ground to the enclosure, A and B. A discharge at the first aperture will probably find a lower impedance path through A, whereas connection B will be more likely to cause the discharge current to flow through the circuitry. Determining the path of lowest impedance is difficult, since the ESD pulse has a large spectral content. Even though the discharge current may not flow directly through the circuitry, the intense fields through path A can couple to the circuitry, causing malfunctions. A single-point ground to the enclosure when properly located serves to divert the ESD current around the sensitive electronics. Again, as with determining lowest-impedance paths on PCBs, it is important to think in terms of the *high-frequency impedance* of paths rather than the dc resistance, since the ESD waveform contains frequencies well into the GHz range. Determining the lowest-impedance path at these very high frequencies is very difficult and not subject to intuition.

In any event, *all circuit grounds internal to a metallic enclosure should be connected to the enclosure to prevent the circuit potential from differing from that of the enclosure.* For example, consider Fig. 12.6, where an external cable

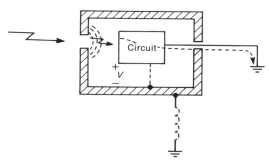

FIGURE 12.6   Illustration of the principle of preventing differences in potential during an ESD discharge by connecting all components to chassis (Green Wire) ground.

penetrates the shield. The Green Wire ground of the power cord will have a rather large inductance. During an ESD event, the passage of the ESD arc discharge current through the inductance of the Green Wire will cause the potential of the enclosure to rise above ground by thousands of volts. If the internal circuitry is not connected to this enclosure, the potential of the circuitry will not be at the same potential as the enclosure, creating the possibility of large potential differences and secondary arcing to the circuitry. Nonmetallic enclosures are also susceptible to this same problem. Arc discharges flowing through the inductances of various paths can create large potential differences between the various electronics, which can cause secondary arcing or dielectric breakdown. Therefore all circuit grounds should be connected (single-point ground) together to prevent these potential differences and to bleed off the accumulated charge. If the product is a three-wire product, the connection should be made to the Green Wire (chassis) ground. It should be pointed out that these single-point connections may cause problems with radiated emissions when the high-frequency potential of the common connection point is nonzero. Nevertheless, if this method is necessary for ESD protection, additional techniques for reducing this radiated emission discussed earlier should be employed.

Once apertures (metal or plastic) enclosures have been properly treated, the cables are the primary mechanisms that allow the effects of the ESD event to penetrate to and affect the internal circuitry. Peripheral cables (including the power cord) are usually quite long, and act as efficient antennas to pick up and couple the fields produced by the ESD event to the interior of the enclosure and thus to the electronic circuitry. Removing all cables (except the power cord) in an ESD test can often determine whether these are the primary points of entry. Shielding all peripheral cables may or may not be an effective prevention measure against coupling the ESD event to the internal electronics. If the cable shield is bonded 360° to the enclosure as illustrated in Fig. 12.7(a), it forms an extension of the metallic enclosure and thus is effective. If the shield is terminated in a pigtail as illustrated in Fig. 12.7(b), the ESD discharge current flowing through the inductance of the pigtail (approximately 15 nH/inch) can create a large voltage difference between the shield and the other enclosure. This couples to the interior wires and subsequently to the internal circuitry.

Use of shielded cables employing 360° bonding is often impractical from a cost standpoint. This is particularly true for plastic enclosures. There remain two basic methods of preventing these signals from entering the circuitry: (1) *blocking* and (2) *diverting* these signals.

Blocking these signals can be accomplished by placing an impedance in the path. ESD events tend to produce both common-mode and differential-mode currents on the peripheral cables. In order to block the induced common-mode currents, we can use (1) common-mode chokes, (2) optical couplers, or (3) place impedances in series with *every wire of the cable, including the ground wires.* Of these three, the simplest and often most effective is the common-mode choke illustrated in Fig. 12.8. A common-mode choke is helpful in preventing

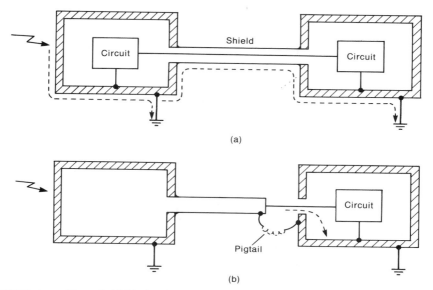

FIGURE 12.7   Use of shielded cables to exclude ESD coupling: (a) circumferentially bonded shields; (b) degradation of shielding due to pigtails.

common-mode currents from entering a product, as well as from exiting the product on these cables. Consequently, common-mode chokes serve a dual role: they tend to prevent radiated emission problems as well as ESD problems due to common-mode currents. In this regard it is important to remember that *all conductors of the cable* (*including the ground wires and the shield pigtail*) *must be routed through the common-mode choke*, as discussed in Chapter 6. Bypassing the common-mode choke with the cable shield connection can effectively defeat the effect of the choke. Common-mode chokes are available in DIP form that allow easy mounting in PCBs where the cables enter.

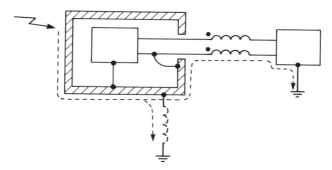

FIGURE 12.8   Use of a common-mode choke to prevent coupling of common-mode ESD signals between subsystems on cables.

The other method of preventing ESD-induced currents from penetrating enclosures via peripheral cables is to *divert them*. This method is critically dependent on *geometry*. For example, placing a capacitor between a signal wire and ground tends to divert the induced current to the enclosure, as shown in Fig. 12.9(a). The effect of the capacitor on the signal current on the cable must also be considered. The capacitor technique will be effective if the spectral content of the ESD current is higher than that of the signal current. Line-to-line and line-to-ground capacitors may be used to divert differential-mode and common-mode ESD currents, respectively, as in power supply filters. In some cases multi-element filters such as Tee or Pi configurations may be necessary. Once again, *geometry* is critical. If the enclosure ground connection is at some distance from this entry point, the currents may find another lower-impedance and undesired path. This is another important reason for *placing all cable connectors at one place on a PCB*. It tends to prevent the formation of lower-impedance paths that are not intended, as illustrated in Fig. 12.9(b). Connecting all ground conductors and diversion devices to the enclosure at one point tends to prevent other less easily observed, lower-impedance paths.

As with other uses of diversion elements such as capacitors, the impedance that they are in parallel with determines their effectiveness. Placing capacitors across low-input-impedance devices will probably be ineffective due to current division, as discussed previously. For low-impedance inputs the technique of *blocking* by placing an impedance in series will be more effective. In this case a ferrite bead will tend to cause less deterioration of the intended signal, while blocking the higher-frequency components of the ESD signal. On the other hand, a capacitor will serve to effectively divert currents from high-input-impedance devices.

Another method of diversion is the use of *clamping devices* such a zener diodes, often referred to as *transient suppressors*. This is illustrated in Fig. 12.10(a). When the input voltage exceeds the threshold of either device, it breaks down. These transient suppressors tend to have activation times inversely proportional to their current-carrying capability: high-current devices have slower response times. Frequently a parallel combination of a low-current,

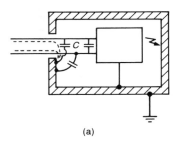

(a)

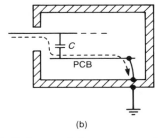

(b)

FIGURE 12.9   Use of parallel capacitors to divert ESD discharges: (a) proper capacitor placement; (b) effect of nonlocal grounding of capacitor forcing discharge through the PCB ground.

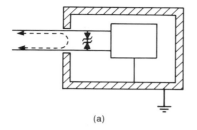

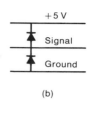

(b)

(a)

FIGURE 12.10 Use of diodes to clamp ESD-induced voltages to safe levels: (a) a zener diode; (b) back-to-back diodes at circuit inputs to prevent overvoltages.

fast-response device and a high-current, slow-response device is employed. Diodes can be used as shown in Fig. 12.10(b) to ensure that input voltages to components remain within safe levels. If the signal line voltage exceeds $+5$ V, the upper diode turns on. Conversely, if the signal voltage attempts to go negative, the lower diode turns on.

The parasitic lead inductance of both the capacitor and the transient absorber are again critical to their success in diverting the ESD signal. It is important to remember that the ESD signal spectral content extends well into the GHz range, so that a small amount of lead inductance can prevent the low-impedance path that these devices are intended to provide.

The previous discussions have concentrated on products that have metallic enclosures where the apertures have been properly treated. Products that are housed in plastic enclosures can utilize these techniques for preventing the penetration of the ESD signal on peripheral cables, but more care is required. In this case we do not have the advantage of a large metallic path. It is of benefit to have a large metallic plane beneath the product, to be used as a diversion for the ESD current and to provide bypass capacitance. It is also important that the plane be connected to all metal parts, including the Green Wire ground.

It is also important that (1) *the electronic grounds of all peripheral cable connectors be connected to this ground plane where the connector enters the PCB* and (2) *all PCBs be placed close to and parallel to this ground plane.* The first principle is important, because we wish to avoid creating large voltage differences between the cable wires and the ground plane, as illustrated in Fig. 12.11. The reason for the second principle is somewhat subtle but effective. Consider an ESD wave propagating across the table on which a product is placed, as illustrated in Fig. 12.12(a). If the table is metallic, the electric field of the wave

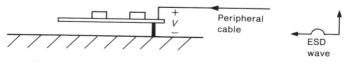

FIGURE 12.11 Illustration of the importance of a local ground where peripheral cables enter the product.

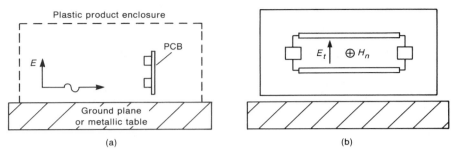

FIGURE 12.12   Illustration of the effect of PCB orientation on their susceptibility to ESD fields: (a) problem dimensions; (b) the ESD fields are oriented with respect to the circuit loop for maximum susceptibility.

near the surface of the table must be perpendicular to the table surface in order to satisfy the boundary conditions of zero tangential electric field on the surface of a perfect conductor. Placing the PCB in the product in the vertical direction as illustrated in Fig. 12.12(b) will mean that the incident electric field will tend to be in the plane of any circuit loops on the PCB (transverse to the transmission line axis). The magnetic field of the ESD wave will possibly be perpendicular to the plane of the circuit loops on the PCB. In Chapter 8 we showed that transmission lines pick up incident field most effectively when the electric field is transverse to the line axis, and the magnetic field is perpendicular to the plane of the loop. This is precisely the situation when the PCB is oriented vertically to the bottom of the product. A better placement of PCBs is horizontal and near the bottom and ground plane of the product. For this placement the electric field will be perpendicular to the plane of the circuit loops, and the magnetic field will be parallel to the plane of the circuit loops. Thus, theoretically, the ESD wave will not couple to the circuit loops on the PCB, since the ground plane will force the electric field lines to be perpendicular to the ground plane (at least close to the ground plane). Placing the PCB very close to the ground plane will insure that any bending of the electric field lines from vertical will be minimized in the vicinity of the circuit loops. This also increases the bypass capacitance to the ground plane. The same holds true for the magnetic field: it will tend to be parallel to the ground plane (near the ground plane).

Proper layout of PCBs in metallic as well as nonmetallic enclosures helps to prevent electric field (capacitive) and magnetic field (inductive) coupling to those circuits from any ESD-induced fields interior to the product. As in the minimization of radiated emissions from a PCB, *loop areas should be kept small to reduce the pickup. Conductor lengths should also be minimized for the same reason.* [3, 4, 7] Ground grids on the PCBs (discussed in Chapter 13) also tend to help. These techniques are very similar to those used to minimize pickup of incident fields as well as radiated emissions from the currents on the conductors, as discussed in Chapter 8. Providing adjacent returns in close proximity to signal lines in a Ground–Signal–Ground–··· or Ground–Signal–Signal–

Ground—··· arrangement tends to minimize the loop areas and associated reception of the fields.

Figure 12.13 illustrates the minimization of loop areas with proper PCB layout. In Chapter 8 we found that the emf induced in a loop by a magnetic field could be represented as a voltage source in that loop that is proportional to the area of the loop. In order to minimize this magnetic field coupling, we need to minimize the loop area. Electric field reception is also dependent on loop area. The electric field induces a current source between the two conductors that is proportional to the loop area. One of the common situations where large loop areas can result is in the $+5$ V–Ground power distribution. Figure 12.13(a) shows a poor routing of the $+5$ V and Ground lands on a PCB. A large loop area that is susceptible to the pickup of ESD-induced fields results. Figure 12.13(b) shows a *no-cost reduction* of this loop area and pickup. Innerplane or multilayer PCBs can be used to further reduce this loop area, as shown in Fig. 12.13(c). The $+5$ V and Ground is distributed around the board on planes that are buried at various depths in the PCB. Connections are made between the IC pins and these planes with "vias." This method permits keeping

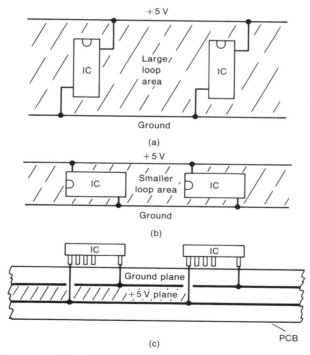

FIGURE 12.13   Reduction of loop area in power distribution circuits to reduce pickup of the ESD field: (a) large loop area; (b) smaller loop area; (c) use of an innerplane (multilayer) board to reduce the loop area and the characteristic impedance of the power distribution circuit.

the area between +5 V and Ground small (between the two planes). It also reduces the inductance of the +5 V and Ground paths, since these planes form a parallel-plate transmission line, as discussed in Chapter 6. This lowers the characteristic impedance of the +5 V–Ground transmission line and hence the voltage drop across these paths. The proximity of the planes also increases the capacitance between the two planes, which also reduces the effect of any induced voltages, as with the decoupling capacitors to be discussed in the next chapter.

Large loop areas may also be formed in like fashion with the signal lines, as illustrated in Fig. 12.14(a). Usually these loop areas are more troublesome than the +5 V–Ground loop areas from the standpoint of ESD susceptibility, since the signals induced in this loop are directly applied to the inputs of the modules. The technique for reducing this loop is to route a ground land adjacent to the signal land as shown in Fig. 12.14(b). With a large number of signal lands on a PCB, this use of dedicated returns may not be practical. A ground

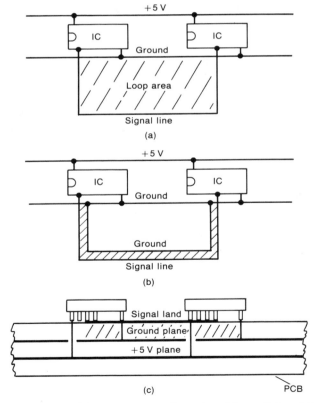

FIGURE 12.14   Reduction of loop areas to reduce the pickup of signal lines: (a) large signal-return area; (b) smaller loop area using a dedicated return; (c) use of an innerplane (multilayer) board to reduce the loop area of the signal-return path.

grid, discussed in Chapter 13, will accomplish the same result, and require less board "real estate." An even better way of reducing the Signal–Ground loop area is with an innerplane or multilayer board, as illustrated in Fig. 12.14(c). Innerplane boards are more expensive than double-sided boards, and so may not be feasible in low-cost products. A well-designed ground grid on a double-sided board can perform as well as an innerplane board, and is much less expensive [1]. Nevertheless, *loop areas must be minimized to prevent ESD susceptibility as well as radiated emissions.*

12.3.3   Software Immunity

Software should be designed to continually verify that faults are detected and recovery action taken. Unlimited wait states should not be used, otherwise an ESD event can cause the product to "lockup." Software should include "watchdog" routines that periodically check whether program flow is correct. If an ESD event has interrupted program flow, a recovery routine may be initiated before any damage is done.

The use of parity bits, checksums and error-correcting codes can prevent the recording of ESD-corrupted data. They cannot restore data, but if correction checks are initiated frequently, much of the stored data will not need to be retransmitted.

Although not properly classified as a software immunity technique, all unused module inputs should be tied to ground or $+5$ V to prevent false triggering by an ESD event. Also, edge-triggered inputs are very susceptible to "spikes" caused by an ESD event. Thus all inputs should be latched and strobed. This reduces the likelihood that the occurence of an ESD event will happen simultaneously with the latching of data and be incorrectly recorded as valid data.

REFERENCES

[1]   H.W. Ott, *Noise Reduction Techniques in Electronic Systems*, second edition, John Wiley Interscience, New York (1988).

[2]   C.R. Paul and S.A. Nasar, *Introduction to Electromagnetic Fields*, second edition, McGraw-Hill, New York (1987).

[3]   W. Boxleitner, *Electrostatic Discharge and Electronic Equipment*, IEEE Press, New York (1989).

[4]   W. Boxleitner, How to defeat electrostatic discharge, *IEEE Spectrum*, pp. 36–40 (August 1989).

[5]   R.J. Calcavecchio and D.J. Pratt, A standard test to determine the susceptibility of a machine to electrostatic discharge, *IEEE International Symposium on Electromagnetic Compatibility*, San Diego, CA, September 1986.

[6]   D. Tran, Some misconceptions in ESD design and testing, *EMC Technology*, pp. 42–48 (November/December 1989).

[7] M. Mardiguian, *Electrostatic Discharge*, Interference Control Technologies, Inc., Gainesville, VA (1986).

PROBLEMS

12.1 If wool is rubbed against Teflon and the two materials separated, what will be the likely charge on the Teflon? [Negative]

12.2 A metallic plate beneath a compact typewriter keyboard has dimensions of 6 inches × 18 inches and is situated an average distance of 1 inch above a large ground plane beneath the typewriter. Determine the capacitance between the two metallic planes. [24 pF]

12.3 Estimate the capacitance of a typical passenger automobile. [333 pF, assuming a sphere of radius 3 m]

12.4 Use SPICE to provide simulations of the ESD current using the R–L–C circuit of Fig. 12.2 with an initial capacitor voltage of 20 kV and (a) $R = 300\ \Omega$, $C = 50$ pF, $L = 0.1\ \mu$H, (b) $R = 1$ kΩ, $C = 250$ pF, $L = 0.1\ \mu$H, and (c) $R = 50\ \Omega$, $C = 100$ pF, $L = 0.5\ \mu$H.

12.5 Investigate the discharge time of a wrist strap having 1 MΩ resistance and length 1 m (15 nH/inch) using SPICE. Model the body as a 50 pF capacitance with a stored voltage of 10 kV.

12.6 Disassemble (carefully) a personal computer, an electronic typewriter and a printer and note the ESD protections that are built into the product as well as those protections that are not provided. How would you "repackage" the product to minimize susceptibility to ESD. What suppression components would you add? Pay close attention to geometry.

12.7 Comment on the ESD protection for the products investigated in Problem 12.6 as to the treatment of peripheral cables.

12.8 In order to investigate the spectral content of ESD waveforms, calculate the Fourier transform of the ESD current pulses of Problem 12.4.

$$\left[(a)|\hat{I}(j\omega)| = \frac{10^{-6}}{\sqrt{(\omega^2/(2 \times 10^{17}) - 1)^2 + 2.25 \times 10^{-16}\omega^2}}, \right.$$

$$(b)|\hat{I}(j\omega)| = \frac{5 \times 10^{-6}}{\sqrt{(\omega^2/(4 \times 10^{16}) - 1)^2 + 6.25 \times 10^{-14}\omega^2}},$$

$$\left. (c)|\hat{I}(j\omega)| = \frac{2 \times 10^{-6}}{\sqrt{(\omega^2/(2 \times 10^{16}) - 1)^2 + 2.5 \times 10^{-17}\omega^2}} \right]$$

Plot these spectra and determine the frequencies where the spectrum is reduced by 6 dB from its dc value. [(a) 100.538 MHz, (b) 44.96 MHz, (c) 31.8 MHz] Determine the energy in each pulse. [(a) 3.333×10^{-5} J, (b) 5×10^{-5} J, (c) 4×10^{-4} J] Compare this with the initial energy stored in the capacitor. [(a) 0.01 J, (b) 0.05 J, (c) 0.02 J] Where does the energy go? [Dissipated in the resistor]

12.9 In order to investigate the coupling of ESD arc currents into electronic circuitry, assume a transmission line 10 cm in length with wire separations of 1 cm and $R_{NE} = R_{FE} = 1$ kΩ lies parallel to one side of the $R-L-C$ circuit of Problem 12.4, as shown in Fig. P12.9. One wire is parallel to and 1 cm from the circuit, and the other wire is parallel to and 2 cm from the circuit. Compute the mutual inductance between the $R-L-C$ circuit and this loop. [13.9 nH] Compute the maximum near-end and far-end voltages induced by magnetic field coupling for the $R-L-C$ parameters of Problem 12.4. [(a), (b) 1387 V, (c) 278 V] Use SPICE to simulate this coupling. Repeat this if the loop sides are moved further away to 10 cm and 11 cm. [1.906 nH, (a), (b) 191 V, (c) 38.1 V]

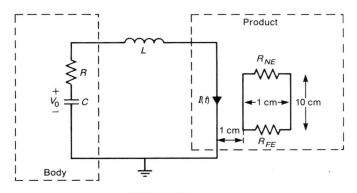

FIGURE P12.9

CHAPTER THIRTEEN

System Design for EMC

The purpose of this chapter is to provide an application of the previously discussed principles in the design of an electronic system to minimize its (1) potential for interfering with other electronic systems, (2) susceptibility to interference from other electronic systems, and (3) potential for interfering with itself. The word "minimized" is used rather than "eliminated" since it is virtually impossible to guarantee that the system will satisfy these goals in every installation. There tends to be a *reciprocity* between interference and susceptibility. If one designs a system to minimize its interference with other electronic systems, its susceptibility to interference from other electronic systems also tends to be reduced, and vice versa.

A product must comply with various governmental regulations that place limits on radiated and conducted emissions in order to be sold. However, if the system complies with these requirements but is susceptible to external sources such as ESD then compliance is a moot point, because customers will soon cease to purchase the product. Therefore all three basic design objectives should be kept in mind to enable the designer to identify any aspect of the design that would adversely impact those goals.

EMC design depends on the *early and continuous* application of the principles and techniques detailed in this text. Electrical engineers may be accustomed to thinking in terms of *ideal behavior* of circuit components. For example, "grounding two points together" by connecting the two points with a long wire may be suitable for placing those two points at approximately the same potential for very low frequencies and/or small currents flowing through that ground connection. However, if we are interested in making sure that these two points are at the same potential at 100 MHz, a 10 inch wire will have a significant impedance (at least 100 Ω) that is primarily due to the wire inductance and *not* its resistance. In the context of providing a low-impedance path for high-frequency currents with this wire, the two points are effectively *unconnected* at this frequency. The problem here is that we tend to think of wires as well as other components of the system in terms of their ideal and/or dc performance and not in terms of how they will behave at the high frequencies that are of interest in EMC.

As another example of simply being vigilant to the potential EMC ramifications of some aspect of the design, consider the routing of cables in the product. This is usually done from a convenience standpoint. Suppose a connector is placed on a PCB so that the cable that attaches to this connector must be routed over the surface of the PCB in order to terminate at its intended destination, e.g., a disk drive connector, as shown in Fig. 13.1. If that cable passes closely over the oscillator of the system (typically > 10 MHz) then the oscillator will most likely couple to the cable, which will then provide an effective path for the oscillator signal to pass to other unintended parts of the system and also to radiate from the cable. Coupling from other components in the vicinity of this cable such as switching transformers will also occur. Designing to avoid this potentially serious problem results in *no additional cost to the system* (or delays). For example, placing the connector at a different location on the board or rotating the board in the product so that the cable need not pass over the board eliminates this problem. Quite frequently the design of the system enclosure (and therefore the location of the other components such as disk drives and connectors to other peripherals) is done independently of and before the design and placement of the PCBs in that system. Therefore it is important that the designers of the subsystems discuss potential EMC problems at a very early stage in the design and frequently throughout the process. A critical aspect of incorporating good EMC design is an awareness of these nonideal effects throughout the functional design process.

Another critical aspect in successful EMC design of a system is to not place reliance on "brute force fixes" such as "shielding" and "grounding." The notions of shielding and grounding are perhaps the most frequently misunderstood and misused aspects of EMC. Their use is often unsuccessful, and may intensify the problem because they cannot be implemented in the ideal sense. The instances

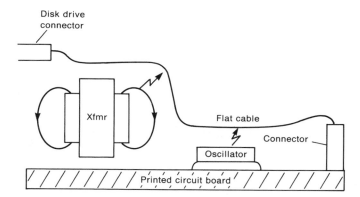

FIGURE 13.1 Packaging considerations that affect radiated and conducted emissions of a product.

where shielding can be successful are discussed in Chapter 11, but these are quite limited and must be used with care appropriate to the situation. Also, many products or their subsystems do not allow the ideal implementation of shields. Appearance and cost constraints frequently require that the product be housed in a plastic enclosure, which provides essentially no shielding. The interior of the enclosure can be sprayed with conductive paint and/or metallic strips can be imbedded in the plastic during molding. However, neither of these remedies will be as effective as expected if holes are present in the enclosure or cables pass from this product to a peripheral device. This also applies to subsystems of the product. For example, enclosing a PCB within a metallic shield can be a waste of metal if conductors that exit the PCB and this "shield" carry high-frequency signals, as is usually the case. Remember that *any metallic conductor is capable of radiating if currents exist on that conductor*; a shield or a grounding conductor is a metallic conductor. If the shield is attached to a point whose high-frequency potential is of the order of millivolts, the shield will become a radiator of that signal and could cause radiated emission as well as conducted emission problems. Once again *it is important to think of the EMC impact of the design at the frequencies of the regulatory limits, e.g., 30 MHz–1 GHz, and not at dc or low frequencies.* "Grounding" is a typical example of the failure to think of performance at high frequencies. We have been conditioned to think of "ground" as a zero-impedance, equipotential surface (a perfect conductor). Currents with frequency components from dc to well above 100 MHz typically pass through "ground." At frequencies in the MHz range resistance of the conductor, even including skin effect, is neglible compared with the impedance due to the ground conductor *inductance*. Therefore it is imperative to think of "ground" as *a path for current to flow* instead of an equipotential surface [1].

When the product is tested to determine its compliance to the regulatory limits (usually toward the end of the design cycle), problems will generally surface. Electronic products are so complex today that it is virtually impossible to consider all EMC issues in the design. However, if attention is paid to EMC throughout the design, the major problems will have been averted, and it will be easier to remedy those problems that do surface without the necessity of adding expensive and difficult-to-implement "fixes." If, for example, the system clock oscillator were placed on the PCB where an input/output (I/O) cable exits the board, coupling from the oscillator to the cable could result in significant EMC problems. During testing, there is little that can be done other than to re-layout the PCB, which is expensive and requires significant delays in the product development schedule. Vigilance to EMC throughout the product design cycle is an important aspect of preventing costly and inconvenient EMC problems.

13.1 GROUNDING

The conventional notion of "ground" is a zero-impedance, equipotential surface and often it is only considered from the standpoint of its dc performance. Neither

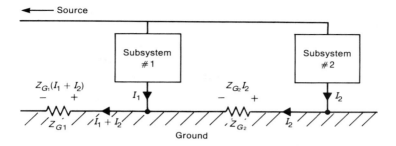

FIGURE 13.2 Illustration of common-impedance coupling.

of these aspects are applicable to "ground" with regard to its application in EMC. *All conductors have a certain amount of impedance*; consequently, *any currents that pass through that "ground" will cause points on its surface to be at different potentials due to the voltage drop across this impedance!* For example, consider Fig. 13.2, which shows two subsystems such as PCBs that are connected to ground (a metallic plane, wire, or PCB land).These subsystems may be either digital or analog or a combination. In digital subsystems, the $+5$ V current returns to its source (the dc power supply) through this ground and this current is constantly changing state when the logic devices switch. In analog subsystems this current may consist of low-frequency or high-frequency narrowband signals as well as broadband signals such as are produced by arcing at dc motor brushes. The analog signals also return to their source through a return. The analog signals often have dedicated returns or "grounds" that are different from the digital returns, although this is not always the case. Nevertheless, for the purposes of illustration, let us assume that the ground or return of subsystem $\#2$ is attached to the ground of subsystem $\#1$ as shown, and both subsystems subsequently share the same return thereafter. The return current I_2 for subsystem $\#2$ combines with that of subsystem $\#1$, and both pass through the common ground impedance Z_{G1}, developing a voltage drop across that part of the return of $Z_{G1}(I_1 + I_2)$. Observe that the signal fluctuations that are unique to subsystem $\#2$ are contained in I_2, and are therefore included in the voltage drop $Z_{G1}I_2$. Thus the ground point of subsystem $\#1$ is varying at a rate that is proportional to the signals in subsystem $\#2$. Therefore the signals in subsystem $\#2$ will couple to subsystem $\#1$ by virtue of this nonzero impedance of the ground and the sharing of the ground return by both signals. Similarly, the voltage of the ground point for subsystem $\#2$ is $Z_{G1}I_1 + (Z_{G1} + Z_{G2})I_2$. Thus the ground point for subsystem $\#2$ has the signals of subsystem $\#1$ imposed on it through Z_{G1}. This is often referred to as *common-impedance coupling*, and was discussed in Chapter 10 with regard to crosstalk. This illustrates the importance of the nonideal effect of ground impedance.

Another important misconception is that the ground impedance is its dc or low-frequency resistance. At the frequencies of the radiated emission limit,

30 MHz–1 GHz, the resistance of conductors, even including skin effect, is negligible compared with the inductance of the conductor! For example, consider a #28 gauge solid wire (radius of 6.3 mils), whose dc resistance is $5.4 \times 10^{-3} \, \Omega/\text{inch}$ and whose resistance at 100 MHz is $65.9 \times 10^{-3} \, \Omega/\text{inch}$. Increasing the diameter of the wire to #20 gauge (radius of 16 mils) decreases these only slightly (to $8.44 \times 10^{-4} \, \Omega/\text{inch}$ at dc and to $25.9 \times 10^{-3} \, \Omega/\text{inch}$ at 100 MHz). Therefore the wire size does not significantly decrease the high-frequency resistance. However, the inductance is on the order of some 15 nH/inch. This gives an impedance at 100 MHz of $9.43 \, \Omega/\text{inch}$, which is significantly larger than the portion due to resistance. Now consider the effect of this return inductance on digital signals. Consider a typical "totem-pole" output of a TTL gate shown in Fig. 13.3. The fanout of this gate, including any interconnection wiring capacitance, is represented by the lumped capacitance C_{LOAD}. With the output in the high state, transistor Q_1 is "on" and Q_2 is "off." With the output in the low state, the reverse is true. During the transition from low to high, C_{LOAD} charges up as shown in the figure. When the gate switches off, Q_1 switches "off" and Q_2 switches "on," so that C_{LOAD} discharges through Q_2. This illustrates why the rise time of TTL totem-pole outputs is typically slower than the falltime; the time constant of the charge path is $(R + R_{\text{LOAD}})C_{\text{LOAD}}$, whereas the time constant of the discharge is $R_{\text{LOAD}}C_{\text{LOAD}}$, where R_{LOAD} represents the sum of the input resistance of the load and the interconnect wiring resistance. (The saturation resistances of the transistors enter into this, but we will disregard them here since they are usually small.) During the low-to-high transition, current is drawn from the dc supply through the inductance of the +5 V supply line and returns through the inductance of the return line to the dc supply. During the transition from high to low, the discharge current of the capacitor passes through the inductance of the "ground connection" between the gate and the load. These sudden changes in the current through those inductances create voltage drops across them. There is also another particularly troublesome current involved in this process. During the transition from "off" to "on" and vice versa, there is a brief time during which both Q_1 and Q_2 are "on," resulting in the so-called "crossover current" that flows from the supply through both transistors and back to the supply through the ground conductor. This crossover current is limited only by the impedance through the Q_1–Q_2 path, and can be quite large (of order 50 mA) with very fast rise/falltimes. Let us consider the voltage developed across the inductance of the return or ground conductor, L_{GND}, between the gate and its load. The current provided is essentially $I_{\text{GND}} = C_{\text{LOAD}} \, dV_{\text{LOAD}}/dt$, with the waveform shown. Current spikes related to the slope of the load voltage occur during the state transitions as shown in the figure. The voltage developed across the ground conductor is essentially $V_{\text{GND}} = L_{\text{GND}} \, dI_{\text{GND}}/dt = C_{\text{LOAD}}L_{\text{GND}} \, d^2V_{\text{LOAD}}/dt^2$. This results in the voltage waveform developed across the ground conductor as shown in the figure, which is related to the slope of the current. For illustration, let us assume a load capacitance of 10 pF and a voltage transition of 3 V in 5 ns. This produces a current through the ground of 6 mA, which will have a rise/falltime less than

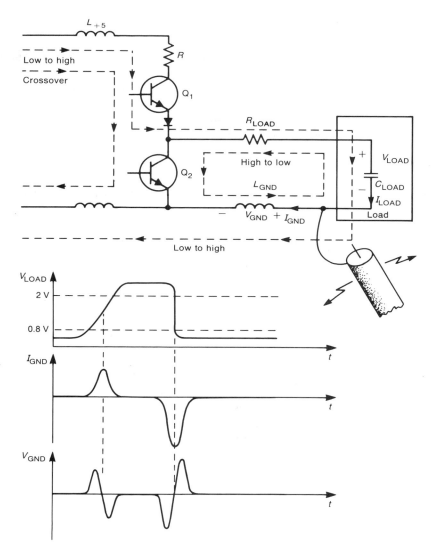

FIGURE 13.3   Illustration of the effect of conductor inductance on ground voltage.

that of the original voltage, e.g., 1 ns. Assume that the total length of the ground or return conductor is 5 inches, with a typical value of 15 nH/inch, giving a net inductance of the return land of some 75 nH. This will give a voltage drop across the ground conductor of 0.45 V. Doubling the length of the return conductor gives a voltage drop across it approaching the noise margin of TTL. When several gates are switching in this manner, it becomes clear that the inductance of the return conductors may create false logic switching because the reference voltages of two gates may differ by the noise margin. The similar effect occurs along the $+5$ V supply conductor. If the shield of a shielded conductor is attached to this assumed "quiet ground" point, it is quite obvious that it will most likely radiate as a very efficient antenna. Clearly something must be done to mitigate this effect of conductor inductance if digital logic circuits are to operate reliably and systems containing them are to comply with the governmental requirements on radiated emissions. We will find in Section 13.3 that a *ground grid* (or a ground plane as with an innerplane (multilayer) PCB) will tend to reduce the inductance of the return path, whereas *decoupling capacitors* will reduce the *effect* of the $+5$ V supply conductor inductance. Observe that if the switch rate is at the system clock frequency, e.g., 10 MHz, then the currents through the return conductor will consist of pulses at twice this frequency (20 MHz), having rise/falltimes less than 5 ns. Is there any doubt as to the radiated emission potential of even small loops of these currents?

Another common misconception is illustrated in Fig. 13.4(a). It is frequently

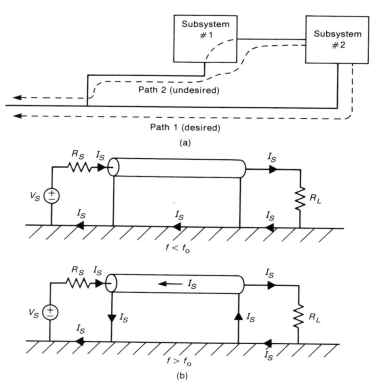

FIGURE 13.4 Illustration of the principle that signals may not return through their desired paths: (a) illustration of the principle; (b) a practical example—a shielded wire above a ground plane.

assumed that currents will return to their source via the paths provided for this purpose. At low frequencies and certainly dc this is usually the case. However, *at high frequencies, and certainly those of the regulatory limits, we cannot assume that currents will return along the intended paths!* A simple example of this is the case of crosstalk to a shielded wire discussed in Chapter 10 and illustrated in Fig. 13.4(b). Above the "cutoff frequency" of the shield–ground plane circuit, f_0, the current will find a lower-impedance path back through the shield rather than through the ground plane, no matter how massive or low impedance the ground plane. So an "intuitive" notion of the lowest-impedance path is false in this case. In fact, if the current has a spectral content that covers a wide frequency range, some of its frequency components will return along the ground plane, while other frequency components *of this very same pulse* will return along a different path: the shield. Therefore *some frequency components of a signal will return along one path, while other frequency components of the same signal may return along other paths.* It is true that *currents will return on the path of lowest impedance.* However, it is often difficult to determine the lowest-impedance path simply by visual inspection or to "designate" a return path on the circuit schematic. Although these paths may seem to violate intuition, there is a logical explanation as evidenced by the case of a shielded wire over a ground plane. We will see in this chapter that to prevent the current taking an undesired return path, we can use a *ground grid* to reduce the impedance (inductance) of the return path whatever that return path may be [1–3].

13.1.1   Safety Ground versus Signal Ground

It is important to realize that *there are several purposes of a ground system.* The concept of a ground as being a zero-potential surface may be appropriate at dc or low frequencies, but is never true at higher frequencies, since conductors have significant impedance (inductance) and high-frequency currents flow through these impedances, resulting in points on the ground having different high-frequency potentials. This highlights the distinction between the two types of grounds: safety ground and signal ground. As discussed in Chapter 9, commercial power is utilized as 120 V, 60 Hz voltage in the US (240 V, 50 Hz in Europe). *A safety ground is normally required in order to provide protection against shock hazard.* A typical commercial power distribution for residences and commercial buildings in the US is illustrated in Fig. 13.5. The commercial power is provided to the residence as 240 V, 60 Hz between one wire (the "red" wire) and another wire (the "black" wire) from the external power distribution system. This enters the residence through the *service entrance panel*, and is distributed on three busses in the service entrance panel. The center bus is referred to as ground, and is connected to physical earth ground at the service entrance panel via a ground rod inserted into the earth. This is to provide shock and fire hazard protection in the event of a fault, as discussed later. The voltage between the two outer busses (red and black) is 240 V, and the voltage between

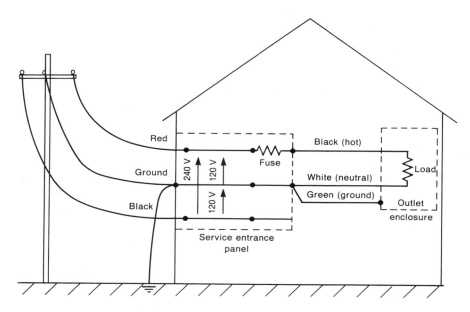

FIGURE 13.5   Residential power distribution in the United States.

each of the outer busses and the center ground bus is 120 V. A 240 V service required by ovens and clothes dryers is obtained by connecting to the outer red and black busses. A fuse or circuit breaker is inserted in each of the leads that are connected to that load. Conventional 120 V service is obtained by connecting to either of the outer busses and the center or ground bus. The lead connected to the red (or black) bus is referred to as the Black or Hot Wire in reference to the conventional color of that wire's insulation in typical cables. It is also at a voltage of 120 V *with respect to earth*. The lead connected to the center ground bus is referred to as the White or Neutral Wire in reference to its insulation color and is at earth potential. In addition to these two wires, there is another wire supplied in typical residential wiring cables: the Green or Ground Wire in reference to its insulation color. The Green Wire is often referred to as the *safety wire* for the following reason. As an outlet, the Black and White Wires are connected to the two outlet terminals. Commercial 120 V power is obtained by plugging the power cord of the device into these two holes in the outlet socket. In addition, the Green Wire that is carried through the residence along with the Black and White Wires is connected to a third hole in the outlet as well as to the metallic case of the outlet. This is to provide a path for fault currents to flow back to the service entrance panel in order to blow the fuse or open the circuit breaker for that circuit. Suppose the Black Wire becomes disconnected inside the outlet and accidentally comes in contact with the metallic enclosure. A path will be provided back to the service entrance panel via the Green Wire for current to flow, thereby opening the circuit breaker

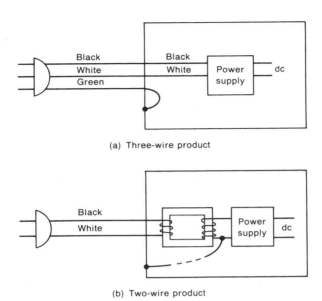

(a) Three-wire product

(b) Two-wire product

FIGURE 13.6 Illustration of power cord connections for (a) a three-wire product and (b) a two-wire product.

for this circuit. If the Green Wire were not present or were not connected to the outlet enclosure, the outlet enclosure would be placed at a 120 V with respect to earth ground which would pose a potential shock hazard to anyone who might touch the outlet enclosure. The only time the Green Wire carries current is during such a fault. The current normally returns via the White Wire. Because of the rather high voltages, a small voltage drop of several hundred millivolts across any of these conductors due to several amps of current flowing through the wires is inconsequential. Thus these conductors are essentially equipotential surfaces with regard to the commercial 120 V, 60 Hz power. The concept of ground in this application is that of a *safety ground*.

It is also worth noting the method by which products that receive power from this outlet utilize this ground. The two common methods, referred to as *three-wire products* and *two-wire products*, are illustrated in Fig. 13.6. The power cord of a three-wire product contains three wires: Black, White and Green. The Green Wire is connected directly to the metallic frame of the product in order to provide shock hazard protection in the same manner as within the power outlet enclosure. The Black and White Wires go to the power supply of the product, where this 120 V, 60 Hz ac voltage is converted to the dc voltages (+ 5 V, + 12 V, − 12 V, + 38 V, etc.) required to power the product's electronics as well as drive motors and other components. A two-wire product illustrated in Fig. 13.6(b) uses only the Black and White Wires. Remember the Black Wire is at 120 V with respect to earth ground and the White Wire is at earth potential.

It would not be feasible to connect the White Wire to the product frame for shock hazard protection, because the user could insert the plug improperly in the outlet. Most two-wire plugs are polarized (one blade is larger than the other), so only one insertion configuration is possible in the outlet holes. However, in order to safeguard against shock, the Black and White Wires are first passed through a transformer in the product, and one of the output wires of the transformer may be connected to the frame of the product. The transformer essentially removes the distinction between which of the two wires is "hot" with respect to earth on the secondary side. Any fault that occurs to the product frame (the overall metallic enclosure or metallic substructure of a plastic enclosure) will draw a large current, which will trip the circuit breaker for that circuit. As discussed in Chapter 9, the elimination of the Green Wire in two-wire products is thought to remove conducted emission problems due to common-mode currents because there is no *physical circuit* for these components of the noise currents through the 50 Ω impedances of the LISN. This assumption is false, since there remains a path via displacement current between the product and the frame of the LISN (which is connected to the Green Wire). There may also be an alternate path through the ground wire in a peripheral cable and through the Green Wire of that peripheral device to the LISN, as illustrated in Fig. 13.4(a). Two-wire products may lessen conducted emission problems, but they typically do not eliminate them.

The other type of ground is the *signal ground*, which permits signal currents to return to their source. It is important to emphasize that *although it may be the designer's intent for the signals to return through these designated paths, there is no guarantee this will occur*. In fact, *some frequency components of a signal may return through one path, while other frequency components of that same signal may return through another path*. A shielded cable above a ground plane is a good example. The frequency components that are below the cutoff frequency of the shield–ground plane circuit will return along the ground plane, however, those above this cutoff frequency will return along the shield and not the ground plane. In the case of signal grounds it is important to think of these as *paths for current to flow*, as was pointed out by Ott [1]. Again, high-frequency impedance of these paths is primarily inductive, and it is this inductive impedance that must be minimized if this "ground" is to have small voltage drops across it. The *ground grid* (discussed later) is an effective, low-cost method of accomplishing this.

13.1.2 Single-Point versus Multipoint Grounding

There are basically two philosophies regarding signal ground schemes: single-point ground systems and multipoint ground systems. A *single-point ground system* is one in which subsystem ground returns are tied to a single point within that subsystem. The intent in using a single-point ground system is to *prevent currents of two different subsystems from sharing the same return path and producing common-impedance coupling*. Figure 13.7 shows typical

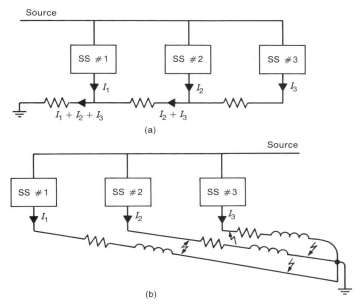

FIGURE 13.7 Illustration of the problems in single-point grounds: (a) common-impedance coupling in a "daisy-chain" connection; (b) unintentional coupling between ground wires in a single-point ground system.

implementations of a single-point ground philosophy. Three subsystems have the same source. The method shown in Fig. 13.7(a) is referred to as the "daisy-chain" or *series connection method*. This technique has the obvious problem of permitting common-impedance coupling between the grounds of the two subsystems. The connection in Fig. 13.7(a) will have the signals of SS # 2 and SS # 3 impressed on SS # 1 as discussed previously. This underscores that we must be cognizant of the return paths for the currents where they are possible to determine. The *parallel connection* shown in Fig. 13.7(b) is the ideal single-point ground connection. However, it too suffers from the disadvantage that the individual ground conductors will have a certain impedance dependent upon the length of these connections. In a distributed system these connection wires may need to be long if we strictly adhere to the single-point ground system philosophy. The ground wires will then possess a possibly large impedance that may negate their positive effect. Also, the return currents flowing through these wires may radiate efficiently to other ground wires and produce coupling between the subsystems in a fashion similar to crosstalk, thereby creating radiated emissions compliance problems. The degree to which this occurs depends upon the spectral content of these return signals: higher-frequency components will radiate and couple more efficiently than lower-frequency components. Therefore a single-point ground philosophy is not a universally ideal ground system philosophy, since it works best for low-frequency subsystems.

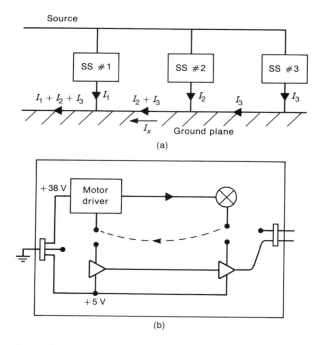

FIGURE 13.8   Illustration of multipoint grounding: (a) the ideal case; (b) illustration of problems that occur in multipoint ground schemes.

The other type of ground system philosophy is the *multipoint ground system* illustrated in Fig. 13.8(a). Typically, a large conductor (often a ground plane) serves as the return in a multipoint ground system. In a multipoint ground system the individual grounds of the subsystems are connected at different points to the ground conductor. In using a multipoint ground system *it is assumed that the ground return to which the individual grounds are terminated have a very low impedance between any two points at the frequency of interest.* Otherwise, there would be no technical distinction between this and the series connection, single-point ground system in Fig. 13.7(a). The advantage a multipoint ground system is thought to have over a single-point ground system is that the connection lead lengths can be shorter, since there is a closer available ground point. But this again *presumes* that the ground has zero, or at least very low, impedance *between the ground connection points at the frequency of interest,* which is not necessarily true. If the ground plane in Fig. 13.8(a) were replaced by a long, narrow land on a PCB, we might believe we were implementing a multipoint ground system if we attached the subsystem grounds at points along this land, when, in fact, this would more closely resemble the series connection, single-point system of Fig. 13.7(a). Quite often, these "semantics" create confusion and misunderstanding. *Simply connecting subsystems to different points on a conductor does not constitute a multipoint ground system unless the spirit of such a system is preserved; that is, the impedance between these connection points along the ground conductor is small at the particular frequency of interest.*

Another problem with a multipoint ground system may be that too little attention is paid to other currents that flow through the ground conductor. For example, suppose the "ground plane" (to which the subsystems are multipoint grounded) has other currents intentionally or otherwise routed through it. An example is illustrated in Fig. 13.8(b), where a dc motor drive circuit is contained on the same PCB as other digital electronic circuits. The $+38$ V dc required to drive the dc motor and the $+5$ V dc required to power the digital electronics are provided to the PCB via a connector. Suppose these circuits are grounded to a common ground net on the PCB. The high-current levels of the motor circuit will pass through this ground, developing potentially large, high-frequency voltages between two points on that ground net as the motor driver switches. If the digital logic circuitry is also terminated to that ground net in a multipoint fashion, these voltages developed across the ground net by the motor return currents may couple into the digital logic circuit, creating problems in its desired performance. In addition, suppose a signal is routed off the PCB via a connector at the opposite side of the PCB from the power connector. The ground wire in that signal cable will be driven at the varying potential of the noisy ground system, and may radiate creating radiated (or conducted) emission problems.

Typically, single-point ground systems are used in analog subsystems, where low-level signals are involved. In these cases, millivolt and even microvolt ground drops can create significant common-impedance coupling interference problems for those circuits. Single-point ground systems are also typically employed in high-level subsystems such as motor drivers, where the intent is to prevent these high-level return currents from developing large voltage drops across a common ground net. Digital subsystems, on the other hand, are inherently "immune" to noise from external sources; however, they are quite susceptible to internal noise. They are said to "shoot themselves in the foot" by internal interference via common-impedance coupling, as illustrated in Fig. 13.3. *In order to minimize this common-impedance coupling, the ground system in digital subsystems tends to be multipoint*, using a large ground plane such as in innerplane board or placing numerous alternate ground paths in parallel such as with a ground grid, thus reducing the impedance of the return path. It is also important to route the signal conductors in close proximity to the ground returns, since this will also reduce the impedance of the return, as we will see in Section 13.3.

There are other types of ground systems that are used less frequently than the previous ones in special circumstances. These are referred to as *hybrid ground systems*, and are a combination of the previous two systems over different frequency ranges. As an example, consider a shielded wire above a ground plane, as shown in Fig. 13.9. We discussed in Chapter 10 the concept that a shielded cable will eliminate inductive coupling to the interior, shielded wire *only if the shield is connected to the ground plane or reference conductor at both ends*. We also pointed out that this permits the possibility of common-impedance

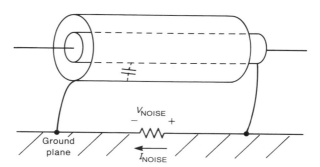

FIGURE 13.9   A way of creating a single-end grounded shield at low frequencies and a shield grounded at both ends at high frequencies to avoid "ground loops."

coupling due to noise currents flowing through the reference conductor generating a voltage across the shield that is coupled to the interior wire. This commonly occurs when low-frequency power currents flow through the reference conductor. A way of selectively implementing the shield grounding and avoiding this low-frequency coupling is illustrated in Fig. 13.9. If the cable has two shields with the inner shield attached to the reference conductor at one end and the outer shield connected to the reference conductor at the other end, no low-frequency connection exists between the two shields, thus avoiding the common-impedance coupling problem due to I_{NOISE} flowing through the reference conductor. However, the parasitic capacitance between the two shields (which is quite large due to the concentric nature of the two shields) provides a high-frequency connection between the two shields, so that the shield is effectively connected to the reference conductor at both ends. This represents the *frequency-selective grounding of a hybrid ground scheme*. A single shield can implement this if we attach one end of the shield to the return conductor *via a capacitor*. At low frequencies the shield will be single-end grounded; whereas at high frequencies the capacitor will present a low impedance, and the shield will be double-end grounded. Typically, this requires a fairly large capacitance. Figure 13.10 depicts two other implementations of a hybrid ground system. The capacitors shown in Fig. 13.10(a) provide a single-point ground system at low frequencies and a multipoint ground system at high-frequencies. The inductors in Fig. 13.10(b) provide just the opposite. The grounding schemes in Fig. 13.10(b) is useful when it is necessary to connect the subsystems to Green Wire ground for safety purposes and to have a single-point ground system at higher frequencies.

   Typical systems require three separate ground systems, as shown in Fig. 13.11(a). Low-signal-level (voltage, current, power) subsystems should be tied to a single dedicated ground point. This is referred to as *signal ground*. Within this signal-ground subsystem, the circuits may utilize single-point ground systems, multipoint ground systems or a combination. The second type of

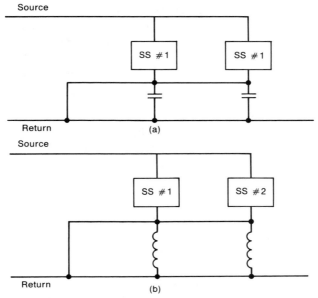

FIGURE 13.10 Hybrid ground schemes: (a) single point at low frequencies and multipoint at high frequencies; (b) single point at high frequencies and multipoint at low frequencies.

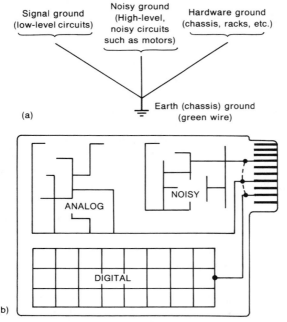

FIGURE 13.11 Segregation of grounds: (a) the ideal arrangement; (b) PCB layout of the ground system.

ground system is referred to as the *noisy ground system.* The noisy ground system represents circuits that operate at high levels and/or generate noise-type signals. A signal may be considered noise in one instance and not in another. For example, the high-frequency spectral content of digital clock signals may be considered noise in complying with the regulatory limits or interfering with other subsystems, yet they are necessary spectral components of the functional signal. On the other hand, arcing at brushes of dc motors is truly noise, and is not necessary for the functional performance of the motor. (Arcing can be suppressed as discussed in Chapter 6 and not impede the motor's performance.) For example, Fig. 13.11(b) shows a PCB that contains digital circuitry, analog circuitry, and noisy, motor driver circuitry. The noisy circuitry ground has a dedicated connection to the board connector that prevents these high-level return currents from passing through the analog or digital ground systems. Similarly, the digital and analog circuitry have dedicated ground returns back to the connector. Note that the ground system within the analog ground system (a signal ground) is substantially a single-point ground system, whereas the ground system within the digital ground system (another signal ground) is substantially a multipoint ground system.

The third type of ground is the *hardware ground* that is connected to chassis, frame, cabinets, equipment racks, etc. This hardware ground is not *intended* to carry current except in the case of a fault or for diversion of ESD signals.

The key to understanding why these different and distinct ground systems are required lies in the fact that *they are intended to prevent common-impedance coupling.* If we allow high-level noise from a motor-driver circuit to pass through a conductor that also serves as the return for a digital circuit, these high-level currents will generate voltage drops across this common return that will be fed into the digital circuit, creating possible functional problems in the digital circuit. It is important to separate low-level and high-level returns, since the larger the magnitude of the return current, the larger the common-impedance voltage drop. Several different low-level circuits may share the same return and not cause interference with each other, since the common-impedance coupling voltage drops generated across the common ground net may not be large enough to cause interference. Not only are the signal levels important in separating ground systems, but their spectral content is also important. Some subcircuits contain inherent filtering at their inputs. Thus high-level noise signals that are presented to their inputs will not create interference problems if the spectral content of that noise is outside the passband of the circuit's input filtering. Digital circuits tend to have very wide bandwidth inputs, so the frequency-selective protection is not present. On the other hand, analog circuits such as comparators tend to have a degree of high-frequency filtering due to the response time of the OpAmp. Parasitics can, however, negate this [4]. The hardware ground is usually separate from the other grounds in order to also avoid the common-impedance coupling problem. High-level 60 Hz power signals as well as ESD signals may pass through this ground. It is important to not provide a connection between hardware ground and the other grounds, in particular

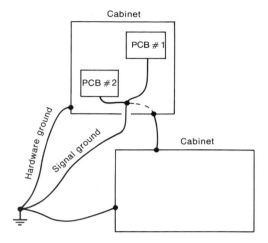

FIGURE 13.12   Illustration of grounding for systems with several cabinets.

the signal ground, so that voltage drops created by, for example, ESD signal diversion will not cause points within the signal ground system to be varying at the noise rate.

As an example where the connection of separate ground systems may create problems, consider the situation shown in Fig. 13.12. Two cabinets or racks constitute the system, and have their hardware grounds connected to a common point. These cabinets are also connected together to prevent potentials being built up between them due, for example, to ESD currents discharging through them. The two PCBs in one cabinet may have their signal grounds connected together, but it would, in general, not be correct to tie this signal ground to the cabinet, since the aforementioned ESD discharge may then cause this signal ground point to vary in accordance with the ESD discharge. However, in some cases it may be necessary to tie these grounds together to prevent ESD problems. Parasitic capacitances and inductances between the cabinet and the interior circuits may cause the high-frequency behavior of this ground system to deviate considerably from the ideal. So circuits should be separated physically as much as possible from each other and the enclosure in order to avoid this high-frequency coupling.

These rules are not inviolate, and may be modified under certain circumstances *as long as the designer considers the potential consequences and the objective that the segregated ground system philosophy is intended to achieve*: eliminate common-impedance coupling. It may be possible to tie all grounds together at the entrance of the connector to the PCB, as indicated in Fig. 13.11(b). Whether this is permissible often depends on the voltage drop (high-frequency) along the ground wires of the connection cable that attaches to this point. If the voltages (high-frequency) developed across the individual ground conductors of the cable are quite different, an indirect path from one ground system to another on the board may be established due to parasitics between the ground systems. It may also be possible to physically connect the ground systems on a board. An example is shown in Fig. 13.13(a). A PCB contains both digital and analog circuitry. The power cable supplying $+5$ V to the digital circuit and $+38$ V to the analog circuit (motor driver) enters at one end of the board via a common power connector from the power supply. In this case it may be permissible to connect the grounds of the digital and the analog high-level circuits where the two areas are adjacent at point P_1. The reason that this may be feasible is that the high-level motor driver currents will most likely take a direct return path to their source, the connector, and so will not pass through the ground net of the digital circuitry. Passing the $+5$ V return currents of the digital circuit through the ground net of the analog circuitry will not cause interference with this circuit, since the common-impedance voltage drops generated across the motor-driver ground will not be significant *with respect*

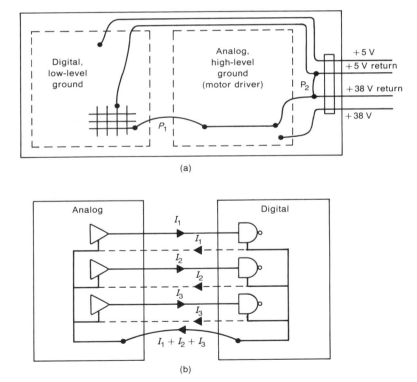

FIGURE 13.13 Ground problems between analog and digital grounds: (a) a case where, because of the location of the connector, the noisy analog signals will not return through the digital ground; (b) an analog system driving a digital system. Use of dedicated returns in close proximity to the signal conductors prevents large loop areas in (b).

to the operating levels of this circuit. The $+5$ V and $+38$ V returns may also be connected at the power connector P_2 for the same reason. On the other hand, if the power connector were placed on the left side of the board, connecting the two ground nets at P_1 would likely cause serious functional problems in the digital circuitry due to passage of the high-level motor-driver currents through the impedance of the digital ground. This further illustrates that *understanding the underlying objectives and theory are the keys to good EMC design.*

A converse problem is illustrated in Fig. 13.13(b). An analog subsystem provides signals to a digital subsystem. If one ground or return is used to connect the two subsystems, potentially large current loops are created that radiate efficiently. A more preferable implementation would be to provide individual returns shown in dashed lines *and* to route each return close to the associated signal line. The lowest impedance for each signal return would most likely be through the intended return because it has the lowest loop impedance due to the proximity between the signal and return wire. For further discussions of "grounding" the reader is referred to [2, 3, 5].

13.1.3 Ground Loops

It is frequently necessary to pass signals between subsystems whose ground systems are intended to be distinct. An example is the motor driver/digital circuit shown in Fig. 13.13(a). The digital circuitry generates pre-drive signals that control the motor drivers located on the analog section of the board. The common ground connection between these two circuits at P_1 may or may not be necessary for functional compatibility between the two circuits. Suppose that the common ground connection at P_1 is left unconnected. The voltage of the digital ground system, V_{GD}, with respect to some reference point and the voltage of the analog ground system, V_{GA}, with respect to this same reference point may be different (and generally will be different). This will create a voltage difference $V_G = V_{GD} - V_{GA}$ between the two ground systems that may be of the order of several hundred millivolts. Connecting the two ground systems together at P_1 will tend to make this voltage difference between the two subsystems inconsequential, but may create common-impedance coupling problems if the high-level analog currents return through the digital ground. Additionally, large current loop areas are illustrated in Fig. 13.13(b) may be created. The question now becomes one of when we can leave these ground systems unconnected and avoid functional problems due to the voltage difference between the two unconnected ground systems. If the input to the analog board is in the form of a Darlington pair as illustrated in Fig. 13.14(a), the noise immunity is 0.6 V $+ 0.6$ V $- 0.5$ V $= 0.7$ V. The two 0.6 V terms are due to the base-emitter voltages of each transistor, while the 0.5 V term is the noise margin voltage of TTL signals. So for this type of input we can leave the grounds unconnected and be able to tolerate a noise voltage difference between the two ground

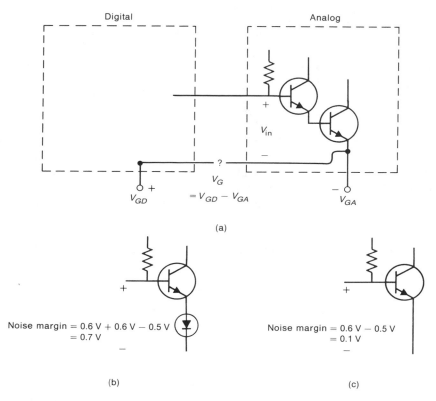

(a)

(b)

(c)

FIGURE 13.14 Illustration of when it may be permissible to disconnect the digital/analog ground connection to prevent common-impedance coupling: (a) the voltage between the circuit grounds is the difference between the ground voltages of each circuit; (b) the noise margin is increased by using a diole in the transistor emitter lead; (c) the noise margin of an unprotected transistor is very small.

systems of as much as 700 mV before functional problems begin to occur. This is a rather large noise margin, which can often be tolerated, so the grounds may not need to be connected. A single transistor would give a noise immunity of 0.6 V − 0.5 V = 0.1 V, which is typically too small to give reliable operation. A diode can be inserted in the emitter of a single transistor to give a noise immunity of 0.7 V as shown in Fig. 13.14(b).

The difference in voltage between two ground systems as discussed in the previous paragraph can result in another potentially serious interference problem, which is referred to as a *ground loop*. This is illustrated in Fig. 13.15(a), where the two subsystems are connected to different ground nets that are at different voltages, or are connected to the same ground system where the two connection points are at different voltages due to the impedance of the ground system. The voltage difference between the two connection points, V_G, acts like

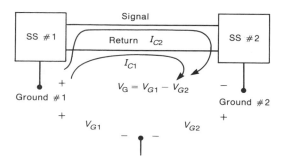

FIGURE 13.15   Illustration of the generation of common-mode currents on interconnect cables due to differences in ground voltages.

a voltage source, and will drive *common-mode currents* I_{C1} and I_{C2} through the signal and return wires between the two systems and between the two connection points. Even if one of the subsystems is not physically connected to a ground point, parasitic capacitance between the subsystem and the ground system can effectively complete the circuit. This is common in small motors, in which the large parasitic capacitance between the motor wiring and the motor frame (which is usually connected to large metallic portions of the frame for thermal considerations) provides a path from the motor input wires through the motor case to the product frame. (See Fig. 6.46(b).) These two common-mode currents flow around potentially large loops: the signal wire/ground return loop and the return wire/ground return loop. These common-mode currents then act like two differential-mode currents from the standpoint of generating radiated emissions. The levels of radiated emissions are proportional to the loop areas, as discussed in Chapter 8.

There are several methods for blocking this path. One of the more common and easily implemented methods is to insert a common-mode choke in the signal/return wires as shown in Fig. 13.16. The common-mode choke was discussed in Chapter 6, and can be represented as a pair of coupled inductors as shown in Fig. 13.16. The magnetic fluxes due to the differential-mode (functional) current tend to subtract in the core, so that the common-mode choke *ideally* presents no impedance to these functional signals (if the wires are wound properly on the core). Leakage inductance and parasitic capacitance between the input and output wires tend to degrade the choke's differential-mode performance. The magnetic fluxes due to the common-mode component of the currents (which returns through the connection between the two ground systems) tend to add in the core, and so an inductive impedance is presented in series with the common-mode currents.

Another method of blocking this common-mode current is with the use of an *optical coupler*, as illustrated in Fig. 13.17(a), which breaks the direct, metallic path. The ground voltage is between the input and the output of the coupler

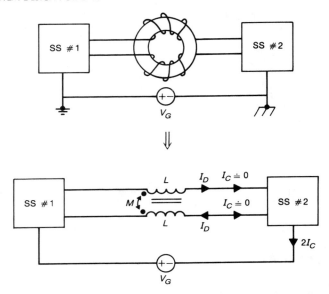

FIGURE 13.16 Use of a common-mode choke to block common-mode currents on interconnect cables.

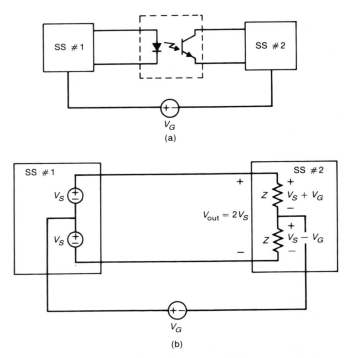

FIGURE 13.17 Methods for decoupling subsystems: (a) the optical isolator; (b) use of balanced terminations (and twisted pairs).

rather than between the two input terminals, and therefore does not create a common-mode current. This method is particularly suitable where rather large differences of voltage occur between ground systems such as the input to a pulse-width modulator in a switching power supply. A balanced system, as shown in Fig. 13.17(b), also tends to provide immunity to ground drops. The output of subsystem #1 is driven in a balanced mode *with respect to the ground of that subsystem*, so that the voltage of the signal wire and the voltage of its return *with respect to the common ground point* are 180° out of phase. The input to subsystem #2 is also balanced in that the impedance between the signal wire and the common ground for that subsystem equals the impedance between the signal return and this common point. A simple analysis of this circuit shows that the output voltage of subsystem #2, which is the *difference* of the two voltages across the impedances with respect to the common, is $V_{out} = (V_S + V_G) - (V_S + V_G) = 2V_S$. Thus the ground noise is subtracted out. This is commonly implemented either with center-tapped transformers (often called *baluns*) or with *differential line drivers and line receivers*. These line drivers and receivers utilize operational amplifiers and rely on their *common-mode rejection ratio* when operated in a balanced mode. This is a common method of long-distance communication of digital data over wire lines. Balancing also aids elimination of capacitive crosstalk coupling, as discussed in Chapter 10. Twisting the two wires together eliminates the pickup of magnetic fields by the signal wire/return wire loop.

The techniques of using common-mode chokes, optical couplers, or balanced transmission are examples of methods of *decoupling subsystems*. We will find numerous other instances where it is important to prevent fluctuations in one subsystem from affecting other subsystems. A common example is the use of *decoupling capacitors* to prevent inductive voltage drops along the power and return conductors supplying subsystems from contaminating those subsystems. This important technique will be considered in Section 13.3.6.

13.2   SYSTEM CONFIGURATION

Perhaps one of the more important aspects of the EMC design of a system is the configuration of the system and its subsystems (location and orientation of the PCBs, cable connectors, power supply and filter, etc.). Quite frequently, improper configuration is the major cause of functional performance problems or failure to comply with regulatory limits. The purpose of this section is to bring attention to this important aspect of EMC design.

13.2.1   System Enclosures

One of the decisions made in the first stage of product development is the external "packaging." The shape and external appearance are typically driven by customer preferences and ergonomic considerations, e.g., location and shape of a keyboard, and location of the on/off switch. This severely constrains the EMC engineer's options with regard to subsystem placement, internal cable routing, etc. It is therefore important that the EMC engineer be actively involved in the design process from the beginning.

One of the first decisions that must be made has to do with the type of system enclosure: metal or plastic. EMC designers prefer metal enclosures, since shielding of the interior electronics can be more easily achieved. However, metal enclosures are difficult to form into smooth shapes, which gives them a "boxy" appearance that has less appeal to the consumer. Enclosures for large computers and Class A devices such as point-of-sale terminals are usually metallic, since consumer preferences are not as significant for these devicees. Seams can be closed quite effectively, reducing radiation from these "slot antennas." Metallic fingers along the edges of doors reduce the leakage (and points of entry for external signals such as ESD). Knitted-wire gaskets can also be placed along the edges of the seam to close this aperature. It is very important to remember that *any cable that penetrates this shielded enclosure will drastically reduce the shielding effectiveness of the enclosure to internal as well as external signals unless these are treated properly*. For example, peripheral cables that exit the product may require filtering in the form of $R-C$ packs on each wire or a common-mode choke through which all of the cable wires, including the shield pigtail wire, must pass. It is difficult to determine prior to construction of a prototype whether this treatment of all peripheral cables will be necessary. It is therefore wise to allow for the later addition of these components if they are required. For example, pin-through holes that allow insertion of an $R-C$ pack or a DIP toroid can be provided on a PCB where a peripheral cable exits the board. These connections can be "wired across" in the initial design. If EMC testing determines that an $R-C$ pack or a DIP toroid is needed to reduce the signals that exit the board via this cable, the wires can be cut and the component mounted without a re-layout of the board. It is quite important to allow space for these components in the initial layout, since board space is often unavailable once the design is completed.

Plastic enclosures are easily molded into any desired shape and present an aesthetically pleasing system enclosure. Such enclosures are cheaper to produce than metal ones (at least before any interior conductive sprays or metal impregnation required for EMC is applied). However, a plastic enclosure provides no shielding, so careful attention to EMC design of the interior electronics is more critical. It is possible to provide some degree of shielding by either coating the interior with a conductive material or by impregnating conductive fibers in the plastic during molding. Conductive paints can be applied, and nickel is a common spray material. There are many other ways of applying this conductive coating, such as flame/arc spraying, vacuum metallization, electroless plating, and metal foil linings. These techniques vary in cost and effectiveness. Conductive fillers can be impregnated in the plastic during molding. A comparison of the shielding effectiveness of these various types of treatment of plastic enclosures is given in [6]. *Even with conductive plastic enclosures, it is important to realize that any penetration of the enclosure such as a peripheral cable or an aperature must be properly treated, or else the ideal shielding effectiveness of the enclosure may be essentially nullified!*

Apertures such as fan vents or other vents must be closed electromagnetically. Screens with many small holes have much better shielding effectiveness than a few large holes or several long slots, as discussed in Chapter 11. Cathode ray tube (CRT) faces also provide apertures for the internal fields to exist the enclosure. Wire mesh screens consisting of many small holes can be laminated between two glass sheets to shield the CRT face while only marginally restricting the view.

13.2.2   Power Line Filter Placement

An example of good EMC design is the placement of the power supply filter. The point at which the commercial ac power enters the product is usually chosen for aesthetics; namely, at the rear of the product. The on/off switch is placed at the front of the product with the consumer in mind. Routing the Phase and Neutral power wires from the rear entrance of the power cord to the front of the product where the switch is located and then back to the rear of the product to pass through the power supply filter and the power supply creates obvious problems. The exposed section of the wires from the rear entrance to the front switch will pick up signals internal to the product. These signals will then pass *unimpeded* out through the ac power cord, where they are efficiently radiated from the cord or conducted into the LISN in a conducted emissions test. Either of these situations will likely cause serious problems in complying with the regulatory limits. The actual switch may be placed at the rear of the product where the power cord enters, and a mechanical linkage can be connected between the switch and the on/off switch lever at the front of the product.

In some designs the power supply filter circuitry may be mounted on the same PCB as the power supply. This also has the disadvantage of placing the filter at some distance from the point of entrance of the power cord, as shown in Fig. 13.18(a), so that internal noise bypasses the filter. A more desirable placement of the filter is directly at the point where the ac power enters the product, as illustrated in Fig. 13.18(b). There exist pre-designed, pre-packaged integral power sockets and power supply filters. These configurations are the most effective in preventing this bypassing of the filter.

13.2.3   Interconnection and Number of Printed Circuit Boards

The next important decision concerns *the number and interconnection of the printed circuit boards in the product*. As a general rule, *it is preferable to have only one system PCB rather than several smaller PCBs* interconnected by cables. The reason for this is that it is easier to limit the voltage drops developed between the grounds of the parts of the system when they are placed on the same board than when they are placed on separate boards and interconnected

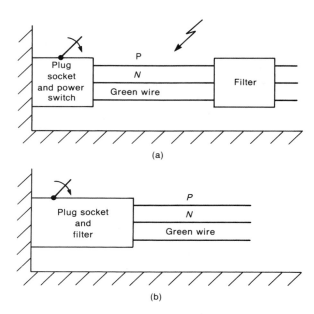

FIGURE 13.18 Effect of power supply filter placement on product emissions: (a) poor placement allowing coupling to the power cord and bypassing of the filter; (b) proper filter placement.

by cables. Cables introduce impedance that affects the voltage drop between the two ends of the cable. The impedance between the two subsystems can also be reduced with a *ground grid* if all subsystems are placed on the same board. Trying to achieve the same result in an interconnection cable between PCBs is difficult, since many interspersed ground wires will be required in a Ground–Signal–Ground–Signal–···configuration in that cable, and these ground wires cannot be as closely spaced to their adjacent signal wires as can lands on a PCB. *High-frequency voltages developed between PCBs create common-mode currents flowing between the boards that accentuate radiated and conducted emissions,* as illustrated in Fig. 13.19. Placing all electronics on one board tends to reduce this voltage difference between subsystems on the board.

High-speed signals that pass between PCBs should be buffered where they *enter* the PCB to reduce fan-out problems. For example, suppose a signal passing from one PCB to another fans out to four other devices. Placing a buffer at the input to the board reduces the current in the interconnect line by a factor of four over placing the buffer at the output of the other board, and hence reduces its radiated emission potential as well as its common-mode current generation potential.

PCBs that must be separated should be connected with low-impedance (*high-frequency impedance*) connections such as short lands on a motherboard or backplane. The PCBs are interconnected by plugging their edge connectors

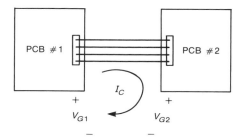

FIGURE 13.19 Illustration that multiple PCBs tend to promote common-mode currents on interconnect cables.

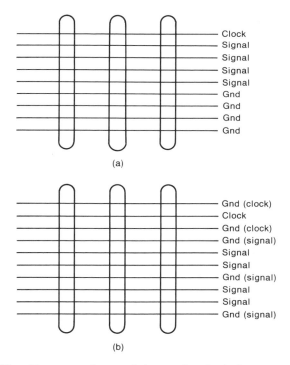

FIGURE 13.20 Use of interspersed grounds (returns) on backplanes to reduce loop areas: (a) large loop areas; (b) reduced loop areas.

into this motherboard. Figure 13.20 illustrates some important considerations in the design of these motherboards. Although it is not always strictly necessary, one clock signal may be passed between PCBs along this motherboard. Where possible, this should be avoided, as in asynchronous communication channels. Also various lower-frequency signals are passed between the PCBs on the conductors of this motherboard. Ground conductors should be placed adjacent to these conductors in order to *reduce the radiated emission loop area as well as the inductance and corresponding voltage drop (high-frequency) along these conductors.* Figure 13.20(a) shows a *poor placement* of these ground lands. The return path for the clock signal is at the opposite end of the board, resulting in a large loop area and inductance. Figure 13.20(b) depicts a *better placement* of these ground lands. Ground or return lands are placed on both sides of and in proximity to the clock signal. Thus the clock signal should return on these lands, reducing the radiated emission potential as well as the inductance of the path of this important signal. Also, ground or return lands are placed around the signal lands for similar purposes. Since these signals (data, etc.) are usually at a much lower repetition rate than the clock, several signal lands can be contained between ground lands. Observe that the key to this scheme being effective is that the signal is *forced (presumed) to return on the designated or desired ground return.* Therefore it is equally important to pay attention to where these return lands are connected on the individual PCBs.

13.2.4   Internal Cable Routing and Connector Placement

The location of cable connectors on PCBs is another critical design consideration. In the next section we will discuss the layout of PCBs. An important aspect of that layout will be to *place the highest-frequency electronics toward the center of the PCB.* The intent is to take advantage of the natural filtering of the signals by the board as the signals pass to the cable connectors. Suppose the system clock oscillator is placed at the center of the board along with the processor it serves. This high-frequency signal will inadvertently couple to other lands, allowing the possibility of passing out through the cables that are attached to the board, where the signal will contaminate other parts of the system and radiate more efficiently. Maintaining the largest distance between the high-frequency signals and points of exit of the board will place filtering (unintentional but useful filtering) such as parasitic capacitance and inductance in their path. Even though this filtering is not intentionally designed and therefore not as efficient as it can be, it nevertheless should be used to advantage. Placing the connectors at the outer edge of the board and placing the high-frequency electronics at the center will maximize this filtering.

In addition, all connectors should be placed on the same edge of the board. The reason for doing this is to prevent the development of voltage drops between the wires in one cable and another cable that will drive common-mode currents out those cables. The voltage drop across two diagonally opposite corners of a PCB will generally be significant no matter how well laid out the board. Also, if the connectors are all placed on one edge of the PCB, it is easier to provide a "quiet ground" along this local area of the board to which cable shields can be connected. If the point of attachment of the shields is noisy, the shields are then driven as effective antennas and so do not achieve their objective. The construction of a "quiet ground" for off-board cables will be discussed in more detail in the next section.

Probably one of the simplest yet potentially serious problems to take care of is the internal routing of cables within the product. Figure 13.1 illustrates that the potential for indirect but highly efficient coupling to cables can be considerable if those cables are allowed to pass near noisy components of the system. Such inadvertent coupling can virtually invalidate any of the conscientious thought given to PCB layout. This again points out that overall system EMC cannot be accomplished unless all participants in the design discuss the potential for EMC problems in every phase of development.

13.2.5 PCB and Subsystem Placement

Attention should be paid to the *placement and orientation of the PCB(s) in the system.* For example, placing two PCBs vertically and near each other as shown in Fig. 13.21 allows coupling of signals from one board to another. Not only may this present functional problems, but it may also allow for a more efficient exiting of these high-frequency signals onto cables or other parts of the system where the signal may radiate more efficiently. It is a waste of time to take care in the placement of the high-frequency electronics at the center of the board and the connectors at the edge in order to impeded the exiting of a high-frequency signal and yet have it couple to another board that is placed nearby. The careful attention to layout in the first board is severely compromised or possibly negated by this board placement. Also, the routing of cables that exit this board should be given consideration to avoid problems of coupling to the cable illustrated in Fig. 13.1. Thus it is essential that the board designers and the packaging personnel confer in the early stages of the design before the system configuration is "frozen."

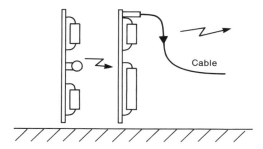

FIGURE 13.21 Illustration of the inadvertent coupling between PCBs due to their close proximity.

13.2.6 Decoupling Subsystems

Once again, it is important to remember that *noise signals should be confined to their known and desired location and not be allowed to propagate to other parts of the system where they may radiate more efficiently and/or cause functional problems.* These noise signals can couple either by radiation from PCB to PCB or via conducting paths such as interconnect cables or backplanes. Eliminating this coupling is referred to as *decoupling the subsystems.* Common-mode currents flowing between subsystems (PCBs, etc.) can be effectively blocked with ferrite, common-mode chokes, as illustrated in Fig. 13.16(a). These are available in DIP packages suitable for automatic insertion as with other components of the PCB.

Another method of decoupling subsystems is to insert a filter in the connection wires or lands between the subsystems. This filter can be in the form of $R-C$ packs, ferrite beads, or a combination. The significant difference between this method of decoupling and the use of a common-mode choke is that a filter will affect differential-mode (functional) signals as well as common-mode (nonfunctional) signals, and thereby may affect the functional performance of the product. Filtering must be used with care.

High-frequency signals on the power distribution system between subsystems can be reduced by the use of decoupling capacitors, as described in the next section. In essence these serve as local reservoirs of charge for the switching signals.

13.3 PRINTED CIRCUIT BOARD DESIGN

The layout of the PCB(s) of the system is probably the most important aspect of the ability of the product to meet the regulatory requirements on conducted and radiated emissions as well as its ability to function reliably in the presence of external interference. This is again an application of the basic EMC design principles discussed previously.

13.3.1 Component Selection

The functional performance specifications of the electronic components are often in direct opposition to the desired EMC performance specifications. For example, from a functional standpoint, the faster (smaller) the rise/falltimes of the digital signals processed by gates, the better. A manufacturer may guarantee that, for example, all gates of a specific type will have rise/falltimes that will not exceed 10 ns. In fact, many of these gates may produce signals with rise/falltimes that are significantly shorter than this maximum, e.g., 1 ns. From an EMC standpoint, we would prefer that the gates have a *minimum* guaranteed rise/falltime, since the shorter the rise/falltime, the more significant is the high-frequency spectral content of the pulses. The problem of having "equivalent parts" from different manufacturers that give equivalent functional performance but drastically different EMC performance is a critical aspect of the EMC design. The regulatory agencies require that *all product units that are manufactured conform to the regulatory limits*. It is therefore not sufficient to construct a "fine-tuned" prototype that meets the regulatory limits. If any product unit fails to meet the regulatory limit, the problem must be fixed. The design must be insensitive to changes in "equivalent components" that typically occur throughout the life cycle of the product. Passing the regulatory tests with a decent margin of safety, say, 3 dB, at all frequencies in the frequency range of the limit is difficult to achieve, but will lessen the impact of changes in parts vendors. A good example of this variability is given in [7]. Therefore it is important to *use the slowest-speed logic possible*.

13.3.2   Splitting Crystal/Oscillator Frequencies

Most digital systems do not need to be operated from a single system clock oscillator/crystal. Asynchronous communication busses can operate with different clock/oscillator frequencies. Where different clock/oscillator frequencies can be tolerated, they should be used to prevent the emissions from different parts of the system from adding in the bandwidth of the test receiver in the course of testing for compliance to the regulatory limits as discussed in Section 7.3.1 and illustrated in Fig. 7.26. Suppose emissions from three different points on the PCB or within the system are due to three different crystals/oscillators, and these signals arrive at the test antenna in phase in the course of testing the product for compliance. If these three signals are within the bandwidth of the receiver and of equal amplitude, their combined effect that appears at the center of the bandpass of the receiver will be 9.54 dB larger than if one signal were present in the bandwidth of the test receiver! The key to avoiding this problem is to *split the crystal/oscillator frequencies* where they do not need to be identical [8]. The key to determining these new values is to realize that *the crystal/oscillator frequencies need to be chosen such that none of their harmonics are closer in frequency than the bandwidth of the receiver being used to measure the emissions*. The minimum 6 dB bandwidth of the test receiver for radiated emissions is 100 kHz. Thus we should separate the crystal/oscillator frequencies *slightly*, so that none of their harmonics are within 100 kHz of each other. For example, suppose two processors are initially intended to use 12 MHz clock frequencies. We could select the crystal/oscillator feeding one processor to be 12 MHz and that feeding the other processor to be 11.90 MHz with little degradation in the system processing speed. This would separate the harmonics by a minimum of 100 kHz within certain frequency bands, as we will show. *This would also allow placement of each crystal/oscillator immediately adjacent to the processor it serves, thus avoiding the perhaps lengthy leads from one crystal/ oscillator to each processor*. This underscores a fundamental point in component placement on the PCB: *keep the leads carrying clock signals as short as possible to avoid radiation from these lands*.

In splitting the frequencies we should realize that the clock frequencies are not necessarily the lowest operating frequencies. Many processors "divide down" this clock frequency to provide lower operating frequencies, which we refer to as *basic frequencies of the processor*. For example the 12 MHz operating frequency of an 8051 processor is divided down to give the 2 MHz ALE (address-latch enable) signal. We do not see the 12 MHz clock signal exiting any of the pins of the processor, but the 2 MHz frequency exits the ALE, Read, Write, etc. pins. Thus the method for splitting the crystal/oscillator frequencies is to determine the lowest operating frequency (the basic frequency), set that to give the required frequency separation, and then determine the resulting crystal/oscillator frequencies. In doing this we must be aware of *foldover* of the harmonics. This is best illustrated by tabulating the frequencies of an example. Consider setting the basic frequencies at 2.000 MHz and at 1.981 MHz. Table 13.1 shows the resulting harmonics.

Observe that at the beginning frequency of the radiated emission frequency range, 30 MHz, the 16th harmonics (32.000 MHz and 31.696 MHz) are separated by 304 kHz, so that they will not add in the bandwidth of the receiver. However, the 99th harmonic of oscillator #1 and 100th harmonic of oscillator #2 are separated by exactly 100 kHz. Thus the 100th harmonic of the 2 MHz signal (200.000 MHz) and the 101st harmonic of the 1.981 MHz signal (200.081 MHz) will be within 81 kHz of each other and *will add in the bandwidth of the receiver*! This separation of the basic frequencies will prevent harmonic additions only over the frequency range of 30–200 MHz. If the radiated emissions above 200 MHz are not a significant problem then this splitting of the basic frequencies will be sufficient. Suppose the two processors divide their crystal/oscillator frequencies by a factor of 6 to produce the basic operating frequencies of 2.000 MHz and 1.981 MHz above. The crystal/oscillator frequency of oscillator

TABLE 13.1

Harmonic	Osc #1 (MHz)	Osc #2 (MHz)	Δf (kHz)
1	2.000	1.981	19
2	4.000	3.962	38
3	6.000	5.943	57
4	8.000	7.942	76
5	10.000	9.905	114
⋮	⋮	⋮	⋮
15	30.000	29.715	285
16	32.000	31.969	304
17	34.000	33.677	323
⋮	⋮	⋮	⋮
98	196.000	194.138	1862
99	198.000	196.119	1881
100	200.000	198.100	1900
101	202.000	200.081	1919

#1 would be 2.000 MHz × 6 = 12.000 MHz and that of oscillator #2 would be 1.981 MHz × 6 = 11.886 MHz. Simple formulae can be developed for choosing these frequencies for a given frequency range of separation to avoid foldover [8]. A spreadsheet program can also be used to quickly calculate the result for more than two crystal/oscillator frequencies.

Another important advantage to splitting crystal/oscillator frequencies in the above manner is *diagnostics*. In the course of testing for regulatory compliance, certain problems will surface, no matter how diligent the engineer has been in applying EMC design principles. It is important at this point to *quickly and correctly diagnose the problem*. It may be evident that the radiated emission is coming from a specific PCB in the system. The question then becomes one of where the emission is being developed on that PCB. In order to suppress this emission, we must be able to determine its point of emission or its source. If we split the crystal/oscillator frequencies, we can use the spectrum analyzer to determine the precise crystal/oscillator that is creating this frequency. This allows us to pinpoint the particular portion of the system that is generating the problem frequency, so that a "fix" may be implemented.

13.3.3 Component Placement

With regard to minimizing the radiated and conducted emission potential of a PCB board design, there are two key goals in component placement on a PCB: (*1*) *minimize the radiated and conducted emission of the traces on the PCB, and* (*2*) *minimize the radiated and conducted emissions of the off-board cables or motherboards* (*backplanes*) *that connect the PCB to the system*. It is important to enunciate these rather obvious goals so that the techniques for their accomplishment can be clearly seen [9].

The first goal, minimizing the radiated and conducted emissions of the lands or traces on the PCB, is simple to achieve. We first must recognize that *the radiation potential of lands on a PCB is a direct function of* (*1*) *their length* (*common-mode currents on them*), (*2*) *the loop area between the signal land and its return* (*differential-mode currents on them*), *and* (*3*) *the high-frequency spectral content of the currents they carry*. "High-speed signals" have more significant spectral content than "low-speed signals." *Signal speed* is intimately tied to spectral content. So a measure of the speed of a signal is proportional to (1) base repetition rate or frequency f_o, (2) rise/falltime τ, and (3) current level I_0. Therefore an effective measure of signal speed is

$$\text{Signal Speed} \approx \frac{f_o I_0}{\tau}$$

Signals having large repetition rates, large current levels and small rise/falltimes will have large high-frequency spectral content.

The signals with the highest-frequency content should be identified. In digital systems these signals are the clocks (crystal/oscillators) in the system. Therefore

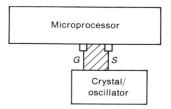

FIGURE 13.22   Placing a crystal/oscillator close to the device it serves tends to reduce the loop area and radiated emissions of the interconnecting lands.

the lands connecting a crystal/oscillator to the device it serves must be kept as short as possible. This is implemented by placing the clocks as close as possible to the devices they serve, as shown in Fig. 13.22, *and* insuring that the interconnecting traces are routed point-to-point rather than by some circuitous route, thereby eliminating the advantage of close module placement. This minimization of the high-frequency trace length can be further enhanced by using dedicated clocks for each module, as described in the previous section. Using one clock to serve several modules virtually guarantees that one set of clock connection lands will be longer than another. *Minimization of the loop area of the signal and return lands of the clock can be insured by routing "ground" or return lands parallel and close to the clock signal land.* In most cases it is advantageous to place returns on both sides of the clock signal land. *Every crystal/oscillator signal land should have a return land routed as close as possible, and generally on both sides.*

The signals having the next lower spectral content must be identified. Typically, in a digital system these are the address/data, read/write signals, etc. Although they have less potential for radiated and conducted emissions, their interconnection traces should be of the shortest possible length, and the loop areas between the signal and return lands should be as small as possible. Examples of these are the signals between the processor and the associated read-only-memory (ROM) and random-access-memory (RAM) components. In a typical 8088 processor system operating at a clock frequency of 8 MHz, the ALE (address latch enable) and data signals operate at 2 MHz. The ROM and RAM modules *must be placed as close as possible to the processors they serve,* and their interconnection lands must be kept as short as possible. Figure 13.23 shows some typical bad and good placements. Figure 13.23(a) shows four RAM modules and their associated address/data latch arranged in a series connection (possibly for aesthetic but certainly not for EMC reasons). The address/data bus conductors are longer than necessary. Figure 13.23(b) shows a more desirable arrangement that minimizes this bus length and associated radiation potential. In addition, since these 2 MHz signals are potentially troublesome, it is sensible to include ground lands next to each land of the bus in a Ground–Signal–Ground–Signal–Ground–... configuration or at least a Ground–Signal–Signal–Ground–... configuration in order to minimize the loop area of these signals.

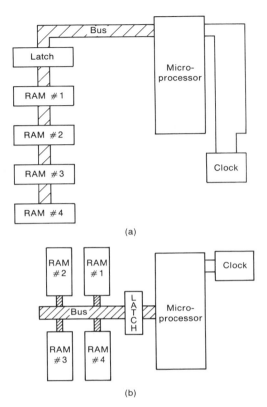

(a)

(b)

FIGURE 13.23   Illustration of (a) poor PCB layout resulting in large loop areas and (b) proper PCB layout.

Once the signals that do not pass off the PCB are treated in this manner, and their spectral content has been reduced as much as possible by slowing their rise/falltimes, the next task is to *prevent them from exiting the PCB via attached cables or motherboards.* Some of these signals are intended to be passed between PCBs, such as with clock signals via motherboards. Where this can be avoided, as with dedicated clocks on each PCB, it should be done in order to prevent the potentially serious radiated emissions from the long traces on the motherboard. Ground traces should be placed parallel and close to clock as well as data signals on the motherboard, as illustrated in Fig. 13.20(b).

To prevent those high-frequency signals that are not intended to exit the PCB from doing so, *place components on a PCB in regions where the distance between the region and any off-board connectors is commensurate with the speed of the signals in that region,* as illustrated in Fig. 13.24. This is to take advantage of the *natural filtering of the intervening board space.* Although this filtering is *unintentional,* it is nevertheless quite effective! Therefore place the highest-speed components (clocks, microprocessors, ASICs) in the center of the board, well

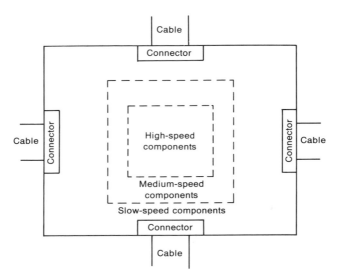

FIGURE 13.24   Illustration of proper component placement on a PCB according to speed of the components. Component speed is directly proportional to the current level and the fundamental frequency of its signal output, and is inversely proportional to the rise/falltime of the signal.

away from the connectors. The next fastest components, such as RAM and ROM modules and their associated latches, may be placed around this region, but must still be far from the cable connectors (as well as edge connectors to motherboards). And finally the slowest speed components that must intentionally interface with other parts of the system via cables (such as motor-driver signals, keyboard, and CRT signals) are placed around the outer edges of the board. The intent here is *to place the signals with the highest spectral content as far as possible from the off-board connectors.*

Figure 13.25 illustrates this implementation for a typical microprocessor/ controller board. An ASIC accepts and queries the keyboard inputs, and drives the CRT diaplsy. A main system microprocessor controls computation and system sequencing. Another controller processor drives the disk drives and any motors of the system. The three processors along with their separate crystals/ oscillators are placed in the center of the board, well away from the cable connectors. Also, the RAM modules and the ROM modules that are accessed by the system processor are placed near that processor to minimize the radiated emissions from their busses. The signals from or to the keyboard, the CRT, and the disk drives must exit the board for proper system function. Although they are at a lower frequency than the system clock, they still have the potential for radiated emission problems once they are present on the off-board cables. The higher-speed signals such as the read/write signals to the hard disk have as much potential for radiated emissions as the ROM/RAM signals. Therefore

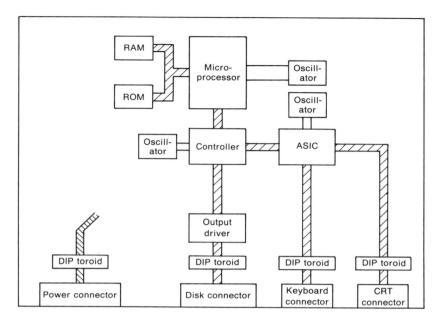

FIGURE 13.25   A good PCB layout for a typical digital system.

these should be passed off-board, where possible, with a motherboard. Nevertheleess, these remaining off-board signals may still cause radiated emission problems, which can be prevented by shielding the cables and/or providing filtering where they exit the PCB.

This brings us to the next topic: *prevention of radiated emissions from off-board cables.* Perhaps the most effective way of controlling off-board cable radiated emissions is to *place all connectors at one edge of the PCB and provide a "quiet" I/O ground at that point,* as illustrated in Fig. 13.26. As discussed earlier, the effectiveness of a shield around any cable is determined by the availability of a "quiet" ground to which the shield is attached. If the shield attachment point is "noisy," the shield will radiate effectively as an antenna. A quiet ground is provided by segmenting the I/O ground and the "noisy" board ground into two segments connected at only one point via a thin land as shown in Fig. 13.26. This prevents noise signals on the noisy ground from inadvertently flowing through the quiet I/O ground. Thus shields can be attached to this quiet I/O ground and be effective. Filtering such as resistor/capacitor ($R-C$) packs between each signal line and the quiet ground can be used to filter high-frequency signals off the cables. In the case of common-mode signals $R-C$ packs and other differential-mode filters such as ferrite beads are not much help. However, provision should be made for DIP toroids. These connection holes can be "wired accross," or 0 Ω surface mount resistors can be inserted in the preliminary design. If compliance testing uncovers significant common-mode

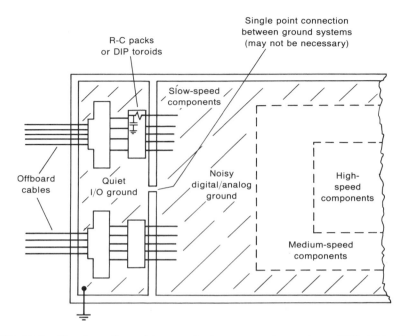

FIGURE 13.26   Creation of a quiet ground where connectors enter a PCB. This allows effective termination of shields and also filtering of off-board cables.

current emissions from one of these cables, the wires can be cut, or the $0\,\Omega$ surface mount resistors can be removed and the DIP toroid inserted without the necessity of a re-layout of the board. $R-C$ packs can be used in the same fashion. In some cases it may be advantageous to omit the single-point connection between the quiet ground and the noisy ground, with the quiet ground being connected to the hardware ground. Provision can be made for connecting the two grounds with a $0\,\Omega$ surface mount resistor in the initial design. If it turns out in testing that the grounds should be isolated, the $0\,\Omega$ resistor can be omitted from the parts list, and the board layout will remain the same. The power connector should also be treated with equal care. If high-frequency signals should couple to this cable, they will radiate as effectively as if they were on the keyboard or CRT cable. It should not be *assumed* that only the low-frequency power signals will be present on the power cable.

If the board must contain both digital and analog circuitry, these two circuit areas should be kept separate, as shown in Fig. 13.13(a), in order to prevent the high-level analog signals from flowing through the ground system of the digital circuitry and creating common-impedance voltage shifts. The common ground connection at point P_1 in Fig. 13.13(a) may or may not be required, depending on the noise immunity of the input circuitry of the analog circuit at the digital/analog interface.

13.3.4   Miscellaneous Considerations

The previous section has presented the most important considerations in the design of a PCB: *topology with respect to component speed.* However, there are a number of seemingly minor points that should be addressed.

An important consideration is the drive capability of the off-chip drivers in ASICs. A common mistake is to give unnecessarily large drive capability to these drivers. If the drive capability that is needed is of order 10 mA, there is no need to provide drive capability of 50 mA. This will only exascerbate the emission levels of the currents on the pins of the ASIC. It is also important to be aware of the phenomenon illustrated in Fig. 13.27. We are now aware of the fact that fast rist/falltime signals will have significant high-frequency spectral content; the faster (shorter) these rise/falltimes, the higher-frequency the spectral content. In order to limit this, we may place a filter at the source output to "round off" the edges of the pulse. If, however, we follow this with a high-speed driver, the previous rounding of the signal will be nullified. Thus placement of filtering is important. *It is also important to use the slowest-speed logic possible* so that unnecessarily fast rist/falltimes will be avoided.

A final consideration that is often critically important is *internal coupling within modules.* The DIP packages that house the component have bonding wires that connect the external pins of the DIP to the small chip at the center of the module. These internal bonding wires can be quite long, in order to accommodate a large number of pins of the chip. Internal coupling between bonding wires can cause pins of the DIP that are immediately adjacent to a high-speed pin to carry those high-frequency signals due to this internal crosstalk, as illustrated in Fig. 13.28. Consequently, we must not allow a supposedly "quiet pin" of a module (such as the reset pin of a processor that normally is activated infrequently) to be routed around the PCB if it is adjacent to a high-frequency pin, because it may carry this high-frequency signal. Read/write (RW) pins also carry high-frequency signals, and could couple to any adjacent "quiet" pins. To correct this problem where the land attached to this "quiet" pin must be routed long distances (as with the reset signal), place a small inductor in series with that pin to block the high-frequency signal from passing along this lond lead. Placement of an inductor in series with a RW or oscillator pin would create obvious functional problems, and could not be tolerated. However, the routing and lengths of these traces will have already

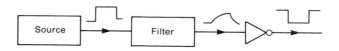

FIGURE 13.27 Illustration of the effect of negating the filtering of a signal with "downstream" components that restore the fast rise/falltimes and radiated emission potential of the signal.

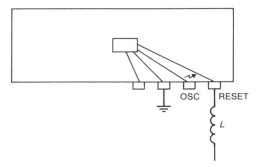

FIGURE 13.28   Illustration of the unintentional coupling of signals between chip bonding wires, causing supposedly quiet module pins to have high-frequency spectral content. Series inductors and/or shunt capacitors can prevent these signals from exiting the module.

been treated with care. But placement of an inductor in series with a reset pin would not create functional problems, because of infrequent activation and slow response required of this signal. A 1 μH inductor would give an impedance at 100 MHz of 628 Ω, which should provide sufficient blockage (assuming that the impedance of the load of this land is sufficiently less than the impedance of the inductor). Ferrite beads could also be used, but their impedance is typically limited to a few hundred ohms. Series resistors are also a possibility, but they affect both low- and high-frequency signals equally and may create functional problems. *Provision should be made for placement of blocking inductors/diversion capacitors for all pins that are adjacent to high-speed pins (pins that contain fast-rise/falltime signals).* If the inadvertent coupling to these "quiet" pins proves to be a problem during compliance testing, the component can be inserted without relayout of the board. Surface-mount and discrete inductors are available in the 1 μH range, and their physical size is no bigger than small ceramic capacitors. Capacitors can be simply soldered across terminals if needed, whereas inductors, ferrite beads, and series resistors can only be included by cutting lands. So points for mounting possible *series elements* must be provided during layout. There exist 0 Ω surface mount resistors that can be used to "wire across" these points in the initial design. If a ferrite bead, resistor, or inductor is needed in order to fix a compliance problem, these 0 Ω resistors can be removed and replaced with surface mount resistors, inductors, or ferrite beads, and only the parts list need be changed. *Preliminary planning such as this prevents serious bottlenecks in the production schedule that a board relayout would cause.*

13.3.5   The Important PCB Ground Grid

We showed previously that the *inductance of PCB lands* provides considerably more *high-frequency impedance* than the resistance of these lands. This inductance of PCB lands and wires can be responsible for generating high-frequency voltage differences between two points on a supposedly equipotential conductor surface, which may result in functional as well as radiated and conducted emission problems. If a shield of an off-board cable is attached to an assumed "quiet ground" point, the shield will be driven at the voltage of the attachment point and will act as an effective antenna. This also applies to "ground" wires in an unshielded cable. In order to avoid this and other potential problems resulting from voltage differences between two points on a conductor or interconnection of conductors, we must have a way of reducing the inductance of the conductors or bypassing their effects. A *ground grid* reduces the inductance of conductors whereas *decoupling capacitors* (discussed in the next section) bypass or reduce the effect of the inductance. *An effective and well-designed ground grid is one of the most important aspects in the ability of the product to meet the regulatory limits and avoid functional problems* [9]. Obvious problems such as long clock leads and placement of high-frequency components adjacent to connectors are easily solved. A good ground grid remedies the remaining ones.

Although not immediately apparent, there are data that indicate a correlation between reduced *ground drop* on a PCB (high-frequency voltage differences between two points on the ground conductor) and a reduction in the radiated emissions of that PCB [10, 11]. At present there is no quantitative correlation. However, experience in product development supports the idea that a reduction of the voltage difference between two points on the ground system of a PCB (ground drop) will result in a reduction in the radiated (and conducted) emissions from that PCB and the attached peripheral cables. Consequently, the design of an effective ground grid on a PCB is a critical aspect to the regulatory compliance of the PCB and its host system. Innerplane (multilayer) boards implement this to a similar degree, but are significantly more expensive than double-sided boards. Even if an innerplane board is used, it is recommended that a ground grid be implemented on the two outer signal layers of that board.

13.3.5.1 *Partial Inductances of Wires and PCB Lands*   The inductance of an isolated wire or PCB land is frequently misunderstood, as was pointed out in Chapter 6. Inductance is associated with a *closed loop*. However, it is possible to uniquely ascribe an inductance to portions of this loop with the concept of *partial self-inductance*. This is usually mistakenly thought to be the internal inductance of the conductor due to magnetic flux internal to the conductor. This internal inductance decreases with the square root of frequency due to skin effect. However, this internal inductance is not the dominant inductance of the conductor. It is dominated by the partial inductance, which is for practical purposes frequency-independent. Similarly, the mutual inductance between parallel conductors of the loop can be uniquely determined with the concepts of *partial mutual inductance.*

The concept of partial inductance is such an important concept it is worthwhile to review it. Consider an isolated conductor (wire or PCB land) of length $\mathcal{L}$ carrying a current I, as shown in Fig. 13.29(a). The *self partial inductance L_p* of this conductor is *the ratio of the magnetic flux generated by*

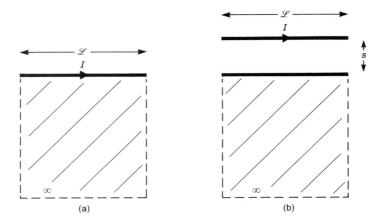

FIGURE 13.29   Illustration of the meaning of partial inductance in terms of the magnetic flux between the conductor and infinity: (a) self partial inductance; (b) mutual partial inductance.

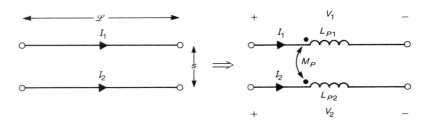

FIGURE 13.30   The partial inductance model of a pair of parallel conductors for computing the voltage drop along one conductor.

current I that passes between the conductor and infinity divided by the current I that produced it. The *mutual partial inductance* M_p between the two conductors that are parallel and separated by a distance s is the *ratio of the magnetic flux due to the current in the first conductor that passes between the second conductor and infinity to the current in the first conductor that produced it,* as shown in Fig. 13.29(b). The equivalent circuit for two parallel conductors is given in Fig. 13.30 in terms of these partial inductances. The voltages developed across the conductors are easily obtained from this equivalent circuit as

$$V_1 = L_{p1}\frac{dI_1}{dt} + M_p\frac{dI_2}{dt} \tag{13.1a}$$

$$V_2 = M_p\frac{dI_1}{dt} + L_{p2}\frac{dI_2}{dt} \tag{13.1b}$$

Let us suppose these two conductors constitute a signal path and its associated return, so that $I_1 = I$ *and* $I_2 = -I_1 = -I$. (If the two conductors do not constitute a signal and associated return, but the return is elsewhere, then there will be mutual couplings between these two conductors and this return conductor that must be included in the model of Fig. 13.30. Otherwise this does not make sense, because the concept of partial inductance requires a closed path.) The above voltage drops become

$$V_1 = (L_{p1} - M_p)\frac{dI}{dt} \qquad (13.2a)$$

$$V_2 = -(L_{p2} - M_p)\frac{dI}{dt} \qquad (13.2b)$$

This shows that *in order to reduce the voltage drop across a conductor, we must maximize the mutual partial inductance between that conductor and its other associated conductor that contains the same current.* This mutual partial inductance can be increased as shown in Chapter 6 in a rather obvious fashion by *placing the two conductors as close as possible.* This is an extremely important concept and corresponds to the case of a shielded wire above a ground plane. Therefore *placing a signal conductor and its associated return close together reduces the effect of the individual conductor inductances.* For example, in Fig. 13.3 the *effects* of the inductances of the +5 V and ground lands are minimized due to the mutual partial inductance between them (not shown in that figure) if we place these two lands as close as possible. The individual partial self-inductances are still present, but their effect is minimized by the mutual partial inductances. Remember that for this to occur the currents in the two conductors must be equal in magnitude and opposite directed; that is, one conductor is truly the return for the other conductor. We will see that this is the reason a *gridded ground system* works well to reduce high-frequency voltage differences between two points on it.

To restate the self and mutual partial inductances between wires and PCB lands that were developed in Chapter 6, the self partial inductance of a wire of radius r_w and length $\mathscr{L}$ is given by

$$L_p = \frac{\mu_0}{2\pi}\mathscr{L}\left\{\ln\left[\frac{\mathscr{L}}{r_w} + \sqrt{\left(\frac{\mathscr{L}}{r_w}\right)^2 + 1}\right] + \frac{r_w}{\mathscr{L}} - \sqrt{\left(\frac{r_w}{\mathscr{L}}\right)^2 + 1}\right\} \qquad (13.3a)$$

or

$$L_p \cong 2 \times 10^{-7}\mathscr{L}\left[\ln\left(\frac{2\mathscr{L}}{r_w}\right) - 1\right] \quad \text{for } \frac{r_w}{\mathscr{L}} \ll 1 \qquad (13.3b)$$

Similarly, for a PCB land of zero thickness (a thin "tape"), width w, and length

$\mathscr{L}$, the self partial inductance is

$$L_{p\,\text{tape}} = \frac{\mu_0}{2\pi}\mathscr{L}\left\{\ln(u + \sqrt{u^2+1}) + u\ln\left[\frac{1}{u} + \sqrt{\left(\frac{1}{u}\right)^2+1}\right]\right. \qquad (13.4\text{a})$$

$$\left. + \frac{u^2}{3} + \frac{1}{3u} - \frac{1}{3u}(u^2+1)^{3/2}\right\}$$

where $u = \mathscr{L}/w$. This result can be simplified for land lengths $\mathscr{L}$ that are long compared with their width, *i.e.*, $u = \mathscr{L}/w \gg 1$, giving

$$L_{p\,\text{tape}} \cong 2\times10^{-7}\mathscr{L}\left[\ln\left(\frac{2\mathscr{L}}{w}\right) + \frac{1}{2}\right] \quad \text{for } \frac{\mathscr{L}}{w} \gg 1 \qquad (13.4\text{b})$$

The mutual partial inductance between two wires of length $\mathscr{L}$ and separation s that are parallel and aligned as in Fig. 13.30 is

$$M_p = \frac{\mu_0}{2\pi}\mathscr{L}\left\{\ln\left[\frac{\mathscr{L}}{s} + \sqrt{\left(\frac{\mathscr{L}}{s}\right)^2+1}\right] + \frac{s}{\mathscr{L}} - \sqrt{\left(\frac{s}{\mathscr{L}}\right)^2+1}\right\} \qquad (13.5\text{a})$$

This can be approximated for conductors whose length is much longer than the separation, $\mathscr{L} \gg s$, to give

$$M_p \cong 2\times10^{-7}\mathscr{L}\left[\ln\left(\frac{2\mathscr{L}}{s}\right) - 1 + \frac{s}{2\mathscr{L}}\right] \qquad (13.5\text{b})$$

$$\cong 2\times10^{-7}\mathscr{L}\left[\ln\left(\frac{2\mathscr{L}}{s}\right) - 1\right] \quad \text{for } \frac{\mathscr{L}}{s} \gg 1$$

The reader is referred to Section 6.1.4 for typical values of these self and mutual partial inductances for wires and PCB lands. Although it is not possible to give a per-unit-length value (since the conductor lengths are involved in natural logarithmic terms), for typical wire gauges the self partial inductances are of order 15–40 nH/inch while the mutual partial inductances range from 1 nH/inch to over 30 nH/inch. Typical PCB lands have self partial inductances of order 30–40 NH/inch and mutual partial inductances ranging from 5 nH/inch to over 30 nH/inch.

Voltage drops along the power distribution conductors are frequently reduced by using lands providing +5 V and power return separated only by a thin insulation (perhaps the PCB thickness). Not only does this reduce the net land inductances, but it also *increases the capacitance between the lands*, a feature we have so far ignored. When viewed in this context, the two lands constitute a transmission line having a characteristic impedance $Z_C = \sqrt{l/c}$, where l is the per-unit-length inductance and c is the per-unit-length capacitance.

As another illustration of the importance of the concept of partial inductances, consider the use of image planes to reduce radiated emissions (of both differential-mode and common-mode currents) [10]. Consider a pair of parallel lands on one side of a PCB, as depicted in Fig. 13.31(a). As we know, there will be two types of currents on these lands: common-mode, I_C, and differential-mode, I_D. Also, the radiated emissions of the differential-mode currents tend to cancel, but those of the common-mode currents tend to add and can provide the dominant radiated emissions. Suppose we place a conducting plane beneath the PCB, as in Fig. 13.31(b). Image currents, equal and opposite to those in the original lands, will be generated in the image plane, as shown in Fig. 13.31(c).

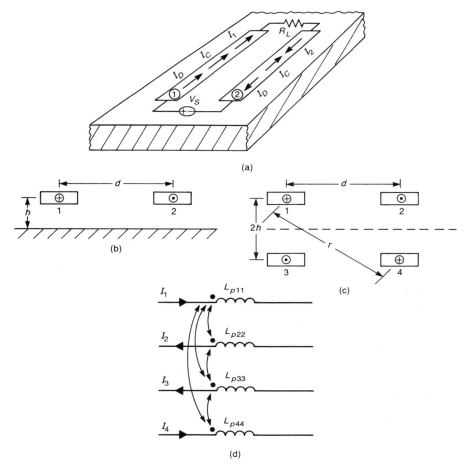

FIGURE 13.31 Use of an image plane to reduce the radiated emissions and ground drop of a PCB: (a) the board schematic; (b) the board cross section; (c) replacement of the image plane with images; (d) the partial inductance model.

It is not necessary to "ground" this plane or otherwise connect it to anything in order to generate these image currents. In essence, the image plane has generated an associated differential-mode current for each of the four original currents in the lands, I_D, I_D, I_C, and I_C. Even though the two original common-mode currents together give increased radiated emissions, we may view each one and its associated image as a differential-mode pair separated only by the board thickness; hence the net radiated emissions from the original common-mode currents will be reduced. The same results apply to the original differential-mode currents. Widely spaced lands carrying differential-mode currents may radiate significantly, but the images associated with these will tend to reduce the radiated emissions from these differential-mode currents since each pair (the original current and its image) is now more closely spaced (separated only by the board thickness). Theoretically, this replacement of the ground plane with images requires that the ground plane be perfectly conducting and infinite in extent. However, the method seems to work well so long as the height of the conductors is small compared with the extent of the ground plane [10].

The image plane can also reduce the voltage drop across each conductor. With reference to Fig. 13.31(c), the voltage drops along each of the original lands are obtained from the equivalent circuit shown in Fig. 13.31(d) as

$$V_1 = sL_{p11}I_1 - sL_{p12}I_2 - sL_{p13}I_3 + sL_{p14}I_4 \tag{13.6a}$$

$$V_2 = -sL_{p12}I_1 + sL_{p22}I_2 + sL_{p23}I_3 - sL_{p24}I_4 \tag{13.6b}$$

Denoting $I = I_1 = I_2 = I_3 = I_4$, the *net partial inductance of each of the original lands* becomes

$$L_{p1} = L_{p11} - L_{p12} - L_{p13} + L_{p14} \tag{13.7a}$$

$$L_{p2} = L_{p22} - L_{p12} + L_{p23} - L_{p24} \tag{13.7b}$$

Assuming identical lands of width w and length $\mathscr{L}$, a board thickness h, and land separation d, the previous results give these inductances (see (13.5b) for self-inductances and (13.4b) for mutual inductances as

$$L_{p11} = L_{p22} = 2 \times 10^{-7}\mathscr{L}\left[\ln\left(\frac{2\mathscr{L}}{w}\right) + \frac{1}{2}\right] \tag{13.8a}$$

$$L_{p12} = L_{p34} = 2 \times 10^{-7}\mathscr{L}\left[\ln\left(\frac{2\mathscr{L}}{d}\right) - 1 + \frac{d}{2\mathscr{L}}\right] \tag{13.8b}$$

$$L_{p13} = L_{p24} = 2 \times 10^{-7}\mathscr{L}\left[\ln\left(\frac{2\mathscr{L}}{2h}\right) - 1 + \frac{2h}{2\mathscr{L}}\right] \tag{13.8c}$$

$$L_{p14} = L_{p23} = 2 \times 10^{-7}\mathscr{L}\left[\ln\left(\frac{2\mathscr{L}}{r}\right) - 1 + \frac{r}{2\mathscr{L}}\right] \tag{13.8d}$$

where

$$r = \sqrt{d^2 + 4h^2} \tag{13.8e}$$

The *net partial inductance* of either of the two identical original lands becomes

$$L_{p1} = L_{p2} = 2 \times 10^{-7} \mathscr{L} \left[\ln\left(\frac{2h}{w}\right) + \ln\left(\frac{d}{r}\right) + \frac{3}{2} + \frac{r - 2h - d}{2\mathscr{L}} \right] \tag{13.9}$$

If the original lands are widely spaced with respect to the board thickness, $d \gg h$, this reduces to

$$L_{p1} = L_{p2} \cong 2 \times 10^{-7} \mathscr{L} \left[\ln\left(\frac{2h}{w}\right) + \frac{3}{2} \right] \quad \text{for } d \gg h, \ \mathscr{L} \gg h \tag{13.10}$$

The net partial inductance of one land without the image plane is

$$L_{p1} = L_{p2} = L_{p11} - L_{p12} \tag{13.11}$$

$$= 2 \times 10^{-7} \mathscr{L} \left[\ln\left(\frac{d}{w}\right) + \frac{3}{2} - \frac{d}{2\mathscr{L}} \right]$$

Comparing (13.10) and (13.11) shows that *a poorly placed return with large separation d between it and the signal conductor can be corrected with regard to voltage drop across that conductor by placing an image plane beneath the PCB*. The net partial inductance of the return is equivalent to an original land spacing of 2h! Reference [10] gives measured and computed results for the voltage drop across one land for an actual board with and without the image plane. Measured radiated emissions without the image plane are dramatically reduced by the addition of the image plane [10].

13.3.5.2 *Ground Drop*

13.3.5.2 *Ground Drop*   The voltage developed across a return conductor is referred to as *ground drop* [2, 10, 11, 12]. There is qualitative evidence to suggest that *the lower the ground drop between two points on a PCB ground net, the lower will be the radiated emissions from that board* [10, 11]. As discussed previously, a ground plane through which currents pass will not be an equipotential surface particularly at the high frequencies of the radiated emission regulatory limit, 30 MHz–40 GHz. We will discuss a method to approximate a ground plane that is less costly to implement that an innerplane (multilayer) board: the ground grid. It is as effective as a ground plane in reducing ground drop and is less expensive.

A common problem that results from radiated emissions being due to ground drop is illustrated in Fig. 13.32. A pair of lands on a PCB connect a source and a load; an off-board shielded cable is to carry this signal to a peripheral device. As the signal changes state, a voltage that is related to the derivative

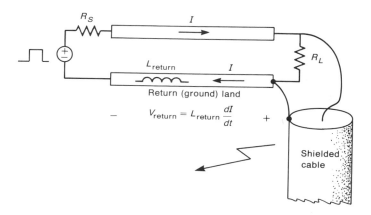

FIGURE 13.32 Illustration of the effect of connecting a cable shield to a noisy ground, causing it to radiate.

of this signal current is developed across the return land. The shield of the cable will probably be connected to the far end of the return land, because it is assumed to be a "quiet ground point." However, the voltage of this point is that which is developed across the return land and drives the shield as an antenna. This results in potentially significant radiated emissions from the shield. This is why good EMC designs will endeavor to provide a quiet ground area where off-board cables exit the PCB. Again, there are numerous cases where the attachment of the shield to the "ground" of a PCB can actually *increase the radiated emissions*, which are due to antenna currents on the outer surface of the shield.

In addition to radiated emissions, inductance of the return path can cause functional problems. An example was shown in Fig. 13.3, where the rapidly changing digital currents create voltage $L_{return}\, dI/dt$ between two points on the return conductor. There are two methods of reducing this ground drop: implementing a ground grid to reduce the inductance of the return path or using decoupling capacitors to lessen the effect of the inductance of the return (and signal) land inductance.

13.3.5.3 *Measurement of Ground Drop*

The correct measurement of ground drop (voltage difference between two points on the return conductor) is critical to a determination of the figure of merit of a particular PCB layout. The *correct method* for the measurement of ground drop is not obvious. In this section we will demonstrate the correct measurement method and the demonstration will depend crucially on the notions of *partial inductances* developed earlier.

Consider two parallel lands connecting a source to a load, as Fig. 13.33(a) illustrates. Suppose we attach wires to each end of the return ("ground") land and measure the voltage generated across a small gap in this wire, V_{meas}. The essential question here is *what distance d between the measurement wire and the*

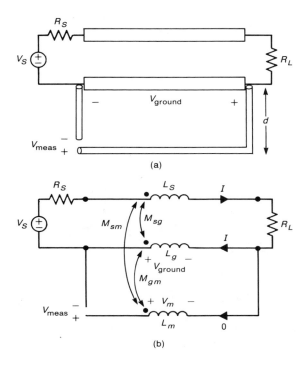

FIGURE 13.33   The correct method of measuring ground drop along a conductor: (a) the physical configuration; (b) the partial inductance equivalent circuit.

two lands should be maintained in order that $V_{meas} \cong V_{ground}$. The answer is is to *make d as large as possible by placing the measurement wire as far away from both lands as practical.* At first it may seem that just the opposite would be true, since magnetic flux from the currents in the lands will thread this measurement loop and will appear to introduce errors in the measurement of V_{ground}. However, it is important to realize that we are not measuring V_{ground} directly but are measuring V_{meas}, and *we want* $V_{meas} \cong V_{ground}$.

In order to demonstrate this important result, consider the equivalent circuit of partial inductances shown in Fig. 13.33(b). (There is no logical reason to omit the capacitances that will occur, but we will ignore them only to simplify the calculation.) The measurement voltage is obtained from this equivalent circuit as

$$V_{meas} = V_m - V_{ground} \tag{13.12}$$
$$= (sM_{sm} - sM_{gm})I - (-sL_g + sM_{sg})I$$

(Observe that the current through the measurement wire is assumed to be zero, so that a high-impedance voltmeter must be used to measure V_{meas}.) In order that the measurement voltage and the ground voltage be identical, the voltage developed across L_m, V_m, must be made zero. This is given by

$$V_m = V_{\text{meas}} - V_{\text{ground}} \qquad (13.13)$$
$$= s(M_{sm} - M_{gm})I$$

In order for this is to be zero so that $V_{\text{meas}} = V_{\text{ground}}$, we must have

$$M_{sm} = M_{gm} \qquad (13.14)$$

The partial mutual inductance M_{sm} is the ratio of the magnetic flux *between the measurement wire and infinity and the current on the signal land*. The partial mutual inductance M_{gm} is similarly the ratio of the magnetic flux *between the measurement wire and infinity and the current on the ground land*. In order for (13.14) to be true, the distance from the signal land to the measurement wire must equal the distance from the ground land to the measurement wire. Placing the measurement wire above the center line between the two lands would accomplish this. Another way to accomplish is to *move the measurement wire away from both lands*. In a more complicated PCB with many lands placing the measurement wire above the PCB and at a distance will aid this accurate measurement of the ground drop voltage. It should be noted that there is a practical limitation to how far the measurement wire can be moved away from the PCB. The combination of the inductance of the measurement loop and the capacitance of the high-impedance measurement voltmeter will create a resonance condition [12]. The resonant frequency will be a direct function of the measurement wire height, an increase in which increases the inductance of the measurement loop. So a compromise may be required in how high above the PCB the measurement wire can be placed and achieve accurate measurement of the ground drop voltage. Skilling gives an alternate proof of this measurement scheme [13].

13.3.5.4 *The Important Gridded Ground System*   The simplest type of PCB is single-sided and has lands on only one side. This is rarely used, because of difficulties in wiring a dense board. The more common double-sided board has lands placed on both sides. As mentioned earlier, the lands on one side are orthogonal to the lands on the other side in order to facilitate wiring. A path can be created by following a land on one side and when a bottleneck occurs jumping to the other side through a "via" and proceeding in the orthogonal direction on that side. The signal can be "viaed" back to the previous side of the board, thus allowing passage across a bottleneck created by a module pin or other lands. The next level of PCB is the *innerplane or multilayer board*, wherein signal planes are buried at various depths in the PCB substrate. Vias are used to make connection between the two outer planes and innerplanes. A common type of innerplane board uses two internal planes: one for $+5$ V distribution and one for the ground or $+5$ V return. The proximity of these

large planes lowers the characteristic impedance and inductance of the power distribution system. It also increases the capacitance between the signal and return, which acts as a decoupling capacitor, a critical component for any digital system as described in the next section. The innerplane board makes wiring much easier than a double-sided board, but is more expensive. Where low cost is an important factor in the success of the product, innerplane boards can be prohibitively expensive, and one has no other resource but to use only double-sided boards. The essential question is how to achieve the benefits of an innerplane board with a double-sided board. The answer is the use of a *gridded ground system and a generous number of decoupling capacitors.*

Recall from Section 13.3.5.1 that *the net inductance of a return land (which is directly related to the voltage drop across that land) can be reduced by placing the signal land near the return land.* Obviously, on a complicated and dense PCB it is not practical to place a return land next to each signal land. The board simply cannot be wired, due to the enormous number of return lands required. Another way of providing return paths close to all signal lands is to create a interconnected grid of lands as in a screen door or panes of a window. An example is shown in Fig. 13.34(a). Suppose a signal land traverses the board

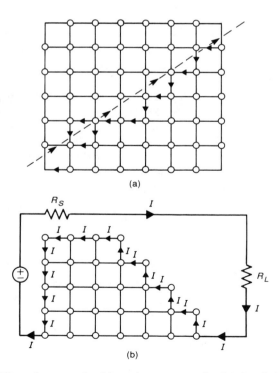

FIGURE 13.34   Effect of a ground grid on the return path of a signal: (a) a signal above the grid; (b) a return path around the grid.

in a diagonal direction as illustrated. It is true that *the current will return to its source along the path of lowest impedance.* Thus the current will return on the portions of the ground grid that are closest to the signal land in a zig-zag pattern as shown. (We are obviously referring to the majority of the return current, since a type of "current division" effect takes place where the current divides through parallel paths in inverse proportion to the relative impedances of those paths. Hence the majority of the current follows the indicated path or a region around that path.) The ideal case would be a solid ground plane, where the current will return along a path directly beneath the signal land. Unless the signal land can be placed closer to the ground plane, this path will have lowest impedance possible. Another example is shown in Fig. 13.34(b). In order for the ground grid to be effective, *the grid must be connected at all points denoted by dots using vias.* In other words, the grids on the bottom and the top *must be "stitched together,"* otherwise abundant alternative paths will not exist. *A well-designed ground grid system will have a large number of possible return paths, to allow each signal to dynamically select a return path as close as possible to the signal path.* The current will return on the path closest to the signal wire (the lowest-impedance path). The problem in many PCB designs is that they give the current no choice but to return at a great distance from the signal land, since that is the "closest" return path. German gives a good example of this return path blockage in [11]. Therefore the key to a well-designed gridded ground system is to provide *many possible return paths between any two points on the PCB.* This will allow the signal current to "choose" the path closest to the signal land, and will thus minimize the net inductance and therefore the impedance of the return path. This will in turn minimize the ground drop. It is recommended that the first step in PCB layout is to put down a sufficiently dense ground grid. The effect of grid size is investigated in [14].

At first glance it may appear that the ground grid will severely limit the number of wiring channels, and thus prevent wiring the board. The key to avoiding this type of bottleneck is to take maximum advantage of the vias to avoid blocking wiring channels. An example is shown in Fig. 13.35. The ground grid lands on the bottom side of the board are indicated in dashed lines, while on the top side of the board are indicated by solid lines. Once again, it is critical that the ground grid be connected with vias at all crossover points indicated with dots. A signal land enters the board from one side. A way of traversing the grid without bottlenecks by "jumping from one side to the other" using vias is shown. Most, but not all, boards can be wired today using only a double-sided board with no innerplanes. Even if an innerplane board is used, it is recommended that the two outer planes employ a gridded ground that is well stitched together and to the innerplane grounds. It is easy to become complacent with innerplane boards. Their complexity tends to give a false sense of security. *A poorly designed innerplane board will perform worse than a well-designed double-sided board.*

The necessity of a good gridded system cannot be overemphasized; it is one of the more important aspects in complying with the regulatory limits.

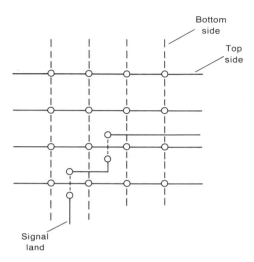

FIGURE 13.35   Illustration of the use of vias on a double-sided PCB to prevent path blockage by the ground grid.

Component placement, short clock leads, etc. are other critically important features of good EMC design. If it is necessary to route lands carrying high-frequency signals such as address/data bus lands long distances on the PCB, ground lands should be interspersed immediately adjacent to each signal land in a so-called "Ground–Signal–Ground–..." pattern, as illustrated in Fig. 13.36(a). This accomplishes the objective of providing a return path close to the signal, which will lower the return inductance, ground drop, loop area, etc. The ground grid will tend to accomplish this to some degree, but it is usually too coarse to create the same effect as closely spaced returns. For "critical signals" (critical from the standpoint of their potential for creating radiated emission problems due to their high-frequency content) such as high-speed signals, a ground or return land on both sides of the signal land is important. In some cases board space does not allow a return to be placed on both sides of each signal. In this case we can locate returns adjacent to every other signal in a "Ground–Signal–Signal–Ground–Signal–..." pattern, as shown in Fig. 13.36(b). These dedicated adjacent ground returns can be an important complement to a well-designed gridded system.

It is important to reemphasize the key feature of a ground grid: *a ground grid provides a large number of alternate paths, so that the signal can choose a return path closest to the signal path, thereby reducing the net inductance of the return path. The ground grid reduces the net inductance of the return path (as well as the signal path) simply by providing an alternate path for the signal current to return on that is close to the signal conductor, hence effectively increasing the mutual partial inductance between the grid and the signal conductor and thereby reducing the net partial inductance of that return path:* $L_{p\,\text{net}} = L_p - M$. If the

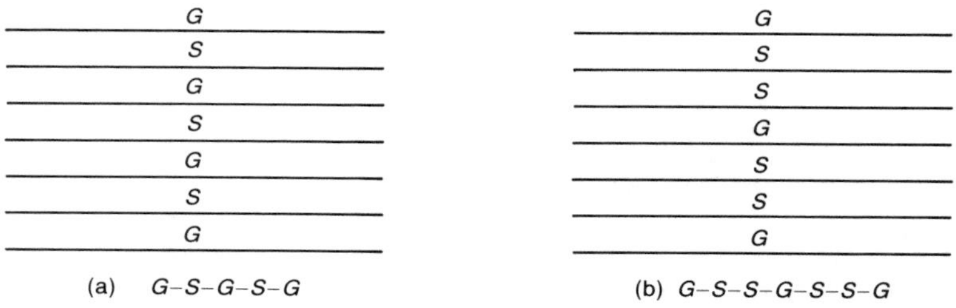

FIGURE 13.36   Illustration of the use of adjacent grounds (returns) to minimize the loop area of large spectral content signals: (a) Ground–Signal–Ground–...; (b) Ground–Signal–Signal–Ground–....

current were constrained to follow a return path that was distant from the signal conductor, virtually no interaction between the conductors would occur, and little subtractive effect would be gained. It is true that placing signal and return paths *very close together* reduces the net inductance of either path. On a particularly dense PCB it is difficult to place return conductors in close proximity to every signal conductor. The ground grid selectively accomplishes this. Reference [14] provides data that show the effect of conductor spacing on the reduction of grid inductances.

13.3.6   Power Distribution and Decoupling Capacitors

A well-designed gridded ground system will have the advantages of reducing (1) the inductance of the return path, (2) the loop area traversed by the signal current, and hence its radiated emissions, and (3) the tendency for currents to be generated on ground conductors such as cable shields that are attached to this ground, thereby producing a quiet ground system. However, the remaining inductance of either path may still be substantial, and the lengths of the conductors carrying these high-frequency signals may be significant even though the loop area may have been reduced by the ground grid. The technique of interspersing ground or return conductors adjacent to all signal conductors that carry high-frequency signals will serve to lower the net inductance of both the signal and the return paths, as we illustrated with the concept of partial inductance. Where possible, the signal conductor should be routed as close as possible to the return conductor in order to minimize the impedance of the path as well as the loop are traversed by the current and hence the radiated emissions of that current. An important example that illustrates this problem is the distribution of the $+5$ V that powers the digital modules. This is illustrated in Fig. 13.3(a), where the rapidly changing currents are drawn through the inductance of the $+5$ V supply conductor. Therefore the voltage drop along the $+5$ V conductor will create "voltage sag" and possible functional problems at the modules. In addition, no matter how close the $+5$ V conductor and its return, the length of this path between the power supply and the module it serves is likely to be quite long. Although the conductors may be close together, their length may still produce a significant loop area. The answer to the elimination of the *effect* of the potentially large loop area between the power source and the module it serves is to use *decoupling capacitors* at each module between the +5 V and ground pins.

The effect of a decoupling capacitor is illustrated in Figs. 13.37 and 13.38. A pair of gates is shown, with the output of one gate feeding the input to the other gate. The paths of the currents for the gates in the two possible states are shown. It is assumed that the ground grid has been implemented and the signal conductor has been placed close to the return or the ground grid so that the inductances L_S and L_G have been minimized. As the gates change state, the current through the +5 V conductor and return land changes value abruptly, creating high-frequency currents that circulate around that potentially large loop. Suppose we place capacitors between the +5 V and ground pins of each gate as shown in Fig. 13.38. These capacitors function as *local sources of charge*, so that during a change in state of the two gates, current is momentarily drawn out of them rather than through the +5 V supply conductor. During the quiescent period between state changes, charge is drawn from the bulk decoupling capacitor on the PCB to recharge these capacitors. However, the current required to replenish the capacitors is drawn at a slower rate than it is removed from them during a state change. Thus the effect of the large loop areas is reduced.

It is important to realize that the currents provided by the decoupling

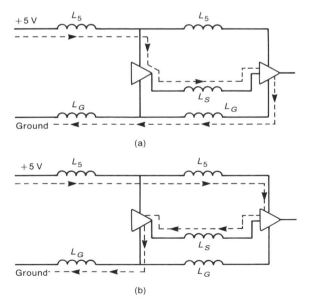

FIGURE 13.37   Loop areas formed when gates switch state: (a) high-to-low; (b) low-to-high.

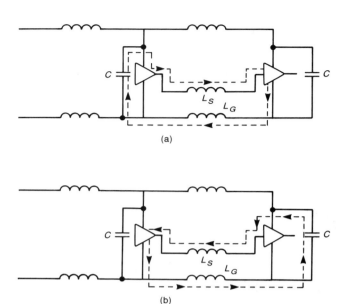

FIGURE 13.38   Use of decoupling capacitors to minimize loop areas when gates switch state: (a) high-to-low; (b) low-to-high.

capacitors have, by design, very high-frequency spectral content, over 100 MHz. Thus, in order for the capacitor to be effective in supplying these high-frequency currents and not have them drawn through the long $+5$ V supply conductor, it is imperative that *they behave as capacitors at these high frequencies.* Typically, this requires small ceramic capacitors. In addition, *it is critical that the net effective lead lengths between the capacitor body and the pins of the module be kept as short as practical in order to produce a self-resonant frequency of the combination as high as possible.* Recall that the self-resonant frequency of the capacitor is a direct result of the *inductance of the connection leads*, as discussed in Chapter 6. The traditional placement of the $+5$ V and ground pins on digital modules has been criticized because they are at diagonally opposite corners of the module and require the longest possible connection lead length for the decoupling capacitor. Nevertheless, there is nothing we can do about this and so must design around it. A preferred placement is shown in Fig. 13.39(a). The capacitor is placed close to the $+5$ V pin of the module and connected to the ground grid. The inductance of the lead to the $+5$ V pin is minimized since its length is kept short. *It is important to remember that the inductance of the capacitor connection leads is that of the total conductor between the capacitor body and the module pins.* A common misconception is that if the visible wires of the decoupling capacitor are kept short, the design is satisfactory. This fails to take into consideration the potentially long *land lengths* that may be required

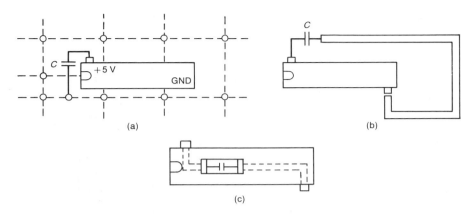

FIGURE 13.39   Decoupling capacitor placement: (a) use of the ground grid to minimize the ground return inductance; (b) the capacitor lead length includes the length of the PCB land required to connect it to the module pin; (c) a surface-mount capacitor beneath the module minimizes the lead length.

to connect the capacitor lead wires to the module pins. An example of this is illustrated in Fig. 13.39(b). The fact that the wires of the capacitor are kept short is of no consequence; the long connection land acts like a long lead wire as far as inductance is concerned.

Every module should have a decoupling capacitor. Capacitors are relatively inexpensive, and, so long as there is sufficient board space, they can and should be placed on every module. Present technology allows the use of *surface-mount technology (SMT) components.* These components are soldered directly to the board and have essentially no lead lengths. They can be placed on the opposite side of the board from the module and directly beneath it, as shown in Fig. 13.39(c). This permits using the shortest possible connection lands. It also minimizes the loop area between these connection lands and the bonding wires in the module, which reduces the radiated emissions from this seemingly small local loop. The ability to place SMT components on the other side of the board allows an enormous component density that would not be possible if components were restricted to one side of the board only. SMT resistors also exist, and $0\,\Omega$ SMT resistors are useful in wiring across connection points that may need to have components inserted for EMC suppression.

The suitable value of the decoupling capacitor is a subject of much current debate. Some work has been done in this area [15]. Traditionally, the value of the decoupling capacitor is chosen based on the allowed value of voltage drop at the $+5\,V$ pin of the module. The current supplied is related to the capacitance and rate-of-change of the voltage across its terminals as

$$I = C\frac{dV}{dt} \tag{13.15}$$

It is from this relation that the usual equation for the capacitance value is taken:

$$C = \frac{I\,dt}{dV} \tag{13.16}$$

In this equation I is the current that is required when the gate switches, dt is the rise/falltime of this change, and dV is the allowed drop in supply voltage. For example, assume that the current transitions 50 mA in 5 ns and the allowed supply voltage is 0.1 V. Equation (13.16) gives a value of 0.0025 μF. This equation, although appearing to be relevant, is not the whole picture, since it confuses a combination of events, and the circuit to which it applies is not clear. Nevertheless, the proper selection of the necessary value of decoupling capacitor is not as simple as (13.16) implies [15]. However, experience has shown that values of order 0.01 μF–470 pF are usually adequate. It is a common mistake to use too large a decoupling capacitor. The self-resonant frequency of the capacitor is dependent on the inductance of the connection leads, L_{lead}, as

$$f_0 = \frac{1}{2\pi\sqrt{L_{\text{lead}}C}} \tag{13.17}$$

as discussed in Chapter 6. For a fixed lead length the self-resonant frequency decreases as the capacitance value is increased. For example, for a lead inductance of 20 nH, a 0.001 μF capacitor will have a self-resonant frequency of 36 MHz. A 0.01 μF capacitor will have a self-resonant frequency factor of $\sqrt{10}$ less, or 11 MHz. If we intend this capacitor to supply currents in the radiated emission regulatory limit, both capacitors may be severe disappointments.

Another method for reducing the effect of connection lead inductance on the self-resonant frequency of decoupling capacitors is to use *distributed decoupling capacitors*. These are available in a form that allows their insertion immediately under the module, as shown in Fig. 13.40, and are constructed from two long metallic strips separated by a thin insulation. Pins are provided that sandwich along with the +5 V and ground pins of the module into the board socket.

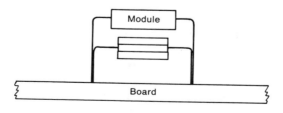

FIGURE 13.40   Distributed (parallel-plate) decoupling capacitors reduce the effect of lead inductance.

The proximity of the two conductors gives a large capacitance and low inductance, as discussed earlier. When viewed from the standpoint of a transmission line, the distributed decoupling capacitor gives low characteristic impedance $Z_C = \sqrt{l/c}$, where l is the per-unit-length inductance and c is the per-unit-length capacitance. Placing wide strips close together evidently reduces the inductance and also increases the capacitance.

Finally, *bulk capacitors* should be placed where power is delivered to the board and at strategic points on the board. The purpose of the bulk capacitors is to serve as charge storage points to replenish the decoupling capacitors of the modules. Generally, the value of the bulk capacitor should be at least ten times the sum of the decoupling capacitor values on the board. For example, if the board contains 12 modules and each is served by a 0.001 μF decoupler, the bulk capacitor should have a value of 0.15 μF. Tantalum electrolytic capacitors can provide these large values of bulk capacitance. Bulk capacitors can also be placed at strategic points on large boards to serve as more local points of charge storage for replenishing the nearby decoupling capacitors.

13.3.7   Reduction of Loop Areas

In Chapter 8 we found that the differential-mode current radiated emissions are proportional to the loop area between the currents. Reducing these loop areas lowers radiated emissions. Similarly, we found that the susceptibility to reception of incident fields is also directly related to loop area. The incident magnetic field induces an emf representable as a voltage source in the loop, and that induced voltage is proportional to the loop area. Also, the incident electric field induces a current source between the two conductors that is also proportional to the loop area between the two conductors. This incident field may be due to some external source such as an AM or FM transmitter or radar, or may be due to ESD-induced fields, as discussed in Chapter 12. In any event, *in order to reduce the radiated emissions of a PCB as well as to reduce the reception of incident fields, loop areas must be minimized.* This minimization of loop areas is a simple application of methods of routing conductors on the PCB.

Radiated emissions from the $+5$ V–Ground lands on a PCB can be a significant part of the radiated emissions of the product. It is often *assumed* that the currents on the $+5$ V land are dc, which usually false. As the ICs switch state, current through the $+5$ V and return lands are rapidly changing, and these currents have high-frequency spectral content due to the fast (ns) transition times. A common problem in the layout of digital PCBs is to route the $+5$ V and Ground distribution lands such that large loops are inadvertently formed, as shown in Fig. 13.41(a). Rearrangement of the ICs to minimize this large loop area is shown in Fig. 13.41(b). Decoupling capacitors can also be better placed in the routing of Fig. 13.41(b) to minimize their lead lengths. Use of an innerplane or multilayer board as in Fig. 13.41(c) is a better but more costly alternative. The $+5$ V and Ground planes are buried at various depths within the PCB, and connections are made from the IC pins to these planes

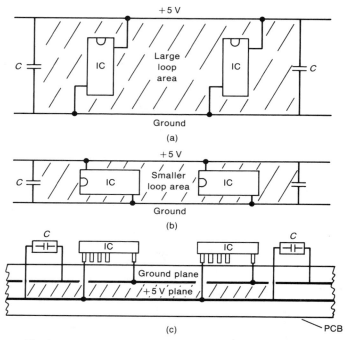

FIGURE 13.41   The important consideration of loop area in minimizing radiated emissions of the power distribution circuits: (a) large loop areas; (b) smaller loop areas; (c) use of an innerplane (multilayer) PCB to reduce the loop area and the characteristic impedance of the power distribution circuit.

with "vias." The innerplane board has additional advantages in that the +5 V and Ground planes form a parallel-plate transmission line that has a small characteristic impedance, as discussed in Chapter 6. Thus the voltage drop along the planes due to the inductance of the paths is reduced. Similarly, the proximity of the planes increases the capacitance between them, which acts as additional decoupling capacitors.

Another troublesome result of large loop areas occurs with the various signal lands. Figure 13.42(a) illustrates the large loop area resulting from the routing of a signal land with respect to the Gound land. Routing a dedicated Ground land close to the signal land as shown in Fig. 13.42(b) tends to reduce that loop area, since the majority of the signal current will return on the land closest to the signal land. Because of the large number of signal lands on a board, use of closely spaced, dedicated lands for every signal land may not be feasible. A well-designed ground grid can accomplish the same result. In the case of "critical signal lands" (critical from the standpoint of having large high-frequency spectral content) such as Address/Data bus lands and clock lands, one generally needs to place adjacent Ground lands on both sides and close to these lands. It is also important to remember when placing Ground lands adjacent to signal lands that the distance the signal must travel to get on these Ground lands at

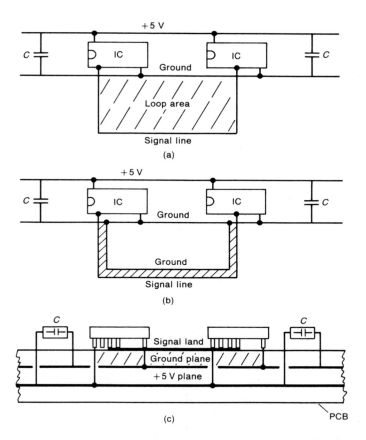

FIGURE 13.42   Large loop areas of the signal-return path should be minimized to reduce radiated emissions: (a) large loop areas; (b) smaller loop areas; and (c) use of an innerplane (multilayer) PCB to reduce the loop area and the characteristic impedance of the signal-return circuit.

the ends of this parallel run must be kept small. It does little good to route a Ground land close to a long signal land while at the endpoints of this run the signal must follow some circuitous route to enter the Ground land. An innerplane or multilayer board, where the lesser cost of a double-sided board is not a factor, can be a better alternative. It is recommended that, even with an innerplane board, one should install a ground grid on both sides and interconnect it to the ground plane with a generous use of vias. Typical board thicknesses are about 47 mils. Therefore the separation between a signal land on one side and the Ground plane may be as "large" as 15 mils. Present technology permits placing a ground land on the board surface as close as 8 mils to the signal land.

Once again it is critical to emphasize that loop areas on a PCB must be minimized to reduce (1) radiated emissions, (2) reception of external fields such as AM and FM transmitters and radar transmitters, and (3) pickup of ESD fields and arc discharge currents. The major problems associated with PCBs are often associated with large loop areas. Remember that "large" can be as small as an inch. From the results of Chapter 8 one can calculate that a 30 MHz current of 50 mA flowing in a loop of 2.25 cm × 2.25 cm (0.886 inch × 0.886 inch) will create a radiated emission at the FCC Class B limit!

13.4   SUMMARY

In this chapter we have brought together all the principles of the previous chapters to provide a coherent and effective EMC design of an electronic system. We have concentrated on digital systems, since these represent the majority of present-day electronic systems. Although different techniques are employed, digital circuit design is little different from the design of high-frequency analog circuits in so far as EMC is concerned. Attention to component selection and placement, connector placement and filtering, the ground grid, Ground–Signal–Ground treatment of long, high-frequency connections, and decoupling capacitor placement and selection in the layout of PCBs are critical and effective measures in the design of a PCB such that it will not create emission as well as susceptibility problems. Attention to board and other subsystem placement within the system, cable routing, treatment of apertures in enclosures, and intelligent segregation of the various ground systems of the system are the final keys to making a product satisfy the objectives of a successful EMC design: the product (1) will not create emissions that cause interference in other products (and will pass the regulatory requirements), (2) will not be susceptible to emissions from other products or electromagnetic phenomena such as ESD, and (3) will not cause interference with itself. Effective EMC design is a matter of applying the basic principles in this text.

REFERENCES

[1]   H.W. Ott, Ground—a path for current flow, *Proceedings 1979 IEEE International Symposium on Electromagnetic Compatibility*, San Diego, CA, October 1979.

[2]   H.W. Ott, *Noise Reduction Techniques in Electronic Systems*, second edition, John Wiley Interscience, New York (1988).

[3]   H.W. Ott, Digital circuit grounding and interconnection, *Proceedings 1981 IEEE International Symposium on Electromagnetic Compatibility*, Boulder, CO, August 1981.

[4]   G.K.C. Chen and J.J. Whalen, Comparative RFI performance of bipolar operational amplifiers, *1981 IEEE International Symposium on Electromagnetic Compatibility*, Boulder, CO, August 1981.

[5]   R. Morrison, *Grounding and Shielding Techniques in Instrumentation*, John Wiley, New York (1967).

[6]   D.R. Bush, A simple way of evaluating the shielding effectiveness of small enclosures, *8th International Symposium and Technical Exhibition on Electromagnetic Compatibility*, Zurich, Switzerland, March 1989.

[7]   D.R. Kerns, Integrated circuit construction and its effects on EMC performance, *Proceedings of the 1984 IEEE International Symposium on Electromagnetic Compatibility*, San Antonio, TX, April 1984.

[8]   K.B. Hardin, Private communication.

[9]   H.W. Ott, Controlling EMI by proper printed wiring board layout, *Proceedings of the 1985 International Symposium on Technical Exhibition on Electromagnetic Compatibility*, Zurich, Switzerland, March 1985.

[10] R.F. German, H.W. Ott, and C.R. Paul, Effect of an image plane on printed circuit board radiation, *IEEE International Symposium on Electromagnetic Compatibility*, Washington, DC, August 1990.

[11] R.F. German, Use of a ground grid to reduce printed circuit board radiation, *Proceedings of the 1985 International Symposium and Technical Exhibition on Electromagnetic Compatibility*, Zurich, Switzerland, March 1985.

[12] C.R. Paul, Modeling electromagnetic interference properties of printed circuit boards, *IBM J. Research and Development*, **33**, 33–50 (1989).

[13] H.H. Skilling, *Electric Transmission Lines*, McGraw-Hill, NY (1951).

[14] T.S. Smith and C.R. Paul, Effect of grid spacing on the inductance of ground grids, *Proceedings of the 1991 IEEE International Symposium on Electromagnetic Compatibility*, Cherry Hill, NJ, August 1991.

[15] J.C. Engelbrecht and K. Hermes, A study of decoupling capacitors for EMI reduction, *Technical Report, International Business Machines*, TR-51.0152 (May 1984).

PROBLEMS

13.1   Remove the cover from a typical personal computer and locate the crystals/oscillators of the system. Identify the modules to which these provide the clock. Do any cables pass near these crystals/oscillators? Identify the entrance points of peripheral cables on the PCB's.

13.2   In order to investigate the effect of common impedance (inductance) coupling, use SPICE to simulate the circuit of Fig. 13.3. Use BJTs having $\beta = 10$. Assume an inductance of the interconnection wires of 15 nH/inch. Let the separation between T1 and T2 be 1 inch and that between the $+5$ V battery and T1 be 10 inches. Assume the connection wires are #28 gauge solid and include the dc resistance in series with the inductance. Assume a signal at the input to T1 as a trapezoidal pulse train of frequency 10 MHz, 50% duty cycle and rise/falltimes of 10 ns. The resulting circuit is shown in Fig. P13.2. Plot the collector voltages of the BJTs without the wiring resistance/inductance included to observe the ideal behavior. Replot these collector voltages with the wiring resistance/inductance included. Plot the voltage between the two emitters of the transistors and the voltage drop across the $+5$ V wire between T1 and T2. Plot similar voltages between the $+5$ V battery and T1. Separate the voltage drop across the interconnection resistance/inductance into the portions due to inductance and due to resistance.

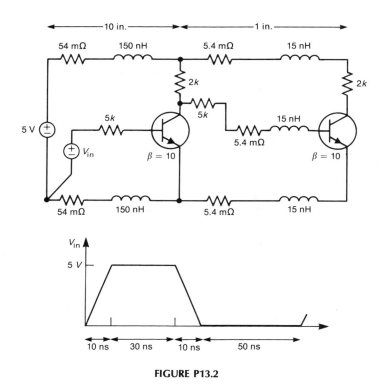

FIGURE P13.2

13.3   In order to investigate the effect of single-point ground systems versus multipoint ground systems, consider the two subsystems fed by a trapezoidal waveform shown in Fig. P13.3. Assume V_{in} to be the waveform of the previous problem and use SPICE to compute the voltage drop waveform across the connection inductances for the two grounding schemes shown.

13.4   Placing signal and return wires close together will force the return current to return on the return wire. This is similar to the case of a shielded wire above a ground plane considered in Chapter 10. In order to illustrate this important point, consider a source and load connected by #28 gauge, solid wires that are 10 inches in length, as shown in Fig. P13.4. Assume that V_{in} is the trapezoidal waveform of Problem 13.3. Another similar wire is used as a common return for other circuits, but is placed

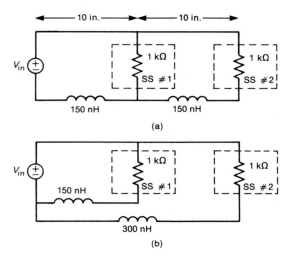

(a)

(b)

FIGURE P13.3

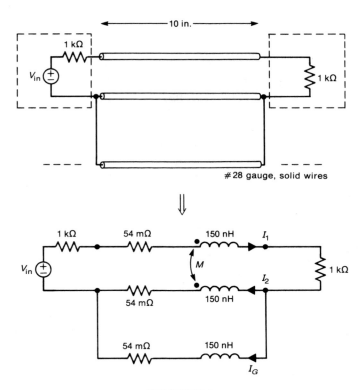

#28 gauge, solid wires

FIGURE P13.4

further from the original two wires. Placing the signal and return wires in close proximity introduces a mutual inductance between them; the closer the wires, the larger the coupling coefficient. Model the circuit as shown in Fig. P13.4 and plot the currents I_1, I_2, and I_G for values of mutual inductance of $M = 140$ nH (close spacing), $M = 100$ nH (moderate spacing), $M = 50$ nH (wide spacing), and $M = 0$ (very wide spacing). Also plot the common-mode current given by $\frac{1}{2}(I_1 - I_2)$ for these various couplings. This represents the effect of "ground drop" illustrated in Fig. 13.15, which can be reduced with the common-mode choke of Fig. 13.16. Careful attention to routing wires closely tends to obviate the need for these "fixes."

13.5   Repeat the previous problem for a balanced configuration illustrated in Fig. 13.17(b).

13.6   Remove the covers from a personal computer, an electronic typewriter, and a printer. Diagram the location of the system microprocessor(s), ASICs, other controllers, and the power and off-board connectors of each PCB. Identify the good and bad points of each topology. How would you re-layout the PCBs and/or consolidate some of the PCBs? Identify the location of the power supply, filter, and on-off switch. Are there any other general system configuration changes you would make with regard to EMC?

13.7   A digital system employs two microprocessors and one ASIC. The first microprocessor divides its nominal operating oscillator frequency of 8 MHz by 4. The second microprocessor divides its nominal operating oscillator of 12 MHz by 6, while the ASIC divides its nominal operating oscillator frequency of 12 MHz by 6. Communication between the two processors and the ASIC is via an asynchronous bus. Determine the oscillator frequencies of the two microprocessors and the ASIC such that none of their harmonics add in the 100 kHz bandwidth of the spectrum analyzer over the frequency band of 30–200 MHz. [$f_1 = 8$ MHz, $f_2 = 11.946$ MHz, $f_3 = 11.892$ MHz]

13.8   Examine the "pinouts" of a typical microprocessor. Identify the most likely "noisy" pins and those "quiet" pins that are adjacent to these noisy pins. Based on the function of the quiet pins, which ones may have an inductor inserted in series to prevent possible, unintentional crosscoupling within the module from exiting the module.

13.9   Compute the partial self and mutual inductances of a 5 inch, parallel pair of #28 gauge, solid wires that are separated by a distance of $\frac{1}{2}$ inch. [161.8 nH, 53.2 nH] Repeat for a length of 10 inches. [358.8 nH, 139.1 nH] Repeat for a length of 20 inches. [788.1 nH, 346.1 nH] Determine an average per-unit-length self and mutual partial inductance from these results. [35.88 nH/inch, 13.95 nH/inch]

13.10 Determine the loop inductances of the configurations of the previous problem (neglect the contributions due to the ends). [217.2 nH, 439.4 nH, 884.0 nH]

13.11 Compare the approximations for self and mutual partial inductances given in (13.3b) and (13.5b) for Problem 13.9. [161.8 nH, 50.7 nH, 358.8 nH, 136.6 nH, 788.0 nH, 343.6 nH]

13.12 In order to further investigate the effect of placing wires/lands in close proximity in order to reduce the net inductance of a return path, compute the net inductance of the three configurations shown in Fig. P13.12. (Assume the current divides among the wires in an ideal way.) [358.8 nH, $L_{eq} = \frac{1}{2}(L + M) = 323.9$ nH, $L_{eq} = \frac{1}{3}(L + M_{12} + M_{13}) = 300.5$ nH]

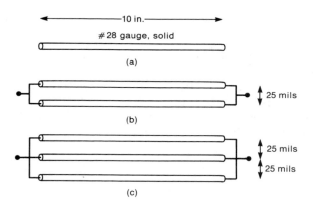

FIGURE P13.12

13.13 Consider two parallel, 10 inch lands whose widths are 15 mils, thicknesses 1.4 mils, and separation 50 mils. Compute the partial self and mutual inductances. [391 nH, 240.6 nH]

13.14 The relation for choosing the value of a decoupling capacitor given in (13.16) is somewhat simplistic, and disregards a number of important factors. In order to illustrate this, consider the circuit of Fig. P13.14, which models a dc voltage source to a switching element such as BJT. The inductance of the interconnecting wires is modeled as L and the on resistance of the switching device is modeled as R. Derive the equation for the capacitor voltage v_C after the switch is closed, assuming that R is very small so that the transient solution is underdamped. [$v_C(t) = (2\alpha V/\beta)e^{\alpha t} \sin \beta t + V$ with $\alpha = -1/2RC$, $\beta = \alpha\sqrt{4R^2C/L - 1}$] Use SPICE to model the circuit shown in Fig. P13.14 with $V = 5$ V, $L = 300$ nH, $R = 2050$ Ω, and $C = 0.01$ μF.

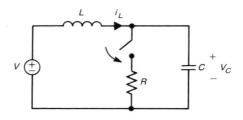

FIGURE P13.14

13.15 In order to investigate the effects of decoupling capacitors on actual circuits, add 0.01 μF decoupling capacitors across the 2 kΩ resistors and BJTs in the circuit of Fig. P13.2 and use SPICE to determine the collector voltages of the two transistors. Compare this result with that obtained in Problem 13.2.

13.16 Choose several simple digital circuits from applications manuals, each of which require a clock. Show a good EMC part layout for the ICs along with connector locations to provide power and I/O cables. Sketch the locations of the lands that interconnect the IC and connector pins. Show your design for a double-sided board and a multilayer board that has $+5$ V and Ground innerplanes.

Index